STUDENT STUDY GUIDE

to accompany

FUNDAMENTALS OF PHYSICS

SECOND EDITION SECOND EDITION EXTENDED

and

PHYSICS • PARTS 1 AND 2

THIRD EDITION

STANLEY A. WILLIAMS
Iowa State University

KENNETH R. BROWNSTEIN
University of Maine

ROBERT L. GRAY
University of Massachusetts

ROBERT RESNICK
Rensselaer Polytechnic Institute

DAVID HALLIDAY
University of Pittsburgh

JOHN WILEY & SONS NEW YORK CHICHESTER BRISBANE TORONTO

ISBN 0 471 06465 3

Printed in the United States of America

10 9 8 7 6 5 4 3

TO THE STUDENT

A complaint, quite frequently uttered by physics students (usually after an examination), is:

> "I understand all of the theory but I cannot seem to solve any problems".

In reality, the truth of the matter is more nearly:

> "I have memorized all the formulas which I can possibly store in my brain but I never seem able to apply them correctly to the solution of problems".

Here the student has confused "memorizing" with "understanding". What can be done in order to improve this state of affairs? Of course there are no magic words which can be said in a study guide (much less here) which will replace the required hours of study and problem solving practice. The objective of this study guide is to make these hours more <u>efficient</u>. It does this by focusing your attention on the most essential parts of the chapters and by providing a re-phrased version of those parts.

This study guide is meant to serve as a companion to either of the textbooks: "Physics" (3rd edition) and "Fundamentals of Physics" (2nd edition, including extended version), both by D. Halliday and R. Resnick. It cannot serve as a substitute for the textbook or for your own efforts. Before we tell you what you can expect to find in the study guide, we would like to make some comments about studying physics in general.

There is a certain strategy of studying physics which you should keep in mind: What concepts, ideas, formulas should you commit to memory? What should you read for "exposure" without memorizing? There are really only two sorts of things which you must memorize.

1. Definitions. In science and engineering, quantities are given very precise definitions. After studying a certain topic, you should ask yourself:
   (a) What new quantities were defined and what mathematical symbols were used to represent them?
   (b) What is the actual definition of these quantities? Usually this will be a mathematical definition.
   (c) What units are used for these quantities?

2. First-principles. These are the fundamental formulas which relate various defined quantities to one another. Some first-principles are mathematically derived, others represent experimental facts. You will encounter many other formulas which represent specific applications of first -principles. These other formulas should <u>not</u> be memorized; rather you should understand <u>how</u> the first-principles were used to derive them.

Most chapters of the study guide we start with a "Review and Preview." This relates the material in the chapter with those of preceding chapters. This is followed by "Goals and Guidelines" which tells you the objectives of the chapter. In the main part of each chapter we emphasize the essential parts of the corresponding chapter(s) of your text. Examples, illustrating these parts, are presented in thorough detail. Finally, most study guide chapters contain a section of "programmed problems"; these have the special feature of encouraging you to actively participate in the solution of problems.*

---

*Special instructions for using the programmed problems are given on page vi.

There are many different ways to use textbooks and study guides. You will probably want to experiment with several to find the method which suits you best. The following method is suggested as a starting point.

1. Locate the section(s) in the study guide which correspond to the assigned textbook material. The study guide's "Table of Contents" and "Index" will be useful in this regard. Look through this study guide material exclusive of examples and programmed problems. This will serve to outline the important parts of the material. You should not expect to understand everything now.
2. Read all the material assigned in the textbook very carefully looking especially for the definitions and first-principles which were emphasized in the study guide.
3. Carefully work through the examples and programmed problems in the study guide. You should also try to redo the examples in your textbook without looking at the solutions.

If you follow these steps conscientiously, you will be able to tackle your assigned homework problems efficiently. You will also find that the study guide serves as a very useful took in reviewing large blocks of material so that you may approach your examinations with confidence.

PREFACE TO THE THIRD EDITION

This third edition of the Study Guide has been extensively revised. The major changes are:

1. With a few exceptions, there has been complete conversion to SI units and abbreviations.
2. A new chapter (34) on alternating currents has been written.
3. The concept of a virtual object has been included (chapter 38).
4. The concept of mutual inductance has been included (chapter 31).
5. An index has been added.

Acknowledgement

We wish to thank Shirley Williams for typing the first and second editions. Andrea-Matilda Brownstein typed the third edition and furnished technical writing assistance. Steven Macomber made several useful suggestions. Finally, one of us (KB) acknowledges Rensselaer Polytechnic Institute where, during his sabbatical leave, most of this third edition was written.

PREFACE TO THE FOURTH EDITION

The major difference between this and the previous edition is in the addition of four new chapters:

Chapter 45: The Atom
Chapter 46: Electrical Conductivity in Solids
Chapter 47: Nuclear Physics
Chapter 48: Special Relativity

These are intended to accompany the "extended version" of "Fundamentals of Physics."

The Index to the entire sutdy guide has been made more complete and a separate Supplementary Index has been added to cover the four new chapters.

ABOUT THE PROGRAMMED MATERIAL

The programmed problems consist of a sequence of numbered frames separated by horizontal lines drawn across the page. Each frame consists of a question part on the left side and an answer part on the right side of the page. The question part of the frame calls for a written response from you. The answer part of the frame allows you to check your answer.

Typically a given problem consists of several frames. This format guides you through the problem step-by-step, with the opportunity to check your work at each step.

In using the programmed problems you should cover the answer part and write your answer (in pencil) in the space provided in the question part. Do not "cheat yourself" by looking at either the answer part or succeeding frames. After you have written your answer to a given frame, expose the answer part to check your work.

Here is an example of the use of two frames of programmed problems. First, we show the frames as they would appear to you with the answer parts covered and before you have written in your response:

1. 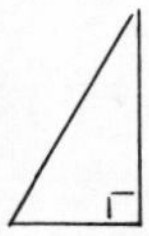 Label the sides of the right triangle shown by A, B, and C with C being the hypotenuse.	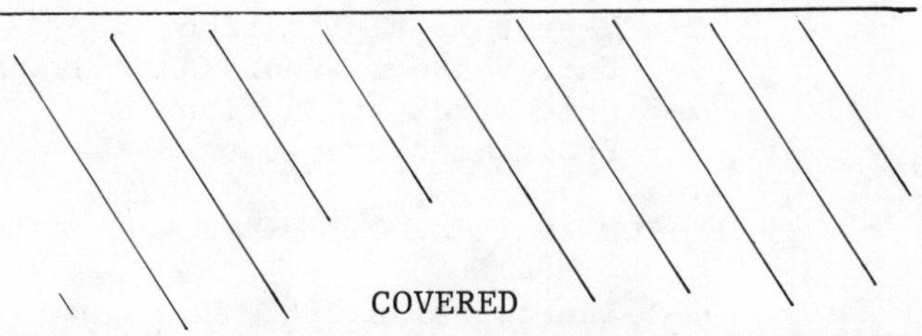
2. What is the mathematical relationship between the lengths A, B and C?	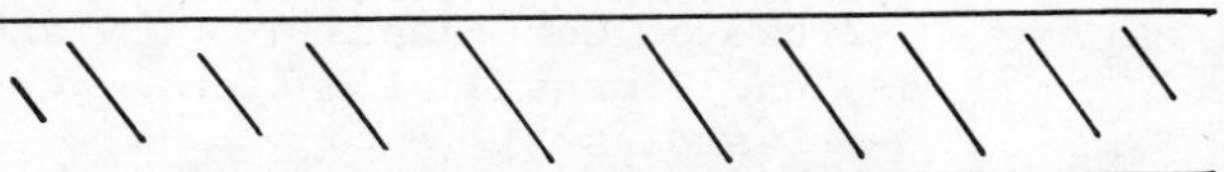

Next we show frame 1 after a student has written his response and exposed the answer part:

1. 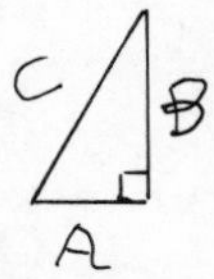 Label the sides of the right triangle shown by A, B, and C with C being the hypotenuse.	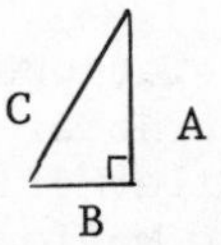

Notice that this student did not label sides A and B in the same order as the printed answer but the hypotenuse C is correctly labeled. This is an example of a correct answer which does not look exactly the same as the printed answer.

Finally we show the second frame after the student has written his response and uncovered the answer part:

2. What is the mathematical relationship between the lengths A, B and C? $C = \sqrt{A + B}$	$C^2 = A^2 + B^2$ (Pythagorean theorem) or $C = \sqrt{A^2 + B^2}$.

The student's answer is incorrect; it is in no way equivalent to the printed answer. At this point the student should try to understand why he has made this mistake after which he should continue with the succeeding frames.

CONTENTS

PART I

*Note: Chapters 45-48 as well as the Supplementary Index pertain to the "extended version" of "Fundamentals of Physics".

PART I

Chapter 1: MEASUREMENT

1-1 Why Formulas?

In a course in physics you are often asked to remember a number of relationships among physical quantities which are often called formulas (more properly formulae). Why is this? Well, first of all, remembering a formula of physics is in a sense no different from remembering a fact of history or an economic principle. Formulas allow you to reproduce some fact of nature without the necessity of repeating some series of laboratory experiments, just as it is not necessary to re-do the War of 1812 or the 1929 market crash. In history or economics one can combine facts to draw conclusions and anticipate "new facts" but, because the number of variables is very large and the relationships quite complex, this usually requires great experience and the "new facts" are not usually certainties. In physics, by contrast, the formulas which you are asked to remember are really very simple and they may be combined to give an unambiguous prediction of other facts.

There are two sorts of formulas you will be asked to learn. First of all there are definition formulas; these state what we mean by a certain physical quantity. For example, if an object travels a certain distance L in a time interval T, its average speed, by definition, is $v = L/T$. The other types of formulas express the "laws of nature" as we understand them. At first it may seem that there are very many formulas but actually there are very few basic or first principle relationships. Your teacher, the text, and this study guide all try to indicate which are basic.

One of the very beautiful things about physics is that formulas for different physical quantities are often very similar in appearance. For example, the electric force between a pair of point charges q_1 and q_2 separated by a distance r is $F = \text{constant} \times q_1q_2/r^2$ and the gravitational force between a pair of point masses m_1 and m_2 separated by a distance r is $F = \text{constant} \times m_1m_2/r^2$. The constants are different, of course, and the origin of the forces are different but the equations are essentially the same. You should be alert to these similarities and exploit them as an aid in remembering the formulas.

1-2 Units

Every physical quantity is expressed as a number of its units. In your text and this study guide we use the SI system of units. In this system the unit of length is the meter (m), that of mass is the kilogram (kg) and that of time is the second (s). Another common system of units is the British Engineering system in which the unit of length is the foot, that of time is the second, and instead of a unit of mass a unit of force, the pound, is used. Sometimes you will need to convert from one system to another. The conversion factors are given as an appendix to your text. You can do the conversion yourself if you treat the units as algebraic quantities. For example, one mile is 1760 yards, and one yard is in turn 0.914 meters. Therefore a velocity of one mile per hour is

$$1\,\frac{\text{mile}}{\text{hour}} = 1\,\frac{\text{mile}}{\text{hour}} \times 1760\,\frac{\text{yard}}{\text{mile}} \times 0.914\,\frac{\text{meter}}{\text{yard}} \times \frac{1}{3600}\,\frac{\text{hour}}{\text{second}} = 0.447\,\frac{\text{meter}}{\text{second}}$$

Notice how the units cancel and combine.

Since you will work mostly in the SI system, you really will not have to carry the units in an equation provided everything in the equation is expressed in SI units. For example, one formula is: Work = Force × Distance. In the SI system the unit of force is the newton and that of distance is the meter.

The unit of work is the joule, so 1 joule = 1 newton-meter. If you express the force in newtons and the distance in meters you know the answer must be in joules. At the beginning, however, it is best to carry along the units and treat them algebraically because this will serve as a check on your use of the equation in question. We will follow this practice in this study guide.

Some fractions and powers of units have names as for example 10^{-2} meter = 1 cm and 10^3 meter = 1 kilometer. In using such lengths in a formula such as Work = Force × Distance, if you want the work in joules you will need to express the distance in meters.

1-3 Use of Powers of 10 and Significant Figures

Often one may need to express a physical quantity as a small fraction of its unit or as a huge number of its unit. We do this through powers of 10. For example: 1½ thousandths of a meter is written as .0015 meter or 1.5×10^{-3} meter. Similarly, fifteen thousand meters is 15 000 meters or 1.5×10^4 meters. It is good practice in numerical problems to convert all numbers to the form a.bc... × $10^{\text{some power}}$. Then all powers of 10 can be collected together. If a number is written as a.bc... × 10^p with $a \neq 0$, then the number of significant figures is the number of figures to the right of the decimal plus one. For example 3.14 is three significant figures. In any given problem each piece of data may be given to a different number of significant figures. When multiplying (or dividing) such numbers, the answer can sensibly only be valid to the least of these. You might be able to grind out more figures but they will have no significance.

The following example illustrates the ideas of this chapter.

>>> Example 1. A certain man's beard grows at the rate of .031 inch per day (about 1/32 inch per day). What is the growth rate, G, expressed in Angstroms per minute? (The Angstron, Å, is a unit of length: 1 Å = 10^{-10} m.)

We must convert from inches to Angstroms and from days to minutes. Thus

$$G = \left(0.031\ \frac{\text{inch}}{\text{day}}\right)\left(2.54 \times 10^{-10}\ \frac{\text{m}}{\text{inch}}\right)\left(\frac{1\ \text{Å}}{10^{-10}\ \text{m}}\right)\left(\frac{1\ \text{day}}{24\ \text{hr}}\right)\left(\frac{1\ \text{hr}}{60\ \text{min}}\right)$$

$$= 5.5 \times 10^3\ \text{Å/min}\quad .$$

Interestingly 5500 Å is a typical wavelength of light in the green part of the visible spectrum!

Notice that each conversion factor (they appear in parentheses in the above formula) is equal to one. For example, since 1 hour = 60 minutes, (1 hr/60 min) = 1. How do we know to multiply by (1 hr/60 min) = 1 instead of say (60 min/1 hr) = 1? It's simple! We want the "hr" units to cancel out in the final result. We retain only two significant figures in the answer since the data (0.031 inch/day) is given to only two significant figures. In this regard, the conversion factors (such as 1 hr = 60 min) are to be regarded as infinitely accurate (i.e. 1.000... hr = 60.000... min). <<<

Chapter 2: VECTORS

REVIEW AND PREVIEW

In your past experience you have dealt with equations in which the quantities represented numbers (called scalars). In physics, many quantities have both a magnitude (numerical value) and a direction; these quantities are called vectors. In this chapter you will learn how to algebraically manipulate vector quantities.

GOALS AND GUIDELINES

In this chapter you have two major goals:

1. Learning the definitions
   a. multiplication of a vector by a scalar (Section 2-4),
   b. vector addition and subtraction (Section 2-5),
   c. the two types of vector multiplication (Section 2-6).
2. Learning to use unit vectors, particularly those for a Cartesian coordinate system (Section 2-7).

Although goal number 2 is more important for problem application, it is not possible to have a sound understanding of vectors without mastering goal number 1 first.

2-1 Introduction

In this chapter we shall give a very brief outline of vectors and some of their properties together with some examples and programmed problems. If after reading your text and working this chapter you feel need of additional instruction we recommend "A Programmed Introduction to Vectors" by Robert A. Carman, John Wiley and Sons, Inc.

2-2 Scalars

Quantities which may be represented by a number, a sign, and a unit are called scalars. Some examples are mass, density, and energy.

2-3 Vectors

Quantities which require the specification of a magnitude (how much) and a direction (which way) are called vectors; some examples are displacement, force, and velocity. Graphically a vector $\underline{\underline{A}}$ is represented by an arrow whose direction is that of the vector it represents and whose length corresponds to the magnitude, $|\underline{\underline{A}}|$, of $\underline{\underline{A}}$.* The magnitude of a vector is by definition an intrinsically positive quantity and need not have the dimensions of length. When we say the length 'corresponds to the magnitude' we mean to some chosen scale. Example: A one centimeter arrow representing a velocity might correspond to a velocity magnitude of 7 meters per second. If $\underline{\underline{A}}$ points into the plane of the paper it is denoted by a cross (×) symbolizing the arrow tail and if $\underline{\underline{A}}$ points out of the paper plane by a dot (·) to symbolize the arrow head. In these cases its length cannot be indicated.

---

*In this Study Guide, vectors are indicated by a double underline: $\underline{\underline{A}}$. In the text, vectors are indicated by bold-face printing. You may choose to indicate a vector with an overarrow: $\vec{A}$. This latter notation is convenient for hand written work.

2-4 Multiplication of a Vector by a Scalar, Negative of a Vector

A vector may be multiplied by a scalar. For example, $\underline{\underline{v}}t = \underline{\underline{s}}$ [$\underline{\underline{v}}$ is velocity, t is time, $\underline{\underline{s}}$ is displacement] is such a product. The magnitude of $\underline{\underline{s}}$ is

$$|\underline{\underline{s}}| = |\underline{\underline{v}}t| = |\underline{\underline{v}}|\,|t|$$

which means the absolute magnitude of the scalar t is taken -- i.e. without sign. The dimensions and units of $\underline{\underline{s}}$ are those of $\underline{\underline{v}}$ times those of t. If t is (positive/negative) then $\underline{\underline{s}}$ points in the (same/opposite) direction as $\underline{\underline{v}}$. For example, let $\underline{\underline{v}}$ be a velocity of 10 meters/second in the direction east, and let t be 5 seconds. Then

$$|\underline{\underline{s}}| = 10\ \frac{\text{meters}}{\text{second}}\ 5\ \text{seconds} = 50\ \text{meters}$$

and the direction of $\underline{\underline{s}}$ is east.

The special case of multiplication by -1 forms the negative of a vector. That is $-\underline{\underline{A}}$ has the same magnitude as $\underline{\underline{A}}$ but is oppositely directed. Unlike scalars, vectors are neither positive nor negative. If $\underline{\underline{A}} = -\underline{\underline{B}}$ then $\underline{\underline{B}}$ is the negative of $\underline{\underline{A}}$ and $\underline{\underline{A}}$ is the negative of $\underline{\underline{B}}$, but neither is in itself positive nor negative.

2-5 Vector Addition and Subtraction

Vectors may be added graphically as illustrated in Fig. 2-1a where the vectors to be added ($\underline{\underline{A}}$ and $\underline{\underline{B}}$) are drawn head to tail and their sum ($\underline{\underline{A}} + \underline{\underline{B}}$) represented by the arrow drawn from the tail of the first to the head of the second. The order of addition is immaterial; that is $\underline{\underline{A}} + \underline{\underline{B}} = \underline{\underline{B}} + \underline{\underline{A}}$. For more than two vectors the procedure is illustrated in Fig. 2-1b.

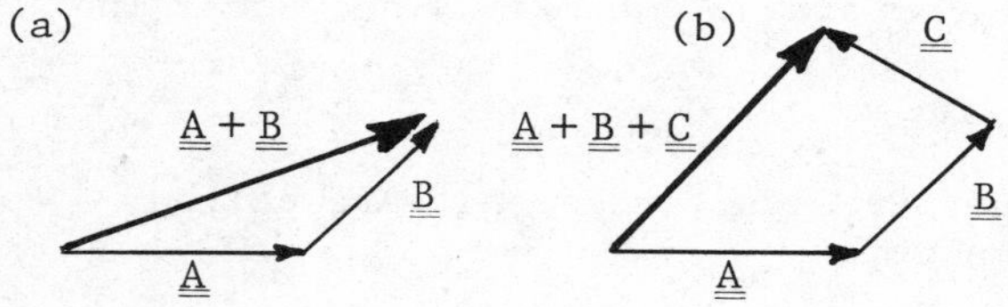

Figure 2-1. (a) The addition of two vectors $\underline{\underline{A}}$ and $\underline{\underline{B}}$. (b) The addition of three vectors $\underline{\underline{A}}$, $\underline{\underline{B}}$, and $\underline{\underline{C}}$.

Vector subtraction is expressed in terms of vector addition and the negative of a vector. The operation $\underline{\underline{A}} - \underline{\underline{B}}$ means $\underline{\underline{A}} + (-\underline{\underline{B}}) = \underline{\underline{A}} + (-1)\underline{\underline{B}}$. The subtraction process is shown in Fig. 2-2. Start with $\underline{\underline{A}}$ and $\underline{\underline{B}}$, form $-\underline{\underline{B}}$ and add it to $\underline{\underline{A}}$.

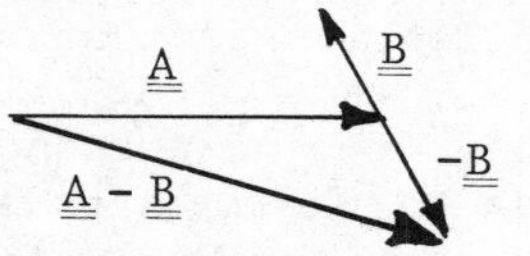

Figure 2-2. Vector subtraction.

2-6 Vector Multiplication

Two kinds of vector multiplication are of interest. The first is the <u>scalar</u> or <u>dot product</u>, and as the name implies the result is a scalar. This product is denoted by $\underline{\underline{A}}\cdot\underline{\underline{B}}$ and is defined by

$$\underline{\underline{A}}\cdot\underline{\underline{B}} = |\underline{\underline{A}}|\,|\underline{\underline{B}}|\cos\theta$$

where θ is the angle between $\underline{\underline{A}}$ and $\underline{\underline{B}}$ as illustrated in Fig. 2-3. Notice that $|\underline{\underline{A}}|\cos\theta$ is the projection of the length of the arrow representing $\underline{\underline{A}}$ onto the direction of $\underline{\underline{B}}$ and similarly $|\underline{\underline{B}}|\cos\theta$ is the projection of the length of the arrow of $\underline{\underline{B}}$ onto the direction of $\underline{\underline{A}}$.

If $\underline{A}$ and $\underline{B}$ are perpendicular ($\theta = 90^o$), then $\underline{A} \cdot \underline{B} = 0$, and such vectors are said to be orthogonal. Also $\underline{A} \cdot \underline{A} = |\underline{A}|^2$ gives the square of the magnitude of $\underline{A}$; it is commonly denoted also by $\underline{A}^2$ or A^2. Often for convenience, the magnitude of a vector $\underline{A}$ will be denoted merely by A.

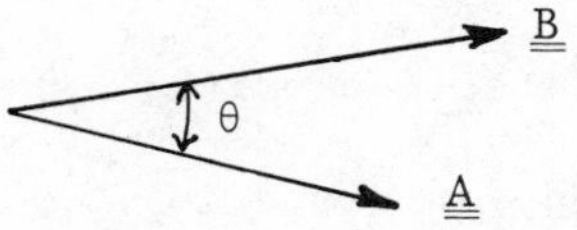

Figure 2-3. Angle between two vectors.

The other kind of product is the <u>vector</u> or <u>cross</u> product denoted by $\underline{A} \times \underline{B}$ and as the name implies the product is itself a vector. By definition if

$$\underline{C} = \underline{A} \times \underline{B}$$

then

$$|\underline{C}| = |\underline{A}|\,|\underline{B}| \sin \theta$$

and the direction of $\underline{C}$ is perpendicular to both $\underline{A}$ and $\underline{B}$ (hence to the plane they define) with its sense along this perpendicular that of a <u>right hand screw</u> when $\underline{A}$ is rotated into $\underline{B}$. This is illustrated in Fig. 2-4. Notice that if $\underline{A}$ and $\underline{B}$ are parallel $\underline{A} \times \underline{B} = 0$.

Figure 2-4. Cross product; $\underline{A} \times \underline{B}$ is directed into the plane of the paper.*

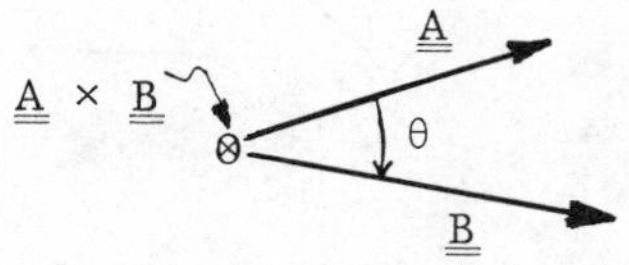

Some useful properties are that

$$\underline{A} \cdot \underline{B} = \underline{B} \cdot \underline{A}$$

and

$$\underline{A} \times \underline{B} = -(\underline{B} \times \underline{A})$$

each of which follows from the definitions. Only the vector product is defined for three or more vectors. For example $\underline{A} \cdot \underline{B} \cdot \underline{C}$ is meaningless, and $\underline{A} \times \underline{B} \times \underline{C}$ is defined only if one specifies by brackets which cross product is to be done first. To see this let us suppose that $\underline{A}$ and $\underline{B}$ have unit length each and are perpendicular (perpendicular vectors are also called orthogonal). Then $\underline{A} \times (\underline{A} \times \underline{B}) = -\underline{B}$ may be seen from Fig. 2-5, but $(\underline{A} \times \underline{A}) \times \underline{B} = 0$.

Figure 2-5. Triple cross product $\underline{A} \times (\underline{A} \times \underline{B})$ where $|\underline{A}| = |\underline{B}| = 1$, and $\underline{A}$ is perpendicular to $\underline{B}$; then $\underline{A} \times (\underline{A} \times \underline{B}) = -\underline{B}$. The $\underline{A} \times \underline{B}$ vector is directed into the page.

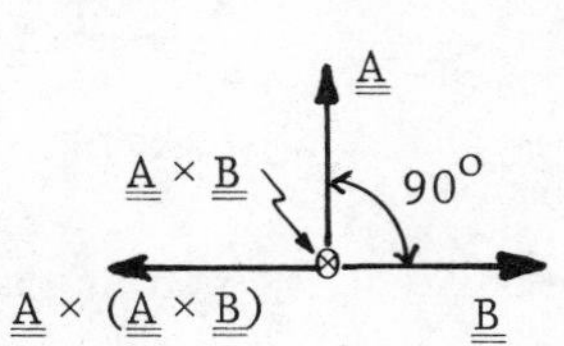

---

* The symbol ⊗ denotes a vector directed <u>into</u> the page ("tail" of the vector arrow), similarly the symbol ⊙ denotes a vector directed out of the page ("head" of the vector arrow).

2-7 Unit Vectors, Coordinate System, Vector Resolution

If $\underline{\underline{A}}$ is not a null vector, i.e. if $|\underline{\underline{A}}| \neq 0$, then the vector $\underline{\underline{A}}/|\underline{\underline{A}}|$ is a vector of unit magnitude whose direction is the same as that of $\underline{\underline{A}}$; it is denoted by $\underline{\underline{e}}_A$. A unit vector does not carry physical units.

>>> Example 1. Let $\underline{\underline{v}}$ denote 10 miles per hour in a direction north; express $\underline{\underline{v}}$ in terms of a unit north vector denoted by $\underline{\underline{e}}$; find $\underline{\underline{e}}$ in terms of $\underline{\underline{v}}$.

Solution: We write $\underline{\underline{v}} = v\underline{\underline{e}}$ where v is the magnitude of $\underline{\underline{v}}$ and is 10 miles per hour. We may find $\underline{\underline{e}}$ as a unit vector in the direction north by

$$\underline{\underline{e}} = \underline{\underline{v}}/v \quad . \qquad <<<$$

Now, any two non-parallel vectors (say $\underline{\underline{A}}$ and $\underline{\underline{B}}$) determine a plane and any vector in that plane can be written in terms of $\underline{\underline{A}}$ and $\underline{\underline{B}}$ as

$$\underline{\underline{C}} = a\underline{\underline{A}} + b\underline{\underline{B}} \quad .$$

This is proved by simple geometry. Since $\underline{\underline{A}}$ and $\underline{\underline{B}}$ are not parallel a parallelogram with $\underline{\underline{C}}$ as its diagonal and with sides parallel to $\underline{\underline{A}}$ and $\underline{\underline{B}}$ can be constructed as in Fig. 2-6.

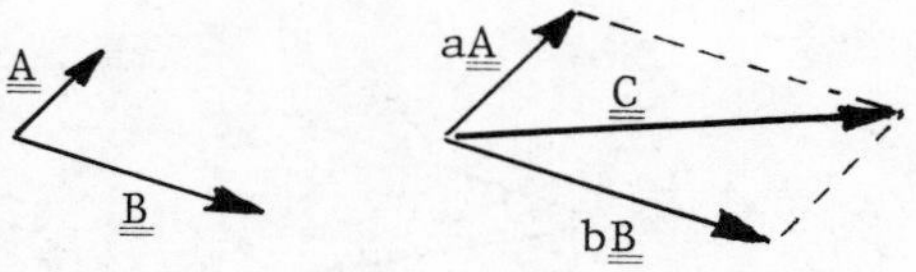

Figure 2-6. An arbitrary vector in a plane expressed in terms of two vectors that determine that plane; a and b are scalars (numbers).

Similarly in three dimensions, our common experience space, any vector can be written in terms of any three non-coplanar vectors. The three constitute a coordinate system. Obviously, unit vectors are of use here.

One of the most useful coordinate systems is the Cartesian system with unit vectors along the x, y and z axes as shown in Fig. 2-7. For these unit vectors the special symbols $\underline{\underline{i}}$, $\underline{\underline{j}}$ and $\underline{\underline{k}}$ are used and they have very beautiful properties

$$\underline{\underline{i}}\cdot\underline{\underline{i}} = \underline{\underline{j}}\cdot\underline{\underline{j}} = \underline{\underline{k}}\cdot\underline{\underline{k}} = 1 \quad ; \quad \underline{\underline{i}}\cdot\underline{\underline{j}} = \underline{\underline{i}}\cdot\underline{\underline{k}} = \underline{\underline{j}}\cdot\underline{\underline{k}} = 0$$

$$\underline{\underline{i}} \times \underline{\underline{j}} = \underline{\underline{k}} \quad , \quad \underline{\underline{j}} \times \underline{\underline{k}} = \underline{\underline{i}} \quad , \quad \underline{\underline{k}} \times \underline{\underline{i}} = \underline{\underline{j}} \quad .$$

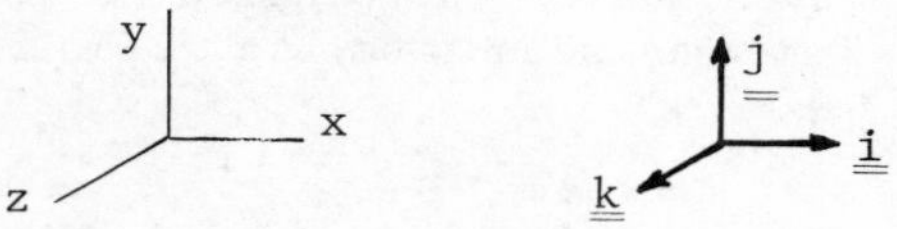

Figure 2-7. Cartesian coordinate system and unit vectors $\underline{\underline{i}}$, $\underline{\underline{j}}$, $\underline{\underline{k}}$.

Any vector $\underline{\underline{v}}$ in three dimensions may be written as

$$\underline{\underline{v}} = v_x\underline{\underline{i}} + v_y\underline{\underline{j}} + v_z\underline{\underline{k}}$$

where the components of v are given by

$$v_x = \underline{\underline{v}}\cdot\underline{\underline{i}} \quad , \quad v_y = \underline{\underline{v}}\cdot\underline{\underline{j}} \quad , \quad v_z = \underline{\underline{v}}\cdot\underline{\underline{k}} \quad .$$

We say that $\underline{v}$ has been resolved into components along the three coordinate axes. Depending upon which way $\underline{v}$ is directed the components may be either positive or negative.

In physics many of the equations you will encounter are written in vector form and as such are independent of your choice of coordinate system. The solutions are more often most easily obtained by selecting a coordinate system and working with the three components.

>>> Example 2. Given two vectors $\underline{\underline{A}} = 4\underline{\underline{i}} + 3\underline{\underline{j}}$, $\underline{\underline{B}} = -2\underline{\underline{i}} + 6\underline{\underline{j}}$:

a) Find the vector which is three times as long as $\underline{\underline{A}}$,
b) Find $\underline{\underline{A}} + \underline{\underline{B}}$ and $\underline{\underline{A}} - \underline{\underline{B}}$,
c) Find $\underline{\underline{A}}\cdot\underline{\underline{B}}$,
d) Find the magnitude of $\underline{\underline{A}}$,
e) Find $\underline{\underline{A}} \times \underline{\underline{B}}$,
f) Find a unit vector in the direction of $\underline{\underline{B}}$,
g) Express $\underline{\underline{A}} \times (\underline{\underline{A}} \times \underline{\underline{B}})$ in terms of $\underline{\underline{A}}$ and $\underline{\underline{B}}$.

Solution:

a) To do so we multiply $\underline{\underline{A}}$ by 3 to obtain

$$3\underline{\underline{A}} = 12\underline{\underline{i}} + 9\underline{\underline{j}} \quad .$$

b)
$$\underline{\underline{A}} + \underline{\underline{B}} = 4\underline{\underline{i}} + 3\underline{\underline{j}} - 2\underline{\underline{i}} + 6\underline{\underline{j}} = 2\underline{\underline{i}} + 9\underline{\underline{j}} \quad .$$

$$\underline{\underline{A}} - \underline{\underline{B}} = 4\underline{\underline{i}} + 3\underline{\underline{j}} + 2\underline{\underline{i}} - 6\underline{\underline{j}} = 6\underline{\underline{i}} - 3\underline{\underline{j}} \quad .$$

c)
$$\underline{\underline{A}}\cdot\underline{\underline{B}} = (4\underline{\underline{i}} + 3\underline{\underline{j}})\cdot(-\,2\underline{\underline{i}} + 6\underline{\underline{j}})$$

$$= (4)(-2)\ \underline{\underline{i}}\cdot\underline{\underline{i}} + (4)(6)\ \underline{\underline{i}}\cdot\underline{\underline{j}} + (3)(-2)\ \underline{\underline{j}}\cdot\underline{\underline{i}} + (3)(6)\ \underline{\underline{j}}\cdot\underline{\underline{j}}$$

$$= -\,8 + 0 + 0 + 18 = 10 \quad .$$

d)
$$\underline{\underline{A}}\cdot\underline{\underline{A}} = |\underline{\underline{A}}|^2 = 16 + 9 = 25 \quad ; \qquad |\underline{\underline{A}}| = \sqrt{25} = 5 \quad .$$

Notice that the positive square root is taken.

e)
$$\underline{\underline{A}} \times \underline{\underline{B}} = (4\underline{\underline{i}} + 3\underline{\underline{j}}) \times (-\,2\underline{\underline{i}} + 6\underline{\underline{j}})$$

$$= -\,8\ (\underline{\underline{i}} \times \underline{\underline{i}}) + 24\ (\underline{\underline{i}} \times \underline{\underline{j}}) - 6\ (\underline{\underline{j}} \times \underline{\underline{i}}) + 18\ (\underline{\underline{j}} \times \underline{\underline{j}})$$

$$= 24\underline{\underline{k}} - 6(-\underline{\underline{k}}) = 30\underline{\underline{k}} \quad .$$

f)
$$|\underline{\underline{B}}|^2 = 4 + 36 = 40 \quad ; \qquad |\underline{\underline{B}}| = \sqrt{40}$$

$$\underline{\underline{e}}_B = \underline{\underline{B}}/|\underline{\underline{B}}| = -\,\frac{2}{\sqrt{40}}\,\underline{\underline{i}} + \frac{6}{\sqrt{40}}\,\underline{\underline{j}} \quad .$$

g) From e) $\underline{\underline{A}} \times \underline{\underline{B}}$ (which is to be done first) is $\underline{\underline{A}} \times \underline{\underline{B}} = 30\underline{\underline{k}}$. Therefore

$$\underline{\underline{A}} \times (\underline{\underline{A}} \times \underline{\underline{B}}) = (4\underline{\underline{i}} + 3\underline{\underline{j}}) \times 30\underline{\underline{k}}$$

$$= -\,120\underline{\underline{j}} + 90\underline{\underline{i}} \quad .$$

This is to be written as

$$\underline{\underline{A}} \times (\underline{\underline{A}} \times \underline{\underline{B}}) = \alpha\underline{\underline{A}} + \beta\underline{\underline{B}} = \alpha(4\underline{\underline{i}} + 3\underline{\underline{j}}) + \beta(-2\underline{\underline{i}} + 6\underline{\underline{j}})$$

where α and β are to be found. Thus, we have

$$(4\alpha - 2\beta)\ \underline{\underline{i}} + (3\alpha + 6\beta)\ \underline{\underline{j}} = -\ 120\underline{\underline{j}} + 90\underline{\underline{i}} \quad .$$

Because $\underline{\underline{i}}$ and $\underline{\underline{j}}$ are not parallel the components may be set equal and we have

$$4\alpha - 2\beta = 90 \qquad [\text{coefficients of } \underline{\underline{i}}]$$

$$3\alpha + 6\beta = -\ 120 \qquad [\text{coefficients of } \underline{\underline{j}}]$$

from which $\alpha = 10$, $\beta = -\ 25$. Thus $\underline{\underline{A}} \times (\underline{\underline{A}} \times \underline{\underline{B}}) = 10\underline{\underline{A}} - 25\underline{\underline{B}}$. <<<

2-8 Some Words of Caution

One must not jump to erroneous conclusions when using vectors and in particular vector products. At first it is best to return to first principles (the definitions and simple conclusions) to check. For example if $\underline{\underline{A}}\cdot\underline{\underline{B}} = 0$, it does not follow that $\underline{\underline{A}}$ and $\underline{\underline{B}}$ are necessarily perpendicular although the converse is true. The possibilities are

i) $|\underline{\underline{A}}| = 0$

ii) $|\underline{\underline{B}}| = 0$ or

iii) $\underline{\underline{A}}$ is perpendicular to $\underline{\underline{B}}$.

Similarly, just because $\underline{\underline{A}}\cdot\underline{\underline{B}} = \underline{\underline{A}}\cdot\underline{\underline{C}}$, do not conclude that $\underline{\underline{C}}$ is necessarily equal to $\underline{\underline{B}}$. The equation merely asserts that the projection of $\underline{\underline{C}}$ onto the direction $\underline{\underline{e}}_A$ is the same as that of $\underline{\underline{B}}$.

2-9 Programmed Problems

1.

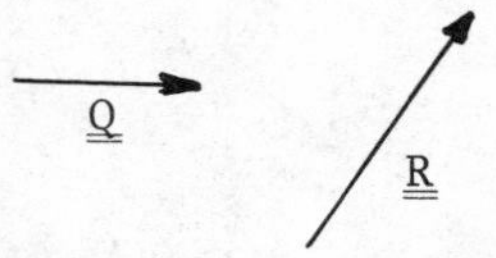

Consider the two vectors represented in the diagram. How do they differ?

(a) Different magnitude (shown as different length).
(b) Different direction.

---

2. The pictorial representation of the vector $\underline{\underline{R}}$ completely specifies the ________ and ________ of $\underline{\underline{R}}$ once you have selected a suitable scale and reference direction.

Magnitude

Direction

3.

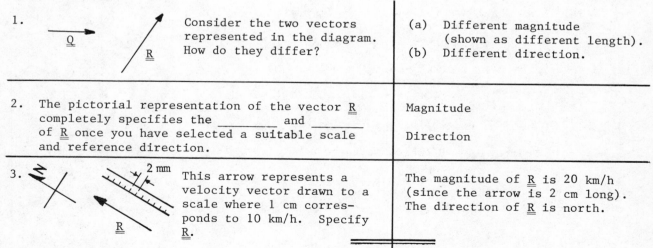

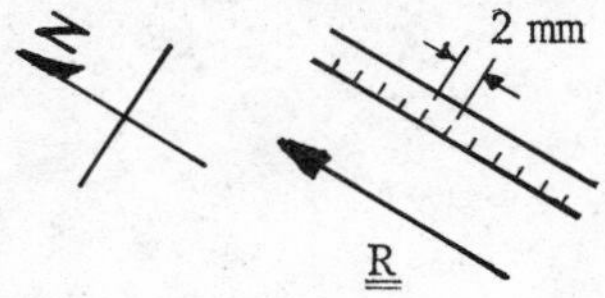

This arrow represents a velocity vector drawn to a scale where 1 cm corresponds to 10 km/h. Specify $\underline{\underline{R}}$.

The magnitude of $\underline{\underline{R}}$ is 20 km/h (since the arrow is 2 cm long). The direction of $\underline{\underline{R}}$ is north.

4. 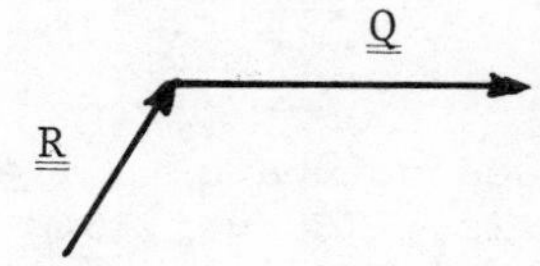

Vectors may be added by the "tail to head" method. Draw on the diagram to the left the vector (R+Q).

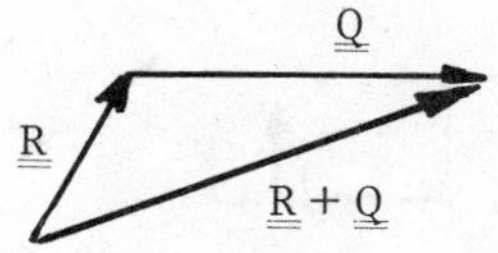

---

5. 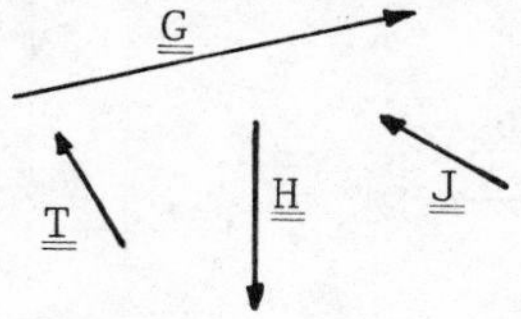

Graphically determine the vector (T+G+H+J). Use the same scale and orientation as shown.

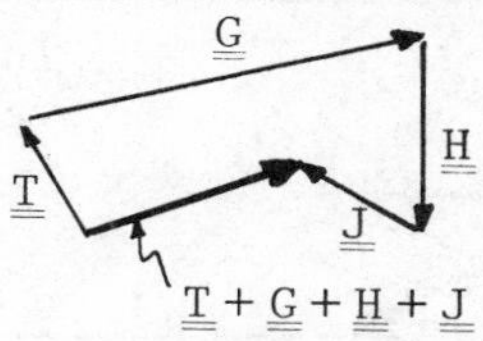

---

6. 

The vector -Q is the same as Q, except oppositely directed. Draw the vector -Q.

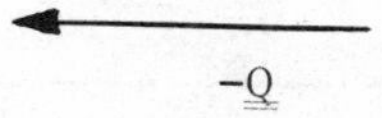

---

7. 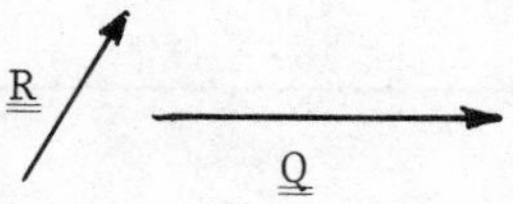

In frame 4 above you graphically determined R + Q. To subtract Q from R we add -Q to R.

Find R + (-Q) graphically.

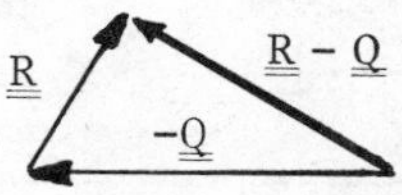

Compare with the answer in frame 4.

---

8. Having reviewed the geometrical method of addition and subtraction we can now look at the analytical method. Here the idea is to resolve vectors into their components with respect to some suitable reference frame. The utility of this method is that one can work with vectors using the ordinary rules of algebra.

What are the unit vectors associated with the x, y and z coordinates respectively of a Cartesian coordinate system?

i, j, k

---

9. 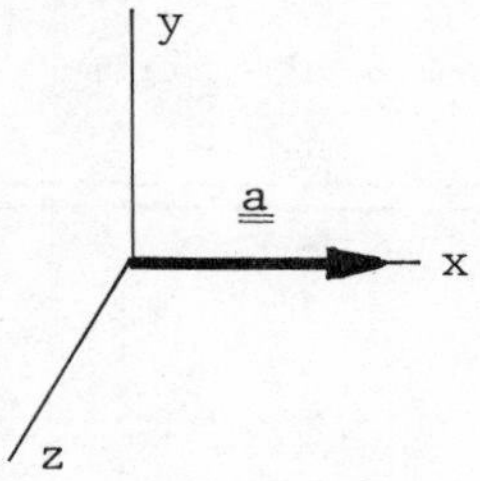

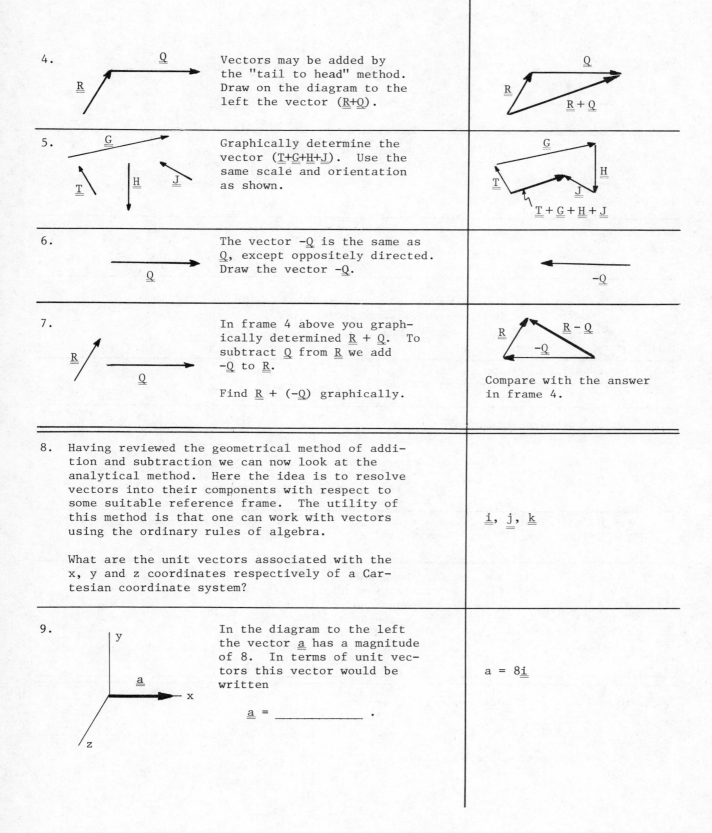

In the diagram to the left the vector a has a magnitude of 8. In terms of unit vectors this vector would be written

a = ____________ .

a = 8i

10. 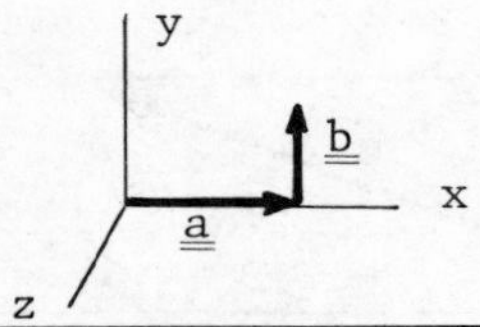

Here we add a new vector $\underline{\underline{b}}$ of magnitude 6.

$\underline{\underline{b}}$ = ________ .

$\underline{\underline{b}} = 6\underline{\underline{j}}$

Note that $\underline{\underline{b}}$ does not have to lie along the y-axis. It must, of course, be parallel to the y axis.

---

11. 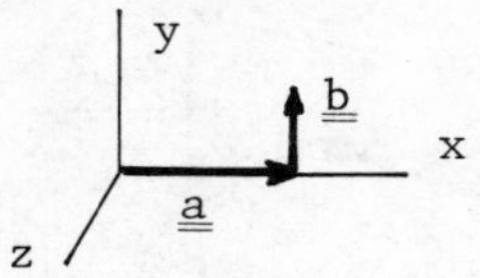

In the diagram show the vector $\underline{\underline{c}} = \underline{\underline{a}} + \underline{\underline{b}}$.

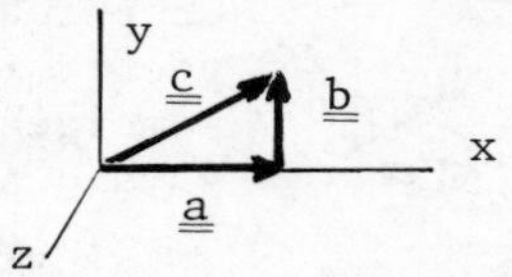

---

12. Since $\underline{\underline{a}} = 8\underline{\underline{i}}$ and $\underline{\underline{b}} = 6\underline{\underline{j}}$

$\underline{\underline{c}}$ = ________ ?

$\underline{\underline{c}} = 8\underline{\underline{i}} + 6\underline{\underline{j}}$

---

13. 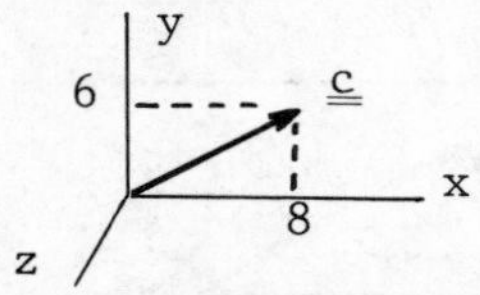

Turning this problem around we can say that $\underline{\underline{c}}$ has the vector components ________ and ________.

$8\underline{\underline{i}}$, $6\underline{\underline{j}}$

---

14. In our particular example of $\underline{\underline{c}} = 8\underline{\underline{i}} + 6\underline{\underline{j}}$ what are the scalar components of $\underline{\underline{c}}$?

8 , 6

---

15. What is the magnitude of $\underline{\underline{c}}$?

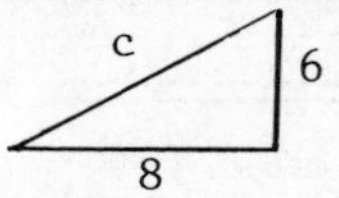

From theorem of Pythagoras.

$c = \sqrt{8^2 + 6^2} = 10$

---

16. Have we completely specified the vector $\underline{\underline{c}}$ now that we know its magnitude is 10?

No. Its direction must also be specified.

---

17. 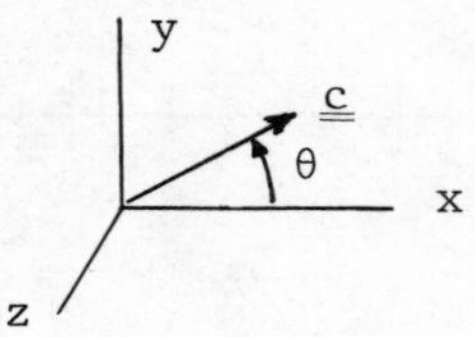

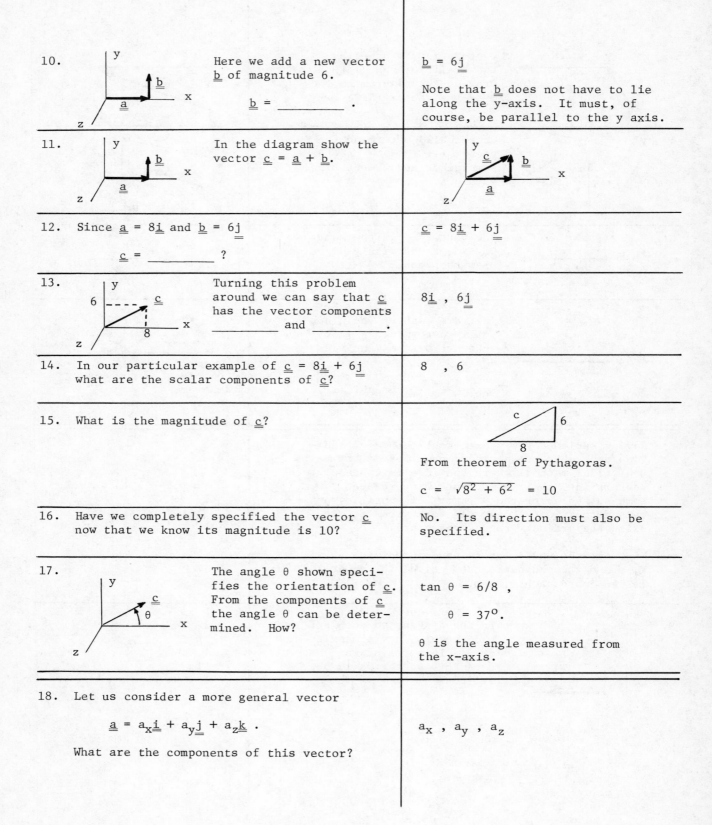

The angle θ shown specifies the orientation of $\underline{\underline{c}}$. From the components of $\underline{\underline{c}}$ the angle θ can be determined. How?

$\tan\theta = 6/8$,

$\theta = 37^{\circ}$.

θ is the angle measured from the x-axis.

---

18. Let us consider a more general vector

$$\underline{\underline{a}} = a_x\underline{\underline{i}} + a_y\underline{\underline{j}} + a_z\underline{\underline{k}} .$$

What are the components of this vector?

a_x , a_y , a_z

19. What is the magnitude of $\underline{a}$?

$|\underline{a}| = \sqrt{a_x^2 + a_y^2 + a_z^2}$

---

20. 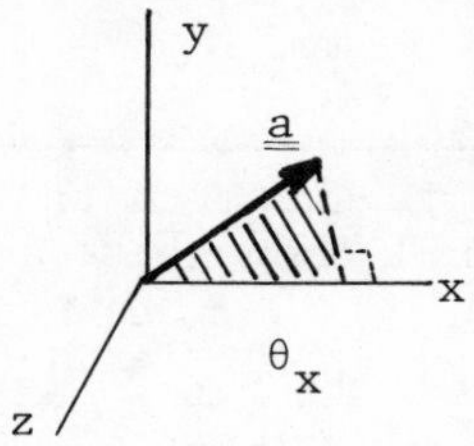

For this vector it is a little more complicated to specify the direction but do not give up yet. The idea of giving an angle between the vector and the coordinate axis will be retained.

In the diagram the vector $\underline{a}$ is not in the xy plane. The dotted line is perpendicular to the x-axis. How can a_x be determined from $|\underline{a}|$ and θ_x?

$a_x = |\underline{a}| \cos \theta_x$

Note that the angle θ_x lies in shaded plane in the illustration.

---

21. We write

$$\cos \theta_x = \frac{a_x}{|\underline{a}|} = \frac{a_x}{\sqrt{(a_x^2 + a_y^2 + a_z^2)}} .$$

This gives the orientation of $\underline{a}$ with respect to the x-axis. Write the expression for the angle θ_y with respect to the y-axis.

$$\cos \theta_y = \frac{a_y}{\sqrt{(a_x^2 + a_y^2 + a_z^2)}}$$

Similarly we could find $\cos \theta_z$.

---

22. 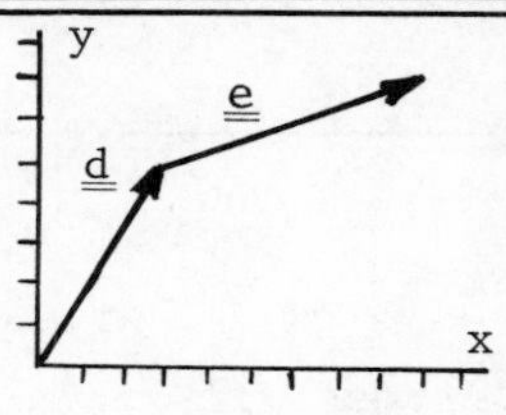

In the diagram to the left determine the components d_x, d_y, e_x and e_y by constructing suitable perpendiculars to the axes. Obtain a number in terms of the axes divisions.

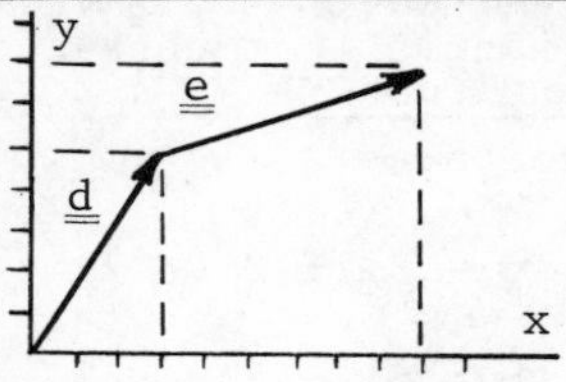

$d_x = 3$, $d_y = 5$

$e_x = 6$, $e_y = 2$

---

23. Write the vector equations for $\underline{d}$ and $\underline{e}$ using the vectors $\underline{i}$ and $\underline{j}$.

$\underline{d} = 3\underline{i} + 5\underline{j}$

$\underline{e} = 6\underline{i} + 2\underline{j}$

---

24. 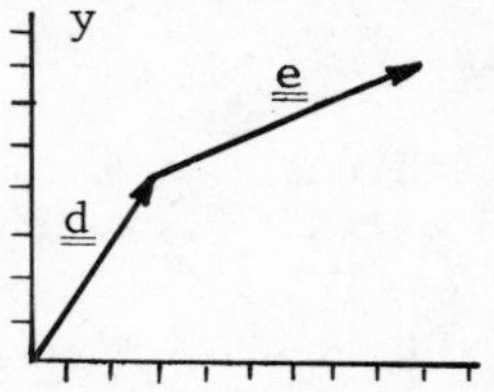

In the figure to the left draw the vector $\underline{f} = \underline{d} + \underline{e}$ and determine the components of $\underline{f}$.

f_x = ______ , f_y = ______ .

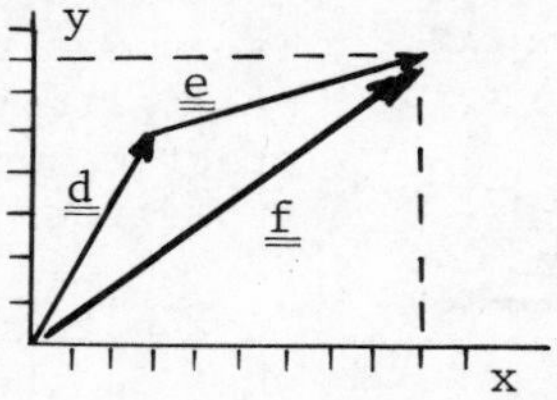

$f_x = 9$, $f_y = 7$

25. Write the vector equations for $\mathbf{d}$, $\mathbf{e}$, and $\mathbf{f}$.

$\mathbf{d} = 3\mathbf{i} + 5\mathbf{j}$

$\mathbf{e} = 6\mathbf{i} + 2\mathbf{j}$

$\mathbf{f} = 9\mathbf{i} + 7\mathbf{j}$

---

26. Looking again at the answer in frame 25 state the rule for adding vectors when they are written in Cartesian component form.

The vector sum is equal to the sum of the individual components.

$6\mathbf{i} + 3\mathbf{i} = 9\mathbf{i}$, $5\mathbf{j} + 2\mathbf{j} = 7\mathbf{j}$

---

27. Vectors when resolved are added by using the ordinary rules of algebra on their scalar components.

Problem: We want to show that the vectors

$$\mathbf{a} = 3\mathbf{i} + 2\mathbf{j} - 7\mathbf{k}$$

$$\mathbf{b} = 5\mathbf{i} + 6\mathbf{j} - 5\mathbf{k}$$

$$\mathbf{c} = 2\mathbf{i} + 4\mathbf{j} + 2\mathbf{k}$$

from a right triangle. First, what is the geometrical requirement for the formation of any triangle from three vectors?

Two of the vectors must have a resultant equal to the third. This ensures a closed figure.

---

28. For the vectors

$$\mathbf{a} = 3\mathbf{i} + 2\mathbf{j} - 7\mathbf{k}$$

$$\mathbf{b} = 5\mathbf{i} + 6\mathbf{j} - 5\mathbf{k}$$

$$\mathbf{c} = 2\mathbf{i} + 4\mathbf{j} + 2\mathbf{k}$$

how is the requirement for a closed figure satisfied?

$\mathbf{a} + \mathbf{c} = \mathbf{b}$

This does not necessarily mean a right triangle.

29.

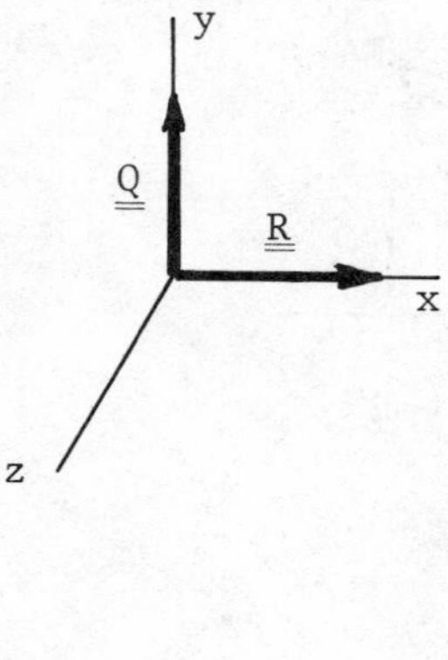

We could look at the sum of the squares of the vector magnitudes, but there is an easier way which will give you additional practice with vector manipulation.

In the figure to the left what is the scalar product of the vectors $\mathbf{Q}$ and $\mathbf{R}$?

Zero.

$\mathbf{Q}\cdot\mathbf{R} = |\mathbf{Q}|\,|\mathbf{R}| \cos\theta$

where $\theta = \pi/2$ and $\cos\theta = 0$.

30. The test then is to find which two vectors have a scalar product of zero. In terms of the scalar components

$$\mathbf{a}\cdot\mathbf{b} = a_x b_x + a_y b_y + a_z b_z \ .$$

For

$$\mathbf{a} = 3\mathbf{i} + 2\mathbf{j} - 7\mathbf{k}$$

$$\mathbf{b} = 5\mathbf{i} + 6\mathbf{j} - 5\mathbf{k}$$

$$\mathbf{c} = 2\mathbf{i} + 4\mathbf{j} + 2\mathbf{k}$$

find $\mathbf{a}\cdot\mathbf{b}$, $\mathbf{b}\cdot\mathbf{c}$, and $\mathbf{a}\cdot\mathbf{c}$.

$$\mathbf{a}\cdot\mathbf{b} = (3 \times 5) + (2 \times 6) + (7 \times 5) = 62 \ .$$

$$\mathbf{b}\cdot\mathbf{c} = (5 \times 2) + (6 \times 4) - (5 \times 2) = 24 \ .$$

$$\mathbf{a}\cdot\mathbf{c} = (3 \times 2) + (2 \times 4) - (7 \times 2) = 0 \ .$$

---

31. The conditions for the formation of the right triangle are

$$\mathbf{a} + \mathbf{c} = \mathbf{b}$$

$$\mathbf{a}\cdot\mathbf{c} = 0 \ .$$

Draw these vectors appropriately to form a right triangle. Choose your own scale.

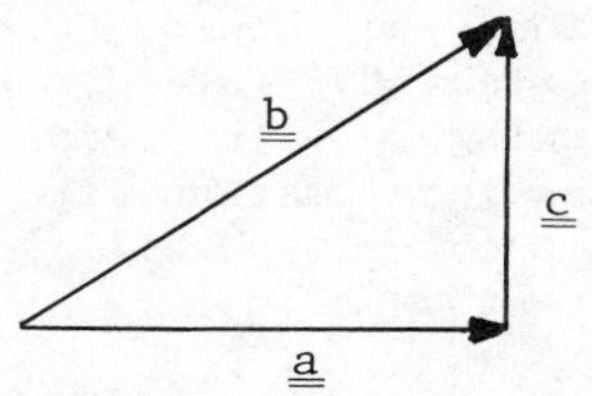

or similarly.

Chapter 3: MOTION IN ONE DIMENSION

REVIEW AND PREVIEW

In the previous chapter you have studied formal properties of vectors and their manipulation. Now, several vector physical quantities dealing with the motion of a particle will be defined. Examples are the position, velocity, and acceleration of a particle. This chapter is concerned with a description of particle motion (kinematics) by which the above physical quantities are related to one another.

GOALS AND GUIDELINES

In this chapter you have two major goals:

1. Learning and understanding the definitions pertaining to particle motion (Section 3-1). Of prime importance are those of position, velocity, (both average and instantaneous) and acceleration (both average and instantaneous). Although the main concern of the rest of this chapter is one-dimensional motion, you should learn the definitions in their full vector form. This is important for future study (e.g. motion in a plane).
2. Learning well the meaning and use of the equations for the case of one-dimensional motion with constant acceleration (Eq. 3-7). A large number of the problems you will solve turn out to involve this special but very important case.

3-1 Introduction

The following terms and definitions must be learned and understood before you can profitably undertake the problems, i.e. without being guilty of "formula plugging"!

Kinematics --- The description of the motions of physical objects without regard to that which causes the motion.

Position vector --- The vector from some chosen origin to a point of interest. It is denoted by $\mathbf{r}$ and in terms of the Cartesian coordinates x, y and z of the point and the Cartesian unit vectors $\mathbf{i}$, $\mathbf{j}$ and $\mathbf{k}$ may be written as

$$\mathbf{r} = x\mathbf{i} + y\mathbf{j} + z\mathbf{k} \quad . \qquad (3\text{-}1)$$

Displacement --- The change in the position vector of a point fixed in a physical object as illustrated in Fig. 3-1.

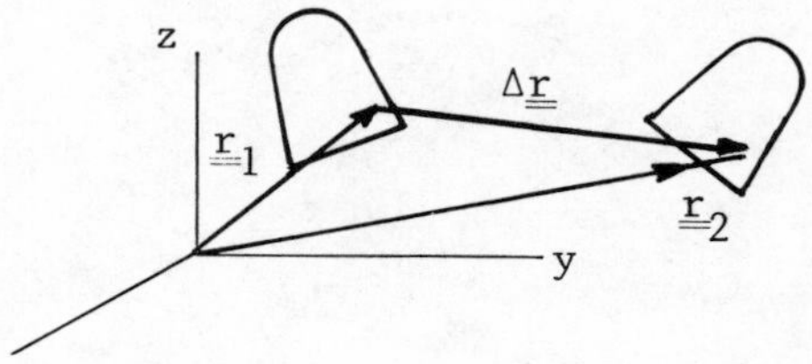

Figure 3-1. Illustration of the displacement vector $\Delta\mathbf{r}$ of a point in a physical object.

Path length --- The actual distance travelled by the particle along its path of motion.

Translational motion --- That motion of a physical object in which every point in the object undergoes precisely the same displacement in a given interval of time.

Particle --- A physical object whose extent is negligible in comparison with the relevant lengths of its environment; mathematically an object which has only one physically distinct point. For some purposes even a large object may be treated as a particle.

Average velocity --- The ratio of the displacement $\Delta\underline{r}$ of a particle to the time interval Δt over which the displacement occurs. It is denoted by $\bar{\underline{v}}$ and is expressed

$$\bar{\underline{v}} = \frac{\Delta\underline{r}}{\Delta t} = \frac{\underline{r}_2 - \underline{r}_1}{t_2 - t_1} \quad ; \qquad t_2 > t_1 \tag{3-2}$$

where 1 and 2 label corresponding position vectors and times. Several important properties are

a) $\bar{\underline{v}}$ is a vector since $1/\Delta t$ is a scalar and $\Delta\underline{r}$ is a vector; its direction is that of $\Delta\underline{r}$;
b) $\Delta\underline{r}$ is independent of the choice of origin although $\underline{r}_2$ and $\underline{r}_1$ are not;
c) nothing is said of the actual path of the particle from 1 to 2;
d) $\bar{\underline{v}}$ is not completely specified until the time interval over which the average is to be taken is specified or understood;
e) the dimensions of $\bar{\underline{v}}$ are those of length/time (L/T) and some common units are meters/second (m/s), feet/second (ft/s), centimeters per second (cm/s), and miles per hour (mph);
f) $\bar{\underline{v}}$ may be zero even though the particle is moving continuously! All that is required is that $\Delta\underline{r}$ should be zero.

Average speed --- The path length divided by the time required to traverse the path. It is often denoted by $\bar{v}$.

Instantaneous velocity --- The limit of the average velocity as $\Delta t \to 0$ but keeping the initial time fixed. The word velocity alone means instantaneous velocity. The velocity $\underline{v}_1$ at time t_1 is

$$\underline{v}_1 = \lim_{t_2 \to t_1} \frac{\underline{r}_2 - \underline{r}_1}{t_2 - t_1} \quad .$$

Note that $\underline{v}_1$ depends upon t_1 but not on t_2. Mathematically the velocity is the derivative of the position vector $\underline{r}$ with respect to the time t. One then writes

$$\underline{v} = \lim_{\Delta t \to 0} \frac{\Delta\underline{r}}{\Delta t} = \frac{d\underline{r}}{dt} \tag{3-3}$$

which means the velocity at some general time t. Two important points are

a) the direction of $\underline{v}$ is tangent to the actual path of the particle (as shown in Fig. 3-2);
b) $\underline{v}$ has the same dimensions and common units as does $\bar{\underline{v}}$.

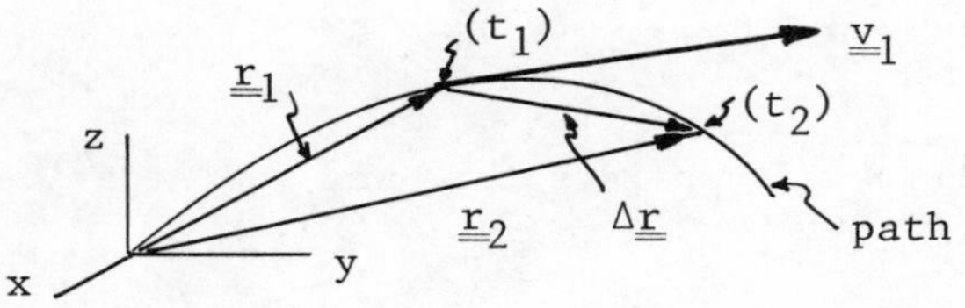

Figure 3-2. The velocity $\underline{v}_1$ at time t_1 is illustrated; $\underline{v}_1$ is tangent to the path followed by the particle in the direction indicated. The average velocity $\bar{\underline{v}}$ has the same direction as $\Delta\underline{r}$.

<u>Speed</u> --- The magnitude of the velocity; it is often denoted as v and has the same dimensions and units as $\underline{\underline{v}}$.

<u>Average acceleration</u> --- The ratio of the change in $\underline{\underline{v}}$ to the time interval over which this change occurs. That is

$$\bar{\underline{\underline{a}}} = \frac{\Delta \underline{\underline{v}}}{\Delta t} = \frac{\underline{\underline{v}}_2 - \underline{\underline{v}}_1}{t_2 - t_1} \ . \qquad (3\text{-}4)$$

a) $\bar{\underline{\underline{a}}}$ is a vector whose dimensions are those of length per time squared (L/T^2).

b) This acceleration may arise because of a change in the <u>magnitude</u> of $\underline{\underline{v}}$, a change in the <u>direction</u> of $\underline{\underline{v}}$, or both.

<u>Acceleration</u> --- The instantaneous acceleration which is the limit of the average accelerations as $\Delta t \to 0$. Mathematically it is the derivative of $\underline{\underline{v}}$ with respect to t.

$$\underline{\underline{a}} = \lim_{\Delta t \to 0} \frac{\Delta \underline{\underline{v}}}{\Delta t} = \frac{d\underline{\underline{v}}}{dt} \ . \qquad (3\text{-}5)$$

Unless otherwise specified, the time at which this derivative is to be evaluated is some general time t.

3-2 Use of Components of $\underline{\underline{r}}$, $\underline{\underline{v}}$ and $\underline{\underline{a}}$; One Dimensional Motion

By resolving the vectors $\underline{\underline{r}}$, $\underline{\underline{v}}$, and $\underline{\underline{a}}$ into Cartesian components, we may deal with three scalar equations rather than with a vector equation. Furthermore, because the <u>unit vectors</u> $\underline{\underline{i}}$, $\underline{\underline{j}}$ <u>and</u> $\underline{\underline{k}}$ <u>are orthogonal and constant in time, each component of the motion may be treated separately of the others.</u> The scalar equations are

$$\begin{aligned} r_x &= x \ , & v_x &= dx/dt \ , & a_x &= dv_x/dt \\ r_y &= y \ , & v_y &= dy/dt \ , & a_y &= dv_y/dt \\ r_z &= z \ , & v_z &= dz/dt \ , & a_z &= dv_z/dt \ . \end{aligned} \qquad (3\text{-}6)$$

These are three essentially independent sets of equations involving the Cartesian components of position, velocity and acceleration. If x is known as a function of time, for example, we may find v_x (by differentiation) and a_x (by differentiation again) quite independently of what happens in the y and z directions. Conversely, if we know a_x as a function of time we may find v_x and x (by a process of "anti-differentiation", called integration).

One may interpret Eqs. (3-6) as follows: v_x is the slope of the curve x versus t and a_x the slope of the curve v_x versus t; a_x is also therefore the rate of change of the slope of x versus t.

By resolving $\underline{\underline{r}}$, $\underline{\underline{v}}$, and $\underline{\underline{a}}$ into components one has to solve three one dimensional motion problems. A particular case of interest is motion in one dimension, say x. Then

$$\underline{\underline{v}} = v_x \underline{\underline{i}} \quad \text{and} \quad \underline{\underline{a}} = a_x \underline{\underline{i}} \ .$$

Since $\underline{\underline{i}}$ points in the plus x direction, v_x is positive if $\underline{\underline{v}}$ points that way and negative if $\underline{\underline{v}}$ points the other way; similarly for a_x and $\underline{\underline{a}}$.

In this case we do not <u>need</u> the full power of the vector notation since the algebraic signs of x, v_x and a_x indicate the directions of $\underline{r}$, $\underline{v}$ and $\underline{a}$, respectively. Often for brevity the coordinate subscript is omitted from v_x and a_x. One might then see v = - 5 mph. The meaning is that the speed is + 5 mph and the direction of the velocity is opposite that of positive x.

3-3 One Dimensional Motion--Constant Acceleration; Free Fall

If the x component of the acceleration is constant, the kinematic equations are

$$v_x(t) = v_x(0) + a_x t \quad , \qquad \text{for constant } a_x \; ; \qquad (3\text{-}7a)$$

$$x(t) = x(0) + v_x(0)t + \tfrac{1}{2} a_x t^2 \quad , \qquad \text{for constant } a_x \; . \qquad (3\text{-}7b)$$

Here $v_x(t)$ is the x component of velocity at some arbitrary time t and $v_x(0)$ is its value at t = 0. Similarly, x(t) is the x component of position at t and x(0) is its value at t = 0. The x component of displacement is x(t) - x(0).

Two other equations are often used; they may be derived from Eq. (3-7a) and (3-7b) and are therefore hardly worth memorizing. The first is found by solving Eq. (3-7a) for t and substituting into Eq. (3-7b) to yield

$$v_x^2(t) = v_x^2(0) + 2a_x[x(t) - x(0)] \quad , \qquad \text{for constant } a_x \; . \qquad (3\text{-}8)$$

To derive the second one, Eq. (3-7a) is solved for a_x which is substituted into Eq. (3-7b) to yield

$$x(t) = x(0) + \tfrac{1}{2}[v_x(t) + v_x(0)]t \quad , \qquad \text{for constant } a_x \; . \qquad (3\text{-}9)$$

It is important to notice that there are 6 quantities [x(0), x(t), $v_x(t)$, $v_x(0)$, a_x, t] which can vary from problem to problem; 2 of these are initial conditions, [x(0) and $v_x(0)$]. If one is interested only in the displacement, x(t)-x(0), rather than position, there are but 5 quantities. These 5 are related in constant acceleration kinematics by only 2 independent equations, (3-7) so 3 of them must be specified <u>in one way or another</u> if a given problem is to be solved. The trick, if it can be called that, to solving the one dimensional, constant acceleration, kinematic problems is to determine which have been specified. We will try to make this clear in the examples.

A very important example of approximately constant acceleration is the acceleration of gravity in the vicinity of the earth's surface which for such problems is treated as a plane. Then the acceleration of gravity, $\underline{g}$, is a vector pointing toward the earth's surface. If the vertical direction is taken as the y axis then a freely falling particle is a one dimensional problem involving the y component only and $a_y = \pm$ g; the plus sign is taken if the positive y axis is down and the minus sign if it is up. Often, the origin is taken at the initial position.

3-4 Examples

We first consider a straightforward problem in which the coordinate is given as a function of time and various quantities are computed.

>>> Example 1. The position of a particle constrained to the x axis is $x = 2t + 5t^2 - 4t^3$. Here x is understood to be in cm and t is the time in seconds. Find a) v_x as a function of time, b) a_x as a function of time, c) the average velocity over the interval 0 to 1 second and d) the average acceleration over the same interval.

Solution

(a) Since $v_x = \frac{dx}{dt}$ we have $v_x = \frac{d}{dt}(2t + 5t^2 - 4t^3)$ or $v_x = 2 + 10t - 12t^2$.

The units of v_x are cm/s.

(b) $a_x = dv_x/dt$, so $a_x = 10 - 24t$. The units of a_x are cm/s^2 .

(c) x(1) = 3 cm, x(0) = 0 cm so that $\bar{v}_x = \frac{x(1) - x(0)}{1\ s} = 3$ cm/s .

(d) $v_x(1) = 0$ cm/s, $v_x(0) = 2$ cm/s, $\bar{a}_x = \frac{0 - 2}{1}$ cm/s $= -\ 2$ cm/s . <<<

Now, we consider a series of problems in which 3 kinematic quantities are specified and the other two must be found.

>>> Example 2. An automobile is brought to a stop with constant acceleration from a speed of 40 mph in 120 feet. What is the acceleration and how long does it take the car to stop?

Now we are to find a_x and t, so $v_x(t)$, $v_x(0)$ and x(t) - x(0) must be given. We choose the positive x axis in the initial direction of motion. Thus the displacement is x(t) - x(0) = 120 ft, and $v_x(0) = 40$ mph; both are positive. At the time of interest, $v_x(t) = 0$ because the car is stopped. So, we have found the specified quantities. From Eq. (3-9) we have

$$x(t) - x(0) = \tfrac{1}{2}\,[v_x(t) + v_x(0)]t$$

so

$$120\ \text{ft} = (\tfrac{1}{2}[0 + 40]\ \text{mph})t$$

hence

$$t = \frac{120\ \text{ft}}{20\ \text{mph}} \;.$$

A convenient number to remember is 60 mph = 88 ft/s. Thus

$$t = \frac{120\ \text{ft}}{20 \times \dfrac{88}{60}\dfrac{\text{ft}}{\text{s}}} \;\; + \;\; \frac{120 \times 60}{20 \times 88}\ \text{s} = 4.09\ \text{s} \;.$$

Then from Eq. (3-7a)

$$a_x = \frac{v_x(t) - v_x(0)}{t} = \frac{0 - \dfrac{40}{60} \times 88\ \dfrac{\text{ft}}{\text{s}}}{4.09\ \text{s}} = -\ 14.3\ \text{ft/s}^2 \;.$$

The minus sign indicates acceleration in the negative x direction. This is opposite to the direction of the chosen positive x axis and the initial velocity. This minus sign automatically resulted from a consistent algebraic treatment of the problem.

>>> Example 3. How much distance is required to stop an automobile going at 108 km/h if the driver's reaction time is 0.1 second and if the maximum possible uniform acceleration is the acceleration of gravity? How long does it take?

We are asked to find the displacement and are given the value of a_x and $v_x(0)$. The acceleration is 0 for the first 0.1 s and then has the value $-\ 9.8\ m/s^2$. We choose $v_x(0)$ to be positive; the minus sign for the acceleration is consistent with this choice. We need one more specified quantity and it is implicit here since at the end the car is stopped, so the final velocity is zero.

It is convenient to divide this problem into two parts: the 0.1 sec for the driver's reaction and T which is the length of time required to stop after the brakes are applied. Since T is unknown we use Eq. (3-7a) taking $v_x(T) = 0$, $v_x(0) = 108$ km/h = 30 m/s. Then

$$v_x(T) = 0 = v_x(0) + a_xT = 30 \text{ m/s} - (9.8 \text{ m/s}^2)T$$

so

$$T = 3.06 \text{ s} \quad .$$

This is the time required to stop after the brakes are applied, so the total time required to stop is

$$3.06 \text{ s} + 0.1 \text{ s} = 3.16 \text{ s} \quad .$$

We may now use Eq. (3-7b) to find out how far the car travels while the brakes are applied.

$$x(T) - x(0) = v_x(0)T + \tfrac{1}{2} a_xT^2$$

so

$$x(T) - x(0) = (30 \text{ m/s})(3.06 \text{ s}) - (4.9 \text{ m/s}^2)(3.06 \text{ s})^2$$

or

$$x(T) - x(0) = 45.9 \text{ m} \quad .$$

To this we must add the distance travelled during the reaction time which is

$$\Delta x = (30 \text{ m/s})(0.1 \text{ s}) = 3.0 \text{ m}$$

so that the total distance required is 48.9 m. <<<

>>> Example 4. A rifle is fired vertically upward and the bullet is in the air for one minute. Find the muzzle velocity.

At first glance it appears that only one quantity is specified. However, we know that the acceleration has magnitude $g = 32$ ft/sec^2. Also, at $t = 1$ minute the displacement is zero since the bullet rises up to a maximum height and then falls back to earth. Thus we know the displacement, acceleration and time. Take the y axis positive upward so that $a_y = -g$, and also take the origin at the earth's surface. Then from Eq. (3-7b) we have

$$y(t) = v_y(0)t - \tfrac{1}{2} gt^2 \quad .$$

The displacement y is zero both at $t = 0$ and at

$$t = T = \frac{2v_y(0)}{g} \quad .$$

The latter time T is the total time the bullet is in the air and is one minute. Thus from the above equation,

$$v_y(0) = \tfrac{1}{2} gT = (0.5)(32 \tfrac{\text{ft}}{\text{s}^2})(60 \text{ s})$$

or

$$v_y(0) = 960 \text{ ft/s} \quad .$$

<<<

3-5 Alternate Notation

Often for brevity the notation for displacement, velocity, etc. is simplified. That is, instead of $v_x(t)$ we might write merely v_x where it is understood that this is the x component of velocity at some time t. Also $v_x(0)$ is abbreviated as v_{xo}. For example, instead of

$$x(t) = x(0) + v_x(0)t + \tfrac{1}{2} a_x t^2$$

one often writes

$$x = x_o + v_{xo}t + \tfrac{1}{2} a_x t^2 \quad .$$

3-6 Programmed Problems

1. The position x of an object is given as a function of time t as

$$x = a + bt + ct^2 + dt^3$$

where a, b, c and d are constants. What is the corresponding expression for the velocity v_x as a function of time?

$$v_x = dx/dt = b + 2ct + 3dt^2$$

2. At t = 0 the velocity is b while at t = 1, the velocity is b + 2c + 3d. What is the average velocity $\bar{v}_x$ for the interval t = 0 to t = 1?

From Eq. (3-2), the definition

$$\bar{v}_x = \frac{x(1) - x(0)}{1 - 0}$$

$$= \frac{a + b + c + d - a}{1}$$

$$= b + c + d$$

Note: $\bar{v}_x$ is <u>not</u> the mean value

$$\tfrac{1}{2}[v(1) + v(0)] = b + c + (3/2)\,d$$

3. A portion of the curve x vs. t is shown

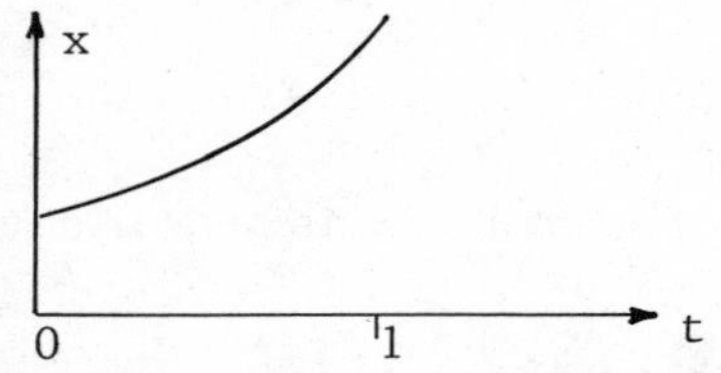

Indicate the geometrical meaning of $v_x(0)$, $v_x(1)$ and $\bar{v}_x$ for the interval 0 to 1.

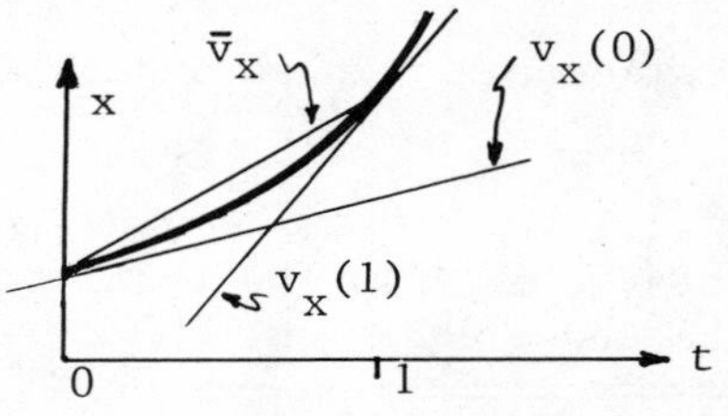

$v_x(0)$ is slope of curve at t = 0,
$v_x(1)$ is slope of curve at t = 1,
$\bar{v}_x$ is slope of chord between t = 0,1.

4. For the given function x(t) what is the acceleration a_x as a function of time?

$$a_x = dv_x/dt = 2c + 6d\,t$$

5. We have

$$a_x(0) = 2c$$

$$a_x(1) = 2c + 6d$$

What is the average acceleration $\bar{a}_x$ for the interval $t = 0$ to $t = 1$?

From Eq. (3-4), the definition,

$$\bar{a}_x = \frac{v_x(1) - v_x(0)}{1 - 0}$$

$$= 2c + 3d$$

Notice that here $\bar{a}_x$ is the same as the mean values

$$\tfrac{1}{2}[a_x(1) + a_x(0)]$$

This occurs if and only if the rate of change of a quantity is itself constant. Here

$$da_x/dt = 6d = \text{constant} .$$

This can be a useful shortcut, but you will always be right to return to the definition.

---

6. Suppose that all one knows is that the acceleration a_x is a constant, say 5 m/s^2. What can you say about the value of the velocity v_x at $t = 2$ seconds?

Not much!

$$v_x(t) = v_x(0) + 5t$$

but, unless $v_x(0)$ is known, one cannot determine $v_x(2) = v_x(0) + 10$. Hence one can only say $v_x(2) - v_x(0) = 10$ m/s. The initial condition information $v_x(0)$ must be given. Suppose $v_x(0) = 8$ m/s .

---

7. So we have

$$v_x(t) = (8 + 5t) \text{ m/s}$$

with t in seconds. What is the displacement of the object from t = 1 s to t = 2 s?

We have

$$x(t) = x(0) + 8t + 2.5t^2$$

so $\quad x(2) = x(0) + 26$

and $\quad x(1) = x(0) + 10.5$.

Thus $x(2) - x(1) = 15.5$ m. Note that the displacement can be determined without knowing $x(0)$. Only if one wants the position at some time does one need the initial condition $x(0)$.

Very often problems cannot be solved by simply plugging the kinematic formulas. The following is an example:

> Ball 1 is released from rest from the top of a building. It strikes the pavement below and rebounds with the same speed it had upon impact. At the same instant that ball 1 rebounds, ball 2 is released from rest from the building top. At what fraction of the building height do they meet?

The first step in doing such two part problems is to break them into simpler parts.

8. Paraphrase the problem assuming that you are given the velocity v_{up} with which ball 1 starts back upward.	Ball 1 starts upward from the ground at a speed v_{up} at the same time as ball 2 is released from rest from the top of a building of height H. Where do they meet in terms of H?
9. Let y_1 be the position of ball 1 above the ground and let y_2 be the position of ball 2 above the ground. Express each as a function of time.	$y_1 = v_{up}t - \frac{1}{2} gt^2$ $y_2 = H - \frac{1}{2} gt^2$ Note that at $t = 0$, $y_1 = 0$ and $y_2 = H$ as required.
10. (a) Express the criterion that the balls meet in terms of y_1 and y_2. (b) Find the time at which they meet, and (c) what else do you need to know!	(a) $y_1 = y_2$ at the same time. Hence (b) they meet at $t = H/v_{up}$ at $y_1 = y_2 = H - \frac{1}{2} g\, H^2/v_{up}^2$. (c) Since y_1 and y_2 are to be found as a fraction of H, we only need to know v_{up}.
11. We've divided the problem into two parts and the first part was solved formally. State the second part of the problem and solve it.	The second part is: A ball is released from rest from the top of a building of height H. It strikes the ground and rebounds with the same speed it had at impact. What is this speed? We choose the origin at the top; the + y axis down. Since the ball falls down with acceleration + g, its velocity is $v = v_o + gt = gt$ since $v_o = 0$. The time t is obtained from $y = H = \frac{1}{2} gt^2$. So $v = v_{up} = \sqrt{2gH}$.
12. Finish the original problem.	Now we need merely use the result from frame 11 in frame 10 to find $y_1 = y_2 = H - \frac{1}{2} g\, H^2/(2gH) = (3/4)\, H$ Would you have been tempted to guess $\frac{1}{2}$ H initially? Why isn't the answer $\frac{1}{2}$ H?

Chapter 4: MOTION IN A PLANE

REVIEW AND PREVIEW

In the previous chapter you learned the definitions of the vector quantities: position, velocity and acceleration. However, you have applied these only to one-dimensional motion. Now you will study two-dimensional motion (i.e., motion in a plane).

GOALS AND GUIDELINES

In this chapter there are three major goals:

1. Learning to apply the previous equations (3-7) to each of the two components of the motion for the case of an acceleration which is constant (both in magnitude and direction). No new formulas are required. Of extreme physical importance is the case of projectile motion. You should strive to understand this as two simultaneous motions (one in the x and one in the y direction).

2. Learning the somewhat more difficult case of uniform circular motion in which the acceleration is constant in magnitude but not in direction. Although you may find the non-Cartesian unit vectors (radial and tangential) difficult, you will gain a much greater understanding of circular motion through their use.

3. Learning to relate the description of the motion of a particle with respect to a second (moving) coordinate system. Of particular importance is the relation between the two velocity descriptions (Eq. 4-24b). Physical applications of these are many including navigation of ships and aircraft.

4-1 Basic Relationships

For motion in one dimension the full power of the vector method was not required, but it is for motion in a plane. The basic relationships are those already given in Chapter 3 but specialized to two dimensions. Using the unit vectors $\underline{\underline{i}}$ and $\underline{\underline{j}}$, in terms of components we have

(position vector) $\quad \underline{\underline{r}} = x\underline{\underline{i}} + y\underline{\underline{j}}$ (4-1)

(velocity) $\quad \underline{\underline{v}} = d\underline{\underline{r}}/dt = v_x\underline{\underline{i}} + v_y\underline{\underline{j}}$ (4-2)

(acceleration) $\quad \underline{\underline{a}} = d\underline{\underline{v}}/dt = a_x\underline{\underline{i}} + a_y\underline{\underline{j}}$ (4-3)

Each component of the motion may be considered separately.

4-2 Constant Acceleration in Plane Motion

If the acceleration is constant, it must be constant in magnitude as well as direction. In terms of components then, a_x = constant and a_y = constant. The equations of motion are those given in Eqs. (3-7), (3-8) and (3-9) for each component.

Thus

$$a_x = \text{constant} \; , \qquad a_y = \text{constant} \; , \qquad (4\text{-}4a)$$

$$v_x(t) = v_x(0) + a_x t \; , \qquad v_y(t) = v_y(0) + a_y t \; , \qquad (4\text{-}4b)$$

$$x(t) = x(0) + v_x(0)t + \tfrac{1}{2}\, a_x t^2 \; , \qquad y(t) = y(0) + v_y(0)t + \tfrac{1}{2}\, a_y t^2 \; , \qquad (4\text{-}4c)$$

$$v_x{}^2(t) = v_x{}^2(0) + 2a_x[x(t) - x(0)] \; , \qquad v_y{}^2(t) = v_y{}^2(0) + 2a_y[y(t) - y(0)] \; , \qquad (4\text{-}4d)$$

$$x(t) = x(0) + \tfrac{1}{2}\, [v_x(t) + v_x(0)]t \; , \qquad y(t) = y(0) + \tfrac{1}{2}\, [v_y(t) + v_y(0)]t \; . \qquad (4\text{-}4e)$$

In vector form these are

$$\underline{\underline{a}} = \text{constant} \; , \qquad (4\text{-}5a)$$

$$\underline{\underline{v}}(t) = \underline{\underline{v}}(0) + \underline{\underline{a}}t \; , \qquad (4\text{-}5b)$$

$$\underline{\underline{r}}(t) = \underline{\underline{r}}(0) + \underline{\underline{v}}(0)t + \tfrac{1}{2}\, \underline{\underline{a}}t^2 \; , \qquad (4\text{-}5c)$$

$$\underline{\underline{v}}(t)\cdot\underline{\underline{v}}(t) = \underline{\underline{v}}(0)\cdot\underline{\underline{v}}(0) + 2\underline{\underline{a}}\cdot[\underline{\underline{r}}(t) - \underline{\underline{r}}(0)] \; , \qquad (4\text{-}5d)$$

$$\underline{\underline{r}}(t) = \underline{\underline{r}}(0) + \tfrac{1}{2}\, [\underline{\underline{v}}(t) + \underline{\underline{v}}(0)]t \; . \qquad (4\text{-}5e)$$

4-3 Projectile Motion

A particle projected with some initial velocity near the earth's surface undergoes motion whose ideal is projectile motion. The situation is idealized by ignoring air resistance and treating the acceleration of gravity as a constant vector pointed perpendicularly downward toward the earth's surface which itself is regarded as a plane. Projectile motion is an example of constant acceleration in plane motion; the plane in this case is perpendicular to the earth's surface and determined by the direction of the initial velocity.

The x and y axes are chosen as indicated in Fig. 4-1. The origin is <u>chosen</u> at the initial position, and the path of the projectile is a portion of a <u>parabola</u> as shown in Fig. 4-2.

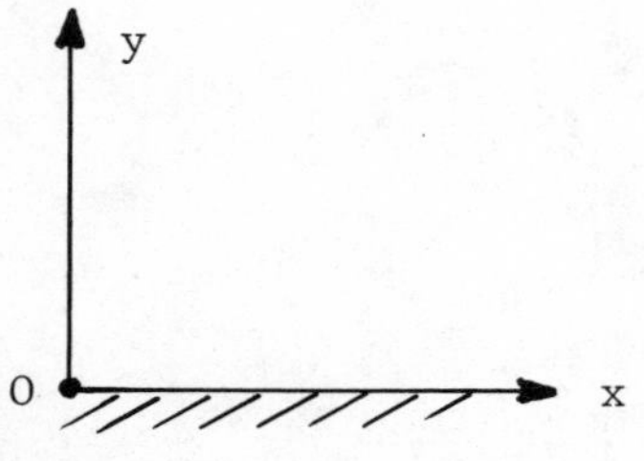

Figure 4-1. Choice of coordinate system for projectile motion.

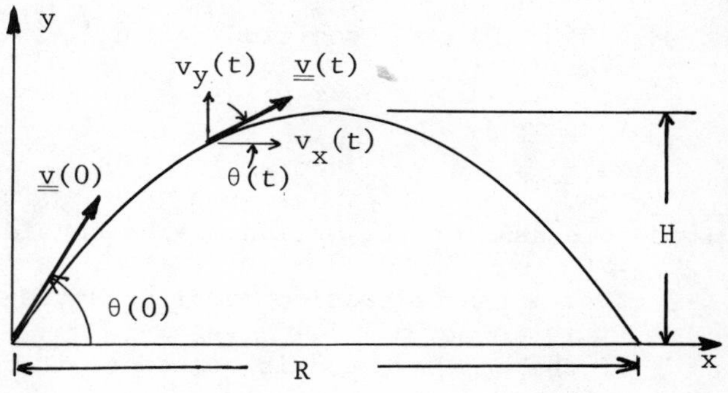

Figure 4-2. Parabolic trajectory of projectile motion.

The velocity is at every point tangent to this trajectory, and is given by

$$\underline{\underline{v}}(t) = v_x(t)\underline{\underline{i}} + v_y(t)\underline{\underline{j}} \quad (4\text{-}6a)$$

or

$$v_x(t) = v(t)\cos\theta(t) \quad (4\text{-}6b)$$

$$v_y(t) = v(t)\sin\theta(t) \quad (4\text{-}6c)$$

where $\theta(t)$ is the angle between $\underline{v}$ and the x axis at time t and the speed, $v(t)$, is given by

$$v(t) = \sqrt{v_x^2(t) + v_y^2(t)} \quad . \quad (4\text{-}7)$$

Since

$$a_x = 0 \quad (4\text{-}8a)$$

$$a_y = -g \qquad \text{(the +y axis is upward)} \quad (4\text{-}8b)$$

the velocity components are given by Eqs. (4-4b) as

$$v_x(t) = v_x(0) \quad (4\text{-}9a)$$

$$v_y(t) = v_y(0) - gt \quad . \quad (4\text{-}9b)$$

The equations of the trajectory of the projectile are Eqs. (4-4c) with $x(0) = y(0) = 0$. Thus

$$x(t) = v_x(0)t \quad (4\text{-}10a)$$

$$y(t) = v_y(0)t - \tfrac{1}{2}gt^2 \quad . \quad (4\text{-}10b)$$

If Eq. (4-10a) is solved for t and this substituted into Eq. (4-10b) then y as a function of x is obtained as

$$y = [v_y(0)/v_x(0)]x - [\tfrac{1}{2}g/v_x^2(0)]x^2 \quad (4\text{-}11)$$

which is a parabola.

The top (vertex) of the parabola corresponds to the maximum height reached by the projectile and its condition is $v_y(t_{top}) = 0$. Therefore from Eq. (4-9b)

$$t_{top} = v_y(0)/g \quad . \quad (4\text{-}12)$$

At this time the height is

$$H = y(t_{top}) = \tfrac{1}{2}v_y^2(0)/g \quad . \quad (4\text{-}13)$$

The <u>range</u> R is the horizontal distance travelled by the projectile when $y(t_R) = y(0)$ which here is zero. Then from Eq. (4-10b)

$$t_R = 2v_y(0)/g = 2t_{top} \quad (4\text{-}14)$$

at which time

$$R = x(t_R) = v_x(0)\,2v_y(0)/g \quad (4\text{-}15a)$$

From Eqs. (4-6) with t = 0 the range may be expressed in terms of the initial speed and angle as

$$R = \frac{v^2(0)\ \sin\ 2\theta(0)}{g} \quad . \tag{4-15b}$$

Notice that the projectile motion is completely specified if $v_x(0)$ and $v_y(0)$ are known. In projectile motion problems these may be given and you will be asked to compute other quantities such as H, R, etc. or the related quantities may be specified and you will be asked to find one or both of the initial conditions. Often, the relevant information is implicit, just to make life interesting!

>>> Example 1. A projectile is fired from a gun at ground level with an initial speed of 300 m/s at an angle 30° to the horizontal. (a) What is the range of the projectile, (b) how high does it go, and (c) how long is it in the air?

(a) The initial conditions $v_x(0)$ and $v_y(0)$ are specified since

$$v_x(0) = v(0)\ \cos\ \theta(0) = (300\ \text{m/s})(0.866)$$

or

$$v_x(0) = 260\ \text{m/s}$$

and

$$v_y(0) = v(0)\ \sin\ \theta(0) = (300\ \text{m/s})(0.500) = 150\ \text{m/s} \quad .$$

Therefore the range, R, is from (4-15a)

$$R = \frac{(2.0)(260\ \text{m/s})(150\ \text{m/s})}{9.8\ \text{m/s}^2} = 7960\ \text{m} = 7.96\ \text{km} \quad .$$

(b) The maximum height is given by (4-13),

$$H = \tfrac{1}{2}\,\frac{v_y{}^2(0)}{g} = (0.5)\,\frac{(150\ \text{m/s})^2}{9.8\ \text{m/s}^2} = 1148\ \text{m} \quad .$$

(c) The projectile is in the air a time interval given by (4-14),

$$t_R = \frac{2v_y(0)}{g} = 31\ \text{s} \quad .$$

<<<

>>> Example 2. A short basketball player can execute a "jump shot" releasing the ball when his hands are 7 feet above the floor and at the moment when he is neither moving up nor down. The basket is 10 feet above the floor. Assume that he is 30 feet from the basket and must start the ball at an angle of 30° to the horizontal; with what speed must he release the ball?

When x is 30 feet, y must be 3 feet. (What does this imply about location of the origin?) From Eq. (4-11) we have

$$y = 3\ \text{ft} = \frac{v(0)\ \sin\ \theta(0)}{v(0)\ \cos\ \theta(0)}\, x - \tfrac{1}{2}\, g\, \frac{x^2}{v^2(0)\ \cos^2\ \theta(0)}$$

$$= (.577)(30\ \text{ft}) - (16\ \text{ft/s}^2)\,\frac{(30\ \text{ft})^2}{v^2(0)(0.75)} \quad .$$

Hence

$$v^2(0) = 1342\ \text{ft}^2/\text{s}^2$$

so

$$v(0) = 37\ \text{ft/s} \quad .$$

<<<

4-4 Circular Motion

If a particle is constrained to move in a circle whose center is the origin, the magnitude r of the position vector $\underline{\underline{r}}$ is constant. For circular motion problems the plane polar coordinates r and θ are more convenient than Cartesian ones. These are the magnitude of $\underline{\underline{r}}$ and its angle of inclination with the x axis respectively. As shown in Fig. 4-3

$$x = r \cos \theta \tag{4-16a}$$

$$y = r \sin \theta \quad . \tag{4-16b}$$

The inverse relationships are

$$r = \sqrt{(x^2 + y^2)} \tag{4-17a}$$

$$\theta = \tan^{-1}(y/x) \quad . \tag{4-17b}$$

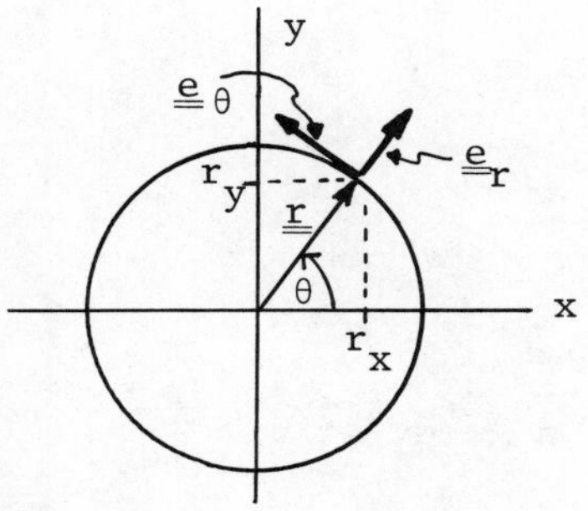

Figure 4-3. Polar coordinates $r_x = r \cos \theta$, $r_y = r \sin \theta$; and unit vectors $\underline{\underline{e}}_r$ and $\underline{\underline{e}}_\theta$.

Corresponding to the coordinates r and θ the unit vectors $\underline{\underline{e}}_r$ and $\underline{\underline{e}}_\theta$ are used. These unit vectors are not constant in direction; they are also shown in Fig. 4-3. The unit vector $\underline{\underline{e}}_r$ is directed radially outward and $\underline{\underline{e}}_\theta$ is tangent to the circle of radius r and points counterclockwise in the direction of increasing θ. In terms of Cartesian unit vectors

$$\underline{\underline{e}}_r = \underline{\underline{r}}/r = \cos \theta \, \underline{\underline{i}} + \sin \theta \, \underline{\underline{j}} \tag{4-18a}$$

and

$$\underline{\underline{e}}_\theta = d\underline{\underline{e}}_r/d\theta = - \sin \theta \, \underline{\underline{i}} + \cos \theta \, \underline{\underline{j}} \quad . \tag{4-18b}$$

Here we have used $d(\cos \theta)/d\theta = - \sin \theta$ and $d(\sin \theta)/d\theta = \cos \theta$.

Now, the position vector $\underline{\underline{r}}$ is

$$\underline{\underline{r}} = r\underline{\underline{e}}_r$$

. (Note that $\underline{\underline{r}}$ has no component in the $\underline{\underline{e}}_\theta$ direction.)

The velocity $\underline{\underline{v}}$ is tangent to the circle so

$$\underline{\underline{v}} = v\underline{\underline{e}}_\theta \tag{4-19}$$

. (Note that $\underline{\underline{v}}$ has no component in the $\underline{\underline{e}}_r$ direction.)

The acceleration $\underline{\underline{a}}$ has in general both components so we can write

$$\underline{\underline{a}} = a_r\underline{\underline{e}}_r + a_\theta\underline{\underline{e}}_\theta \quad . \tag{4-20}$$

We may now find v_θ by first obtaining $\underline{\underline{v}}$ by differentiating $\underline{\underline{r}}$ with respect to time. Thus, since r is constant

$$\underline{\underline{v}} = d\underline{\underline{r}}/dt = r(d\underline{\underline{e}}_r/dt) = r(d\underline{\underline{e}}_r/d\theta)(d\theta/dt)$$

$$\underline{\underline{v}} = r(d\theta/dt)\ \underline{\underline{e}}_\theta\ .$$

From Eq. (4-19) we see that

$$v = r(d\theta/dt) \tag{4-21}$$

which may be positive or negative. The acceleration is

$$\underline{\underline{a}} = d\underline{\underline{v}}/dt = (dv/dt)\underline{\underline{e}}_\theta + v(d\underline{\underline{e}}_\theta/dt)\ .$$

From Eqs. (4-18b) and (4-21)

$$d\underline{\underline{e}}_\theta/dt = (d\underline{\underline{e}}_\theta/d\theta)(d\theta/dt) = -\ \underline{\underline{e}}_r\ v/r\ .$$

Therefore

$$\underline{\underline{a}} = -\ (v^2/r)\underline{\underline{e}}_r + (dv/dt)\underline{\underline{e}}_\theta\ .$$

So, we see that the acceleration components are $a_r = -\ v^2/r$, $a_\theta = dv/dt$.

Let

$$a_T = a_\theta = dv/dt \tag{4-22a}$$

and

$$a_R = -\ a_r = v^2/r\ . \tag{4-22b}$$

The term a_T is the component of $\underline{\underline{a}}$ which is tangent to the particle's path and arises from a change in the magnitude of the velocity; it is called the tangential acceleration. The term a_R is the component at $\underline{\underline{a}}$ along the radially inward direction and arises from the change in direction of $\underline{\underline{a}}$; it is called the centripetal acceleration; it is always positive. Thus we have

$$\underline{\underline{a}} = a_T\underline{\underline{e}}_\theta + (-\ a_R\underline{\underline{e}}_r)\quad .$$

Note that the radial part of $\underline{\underline{a}}$ points toward the center of the circle (since $a_R > 0$).

In uniform circular motion v is constant, so only the centripetal acceleration arises. In this case $\underline{\underline{a}}$ has the constant magnitude v^2/r, but continually changes direction.

Care must be taken in using Eq. (4-21) in that often $d\theta/dt$ is given in revolutions per unit time. This must be expressed in radians per unit time so that v has the dimensions L/T and r those of L; one revolution is 2π radians. This is because an arc length on a circle is given by $s = r\theta$ with θ in radians; then $v = ds/dt = r(d\theta/dt)$.

Problems involving circular motion are generally straightforward. Typically we have --

>>> Example 3. An airplane propeller of length 2 meters is rotating at 1800 rpm with the airplane sitting still. What is the speed and centripetal acceleration of the propeller tip? Express the latter in terms of the acceleration of gravity.

We have $d\theta/dt = 1800$ rev/min and we must express this as radians per minute or per second. Thus

$$\frac{d\theta}{dt} = 1800\,\frac{\text{rev}}{\text{min}} = \frac{(1800)(2\pi)}{60}\,\frac{\text{radians}}{\text{second}} \quad .$$

Then $$v = r\,d\theta/dt = 1 \text{ meter } (60\pi\ \text{s}^{-1}) = 188.5 \text{ m/s} \quad .$$

The centripetal acceleration is

$$a = \frac{v^2}{r} = \frac{(188.5)^2\ \text{m}^2/\text{s}^2}{1\ \text{m}} = 3.55 \times 10^4\ \text{m/s}^2 \quad .$$

Since the acceleration of gravity g is 9.8 m/s^2 we have

$$a = 3.6 \times 10^3\ g \quad .$$ <<<

4-5 Relative Velocity and Acceleration

Consider two coordinate systems denoted S and S'. Let the position vector, velocity and acceleration of a particle <u>relative</u> to (i.e., as measured from) S be denoted respectively by $\underline{r}$, $\underline{v}$ and $\underline{a}$. Similarly let the corresponding quantities relative to S' be $\underline{r}'$, $\underline{v}'$, and $\underline{a}'$. Also let $\underline{R}$ denote the position vector of the origin of S' from S, as shown in Fig. 4-4. The velocity of S' relative to S is $\underline{u}$ and

$$\underline{u} = d\underline{R}/dt \quad . \tag{4-23a}$$

The acceleration of S' relative to S is $\underline{a}_{S'}$ and is given by

$$\underline{a}_{S'} = d\underline{u}/dt \quad . \tag{4-23b}$$

Note that as seen from S' the position, velocity and acceleration of S are $-\underline{R}$, $-\underline{u}$ and $-\underline{a}_{S'}$.

Figure 4-4. Relative coordinates.

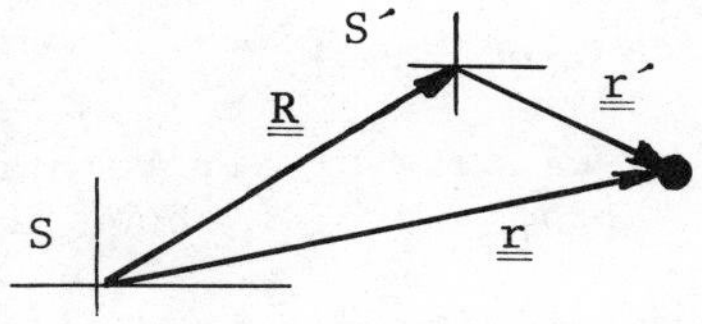

The kinematic quantities in the two coordinate systems are related by

$$\underline{r} = \underline{R} + \underline{r}' \ , \tag{4-24a}$$

$$\underline{v} = d\underline{r}/dt = \underline{u} + \underline{v}' \ , \tag{4-24b}$$

and $$\underline{a} = d\underline{v}/dt = \underline{a}_{S'} + \underline{a}' \ . \tag{4-24c}$$

Often we consider the case where $\underline{u}$ is constant, so $\underline{a}_{S'} = 0$ and $\underline{a} = \underline{a}'$.

The only difficulty encountered with relative velocity problems is in determining <u>consistently</u> which coordinate system is S and which is S'. One must remember that for Eqs. (4-24) to apply, $\underline{R}$ and $\underline{u}$ must be respectively the position and velocity of the S' origin <u>as seen from</u> S.

>>> Example 4. An airplane is heading due north at an airspeed of 100 mph and the wind is <u>from</u> the west at 20 mph. What is the airplane's velocity relative to the ground?

The airplane is treated as a particle and the earth reference from is S. A coordinate frame fixed with respect to the air is S´. Thus

$\underline{\underline{u}}$ = velocity of air with respect to the ground [20 mph, east];

$\underline{\underline{v}}'$ = velocity of the airplane with respect to the air [100 mph, north];

$\underline{\underline{v}}$ = velocity of the airplane relative to the ground.

From Eq. (4-24b) we have

$$\underline{\underline{v}} = \underline{\underline{u}} + \underline{\underline{v}}' \quad .$$

Let north be the positive y axis and east the positive x axis. Then

$$\underline{\underline{v}} = 20\underline{\underline{i}} + 100\underline{\underline{j}} \quad .$$

We may also write for the "ground speed"

$$v = \sqrt{(100)^2 + (20)^2} = 102 \text{ mph} \quad .$$

The direction of $\underline{\underline{v}}$ is given by

$$\theta = \tan^{-1}(v_y/v_x) = \tan^{-1}(100/20) = 78.7^\circ \quad .$$ <<<

A variation in the same problem is a common light aircraft navigation problem.

>>> Example 5. The wind is <u>from</u> a magnetic direction of 300° (north is 360° or 0°) at 25 mph and an airplane has an airspeed of 145 mph. In what direction must the pilot head the aircraft to have a ground velocity that is due north? What is the airplane's ground speed?

In this case we have:

$\underline{\underline{u}}$ = velocity of air relative to the ground [25 mph in a direction 120° from north as shown in Fig. 4-5];

$\underline{\underline{v}}$ = velocity of aircraft relative to the ground; we know its direction is to be north, but its magnitude is not known;

$\underline{\underline{v}}'$ = velocity of aircraft relative to the air; we know its magnitude is 145 mph but its direction is unknown.

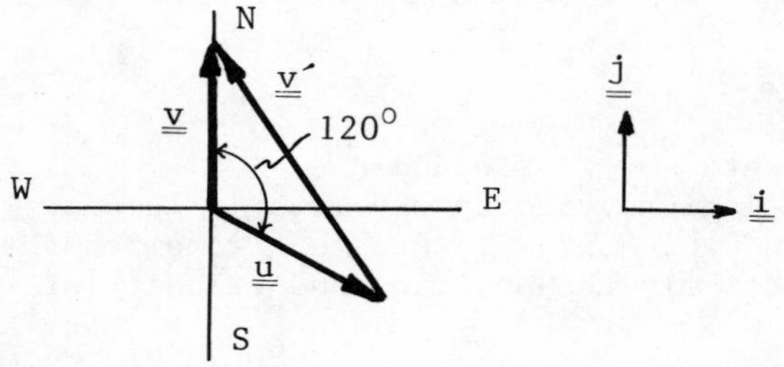

Figure 4-5. Example 5. A light aircraft navigation system.

Let $\underline{\underline{i}}$ denote the direction east and $\underline{\underline{j}}$ the direction north. Then from Fig. 4-5

$$\underline{\underline{u}} = u \cos 30^{o}\ \underline{\underline{i}} - u \sin 30^{o}\ \underline{\underline{j}}$$

or

$$\underline{\underline{u}} = (21.7\underline{\underline{i}} - 12.5\underline{\underline{j}})\ \text{mph} \quad .$$

Also

$$\underline{\underline{v}} = v\underline{\underline{j}} \quad ,$$

and

$$\underline{\underline{v}}' = v'_x\underline{\underline{i}} + v'_y\underline{\underline{j}} \quad .$$

Then from Eq. (4-24b)

$$\underline{\underline{v}} = v\underline{\underline{j}} = \underline{\underline{u}} + \underline{\underline{v}}' = (21.7 + v_x')\underline{\underline{i}} + (v_y' - 12.5)\underline{\underline{j}} \quad .$$

Thus

$$v_x' = -\ 21.7\ \text{mph}$$

$$v_y' - 12.5 = v \quad .$$

We know that $\sqrt{v_x'^2 + v_y'^2} = 145$ mph. Let ϕ denote the angle between $\underline{\underline{v}}'$ and the <u>negative</u> x axis. Then

$$v_x' = -\ 145 \cos \phi\ \text{mph}$$

$$v_y' = 145 \sin \phi\ \text{mph} \quad .$$

Therefore

$$145 \cos \phi = 21.7$$

or

$$\cos \phi = 21.7/145 = .1497 \quad .$$

Hence $\phi = 81.4^{o}$ and $\sin \phi = .9888$, so $v_y' = 143.4$ mph and then

$$v = 131\ \text{mph} \quad .$$

Thus the aircraft heading should be approximately 351° and the ground speed will be 131 mph.

<<<

4-6 Programmed Problems

A ball is thrown with an initial velocity $\underline{v}$ at an angle θ with the horizontal. The ball is released a distance d above the lower level. The problem is to determine when and where the ball will land. We will use numbers later, but for now a few general ideas about projectile motion are in order.

1.

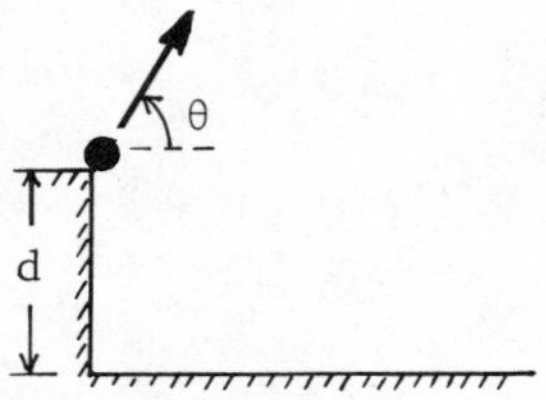

On the diagram sketch the trajectory of the ball. Ignore air resistance.

Answer:

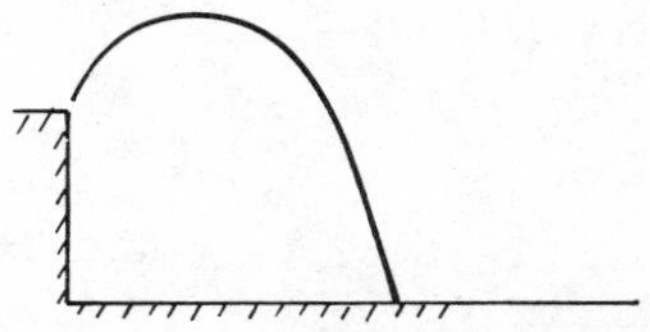

The curve is a portion of a parabola.

2.

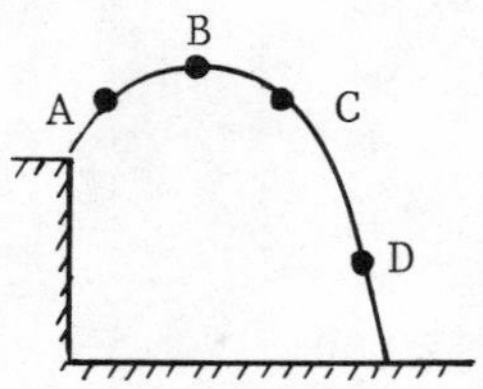

Sketch in vectors representing the <u>horizontal</u> velocity of the ball at the points indicated by A, B, C and D. Sketch vectors carefully with respect to length (magnitude) and orientation (direction).

Answer:

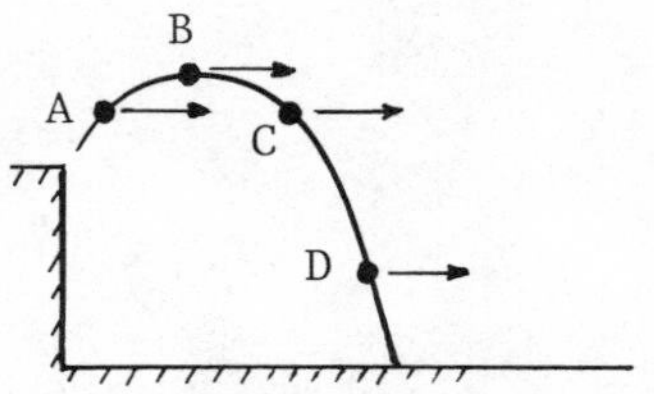

All vectors have the same length.

---

3. What is the horizontal acceleration of the ball?

Answer: Zero.
No change in velocity means no acceleration.

4.

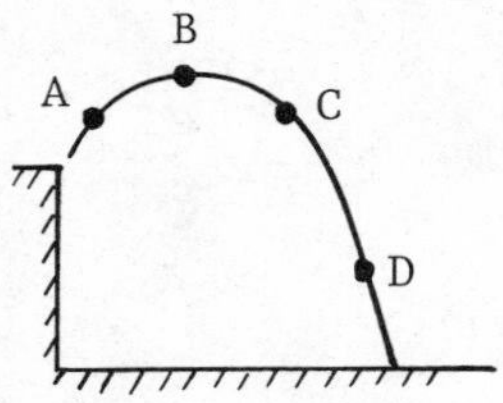

Sketch in vectors representing the <u>vertical</u> velocity of the ball at the points indicated.

Answer:

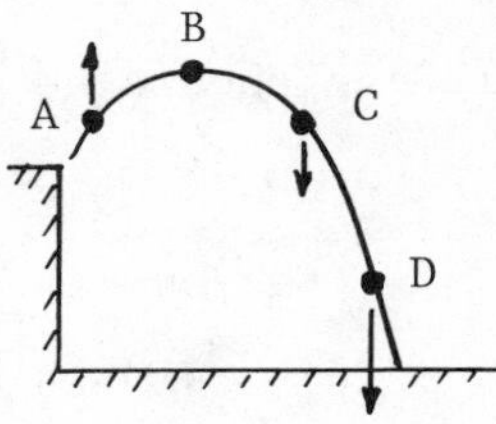

Note: Zero vertical velocity at B.

A correct answer to the last frame is very important. Make certain you understand the answer and then we will look more closely at the reason. (a) At A the ball has a vertical velocity which is "up". (b) At B the ball has <u>no</u> vertical velocity. (c) At C the ball is going "down" and has a velocity "down". (d) D is similar to C but the speed is greater.

5. The previous frames constitute what might be called a "physical feel" for the motion. Now look at the physics.

Is the vertical velocity constant?

No. First it's up, then it's down.

---

6. Is the vertical motion of the ball subject to acceleration?

Yes. Change in velocity means acceleration.

7.

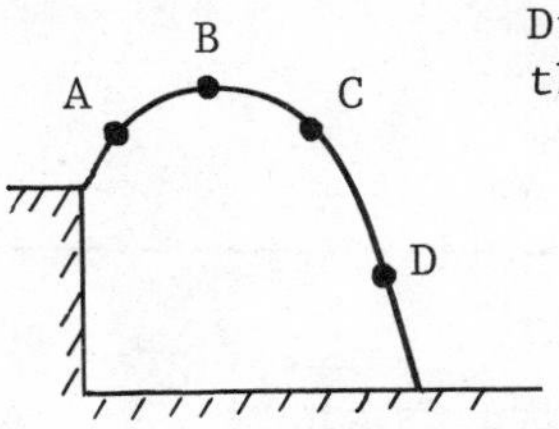

Draw vectors representing the vertical acceleration.

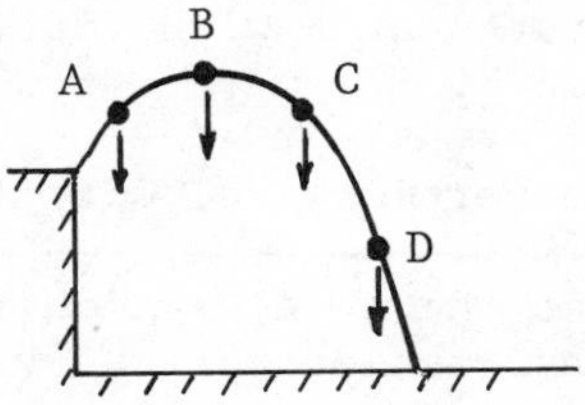

All vectors are the same in magnitude ($g = 9.8\ m/s^2$) and in direction (downward).

8.

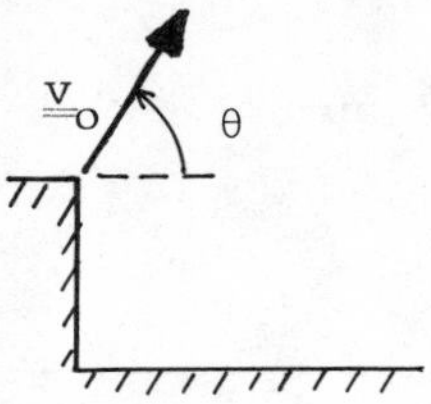

So far we have seen that the motion consists of a horizontal part with constant velocity and a vertical part with constant acceleration.

The initial velocity is $\underline{v}_o$ and the launching angle is θ.

What is the initial vertical velocity v_{yo} and the initial horizontal velocity v_{xo} in terms of these quantities?

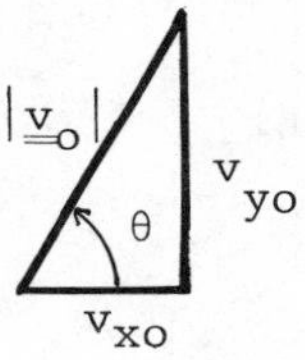

$$v_{yo} = |\underline{v}_o| \sin\theta$$

$$v_{xo} = |\underline{v}_o| \cos\theta$$

9.

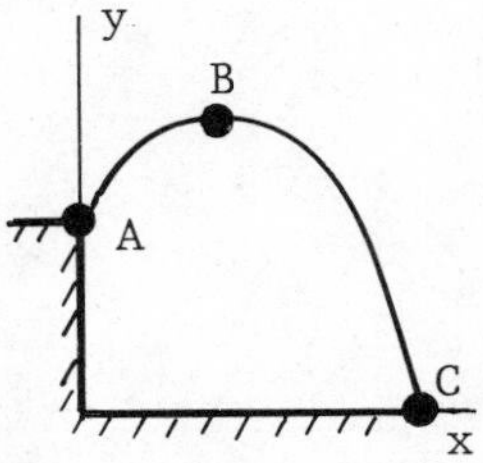

A reference frame has been superimposed on the diagram. In this system "up" and "right" will be positive.

If the projectile is in flight for a time T, what is the expression for its x position on landing?

[In this and in the frames to follow, note that A is now the launching point, B is the point at the top of the parabola, C is the landing point.]

$$x = x_o + (v_o \cos\theta)\,T$$

where x_o will be chosen to be zero [note that initial value of x (at A) is zero].

(The symbol v_o denotes the magnitude of the initial velocity: $v_o = |\underline{v}_o|$.)

10. To determine the flight time T we must consider the vertical motion. The convenient method of analyzing the vertical component of the motion is to break it up into pieces. Concentrate first on the path from A to B. Rather than trying to remember just the right kinematic formula, let us concentrate on what is happening.

The ball starts at A with vertical velocity $v_{yo} = v_o \sin\theta$.

What is the vertical velocity at B?

Zero.

---

11. We know what the acceleration is so we can write the expression for the velocity at B (which happens to be zero) as

$$v_y = v_{yo} - \underline{\qquad\qquad}.$$

vertical velocity at B		vertical velocity at A

Solve this for the time for the projectile to go from A to B.

gt_{up}

The minus sign is there because $\underline{g}$ points in the minus y direction.

t_{up} is the time required to go from A to B.

$0 = v_o \sin\theta - gt_{up}$

$$t_{up} = \frac{v_o \sin\theta}{g}$$

---

12. 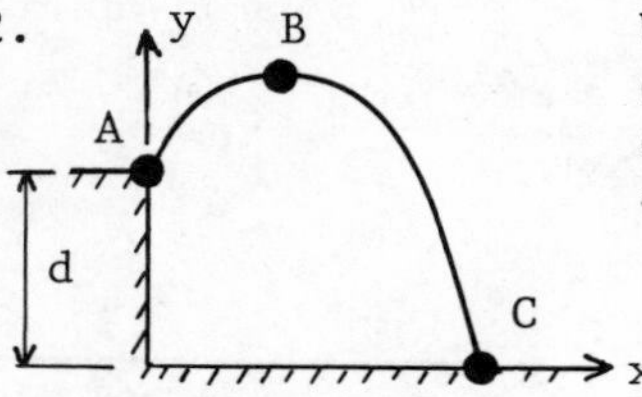

Write the kinematic equation for the y coordinate of the ball as it reaches B.

$y = y_o + v_{yo}t_{up} - \frac{1}{2} gt_{up}^2$,

$v_{yo} = v_o \sin\theta$ and

$$t_{up} = \frac{v_o \sin\theta}{g}$$

---

13. Now is a good time to put in some numbers. Let $|\underline{v}_o| = 98$ m/sec, $\theta = 30^o$, and $d = 100$ m. Find t_{up} and y at B. Remember $g = 9.8$ m/s^2 .

$t_{up} = 5$ sec

$y = 222.5$ m

[Note that $y_o = d = 100$ m]

14.

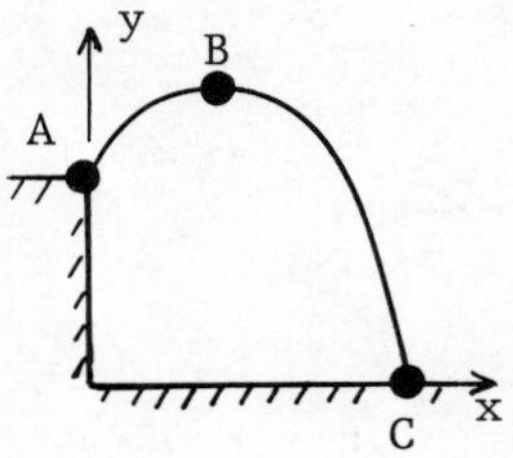

The ball is at B at the end of 5 seconds. The vertical motion is momentarily halted. The vertical motion from B to C will now be considered <u>with B taking the role of the initial position</u>.

The initial vertical velocity in going from B to C is ________ .
The initial position at B is y = ________ .
The final position at C is y = ______ .
The acceleration in the y direction during the motion from B to C is ______ .

Zero

222.5 m

Zero

$-g = -\ 9.8\ m/s^2$.

---

15. Write the kinematic equation, which relates the four terms in the previous frame with the time for the motion from B to C, t_{down}.

$y = y_o + v_{yo}t_{down} - \frac{1}{2}\ gt^2_{down}$

Note that <u>g</u> is still in the minus y direction.

For our case

$0 = 222.5 + 0 - \frac{1}{2}\ (9.8)t^2_{down}$

---

16. Solve the last equation of the previous answer for t_d.

$t_{down} = 6.7\ s$

17.

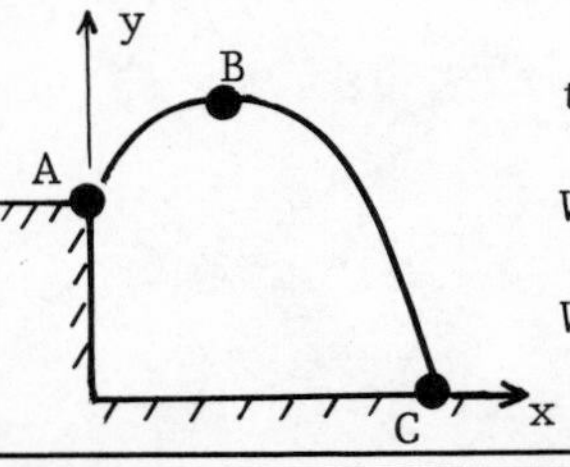

$v_o = 98\ m/s\ ,\ \theta = 30^o$.

$t_{up} = 5\ s\ ,\ t_{down} = 6.7\ s$.

What is the flight time T?

Where will the ball hit?

$T = t_{up} + t_{down} = 11.7\ s$.

$x = v_o \cos\theta\ T$

$x = 84.9\ m/s \times 11.7\ s = 993\ m$.

Note: x is <u>not</u> the range as given by Eq. (4-15b).

18.

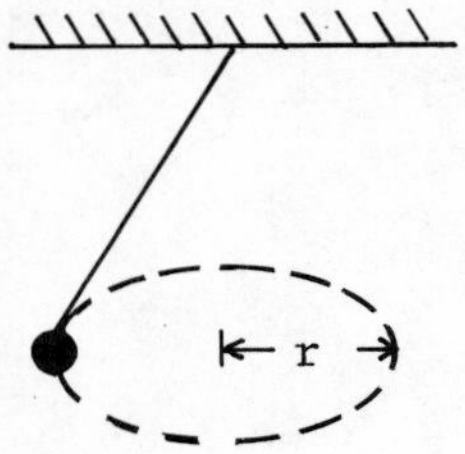

A ball suspended by a string moves in a circle in the plane perpendicular to the diagram. The speed of the ball is constant.

Is the velocity constant?

No.
The magnitude (speed) is, but the direction of the velocity vector is continually changing.

19. 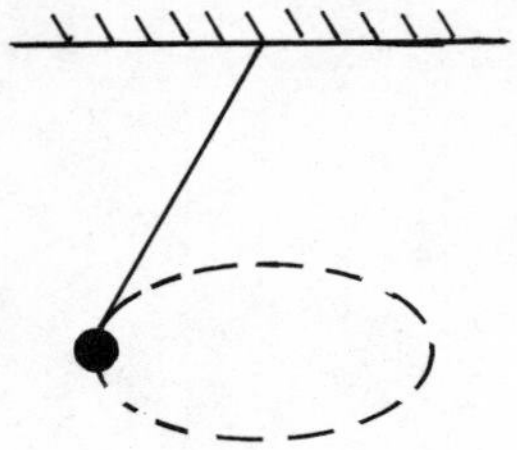

Draw a vector on the ball indicating the direction of the acceleration of the ball.

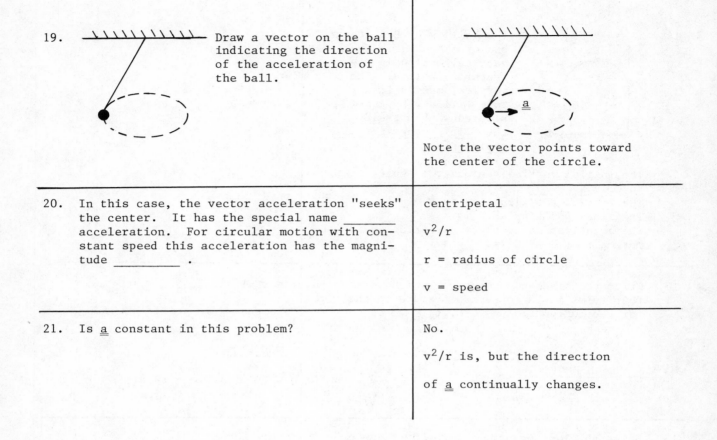

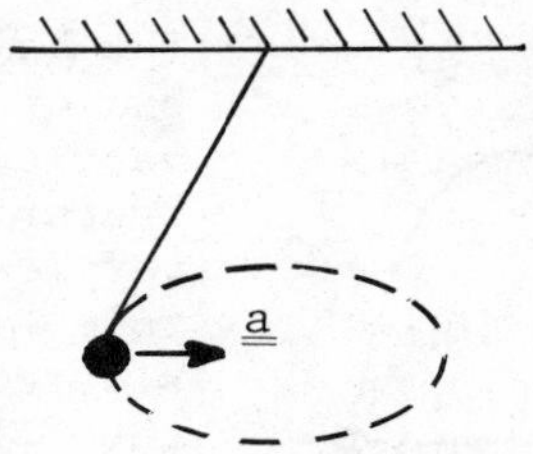

Note the vector points toward the center of the circle.

---

20. In this case, the vector acceleration "seeks" the center. It has the special name _______ acceleration. For circular motion with constant speed this acceleration has the magnitude _________ .

centripetal

v^2/r

r = radius of circle

v = speed

---

21. Is $\underline{\underline{a}}$ constant in this problem?

No.

v^2/r is, but the direction

of $\underline{\underline{a}}$ continually changes.

Chapter 5: PARTICLE DYNAMICS

REVIEW AND PREVIEW

So far, you have studied kinematics (i.e., the description of the motion of a particle). This involved the inter-relationship among the position ($\underline{\underline{r}}$), the velocity ($\underline{\underline{v}}$) and the acceleration ($\underline{\underline{a}}$). For example, in the case of one-dimensional motion with constant acceleration, one had

$$v = v_o + at$$

$$x = x_o + v_o x + \tfrac{1}{2} at^2 \ .$$

Another example studied was that of uniform circular motion in which the (vector) acceleration was not constant. At the heart of kinematics is the fact that given the acceleration, one can find velocity and the position at any time (past, present, future). Now you will learn how to deduce the acceleration of a particle. This is accomplished by means of the concepts of

1. force (caused by the particle's environment)
2. mass (a property of the particle).

The mass governs the response of particle to the force.

GOALS AND GUIDELINES

In this chapter you have three major goals:

1. Understanding Newton's three laws (Section 5-2).
   a. First law: Motion in the absence of forces.
   b. Second law: Relationship among force, mass and acceleration.
   c. Third law: Action and reaction force pairs.
2. Learning the techniques of applying Newton's laws to problems (Section 5-5): A procedural recipe is given which you should follow in all respects. You should resist the temptation to take shortcuts with simple problems; these shortcuts frequently lead to disaster in more complicated problems. Of particular importance is the technique of drawing a clearly labelled free body diagram, as shown in the examples.
3. Learning the properties of the two types of friction: static friction and kinetic friction (Section 5-6). You should pay particular attention to their similarities and differences in addition to learning the two simple equations which they obey.

5-1 Basic Ideas

Dynamics is that aspect of mechanics in which the kinematics of the motion of a body are related to the causes of that motion. The causes in turn are related to the relevant parts of the environment of the body. The central problem of classical mechanics is

(a) given a particle whose intrinsic characteristics (mass, electrical charge, etc.) are known,

(b) given the initial conditions of its motion, and
(c) given a complete description of the relevant parts of its environment,
(d) what is the subsequent motion of the particle?

The influence of the relevant parts of a particle's environment is expressed in terms of the concept of force which the environment exerts on the particle. The problem has two aspects

(1) the force law which relates a force acting upon a given particle to certain properties of its relevant environment, and
(2) the laws of motion which determine the motion of the particle under the action of a given force.

5-2 Newton's Laws; Force; Mass

Newton's laws of motion apply in inertial reference frames. The first two laws of motion are

(1) Every body persists in its state of rest or of uniform motion in a straight line unless compelled to change that state by forces impressed upon it.

(2) A force acting upon a body produces an acceleration whose direction is that of the applied force and whose magnitude is proportional to the magnitude of the force and inversely proportional to the mass of the body.

If several forces act, each produces its own acceleration independently; the resultant force is the vector sum of the several forces and the resultant acceleration is the vector sum of the corresponding accelerations. Mathematically Newton's second law is written

$$\underline{\underline{F}} = m\underline{\underline{a}} \tag{5-1}$$

where $\underline{\underline{F}}$ is the applied force, m the mass of the body, and $\underline{\underline{a}}$ its acceleration. If N forces act, then for each force $\underline{\underline{F}}_j$

$$\underline{\underline{F}}_j = m\underline{\underline{a}}_j \tag{5-2}$$

and

$$\underline{\underline{F}} = \sum_{j=1}^{N} \underline{\underline{F}}_j = \sum_{j=1}^{N} m\underline{\underline{a}}_j = m\underline{\underline{a}} \ . \tag{5-3}$$

The notation $\sum_{j=1}^{N} \underline{\underline{F}}_j$ means $\underline{\underline{F}}_1 + \underline{\underline{F}}_2 + \ldots\ldots\ldots + \underline{\underline{F}}_N$.

Note that the first law can be considered as a special case of the second law in that if the resultant force acting upon a body is zero, then its acceleration is zero.*

*The second law is in part really just the operational definition of force. That is, one defines the concept of force by using a standard of mass. In the SI system the mass standard is defined to be one kilogram and the unit of force is one newton. If the mass standard (one kilogram) undergoes an acceleration of $\underline{\underline{a}}$ (m/s^2) in a given environment then the environment has exerted a force of $\underline{\underline{F}}$ (newtons) where $|\underline{\underline{F}}|$ is numerically equivalent to $|\underline{\underline{a}}|$.

(3) To every action there is an equal but opposite reaction. That is, if a particle A exerts a force $\mathbf{F}_{BA}$ on another particle B, then B in turn exerts a force $\mathbf{F}_{AB}$ on A such that

$$\mathbf{F}_{BA} = -\mathbf{F}_{AB} \quad . \tag{5-4}$$

Either of these two forces is called action, the other reaction. The third law asserts two important points:

(i) Forces occur in pairs, and
(ii) That force called action acts on one body while the reaction force acts on the other body. *Action and reaction pairs never act on the same body.* The third law states that not only do these forces obey Eq. (5-4), but further each force acts along a line joining the two particles.

5-3 Mechanical Units

The common mechanical units are summarized in Table 5-1.

Table 5-1

SYSTEM	FORCE UNIT	MASS UNIT	ACCELERATION UNIT	FUNDAMENTAL UNITS	DERIVED UNITS
SI	newton (N)	kilogram (kg)	m/s^2	m, kg, s	$N = \frac{kg \cdot m}{s^2}$
cgs	dyne (dyn)	gram (g)	cm/s^2	cm, g, s	$dyne = \frac{g \cdot cm}{s^2}$
BE	pound (lb)	slug	ft/s^2	ft, lb, s	$slug = \frac{lb}{ft/s^2}$

5-4 Weight and Mass

The *weight* of a body is the gravitational force exerted on it by the earth and as such is a vector directed toward the earth's surface. The magnitude of the weight is the product of the mass of the body and the acceleration of gravity. In vector form

$$\mathbf{W} = m\mathbf{g} \quad . \tag{5-5}$$

Since $\mathbf{g}$ varies slightly from point to point on the earth's surface the weight of a body depends upon its location. Its mass, however, is an intrinsic property independent of location.

5-5 Applications of Newton's Laws

We shall now consider by examples and problems applications of Newton's laws. Mechanics problems of this sort may be broken down into 5 parts.

(1) Identify the body whose motion is to be considered.
(2) Determine from its environment just what forces act *on* the body. The magnitude and direction of these will often be the unknowns which you are trying to find; in these cases you must know of their existence.

(3) Draw a separate diagram of the body alone together with all the forces acting on it. This is called a free body diagram and is a very essential guide to formulating the problem. More often than not students who fail to master mechanics problems do so because they fail to learn to draw free body diagrams; even the "old pros" use them!

(4) Choose a suitable inertial (non-accelerated) reference frame. The location of the origin and the orientation of the axes are at your disposal and you should try to choose them as conveniently as possible. Do not worry at first about how shrewdly you can choose them.

(5) Apply Newton's second law

$$\underline{\underline{F}} = m\underline{\underline{a}} \quad ; \quad \underline{\underline{F}} = \sum_{j=1}^{N} \underline{\underline{F}}_j$$

in component form to each component of $\underline{\underline{F}}$ and $\underline{\underline{a}}$. That is, use

$$F_x = ma_x \tag{5-6a}$$

$$F_y = ma_y \tag{5-6b}$$

$$F_z = ma_z \tag{5-6c}$$

where F_x is the x component of the resultant force and a_x is the x component of the acceleration, etc.

Finally a word of advice; do not be in a hurry to substitute the numerical values into the equations. Work out the answer algebraically. For numerical problems, substitute the numbers as the very last step.

We shall consider a series of problems involving a block on an inclined plane.

>>> Example 1. Consider a block of mass m resting on a smooth inclined plane whose surface makes an angle θ with the horizontal as shown in Fig. 5-1a.

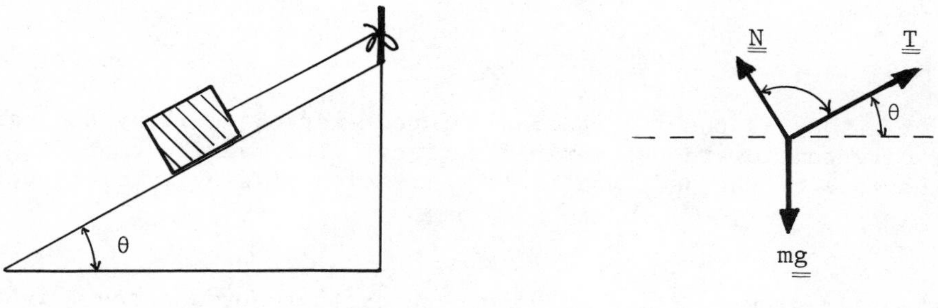

(a) (b)

Figure 5-1. Example 1. (a) A block of mass m is held at rest on a smooth incline by a string. (b) Free body diagram.

(a) Suppose that a light inextensible string is attached to the block and pulls on it parallel to the plane surface. What must be the tension in the string (i.e. the force it exerts on the block) if the block remains at rest?

The block is the body to be considered and the forces acting on it are due to gravity, the plane, and the string; these three constitute the relevant environment of the block. The force of gravity is the block's weight $\mathbf{mg}$ which acts vertically downward. The plane is smooth (frictionless) and therefore can exert a force $\mathbf{N}$ normal to its surface only. Finally the string exerts a pull $\mathbf{T}$, its tension, directed along its length. These are shown in Fig. 5-1b on the free body diagram. Notice that in this case we know the directions of $\mathbf{N}$ and $\mathbf{T}$ but not their magnitudes. It can be that neither the direction nor magnitude of a force are known.

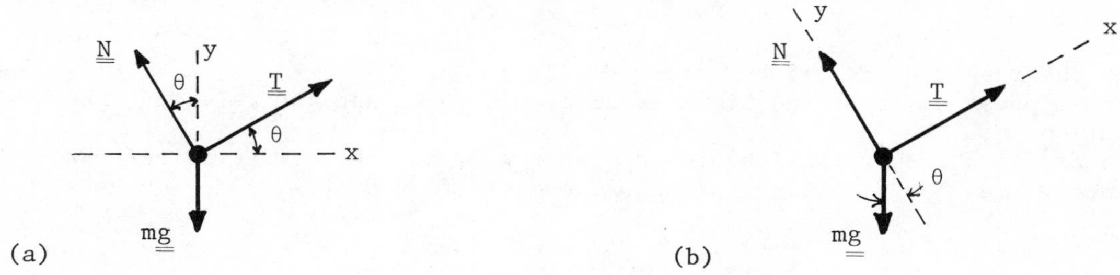

Figure 5-2. Example 1. Free body diagrams with (a) one choice of axes and (b) a better choice of axes.

In Fig. 5-2a we show again the free body diagram with one possible choice of inertial axes. The resultant force is

$$\sum \mathbf{F} = \mathbf{mg} + \mathbf{N} + \mathbf{T} \quad .$$

Since the body remains at rest

$$m\mathbf{a} = 0 \quad ,$$

and therefore Newton's second law says

$$\mathbf{mg} + \mathbf{N} + \mathbf{T} = 0 \quad .$$

With our choice of axes the x component of this equation is

$$- N \sin\theta + T \cos\theta = 0 \quad . \tag{5-7}$$

For the y component we have

$$- mg + N \cos\theta + T \sin\theta = 0 \quad . \tag{5-8}$$

To eliminate one unknown, say N, we multiply Eq. (5-7) by cos θ and Eq. (5-8) by sin θ and add. Then

$$- mg \sin\theta + T \sin^2\theta + T \cos^2\theta = 0$$

or

$$- mg \sin\theta + T (\sin^2\theta + \cos^2\theta) = 0 \quad .$$

Since $\cos^2 \theta + \sin^2 \theta = 1$, we find

$$T = mg \sin \theta \quad . \tag{5-9}$$

A more shrewdly chosen set of axes is shown in Fig. 5-2b. With this choice $\underline{\underline{T}}$ has only an x component and $\underline{\underline{N}}$ only a y component. Since we are interested in $\underline{\underline{T}}$ we consider the x component only and immediately write

$$\sum F_x = 0 = - mg \sin \theta + T$$

which is the result of Eq. (5-9).

(b) Suppose now that the block is placed on the plane and released; what is its acceleration?

The block no longer remains at rest, so $\underline{\underline{a}} \neq 0$; now the string is not part of the environment, i.e. $\underline{\underline{T}} = 0$. Using the second choice of axes (Fig. 5-2b) we have

$$\sum F_x = - mg \sin \theta = ma_x \quad .$$

Therefore

$$a_x = - g \sin \theta \tag{5-10}$$

and the negative sign <u>together with our choice of the positive</u> x <u>direction</u> tells us that the block accelerates down the plane. Here, that is intuitively obvious, but in more complex problems this is not always the case. If $\theta = 0$, i.e. the plane is horizontal, we see that $a_x = 0$ which makes good physical sense. If $\theta = 90^o$ then $a_x = - g$ and the block falls as if the plane were not present. By obtaining an <u>algebraic answer</u> we have easily checked that our solution makes sense in these two special cases.

<<<

>>> Example 2. Suppose that a block of mass m rests on a plane inclined at an angle θ and is attached to a string which passes over a pulley at the top of the incline as in Fig. 5-3a. Another mass M is attached to the other end of the string. Ignore the mass of the string and the mass and any friction in the pulley. Find the acceleration of m and M.

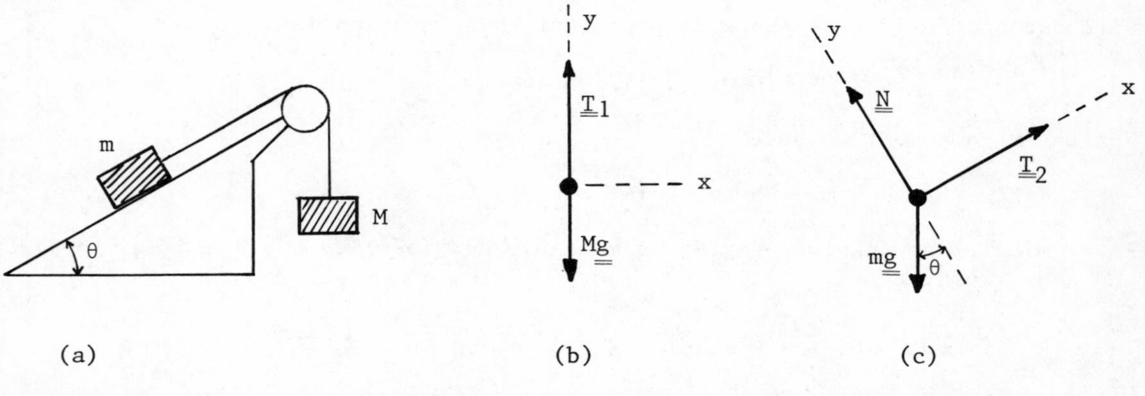

Figure 5-3. Example 2. (a) A block of mass m on an incline and connected via a string and pulley to another block of mass M. (b) Free body diagram and axes for M. (c) Free body diagram and axes for m.

In Fig. 5-3b we have a free body diagram for M together with a choice of axes for its motion. Since the x and z components of motion of M are zero we need consider only the y component. Let $\underline{\underline{T}}_1$ be the upward pull of the string on M. Then

$$\sum F_y = T_1 - Mg = MA_y \qquad (5\text{-}11)$$

where A_y is the acceleration of M.

In Fig. 5-3c we have a free body diagram for m together with a choice of axes for its motion. Notice that they need not be the same axes as those chosen for M. Let the pull of the string on m be $\underline{\underline{T}}_2$. Then for the x component of the motion m we have

$$\sum F_x = T_2 - mg \sin \theta = ma_x \quad . \qquad (5\text{-}12)$$

In this sort of situation where the mass of the string is ignored, where the pulley is massless and frictionless and serves only to change the direction of the string, the magnitude of the tension is the same throughout the string. Thus

$$T_2 = T_1 \quad . \qquad (5\text{-}13)$$

Now, to eliminate T_1 and T_2 we subtract Eqs. (5-11) and (5-12) and use Eq. (5-13) to find

$$Mg - mg \sin \theta = ma_x - MA_y \quad .$$

Because the string is inextensible, if m goes up the plane (along its positive x axis) M must move down (along its negative y axis) by the same amount. Therefore their velocities and accelerations have the same magnitude (but different directions). As a result

$$a_x = - A_y \quad .$$

Thus

$$Mg - mg \sin \theta = (m + M)a_x$$

so

$$a_x = \frac{g(M - m \sin \theta)}{M + m} = - A_y \quad . \qquad (5\text{-}14)$$

Notice that if $M > m \sin \theta$, a_x is positive and m moves up the plane while M goes downward; if $M < m \sin \theta$, a_x is negative and m moves down the plane and M moves upward. The sign automatically comes out right. In such problems as far as the signs of components are concerned, only logical consistency is required. Here that is specified by $a_x = - A_y$. [In hindsight it would have been more convenient to take the y axis "down" in Figure 5-3b. Then we would have $a_x = + A_y$].

<<<

5-6 Frictional Forces

Frictional forces arise when one surface is in contact with another. They act parallel to the surface of contact and always oppose the relative motion. Two sorts of frictional forces are usually considered.

(1) Static friction; magnitude denoted f_s

(a) arises between surfaces at rest relative to one another;
(b) variable in magnitude (being always just large enough to cancel components of other forces in the problem) up to its maximum, $f_{s\ max}$;
(c) $f_{s\ max}$ is proportional to the magnitude of the normal force N which is the force exerted between the surfaces and is normal (perpendicular) to each. The constant of proportionality is called the coefficient of static friction and is denoted μ_s; its value depends upon the materials of both surfaces.

In equation form we have

$$f_s \leq \mu_s N \tag{5-15}$$

and the direction of $\underline{f}_s$ is such as to oppose the relative motion.

(2) Kinetic friction; magnitude denoted f_k

(a) arises between surfaces in relative motion;
(b) f_k is independent of velocity for reasonable speeds;
(c) f_k is independent of the contact area;
(d) f_k is proportional to the normal force N: the coefficient of proportionality is called the coefficient of kinetic friction and is denoted μ_k; its value depends upon the materials of both surfaces.

In equation form then

$$f_k = \mu_k N \tag{5-16}$$

and the direction of $\underline{f}_k$ is such as to oppose the relative motion. Both μ_k and μ_s are dimensionless.

>>> Example 3. A block of ice weighing 50 lb rests on a wooden floor. If a force of 5 lb will just start the ice moving, what is the static coefficient of friction? In Fig. 5-4a is shown a free body diagram of the ice block together with a coordinate system choice.

Figure 5-4. Example 3. (a) Free body diagram with static friction and (b) free body diagram with kinetic friction.

Now $f_s = f_{s\ max}$ because the block just starts to move. So we have

$$\sum F_y = N - mg = 0$$

$$\sum F_x = F - f_{s\ max} = F - \mu_s N = 0 \quad .$$

Hence $$\mu_s = \frac{F}{N} = \frac{F}{mg} = \frac{5\ \text{lb}}{50\ \text{lb}} = 0.1 \quad .$$

If this same force now causes the block to accelerate uniformly at 1.6 ft/s^2 what is μ_k? Fig. 5-4b is a free body diagram together with axes. Then

$$\sum F_y = N - mg = 0$$

or

$$N = mg \quad .$$

Also

$$\sum F_x = F - f_k = F - \mu_k N = ma_x \quad .$$

Thus

$$\mu_k = \frac{F}{N} - \frac{ma_x}{N} = \frac{F}{mg} - \frac{a_x}{g} \quad .$$

Hence

$$\mu_k = \frac{5\ \text{lb}}{50\ \text{lb}} - \frac{1.6\ \text{ft/s}^2}{32\ \text{ft/s}^2} = .05 \quad .$$

<<<

5-7 Circular Motion Dynamics

If a particle undergoes uniform or non-uniform circular motion it is subjected to a centripetal acceleration $a_R = v^2/r$ directed radially inward toward the circle center (see Section 4-4). This must be provided by some force which is given the name centripetal force meaning center seeking. This is not a new kind of force; the name merely describes how the force acts. The magnitude of the centripetal force is mv^2/r and its direction is radially inward. If $\underline{\underline{e}}_r$ is a unit vector pointing radially outward as in Fig. 5-5 then from Newton's second law

$$\underline{\underline{F}}_{\text{centripetal}} = - (m\ v^2/r)\ \underline{\underline{e}}_r \quad . \tag{5-17}$$

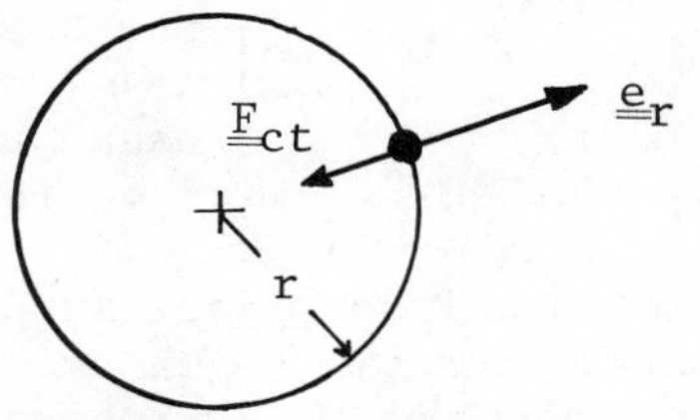

Figure 5-5. Centripetal force $\underline{\underline{F}}_{ct}$ in circular motion of radius r; $\underline{\underline{e}}_r$ is a unit radial vector.

>>> Example 4. Consider a large jet airplane making a level circular turn near the earth's surface at a constant speed of 300 mph. The only forces to be considered are the lift $\underline{\underline{L}}$, which acts perpendicular to the wings, and its weight. The wings make an angle $\theta = 60^\circ$ with the horizontal. Show that the radius of turn (the circle) is approximately 2/3 mile.

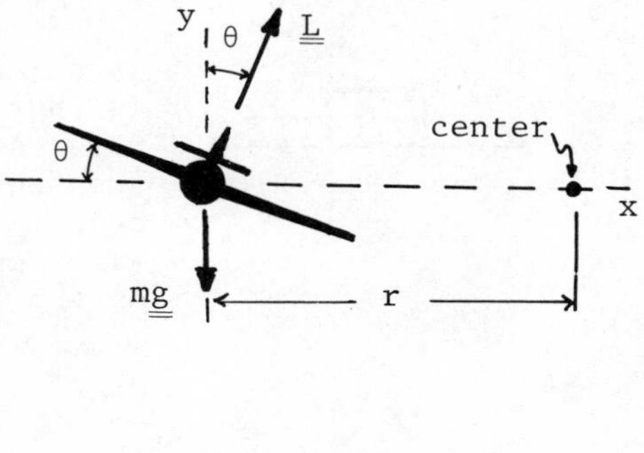

Figure 5-6. Example 4. Free body diagram.

Notice that the weight is not given; it is not needed. A free body diagram is shown in Fig. 5-6. Then

$$L \sin \theta = ma_R = m\,v^2/r$$

$$L \cos \theta = mg \quad .$$

By dividing we have

$$\tan \theta = v^2/rg$$

so

$$r = v^2/(g \tan \theta) \quad .$$

Now 300 mph = (300/60)(88) ft/s , $g = 32$ ft/s^2 and $\tan \theta = \sqrt{3}$. Thus

$$r = \frac{(5)^2(88)^2}{(32)(1.732)}\ \text{ft} = \frac{(5)^2(88)^2}{(32)(1.732)(5280)}\ \text{mile}$$

or

$$r = 0.66 \text{ mile} \simeq 2/3 \text{ mile} \quad .$$

<<<

5-8 Programmed Problems

The emphasis in this first problem will be to guide you through the analysis in a sort of prescriptive form. While your style of solving problems may become more (or less) sophisticated as you gain experience, the essential attack will remain the same. We will follow the process as outlined in Section 5 of this chapter.

1.

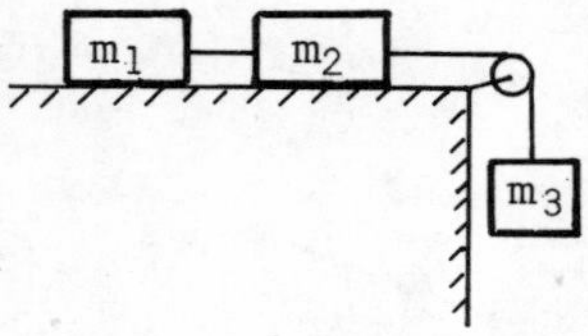

Three masses connected by two strings are arranged as shown. The problem is to determine the acceleration of each mass and the tension in each string. The surface is frictionless.

The motion of an object such as m_1 or m_2 or m_3 is determined by the interaction between these objects and their environment. Thus we must determine the pertinent environment. Isolate the masses by enclosing each in a dashed circle.

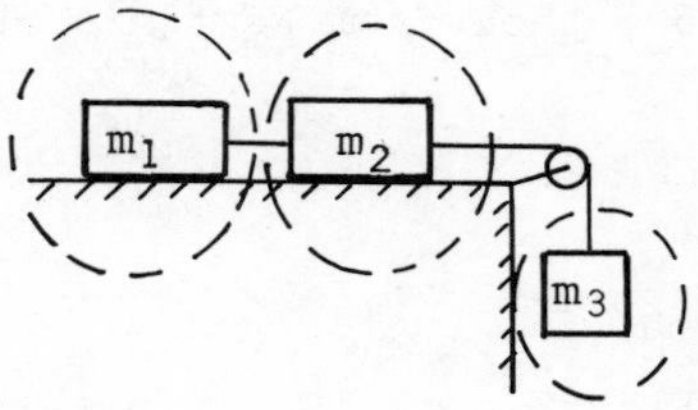

2.

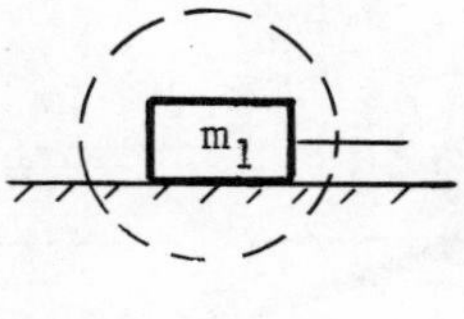

The utility of isolating an object diagrammatically is that the dashed circle focuses your attention on the "environment" of the object. What is the environment of the mass m_1 as shown?

a. The string

b. The surface

c. The earth's gravitational attraction. Do not forget this one.

3. 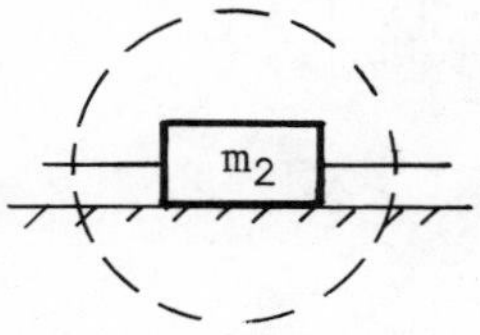

It turns out that the dashed circle always cuts through the environment, either in fact as with the string and surface, or in principle as with the earth's gravitational attraction.

What is the environment of m_2?

a. Two strings.
b. The surface.
c. The earth's gravitational attraction.

---

4. Before drawing free body diagrams for the masses a few words about strings are in order.

(1) Strings under tension always exert forces acting away from the object to which they are attached, i.e. they always "pull". Try pushing a horse with a rope.
(2) The tension is everywhere the same in a string (at least for the problems you will encounter most of the time).

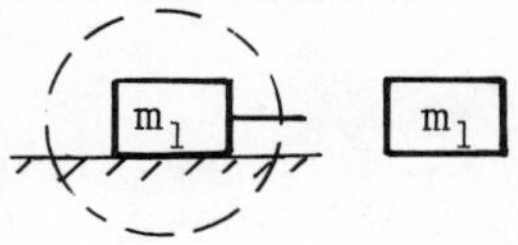

Having identified previously the three elements of the environment of m_1, indicate the nature of their effect on m_1 by vectors.

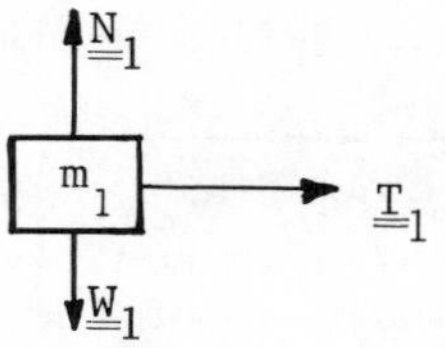

The subscripts on $\mathbf{N}_1$ and $\mathbf{W}_1$ mean that they act on m_1. $\mathbf{T}_1$ means the tension in string number 1.

Always resist the temptation to omit forces like $\mathbf{N}_1$ and $\mathbf{W}_1$ even though you may think or know that they are not important in a given problem.

---

5. a. $\mathbf{N}_1$ is the force of the ________ on m_1.

b. $\mathbf{T}_1$ is the force of the ________ on m_1.

c. $\mathbf{W}_1$ is the force of the ________ on m_1.

a. Surface.
b. String.
c. Earth.

---

6. Note $\mathbf{T}_1$ acts away from m_1, $\mathbf{W}_1$ acts toward the earth, $\mathbf{N}_1$ acts normal to the surface and away from the surface.

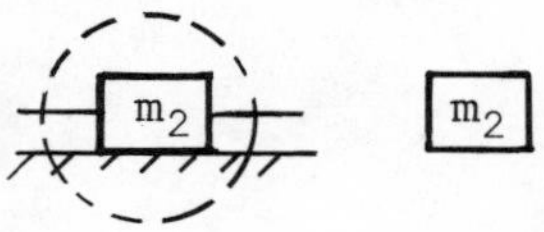

Draw a free body diagram for m_2. Remember the string on the left is the same string which acts on m_1.

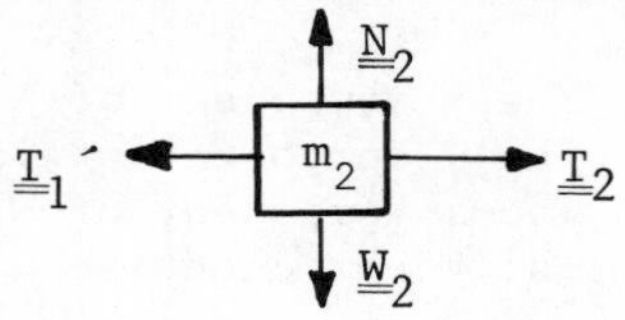

$\mathbf{T}_1' = -\ \mathbf{T}_1$

---

7. 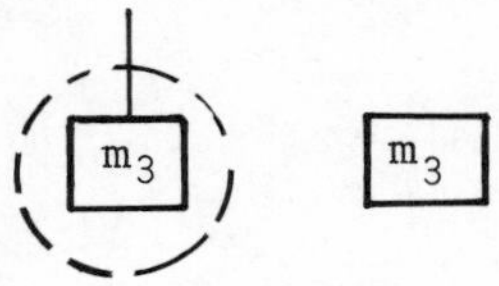

Draw the free body diagram for m_3. The string attached to m_3 is the same as that attached to the right side of m_2.

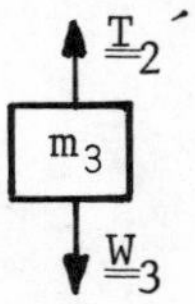

$|\mathbf{T}_2'| = |\mathbf{T}_2|$

8.

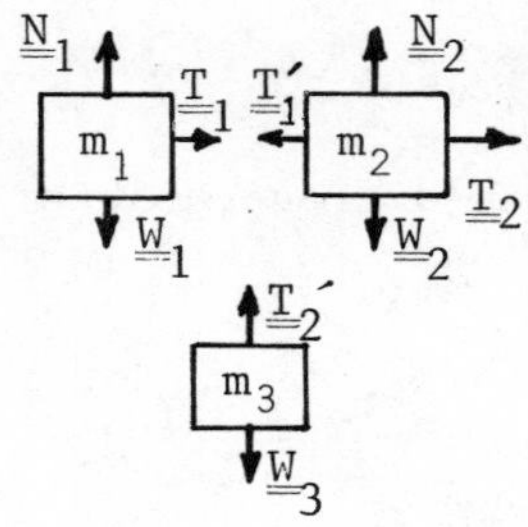

Complete the following concerning the diagram to the left.

a. $\sum \underline{F}$ on m_1 = ____ .

b. $\sum \underline{F}$ on m_2 = ____ .

c. $\sum \underline{F}$ on m_3 = ____ .

d. $\underline{T}_1$ on m_1 (is/is not) equal to $\underline{T}_1'$ on m_2.

e. $|\underline{T}_1|$ on m_1 (is/is not) equal to $|\underline{T}_1'|$ on m_2.

a. $\underline{N}_1 + \underline{T}_1 + \underline{W}_1$

b. $\underline{T}_1' + \underline{N}_2 + \underline{W}_2 + \underline{T}_2$

c. $\underline{T}_2' + \underline{W}_3$

d. Is not (oppositely directed).

e. Is (same string).

---

9. We have now completed the part of the problem which is usually most difficult for beginners. The name of the game in dynamical mechanics is "find the force". Now that we have done that we can solve the problem.

Write down Newton's second law for the motion for mass m_1.

$\underline{F} = m_1\underline{a}$

where $\underline{F} = \underline{N}_1 + \underline{T}_1 + \underline{W}_1$ is the vector sum of the forces acting on m_1. $\underline{a}_1$ is the vector acceleration of m_1.

10.

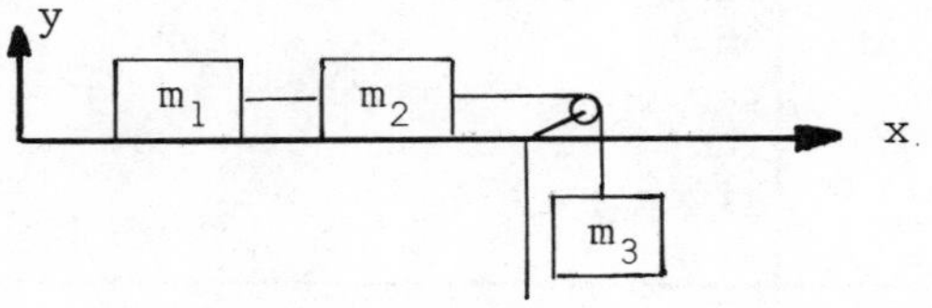

To facilitate solving this problem we will resolve all vectors into components.

Superimposed on the diagram is a two-dimensional coordinate system with sign conventions indicated. Answer the following questions.

a. The y component of $\underline{N}_1$ is _____ .
b. The x component of $\underline{T}_1'$ on m_2 is _____ .
c. The x component of $\underline{W}_1$ is _____ .
d. The y component of $\underline{W}_1$ is _____ .

The force vectors are shown in frame 8.

a. N_1
b. $-T_1$
c. Zero
d. $-W_1$

Note: The resolution is easy in this case because all vectors lie along coordinate axes.

11. From the diagram in the previous frame,

a. $\sum F_x$ on m_1 = ________.

b. $\sum F_y$ on m_1 = ________.

c. $\sum F_x$ on m_2 = ________.

d. $\sum F_y$ on m_2 = ________.

e. $\sum F_x$ on m_3 = ________.

f. $\sum F_y$ on m_3 = ________.

Here $\sum$ means the algebraic sum of the components.

a. T_1

b. $N_1 - W_1$

c. $T_2 - T_1$

d. $N_2 - W_2$

e. 0

f. $T_2 - W_3$

Note that we have used the fact that $|\underline{T}_1'| = |\underline{T}_1|$ and $|\underline{T}_2'| = |\underline{T}_2|$.

---

12. From the character of the motion of the masses in this problem what can you say about $\sum F_y$ acting on m_1 or m_2?

Zero in both cases since $\sum F_y = ma_y$ and a_y is zero for both m_1 and m_2. (They do not crash through the surface.)

---

13. The strings tying the blocks together do not stretch. What does this tell you about the relationship among the accelerations of the three blocks?

Their magnitudes must be the same. If blocks 1 and 2 accelerate to the right (along + x) then block 3 accelerates down (along - y). If we denote a_{1x} simply by a, then the three accelerations are related by:

$$a_{1x} = a_{2x} = a, \quad a_{3y} = -a.$$

---

14. Using the results of frame 11, write Newton's second law for blocks 1, 2 and 3. (Express all accelerations in terms of the common acceleration "a" introduced in the answer to the above frame.)

(1) $T_1 = m_1 a$

(2) $T_2 - T_1 = m_2 a$

(3) $T_2 - W_3 = - m_3 a$ (where $W_3 = m_3 g$)

---

15. Now we have to solve the above three equations for a, T_1 and T_2. Note that there are <u>three equations in three unknowns</u>.

(1) $T_1 = m_1 a$

(2) $T_2 - T_1 = m_2 a$

(3) $T_2 - m_3 g = - m_3 a$

Eliminate T_1 from (1) and (2) and then solve for T_2.

(1) $T_1 = m_1 a$

(2) $T_2 - T_1 = m_2 a$

} add

$T_2 = (m_1 + m_2)a$

16. Substitute the answer to frame 15 into (3) and solve for a

$T_2 - m_3 g = -m_3 a$

$[(m_1 + m_2)a] - m_3 g = -m_3 a$

$$a = \frac{m_3}{(m_1 + m_2 + m_3)}\, g$$

17. Substitute this "a" from the above answer into (1) of frame 15 to determine T_1 and into (3) to determine T_2.

$$T_1 = \frac{m_1 m_3}{(m_1 + m_2 + m_3)}\, g$$

$$T_2 = \frac{m_3 (m_1 + m_2)}{(m_1 + m_2 + m_3)}\, g$$

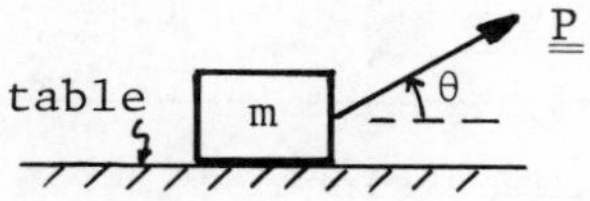

A force $\underline{P}$ acts at an angle θ on an object of mass m. The coefficient of kinetic friction between the object and the surface is μ_k. What is the acceleration of the object?

18. Draw a free body diagram of the object showing all the forces which act on it.

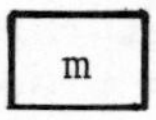

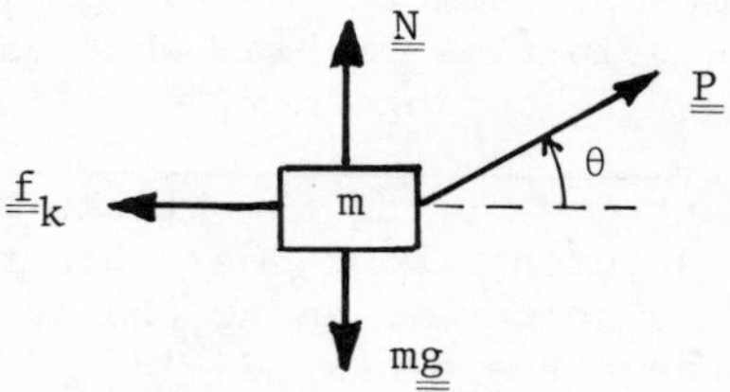

19. Are any of the forces shown in the previous frame acting on the table?

No. The forces shown are all acting on m, not on the table.

20. Newton's third law says that to every action there is an equal and opposite reaction. Describe the force which is the reaction to the force $\underline{N}$ in this problem. Draw a vector representing this reaction force and show the body on which it acts. What is the source of this reaction force?

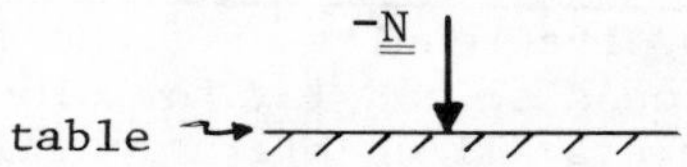

The reaction to $\underline{N}$ is $-\underline{N}$ acting on the table. The object m exerts this force on the table.

21.

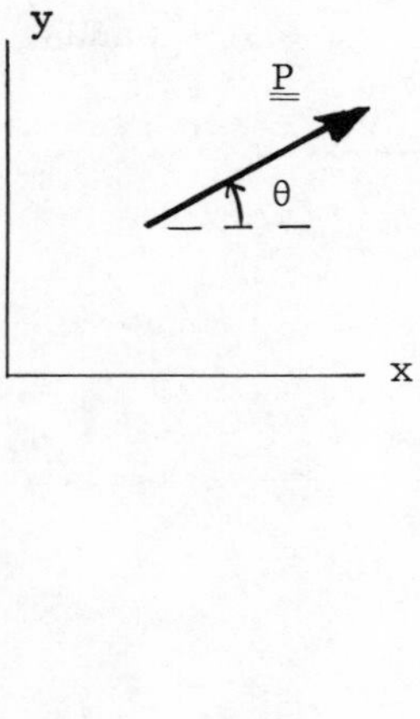

For the coordinate system shown, the components of $\underline{P}$ are:

P_x = ________ ,

P_y = ________ .

$P_x = P \cos\theta$,

$P_y = P \sin\theta$.

22.

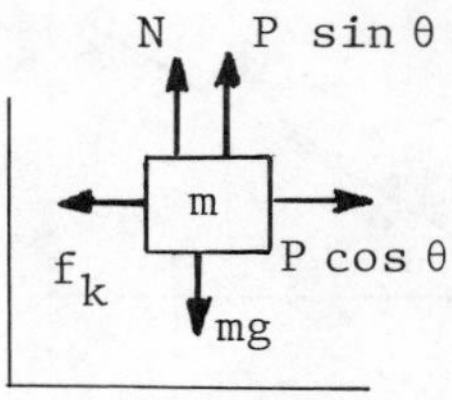

Here we have shown all the forces resolved into their components. Note that by selecting this coordinate system, the forces $\underline{\underline{N}}$, $\underline{\underline{f}}_k$ and $\underline{\underline{mg}}$ were already resolved.

Using this diagram write the equations for the x and y components of the motion of m (i.e. apply Newton's second law).

$\sum F_x = ma_x$:

$$P \cos \theta - f_k = ma_x$$

$\sum F_y = ma_y$:

$$N + P \sin \theta - mg = 0 \quad (\text{since } a_y = 0)$$

---

23. In this problem is the normal force N equal to the weight mg of the object m?

No. From the answer to the above frame, $N = mg - P \sin \theta$.

---

24. In view of the previous frame you should not assume that a normal force exerted by a surface is always equal to the weight of the object it is supporting. It may be less (or in some cases, more). Write down the equation relating the frictional force f_k and the normal force N. Remember that this is not a static friction problem).

$f_k = \mu_k N$.

If we <u>were</u> dealing with static friction then all we could say would be $f_s \leq \mu_s N$.

25.

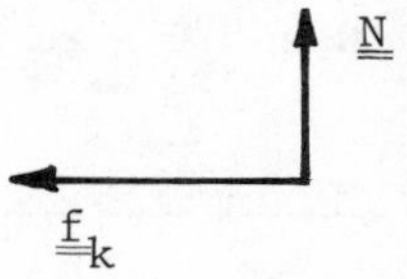

Sketch the resultant of the two forces shown at the left. What environment is the cause of these froces?

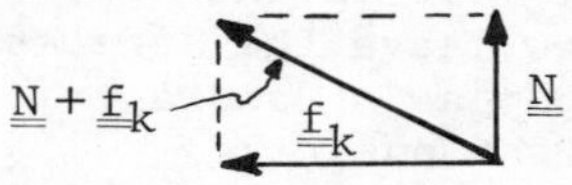

The table exerts these forces on the object. $\underline{\underline{N}} + \underline{\underline{f}}_k$ is called the total contact force. In problems without friction, contact forces have only normal components.

---

26. Returning to the equations

(1) $P \cos \theta - f_k = ma_x$

(2) $N + P \sin \theta - mg = 0$,

solve (1) algebraically for a_x and (2) for N.

$$a_x = \frac{P \cos \theta - f_k}{m}$$

$$N = mg - P \sin \theta$$

---

27. Substitute appropriately for f_k into the above answer for a_x.

$$f_k = \mu_k N = \mu_k(mg - P \sin \theta)$$

$$a_x = \frac{P \cos \theta - \mu_k(mg - P \sin \theta)}{m}$$

---

28. For P = 8 lb, $\theta = 45^\circ$, and $\mu_k = 0.2$ what is the acceleration of the object if it weighs 16 lb?

Noting that mg = 16 lb (given) and that m = mg/g = 0.5 slug, the above formula for a_x evaluates to

$$a_x = \frac{8(0.707) - 0.2[16 - 8(0.707)]}{0.5}$$

$$= 7.2 \text{ ft/s}^2 \ .$$

A racing car travels around a circular track of radius R. The track is banked an an angle θ relative to the horizontal. There is one speed v with which the car can negotiate the curve such that there is no sideways (frictional) force on the car tires. Find this speed.

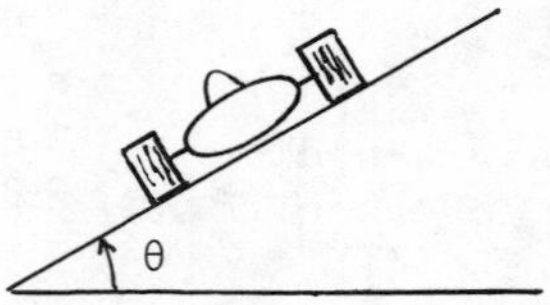

29. Start by drawing a free body diagram of the car showing all the forces acting on it. Label these forces and give the cause of each.

N is the force of the track on the car. It is normal to the track since there is no friction at this speed (given). Of course mg is the weight of the car (i.e. the gravitational attraction of the earth on the car.

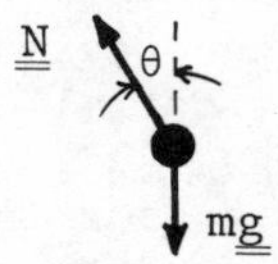

30. Choose the axis system shown and write Newton's second law for the car in component form. Put in the force components but for now leave the acceleration components simply as a_x and a_y. Note that the positive x axis points toward the center of the circle.

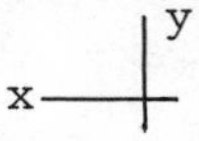

$\sum F_x = ma_x$: $N \sin \theta = ma_x$

$\sum F_y = ma_y$: $N \cos \theta - mg = ma_y$

31. (a) What do you know about a_x in this problem?

(b) What do you know about a_y in this problem?

(c) In view of your answers to (a) and (b), simplify the two equations in the previous answer frame.

(a) For uniform circular motion the acceleration is v^2/R and it is directed toward the center. Thus $a_x = v^2/R$.

This is called the centripetal acceleration.

(b) There is no acceleration in the y direction, i.e. $a_y = 0$.

(c)

$$N \sin \theta = mv^2/R ,$$

$$N \cos \theta = mg .$$

32. Solve for the "correct" speed v.

Dividing the above two equations eliminates N. Solving for v,

$$v = \sqrt{Rg \tan \theta}$$

33. (a) If the car exceeds this speed v, what is the direction of the frictional force on the tires? Which way does the car tend to skid?
(b) If the car goes too slow, what is the direction of the frictional force? Which way does the car tend to skid?

(a) In this case the centripetal acceleration (v^2/R) is larger than at the correct speed. Hence friction must act down the track surface. The car tends to skid outward and friction must hold it in.
(b) Now the centripetal acceleration is too small. The car tends to skid inward and the friction must act back up the track surface.

Now one last exercise involving the ideas associated with uniform circular motion. A circus performer rides a motorcycle around the inside of a rough cylinder of radius R. He does not slide down the side of the cylinder.

34.

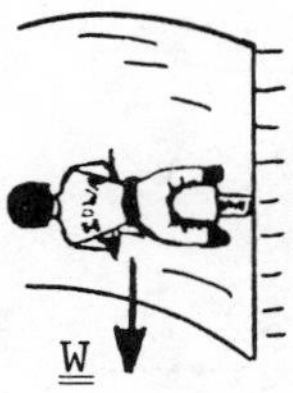

In the diagram we have drawn the gravitational force acting on the cycle and rider. Draw the force which is required in order that they go in a circle.

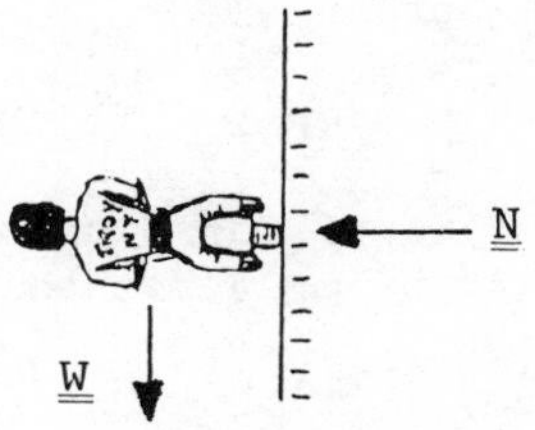

Note: N is the centripetal force. In this particular problem the centripetal force is the normal force exerted on the cycle by the cylinder.

---

35. As shown here the cycle and rider system would accelerate down the cylinder. How is it possible that the system doesn't?

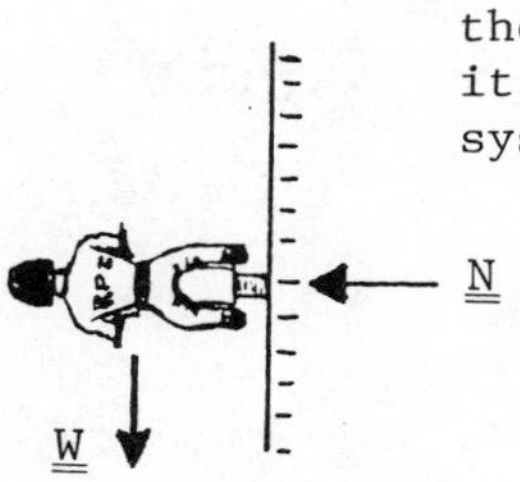

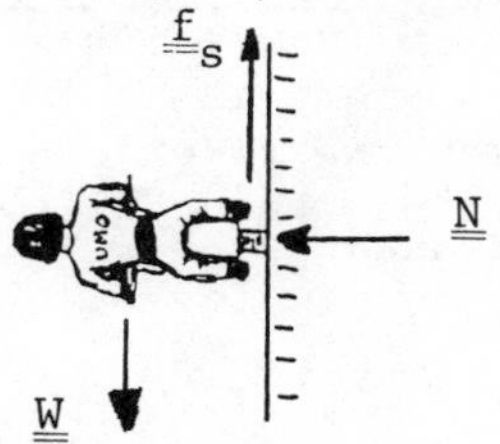

A frictional force f_s, exerted on the tire by the cylinder must oppose W. [Are you surprised that it is the static frictional force? Remember there is no slipping at the point where the tire makes contact with the cylinder.]

---

36. Is there an additional force, called centrifugal, which acts on this system to oppose N?

No. This is a popular misconception. If there were such a force then the body would be in equilibrium and thus go in a straight line rather than a circle. [The centrifugal force is exerted on the cylinder by the tires.]

37. For the motion to be circular without slipping we must have:

(1) ________ $= mv^2/R$

(2) f_s - ______ = 0

(3) $f_s \leq$ ________ $(= f_{s\ max})$

(1) N

(2) W

(3) $\mu_s N$

---

38. The above three equations apply as the rider is going in a circle without slipping. If the rider speeds up a little while maintaining the same circular orbit, what happens to the normal force N?

From (1) we see that N must increase.

---

39. Using W = mg we can substitute (2) into (3) and obtain

$$mg \leq \mu_s N \quad .$$

Substitute for N using (1), divide both sides by mg and obtain

$$1 \leq \_\_\_\_\_\_\_\_\_\_$$

$$1 \leq \frac{\mu_s v^2}{Rg}$$

---

40. The result of the last frame shows that the quantity $\mu_s v^2/Rg$ has a (maximum/minimum) value. What is this value?

minimum, one. The result of the last frame says that this quantity is "greater than or equal to one".

---

41. Thus there is a minimum value of v for which the rider can maintain the circular motion without slipping. Calculate this minimum speed if R = 25 m and $\mu_s = 0.50$.

$$\mu_s v^2/Rg = 1 \quad \text{(at minimum v)}$$

$$v = \sqrt{Rg/\mu_s}$$

$$= \sqrt{(25\ \text{m})(9.8\ \text{m/s}^2)/(0.500)}$$

$$= 22.1\ \text{m/s} \simeq 50\ \text{mph} \quad .$$

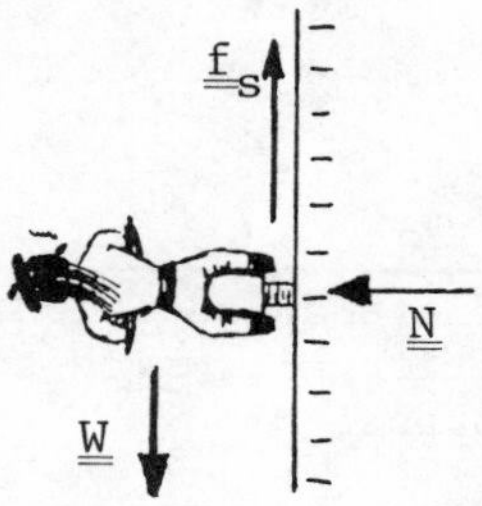

Chapter 6: WORK AND ENERGY

REVIEW AND PREVIEW

Previously you have learned that knowledge of the acceleration at all times allows one to deduce the detailed motion of a particle. The acceleration which you studied is an instantaneous property associated with the force on a particle, that is a property associated with the motion of a particle over an infinitesimal part of its motion. In this chapter you will study a property of the force during a finite motion of the particle (i.e. the *work* done by the force). You will meet a quantity called *kinetic energy* whose change is related to the work.

GOALS AND GUIDELINES

In this chapter you have four major goals:

1. Learning the definition of work (Sections 6-2 and 6-4). It is important to realize that work is a scalar, but we need to know the direction of the force as well as its magnitude to compute the work it does.
2. Learning the definition of the kinetic energy associated with the motion of a particle (Eq. 6-7). Like work, kinetic energy is also a scalar quantity.
3. Understanding the relationship between kinetic energy and the work done on a particle (i.e., the *work energy* theorem, Eq. 6-11). Frequently the use of this theorem allows us to determine the speed of a particle without detailed knowledge of its motion.
4. Learning the definition of power.

6-1 Basic Ideas

Given the resultant force $\mathbf{F}$ on a particle its acceleration is

$$\mathbf{a} = \mathbf{F}/m$$

which is valid whether or not $\mathbf{F}$ is constant. If $\mathbf{F}$ is not constant, then neither is $\mathbf{a}$ and the constant acceleration equations (4-5) do *not apply*. While it is possible to continue applying Newton's laws to problems involving non-constant forces, the mathematics becomes more difficult. In order to progress we will attack these problems from a different approach which will require the use of the concepts *work*, *kinetic energy* and the *work-energy theorem*.

6-2 Work Done by a Constant Force

If $\mathbf{F}$ is a *constant* force acting *on* a particle which undergoes a displacement $\mathbf{d}$, the *work* done *by* $\mathbf{F}$ *on* the particle is *defined* to be the scalar product of $\mathbf{F}$ and $\mathbf{d}$

$$W = \mathbf{F}\cdot\mathbf{d}, \text{ constant force.} \qquad (6\text{-}1)$$

If N constant forces $\mathbf{F}_j$, j = 1, 2, ... N, act on the particle, the work done by each is $W_j = \mathbf{F}_j\cdot\mathbf{d}$ and the *total* work done *on* the particle is

$$W = \sum_{j=1}^{N} W_j = \sum_{j=1}^{N} \mathbf{F}_j\cdot\mathbf{d} = \Big[\sum_{j=1}^{N} \mathbf{F}_j\Big]\cdot\mathbf{d} = \mathbf{F}\cdot\mathbf{d} \ . \qquad (6\text{-}2)$$

where $\mathbf{F}$ is the resultant force acting *on* the particle.

Note that W is a scalar, may be positive or negative, and is zero if $\mathbf{d} = 0$ or if $\mathbf{F}$ has no component along the direction of $\mathbf{d}$.

Let the displacement direction be the x axis and the displacement be Δx. If F_x is the component of $\mathbf{F}$ along this direction, then the work done by $\mathbf{F}$, $W = F_x \Delta x$, may be given a geometrical interpretation as the area under the curve F_x versus x for the interval Δx. This is illustrated in Fig. 6-1.

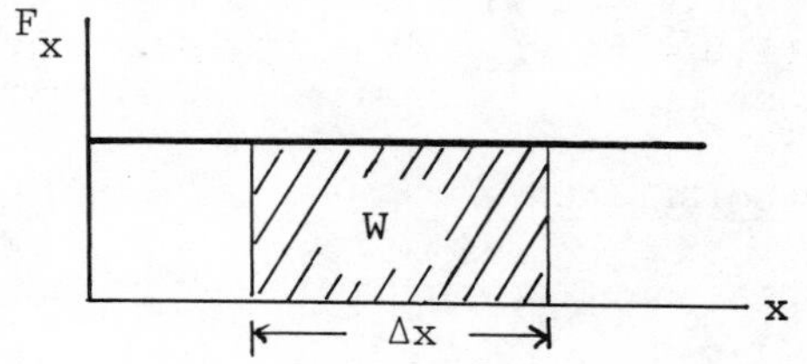

Figure 6-1. The work done by a constant force shown as the area under the curve F_x versus x.

\>>> Example 1. Consider a block of mass m sliding a distance d down an inclined plane of angle θ. The frictional force is f_k. What is the work done by the weight of the block, by friction, and by the normal force of the plane? What is the total work done on the block?

Fig. 6-2 is a free body diagram of the block together with coordinate axes. Notice that the positive x axis has been chosen down the plane.

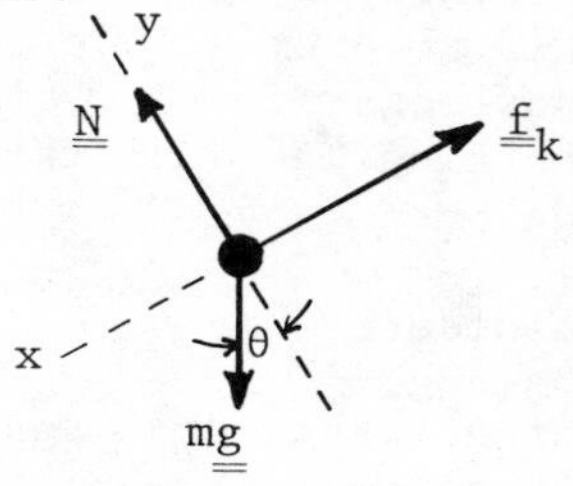

Figure 6-2. Example 1. Free body diagram of a block sliding down an inclined plane with friction f_k.

The forces are all constant and are given by

$$\mathbf{F}_{weight} = m\mathbf{g} = mg \sin\theta \mathbf{i} - mg \cos\theta \mathbf{j} \quad ,$$

$$\mathbf{F}_{friction} = - f_k \mathbf{i} \quad ,$$

$$\mathbf{F}_{normal} = N\mathbf{j} \quad .$$

The displacement is $\mathbf{d} = d\mathbf{i}$ so the work done by the weight of the block (i.e. work done by gravity) is

$$W_{mg} = [mg \sin\theta \mathbf{i} - mg \cos\theta \mathbf{j}] \cdot d\mathbf{i} = mgd \sin\theta \quad ,$$

since $\mathbf{i} \cdot \mathbf{i} = 1$ and $\mathbf{j} \cdot \mathbf{i} = 0$. The work done by friction is

$$W_{f_k} = (- f_k \mathbf{i}) \cdot d\mathbf{i} = - f_k d$$

and the work done by the normal force is

$$W_N = N\mathbf{j} \cdot d\mathbf{i} = 0 \quad .$$

Therefore, the total work done on the block is

$$W_{Total} = W_{mg} + W_{f_k} + W_N = (mg \sin \theta - f_k)d \quad . \qquad <<<$$

The unit of work is the unit of force times the unit of distance. The units of force in the various systems are summarized in Table 6-1.

Table 6-1

WORK UNITS IN VARIOUS SYSTEMS OF UNITS

System	Work Unit	Conversion Factor
SI	joule = N·m	J = 10^7 erg = 0.738 ft·lb
Cgs	erg = dyn·cm	erg = 10^{-7} J = 7.38×10^{-8} ft·lb
BES	ft·lb	ft·lb = 1.36 J = 1.36×10^7 erg

6-3 Primer on Integration

The student who is unfamiliar with integration should see the Appendix.

6-4 Work Done by a Variable Force

First consider a one dimensional case where both the displacement and the force are directed along the x-axis. We will also consider the special case where $\mathbf{F}$ is the known function of x, i.e. $\mathbf{F} = F(x)\mathbf{i}$, where F(x) is known. During a small displacement dx the force does an increment of work dW given by

$$dW = F(x)\, dx \quad . \qquad (6\text{-}3)$$

As x varies from x_1 to x_2, W varies from the value 0 to the value W_{12}. Thus

$$\int_0^{W_{12}} dW = W_{12} = \int_{x_1}^{x_2} F(x)\, dx \quad . \qquad (6\text{-}4)$$

This is the area under the curve F(x) versus x between x_1 and x_2.

In the case where $\mathbf{F}$ varies both in direction and magnitude, we let an infinitesimal displacement be $d\mathbf{r}$; over this interval $\mathbf{F}$ does work

$$dW = \mathbf{F} \cdot d\mathbf{r} = F \cos \theta \, ds \qquad (6\text{-}5)$$

where θ is the angle between $\mathbf{F}$ and $d\mathbf{r}$ and ds is an increment of length along the displacement path (i.e. the magnitude of dr). Then

$$W_{ab} = \int_a^b F \cos \theta \, ds \quad . \qquad (6\text{-}6)$$

Such integrals are called line integrals and can be evaluated only if we know how θ and F vary along the path.

6-5 Kinetic Energy and the Work-Energy Theorem

The kinetic energy of a particle is denoted as K and defined by

$$K = \tfrac{1}{2} mv^2 \tag{6-7}$$

where m is the mass of the particle and v is its speed. Also, since v^2 is the square of the magnitude of $\underline{v}$

$$K = \tfrac{1}{2} m\underline{v}\cdot\underline{v} \quad . \tag{6-8}$$

Differentiating K with respect to t we find

$$dK/dt = \tfrac{1}{2} m\,(d\underline{v}/dt)\cdot\underline{v} + \tfrac{1}{2} m\underline{v}\cdot d\underline{v}/dt = m\underline{v}\cdot d\underline{v}/dt \quad .$$

Now $d\underline{v}/dt = \underline{a}$, so

$$dK/dt = m\underline{a}\cdot\underline{v} \quad . \tag{6-9}$$

Then since $\underline{v} = d\underline{r}/dt$ we have

$$dK = m\underline{a}\cdot d\underline{r} \quad .$$

But, $m\underline{a}$ is the force acting on the particle, so

$$dK = \underline{F}\cdot d\underline{r} \quad . \tag{6-10}$$

Now integrate both sides of Eq. (6-10) from some initial value to some final value. The left hand side is the change in kinetic energy and the right hand side is the (total) work done on the particle. Thus

$$K_f - K_i = W_{if} \tag{6-11}$$

which is the work-energy theorem. Eq. (6-11) merely relates two defined quantities, work and kinetic energy. If the kinetic energy decreases, the work done on the particle is negative; if the particle's speed is constant the force does no work because $K_f = K_i$.

>>> Example 2. A particle of mass m is initially held at rest at a point x_1 relative to some origin. The particle is repelled from the origin by a force $\underline{F}(x) = (k/x^2)\,\underline{i}$. It is now released. What is the velocity of the particle when it is very far from the origin?

The work done on the particle as it moves from x_1 to some value x_2 is

$$W_{12} = \int_1^2 dW = \int_{x_1}^{x_2} \left(\frac{k}{x^2}\,\underline{i}\right)\cdot(dx\underline{i}) = \int_{x_1}^{x_2} \frac{k}{x^2}\,dx = k\left(\frac{1}{x_1} - \frac{1}{x_2}\right) \quad .$$

From the work-energy theorem

$$W_{12} = K_2 - K_1 \quad .$$

Since $K_1 = 0$, $W_{12} = K_2 = \tfrac{1}{2} mv_2{}^2$ and therefore

$$v_2 = \left[\frac{2k}{m}\left(\frac{1}{x_1} - \frac{1}{x_2}\right)\right]^{\frac{1}{2}} \quad .$$

Now, very far from the origin x_2 is very large compared with x_1, so $1/x_2$ is very small compared with $1/x_1$. Thus for large enough values of x_2 the velocity is approximately constant and is given by

$$\mathbf{v} = \left[\frac{2k}{mx_1}\right]^{1/2} \mathbf{i} \quad .$$

The force in this problem represents the kind which occurs between two protons. <<<

6-6 Power

<u>Power</u> is the rate of doing work; the instantaneous power P delivered by a working agent is

$$P = dW/dt \quad . \tag{6-12}$$

The <u>average power</u> $\bar{P}$ is the total work done divided by the time interval over which it is done

$$\bar{P} = W/t \quad . \tag{6-13}$$

From the work-energy theorem, Eq. (6-11), applied to an infinitesimal increment we have

$$dW = dK$$

and from Eq. (6-9) then

$$P = dW/dt = dK/dt = m\mathbf{a}\cdot\mathbf{v} = \mathbf{F}\cdot\mathbf{v} \quad . \tag{6-14}$$

This is another expression for the power delivered by the resultant force acting on a particle.

The unit of power in the SI units is a <u>watt</u> (abbreviated w) = joule/second. In the Cgs system a unit of power is an erg/s which is not given another name; similarly in the British system the natural unit of power is the ft·lb/s. Often a larger unit, the horsepower, is used 1 hp = 550 ft·lb/s = 746 w .

Work can be expressed in units of power times time; you pay for the <u>work</u> done by the electricity supplied to your home, i.e. you pay for so many kilowatt-hours.

6-7 Programmed Problems

1.

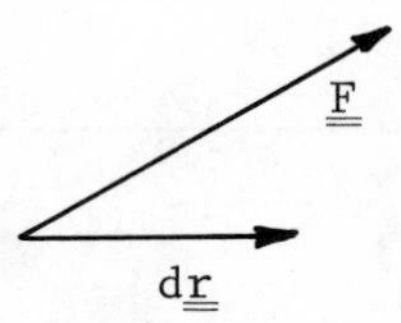

The concept of work in physics is not measured in terms of human physical exertion or perspiration. The increment of work dW done by a force $\mathbf{F}$ in displacing an object a distance $d\mathbf{r}$ is

dW = ______ .

$\mathbf{F}\cdot d\mathbf{r}$

2.

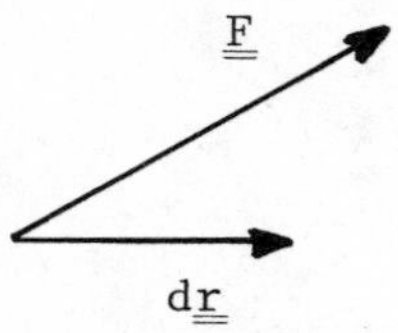

$\mathbf{F} \cdot d\mathbf{r}$ is the scalar or dot product of $\mathbf{F}$ and $d\mathbf{r}$. This can be written as

$$\mathbf{F} \cdot d\mathbf{r} = |\mathbf{F}|\,|d\mathbf{r}| \cos\theta \ .$$

Indicate θ on the diagram.

This is sometimes confusing in that there are two angles "between" $\mathbf{F}$ and $d\mathbf{r}$. The correct one is the smaller of the two angles.

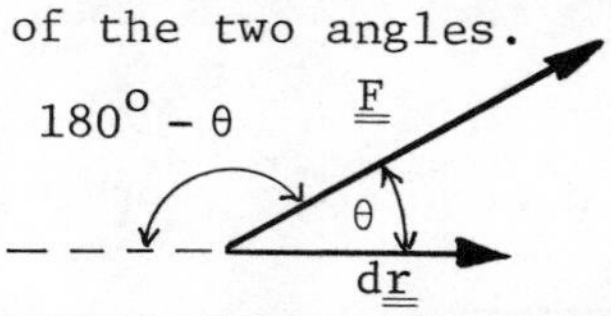

3.

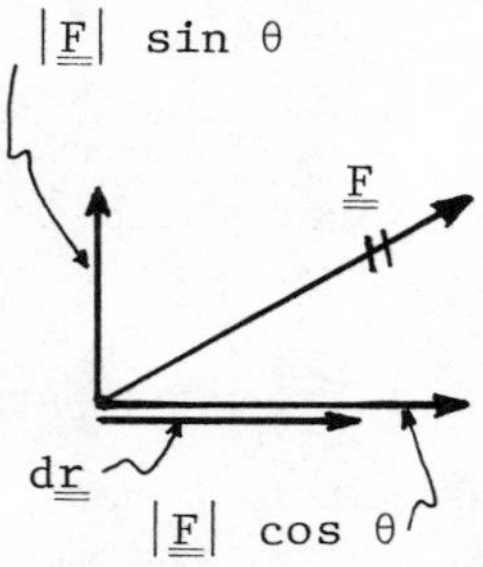

You will note in the expression

$$|\mathbf{F}|\,|d\mathbf{r}| \cos\theta$$

that $|\mathbf{F}| \cos\theta$ is the component of $\mathbf{F}$ in the direction of the displacement $d\mathbf{r}$.

What is the work done by the component $|\mathbf{F}| \sin\theta$?

Zero, the vector component whose magnitude is $|\mathbf{F}| \sin\theta$ is perpendicular to $d\mathbf{r}$ and this component does no work.

---

4. The angle between $\mathbf{F}$ and $d\mathbf{r}$ can vary from 0 to 180°. The cosines of 0°, 90°, and 180° are _____, _____, and _____ respectively.

1, 0, -1.

This implies that the scalar quantity work may be positive, negative, or zero. It is zero whenever $\mathbf{F}$ and $d\mathbf{r}$ are <u>perpendicular</u>.

5.

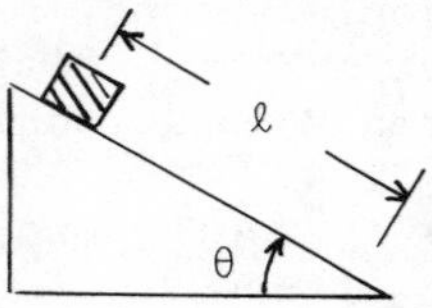

A block of mass m starts from rest and slides down a smooth incline. What is the speed of the block at the bottom of the incline?

Draw a free body diagram of m showing the forces acting on it. Keep the orientation of m as shown.

6.

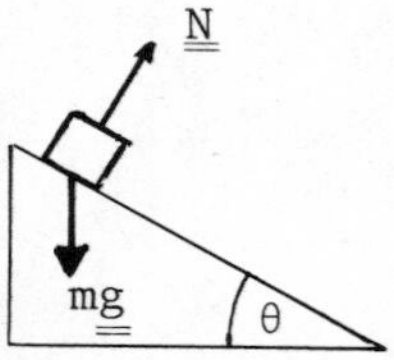

Superimpose a coordinate system on this diagram with the x-axis parallel to (and down) the incline.

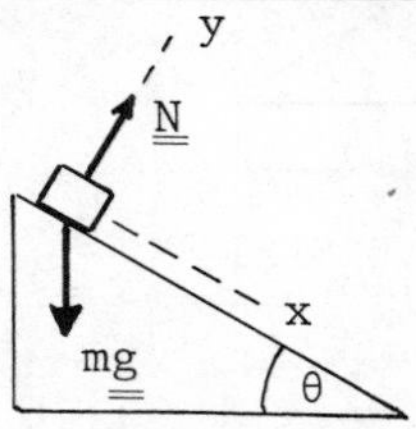

7. 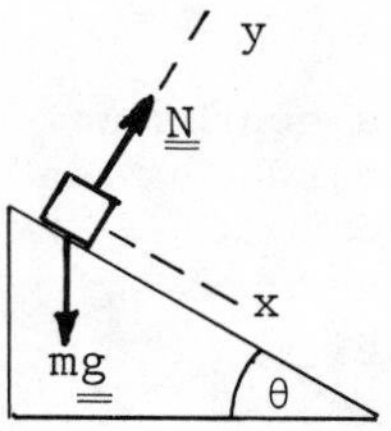

Resolve both forces into components.

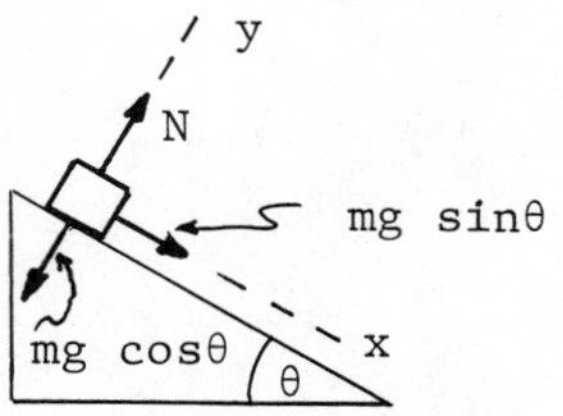

$\underline{\underline{N}}$ is automatically resolved because of the choice of axes.

---

8. Write Newton's second law for the x component of motion. Express forces in the manner of the previous answer.

$F_x = ma_x$

$mg \sin\theta = ma_x$

---

9. What is the acceleration down the plane? (So far this problem is just like those in the chapters on particle dynamics.)

From the previous answer,

$a_x = g \sin\theta.$

---

10. In kinematics you used the expression

$$v_x^2 = v_{xo}^2 - 2a_x(x - x_o)$$

to calculate velocities. Reread frame 5 to refresh your mind about the problem and using the a_x in frame 9, solve for v_x^2 at the bottom of the incline.

$v_x^2 = v_{xo}^2 - 2a_x(x - x_o)$

$v_{xo}^2 = 0$, the block starts from rest.

$a_x = g \sin\theta$

$x - x_o = \ell$

$v_x^2 = 2g\ell \sin\theta$

---

11. What is the kinetic energy of m at the bottom?

$K = \frac{1}{2} mv_x^2$

$K = \frac{1}{2}\, 2\, mg\ell \sin\theta$

$K = mg\ell \sin\theta$

---

12. 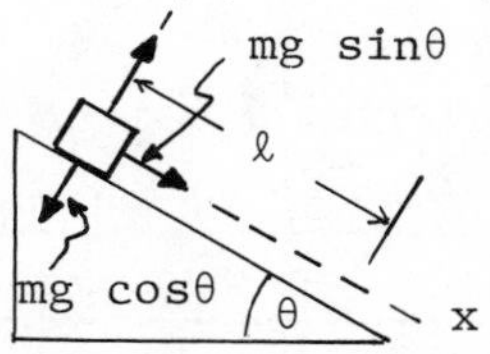

From $x = 0$ to $x = \ell$, the

(a) work done by $\underline{\underline{N}}$ = ____,

(b) work done by $\underline{\underline{mg}}$ = __________ .

a. Zero ($\underline{\underline{N}} \perp$ to displacement).

b. $mg\ell \sin\theta$.

Note that only the component of $\underline{\underline{mg}}$ down the incline contributes to the work in b.

---

13. The work-energy theorem states that $W = \Delta K$ where W is the work of the resultant force and ΔK the change in kinetic energy. In this problem is $mg \sin\theta$ the magnitude of the resultant force?

Yes. The components N and $mg \cos\theta$ add up to zero.

Our calculation shows

$$W = mg\ell \sin\theta \qquad \text{frame 12.}$$

$$\Delta K = mg\ell \sin\theta \qquad \text{frame 11.}$$

We obtained ΔK by finding the force, calculating the acceleration and then using a kinematic equation to find v_x. Calculating the work and setting this equal to the change in kinetic energy is clearly an efficient calculational method. It is particularly useful when the resultant force (and hence acceleration) is not constant.

---

14. A ball is thrown vertically upward near the earth's surface with an initial speed of 50 ft/s. Will there be a resultant force acting on this ball? If so, identify it fully.

Yes.

$\mathbf{W} = m\mathbf{g}$ acting downward.

---

15. 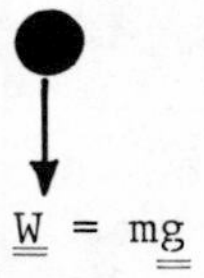

Here we show the ball at a point prior to the top of its trajectory. Draw vectors representing its velocity $\mathbf{v}$ and an infinitesimal displacement $d\mathbf{r}$.

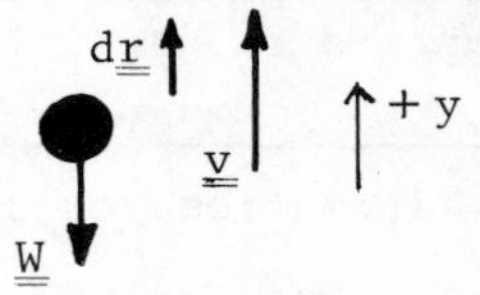

$d\mathbf{r}$ is in the direction of motion; the magnitude of $d\mathbf{r}$ is $|d\mathbf{r}| = dy$.

---

16. The work-energy theorem can be written

$$\int_1^2 \mathbf{F}\cdot d\mathbf{r} = \tfrac{1}{2}\, mv_2^{\,2} - \tfrac{1}{2}\, mv_1^{\,2}$$

where 1 and 2 are the initial and final positions. From the definition of the scalar product this can be written for our problem

$$-\int_1^2 |mg|\,|dy| = \tfrac{1}{2}\, mv_2^{\,2} - \tfrac{1}{2}\, mv_1^{\,2} \ .$$

What is it about the problem that requires the minus sign in front of the integral?

$\mathbf{F}$ and $d\mathbf{r}$ have opposite directions.

$$\mathbf{F}\cdot d\mathbf{r} = |\mathbf{F}|\,|d\mathbf{r}| \cos\theta$$
$$= (mg)(dy)(-1)$$
$$= -\ mg\ dy\ .$$

Note that $\theta = 180^{\circ}$.

17.

$$-\int_1^2 mgdy = \tfrac{1}{2}\, mv_2^{\,2} - mv_1^{\,2}\ .$$

For our problem we wish to know the maximum height that the ball will attain. What will v_2 be at that position?

Zero. The ball will momentarily be at rest.

---

18. Now we have

$$-\int_1^2 mgdy = -\ \tfrac{1}{2}\, mv_1^{\,2}$$

or

$$\int_1^2 mgdy = \tfrac{1}{2}\, mv_1^{\,2}\ .$$

$$\int_1^2 mg\ y\ dy = mg\ y\Big|_1^2$$
$$= mg(y_2 - y_1)$$

Note that m and g are constants and can be factored out of the integral.

19. Finally, cancelling the common factor m, we have

$$gy\Big|_1^2 = \tfrac{1}{2} v_1^2$$

with y at position 1 = 0 and y at position 2 = h.

Substitute appropriately and solve for h. Remember that v_1 = 50 ft/sec. Obtain a numerical result.

$$g[h - 0] = \tfrac{1}{2} v_1^2$$

$$h = \tfrac{1}{2} v_1^2/g$$

$$= \tfrac{1}{2} (50 \text{ ft/s})^2/(32 \text{ ft/s}^2)$$

$$= 39 \text{ ft.}$$

Chapter 7: CONSERVATION OF ENERGY

REVIEW AND PREVIEW

In the previous chapter you studied the important concept of energy associated with the speed of a particle (i.e., its kinetic energy). In this chapter you will meet an energy associated with the position of a particle (called its potential energy). Under certain circumstances the total mechanical energy (i.e., kinetic plus potential) is a constant. Under such conditions, there is conservation of mechanical energy which when it applies is an extremely powerful concept.

GOALS AND GUIDELINES

In this chapter you have four major goals:

1. Understanding the distinction between conservative and non-conservative forces which determine whether or not one can define a potential energy associated with the force (Section 7-1).
2. Learning how to compute the potential energy associated with a given conservative force (Eqs. 7-1a and 7-1b).
3. Learning to use the concept of conservation of mechanical energy in problems (Section 7-2, Example 1 and programmed problems).
4. Learning how to handle non-conservative forces (Section 7-3 and programmed problems).

7-1 Basic Ideas; Conservative and Non-conservative Forces

We will begin this section with a programmed problem.

1. As an exercise in calculating work consider a particle of mass m which is constrained to move in the (x,y) plane. We wish to calculate the work done by the force $\underline{\underline{F}} = (4 - 2y)\underline{\underline{i}}$. 1. What is F_x? ______ 2. What is F_y? ______ 3. What is F_z? ______	1. (4 - 2y) 2. Zero 3. Zero Note that the component F_x happens to depend upon the y coordinate.
2. With the force $\underline{\underline{F}} = (4 - 2y)\underline{\underline{i}}$ 1. $\underline{\underline{F}}$ = ______ when y = 0. 2. $\underline{\underline{F}}$ = ______ when y = 2.	1. $\underline{\underline{F}} = 4\underline{\underline{i}}$ 2. $\underline{\underline{F}} = 0$; $(4 - 2y) = 0$ when y = 2.

3.

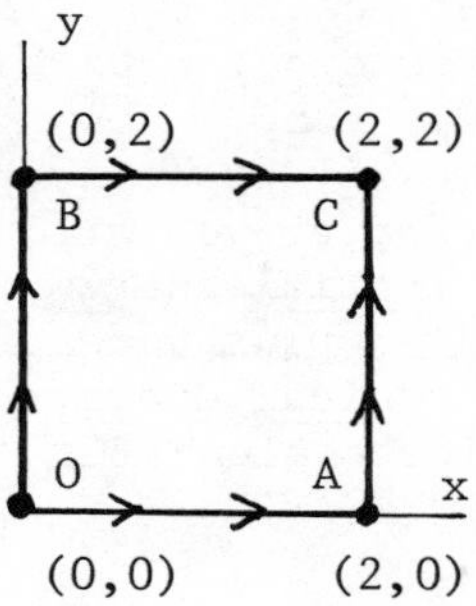

To the left we show two paths, OBC and OAC, which take m from the point O to the point C.

If it occurs to you that a particle starting from rest at O could not get to C under the influence of the force $\underline{\underline{F}} = (4 - 2y)\underline{\underline{i}}$, you are right. There must be other forces acting, but we only want to calculate the work done by this force.

For the force $\underline{\underline{F}} = (4 - 2y)\underline{\underline{i}}$ describe specifically the force on m along the path

1. OB ; 2. BC ; 3. OA ; 4. AC.

Remember that you must give both magnitude and direction to describe a vector fully.

Since the force always points in the direction of the unit vector $\underline{\underline{i}}$ (to the right) each answer below will be understood to contain that fact.

1. The magnitude of $\underline{\underline{F}}$ depends upon the y coordinate along OB.

2. The magnitude is zero along BC because $(4 - 2y)$ is zero for $y = 2$.

3. The magnitude is constant along OA and is equal to 4 when $y = 0$.

4. The magnitude depends upon the y coordinate along AC.

---

4. Referring back to the diagram of frame 3 write the displacement vectors for the path segments as indicated below. Answer 1 is given as an example.

1. Displacement from O to B = $2\underline{\underline{j}}$.
2. Displacement from B to C = ______.
3. Displacement from O to A = ______.
4. Displacement from A to C = ______.

1. $2\underline{\underline{j}}$
2. $2\underline{\underline{i}}$
3. $2\underline{\underline{i}}$
4. $2\underline{\underline{j}}$

Note that all have the same magnitude. The path length is the same for each segment.

5.

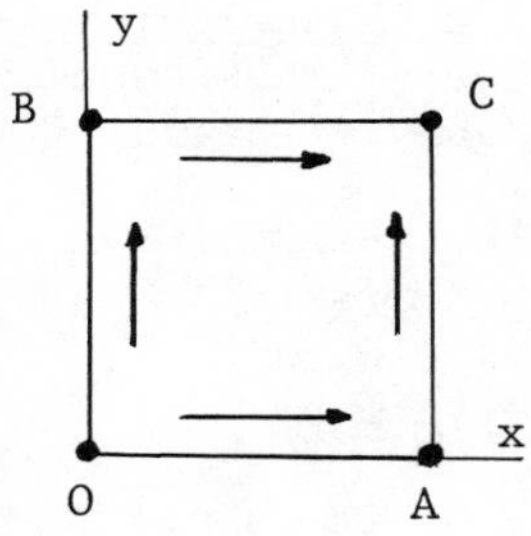

A little qualitative reasoning will make this problem easier. We break the problem into parts and consider each part separately.

On the diagram to the left are drawn vectors which are intended to represent displacements along the four path segments.

For each of the four displacement vectors draw an associated force vector to represent $\underline{\underline{F}} = (4 - 2y)\underline{\underline{i}}$. Don't worry about relative magnitude.

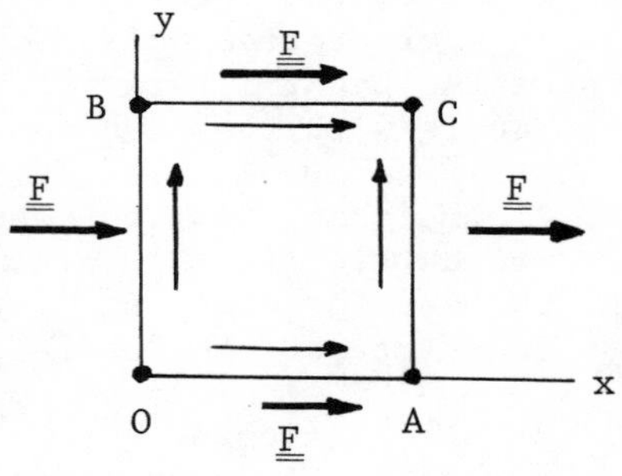

Note that the force vectors are all pointing in the $\underline{\underline{i}}$ direction.

6. Looking at the answer to the previous frame, why is it true that for $\underline{\underline{F}} = (4 - 2y)\underline{\underline{i}}$ the 1. Work to go from 0 to B is zero? 2. Work to go from A to C is zero?	Because in both cases the force vector is perpendicular to the displacement vector.
7. Now we have only the segments OA and BC to deal with. Look at answer 2 of frame 3 and explain why the work along BC is zero.	Because the force along BC is zero.
8. Now the only segment left to consider is OA along which vector $\underline{\underline{F}}$ is constant. $\underline{\underline{d}} = 2\underline{\underline{i}}$ and $\underline{\underline{F}} = 4\underline{\underline{i}}$. $W_{OA} = \underline{\underline{F}} \cdot \underline{\underline{d}} =$ ______. Let d have the units meters and F have the units newtons. Obtain a numerical result.	$\underline{\underline{F}} \cdot \underline{\underline{d}} = 4\text{ N} \times 2\text{ m} = 8\text{ J}$
9. Now we review the result: $W_{OB} = 0$, $W_{OA} = 8$ joules, $W_{BC} = 0$ and $W_{AC} = 0$, so, $W_{0 \to B \to C} = W_{OB} + W_{BC} =$ _____ $W_{0 \to A \to C} = W_{OA} + W_{AC} =$ _____.	 0 8 J

This exercise in calculating work has an important consequence that will be developed in this chapter. In this problem the work done by $\underline{\underline{F}} = (4 - 2y)\underline{\underline{i}}$ on the mass m as it moved from 0 to B to C was zero, while in going from 0 to A to C the work was 8 joules. Although in both cases it finally arrived at C, the work done by the force depended on the path from 0 to C. Because of this result, the force of this problem is called non-conservative.

The force in the foregoing programmed problem was an example of a non-conservative force. There are forces called conservative forces for which the work is the same regardless of the path. There are three equivalent definitions of conservative forces. The first deals with the resultant force (or with a single force if only it acts).

1. If the change in the kinetic energy, ΔK, of a particle moving in any round trip (i.e. returning to its starting point) is zero, the resultant force acting on the particle is conservative. If $\Delta K \neq 0$, at least one of the forces acting on the particle is non-conservative.

2. A force is (conservative/non-conservative) if the work done by it on a particle that moves through any round trip is (zero/not zero).

Definition 2 is shown to be equivalent to 1 by the work-energy theorem.

3. A force is (conservative/non-conservative) if the work done by it on a particle that moves between any two points depends on (only the points and not the path between them/the path taken between the points).

Definition 3 is shown to be equivalent to 2 (and hence to 1) by traversing one segment of some round trip path a → b → a backwards; here a and b are points of interest.

Many of the forces you will encounter such as that of a spring and gravity are conservative. Kinetic friction is another example of a non-conservative force.

7-2 Potential Energy -- Conservation of Mechanical Energy

Potential energy is energy a particle possesses by virtue of its position. More generally, it is the energy of configuration of a system. For particles the system is the particle and its relevant environment. Potential energy is defined in terms of the work done by a force in moving a particle from some reference point to the position of interest. Therefore, the concept only makes sense for conservative forces; in that case the work depends only on the reference point and the position of interest. For example, if friction (a non-conservative force) is considered, the work done by it in going from the reference point to some other position depends upon how the particle gets there (path dependence) and indeed on how many times the particle has been there before.

In one dimension, if F(x) is a conservative force and x_0 is the reference point, the potential energy may be defined by

$$U(x) = -\int_{x_0}^{x} F(x)dx + U(x_0), \tag{7-1a}$$

or if i is the initial point, f the final point and W_{if} the work done by the (conservative) force as the system goes from i to f, then

$$U_f - U_i = -W_{if} \text{ (conservative).} \tag{7-1b}$$

That is, the change in potential energy is the negative of the work done by the conservative force as the particle moves from the reference point. The choice of reference point is arbitrary and is taken for convenience; often x_0 is taken as that point for which F(x) is zero. The choice of the value of $U(x_0)$ is similarly arbitrary.

From Eq. (7-1) it follows that

$$F(x) = -\frac{dU(x)}{dx} \tag{7-2}$$

which says that the potential energy is a function of position whose negative derivative is the conservative force. The potential energy function is defined only up to an arbitrary constant. In other words, it is a particular integral of - F(x) [see the Appendix]. Any constant could be added to U(x) and Eq. (7-2) would still yield the same force. Also the change in potential energy, which is the negative of the work done by F(x), would still be the same. Physically this is a choice in the zero of potential energy.

The total mechanical energy is defined as the sum of the kinetic plus potential energies. That is

$$E = K + \sum U \tag{7-3}$$

where there is one potential energy term in the sum for each conservative force.

If only conservative forces act then the total mechanical energy E is conserved, that is, it is constant. This is the law of conservation of mechanical energy. It follows from Eq. (7-3) since

$$(E_f - E_i) = (K_f - K_i) + \sum (U_f - U_i) \tag{7-4a}$$

but

$$\sum (U_f - U_i) = -\sum W_{if} \text{ (conservative)} \tag{7-4b}$$

and from the work-energy theorem

$$E_f - E_i = K_f - K_i - \sum W_{if} \text{ (conservative)} = 0 \quad . \qquad (7\text{-}4c)$$

This says that $E_f = E_i$ and therefore E is a constant.

Often the conservation of mechanical energy is written in differential form

$$dE = dU + dK = 0 \qquad (7\text{-}5)$$

where U stands for the total potential energy, and dU is an infinitesimal increment. It is related to the infinitesimal work of the resultant conservative force $\underline{\underline{F}}$ by

$$dU = - dW = \underline{\underline{F}} \cdot d\underline{\underline{r}} \quad . \qquad (7\text{-}6)$$

In two or more dimensions the change in potential energy is

$$\Delta U = U_b - U_a = - \int_a^b \underline{\underline{F}} \cdot d\underline{\underline{r}} \quad . \qquad (7\text{-}7)$$

Because $\underline{\underline{F}}$ is conservative, any convenient path may be used to evaluate the integral of Eq. (7-7). The relationship analogous to Eq. (7-2) is

$$\underline{\underline{F}}(\underline{\underline{r}}) = - \underline{\underline{\nabla}} U(\underline{\underline{r}}) \qquad (7\text{-}8)$$

where $\underline{\underline{\nabla}}$ is the symbol for the gradient. One says that $\underline{\underline{F}}$ is the negative gradient of $U(\underline{\underline{r}})$, and that conservative forces are derivable from a potential energy function. In Cartesian coordinates Eq. (7-8) is equivalent to

$$F_x = - \frac{\partial U(x,y,z)}{\partial x} \qquad (7\text{-}9a)$$

$$F_y = - \frac{\partial U(x,y,z)}{\partial y} \qquad (7\text{-}9b)$$

$$F_z = - \frac{\partial U(x,y,z)}{\partial z} \qquad (7\text{-}9c)$$

where the round differentiation symbol, ∂, means for example

$$\frac{\partial U(x,y,z)}{\partial x} \equiv \left[\frac{dU(x,y,z)}{dx}\right]_{y,z \text{ held constant}} \quad . \qquad (7\text{-}10)$$

In other words, the explicit x dependence is differentiated with the other variables (y and z) regarded as constant. For example, if

$$U(x,y,z) = 4x^2 + 2xy + 3yz$$

then

$$\frac{\partial U(x,y,z)}{\partial x} = 8x + 2y + 0 \quad .$$

>>> Example 1. The force due to gravity is a conservative force. The gravitational potential energy of a particle of mass m near the earth's surface (where g is constant) is

$$U(y) - U(0) = - \int_0^y (- mg)\, dy = m\, g\, y$$

The origin is at the earth's surface and the y axis is positive upward. The value of the potential energy function at y = 0, U(0) may be chosen to be any constant value, without change in physical content. It is usually convenient to take U(0) = 0 at the surface of the earth (y = 0).

The conservation of mechanical energy yields

$$E = K + U = \tfrac{1}{2} mv^2 + mgy = \text{constant}.$$

The value of E depends upon the initial conditions. For example if the particle starts at y = 0 with velocity $\underline{\underline{v}} = v_o \underline{\underline{j}}$ then

$$E = \tfrac{1}{2} mv_o{}^2$$

and

$$\tfrac{1}{2} mv^2 + mgy = \tfrac{1}{2} mv_o{}^2$$

or

$$v^2 = v_o{}^2 - 2gy$$

which is the result of Eq. (4-4d) with $v_y(0) = v_o$, $a_y = -g$, and $y(0) = 0$. <<<

The conservation of mechanical energy is used to solve some problems which would otherwise prove to be quite difficult as well as those which can be easily solved by the use of familiar dynamic and kinematic equations.

7-3 Nonconservative Forces -- Conservation of Energy

The work done on an object by nonconservative forces has always been found to be associated with a change in some new form of energy. For example, the work W_f done by friction on an object is found to increase the heat energy of the system (particle plus environment). Formally

$$W_f = -\Delta Q \tag{7-11}$$

where ΔQ is the change in heat energy. Since W_f is negative, Q is positive. With each nonconservative force we associate a change in some form of energy and write the conservation of total energy as

$$\Delta E_{Total} = \Delta K + \Delta U + \Delta Q + \text{changes in other forms of energy} = 0. \tag{7-12}$$

Here Δ denotes "the change in". In words Eq. (7-12) asserts the conservation of energy:

> Energy may be transformed from one kind to another, but it is neither created nor destroyed; total energy is conserved.

The principle of the conservation of energy is one of the most fundamental in physics. The conservation of mechanical energy is a special case which applies only when all forces are conservative.

More Programmed Problems

1. We will consider the problem of frames 5-13 of Chapter 6. You should review these first.

   What work does the normal force $\mathbf{N}$ do as the block slides down the incline?

None. The normal force $\mathbf{N}$ is perpendicular to the displacement so

$$dW = -\mathbf{F}\cdot d\mathbf{r} = -\mathbf{N}\cdot d\mathbf{r}$$

$$= N\ dy \cos 90^{o} = 0\ .$$

---

2. Since $\mathbf{N}$ contributes no energy change in the problem we need consider only the change in potential energy of the gravitational field. Use the positive y axis upward. This is a different choice of axes than was used before; it is convenient here.

   (a) What is the force of gravity (vectorially)?
   (b) How much work would gravity do in taking the block to the top of the incline?
   (c) What is the change in gravitational potential energy?

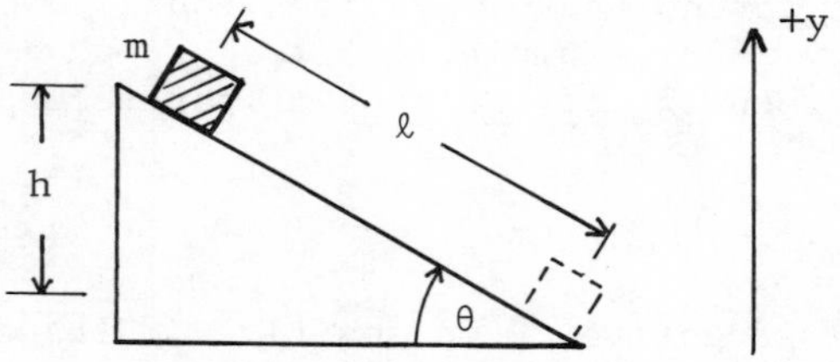

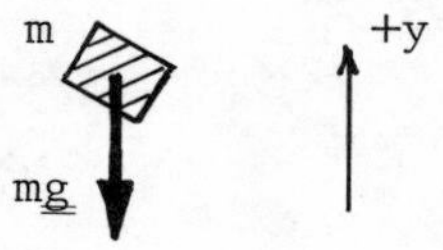

a. $\mathbf{F}_{gravity} = -mg\ \mathbf{j}$

b. Since gravity is a conservative force the work it does is path independent. Therefore we may imagine that the block was lifted (by someone) straight up. Then a little displacement is $d\mathbf{r} = dy\ \mathbf{j}$ so

$$W_{\substack{by\\gravity}} = \int_0^h (-mg\ \mathbf{j})\cdot dy\ \mathbf{j}$$

$$= -mgh = -mg\ell \sin\theta .$$

c.

$$\Delta U_{gravity} = U_{final} - U_{initial}$$

$$= -W_{by\ gravity}$$

$$= mg\ell \sin\theta .$$

---

3. Since the change in potential energy of gravity is $mg\ell \sin\theta$ as the block is raised to the top of the incline, what would be the change in gravitational potential energy as the block comes back down the incline?

$$\Delta U_{gravity} = -mg\ell \sin\theta\ .$$

You could figure this by recomputing the change or by realizing that if the potential energy increases as the block goes up, it must decrease by the same amount as the block goes down the incline.

4. To find the speed of the block at the bottom of the incline write a statement of the conservation of mechanical energy.

There are a couple of ways you may have written this:

i)

$K_{top} + U_{top} = K_{bottom} + U_{bottom}$.

$K_{top} = 0$ because the block starts from rest.

$U_{top} = mgh = mg\ell \sin \theta$.

This corresponds to making the arbitrary (but useful) choice that $U_{gravity} = 0$ at the bottom.

$K_{bottom} = \frac{1}{2} mv^2_{bottom}$.

$U_{bottom} = 0$.

ii) $\Delta K + \Delta U = 0$.

$(\frac{1}{2} mv_b{}^2 - 0) + (- mg\ell \sin \theta) = 0$

Either form is correct and one follows from the other.

---

5. What is the speed of the block at the bottom of the incline?

$\frac{1}{2} mv^2_{bottom} = mg\ell \sin \theta$

$v_b = \sqrt{2 g\ell \sin \theta} = \sqrt{2 gh}$.

Notice that the speed is exactly the same as if the block had fallen freely from a height $h = \ell \sin \theta$ because the normal force, $\underline{\underline{N}}$ does no work as the block comes down the incline.

---

6. If instead of a nice straight incline suppose the block starts from a height h on a frictionless roller-coaster type incline.

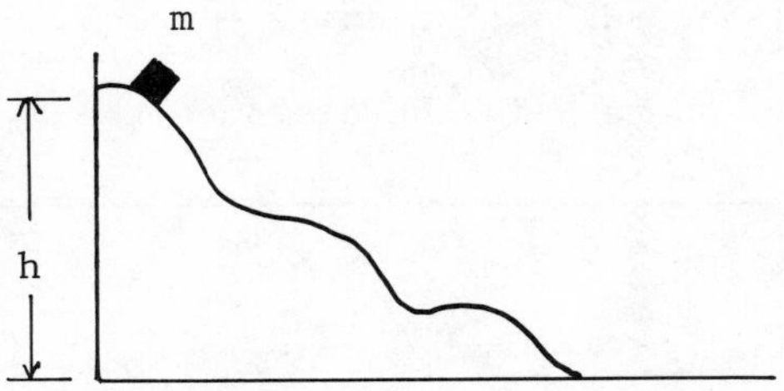

What is the speed of the block at the bottom?

$v_{bottom} = \sqrt{2 gh}$

This is exactly the same answer as before. You may verify this by repeating frames 2-5 for this case. The speed of the block will vary in a complicated way coming down this incline depending upon the shape of the incline, but at the bottom the speed will be as before.

7. Suppose now that the mass is at the bottom of the incline on a horizontal surface and has speed v. Let the horizontal surface have coefficient of kinetic friction μ_k with the block. Take the positive x axis to the right.

(a) What is the frictional force?
(b) Is this force conservative?

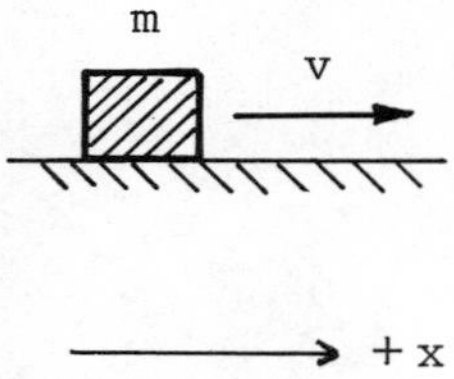

a.

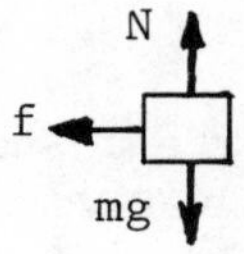

$\underline{\underline{f}} = -\mu_k N\underline{\underline{i}} = -\mu_k mg\underline{\underline{i}}$

The minus sign occurs because friction opposes the motion and hence $\underline{\underline{f}}$ is directed to the left.

b. No! To see this imagine that you slide the block over this horizontal surface from some point back to that same point without giving it any final velocity. The change in kinetic energy around this closed path is zero, but friction does a great deal of work depending upon how long the path is. [Caution: $f = +\mu_k mg\underline{\underline{i}}$ when the block moves to the left]

---

8. Suppose we want to know how far the block slides over this horizontal surface given that its initial speed is v.

Does gravity now play a role?

No. The gravitational force is now normal to any displacement so gravity does no work. Thus there is no change in gravitational potential energy.

---

9. To find how far the block slides write expressions for

(a) The initial total mechanical energy.
(b) The final total mechanical energy.
(c) The work done by friction.

a. $E_{initial} = K_{initial} = \frac{1}{2}mv^2$.

b. $E_{final} = K_{final} = 0$ because the block came to rest.

Notice that gravitational potential energy is ignored in each case because it does not change.

c. $W_f = \int_0^D \underline{\underline{f}} \cdot dx\underline{\underline{i}} = -\mu_k mgD$

where D is the distance the block slides.

---

10. Write an expression for the change in total energy and find how far the block slides.

$$\Delta E = 0 = \Delta K + \Delta U + \Delta Q$$
$$= (K_f - K_i) + (U_f - U_i) + (-W_f)$$
$$= -\frac{1}{2}mv^2 + 0 + \mu_k mgD \ .$$

Therefore $D = \dfrac{v^2}{2g\mu_k}$

11. A more complicated problem in which you would use the conservation of total energy would be the block on the straight incline starting at height $h = \ell \sin\theta$ and sliding to the bottom, but now the incline exerts friction on the block with coefficient μ_k. What is the speed of the block at the bottom of the incline?

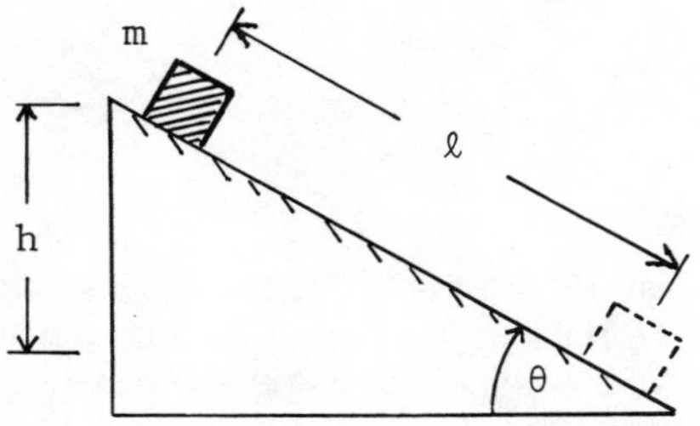

Here you must take into account both the work of friction (which goes into heat) and the change in gravitational potential energy.

$f = \mu_k N = \mu_k mg \cos\theta$.

$W_f = -(\mu_k mg \cos\theta)\ell$.

$K_{initial} = 0$.

$K_{final} = \frac{1}{2} mv^2$.

$U_{initial} = mgh = mg\ell \sin\theta$.

$U_{final} = 0$.

$$\Delta E = 0 = (0 - \tfrac{1}{2} mv^2) + (mg\ell \sin\theta - 0) + (\mu_k \ell mg \cos\theta) .$$

Thus

$$v = \sqrt{2 g\ell \sin\theta - 2\mu_k g\ell \cos\theta}$$

As a qualitative check, notice that the speed is less than it would be if friction were not present, i.e. if $\mu_k = 0$. This is reasonable.

Chapter 8: CONSERVATION OF LINEAR MOMENTUM

REVIEW AND PREVIEW

Previous chapters have dealt with the motion of a single particle. In this sense it is well for you to recall that an object need not necessarily be small for it to behave as a particle, rather each part must simultaneously undergo the same displacement in the same time interval. In this chapter you will learn to deal with a point associated with a rigid body (or with a collection of particles) called the center of mass; you will also learn how to describe its motion via the concept of linear momentum. You will further be introduced to one of the most powerful of all physical laws, the conservation of linear momentum.

GOALS AND GUIDELINES

This chapter sets for you three major goals:

1. Learning and understanding the concept of the center of mass of a collection of particles or a rigid body (Section 8-1) and learning to describe the motion of the center of mass (Section 8-2, Eq. 8-8). You should not be surprised to find that you have already mastered the second part of this goal.
2. Learning the definition of the total linear momentum of a system and understanding the concept of the conservation of this linear momentum (Section 8-3).
3. Learning to apply the conservation of linear momentum principle (Section 8-4 and 8-5).

8-1 Center of Mass

Consider a collection of N particles. Let the mass of the ith particle be m_i and its position vector from some arbitrary origin be $\underline{\underline{r}}_i$ (as shown in Fig. 8-1). The position vector of the center of mass $\underline{\underline{R}}$ is defined to be

$$\underline{\underline{R}} = \frac{1}{M}\sum_{i=1}^{N} m_i \underline{\underline{r}}_i \quad , \quad \text{where} \quad M = \sum_{i=1}^{N} m_i = \text{total mass.} \tag{8-1}$$

Eq. (8-1) is equivalent to three scalar equations for the components of $\underline{\underline{R}}$. These are

$$x_{CM} = \frac{1}{M}\sum_{i=1}^{N} m_i x_i \quad , \tag{8-2a}$$

$$y_{CM} = \frac{1}{M}\sum_{i=1}^{N} m_i y_i \quad , \tag{8-2b}$$

$$z_{CM} = \frac{1}{M}\sum_{i=1}^{N} m_i z_i \quad , \tag{8-2c}$$

where x_i, y_i, and z_i are the Cartesian components of $\underline{\underline{r}}_i$.

Figure 8-1. Position vectors of individual particles in a collection.

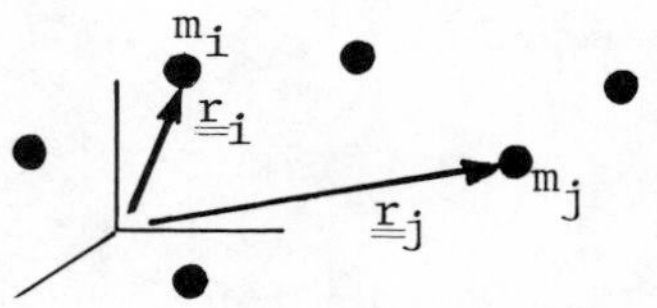

Often a convenient choice for the origin is at the center of mass, i.e. $\underline{\underline{R}} = 0$. In that case

$$\sum_{i=1}^{N} m_i \underline{\underline{r}}_i = 0 \quad , \text{ origin at center of mass.} \qquad (8\text{-}3)$$

In a continuous body the sums in Eqs. (8-1) to (8-3) must be replaced by integrals. If dm is an infinitesimal mass at the position $\underline{\underline{r}}$ then

$$\underline{\underline{R}} = \frac{1}{M} \int \underline{\underline{r}}\, dm \quad , \quad \text{where} \quad M = \int dm = \text{total mass.} \qquad (8\text{-}4)$$

To evaluate such integrals one uses the concept of mass density, ρ, which is mass per unit volume and may vary from point to point in the body. Then $dm = \rho\, dV$ where dV is an infinitesimal volume element. The integrations of Eq. (8-4) range over the volume of the continuous body. They will be considered further in Chapter 11.

If ρ is a constant and therefore independent of position, then $M = \rho V$ and

$$\underline{\underline{R}} = \frac{1}{\rho V} \int \rho \underline{\underline{r}}\, dV = \frac{1}{V} \int \underline{\underline{r}}\, dV \quad . \qquad (8\text{-}5)$$

If a body has a point, line, or plane of symmetry then the center of mass lies at that point, on that line, or in that plane.

>>> Example 1. Find the center of mass of particles of masses $m_1 = m_3 = 1.0$ kg at opposite corners of a square and $m_2 = 2.0$ kg and $m_4 = 3.0$ kg at the other corners. A side of the square is 1.0 meter.

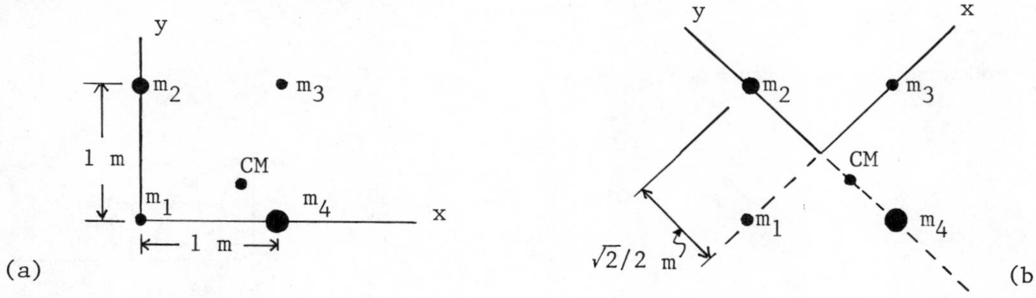

Figure 8-2. Example 1. (a) First choice of axes for finding the center of mass of the system. (b) A better choice which displays the symmetry of the figure.

In Fig. 8-2a is shown the particles and one choice of coordinate axes and origin.

Then

$$x_{CM} = \frac{(1.0\text{ kg})(0.0\text{ m}) + (2.0\text{ kg})(0.0\text{ m}) + (1.0\text{ kg})(1.0\text{ m}) + (3.0\text{ kg})(1.0\text{ m})}{(1.0 + 2.0 + 1.0 + 3.0)\text{ kg}} = \frac{4}{7}\text{ m} \quad .$$

Similarly,

$$y_{CM} = \frac{(1.0)(0.0) + (2.0)(1.0) + (1.0)(1.0) + (3.0)(0.0)}{7.0}\text{ m} = \frac{3}{7}\text{ m} \quad .$$

The location of this point is shown on Fig. 8-2a and denoted C.M.; notice that it lies on the diagonal from m_2 to m_4 at a distance $\sqrt{2}/14$ from the center toward m_4. This diagonal is a line of symmetry.

Fig. 8-2b shows a better choice of origin and axes. Here we've made use of the symmetry of the four particle system about a line drawn from m_2 to m_4; this is the diagonal above. Then we have immediately in this coordinate system

$$x_{CM} = 0$$

and for the y component of $\underline{\underline{R}}$

$$y_{CM} = \frac{m_2 y_2 + m_4 y_4}{m_1 + m_2 + m_3 + m_4} = \frac{(2.0)\sqrt{2}/2 - (3.0)\sqrt{2}/2}{7.0} \text{ m} = -\sqrt{2}/14 \text{ m} \quad .$$

This is the <u>same</u> physical point as we found before, but because we have chosen a different origin and axes the <u>numerical values</u> of X_{CM} and Y_{CM} differ from those found before. <<<

>>> Example 2. Find the center of mass of a uniform circular plate of radius R with a hole of radius r cut a distance D from the center.

If the thickness of the plate is t, then for purposes of finding X_{CM} and Y_{CM}, $dV = t\,dA$ and $V = tA$ where A is the area. Eq. (8-5) becomes

$$\underline{\underline{R}} = \frac{1}{A}\int \underline{\underline{r}}\, dA \quad ; \quad A = \pi R^2 - \pi r^2 \quad .$$

Choose the x and y axes as shown in Fig. 8-3.

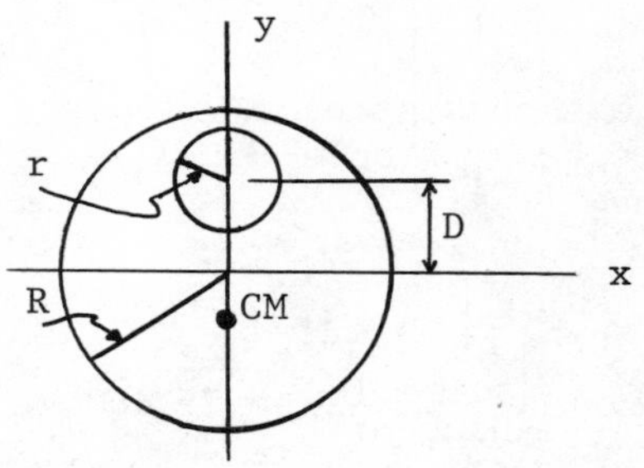

Figure 8-3. Example 2. Choice of axes.

From symmetry, $x_{CM} = 0$, and we need only compute y_{CM}. Then

$$y_{CM} = \frac{1}{A}\int_{\text{plate with hole}} y\, dA \quad .$$

The integration runs over the area of the plate with the hole in it. Let us rewrite this as

$$y_{CM} = \frac{1}{A}\left[\int_{\text{plate}} y\, dA - \int_{\text{hole}} y\, dA\right] \quad .$$

But $\int_{\text{plate}} y\, dA$ is just y_{CM} of the entire plate <u>with the hole filled</u> and is therefore zero for our choice of origin.

The corresponding integral over the hole is the y component of the center of mass of the hole times the area of the hole, i.e.

$$\int_{\text{hole}} y \, dA = (D)(\pi r^2) \quad .$$

Thus

$$y_{CM} = -\frac{D\pi r^2}{\pi R^2 - \pi r^2} = -\frac{r^2 D}{R^2 - r^2} \quad .$$

Notice that y_{CM} lies on the opposite side of the origin from the hole. Where is z_{CM}?
<<<

8-2 Center of Mass Motion

Consider a system of constant* total mass $M = \sum m_i$. If we differentiate Eq. (8-1) with respect to time we have

$$M\underline{\underline{V}}_{CM} = \sum_i m_i \underline{\underline{v}}_i \quad ; \quad \underline{\underline{V}}_{CM} = d\underline{\underline{R}}/dt \quad , \tag{8-6}$$

where $\underline{\underline{v}}_i$ is the velocity of the ith particle and $\underline{\underline{V}}_{CM}$ the velocity of the center of mass. Differentiating again we find

$$M\underline{\underline{A}}_{CM} = \sum_i m_i \underline{\underline{a}}_i \quad ; \quad \underline{\underline{A}}_{CM} = d\underline{\underline{V}}_{CM}/dt \quad . \tag{8-7}$$

Since $m_i \underline{\underline{a}}_i = \underline{\underline{F}}_i$ where $\underline{\underline{F}}_i$ is the resultant force on the ith particle, one has

$$M\underline{\underline{A}}_{CM} = \sum_i \underline{\underline{F}}_i \quad .$$

The $\underline{\underline{F}}_i$ may consist of both internal forces (due to the other particles) and external forces; from Newton's third law the internal forces cancel in pairs when we sum the $\underline{\underline{F}}_i$. Thus

$$M\underline{\underline{A}}_{CM} = \sum \underline{\underline{F}}_{external} = \underline{\underline{F}}_{ext} \quad , \tag{8-8}$$

where $\underline{\underline{F}}_{ext}$ is the resultant external force.

The meaning of Eq. (8-8) is that the center of mass of a fixed total mass system moves as if all the mass were concentrated at that point and as if all the external forces were applied there.

8-3 Linear Momentum -- Conservation

The linear momentum of a particle of mass m moving with velocity $\underline{\underline{v}}$ is defined to be

$$\underline{\underline{p}} = m\underline{\underline{v}} \quad . \tag{8-9}$$

Newton's second law then reads

$$\underline{\underline{F}} = m\underline{\underline{a}} = m \, d\underline{\underline{v}}/dt = d(m\underline{\underline{v}})/dt = d\underline{\underline{p}}/dt \tag{8-10}$$

since m is constant.

---

*This means that no particles enter or leave the system.

For a system of particles of <u>constant total mass</u> the total momentum is given by Eq. (8-6) as

$$M\underline{\underline{V}}_{CM} = \underline{\underline{P}} = \sum_i \underline{\underline{p}}_i \quad . \tag{8-11}$$

From Eq. (8-8) it follows that

$$\underline{\underline{F}}_{ext} = d\underline{\underline{P}}/dt \quad , \quad M = \text{constant}. \tag{8-12}$$

It is important to note that this equation applies only to systems of <u>constant total mass</u>.

If the resultant external force $\underline{\underline{F}}_{ext}$ is zero then $d\underline{\underline{P}}/dt = 0$ and hence $\underline{\underline{P}}$ is a constant. This is the principle of conservation of linear momentum. Formally

$$\text{if } \underline{\underline{F}}_{ext} = 0 \quad , \quad \underline{\underline{P}} = \text{constant}. \tag{8-13}$$

For a system of particles, their individual momenta may change, but in the absence of external forces or zero resultant external force, the total momentum is conserved. Eq. (8-13) is a vector equation. Therefore, if any component of $\underline{\underline{F}}_{ext}$ is zero the corresponding component of $\underline{\underline{P}}$ is conserved even though the other components need not be.

8-4 Applications of the Conservation of Momentum Principle

In applying this principle as discussed so far one must apply it to a system of <u>constant</u> total mass. Variable mass is considered separately in Section 8-5. To apply this principle one must

i) pick a system such that the total mass is constant;
ii) determine $\underline{\underline{F}}_{ext}$;
iii) if $\underline{\underline{F}}_{ext} = 0$, then $\underline{\underline{P}}$ is a constant.

>>> Example 3. A driverless runaway car weighing 3000 lb is moving at a constant speed of 30 miles per hour. An alert truck driver driving a 12,000 lb truck collides with the car head on to stop it. What must be the speed of the truck so that both car and truck come to rest at the collision point? Is mechanical energy conserved or not; if not, how much is lost or gained?

The forces involved in the collision are internal; the net external force, $\underline{\underline{F}}_{ext}$, is zero so Eq. (8-11) applies -- i.e. momentum is conserved. Since $\underline{\underline{P}}_{final}$ is zero so is $\underline{\underline{P}}_{initial}$. Let the mass and velocity of the car be m and $\underline{\underline{v}}$ and those of the truck M and $\underline{\underline{V}}$.

$$\underline{\underline{P}}_{initial} = m\underline{\underline{v}} + M\underline{\underline{V}} = \underline{\underline{P}}_{final} = 0.$$

Therefore $\underline{\underline{V}} = -(m/M)\underline{\underline{v}}$ is the required velocity of the truck. Since $m/M = 1/4$, the speed of the truck is $|\underline{\underline{V}}| = 7.5$ mph. Mechanical energy is not conserved; in fact the total final kinetic energy is zero. The "lost" mechanical energy which goes into deforming the car and truck is K_i. <<<

8-5 Systems of Variable Mass

Systems involving variable mass often seem to give students of physics some difficulty. In a rocket problem, for example, one wants to watch the rocket whose mass is not constant, but the momentum equation, Eq. (8-12) and the momentum conservation equation, Eq. (8-13), apply <u>only</u> to constant mass systems.

All one needs to do then is to consider an overall system whose total mass is constant; for example, in a rocket problem, the rocket plus the ejected material.

Call that part of the system which we want to watch the principal part. Its mass at some time t is M(t) and its velocity relative to some inertial origin 0 is $\underline{\underline{v}}(t)$. The velocity relative to 0 of the mass being ejected (or absorbed) at time t is $\underline{\underline{u}}(t)$. Then if $\underline{\underline{F}}_{ext}$ is the resultant external force on the overall constant mass system, the correct dynamic equation is

$$\underline{\underline{F}}_{ext} = \frac{d}{dt}(M(t)\underline{\underline{v}}(t)) - \underline{\underline{u}}(t)\frac{dM(t)}{dt} \quad . \tag{8-14}$$

This is usually written

$$M(t)\frac{d\underline{\underline{v}}(t)}{dt} = \underline{\underline{F}}_{ext} + \frac{dM(t)}{dt}(\underline{\underline{u}}(t) - \underline{\underline{v}}(t)) \quad . \tag{8-15}$$

The quantity $\underline{\underline{u}}(t) - \underline{\underline{v}}(t)$ is the velocity of the ejected (or absorbed) mass relative to the principal part, again at the time t. We denote it as $\underline{\underline{v}}_{rel}$. Often $\underline{\underline{v}}_{rel}$ will be constant even though $\underline{\underline{u}}(t)$ and $\underline{\underline{v}}(t)$ are not. The second term on the right of equation (8-15) is often denoted

$$\underline{\underline{F}}_{reaction} = \frac{dM(t)}{dt}\underline{\underline{v}}_{rel} \tag{8-16}$$

and for a rocket this is called thrust.

For a rocket as shown in the sketch, $\underline{\underline{v}}_{rel}$ is in the negative x direction but dM(t)/dt is negative (since the mass of the rocket decreases) so the thrust is directed in the positive x direction.

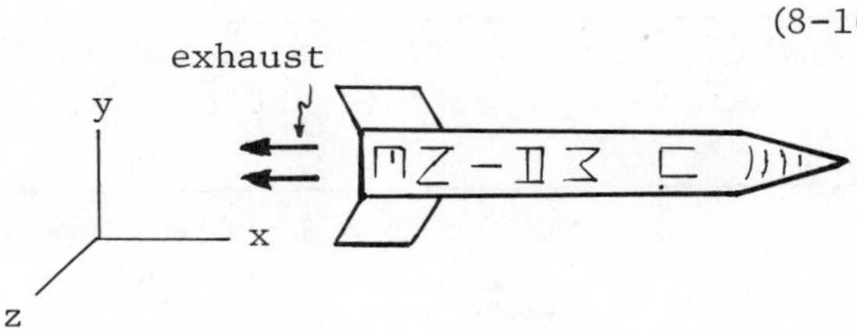

>>> Example 4. To double his car's capability for acceleration when pursued, James Bond discharges oil of density 800 kg/m^3 at the rear of his 2000 kg car. The oil is discharged through a nozzle 6 cm in diameter with a speed relative to the car of 70 m/s. What thrust is obtained from the ejection of the oil? What acceleration is attained at the instant that the valve is opened?

To find the thrust we need to determine dM/dt. Take a small volume element of cross sectional area equal to that of the nozzle and of length dx as shown in Fig. 8-4.

Figure 8-4. Example 4. Element of volume for computation of dM/dt.

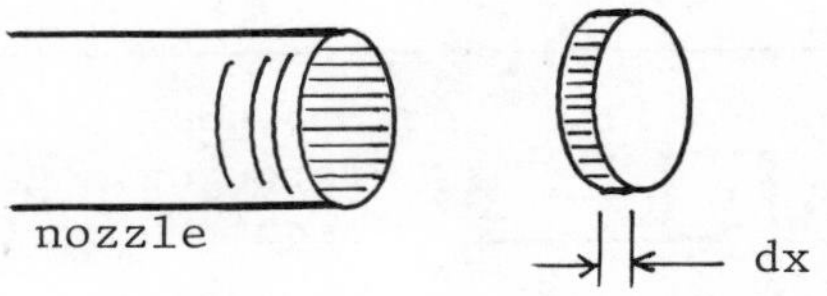

The increment of mass is

$$|dM| = \rho\ dV = \rho\ A\ dx$$

so that

$$|dM/dt| = \rho\ A\ dx/dt = \rho\ A\ v_{rel}$$

where v_{rel} is the relative speed of the ejected oil. Then the thrust is

$$F_{reaction} = |dM/dt|\ v_{rel} = \rho\ A\ v_{rel}^2$$

$$= (800\ kg/m^3)[\pi(6 \times 10^{-2}\ m)^2/4](70\ m/s)^2$$

$$= 1.11 \times 10^4\ N \quad .$$

Since we are told that this "doubles" his car's acceleration capability then the force due to the engine must be also 1.11×10^4 N. From Eq. (8-15) then

$$Ma = F_{ext} + F_{reaction} = 2.22 \times 10^4 \text{ N} \ .$$

At the instant that the nozzle opens M is 2000 kg, so the magnitude of the acceleration is

$$a = (2.22 \times 10^4 \text{ N})/(2 \times 10^3 \text{ kg}) = 11 \text{ m/s}^2 \ . \qquad <<<$$

8-6 Programmed Problems

1. The exercises and problems for this chapter will begin with a brief review of terms and ideas. The linear momentum of a particle of mass m with velocity **v** is **p** = _______.	m**v**
2. To say that the (linear) momentum of a particle is constant is to say that its velocity is ___.	Constant both in magnitude and direction. The mass of a particle is constant.
3. 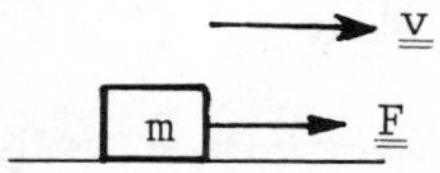 A 4 lb block moves in a straight line with a constant speed of 16 ft/s. The force **F** is constant and equal to 6 lb. What is the momentum of the block?	$p = mv$ $p = (W/g)v \qquad$ [W = weight] $p = \dfrac{4 \text{ lb}}{32 \text{ ft/s}^2} \times 16 \text{ ft/s}$ $p = 2 \text{ slug}\cdot\text{ft/s}$.
4. Is **p** constant for m because **F** is constant?	No. **p** is constant because **v** is constant. The resultant force on m must be zero, thus there must be another force.
5. 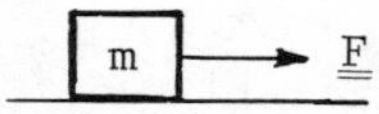 Show how the resultant force on m can be zero as required by the constant momentum.	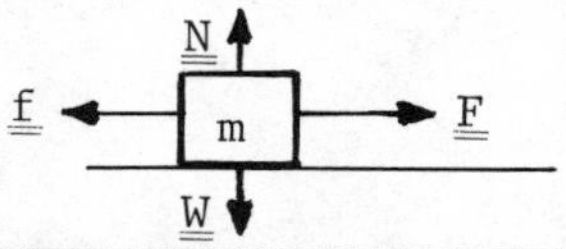 **N** = normal force **f** = frictional force
6. Assume the moon travels in a circular orbit about the earth with constant speed. Is the momentum of the moon a conserved quantity?	No. There must be a resultant force acting to accelerate the moon. Thus **p** is not constant. The orientation of **p** is continually changing.
7. Conservation of mechanical energy requires that the speed of a ball initially thrown vertically upward will be the same when it returns to the starting point. Is the momentum of the ball constant during its motion?	No again. The object still has a resultant force (m**g**) acting. $m\mathbf{g} = d\mathbf{p}/dt$ $m\mathbf{g} \neq 0$; $\mathbf{p} \neq$ constant.

8. 

A massless spring is compressed between two blocks on a frictionless surface. Both blocks are initially at rest. We seek the ratio of the speeds of m_1 and m_2.

When the blocks are released show all the forces on each mass while they are still in contact with the spring.

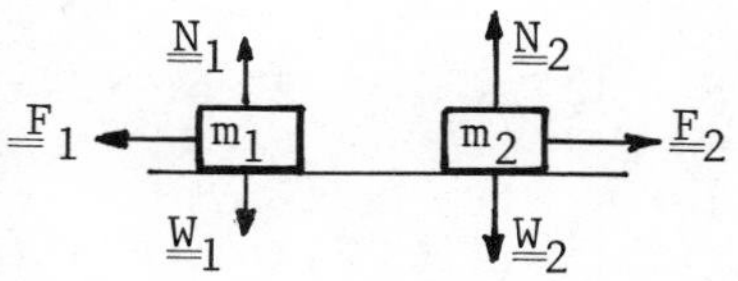

$\mathbf{F}_1$ and $\mathbf{F}_2$ are the forces due to the spring.

---

9. We will ignore $\mathbf{N}_1$, $\mathbf{N}_2$, $\mathbf{W}_1$ and $\mathbf{W}_2$ since they do not influence the horizontal motion. While the spring is still compressed, but expanding:

(a) Is $\mathbf{F}_1 = \mathbf{F}_2$?
(b) Is the resultant force on m_1 zero?
(c) Is $\mathbf{p}_1$ a constant during the motion of m_1?
(d) Is the resultant force on m_2 zero?
(e) Is $\mathbf{p}_2$ a constant during the motion of m_2?

a. No, $|\mathbf{F}_1| = |\mathbf{F}_2|$ but $\mathbf{F}_1 = -\mathbf{F}_2$.

b. No.

c. No, because of answer b.

d. No.

e. No, because of answer d.

---

10. 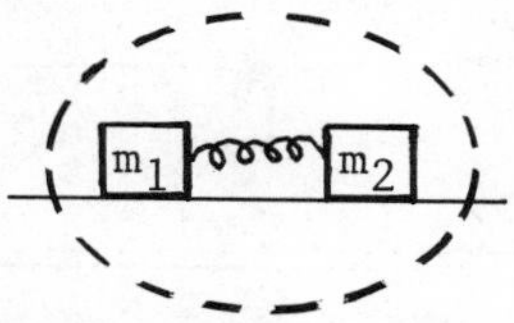

Continuing to ignore the $\mathbf{N}$ and $\mathbf{W}$ forces, what other environment forces "cut" through the isolating dashed line in this problem?

None. The spring is inside.

---

11. 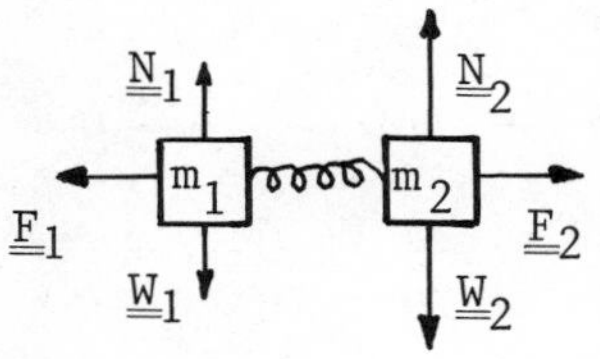

Here we have isolated the system consisting of m_1 and m_2.

(a) Is the resultant force on m_1 zero?
(b) Is the resultant force on m_2 zero?
(c) Is the total external force on the system zero?

a. No.

b. No.

c. Yes.

---

12. For the system then $\mathbf{F}_{ext}$ is zero. Thus $d\mathbf{P}/dt = 0$ for the system as defined by the sketch of frame 11. This means $\mathbf{P}$ of the system is a constant.

$$\mathbf{P}_{system} = \mathbf{p}_1 + \mathbf{p}_2 = \text{constant.}$$

What is $\mathbf{P}_{system}$ initially before the masses are released?

Zero; $\mathbf{v}$ for both m_1 and m_2 is zero.

13. When the spring is released, expanding, and still in contact with m_1 and m_2:

(a) $\mathbf{p}_1$ (is/is not) zero; (is/is not) constant.
(b) $\mathbf{p}_2$ (is/is not) zero; (is/is not) constant.
(c) $(\mathbf{p}_1 + \mathbf{p}_2)$ (is/is not) zero.
(d) $(\mathbf{p}_1 + \mathbf{p}_2)$ (is/is not) constant.

a. Is not; is not.
b. Is not; is not.
c. Is.
d. Is.

14.

Draw vectors representing $\mathbf{p}_1$ and $\mathbf{p}_2$ while m_1 and m_2 are under the influence of the spring.

Write the scalar equation for $(\mathbf{p}_1 + \mathbf{p}_2)$ of the system in terms of the velocity components.

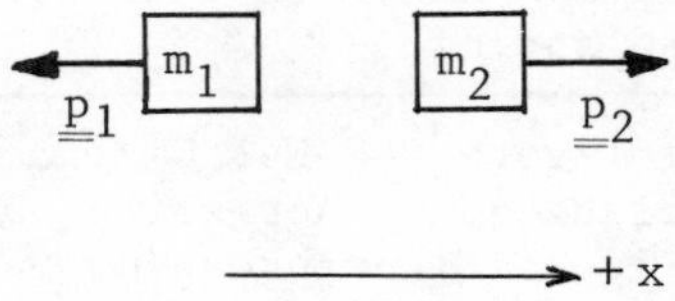

$m_1v_{1x} + m_2v_{2x} = 0$

---

15. What is the ratio of the speeds of m_1 and m_2 after the masses cease to have contact with the spring?

$m_1v_{1x} = -m_2v_{2x}$

$$\frac{v_{1x}}{v_{2x}} = -\frac{m_2}{m_1}$$

Since speed is the magnitude of the velocity, the ratio of the speeds is m_2/m_1.

16.

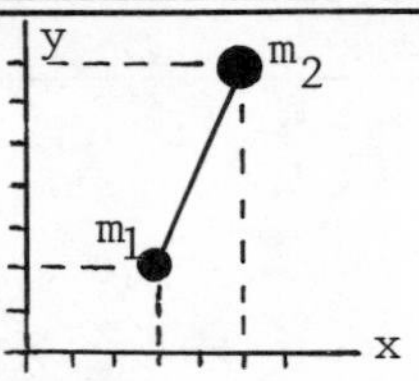

Let us now determine the center of mass of a dumbbell.

Start by drawing the position vectors $\mathbf{r}_1$ and $\mathbf{r}_2$ of m_1 and m_2 with respect to the origin; m_1 and m_2 are connected by a massless rod.

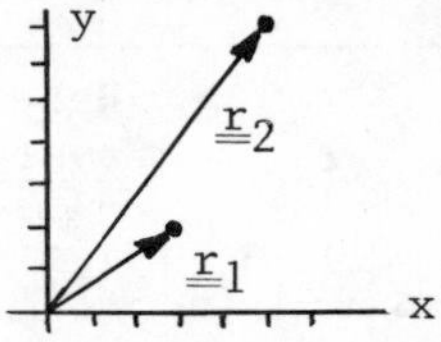

$\mathbf{r}_1 = 3\mathbf{i} + 2\mathbf{j}$

$\mathbf{r}_2 = 5\mathbf{i} + 7\mathbf{j}$

---

17. The center of mass of a collection of particles is defined as

$$\mathbf{R}_{CM} = \frac{1}{M}\sum_{i=1}^{N} m_i\mathbf{r}_i \quad \text{with } M = \sum_{i=1}^{N} m_i .$$

(a) Write the equations for the scalar equivalents X_{CM} and Y_{CM} of $\mathbf{R}_{CM}$ for the coordinate system in frame 16.

(b) What is M for the system of frame 16?

a. $$X_{CM} = \frac{1}{M}\sum_{i=1}^{2} m_i x_i$$

$$Y_{CM} = \frac{1}{M}\sum_{i=1}^{2} m_i y_i$$

b. $$M = \sum_{i=1}^{2} m_i$$

$$M = m_1 + m_2$$

18. The $\mathbf{r}_i$'s for the $\mathbf{R}_{CM}$ formula are shown in the answer of frame 16. From that

(a) x_1 of $\mathbf{r}_1$ = ______ .
(b) y_1 of $\mathbf{r}_1$ = ______ .
(c) x_2 of $\mathbf{r}_2$ = ______ .
(d) y_2 of $\mathbf{r}_2$ = ______ .

a. $x_1 = 3$

b. $y_1 = 2$

c. $x_2 = 5$

d. $y_2 = 7$

---

19. Let $m_1 = 1$ kg and $m_2 = 2$ kg. Using the answers to the frames 17 and 18 calculate

(a) X_{CM}.

(b) Y_{CM}.

a. $X_{CM} = \frac{1}{M} \sum_{i=1}^{2} m_i x_i$

$X_{CM} = \frac{1}{3} [(1)(3) + (2)(5)]$

$= (13/3)$.

b. $Y_{CM} = \frac{1}{M} \sum_{i=1}^{2} m_i y_i$

$Y_{CM} = \frac{1}{3} [(1)(2) + (2)(7)]$

$= (16/3)$.

---

20. $\mathbf{R}_{CM}$ can be written as

$\mathbf{R}_{CM}$ = ______ $\mathbf{i}$ + ______ $\mathbf{j}$.

$\mathbf{R}_{CM} = (13/3)\mathbf{i} + (16/3)\mathbf{j}$.

---

21. 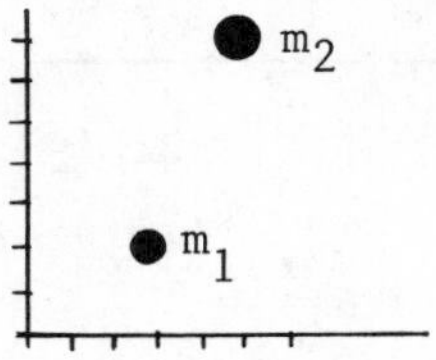

Draw $\mathbf{R}_{CM}$ to scale on this diagram.

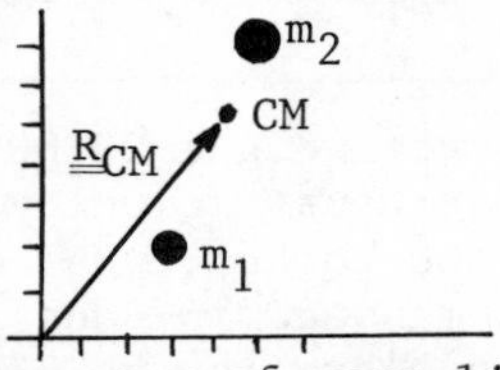

The center of mass lies on the connecting rod, two thirds of the way toward the larger mass.

---

22. 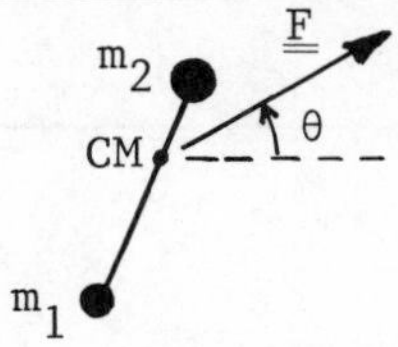

Now let us look at the motion of the dumbbell under the action of an external force.

$m_2 = 2.0$ kg ; $m_1 = 1.0$ kg

If $|\mathbf{F}| = 5$ N and $\theta = 30^o$, what is

(a) A_x of CM?

(b) A_y of CM?

a.

$A_x = F_x/M = 4.3 \text{ N}/3 \text{ kg} = 1.4 \text{ m/s}^2$.

b.

$A_y = F_y/M = 2.5 \text{ N}/3 \text{ kg} = 0.8 \text{ m/s}^2$.

23. 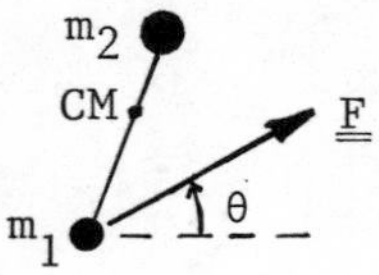

Consider this similar situation.

If $|\underline{F}|$ = 5 N and $\theta = 30^{\circ}$, what is

(a) A_x of CM?

(b) A_y of CM?

Same as in frame 22. The motion of the CM is as if all mass were at the CM and all external forces acted at CM. Note that the <u>individual</u> motions of m_1 and m_2 are different in frame 22 and 23. In frame 23 the masses would rotate as well as translate.

---

24. 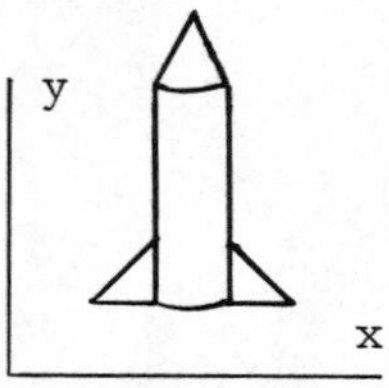

Consider the vertically launched rocket shown at lift off. If the rocket is the "system" whose motion is desired, is this a system of constant mass?

No. The mass of the rocket is continually decreasing due to fuel consumption and exhaust.

---

25. 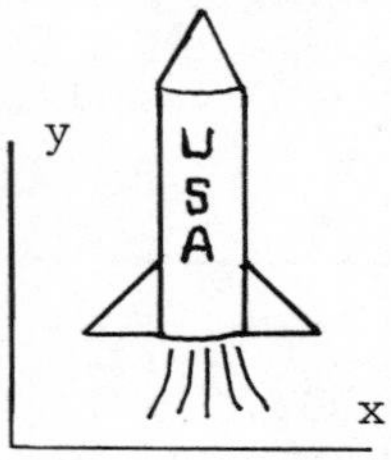

The equation of motion for the rocket can be written as

$$M \frac{d\underline{v}}{dt} = \underline{F}_{ext} + \underline{v}_{rel} \frac{dM}{dt} .$$

Draw vectors representing $\underline{v}$, $\underline{F}_{ext}$, and $\underline{v}_{rel}$ (dM/dt).

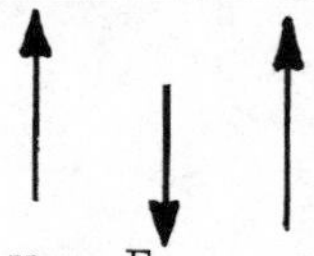

$\underline{v}$ $\underline{F}_{ext}$ $\underline{v}_{rel}$(dM/dt)

$\underline{v}_{rel}$ is in $-\underline{j}$ direction but dM/dt is negative.

---

26. The term $\underline{v}_{rel}$(dM/dt) is called the _______ .

Thrust.

---

27. If the rocket weighs 20,000 lb and has a constant gas exhaust velocity of 7,000 ft/s relative to the rocket, what must be the rate at which its mass diminishes in order that the initial acceleration be g?

Looking again at the answer to frame 25 write the scalar form of Newton's second law for the rocket. Do not put in numbers as yet.

$M\,(dv/dt) = -Mg - |v_{rel}|\,(dM/dt)$

since $\underline{v}_{rel} = -|v_{rel}|\underline{j}$

---

28. For $M\,(dv/dt) = -Mg - |v_{rel}|\,(dM/dt)$ of this problem

(a) M = ______ .

(b) g = ______ .

(c) dv/dt = ______ .

(d) $|v_{rel}|$ = ______ .

(e) dM/dt = _______.

a. 625 slug

b. 32 ft/s^2

c. g (given)

d. 7,000 ft/s .

e. − 5.7 slug/s . [The mass of the rocket decreases at the rate 5.7 slug/s.]

Chapter 9: COLLISIONS

REVIEW AND PREVIEW

In the previous chapter you learned that total linear momentum of a system is conserved in the absence of external forces. In this chapter you will study an important application in which the effect of external forces may be neglected, namely the collision of two particles. You will find that linear momentum is always conserved in such cases but mechanical energy may or may not be conserved.

GOALS AND GUIDELINES

In this chapter you have two major goals:

1. Learning the definition of impluse and its relationship to change in momentum. Learning the definition of a collision and learning the classification of the two types of collisions, namely elastic and inelastic (Section 9-1).
2. Learning to solve and to distinguish between elastic and inelastic collision problems. You should pay particular attention to the recipe suggested at the beginning of Section 9-2. Section 9-3 which deals with the solution of collision problems in center of mass coordinates is worth studying for the insight it provides, but is not necessary for solving collision problems.

9-1 Collisions; Impulse; Momentum Conservation

Broadly speaking, a collision is a physical event in which two or more objects come together or separate. This event must take place over a time interval Δt which is short in comparison with the observation time interval ΔT, i.e. $\Delta t \ll \Delta T$. Other than during the time interval Δt the objects are presumed not to interact with one another. In other words, time can be separated into before the collision, during the collision, and after the collision.

Collision theory concentrates on predicting conditions after the collision given the conditions before but without detailed knowledge of what goes on during. Basically it is nothing more than an application of the conservation of energy and momentum.

Relatively large forces may act during the collision and these may vary with time in a very complicated manner. The impulse, $\underline{\underline{J}}$, of a force acting over a time interval Δt from some initial time t_i to some final time t_f is defined to be

$$\underline{\underline{J}} \equiv \int_{t_i}^{t_f} \underline{\underline{F}}(t)\, dt \quad . \tag{9-1}$$

Each component of the impulse is the area under the curve of the corresponding component of $\underline{\underline{F}}$ versus time, as shown in Fig. 9-1 for the x component. If $\underline{\underline{F}}(t)$ acts on a body for an increment dt the change in momentum $d\underline{\underline{p}}$ of the body is given by

$$d\underline{\underline{p}} = \underline{\underline{F}}(t)\, dt \quad . \tag{9-2}$$

The change in momentum of the body over the time interval ($\Delta t = t_f - t_i$) is the impulse. Formally

$$\underline{\underline{p}}_f - \underline{\underline{p}}_i = \underline{\underline{J}} = \int_{t_i}^{t_f} \underline{\underline{F}}(t)\, dt \quad . \tag{9-3}$$

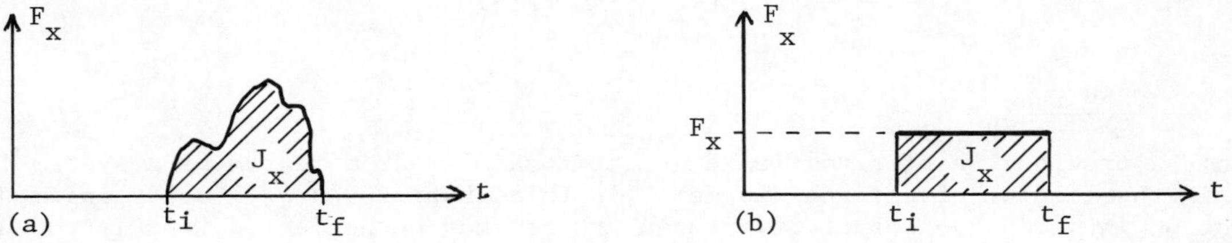

Figure 9-1. (a) The x component of impulse is the area under the curve F_x versus time. (b) The x component of the average force $\bar{F}$.

In a collision between two particles (1 and 2) in the *absence of external forces*, the force on 1 is that exerted by 2, $\mathbf{F}_{12}$ and similarly the force on 2 is that exerted by 1, $\mathbf{F}_{21}$. Since these are internal forces for the two particle system,

$$\mathbf{F}_{12} = -\,\mathbf{F}_{21} \tag{9-4}$$

so no matter how complicated their variation with time

$$\mathbf{J}_{12} = \int_{t_i}^{t_f} \mathbf{F}_{12}\,dt = -\int_{t_i}^{t_f} \mathbf{F}_{21}\,dt = -\,\mathbf{J}_{21} \quad . \tag{9-5}$$

Therefore the change in the total momentum of the system is

$$\Delta\mathbf{P} = \Delta\mathbf{p}_1 + \Delta\mathbf{p}_2 = \mathbf{J}_{12} + \mathbf{J}_{21} = 0 \quad . \tag{9-6}$$

Thus, *in the absence of external forces total momentum is conserved during a collision.*

\>>> Example 1. A golfer strikes a golf ball which weighs 0.1 lb and imparts to it a velocity of 150 ft/s. What is the impulse of the force on the golf ball?

We do not know $\mathbf{F}$ so we cannot compute the impulse from the integral of Eq. (9-1). We can, however, compute the impulse from the change in momentum of the golf ball. Its initial velocity is zero and its final speed is 150 ft/s. Let the final velocity direction be denoted $\mathbf{n}$. Then

$$\mathbf{J} = \mathbf{p}_f - \mathbf{p}_i = mv_f\mathbf{n}$$

shows that the direction of $\mathbf{J}$ is the same as $\mathbf{v}_f$. The magnitude of $\mathbf{J}$ is

$$\begin{aligned} J &= mv_f = wv_f/g \\ &= \frac{(0.1\text{ lb})(150\text{ ft/s})}{32\text{ ft/s}^2} = 0.47\text{ lb}\cdot\text{s} \end{aligned}$$

The duration of contact can be expected to be of the order of one millisecond i.e. 10^{-3} second. What is the average force exerted by the club head?

$$\bar{F} = \frac{J}{\Delta t} = \frac{\Delta p}{\Delta t} = \frac{0.47\text{ lb}\cdot\text{s}}{10^{-3}\text{ s}} = 470\text{ lb} \quad . \qquad <<<$$

This result is called the *average* force exerted. This can be understood if we remember that the impulse of a force has been interpreted as the area under a plot of the force as a function of time. In Fig. (9-1) we show two curves which have the same area (thus the same impulse). In other words we can replace a rather complicated force (Fig. 9-1a) by an average constant force (Fig. 9-1b), in the sense that they have the same impulse.

External forces are rarely completely absent. However, they are generally much weaker than the collision forces, so that

$$|\underline{\underline{J}}_{external}| << |\underline{\underline{J}}_{collision}| \tag{9-7}$$

and the external forces may be neglected during the collision. Eq. (9-7) may then be taken as a condition for the validity of describing an event as a collision. In review then, an event is a collision if

(1) $\Delta t << \Delta T$ -- time can be separated into before, during and after;

(2) $|\underline{\underline{J}}_{external}| << |\underline{\underline{J}}_{collision}|$ -- external forces can be neglected and total momentum is conserved.

Although momentum is conserved in a collision, kinetic energy may or may not be. This leads to a classification of collisions: collisions are

(1) elastic -- kinetic energy is conserved;

(2) inelastic -- kinetic energy is not conserved; in the special case where the colliding particles stick together we have a completely inelastic collision. In a completely inelastic collision the maximum kinetic energy is lost consistent with momentum conservation.

The principles involved in collision problems are really quite simple. The major difficulty is often one of bookkeeping or notation. We shall use the following notation:

m_1, m_2 -- masses of the two particles,

$\underline{\underline{v}}(1,i)$, $\underline{\underline{v}}(2,i)$ -- initial (before collision) velocities of particles 1 and 2 respectively,

$\underline{\underline{v}}(1,f)$, $\underline{\underline{v}}(2,f)$ -- final (after collision) velocities of particles 1 and 2 respectively.

9-2 Examples of Collision Problems

Your success in solving such problems usually depends in large part on drawing a figure. On the figure you should label all the quantities of interest such as $\underline{\underline{v}}(1,f)$, $\underline{\underline{v}}(2,i)$, etc. Then, catalog those which are known and those which are unknown. Write the equations which relate these quantities -- one for each component of momentum and one more (i.e. conservation of kinetic energy) if the collision is elastic. Also, you should write down any subsidiary conditions specified by the problem. Count the number of unknowns. If this number exceeds the number of equations plus the number of subsidiary conditions you can't solve the problem. Generally, this means you've missed something so go back over the process.

>>> Example 2. A particle of mass 3 kg initially has velocity 9 m/s along the positive y direction of some coordinate system. A second particle of mass 5 kg has initial velocity of 3 m/s along the negative x-axis. They collide and stick together. What is their final velocity?

The fact that the two stick together should signal a totally inelastic collision. This means that we'll only have the momentum conservation relationships. Fig. 9-2 illustrates the problem.

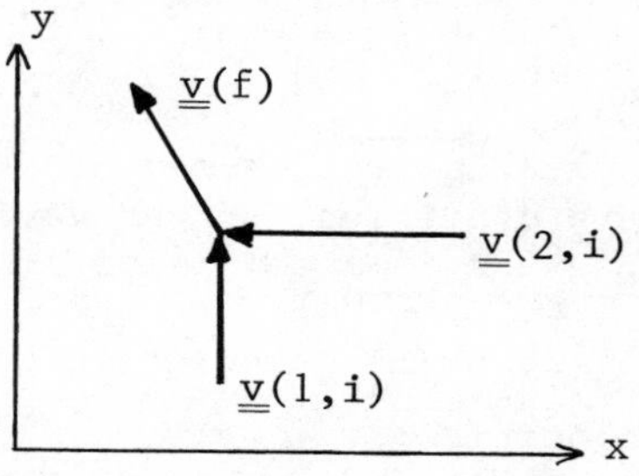

Figure 9-2. Example 2. Illustration of an inelastic collision.

We've let the 3 kg mass be m_1 and the 5 kg mass m_2. Let's tabulate the data:

$$m_1 = 3 \text{ kg} \quad , \qquad m_2 = 5 \text{ kg}$$

$$\underline{\underline{v}}(1,i) = 9\underline{\underline{j}} \text{ m/s} \ , \quad \underline{\underline{v}}(2,i) = -\ 3\underline{\underline{i}} \text{ m/s} \quad .$$

There is only one (vector) unknown (or really <u>three</u> unknown vector components) $\underline{\underline{v}}(f)$. The momentum conservation equation is

$$m_1\underline{\underline{v}}(1,i) + m_2\underline{\underline{v}}(2,i) = (m_1 + m_2)\underline{\underline{v}}(f) \quad .$$

Therefore

$$\underline{\underline{v}}(f) = \frac{m_1}{m_1 + m_2}\,\underline{\underline{v}}(1,i) + \frac{m_2}{m_1 + m_2}\,\underline{\underline{v}}(2,i)$$

so

$$\underline{\underline{v}}(f) = \frac{3}{8}\,(9\underline{\underline{j}}) + \frac{5}{8}\,(-\ 3\underline{\underline{i}}) \text{ m/s} = -\ 1.88\ \underline{\underline{i}} + 3.38\ \underline{\underline{j}} \ \text{ m/s} \ .$$

We can also find the kinetic energy lost in the collision. Since

$$K_i = \tfrac{1}{2}\, m_1 v^2(1,i) + \tfrac{1}{2}\, m_2 v^2(2,i) = 144 \text{ J}$$

and

$$K_f = \tfrac{1}{2}\,(m_1 + m_2)v^2(f) = 59.8 \text{ J} \ ,$$

the loss in kinetic energy is

$$K_i - K_f = 84.2 \text{ J} \ .$$

<<<

>>> Example 3. Now we consider an elastic collision. The data are:

$$m_1 = 15 \text{ kg} \quad , \qquad m_2 = 5 \text{ kg}$$

$$\underline{\underline{v}}(1,i) = 3\underline{\underline{j}} \text{ m/s} \quad , \quad \underline{\underline{v}}(2,i) = -\ 9\underline{\underline{i}} \text{ m/s} \quad .$$

Suppose now that after the collision the 5 kg mass is found to have a velocity along the negative y-axis (see Fig. 9-3). Find the final velocities.

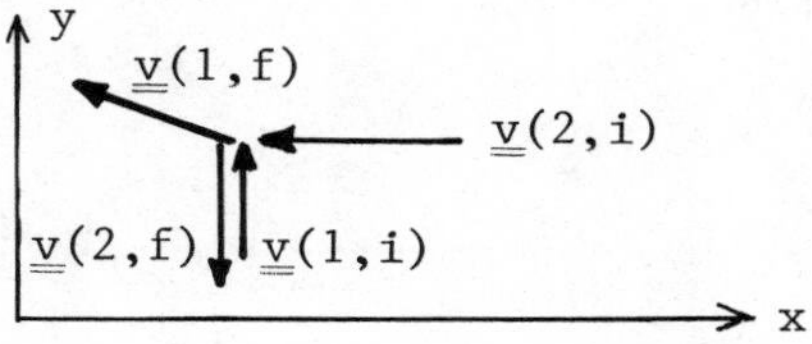

Figure 9-3. Example 3. An elastic collision in two dimensions. For clarity we have displaced $\underline{\underline{v}}(2,f)$ slightly to the left.

There are <u>four</u> unknown quantities, namely the x and y components of the two final velocities: $v_x(1,f)$, $v_y(1,f)$, $v_x(2,f)$, $v_y(2,f)$.

The collision is elastic so we have three equations relating these quantities. These are

$$m_1 v_x(1,i) + m_2 v_x(2,i) = m_1 v_x(1,f) + m_2 v_x(2,f)$$

$$m_1 v_y(1,i) + m_2 v_y(2,i) = m_1 v_y(1,f) + m_2 v_y(2,f)$$

$$\tfrac{1}{2} m_1 v^2(1,i) + \tfrac{1}{2} m_2 v^2(2,i) = \tfrac{1}{2} m_1 v^2(1,f) + \tfrac{1}{2} m_2 v^2(2,f) \quad .$$

We have a <u>necessary</u> subsidiary condition, namely the direction of $\underline{\underline{v}}(2,f)$. Thus

$$v_x(2,f) = 0 \quad ,$$

so there are three unknowns now and three equations. The rest is algebra and the answers are

$$v_x(1,f) = -\ 3\ \text{m/s} \quad , \quad v_y(1,f) = 4.5\ \text{m/s} \quad , \quad v_y(2,f) = -\ 4.5\ \text{m/s} \quad .$$ <<<

9-3 Center of Mass Coordinates -- One Dimension

In a collision between two particles the momentum of the center of mass is constant and it is often convenient to choose a coordinate system fixed at the center of mass, i.e. one in which the center of mass is at rest. Most advanced collision theory, particularly in quantum mechanics, utilizes the center of mass coordinate system. In this coordinate system the notation for the velocities in the x direction is

$u(1,i)$, $u(2,i)$ -- initial velocities of particles 1 and 2 respectively

$u(1,f)$, $u(2,f)$ -- final velocities of particles 1 and 2.

The conservation of linear momentum together with the fact that the coordinate system is the center of mass system yields

$$m_1 u(1,i) + m_2 u(2,i) = 0 \quad , \tag{9-8a}$$

$$m_1 u(1,f) + m_2 u(2,f) = 0 \quad . \tag{9-8b}$$

thus

$$u(2,i) = -\ (m_1/m_2)\ u(1,i) \quad , \tag{9-8c}$$

$$u(2,f) = -\ (m_1/m_2)\ u(1,f) \quad . \tag{9-8d}$$

For an <u>elastic collision</u> the conservation of kinetic energy yields

$$\tfrac{1}{2}\, m_1 u^2(1,i) + \tfrac{1}{2}\, m_2 u^2(2,i) = \tfrac{1}{2}\, m_1 u^2(1,f) + \tfrac{1}{2}\, m_2 u^2(2,f) \quad . \tag{9-9}$$

From Eqs. (9-8) and (9-9) one finds

$$u(1,f) = -\, u(1,i) \tag{9-10a}$$

$$u(2,f) = -\, u(2,i) \quad . \tag{9-10b}$$

The conversion between the laboratory system and the center of mass system is given by

$$v = u + v_{CM} \quad , \tag{9-11}$$

where v is any of the four v's and u is the corresponding one of the four u's and where the center of mass velocity, v_{CM}, is given by

$$m_1 v(1,i) + m_2 v(2,i) = (m_1 + m_2) v_{CM} \quad . \tag{9-12}$$

In Fig. 9-4 we illustrate the elastic collision for the special case $m_1 = m_2$ as seen in the laboratory and as seen in the center of mass system.

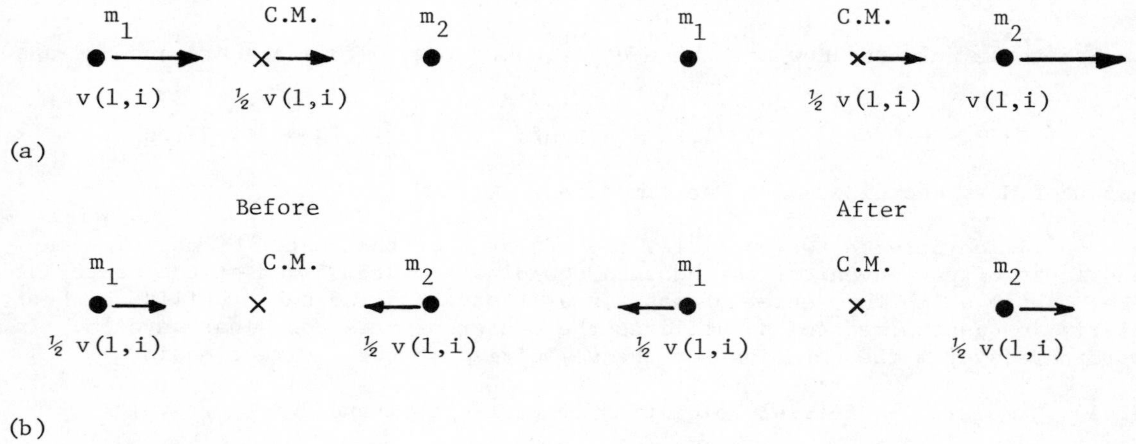

Figure 9-4. A one dimensional elastic collision with $m_1 = m_2$ as seen (a) in the laboratory and (b) in the center of mass system before and after collision.

For a <u>completely inelastic</u> collision $v(1,f) = v(2,f) = v_{CM}$ so $u(1,f) = u(2,f) = 0$. The initial kinetic energy is

$$\begin{aligned} K_i &= \tfrac{1}{2}\, m_1 v(1,i)^2 + \tfrac{1}{2}\, m_2 v(2,i)^2 = \tfrac{1}{2}\, m_1 [u(1,i) + v_{CM}]^2 + \tfrac{1}{2}\, m_2 [u(2,i) + v_{CM}]^2 \\ &= \underbrace{\tfrac{1}{2}\, m_1 u(1,i)^2 + \tfrac{1}{2}\, m_2 u(2,i)^2}_{\text{"K relative to CM"}} + \underbrace{\tfrac{1}{2}\,(m_1 + m_2) v_{CM}{}^2}_{\text{"K of CM"}} + \underbrace{[m_1 u(1,i) + m_2 (u_2,i)] v_{CM}{}^2}_{\text{Zero from (9-8)}} \end{aligned}$$

After the collision

$$K_f = \tfrac{1}{2}(m_1 + m_2)v_{CM}^2 \quad .$$

Thus the kinetic energy lost is that due to the motion of m_1 and m_2 relative to the center of mass. The kinetic energy of the center of mass is conserved. If the collision is only partially inelastic then some fraction of the kinetic energy of m_1 and m_2 relative to the center of mass is lost.

9-4 Programmed Problems

1. 

A ball strikes a rough wall with initial momentum $\mathbf{p}_i$ and rebounds with final momentum $\mathbf{p}_f$. $\mathbf{p}_i$ and $\mathbf{p}_f$ are shown as vectors. Draw the vector $\mathbf{p}_f - \mathbf{p}_i$ on the diagram.

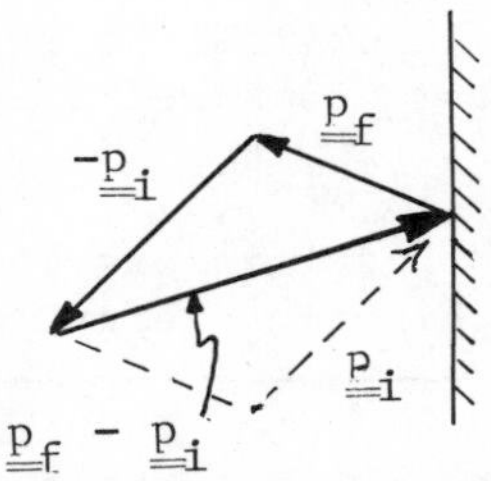

---

2. The vector $\mathbf{p}_f - \mathbf{p}_i$ is the change of momentum of the ball as a result of having collided with the wall. From Newton's second law we may write

$$\mathbf{p}_f - \mathbf{p}_i = \int_{t_i}^{t_f} \mathbf{F}\, dt.$$

What is the name of the right hand side of this equation? What is the source of $\mathbf{F}$?

Impulse, wall.

---

3. Let

$$\mathbf{p}_i = 4\mathbf{i} + 3\mathbf{j}$$

and

$$\mathbf{p}_f = -2\mathbf{i} + 2\mathbf{j}$$

in units kg·m/s .

What is $|\mathbf{p}_f - \mathbf{p}_i|$? Hint: Remember how to "add" vectors when expressed in Cartesian components and how to determine the magnitude of a vector from its Cartesian components.

$$\mathbf{p}_i = -2\mathbf{i} + 2\mathbf{j}$$
$$-\mathbf{p}_f = -4\mathbf{i} - 3\mathbf{j}$$
$$\mathbf{p}_f - \mathbf{p}_i = -6\mathbf{i} - \mathbf{j}$$
$$|\mathbf{p}_f - \mathbf{p}_i| = \sqrt{36 + 1} = 6.1 \text{ kg·m/s} \; .$$

This is the change in momentum of the ball.

---

4. If the ball is in contact with the wall for 0.01 seconds, what is the magnitude of the impulse delivered to the ball by the wall?

If you did anything other than repeat the answer to the last frame, shame on you.

Impulse = change in momentum
= 6.1 kg·m/s .

5. If the ball is in contact with the wall for 0.01 seconds, what is the magnitude of the average force exerted by the wall on the ball? Also, what is the direction of this impulse force?

$$|\bar{\mathbf{F}}| = \frac{\Delta \mathbf{p}}{\Delta t}$$

$$|\bar{\mathbf{F}}| = \frac{6.1\ \text{kg}\cdot\text{m/s}}{0.01\ \text{s}}$$

$$|\bar{\mathbf{F}}| = 610\ \text{N}\ .$$

The direction of $\bar{\mathbf{F}}$ is that of $\mathbf{p}_f - \mathbf{p}_i$.

---

6. Collisions are classified as

(a) ______ collisions in which both momentum and kinetic energy are conserved

or

(b) ______ collisions in which momentum but not kinetic energy is conserved.

a. Elastic

b. Inelastic

7.

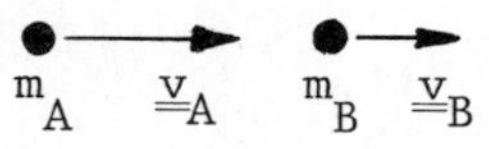

Particle A moving to the right with velocity $\mathbf{v}_A$ collides elastically with particle B which is moving to the right with velocity $\mathbf{v}_B$.

Is it true that conservation of kinetic energy for this system means

K of m_A	=	K of m_A
before collision		after collision

and

K of m_B	=	K of m_B
before collision		after collision?

No, this is false. Conservation of kinetic energy means that the total kinetic energy before the collision equals the total kinetic energy after collision. Here total kinetic energy is

$$\tfrac{1}{2}\, m_A v_A^2 + \tfrac{1}{2}\, m_B v_B^2$$

It is possible for $\mathbf{v}_A$ and $\mathbf{v}_B$ to be different after the collision.

---

8. Below are represented the momentum vectors of two collision partners. $\mathbf{p}_{1i}$ is the momentum of particle 1 before the collision. $\mathbf{p}_{2i}$ is the momentum of particle 2 before the collision.

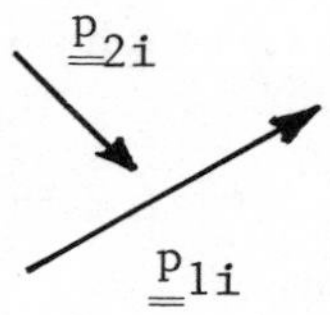

Draw approximately to scale the total initial momentum $\mathbf{p}_i$ of this two particle system.

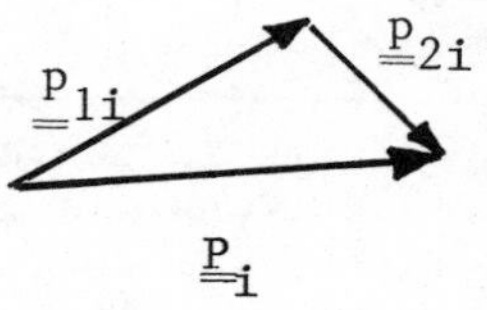

9. Considering the collision to be either elastic or inelastic draw the vector $\mathbf{p}_f$.

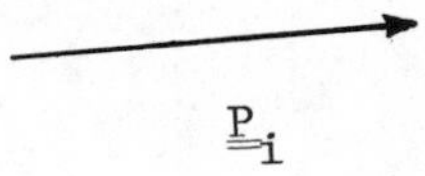

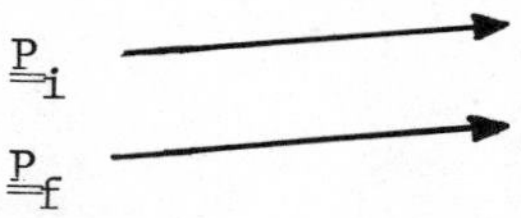

These must be the same to conserve the total momentum in either the elastic or inelastic case.

10.

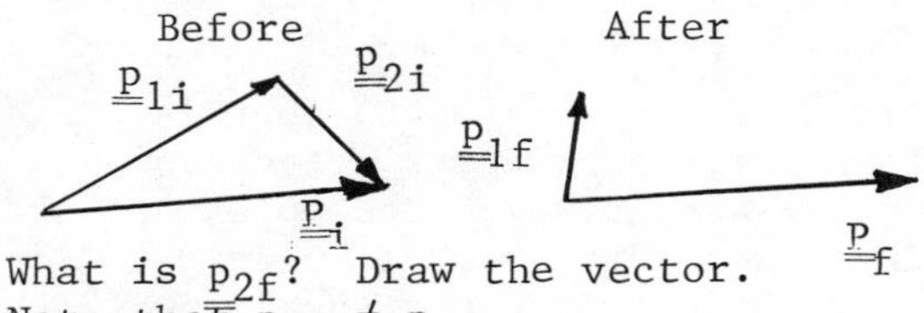

What is p_{2f}? Draw the vector. Note that $\mathbf{p}_{1i} \neq \mathbf{p}_{1f}$.

The point to be learned here is that momentum conservation means

$$\mathbf{p}_{1i} + \mathbf{p}_{2i} = \mathbf{p}_{1f} + \mathbf{p}_{2f}$$

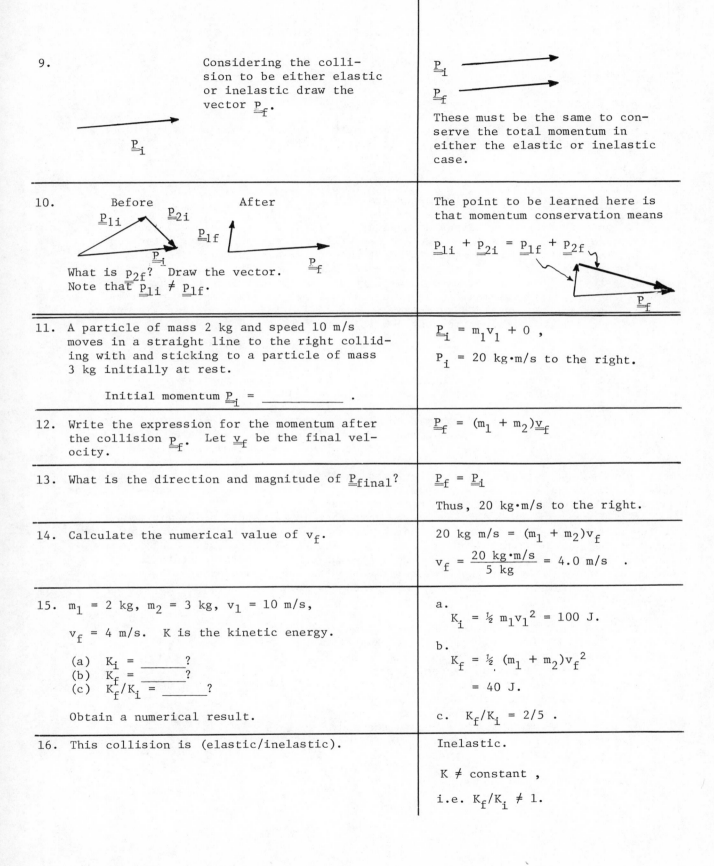

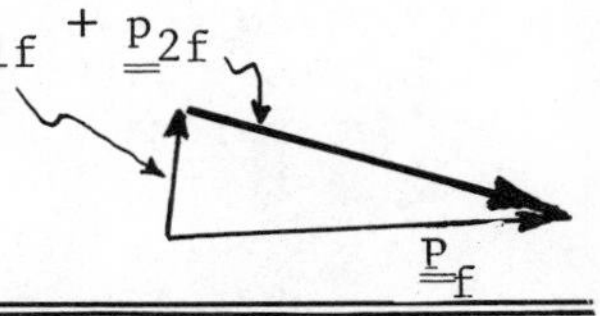

---

11. A particle of mass 2 kg and speed 10 m/s moves in a straight line to the right colliding with and sticking to a particle of mass 3 kg initially at rest.

Initial momentum $\mathbf{p}_i$ = ___________ .

$\mathbf{p}_i = m_1v_1 + 0$,

$p_i = 20$ kg·m/s to the right.

---

12. Write the expression for the momentum after the collision $\mathbf{p}_f$. Let $\mathbf{v}_f$ be the final velocity.

$\mathbf{p}_f = (m_1 + m_2)\mathbf{v}_f$

---

13. What is the direction and magnitude of $\mathbf{p}_{final}$?

$\mathbf{p}_f = \mathbf{p}_i$

Thus, 20 kg·m/s to the right.

---

14. Calculate the numerical value of v_f.

$20 \text{ kg m/s} = (m_1 + m_2)v_f$

$$v_f = \frac{20 \text{ kg·m/s}}{5 \text{ kg}} = 4.0 \text{ m/s} \quad .$$

---

15. $m_1 = 2$ kg, $m_2 = 3$ kg, $v_1 = 10$ m/s,

$v_f = 4$ m/s. K is the kinetic energy.

(a) K_i = ______?
(b) K_f = ______?
(c) K_f/K_i = ______?

Obtain a numerical result.

a.
$K_i = \frac{1}{2} m_1 v_1^2 = 100$ J.

b.
$K_f = \frac{1}{2} (m_1 + m_2)v_f^2$

$= 40$ J.

c. $K_f/K_i = 2/5$.

---

16. This collision is (elastic/inelastic).

Inelastic.

$K \neq$ constant ,

i.e. $K_f/K_i \neq 1$.

Chapter 10: ROTATIONAL KINEMATICS

REVIEW AND PREVIEW

Your first introduction to mechanics began with the kinematics of linear motion; you learned the definitions of position, velocity and acceleration and relationships among them. In this chapter you will meet their rotational motion analogues: angular orientation, angular velocity, and angular acceleration.

GOALS AND GUIDELINES

In this chapter you have two major goals:

1. Learning the definitions of angular position, angular velocity (both average and instantaneous) and angular acceleration. Pay particular attention to Eqs. (10-6) which pertain to constant angular acceleration and point out the analogy with one dimensional constant linear acceleration. Time invested in study of this parallelism is well spent for you can then use the insight you developed in solving linear problems to solve similar angular problems.
2. Learning the vector description of angular motion. You will again encounter centripetal acceleration but now in a new form.

10-1 Rotational Kinematics

Rotational kinematics is a description of the rotational motions of physical objects. One idealizes a physical body as a rigid body which cannot warp, bend, etc. We say then

> a rigid body moves in pure rotation if every point in the body moves in an arc of a circle; the centers of these circles lie on a single straight line called the axis of rotation.

It is not necessary that this axis be fixed in direction for all time, so a point may never complete the circle. In general, the axis of rotation is an instantaneous axis, and its direction may change from moment to moment.

A rigid body has six degrees of freedom; that is, a complete description of its motion requires the specification of six coordinates. Three of these are the components of the position vector of its center of mass relative to a fixed (laboratory) coordinate system. The other three are angles which specify the orientation of a coordinate system fixed in the body relative to the laboratory axes. The concern of rotational kinematics is with these angles and how they change with time. The six variables are illustrated in Fig. 10-1.

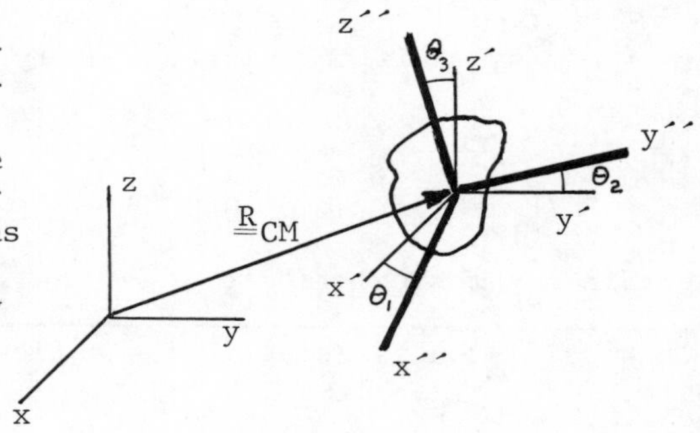

Figure 10-1. Illustration of the six variables used to describe the motions of a rigid body. Three are the components of the center of mass position vector, $\underline{\underline{R}}_{C.M.}$. The (x´, y´, z´) axes translate with the center of mass but maintain the same orientation as the inertial axes (x,y,z). The axes (x´´,y´´,z´´) are rigidly fixed to the body and their orientation angles θ_1, θ_2 and θ_3 are the other three coordinates.

10-2 Angular Variables -- Fixed Axis of Rotation

If a body is constrained to rotate about a fixed axis, all points in the body undergo circular motion; the centers of these circles all lie on the rotation axis. In Fig. 10-2 we show a plane passed through a point of interest, P, and perpendicular to the axis of rotation which is taken to be the z axis. The position vector of P in this plane is $\underline{\underline{r}}$; $|\underline{\underline{r}}| = r$ is constant so that given $|\underline{\underline{r}}|$ the point may be located by specifying θ.

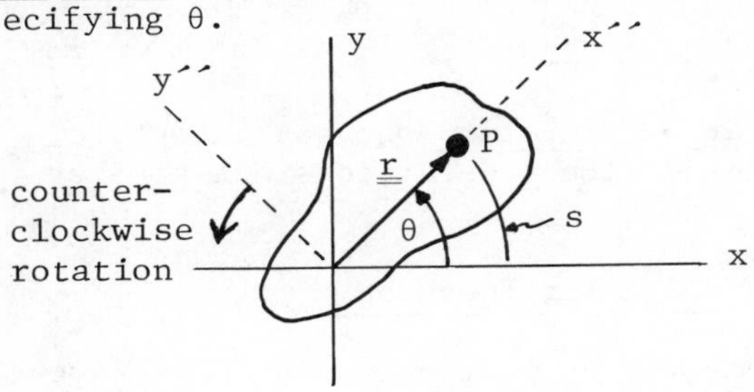

Figure 10-2. Cross sectional view of a body rotating about a fixed axis (the z axis); P is a point in the body and the paper plane. The chosen body fixed (x´´,y´´,z´´) axes are also shown; for simplicity P was taken on the x´´ axis. The body rotates counterclockwise and P describes a circle of radius $|\underline{\underline{r}}|$.

A possible choice of body fixed axes is also shown and in this case the same value of θ specifies the orientation of the body fixed axes relative to those of the laboratory. In general there would be two more angles, say θ_2 and θ_3. Because of the fixed axis constraint however, $\theta_3 = 0$ and $\theta_2 = \theta_1 \equiv \theta$. The quantity s denotes a length of arc along the circle described by P and we adopt the convention that θ increases for counterclockwise rotations. We measure θ in radians so that

$$\theta = s/r \quad . \tag{10-1}$$

The average angular speed $\bar{\omega}$ is defined by

$$\bar{\omega} = \Delta\theta/\Delta t \tag{10-2}$$

and the instantaneous angular speed is

$$\omega = d\theta/dt \quad ; \tag{10-3}$$

these are measured in radians per second. Similarly, the average angular acceleration is denoted

$$\bar{\alpha} = \Delta\omega/\Delta t \tag{10-4}$$

and the instantaneous angular acceleration is

$$\alpha = d\omega/dt \quad ; \tag{10-5}$$

α is measured in radian/second^2. Note that ω as well as α is the same for all points in the body.

Rotation of a particle or rigid body about a fixed axis has a complete formal correspondence to the translational motion along a fixed direction.

The corresponding variables are

Linear	Angular
x	θ
$v = dx/dt$	$\omega = d\theta/dt$
$a = dv/dt$	$\alpha = d\omega/dt$.

For rotation about a fixed axis with constant angular acceleration one has therefore (in correspondence with the linear motion along a fixed direction at constant acceleration)

Linear Equation	Angular Equation	
$v = v_o + at$	$\omega = \omega_o + \alpha t$	(10-6a)
$x = x_o + v_o t + \frac{1}{2} at^2$	$\theta = \theta_o + \omega_o t + \frac{1}{2} \alpha t^2$	(10-6b)
$x = \frac{1}{2} (v + v_o)t + x_o$	$\theta = \frac{1}{2} (\omega + \omega_o)t + \theta_o$	(10-6c)
$v^2 = v_o^2 + 2a (x - x_o)$	$\omega^2 = \omega_o^2 + 2\alpha (\theta - \theta_o)$	(10-6d)

Problems involving rotation about a fixed axis with constant angular acceleration are similar to constant linear acceleration problems. That is, there are six variables (θ, θ_o, ω, ω_o, α and t) related by three equations (Eqs. (10-6a) and (10-6b) plus α = constant). In any given problem then, your job is to determine how three of these quantities have been specified.

>>> Example 1. A brake is applied to a wheel initially rotating at 880 revolutions per minute to provide a constant angular acceleration of -40 radian/s^2 (the minus sign denotes deceleration). Through how many revolutions does the wheel turn before coming to rest?

From Eq. (10-6d) with $\omega = 0$ the angular displacement is

$$\theta - \theta_o = -\omega_o^2/(2\alpha) \quad .$$

Thus

$$-\frac{(880 \text{ rev/min})^2}{2(-40 \text{ rad/s}^2)} \frac{(2\pi \text{ rad})}{(1 \text{ rev})} \frac{(1 \text{ min})^2}{(60 \text{ s})^2} = 16.9 \text{ revolutions} \quad .$$ <<<

In this example ω_o and α were specified and we had the implicit information that $\omega = 0$ at some time t. Since we needed only the displacement there were really only five variables and three were specified. We could have used Eq. (10-6a) to determine the time required for the wheel to stop, but this was not asked.

>>> Example 2. If a rigid body is initially at rest and is subjected to a uniform angular acceleration of 5 revolutions per second^2, when will it have turned through 100 revolutions?

In this case we're given α and $\theta - \theta_o$ and asked to find t; ω_o is implicitly specified as zero.

In Eq. (10-6b) then, since both $\theta - \theta_o$ and α are expressed in revolutions we do not need to convert to radians. Thus

$$t = [2(\theta - \theta_o)/\alpha]^{1/2} = [2(100)/5]^{1/2} = 6.3 \text{ s} \quad . \qquad <<<$$

10-3 Relationship Between Linear and Angular Kinematics -- Fixed Axis

Let P be some point in a body at a perpendicular distance $r_\perp$ from the fixed axis of rotation (P could also be a particle undergoing circular motion). Then, the arc length of the circle traversed by P is given by Eq. (10-1) as

$$s = r_\perp \theta \quad . \qquad (10\text{-}7)$$

Since $r_\perp$ is constant, the speed of P is given by

$$v = ds/dt = r_\perp \, d\theta/dt = r_\perp \omega \quad . \qquad (10\text{-}8)$$

The tangential acceleration is the rate of change of v, and is given by

$$a_T = dv/dt = r_\perp \, d\omega/dt = r_\perp \alpha \qquad (10\text{-}9)$$

where α is the angular acceleration. The centripetal acceleration is

$$a_C = v^2/r_\perp = r_\perp \omega^2 \quad . \qquad (10\text{-}10)$$

From these equations, the economy and simplicity of using angular variables should be apparent, because $r_\perp$ and hence v, a_C, and a_T vary from point to point of the body, but ω and α are the same for all points at any given time. [CAUTION: In equations (10-7,8,9, 10) the angular unit must be the radian.]

>>> Example 3. A racing airplane has an engine designed to operate at 3600 rpm (revolutions per minute) and the airplane speed is to be 450 mph. What is the maximum diameter propeller that may be used if the propeller tip speed is not to exceed the velocity of sound (approximately 740 mph)?

The propeller tip has two components of velocity. One is that of the forward velocity of the airplane and the other is that due to rotation at 3600 rpm. Call the latter $v = \omega r_\perp$ and the former v_p. Since these are at right angles the square of the resultant is

$$v_p^2 + \omega^2 r_\perp^2 \le v^2_{sound} \quad .$$

Therefore the propeller radius can at most be

$$r_\perp = \sqrt{v_s^2 - v_p^2}/\omega \quad .$$

Then since $\sqrt{v_s^2 - v_p^2} = \sqrt{(740)^2 - (450)^2} = 587$ mph , and $\omega = 3600$ rpm $= (3600)(2\pi)$ rad/min ,

$$r_\perp = \frac{587 \text{ miles min}}{(3600)(2\pi) \text{ hr}} = \frac{(587)(5280)}{(3600)(2\pi)(60)} \text{ ft.} = 2.3 \text{ ft.}$$

Therefore the maximum diameter is $d = 2r_\perp = 4.6$ feet . <<<

>>> Example 4. A carborundum grinding wheel has a diameter of 15 cm. If the wheel is spinning at 3000 rpm, what is the acceleration of a point near the rim? Suppose a small chip comes loose; with what speed will it leave the wheel?

The acceleration of a point near the rim is entirely centripetal since the angular speed is constant, i.e. the tangential acceleration is zero as defined in Eq. (10-9). Therefore

$$a_C = r_\perp \omega^2 = (7.5 \text{ cm})(3000 \text{ rev/min})^2(2\pi \text{ rad/rev})^2 \times (1 \text{ min}/60 \text{ s})^2 = 7.4 \times 10^5 \text{ cm/s}^2 \quad .$$

In terms of g, a_C = 755 g. The linear velocity of a point near the rim is

$$v = \omega r_\perp = a_C/\omega = (7.4 \times 10^5 \text{ cm/s}^2)/[(3000)(2\pi/60)\text{rad/s}] = 2360 \text{ cm/s} \quad .$$

This is the speed at which a small chip would leave the wheel. <<<

10-4 Vectorial Properties of Rotation

Consider the simple case of a particle moving in a circle about the z axis as shown in Fig. 10-3. Notice that the origin has been taken to lie on the z axis but <u>not</u> at the center of the circle. The angular speed of the particle relative to the z axis at some instant of time is ω, and its angular acceleration is α. If the circle radius is $r_\perp$, then the linear speed of the particle is given by Eq. (10-8) and the linear <u>velocity</u> is directed tangent to the circle as shown. The tangential acceleration is given by Eq. (10-9) and is directed tangent to the circle, while the centripetal acceleration is given by Eq. (10-10) and is directed toward the center of the circle.

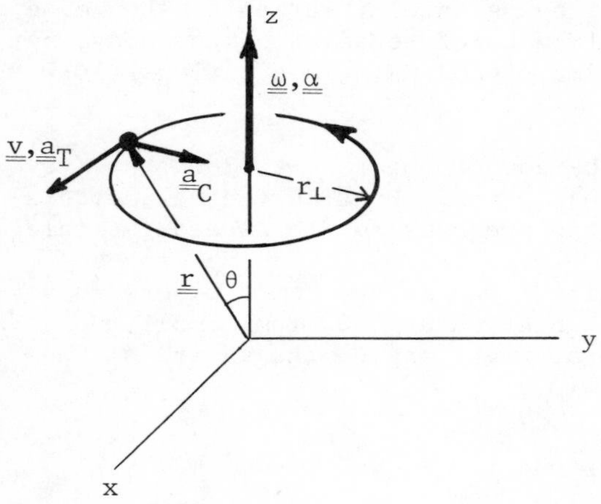

Figure 10-3. A particle moving in a circle of radius $r_\perp$ about the z axis. Also shown are the position vector $\mathbf{r}$, the velocity $\mathbf{v}$, the tangential acceleration $\mathbf{a}_T$, the centripetal acceleration $\mathbf{a}_C$, the angular velocity $\boldsymbol{\omega}$, and angular acceleration $\boldsymbol{\alpha}$. The directions of $\boldsymbol{\alpha}$ and $\mathbf{a}_T$ are for $\omega = |\boldsymbol{\omega}|$ increasing; for ω decreasing the directions of $\boldsymbol{\alpha}$ and $\mathbf{a}_T$ would be reversed.

Now, given that

$$v = \omega r_\perp \quad ,$$

suppose we express v in terms of r rather than $r_\perp$. From Fig. 10-3 then $r_\perp = r \sin\theta$ and we have

$$v = \omega r \sin\theta$$

where θ is the angle between the rotation axis and the position vector $\mathbf{r}$. This looks like the magnitude of a vector cross product between $\boldsymbol{\omega}$ and $\mathbf{r}$.

If we associate the rotation axis with the direction of the angular velocity of magnitude ω, then one is tempted to define a vector angular velocity $\underline{\underline{\omega}}$ which in this case can be written as

$$\underline{\underline{\omega}} = \omega\underline{\underline{k}}$$

since the axis of rotation here is in the z direction. In general then we define the vectorial angular velocity direction as the axis of rotation, and define the magnitude as the angular speed. The sense of $\underline{\underline{\omega}}$ is given by a right hand rule in that if the fingers of the right hand point in the direction of rotation the right thumb points in the direction of $\underline{\underline{\omega}}$. Thus we have

$$\underline{\underline{v}} = \underline{\underline{\omega}} \times \underline{\underline{r}} \quad . \tag{10-11}$$

It is clear that this definition correctly gives the magnitude, v, of $\underline{\underline{v}}$ and further we see that this also gives the correct direction. This does not of course prove that $\underline{\underline{\omega}}$ is truly a vector. Let us see if this definition is consistent with other aspects of the motion with which we are already familiar.

Differentiate Eq. (10-11) with respect to time. This yields

$$\underline{\underline{a}} = d(\underline{\underline{v}})/dt = d(\underline{\underline{\omega}} \times \underline{\underline{r}})/dt = (d\underline{\underline{\omega}}/dt) \times \underline{\underline{r}} + \underline{\underline{\omega}} \times (d\underline{\underline{r}}/dt)$$

$$= (\underline{\underline{\alpha}} \times \underline{\underline{r}}) + (\underline{\underline{\omega}} \times \underline{\underline{v}}) \tag{10-12}$$

where we've used the fact that $\underline{\underline{r}}$ is the position vector, so $\underline{\underline{v}} = d\underline{\underline{r}}/dt$ and we've defined the vectorial angular acceleration $\underline{\underline{\alpha}}$ as

$$\underline{\underline{\alpha}} = d\underline{\underline{\omega}}/dt \quad . \tag{10-13}$$

As previously stated $\underline{\underline{\omega}} = \omega\underline{\underline{k}}$ so

$$\underline{\underline{\alpha}} = d\underline{\underline{\omega}}/dt = \omega \, d\underline{\underline{k}}/dt + (d\omega/dt)\,\underline{\underline{k}} \; .$$

In this special case, $d\underline{\underline{k}}/dt = 0$, i.e. the direction of $\underline{\underline{\omega}}$ is fixed, so we have

$$\underline{\underline{\alpha}} = (d\omega/dt)\,\underline{\underline{k}} = \alpha\underline{\underline{k}} \quad ,$$

and $\underline{\underline{\alpha}}$ is directed along the rotation axis. If α is positive so that ω is increasing, $\underline{\underline{\alpha}}$ is parallel to $\underline{\underline{\omega}}$, but if α is negative so that ω is decreasing, $\underline{\underline{\alpha}}$ is antiparallel to $\underline{\underline{\omega}}$.

We note that Eq. (10-12) indicates that the total acceleration of the object consists of two terms, $\underline{\underline{\alpha}} \times \underline{\underline{r}}$ and $\underline{\underline{\omega}} \times \underline{\underline{v}}$. The first term has magnitude $\alpha r \sin\theta = \alpha r$ which is the tangential acceleration. If α is positive, this tangential acceleration $\underline{\underline{a}}_T$ is parallel to $\underline{\underline{v}}$ while if α is negative $\underline{\underline{a}}_T$ is opposed to (antiparallel) to $\underline{\underline{v}}$. The second term in Eq. (10-12) has magnitude

$$\omega v \sin 90^\circ = \omega v = \omega^2 r = v^2/r$$

and is directed toward the center of the circle; it is the centripetal acceleration, $\underline{\underline{a}}_C$. Note that the magnitude has the familiar form.

So, in the special case of a fixed direction rotation axis we see that the equations

$$\underline{\underline{v}} = \underline{\underline{\omega}} \times \underline{\underline{r}} \tag{10-14a}$$

$$\underline{\underline{a}}_T = \underline{\underline{\alpha}} \times \underline{\underline{r}} \tag{10-14b}$$

$$\underline{\underline{a}}_C = \underline{\underline{\omega}} \times \underline{\underline{v}} \tag{10-14c}$$

correctly give the linear vector quantities; that is, correct both in magnitude and direction. The nice feature of using these vector equations is that the directions come out automatically.

Now to achieve Eqs. (10-14) we ascribed a vectorial property to $\boldsymbol{\omega}$. In order that this be completely valid, it is also necessary that if a body has angular velocity $\boldsymbol{\omega}_1$ about some axis and simultaneously $\boldsymbol{\omega}_2$ about some other axis, the resultant motion should be an angular velocity $\boldsymbol{\omega}$ about some axis such that

$$\boldsymbol{\omega} = \boldsymbol{\omega}_1 + \boldsymbol{\omega}_2 \quad . \tag{10-15}$$

Eq. (10-14) can be verified by experiment.

Both the angular velocity $\boldsymbol{\omega}$ and the angular acceleration $\boldsymbol{\alpha}$ are vectors which means that to each we can ascribe a direction and a magnitude and for simultaneous motions the resultant angular velocity and angular acceleration are each given correctly by the vector addition law. The rotation of a body through an angle θ about some axis could also be ascribed a direction and magnitude. If two such rotations are performed simultaneously or sequentially the resultant rotation is not in general given by vector addition. To show this we need only one counter-example. Consider a π rotation about the x axis followed by a π rotation about the y axis. This is seen to be equivalent to a π rotation about the z axis (see Fig. 10-4). But vector addition would require that the resultant be a rotation of π about an axis directed at 45° to both the x and y axes. This failure of the law of vector addition to correctly give the resultant rotation is the origin of the statement that finite rotations are not vectors. On the other hand angular velocities and angular acceleration are vectors because the vector addition law gives the correct resultant.

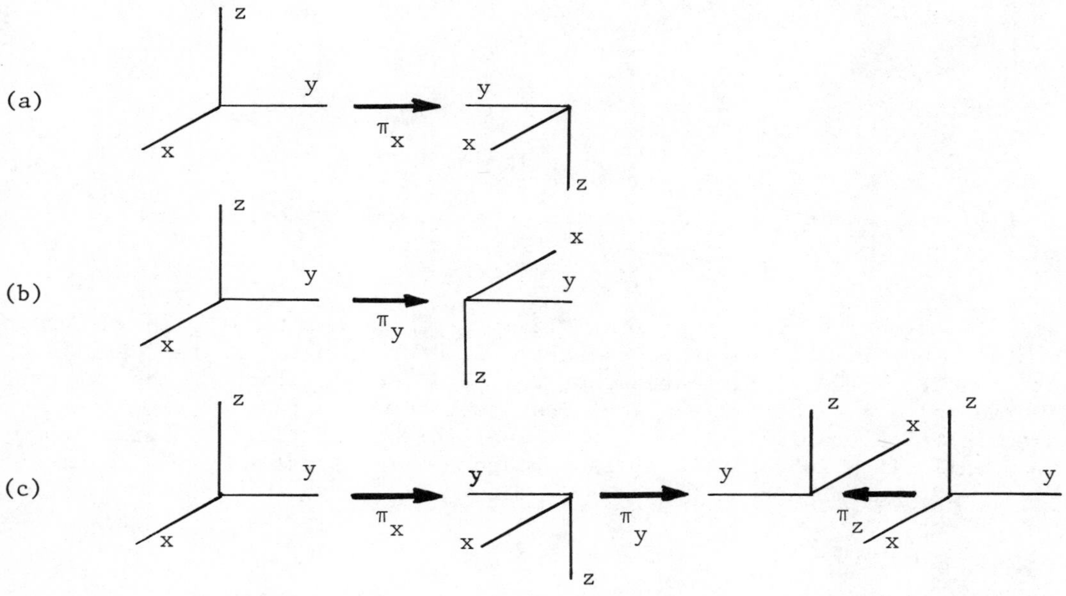

Figure 10-4. Illustration (a) a π rotation about the x axis, (b) a π rotation about the y axis, (c) a π rotation about x followed by a π rotation about y and its equivalence to a π rotation about z.

10-5 Programmed Problems

1.

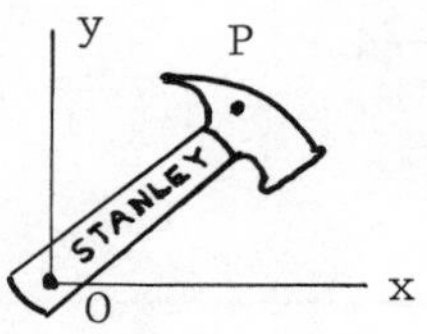

A hammer is pinned at 0 and rotates with constant speed about 0 in the x-y plane. Initially a point P is located at $\underline{r} = 3\underline{i} + 3\underline{j}$. For a rigid body the separation of P and 0 is constant.

a. What is $|\underline{r}|$?
b. What is the position vector of P when P coincides with the positive y axis?

a. $|\underline{r}| = \sqrt{r_x^{\,2} + r_y^{\,2}}$

$|\underline{r}| = 3\sqrt{2}$

b. $\underline{r} = (3\sqrt{2})\underline{j}$

---

2. As the point P moves from

$$\underline{r} = 3\underline{i} + 3\underline{j} \quad \text{to} \quad \underline{r} = 3\sqrt{2}\ \underline{j}$$

$d|\underline{r}|/dt$ has the value ______ .

Zero.
$|\underline{r}|$ is a constant.

---

3. For this uniform circular motion of P we already know "how far from 0" P will always be. Thus to complete the specification of P we need only know what its "orientation" is with respect to some reference. For this case the variable "where" is an angle. In the diagram to the right the line segments labeled "r" are the same and are both equal to the magnitude of $\underline{r}$.

Motion of hammer

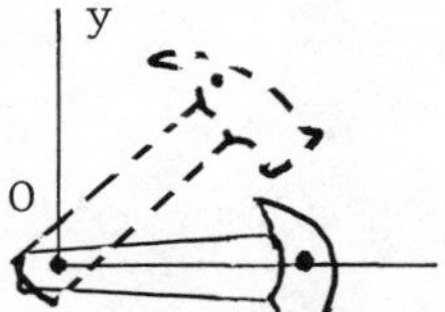

Motion of point P

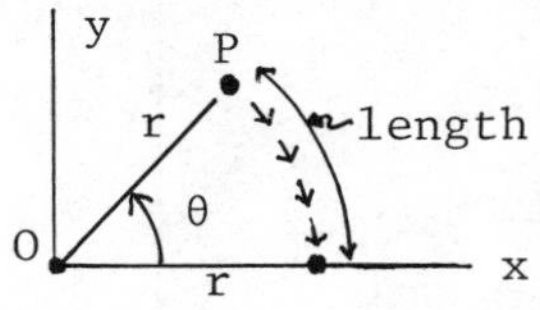

Measuring θ counterclockwise from the x axis we can define θ in radians as

$\theta =$ ______ .

$\theta = s/r$

s is the arc of a circle of radius r.

---

4. $v = ds/dt$ is the tangential (linear) speed of the point P on the head of the hammer. How is ds/dt related to the rate of change of θ?

$ds/dt = d(r\theta)/dt = r\, d\theta/dt$

$d\theta/dt$ has the units rad/s .

---

5. ds/dt is called the ______ of point P on the hammer.
$d\theta/dt$ is called the ______ of point P on the hammer.

tangential or linear velocity (speed)

angular velocity (speed)

6. The relationship between the linear and angular speed of the point P is written $v = r\omega$. Differentiate this expression with respect to time.	$dv/dt = r\, d\omega/dt$ since r = constant .
7. dv/dt is called the ______ of the point P. $d\omega/dt$ is called the ______ of the point P.	tangential (linear) acceleration. angular acceleration.
8. The rotational analogues of x, v, a and t are θ, ω, α and t. Write the analogue equations for the following:	$x \rightarrow \theta$ $v \rightarrow \omega$ $a \rightarrow \alpha$ $t \rightarrow t$
Linear Kinematics (constant a) / Rotational Kinematics (constant α) a. $v = v_o + at$ a. __________ . b. $x = x_o + v_o t + \frac{1}{2} at^2$ b. __________ . c. $v^2 = v_o^2 + 2\,ax$ c. __________ .	a. $\omega = \omega_o + \alpha t$. b. $\theta = \theta_o + \omega_o t + \frac{1}{2}\alpha t^2$ c. $\omega^2 = \omega_o^2 + 2\,\alpha\theta$
9. In Chapter 3 the above linear equations were essentially derived from the definitions of average velocity and acceleration. Write the analogous definitions of average angular velocity and acceleration. Linear / Angular a. $\bar{v} = (x_2 - x_1)/(t_2 - t_1)$ a. $\bar{\omega}$ = ______ . b. $\bar{a} = (v_2 - v_1)/(t_2 - t_1)$ b. $\bar{\alpha}$ = ______ .	a. $\bar{\omega} = (\theta_2 - \theta_1)/(t_2 - t_1)$ b. $\bar{\alpha} = (\omega_2 - \omega_1)/(t_2 - t_1)$
10. The angular velocity of our hammer increases uniformly from 120 rpm to 210 rpm in 3 seconds. ω_i = 120 rpm = _____ rad/s ω_f = 210 rpm = _____ rad/s Leave your answer in terms of π.	$1\,\frac{\text{rev}}{\text{min}} \times \frac{1}{60}\frac{\text{min}}{\text{s}} = \frac{1}{60}\frac{\text{rev}}{\text{s}}$. $\frac{1}{60}\frac{\text{rev}}{\text{s}} \times 2\pi\frac{\text{rad}}{\text{rev}} = \frac{\pi}{30}\frac{\text{rad}}{\text{s}}$. Thus 1 rpm $= \frac{\pi}{30}\frac{\text{rad}}{\text{s}}$. ω_i = 120 rpm = 4π rad/s . ω_f = 210 rpm = 7π rad/s .

11. What is the average acceleration of the hammer during the three seconds? Obtain the answer in rad/s^2.	$\bar{\alpha} = (\omega_f - \omega_i)/(t_f - t_i)$ $\bar{\alpha} = (7\pi\ rad/s - 4\pi\ rad/s)/(3\ s)$ $\bar{\alpha} = \pi\ rad/s^2$. Note: In this example the average and instantaneous angular accelerations are the same (and constant).
12. What is the angular displacement of the hammer during the 3 second time interval?	Using equation (b) from the answer of frame 9 $\theta - \theta_o = \omega_o t + \frac{1}{2}\alpha t^2$ $\theta - \theta_o = (4\pi\ rad/s)(3\ s) + \frac{1}{2}(\pi\ rad/s^2)(9\ s^2)$. $\theta - \theta_o = 16.5\pi$ radians.

A child's record has a sharp radial scratch, consistent with the condition of most children's records. You enter the room as it is playing and in disgust turn off the record player. Your keen scientific observation causes you to note that during the slowdown the scratch makes 2 revolutions in 4 seconds after which you leave the room. We shall assume that the acceleration during slowdown is constant at $\pi/4\ rad/s^2$, and calculate the angular velocity of the record as you leave.

13. To begin with, what is the change in angular velocity during the 4 second interval?	$\omega - \omega_o = \alpha t$ $\omega - \omega_o = (-\pi/4\ rad/s^2)(4\ s)$ $\omega - \omega_o = -\pi\ rad/s$. The minus sign indicates a decrease in angular velocity.
14. Through how many radians did the record turn during the 2 revolutions?	1 rev = 2π radians, so 2 rev = 4π radians.
15. What terms in the following expression do you already know for the time period under discussion? $\theta - \theta_o = \omega_o t + \frac{1}{2}\alpha t^2$.	$\theta - \theta_o = 4\pi$ radians $t = 4\ s$ $\alpha = -\pi/4\ rad/s^2$

16. What was the angular speed of the record when you turned the machine off?

From the previous frame

$$\omega_o = [(\theta - \theta_o) - \tfrac{1}{2}\alpha t^2]/t$$

$$= [4\pi - \tfrac{1}{2}(-\pi/4)(4^2)]/4$$

$$= 3\pi/2 \text{ rad/s} \quad (= 45 \text{ rpm!})$$

---

17. Look at the answers to frame 13 and 16; what was the angular speed as you left?

$$\omega - \omega_o = -\pi \text{ rad/s}$$

$$\omega = -\pi \text{ rad/s} + 3\pi/2 \text{ rad/s}$$

$$= \pi/2 \text{ rad/s} .$$

This problem is one in which you cannot substitute directly into a convenient kinematic formula and solve for ω. The problem requires the calculation of ω_o before you can determine ω.

---

18. 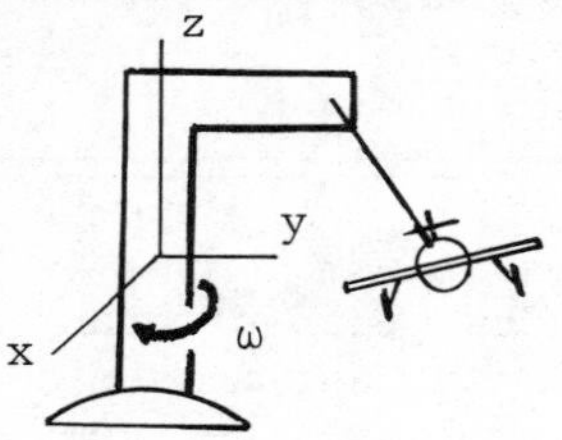

A popular amusement ride consists of an airplane connected by a cable to an inverted L structure which rotates in the direction shown; ω is constant.

For the coordinate system shown, indicate the resultant force on the airplane, the position vector of the airplane, the velocity of the airplane and the angular velocity of the airplane.

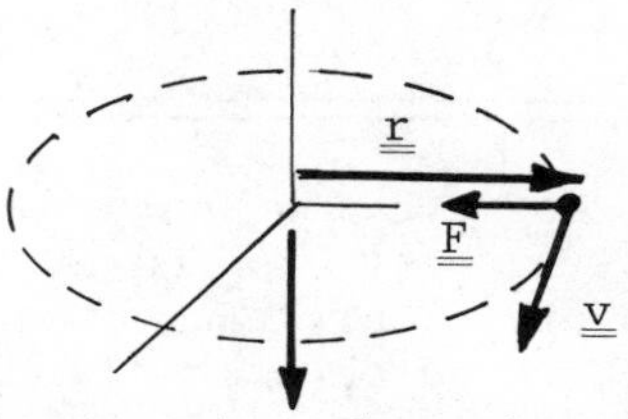

$\underline{F}$ is <u>centripetal</u>, $\underline{v}$ is tangent to a circle in the xy plane and $\underline{\omega}$ is along the negative z axis.

---

19. Write the vector product which relates $\underline{v}$, $\underline{\omega}$ and $\underline{r}$ of the previous answer.

$$\underline{v} = \underline{\omega} \times \underline{r}$$

Note that the directions of $\underline{v}$, $\underline{\omega}$ and $\underline{r}$ of the previous answer satisfy the right hand rule.

---

20. From the definition of the vector product we can write:

$$|\underline{v}| = |\underline{\omega}|\,|\underline{r}| \sin\phi$$

where $\sin\phi$ = _____ in this problem. Look at the answer to frame 18 before answering.

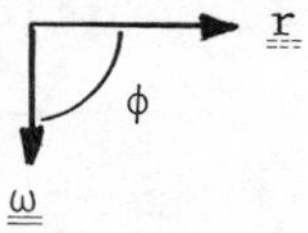

$\phi = 90^o$
$\sin\phi = 1$

---

21. From the previous frame then, we have $v = \omega r$. We can rewrite the resultant force $F = mv^2/r$ as

$$F = mr\,(\_\_\_\_\_)^2 \quad .$$

$F = mr\,\omega^2$

---

22. The term $r\omega^2$ is the ______ acceleration of the airplane.

Radial or centripetal.

Chapter 11: ROTATIONAL DYNAMICS

REVIEW AND PREVIEW

The previous chapter was an introduction to the descriptive nature of angular motion, angular kinematics. In this chapter you will learn to deal with the causes of rotational motion. That is, you will study rotational dynamics. Rotational dynamics is the relationship between rotational kinematic quantities (θ, $\underline{\omega}$, $\underline{\alpha}$) and properties of the body or system undergoing rotational motion together with properties of its relevant environment which cause those rotational motions. To each linear kinematic quantity there is a rotational analogue (see Chapter 10). In the same manner, to each linear dynamic and inertial quantity there is also a rotational analogue. Further the laws of rotation mechanics have their analogues in laws of linear mechanics.

You will be introduced to the concept of torque which is the rotational analogue of force and to angular momentum which is the rotational analogue of linear momentum. You will also meet rotational inertia which plays the mass role in rotational motion. Equations will be introduced which relate these quantities and also you will meet equations which relate the cause of rotational motion to the rotational kinematics; for example, in certain circumstances one may write torque = rotational inertia times angular acceleration. It is important to realize that no new physical laws are introduced to relate these quantities, only Newton's laws are used. You will also meet the third conservation law, conservation of angular momentum.

GOALS AND GUIDELINES

In this chapter you will have four main goals:

1. Learning the definitions of the vectors torque $\underline{\tau}$ and angular momentum $\underline{\ell}$ and the basic equation which relates the two. This is the basic equation of rotational dynamics.
2. Learning the definitions of rotational inertia and rotational kinetic energy and how to calculate these quantities.
3. Learning well the special but important case of fixed axis rotational dynamics. Pay particular attention to Table 11-1 which draws the parallel with one dimensional linear motion. Again time invested here is well spent because you can use your previously developed experience to solve fixed axis rotation problems.
4. Learning to solve problems involving combined rotation and translation. Of particular importance is the case of rolling about a fixed direction axis along a surface without slipping.

11-1 Torque and Angular Momentum of a Particle

If a force $\underline{F}$ acts upon a particle or a system at some point P whose position vector from some origin 0 is $\underline{r}$ then the torque $\underline{\tau}$ of this force with respect to 0 is defined to be

$$\underline{\tau} = \underline{r} \times \underline{F} \quad . \tag{11-1}$$

Some properties of $\underline{\tau}$ are:

1. $\underline{\tau}$ is a vector whose magnitude is $\tau = rF \sin \theta$ where θ is the angle between $\underline{r}$ and $\underline{F}$. The direction of $\underline{\tau}$ is perpendicular to $\underline{r}$ and to $\underline{F}$ (hence to the plane determined by the two) and the sense of $\underline{\tau}$ is given by the right hand rule. As shown in Fig. 11-1a if the fingers of the right hand rotate $\underline{r}$ into $\underline{F}$ then the right thumb points in the direction of $\underline{\tau}$.

(a)

(b)   line of $\underline{F}$

Figure 11-1. (a) Illustration of the torque $\underline{\tau}$ of a force $\underline{F}$ which acts at P whose position vector relative to 0 is $\underline{r}$. (b) Illustration of the line of action of a force $\underline{F}$ and the moment arm $r_\perp$ of the force.

2. $\underline{\tau}$ depends upon the choice of the origin 0 through $\underline{r}$.
3. The dimensions of $\underline{\tau}$ are the <u>same as</u> those of energy, that is ML^2/T^2. In SI units, however, one usually uses the N·m rather than J. Similarly in the Cgs system the common unit is the dyn·cm and in the BES the ft·lb.
4. As shown in Fig. 11-1b, $r \sin \theta \equiv r_\perp$ is the perpendicular distance from 0 to the <u>line of action</u> of $\underline{F}$. One often speaks of $r_\perp$ as the <u>moment arm</u> of $\underline{F}$ and calls $\underline{\tau}$ the <u>moment of the force</u>.

If a particle is at a point P whose position vector with respect to some origin 0 is $\underline{r}$, and if the particle has linear momentum p, its <u>angular momentum $\underline{\ell}$ with respect to 0</u> is defined to be

$$\underline{\ell} = \underline{r} \times \underline{p} \ . \tag{11-2}$$

Some properties of $\underline{\ell}$ are:

1. $\underline{\ell}$ is a vector whose magnitude is $\ell = rp \sin \theta$ where θ is the angle between $\underline{r}$ and $\underline{p}$. The direction of $\underline{\ell}$ is perpendicular to both $\underline{r}$ and $\underline{p}$ with the sense of $\underline{\ell}$ given by the right hand rule as shown in Fig. 11-2a.

(a)

(b)   line of $\underline{p}$

Figure 11-2. (a) Illustration of the angular momentum $\underline{\ell}$ of a particle whose position vector relative to 0 is r and whose momentum is p. (b) Illustration of the line of action of $\underline{p}$ and the moment arm $r_\perp$ of $\underline{p}$.

2. $\underline{\ell}$ depends upon the choice of the origin 0 through $\underline{r}$.
3. The dimensions of $\underline{\ell}$ are those of energy times time, that is ML^2/T. In the SI system one uses the unit J·s and in the Cgs system the erg·s while in the BES one uses the ft·lb·s.

4. As shown in Fig. 11-2b, $r \sin\theta \equiv r_\perp$ is the perpendicular distance from 0 to the line of action of $\underline{\underline{p}}$. One often calls $r_\perp$ the moment arm of $\underline{\underline{p}}$ and $\underline{\underline{\ell}}$ the moment of the momentum.

Now, $\underline{\underline{\tau}}$ is the rotational analogue of $\underline{\underline{F}}$ and $\underline{\underline{\ell}}$ is the rotational analogue of $\underline{\underline{p}}$. The rotational analogue of Newton's second law is

$$\underline{\underline{\tau}} = d\underline{\underline{\ell}}/dt \quad . \tag{11-3}$$

In order that Eq. (11-3) be valid, it is necessary that

i) $\underline{\underline{\tau}}$ and $\underline{\underline{\ell}}$ be referred to the same origin 0;
ii) the coordinate frame in which they are measured must be inertial. This is necessary so that $\underline{\underline{F}} = d\underline{\underline{p}}/dt$, from which Eq. (11-3) follows, be valid.

11-2 Angular Momentum of a System of Particles

For a system of N particles the total angular momentum is defined by the vectorial sum

$$\underline{\underline{L}} = \sum_{j=1}^{N} \underline{\underline{\ell}}_j \tag{11-4}$$

where $\underline{\underline{\ell}}_j$ is the angular momentum of the jth particle. The rotational analogue of

$$\underline{\underline{F}}_{ext} = d\underline{\underline{P}}/dt \tag{11-5}$$

where $\underline{\underline{P}}$ is the total linear momentum of a system is

$$\underline{\underline{\tau}}_{ext} = d\underline{\underline{L}}/dt \;, \quad \text{common origin in an inertial frame} \tag{11-6}$$

where $\underline{\underline{\tau}}_{ext}$ denotes the net external torque on the system. Eq. (11-6) follows from Eqs. (11-1) - (11-5) by noting that the torque on the jth particle is due to both internal and external forces. Provided that the internal forces are not only oppositely directed (Newton's third law) but also act along the same line (strong form of the third law) the internal torques cancel in pairs and Eq. (11-6) follows.

Again, $\underline{\underline{\tau}}_{ext}$ and $\underline{\underline{L}}$ must be referred to the same origin in some inertial frame of reference. Eq. (11-6) is also valid however when $\underline{\underline{\tau}}_{ext}$ and $\underline{\underline{L}}$ are measured with respect to an origin at the center of mass of the system. This fact allows one to separate general motion of a system into motion of the center of mass and motion about the center of mass. Thus Eq. (11-6) may be written as

$$\underline{\underline{\tau}}_{ext} = d\underline{\underline{L}}/dt \;, \quad \text{common origin either in an inertial frame or at the center of mass.} \tag{11-6$'$}$$

11-3 Rotational Inertia -- Fixed Axis, Rotational Kinetic Energy

The moment of inertia, or rotational inertia of a rigid body with respect to some axis is defined by

$$I = \sum m_j d_j^{\,2} \tag{11-7}$$

where d_j is the perpendicular distance of the jth particle of mass m_j from the axis. Eq. (11-7) applies to a rigid body made of mass points and the sum is taken over all the mass points of the body. For a continuous body this sum is replaced by an integral,

$$I = \int r_\perp^2 \, dm \quad , \quad r_\perp = \text{perpendicular distance of dm from the axis.} \qquad (11\text{-}8)$$

The mass element dm under consideration is written as $dm = \rho \, dV$ where ρ is the density (mass per unit volume) and dV is an infinitesimal volume element. In general, ρ would be a function of position; for homogeneous bodies it is not. In that case $\rho = M/V$ where M is the mass of the body and V its volume. Then

$$I = \frac{M}{V} \int r_\perp^2 dV \quad , \quad \text{homogeneous body} \qquad (11\text{-}9)$$

where the integration is over the volume of the body.

>>> Example 1. Find the rotational inertia of a thin homogeneous disc of radius R and mass M about a diameter.

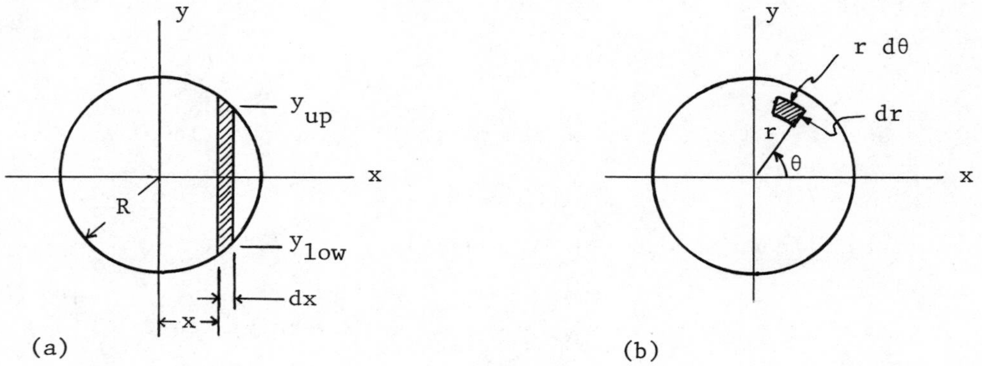

Figure 11-3. Example 1. The rotational inertia of the disc about the y-axis will be calculated. (a) Volume element used in Cartesian coordinates. (b) Volume element used in polar coordinates.

We'll do this two ways. First of all consider Fig. 11-3a in which we've taken a volume element

$$dV = t(y_{up} - y_{low})dx$$

where t is the thickness and y_{up} and y_{low} stand for the upper and lower values of y for a given value of x. The equation for the circular disc is

$$x^2 + y^2 = R^2 \quad ,$$

so,

$$y_{up} = \sqrt{R^2 - x^2}$$

and

$$y_{low} = -\sqrt{R^2 - x^2} \quad .$$

Thus

$$dV = 2t\sqrt{R^2 - x^2}\,dx \quad .$$

Every point in this strip is at a perpendicular distance x from the y-axis and it is precisely for that reason that we divide the body into strips in this direction rather than in some other direction. The disc volume is

$$V = \pi R^2 t \quad .$$

Thus

$$I = (M/\pi R^2 t)\int_{-R}^{R} x^2\, 2t\sqrt{R^2 - x^2}\,dx \quad .$$

This is integrated to

$$I = MR^2/4 \quad .$$

It is relatively easy to set up I in Cartesian coordinates but somewhat difficult to integrate. Let us now use polar coordinates as shown in Fig. 11-3b. Then the volume element is

$$dV = (t)(r\,d\theta)(dr)$$

where r varies from 0 to R and θ from 0 to 2π. The perpendicular distance of this volume element from the y-axis is $r\cos\theta$. Therefore

$$I = \frac{M}{\pi R^2 t}\int_0^R\int_0^{2\pi} (r\cos\theta)^2\, tr\,d\theta\,dr$$

$$= \frac{M}{\pi R^2}\int_0^R r^3 dr\int_0^{2\pi}\cos^2\theta\,d\theta$$

$$= \frac{M}{\pi R^2}\frac{R^4}{4}\left[\frac{\theta}{2} + \frac{1}{4}\sin 2\theta\right]_0^{2\pi} = \frac{M}{4}R^2 \quad .$$ <<<

The point of doing this example two ways is to show that if one chooses coordinates which make as much use of the symmetry of the body as possible the calculation is simplified, but this is not necessary to obtain the result. Don't worry if at first you cannot choose the best possible coordinate system.

A very useful result is the <u>parallel axis theorem</u>. If,

I = rotational inertia about some axis,
I_{CM} = rotational inertia about an axis parallel to the first axis but passing through the center of mass,
d = perpendicular distance between the two axes,
M = mass of the body,

then the parallel axis theorem is

$$I = I_{CM} + Md^2 \quad . \tag{11-10}$$

>>> Example 2. A thin homogeneous equilateral triangular plate has mass m and sides ℓ. Determine the moment of inertia about the axes (a) the perpendicular bisector of one of the angles, (b) a line passing through the center of mass and parallel to one edge, (c) an axis through the center of mass and perpendicular to the plate, and (d) an axis along one edge.

Because the plate is thin, we can use mass per unit area, σ, as our density and because it is homogeneous

$$\sigma = \frac{M}{A} = \frac{M}{\frac{\sqrt{3}}{4}\ell^2} .$$

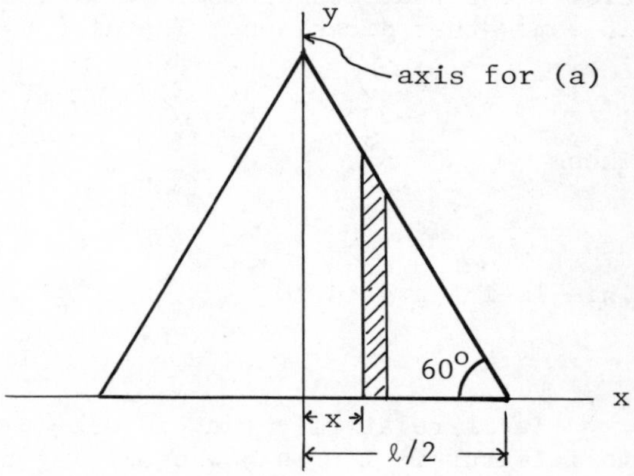

To do part (a) we divide the triangle into strips as shown in the sketch. The height of this strip is y which is given by

$$\frac{y}{\frac{\ell}{2} - x} = \tan 60^\circ = \sqrt{3} .$$

This expression is valid for $x \geq 0$. The area of the strip is $y\,dx = \sqrt{3}(\ell/2 - x)dx$. The moment of inertia of the entire triangle is twice that of the triangle between $x = 0$ and $x = \ell/2$, i.e.

$$I = 2\int_0^{\ell/2} x^2(\sigma\, y\, dx) = 2\,\sigma\,\sqrt{3}\int_0^{\ell/2} x^2(\frac{\ell}{2} - x)dx = \sigma\,\sqrt{3}\,\ell^4/96 ,$$

$$I_{(a)} = \frac{M\ell^2}{24} .$$

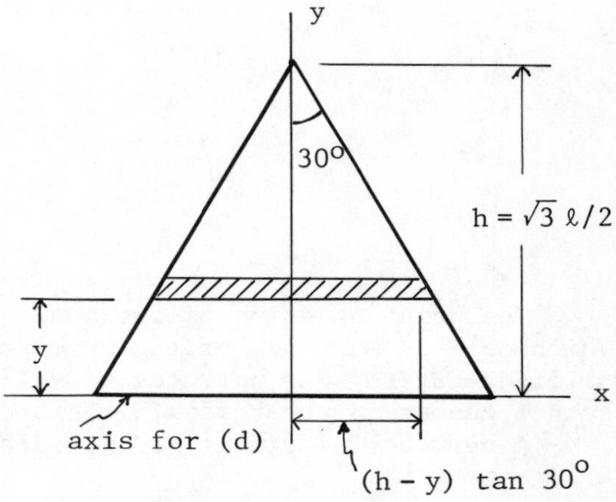

In this case, it is a bit easier to do part (d) first and use the parallel axis theorem for part (b). To do part (d) we divide the triangle into strips as shown in the sketch. This strip is at a distance y from the x axis and has area

$$2(h - y)\tan 30^\circ\, dy = \frac{2}{3}\sqrt{3}\,(h - y)\,dy .$$

Thus this strip has mass

$$dm = \frac{2\sigma}{3}\sqrt{3}(h - y)\,dy$$

and the moment of inertia about the x axis is

$$I_{(d)} = \int_0^h \frac{2\sigma}{3}\sqrt{3}(h - y)y^2dy = \frac{\sigma\sqrt{3}h^4}{18} = \frac{M\ell^2}{8} .$$

Now we can do part (b), because the axis of part (b) will be parallel to the axis of part (d) and at a distance h/3 above it. Thus

$$I_{(b)} = I_{(d)} + M\frac{h^2}{9} = \frac{M\ell^2}{8} + M\frac{\ell^2}{12} = \frac{5M\ell^2}{24} .$$

To do part (c) we use a very nifty theorem! For a plate-like body one can easily show that the sum of the moments of inertia about two perpendicular axes in the plane of the plate and passing through the center of mass is equal to the moment of inertia about an axis perpendicular to the plate through the center of mass. Thus

$$I_{(c)} = I_{(a)} + I_{(b)} = M\ell^2/4 \quad . \qquad <<<$$

The rotational inertia of a rigid body occurs in expressions for the kinetic energy of rotation. We speak of a fixed axis if

(1) the axis of rotation is fixed in an inertial frame, or
(2) the axis of rotation passes through the center of mass of the body and maintains a constant direction in space.

In these cases every point in the body has linear speed $v = \omega d$ due to the rotation where ω is the instantaneous angular speed and d is the perpendicular distance from the point to the rotation axis. The rotational kinetic energy is

$$K = \tfrac{1}{2} I\omega^2 \quad . \qquad (11\text{-}11)$$

In Eq. (11-11), I is the rotational inertia of the body relative to the axis of rotation. For a fixed axis, I is constant in time.

In case (2), the total kinetic energy is the sum of the rotational kinetic energy plus the translational kinetic energy or (see section 11-5)

$$K_{Total} = \tfrac{1}{2} I\omega^2 + \tfrac{1}{2} MV^2 \qquad (11\text{-}12)$$

where M is the mass of the body and V is the speed of its center of mass. Here I is the rotational inertia relative to an axis through the center of mass.

11-4 Rotational Dynamics of a Rigid Body -- Fixed Axis

The equations of rotational dynamics relate the external torques on a rigid body to the kinematic quantities. The formulae of this section apply only to the case of a fixed axis.

If a force $\underline{\underline{F}}$ is applied to a rigid body at a point $\underline{\underline{r}}$ relative to some origin on the rotation axis, the torque of this force is $\underline{\underline{\tau}} = \underline{\underline{r}} \times \underline{\underline{F}}$. If the fixed axis is that of case (2) of Sec. 11-3, the origin must be at the center of mass. The power input of this force to the rigid body, or the rate at which the force does work on the rigid body is given by

$$P = \underline{\underline{F}} \cdot \underline{\underline{V}} \quad ,$$

where $\underline{\underline{V}}$ is the total velocity, i.e.

$$\underline{\underline{V}} = \underline{\underline{V}}_{CM} + (\underline{\underline{\omega}} \times \underline{\underline{r}}) \quad .$$

That part of the power input $\underline{\underline{F}} \cdot \underline{\underline{V}}_{CM}$ changes the translational kinetic energy of the body. The remaining part changes the rotational kinetic energy and is of interest here. Since the velocity of the body at $\underline{\underline{r}}$ due to the rotation is $\underline{\underline{\omega}} \times \underline{\underline{r}}$,

$$\underline{\underline{F}} \cdot (\underline{\underline{\omega}} \times \underline{\underline{r}}) = \underline{\underline{\omega}} \cdot (\underline{\underline{r}} \times \underline{\underline{F}}) = \underline{\underline{\omega}} \cdot \underline{\underline{\tau}} \quad .$$

The rate at which this work is done on the body is equal to the rate at which its kinetic energy of rotation changes. We shall call this rate P_{rot}, and

$$P_{rot} = dW/dt = dK/dt = d(\tfrac{1}{2}\, I\omega^2)/dt \quad .$$

Because the axis is fixed I is constant in time and

$$P_{rot} = \tfrac{1}{2} \cdot 2I\omega d\omega/dt = I\alpha \quad . \tag{11-13}$$

Thus we write $\underline{\underline{\omega}}\cdot\underline{\underline{\tau}} = \omega\tau_\omega$ where τ_ω is the component of $\underline{\underline{\tau}}$ along the $\underline{\underline{\omega}}$ direction and have the important result

$$\tau_\omega = I\alpha \quad , \quad \text{fixed axis} \quad . \tag{11-14}$$

The rotational kinetic energy of the body can also be written as

$$\begin{aligned} K &= \tfrac{1}{2} \sum_j m_j(\omega \times \underline{\underline{r}}_j)\cdot\underline{\underline{v}}_j \\ &= \tfrac{1}{2} \sum_j m_j(\underline{\underline{r}}_j \times \underline{\underline{v}}_j)\cdot\underline{\underline{\omega}} = \tfrac{1}{2}\, \underline{\underline{\omega}}\cdot\underline{\underline{L}} \quad . \end{aligned} \tag{11-15a}$$

Let $L_\omega = \underline{\underline{\omega}}\cdot\underline{\underline{L}}$ denote the component of $\underline{\underline{L}}$ along the $\underline{\underline{\omega}}$ direction. Then

$$K = \tfrac{1}{2}\, \omega L_\omega \quad . \tag{11-15b}$$

Note that even for the fixed axis case, $\underline{\underline{L}}$ is <u>not</u> in general parallel to $\underline{\underline{\omega}}$. However the <u>component</u> of $\underline{\underline{L}}$ along the $\underline{\underline{\omega}}$ direction is related to ω and I by

$$L_\omega = \omega I \quad . \tag{11-16}$$

In Table 11-1 we summarize the fixed axis rotational quantities together with the linear analogues.

Table 11-1

Linear Quantity		Fixed Axis Rotation Quantity	
Displacement	x	Angular displacement	θ
Velocity	$v = dx/dt$	Angular velocity	$\omega = d\theta/dt$
Acceleration	$a = dv/dt$	Angular acceleration	$\alpha = d\omega/dt$
Inertia property mass	m	Inertia property rotational inertia	I, constant
Force	$F = ma$	Torque component	$\tau_\omega = \alpha I$
Kinetic energy	$K = \tfrac{1}{2}\, mv^2$	Kinetic energy	$K = \tfrac{1}{2}\, I\omega^2$
Work	$dW = F\, dx$	Work	$dW = \tau_\omega d\theta$
Power	$P = Fv$	Power	$P_{rot} = \tau_\omega\omega$
Momentum	$p = mv$	Angular momentum component	$L_\omega = I\omega$

>>> Example 3. A uniform disc of radius R and mass M is mounted with a frictionless bearing at its center and is rotating about an axis through the bearing perpendicular to the disc. If the initial angular speed is ω_o and a brake applies a frictional force tangent to the disc edge so as to stop the disc uniformly in θ_o radians, what is this force?

We must first of all find the angular acceleration. Since we're told that it is uniform we may use Eq. (10-6d) with the displacement θ_o. Then since $\omega = 0$,

$$\alpha = -\omega_o^2/2\theta_o \quad .$$

This angular acceleration is related directly to the torque by Eq. (11-14) since this is a fixed axis case. Thus

$$\tau_\omega = I\alpha \quad .$$

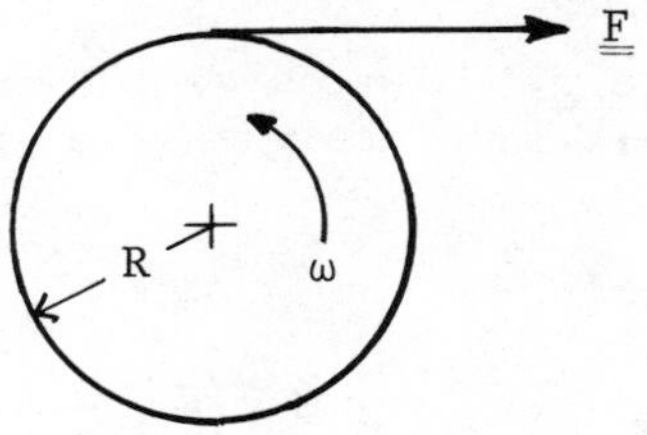

Figure 11-4. Example 3. A tangential force $\underline{\underline{F}}$ applied to stop a rotating disc uniformly.

Since the force is applied tangentially as in Fig. 11-4

$$\tau_\omega = -RF$$

so we have

$$-RF = I\alpha = (MR^2/2)(-\omega_o^2/2\theta_o) \quad .$$

Thus

$$F = MR\omega_o^2/4\theta_o \quad .$$

<<<

>>> Example 4. Find the work done by the torque in Example 3 and show that it is equal to the change in kinetic energy.

For a small angular displacement $d\theta$ the work done by τ_ω is

$$dW = \tau_\omega d\theta \quad .$$

Hence the total work done by τ_ω is

$$W = \int_0^{\theta_o} \tau_\omega d\theta = \tau_\omega \int_0^{\theta_o} d\theta = \tau_\omega \theta_o$$

since τ_ω is constant. Thus

$$W = -MR^2\omega_o^2/4 \quad .$$

The initial kinetic energy is

$$K_i = \tfrac{1}{2} I\omega_o^2 = MR^2\omega_o^2/4 \ ,$$

and the final kinetic energy is 0. Therefore

$$\Delta K = K_f - K_i = -MR^2\omega_o^2/4$$ <<<

11-5 Combined Rotational and Translational Motions

If a rigid body rotates about a fixed direction axis through its center of mass and in addition, the center of mass translates with velocity $\underline{\underline{V}}$ relative to some inertial origin 0, then a mass point m_j of the body has total velocity $\underline{\underline{v}}_j$ relative to 0 of

$$\underline{\underline{v}}_j = \underline{\underline{V}} + \underline{\underline{v}}_j' \quad , \quad \underline{\underline{v}}_j' = d\underline{\underline{r}}_j'/dt = \underline{\underline{\omega}} \times \underline{\underline{r}}_j' \tag{11-17}$$

where $\underline{\underline{v}}_j'$ is its velocity and $\underline{\underline{r}}_j'$ its position vector, <u>both relative to the center of mass</u>. The total kinetic energy of the body is

$$K = \tfrac{1}{2}\sum_j m_j \underline{\underline{v}}_j \cdot \underline{\underline{v}}_j = \tfrac{1}{2}\sum_j m_j (\underline{\underline{V}} + \underline{\underline{v}}_j') \cdot (\underline{\underline{V}} + \underline{\underline{v}}_j')$$

$$= \tfrac{1}{2}\sum_j m_j \underline{\underline{v}}_j' \cdot \underline{\underline{v}}_j' + (\sum_j m_j \underline{\underline{v}}_j') \cdot \underline{\underline{V}} + \tfrac{1}{2}(\sum_j m_j) \underline{\underline{V}} \cdot \underline{\underline{V}} \ .$$

The second term is just the rate of change of $\sum m_j \underline{\underline{r}}_j'$; but $\underline{\underline{r}}_j'$ is the position vector of m_j relative to the center of mass so $\sum m_j \underline{\underline{r}}_j' = 0$ and therefore the second term vanishes. Because $\sum_j m_j = M$, where M is the total mass of the body, the last term is just the translational kinetic energy of the body which we denote as K_{CM};

$$K_{CM} = \tfrac{1}{2} MV^2 \ . \tag{11-18a}$$

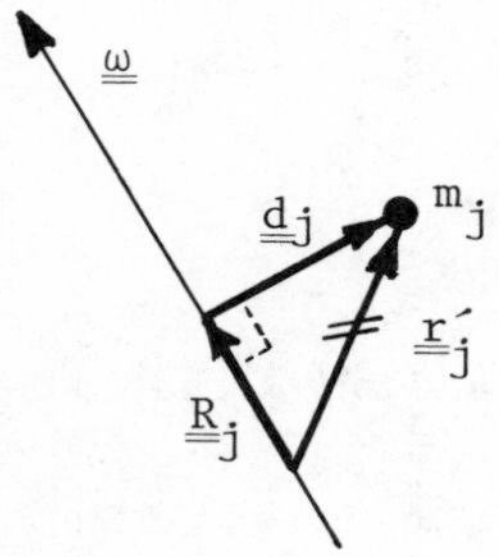

Figure 11-5. Resolution of the center of mass position vector $\underline{\underline{r}}_j'$ of a mass m_j into a vector $\underline{\underline{R}}_j$ along $\underline{\underline{\omega}}$ and a vector $\underline{\underline{d}}_j$ perpendicular to $\underline{\underline{\omega}}$.

The first term is just the rotational kinetic energy. In Fig. (11-5) we show $\underline{\underline{r}}_j'$ resolved into a vector $\underline{\underline{R}}_j$ along $\underline{\underline{\omega}}$ and a vector $\underline{\underline{d}}_j$ perpendicular to $\underline{\underline{\omega}}$. Then

$$\underline{\underline{v}}_j' = \underline{\underline{\omega}} \times \underline{\underline{r}}_j' = \underline{\underline{\omega}} \times (\underline{\underline{R}}_j + \underline{\underline{d}}_j) = \underline{\underline{\omega}} \times \underline{\underline{d}}_j \ .$$

Thus

$$K_{rotation} = \tfrac{1}{2}\sum_j m_j (\underline{\underline{\omega}} \times \underline{\underline{d}}_j) \cdot (\underline{\underline{\omega}} \times \underline{\underline{d}}_j)$$

$$= \tfrac{1}{2} \sum_j m_j[(\underline{\underline{\omega}}\cdot\underline{\underline{\omega}})(\underline{\underline{d}}_j\cdot\underline{\underline{d}}_j) - (\underline{\underline{\omega}}\cdot\underline{\underline{d}}_j)^2] \quad ,$$

but since $\underline{\underline{\omega}}$ is perpendicular to $\underline{\underline{d}}_j$, we have

$$K_{rotation} = \tfrac{1}{2}\,\omega^2 \sum_j m_j d_j^{\,2} = \tfrac{1}{2}\,I\omega^2 \tag{11-18b}$$

where I is the rotational inertia of the body relative to the axis $\underline{\underline{\omega}}$ which passes through the center of mass 0´. The total kinetic energy is

$$K = K_{rotation} + K_{translation} = \tfrac{1}{2}\,I\omega^2 + \tfrac{1}{2}\,MV^2 \quad . \tag{11-19}$$

Eq. (11-19) is completely general for the fixed axis case. For the special case of a body that rolls without slipping as the body rolls forward through an angle $d\theta$, the center of mass moves forward a distance $ds = R\,d\theta$. In this special case, the center of mass velocity is related to the angular velocity by

$$V = ds/dt = R\,d\theta/dt = R\omega \quad . \tag{11-20}$$

Then

$$K = \tfrac{1}{2}\,I\omega^2 + \tfrac{1}{2}\,MR^2\omega^2 = \tfrac{1}{2}\,I_p\omega^2 \tag{11-21}$$

where I_p is the rotational inertia of the body relative to an axis through the point of contact and parallel with the axis through the center of mass. (See Section 11-3, parallel axis theorem.)

Thus, for an object that rolls without slipping, the motion can be considered as a combination of translation of the center of mass plus a rotation about the center of mass or as rotation only with the same angular speed but about the point of contact.

>>> Example 5. Find the translational speed of a uniform sphere of radius R and mass M that rolls without slipping down an incline of angle θ if it starts from some height h.

We shall use the energy method. The final kinetic energy is given by

$$K = \tfrac{1}{2}\,I_p\omega^2$$

and

$$I_p = 2MR^2/5 + MR^2 = 7MR^2/5 \quad .$$

The initial kinetic energy is zero. The initial potential energy is Mgh and the final potential energy is zero. The conservation of total energy yields

$$K_i + U_i = K_f + U_f$$

or

$$0 + Mgh = (7MR^2/5)\omega^2/2 + 0 \quad .$$

Thus

$$R\omega = \sqrt{10\,gh/7} \quad .$$

Now, the translational speed is the speed of the center of mass and is given by

$$V = \omega R = \sqrt{10\,gh/7} \quad .$$

Notice that had the sphere slid without rolling down a frictionless incline V would have been $\sqrt{2gh}$.

Therefore one concludes that

$$V_{\text{rolling without slipping}} < V_{\text{frictionless}} \quad .$$

Physically this occurs because some of the potential energy must be converted into kinetic energy of rotation.

Now, suppose we want to find the acceleration. We shall use dynamic methods. Consider a free body diagram of the sphere as in Fig. 11-6.

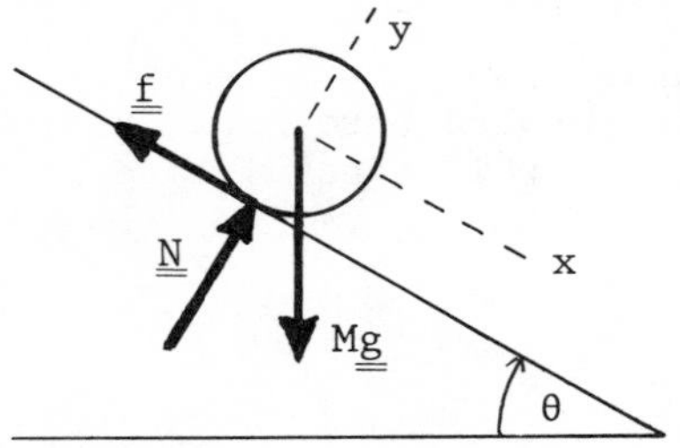

Figure 11-6. Example 5. A sphere rolling without slipping down an incline of angle θ; choice of axes and origin.

The x component of Newton's law, $\sum F_x = Ma_x$ yields $Mg \sin\theta - f \equiv Ma$ where f is the frictional force. The y component yields (since $a_y = 0$), $N - Mg\cos\theta = 0$. Next we take the sum of the z components of the torques about an axis through the center of mass and apply $\sum \tau_z = I\alpha$:

$$fR = I\alpha \quad .$$

Because the sphere does not slip the center of mass velocity V is related to ω by

$$V = R\omega$$

so the acceleration of the center of mass is $a = R\alpha$. The torque equation yields $fR = Ia/R$ or $f = aI/R^2$. This is the value the friction must have to prevent slippage. From the x-component force equation we have

$$Mg \sin\theta - I\, a/R^2 = Ma \quad ,$$

so

$$a = g \sin\theta/(1 + I/MR^2)$$

This equation is valid for any object of circular cross section that rolls without slipping. For all such bodies the rotational inertia will have the form $I = \beta MR^2$ where β is some number. Therefore

$$a = g \sin\theta/(1 + \beta) \quad .$$

Notice that this result is independent of R or M and depends only upon β. As β increases, the acceleration decreases, so the smaller the value of β the larger the value of the acceleration independent of M and R. For a sphere $\beta = 2/5$, for a cylinder $\beta = 1/2$, for a spherical shell $\beta = 2/3$ and for a hoop $\beta = 1$. Therefore if one released a sphere, a cylinder, a spherical shell, and a hoop from the top of an incline at the same instant the sphere would reach the bottom first followed by the cylinder, spherical shell and hoop in that order quite independent of their relative sizes and masses.

Finally, we note that we may place a limit on the angle which the incline may have and yet the object not slip. Since f is the force of static friction it is limited by $f \le \mu_s N = \mu_s mg \cos\theta$ where μ_s is the coefficient of static friction. Then, one has from $Mg \sin\theta - f = Ma$,

$$\tan\theta \le \mu_s[1/\beta + 1]$$

as the condition limiting the angle of the incline. <<<

11-6 Conservation of Angular Momentum

From the general equation $\underline{\tau}_{ext} = d\underline{L}/dt$ one has the conservation of angular momentum. If the net external torque is zero the total angular momentum of a system is conserved -- i.e. is constant. For a fixed axis system (again recall this may be an axis through the center of mass but of constant orientation) if τ_ω is zero then L_ω is constant. This is conveniently expressed by

$$\omega I_\omega = \omega_o I_{\omega_o} = \text{constant, fixed axis} \quad . \qquad (11\text{-}22)$$

Thus if I changes then so must ω so that the product is constant. The rotational kinetic energy is not necessarily conserved however since

$$K_i = \tfrac{1}{2}\,\omega_i L_{\omega_i}$$

and

$$K_f = \tfrac{1}{2}\,\omega_f L_{\omega_f} \quad .$$

Thus

$$K_f/K_i = \omega_f/\omega_i \quad .$$

>>> Example 6. A simple clutch consists of two discs having rotational inertias I_A and I_B as indicated in Fig. 11-7. Suppose A initially rotates with angular speed ω and B is at rest. If they are coupled what is the final angular speed and how much energy is lost?

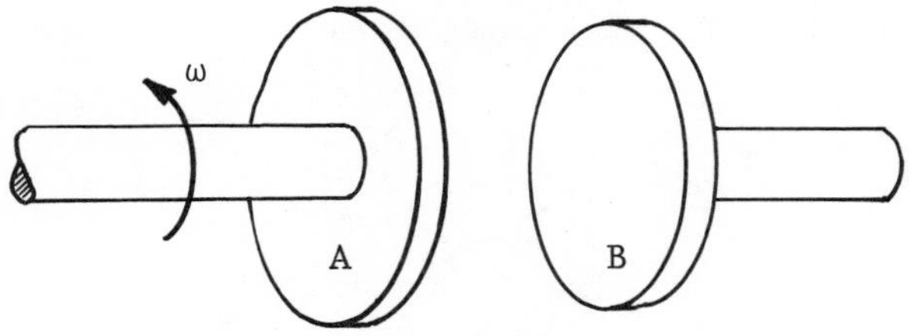

Figure 11-7. Example 6. A simple clutch of discs A and B; A initially rotates with angular speed ω and B is at rest.

Initially we have

$$L_A = \omega I_A \quad , \quad L_B = 0$$

so for the system (both clutch discs) initially

$$L_{system} = I_A\omega \quad .$$

All forces and hence all torques are internal for this choice of system so L_{system} is conserved. Thus since I_{system} finally is $I_A + I_B$ we have

$$\omega_f(I_A + I_B) = \omega I_A$$

so the final angular speed is

$$\omega_f = \omega I_A/(I_A + I_B) \quad .$$

The initial kinetic energy is

$$K_i = \tfrac{1}{2}\,\omega^2 I_A$$

and the final kinetic energy is

$$K_f = \tfrac{1}{2}\,\omega_f^2(I_A + I_B) = \tfrac{1}{2}\,\omega^2 I_A^2/(I_A + I_B) \quad .$$

Therefore the energy lost is $K_i - K_f$ or

$$\Delta K = \tfrac{1}{2}\,\omega^2 I_A I_B/(I_A + I_B) \quad . \qquad <<<$$

11-7 Relationship of $\underline{\underline{L}}$ to $\underline{\underline{\omega}}$ -- Fixed Axis

In general, even for a fixed axis, $\underline{\underline{L}}$ is not parallel to $\underline{\underline{\omega}}$. To see this consider a collection of mass points where mass m_j has position vector $\underline{\underline{r}}_j$ from some origin on the rotation axis. Then, because the axis is fixed, the velocity $\underline{\underline{v}}_j$ of mass m_j is just $\underline{\underline{\omega}} \times \underline{\underline{r}}_j$. Thus

$$\underline{\underline{L}} = \sum_j \underline{\underline{r}}_j \times m_j\, \underline{\underline{v}}_j = \sum_j m_j \underline{\underline{r}}_j \times (\underline{\underline{\omega}} \times \underline{\underline{r}}_j) = \sum_j m_j[\underline{\underline{\omega}} r_j^2 - (\underline{\underline{\omega}}\cdot\underline{\underline{r}}_j)\underline{\underline{r}}_j] \quad .$$

For definiteness, suppose that $\underline{\underline{\omega}} = \omega\underline{\underline{k}}$. Then with $\underline{\underline{r}}_j = x_j\underline{\underline{i}} + y_j\underline{\underline{j}} + z_j\underline{\underline{k}}$

$$\underline{\underline{L}} = \omega\underline{\underline{k}}(\sum_j m_j(x_j^2 + y_j^2)) - \omega\underline{\underline{j}}(\sum_j m_j z_j y_j) - \omega\underline{\underline{i}}(\sum_j m_j z_j x_j) \quad .$$

In general, $\underline{\underline{L}}$ has a component along $\underline{\underline{\omega}}$ (here $\underline{\underline{k}}$) and a component along some direction perpendicular to $\underline{\underline{\omega}}$. We see here that $\underline{\underline{L}}$ will lie along $\underline{\underline{\omega}}$ only if

$$\sum_j m_j z_j y_j = 0 \qquad \text{and} \qquad \sum_j m_j z_j x_j = 0 \quad .$$

These equations require that the z axis (rotation axis) be what is known as a principal axis. A principal axis has the sort of symmetry indicated by these equations. This symmetry may not be obvious if the mass points have different masses. For a uniform continuous body (where the sums are to be replaced by integrals) such axes are usually more evident. One solution is if all the mass points are equal and if there is mass at (x,y,z) there is equal mass at (-x, -y, z).

If $\underline{\underline{L}}$ is not parallel to $\underline{\underline{\omega}}$ then even if $\underline{\underline{\omega}}$ is constant, the direction of $\underline{\underline{L}}$ changes with time. The time rate of change of $\underline{\underline{L}}$ is easily computed in this case. From the original equations above,

$$d\underline{\underline{L}}/dt = \sum_j \underline{\underline{v}}_j \times m_j \underline{\underline{v}}_j + \sum_j m_j \underline{\underline{r}}_j \times (\underline{\underline{\omega}} \times \underline{\underline{v}}_j) \quad .$$

The first term is zero and the second yields

$$d\mathbf{L}/dt = \sum_j m_j[\boldsymbol{\omega}(\mathbf{r}_j \cdot \mathbf{v}_j) - \mathbf{v}_j(\mathbf{r}_j \cdot \boldsymbol{\omega})] \quad .$$

But again, $\mathbf{v}_j = \boldsymbol{\omega} \times \mathbf{r}_j$ is perpendicular to $\mathbf{r}_j$ so the first term here is zero. So

$$d\mathbf{L}/dt = -\sum_j m_j \mathbf{v}_j(\mathbf{r}_j \cdot \boldsymbol{\omega}) \quad .$$

Now, consider $\boldsymbol{\omega} \times \mathbf{L}$. This is given by

$$\boldsymbol{\omega} \times \mathbf{L} = \sum_j m_j\, \boldsymbol{\omega} \times (\mathbf{r}_j \times \mathbf{v}_j) = \sum_j m_j[\mathbf{r}_j(\boldsymbol{\omega} \cdot \mathbf{v}_j) - \mathbf{v}_j(\mathbf{r}_j \cdot \boldsymbol{\omega})] \quad .$$

However, since $\mathbf{v}_j$ is also perpendicular to $\boldsymbol{\omega}$, the first term here vanishes. Thus we see that for constant $\boldsymbol{\omega}$

$$d\mathbf{L}/dt = \boldsymbol{\omega} \times \mathbf{L} \ .$$

One can visualize this as if the tip of the angular momentum vector describes a circle about the rotation axis. The radius of this circle is the magnitude of the component of $\mathbf{L}$ perpendicular to $\boldsymbol{\omega}$. The magnitude of $\mathbf{L}$ remains constant, but the direction of $\mathbf{L}$ constantly changes.

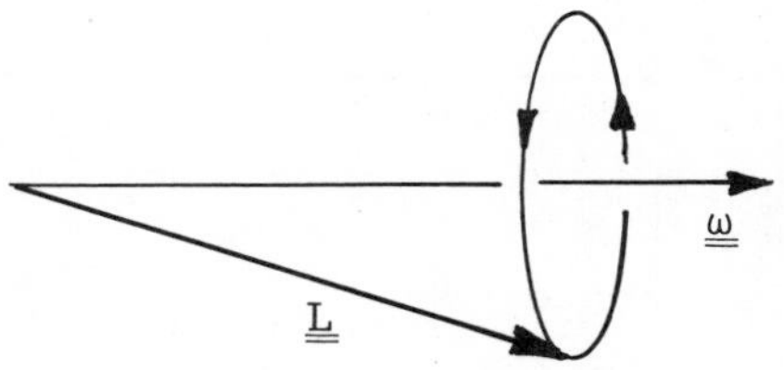

A physical object such as a flywheel is kept rotating about a fixed axle (axis of rotation) by bearings on the axle. If the object is not rotating about a principal axis then these bearings must supply the torque necessary to produce this change in the direction of $\mathbf{L}$. The bearings do so by exerting a force on the axle which in turn exerts an equal (but oppositely directed) force on the bearings and ultimately upon whatever supports these bearings. This force changes direction constantly and is the origin of the vibrations or wobble felt in an imperfectly balanced or slightly out-of-round automobile tire.

11-8 Programmed Problems

1.

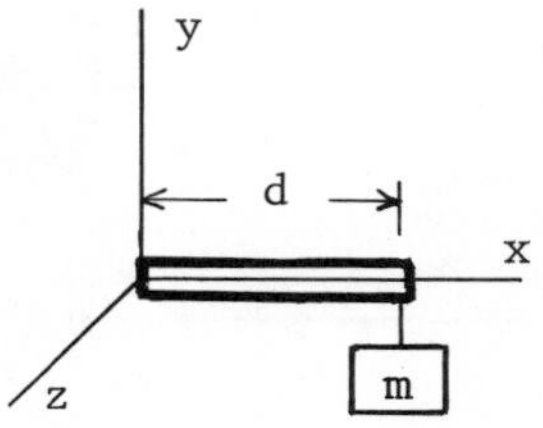

A mass m is attached to a massless bar as shown. Draw vectors representing the force exerted on the bar by the attached mass and draw the position vector of the point of application of that force with respect to the pivot point.

$\mathbf{W} = m\mathbf{g} = -mg\mathbf{j}$

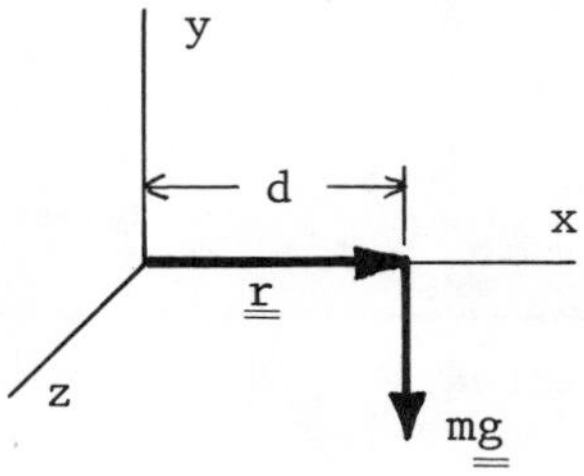

---

2. The torque exerted on the bar by $\mathbf{F}$ acting at $\mathbf{r}$ is defined by the vector product

$\boldsymbol{\tau} =$ ______________ .

$\boldsymbol{\tau} = \mathbf{r} \times \mathbf{F}$

3.

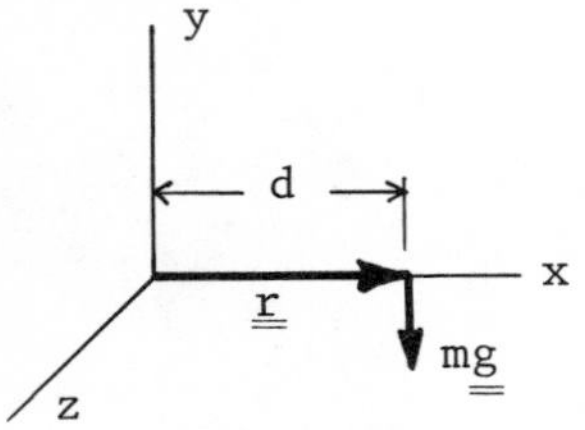

a. Represent the torque by an appropriate vector on the diagram.

b. What is the magnitude of the torque in terms of the variables given?

The direction of $\underline{\tau}$ satisfies the right hand rule.

a.

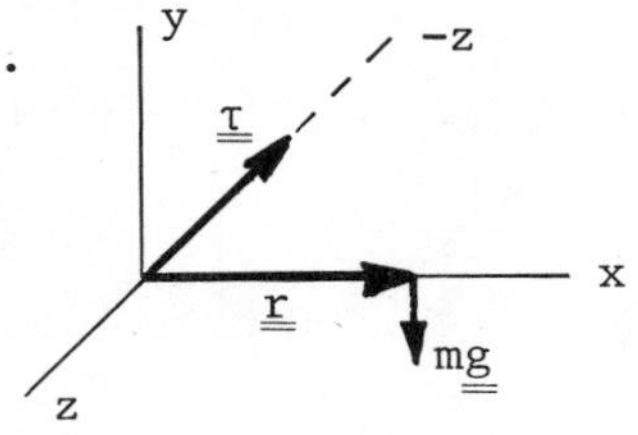

b. $|\underline{\tau}| = |\underline{r}|\,|\underline{mg}| \sin\theta$

$\theta = \pi/2, \sin\theta = 1$

$|\underline{\tau}| = mgd$

4.

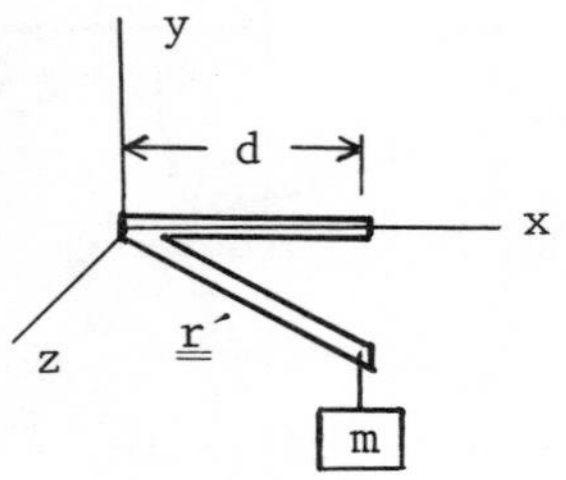

The situation is now modified with the addition of a new bar of length $\underline{r}'$ such that the mass is now suspended directly below its previous position.

Show on the diagram the angle θ which appears in the equation below for $|\underline{\tau}|$.

$$|\underline{\tau}'| = |\underline{r}'|\,|\underline{mg}| \sin\theta\ .$$

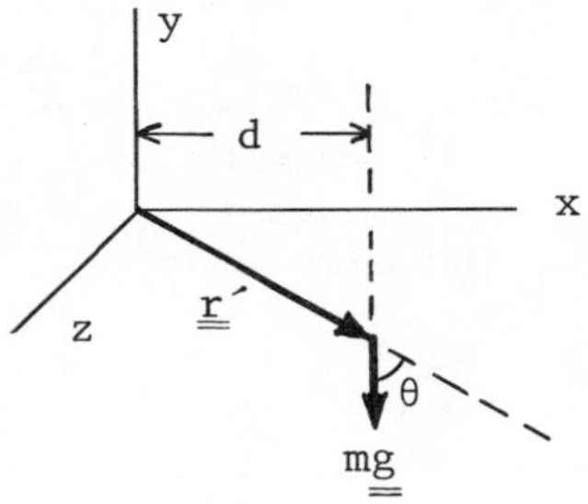

5.

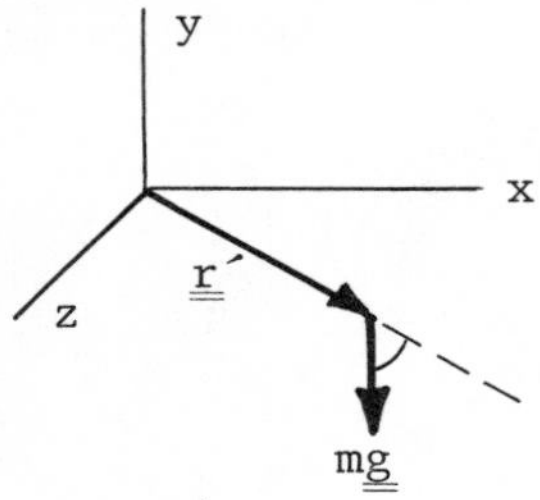

The magnitude of the torque of $\underline{mg}$ is

$$|\underline{\tau}'| = |\underline{mg}|\,|\underline{r}'| \sin\theta\ .$$

Remembering from geometry that angles between intersecting lines are equal, identify the length $|\underline{r}'| \sin\theta$ in the diagram to the left.

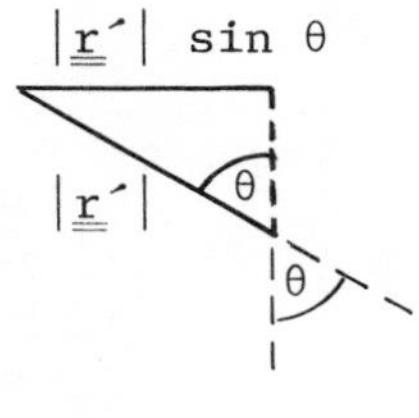

Right triangle.

---

6. The distance $|\underline{r}'| \sin\theta$ is the perpendicular distance to the axis of rotation from the line of action of the force.

a. Since $d = |\underline{r}'| \sin\theta$, then $|\underline{\tau}'| = mgr' \sin\theta =$ _____ .

b. The distance $r' \sin\theta$ has the name _____ .

a. mgd. Same as answer in frame 3.

b. Moment arm.

7. 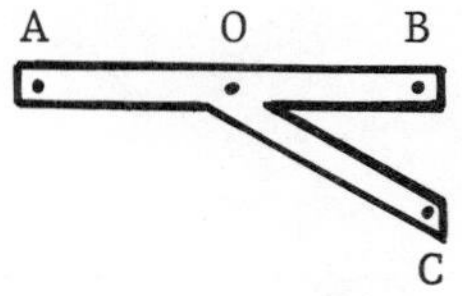

In the diagram

a. If two equal weights were placed at A and B, the bar pinned at 0 (would, would not) rotate.

b. If two equal weights were placed at A and C the bar (would, would not) rotate.

a. Would not - equal and opposite torques.
b. Would not - check frames 6 and 3 again if you missed this one. It is the moment arm that counts. The moment arm is frequently not the magnitude of the position vector of the point of application of the force.

---

8. 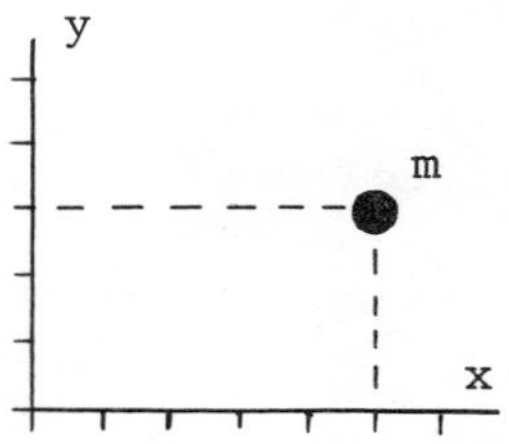

[each unit is one meter]

A particle of mass m = 2 kg is moving with constant velocity $\mathbf{v} = 20\mathbf{i}$ m/sec.

a. Draw the position vector of m.
b. Draw a vector representing the momentum of m.
c. Write $\mathbf{r}$ in terms of the unit vectors $\mathbf{i}$ and $\mathbf{j}$.
d. What is $|\mathbf{p}|$?

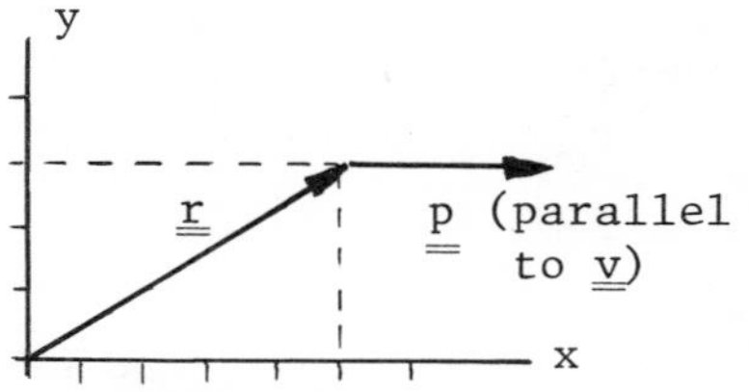

c. $\mathbf{r} = (5\mathbf{i} + 3\mathbf{j})$ meters.
d. $|\mathbf{p}| = m|\mathbf{v}| = 40$ kg·m/s .

---

9. The definition of angular momentum is

$$\boldsymbol{\ell} = \mathbf{r} \times \mathbf{p} \ .$$

Here $\boldsymbol{\ell} = (5\mathbf{i} + 3\mathbf{j}) \times (40\mathbf{i})$ kg m^2/s . Expand this last expression remembering that the distributive property is:

$$(\mathbf{a} + \mathbf{b}) \times (\mathbf{c}) = (\mathbf{a} \times \mathbf{c}) + (\mathbf{b} \times \mathbf{c}) \ .$$

$\boldsymbol{\ell} = [(5\mathbf{i} \times 40\mathbf{i}) + (3\mathbf{j} \times 40\mathbf{i})]$ kg·m^2/s .

---

10. We have

$$\boldsymbol{\ell} = [(5\mathbf{i} \times 40\mathbf{i}) + (3\mathbf{j} \times 40\mathbf{i})] \text{ kg·m}^2/\text{s} \ .$$

Use the diagram to the left, the right hand rule and the definition of vector products to answer the following:

a. $\mathbf{i} \times \mathbf{i} =$ ______ .

b. $\mathbf{j} \times \mathbf{i} =$ ______ .

a. $\mathbf{i} \times \mathbf{i} = 0$.

b. $\mathbf{j} \times \mathbf{i} = -\mathbf{k}$.

---

11. Using the information of frame 10,

$\boldsymbol{\ell} =$ ______.

$\boldsymbol{\ell} = -120\,\mathbf{k}$ kg·m^2/s .

Note $\boldsymbol{\ell}$ is perpendicular to the plane of $\mathbf{r}$ and $\mathbf{p}$ as shown in frame 8. $\boldsymbol{\ell}$ points <u>into</u> the page.

Frame	Answer
12. If the mass is moving with constant velocity, what is the force on m?	Zero, since $\mathbf{v}$ is constant: $\mathbf{F} = m\,d\mathbf{v}/dt$ $= 0$.
13. Since the force on m is zero, what will the torque be with respect to the origin?	Zero. $\boldsymbol{\tau} = \mathbf{r} \times \mathbf{F}$, and $\mathbf{F} = 0$.
14. If the torque is zero, what will be the time rate of change of $\boldsymbol{\ell}$?	Also zero, since $\boldsymbol{\tau} = 0$ and $\boldsymbol{\tau} = d\boldsymbol{\ell}/dt$. Therefore $d\boldsymbol{\ell}/dt = 0$ so that $\boldsymbol{\ell}$ is constant (in both magnitude and direction).

15.

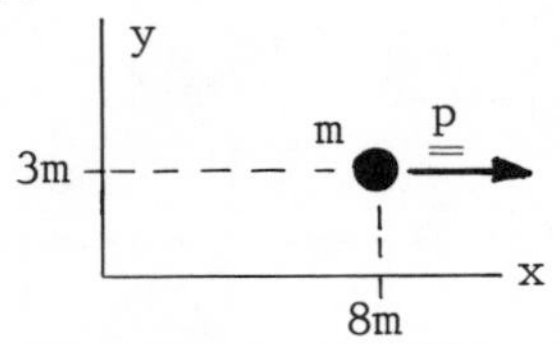

$\mathbf{p} = 20\mathbf{i}$ kg·m²/s.

This diagram represents the situation at some later time. m has moved in a straight line. Now,

a. $\mathbf{r} =$ ______ .

b. $\boldsymbol{\ell} = \mathbf{r} \times \mathbf{p} =$ ______ .

Expand the vector product and obtain an answer as you did previously for $\boldsymbol{\ell}$.

Answer:

a. $\mathbf{r} = (8\mathbf{i} + 3\mathbf{j})$ meters

b. $\boldsymbol{\ell} = \mathbf{r} \times \mathbf{p}$

$\boldsymbol{\ell} = (8\mathbf{i} + 3\mathbf{j}) \times (40\mathbf{i})\ \mathrm{kg\cdot m^2/s}$.

$\boldsymbol{\ell} = (-\ 120\mathbf{k})\ \mathrm{kg\cdot m^2/s}$.

The answer just determined is the identical answer to that of frame 11. You should keep in mind the following ideas from this little exercise. 1. Objects <u>need</u> <u>not</u> move in circles or even curved paths to have angular momentum with respect to some reference point. 2. If the <u>torque</u> on an object relative to some point is zero, then the <u>angular momentum</u> relative to that point is a <u>constant</u> of the motion, i.e. $\boldsymbol{\ell}$ = constant. Although here we have a somewhat trivial case because $\mathbf{F} = 0$, this is actually a powerful conservation law.

16.

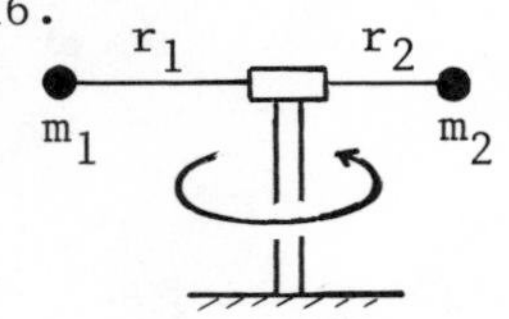

Two masses connected by a massless rod are constrained to move in a horizontal plane at a constant angular velocity of 200 rad/s.

m_1 = 5 kg

m_2 = 10 kg

r_1 = 10 m

r_2 = 4 m

a. $v_1 =$ ____ m/s .

b. $v_2 =$ ____ m/s .

c. $\frac{1}{2} m_1 v_1^{\,2} =$ ____ J .

d. $\frac{1}{2} m_2 v_2^{\,2} =$ ____ J .

Obtain numbers.

Answer:

$v = \omega r$

a. 2,000 m/s .

b. 800 m/s .

c. 10^7 J .

d. 3.2×10^6 J .

17. Treating the masses as point particles, the moment of inertia of the system is defined as

$$I = \sum_{i=1}^{2} m_i r_i^2$$

and with respect to the axis of rotation, I = ______ kg·m² (numerical answer please).

$I = m_1 r_1^2 + m_2 r_2^2$

$I = 5 \cdot 10^2 + 10 \cdot 4^2$

$I = 660\ \text{kg}\cdot\text{m}^2$

---

18. We have from frame 16

$$KE_{Total} = \tfrac{1}{2} m_1 v_1^2 + \tfrac{1}{2} m_2 v_2^2 = 13.2 \times 10^6\ \text{J} .$$

Treating the system as a rotating rigid body we can write

$$KE_T = \tfrac{1}{2} I\omega^2 = \_\_\_\_\_\_\ \text{J} .$$

Use the moment of inertia from frame 17 and $\omega = 200$ rad/s. Do the arithmetic.

$KE_T = \tfrac{1}{2} I\omega^2$

$KE_T = \tfrac{1}{2}\,(660\ \text{kg}\cdot\text{m}^2)(200\ \text{rad/s})^2$

$KE_T = 13.2 \times 10^6\ \text{J}$.

---

19. We get the same answer both ways. When we use the rotational approach the body is characterized by a single quantity, the moment of inertia, rather than a collection of quantities such as m_1, r_1 .... I is the rotational analogue of ______ .

Mass.

---

20. For an object of continuous mass distribution the summation of frame 17 is replaced by an integral.

I = ______.

$I = \int r^2\, dm$

---

21. Calculate the moment of inertia of a homogeneous rod about an axis through one end perpendicular to the length ℓ.

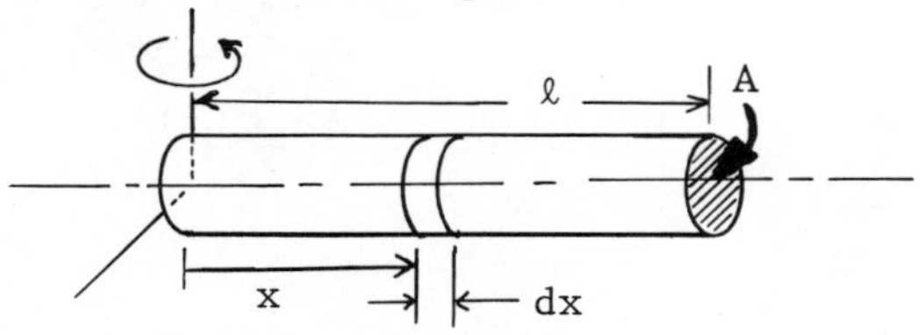

The general technique is to replace dm by a volume density times a volume element.

$$dm = \rho\, dV \quad .$$

a. For the volume element shown of length dx and cross-sectional area A located at position x

dV = ______ .

b. For the homogeneous rod of total mass M and total volume V,

ρ = ______ .

a. $dV = A\, dx$.

b. $\rho = M/V$, constant.

22. Using the information of the previous frame we can write $$I = \int r^2 dm = \_\_\_\_\_\_ .$$	$I = \int x^2 \rho \, dV$ $I = \rho \int Ax^2 dx$
23. So A is the same for every volume element slice. $$I = \rho A \int x^2 dx \quad .$$ a. Looking back at frame 21, what are the limits for x of this integral? b. Evaluate the integral.	x goes from 0 to ℓ. $I = \rho A \int_0^{\ell} x^2 dx = \rho A \ell^3/3$
24. The answer to frame 23 is not yet in the form usually given. We have already that $\rho = M/V$. Express the cross-section area A in terms of the volume and length of the rod.	$V = \ell A$ $A = V/\ell$
25. Now substitute appropriately for ρ and A in the answer to frame 23. $$I = \_\_\_\_\_\_ .$$	$I = \rho A \ell^3/3$ $I = (M/V)(V/\ell)\ell^3/3$ $I = M\ell^2/3$

Chapter 12: EQUILIBRIUM OF RIGID BODIES

REVIEW AND PREVIEW

This chapter should provide a breather for you as you will now learn to solve problems in which there is no motion at all! You will make use of the dynamic equations you learned in Chapters 5 and 11, but with neither linear nor angular acceleration. The only really new concept you will meet is that of center of gravity.

GOALS AND GUIDELINES

In this chapter you have one major goal:

> Learning the basic equations of static equilibrium and how to apply these to static problems. Of particular importance will be the application of the equilibrium equations for plane (two dimensional) rigid bodies. You should pay attention to the recipe given in Section 12-3 for solving equilibrium problems, and to the examples that follow which illustrate this recipe.

12-1 Equilibrium Conditions

A rigid body is defined to be in mechanical equilibrium if with reference to some inertial frame of reference

1. $\underline{\underline{a}}_{CM} = 0$ -- the acceleration of the center of mass is zero

and

2. $\underline{\underline{\alpha}} = 0$ -- the angular acceleration about any fixed axis in the inertial frame is zero.

Note that the body need not be at rest in order to be in equilibrium. In other words there need not be lack of motion; if there is motion, however, it must be uniform: If $V_{CM} \neq 0$, it must be constant (in direction as well as magnitude), and if the body is rotating with angular velocity $\underline{\underline{\omega}} \neq 0$, it must be constant (in direction as well as magnitude).* If $\underline{\underline{\omega}} = 0$ and $\underline{V}_{CM} = 0$ then we speak of static equilibrium.

These conditions imply that

$$\sum_{\substack{\text{external}\\\text{forces}}} \underline{\underline{F}}_j = 0 \tag{12-1a}$$

and

$$\sum_{\substack{\text{external}\\\text{torques with}\\\text{respect to any}\\\text{inertial origin}}} \underline{\underline{\tau}}_j = 0 \quad . \tag{12-1b}$$

---

* It is possible to have $\underline{\underline{\alpha}} = 0$ and yet require a non-zero torque to do this. This situation would occur for a body rotating with constant angular velocity about a non-principle axis, e.g. the case discussed at the end of Section 11-7. This type of situation will not be discussed here.

The choice of origin in the inertial frame from which all the $\underline{\underline{\tau}}_j$ are evaluated is immaterial and may be chosen for convenience.

The equilibrium conditions are 6 independent conditions that must be satisfied -- one for each degree of freedom of the rigid body. Often one deals with plane problems -- all possible motion of interest in a single plane. Then there are but 3 degrees of freedom, two for translation and one for rotation. Similarly, there are but 3 equilibrium conditions

$$\sum F_x(j) = 0 \qquad (12\text{-}2a)$$

$$\sum F_y(j) = 0 \qquad \text{x,y plane equilibrium conditions.} \qquad (12\text{-}2b)$$

$$\sum \tau_z(j) = 0 \qquad (12\text{-}2c)$$

12-2 Center of Gravity

One of the forces encountered in equilibrium problems is that of gravity. The center of gravity is defined to be that position through which the force of gravity can be considered to act. That is, if $\underline{\underline{F}}_g$ is the force of gravity and $\underline{\underline{\tau}}_g$ its torque, then if $\underline{\underline{R}}$ is the position vector of the center of gravity

$$\underline{\underline{\tau}}_g = \underline{\underline{R}} \times \underline{\underline{F}}_g \quad . \qquad (12\text{-}3)$$

In the vicinity of the earth's surface, if an object is sufficiently small, the acceleration of gravity is essentially constant over the body and is denoted by $\underline{\underline{g}}$. That is, at every mass point m_j the force of gravity is $\underline{\underline{F}}_j = m_j\underline{\underline{g}}$. Therefore, the resultant force of gravity on the body is

$$\underline{\underline{F}}_g = \sum_j m_j\underline{\underline{g}} = M\underline{\underline{g}} \qquad (12\text{-}4)$$

where $M = \sum_j m_j$ is the total mass. The torque of the gravitational force at the jth mass point is $\underline{\underline{\tau}} = \underline{\underline{r}}_j \times m_j\underline{\underline{g}}$ where $\underline{\underline{r}}_j$ is the position vector of m_j relative to some inertial origin, O. The net torque is

$$\underline{\underline{\tau}}_g = \sum_j \underline{\underline{r}}_j \times m_j\underline{\underline{g}} = \Big(\sum_j m_j\underline{\underline{r}}_j\Big) \times \underline{\underline{g}} \quad . \qquad (12\text{-}5)$$

Let $\underline{\underline{r}}_j'$ be the position vector of m_j relative to an origin at the center of gravity. Then

$$\underline{\underline{r}}_j = \underline{\underline{R}} + \underline{\underline{r}}_j' \quad ,$$

so

$$\underline{\underline{\tau}}_g = \Big[\sum_j m_j(\underline{\underline{R}} + \underline{\underline{r}}_j')\Big] \times \underline{\underline{g}} = M\underline{\underline{R}} \times \underline{\underline{g}} + \Big(\sum_j m_j\underline{\underline{r}}_j'\Big) \times \underline{\underline{g}}$$

$$= \underline{\underline{R}} \times M\underline{\underline{g}} + \Big(\sum_j m_j\underline{\underline{r}}_j'\Big) \times \underline{\underline{g}} = \underline{\underline{R}} \times \underline{\underline{F}}_g + \Big(\sum_j m_j\underline{\underline{r}}_j'\Big) \times \underline{\underline{g}} \quad .$$

This equation defines $\underline{\underline{R}}$ since we want $\underline{\underline{R}} \times \underline{\underline{F}}_g = \underline{\underline{\tau}}_g$ which requires

$$\Big(\sum_j m_j\underline{\underline{r}}_j'\Big) \times \underline{\underline{g}} = 0 \quad . \qquad (12\text{-}6)$$

This is certainly satisfied if $\underline{\underline{R}}$ is the position vector of the center of mass for then

$$\sum_j m_j \underline{\underline{r}}_j' = 0 \quad .$$

Therefore, for small objects near the earth's surface we use

$$\underline{\underline{F}}_g = m\underline{\underline{g}} \qquad \text{center of gravity} \tag{12-7a}$$

and center of mass coincide

$$\underline{\underline{\tau}}_g = \underline{\underline{R}} \times m\underline{\underline{g}} \tag{12-7b}$$

where $\underline{\underline{R}}$ is the position vector of the center of mass.

12-3 Solving Equilibrium Problems

Most equilibrium problems you will be asked to solve involve static equilibrium. The system under consideration will perhaps consist of several bodies, rods, wires, strings, etc. Depending upon what is asked, the first order of business is to

(1) Determine what body or system of bodies the equilibrium conditions will be applied to.

(2) Isolate the system of interest and draw a free body diagram showing all external forces which act on the body. (Internal forces and torques cancel in pairs and are of no interest.) Often there will be forces where magnitude and direction are both unknown. In this case try to make the best possible guess as to direction. If you guess wrong your solution will yield a negative value for the force, so the procedure is self correcting provided you treat the assumed direction consistently.

(3) Choose a convenient reference frame along whose axes all the external forces will be resolved to apply the first three equilibrium conditions

$$\sum F_x(j) = 0 \quad , \quad \sum F_y(j) = 0 \quad , \quad \sum F_z(j) = 0 \quad .$$

Sometimes the best choice of reference frame will be obvious, sometimes not. If you make less than the best choice your algebra will be a bit more complicated but nothing drastic happens!

(4) Choose a convenient reference frame and origin along whose axes and about which origin the torque components will be evaluated to apply the last three equilibrium conditions

$$\sum \tau_x(j) = 0 \quad , \quad \sum \tau_y(j) = 0 \quad , \quad \sum \tau_z(j) = 0 \quad .$$

This need not be the same reference frame used in (3) although it is often convenient to choose them the same. The choice of origin location is not crucial, and can often be made to advantage. For example, if the origin is chosen at the point of application of one of the forces, its torque is zero.

>>> Example 1. Consider a simple balance as shown in Fig. 12-1a. The uniform balance bar has a known mass M and the right-most mass m_1 is also known. The bar has a length L and the distance from the knife edge to the right end can be measured. Find m_2 if this distance to the right end is x.

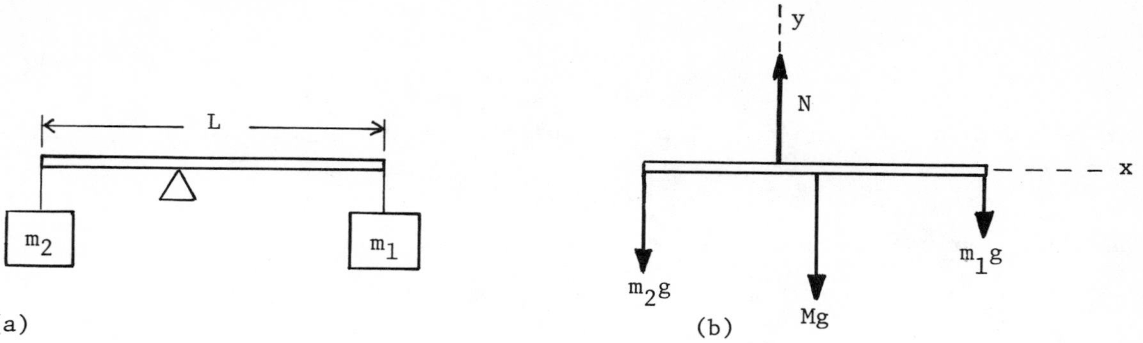

Figure 12-1. Example 1. (a) A simple balance consisting of a uniform bar of mass M, two masses m_1 and m_2 and a knife edge. (b) A free body diagram with choice of axes.

In Fig. 12-1b is shown a free body diagram together with a choice of axes; the z axis points out of the paper plane. The z component of the torque equilibrium condition yields

$$\sum \tau_z = m_2 g(L - x) - Mg(x - L/2) - m_1 gx = 0$$

so

$$m_2 = \frac{M(x - L/2) + m_1 x}{L - x} \quad .$$

Notice we needed only the torque condition. Suppose the origin were taken instead at the center of mass. Then the torque condition would yield

$$\sum \tau_z = m_2 g\ L/2 - N(x - L/2) - m_1 g\ L/2 = 0 \quad .$$

But N is not known. The y component of the force equilibrium conditions yields however

$$\sum F_y = N - m_1 g - M\ g - m_2 g = 0 \quad .$$

Substituting this result into the torque equation we would find the same result for m_2 as before. <<<

>>> Example 2. Consider a sign of weight W and length 2a hung from a light (i.e. massless) rod pivoted at a wall and cable as shown in Fig. 12-2a. The left end of the sign is a distance b from the wall. Find the tension in the cable.

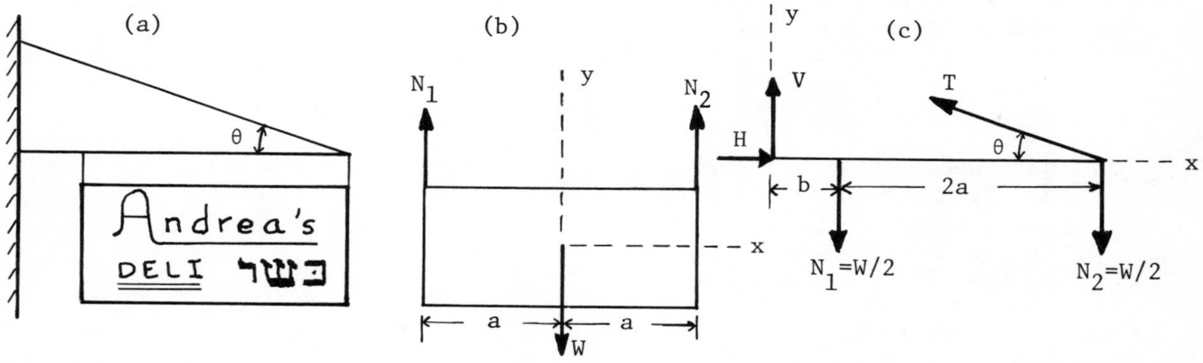

Figure 12-2. Example 2. (a) A sign is hung from a massless rod pivoted at a wall and supported by a cable. (b) Free body diagram of the sign with chosen axes. (c) Free body diagram of the rod with chosen axes.

In Fig. 12-2b we show a free body diagram of the sign only together with a choice of axes and origin. The equilibrium conditions yield

$$\sum F_y = N_1 + N_2 - W = 0$$

$$\sum \tau_z = N_2 a - N_1 a = 0 \quad .$$

Therefore $N_1 = N_2 = W/2$.

In Fig. 12-2c we show a free body diagram of the rod together with a choice of axes. The force at the pivot is unknown and we've designated it by its components H and V. We choose the axis origin at the pivot and use the torque condition

$$\sum \tau_z = -(W/2)\, b - (W/2)\,(b + 2a) + T \sin\theta\,(b + 2a) = 0 \quad .$$

Therefore the tension is

$$T = \frac{W(a + b)}{(b + 2a)\sin\theta} \quad .$$

Also, H and V are easily found. From

$$\sum F_x = -T\cos\theta + H = 0$$

and

$$\sum F_y = T\sin\theta + V - W = 0$$

we have

$$H = T\cos\theta = \frac{W(a + b)}{(b + 2a)}\cot\theta$$

and

$$V = W - T\sin\theta = W - \frac{W(a + b)}{(b + 2a)} = W\,\frac{a}{(b + 2a)} \quad .$$ <<<

>>> Example 3. A cylinder of radius R and mass M rests in equilibrium on an inclined plane of angle θ as shown in Fig. 12-3a. A horizontal cord is attached to the cylinder at its top-most point and to the incline. Find the tension in the cord, the normal force of the plane, the frictional force between the cylinder and the plane and the minimum value of the coefficient of friction for equilibrium.

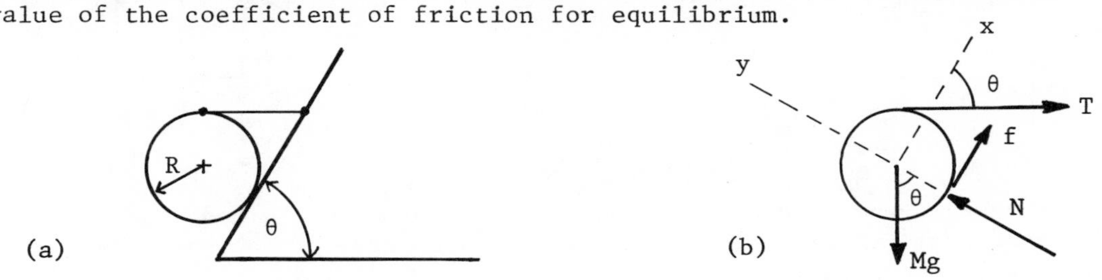

Figure 12-3. Example 3. (a) A cylinder rests on an incline of angle θ and is attached to a horizontal cord. (b) A free body diagram with choice of axes.

In Fig. 12-3b we show a free body diagram of the cylinder together with a choice of axes and origin. The equilibrium conditions are

$$\sum F_x = T\cos\theta + f - Mg\sin\theta = 0$$

$$\sum F_y = N - Mg\cos\theta - T\sin\theta = 0$$

$$\sum \tau_z = Rf - RT = 0 \quad .$$

Thus

$$T = \frac{Mg\sin\theta}{\cos\theta + 1} = f$$

and

$$N = Mg\cos\theta + T\sin\theta = Mg\cos\theta + \frac{Mg\sin^2\theta}{\cos\theta + 1}$$

or

$$N = Mg \quad .$$

<<<

In these three examples there have been the same number of unknowns as equilibrium conditions (3) so that a unique solution exists. Often this is not the case and there will be an infinity of possible solutions. In a true physical situation in which one accounts for the fact that the "rigid bodies" are not truly rigid but actually do bend and warp slightly there will be a unique solution. This, however, depends upon more conditions than just the equilibrium conditions. In the last example we shall consider a case which does not have a unique solution.

>>> Example 4. A rectangular uniform door of weight W and sides a and b is hung by two hinges as shown in Fig. 12-4a. Find the forces at the hinges necessary for equilibrium.

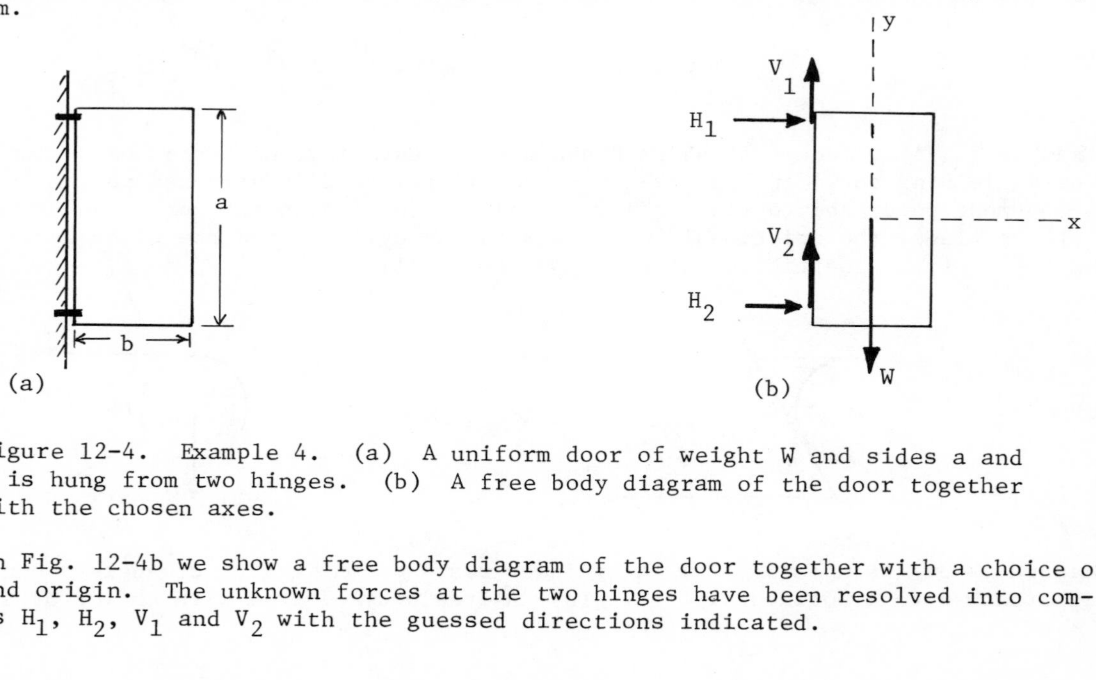

Figure 12-4. Example 4. (a) A uniform door of weight W and sides a and b is hung from two hinges. (b) A free body diagram of the door together with the chosen axes.

In Fig. 12-4b we show a free body diagram of the door together with a choice of axes and origin. The unknown forces at the two hinges have been resolved into components H_1, H_2, V_1 and V_2 with the guessed directions indicated.

The equilibrium conditions are

$$\sum F_x = H_1 + H_2 = 0$$

$$\sum F_y = V_1 + V_2 - W = 0$$

$$\sum \tau_z = -(b/2)V_1 - (b/2)V_2 - (a/2)H_1 + (a/2)H_2 = 0 \quad .$$

One has therefore

$$H_2 = bW/(2a)$$

and

$$H_1 = -bW/(2a) \quad .$$

The minus sign indicates that we guessed wrongly about H_1 and that it points to the left. Now we find

$$V_1 + V_2 = W$$

but we cannot determine V_1 and V_2 separately.

Suppose we tried to take torques about another point as origin, say the lower right hand corner. Then

$$\sum_{\substack{\text{lower right} \\ \text{corner origin}}} \tau = (b/2)\,W - bV_2 - bV_1 - aH_1 = 0 \quad .$$

But this merely yields

$$b(V_1 + V_2) = (b/2)W - a[-bW/(za)] = bW$$

or

$$V_1 + V_2 = W$$

again. The essential point is that there are 4 unknowns (H_1, H_2, V_1, V_2), only 3 equations ($\sum F_x = 0$, $\sum F_y = 0$, $\sum \tau_z = 0$), and this is the best we can do without further conditions such as the bending of the door.

<<<

Chapter 13: OSCILLATIONS

REVIEW AND PREVIEW

Previously you studied the concepts of kinetic energy, potential energy, and conservation of mechanical energy. In the problems you solved it did not matter whether the motion of the body under consideration repeated itself after some time or not. In this chapter you will study the special, but very interesting and most important case in which the body retraces its motion after some period of time, and continues to do so at regular intervals. Such motion is called oscillatory.

GOALS AND GUIDELINES

In this chapter you have two major goals:

1. Learning the vocabulary of oscillatory motion. This may seem tedious, but you must learn the language before you can successfully attempt the problems.
2. Understanding the very special case of simple harmonic motion. Pay particular attention to the forms of the equations involved. These equations apply to any type of simple harmonic motion whether linear or angular or whatever; the variables involved will change with changing physical situations, but the form of the equation is the same.

13-1 Oscillations -- Vocabulary

One of the most interesting phenomena in all of physics is motion which repeats itself at equal or regular intervals of time. Such motion is called periodic motion. The same laws of mechanics with which you are familiar apply here also, but there are some new definitions and vocabulary which you must learn.

Periodic Motion -- Any motion of a system which repeats itself at regular equal intervals of time. Periodic motion is often called harmonic motion.

Periodic Functions -- Functions which repeat their values at equal intervals of their arguments. Some examples are sin ωt, cos ωt, etc. They are used to describe periodic motion.

Oscillatory or Vibratory Motion -- The name given to periodic motion of a particle which moves back and forth over the same path.

Damped Harmonic Motion -- Motion which is harmonic but traces out less and less of its full path due to dissapative or frictional forces.

Period -- The time interval between one configuration of a system undergoing periodic motion and the next identical configuration. A configuration is specified by the coordinates and momenta of all parts of a system. The period is also the interval of repetition; it is usually denoted by T.

Frequency -- The number of times the configuration of a periodic motion system repeats itself per unit time. The reciprocal of the period; it is usually denoted by

$$\nu = 1/T \quad . \qquad (13\text{-}1)$$

Cycle -- A complete repetition from one configuration to the next identical one. The frequency is the number of these cycles per unit time.

Hertz -- The SI unit of frequency, a cycle per second; it is abbreviated Hz.

13-2 A Particle in Harmonic Motion -- One Dimension

For a particle undergoing harmonic motion one needs the vocabulary

Equilibrium -- that position at which no net force acts on the particle;

Displacement -- the distance of the oscillating particle from its equilibrium position.

Since the particle must return to a given location with the same momentum (hence kinetic energy) as it had when it was there before, the force acting on the particle must be conservative. That is, the force must be derivable from a potential energy function, i.e.

$$F = -\,dU/dx \quad . \tag{13-2}$$

The equilibrium position occurs at the minima of the potential energy ($dU/dx = 0$); also, the force must tend to return the particle to the equilibrium position. A possible potential energy is shown in Fig. 13-1; the equilibrium position has been chosen as the origin.

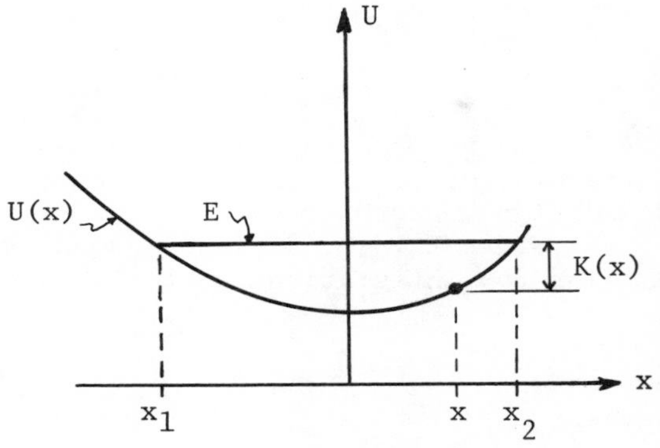

Figure 13-1. Illustration of potential energy U(x) versus x; the total mechanical energy is the horizontal solid line and the kinetic energy K(x) is shown. The origin has been chosen as the equilibrium point; the limits of oscillation are x_1 and x_2.

The possible values which the displacement may have are determined by the total mechanical energy

$$E = K + U(x) = \text{constant} \tag{13-3}$$

where K is the kinetic energy. If we draw a line parallel to the x axis representing the function E(x) = constant as shown in Fig. 13-1 the difference between E(x) and U(x) is K(x) which must be positive. The limits are then given by

$$U(x) = E \tag{13-4}$$

and are indicated by x_1 and x_2. At x_2, F(x) is negative ($dU/dx > 0$) so a particle released there is accelerated toward the origin at which point it has acquired kinetic energy $K = E - U(0)$. Note that U(0) need not be zero. At x = 0 the particle continues against the retarding force ($F > 0$) until it reaches x_1 where again it has zero velocity. It is then accelerated toward the origin and eventually returns to x_2 having completed one cycle of the periodic motion. For a general potential energy function (or briefly potential) apart from the periodicity, the particle motion would be quite complicated. For this reason one studies the simple harmonic oscillator potential.

13-3 Simple Harmonic Oscillator -- Simple Harmonic Motion

The simple harmonic oscillator potential in one dimension is

$$U = \tfrac{1}{2} kx^2 \tag{13-5}$$

so the restoring force is

$$F = - dU/dx = - kx \quad , \tag{13-6}$$

which clearly tends to return the particle toward the equilibrium position which is the origin. A particle moving in this potential is called a <u>simple harmonic oscillator</u> (SHO) and its motion is called <u>simple harmonic motion</u> (SHM). The two limiting positions are equally spaced from the origin at

$$x_{limit} = \pm \sqrt{2E/k} \quad . \tag{13-7}$$

The differential equation of motion of the SHO is obtained from Newton's second law:

$$F = - kx = ma_x = mdv_x/dt = md^2x/dt^2$$

and therefore

$$d^2x/dt^2 + kx/m = 0 \quad . \tag{13-8}$$

<u>In any physical situation governed by an equation of this form one speaks of SHM</u>. Examples occur in acoustics, optics, electrical circuits, atomic and nuclear physics in addition to mechanics.

Since Eq. (13-8) is a second order differential equation its general solution contains two integration constants.* A general solution is written as

$$x = A \cos (\omega t + \delta) \tag{13-9}$$

where

$$\omega = \sqrt{k/m} \quad . \tag{13-10}$$

This solution is verified by substitution into Eq. (13-8). In Eq. (13-9)

A = the amplitude of the SHM

$\omega t + \delta$ = the phase

δ = the phase constant

ω = angular frequency; $\omega = 2\pi/T = 2\pi\nu$ where ν is the frequency.

---

*Recall one dimensional, constant acceleration, kinematics. Then $a_x = \text{constant} = d^2x/dt^2$. In that case $x(0)$ and $v_x(0)$ are the two integration constants.

The amplitude, A, and phase constant, δ, are determined by the initial conditions or by conditions equivalent to them. Note that ω is independent of the amplitude; this is characteristic of SHM. The amplitude may be related to the total mechanical energy. From Eq. (13-9) x_{max} occurs when $\omega t + \delta = 0, \pm 2\pi, \pm 4\pi$, etc. and the minimum value which is $-x_{max}$ occurs at $\omega t + \delta = \pm \pi, \pm 3\pi, \pm 5\pi$, etc. These are the x_{limits} of Eq. (13-7) and since the amplitude is taken as intrinsically positive

$$A = \sqrt{2E/k} \quad . \tag{13-11}$$

Two SHM's may have the same amplitude and frequency and differ in phase constant. If

$$x_1 = A_1 \cos (\omega t + \delta)$$

and

$$x_2 = A_2 \cos (\omega t + \delta + \phi)$$

are two such SHM's, one says that x_2 leads x_1 in phase by ϕ. It is not necessary that the amplitudes be the same for this statement to have meaning. The statement means that x_2 reaches a given value (relative to its maximum) sooner than x_1 does.

>>> Example 1. Consider the SHM $x = A \cos \omega t$. Show that the velocity leads the displacement by $\pi/2$ in phase and that the acceleration leads the velocity by $\pi/2$ in phase.

Since $x = A \cos \omega t$, the velocity is

$$v = dx/dt = -\omega A \sin \omega t = \omega A \cos (\omega t + \pi/2)$$

so it leads x by $\pi/2$ in phase. Next,

$$a = dv/dt = -\omega^2 A \cos \omega t = \omega^2 A \cos (\omega t + \pi)$$

which leads v by $\pi/2$ and x by π. An easy way to see this phase relationship is to graph x/A, $v/\omega A$ and $a/\omega^2 A$ versus ωt and watch the location of the first maxima past the origin. This is done in Fig. 13-2.

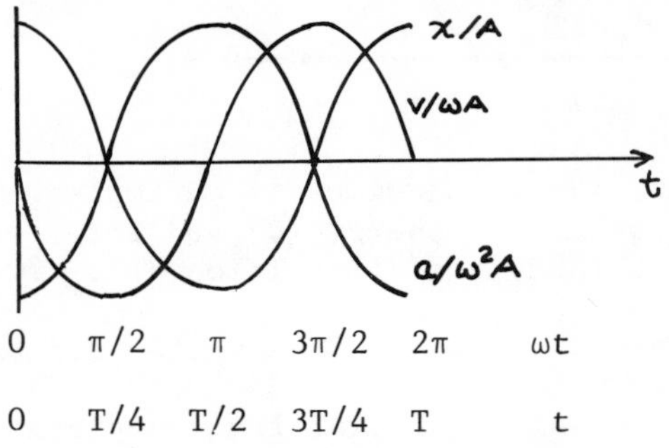

Figure 13-2. A graph of the displacement (x), velocity (v) and acceleration (a) divided by their respective maximum values for SHM. This illustrates that "a" leads "v" by $\pi/2$ and that "v" leads "x" by $\pi/2$.

The values of ωt are indicated as well as the values of t. The introduction of a phase constant would merely shift the time origin one way or the other, but would not alter the above phase relationships. <<<

13-4 Energy Considerations in SHM

If the displacement is given by

$$x = A \cos (\omega t + \delta)$$

the velocity is given by

$$v_x = dx/dt = -\omega A \sin(\omega t + \delta) \quad .$$

The kinetic energy may be expressed as a function of time as

$$K = \tfrac{1}{2} m v_x{}^2 = \tfrac{1}{2} m\omega^2 A^2 \sin^2(\omega t + \delta) = \tfrac{1}{2} kA^2 \sin^2(\omega t + \delta) \qquad (13\text{-}12a)$$

or as a function of position as

$$K = \tfrac{1}{2} m\omega^2 (A^2 - x^2) = \tfrac{1}{2} k(A^2 - x^2) \quad . \qquad (13\text{-}12b)$$

Similarly the potential energy as a function of position is

$$U = \tfrac{1}{2} kx^2 \qquad (13\text{-}13a)$$

and as a function of time is

$$U = \tfrac{1}{2} kA^2 \cos^2(\omega t + \delta) \quad . \qquad (13\text{-}13b)$$

The total mechanical energy is given by

$$E = K + U = \tfrac{1}{2} kA^2 \qquad (13\text{-}14)$$

as in Eq. (13-11). Fig. 13-3a shows K, U and E as functions of t and in Fig. 13-3b they are shown as functions of x. Here, for simplicity, δ was taken to be zero.

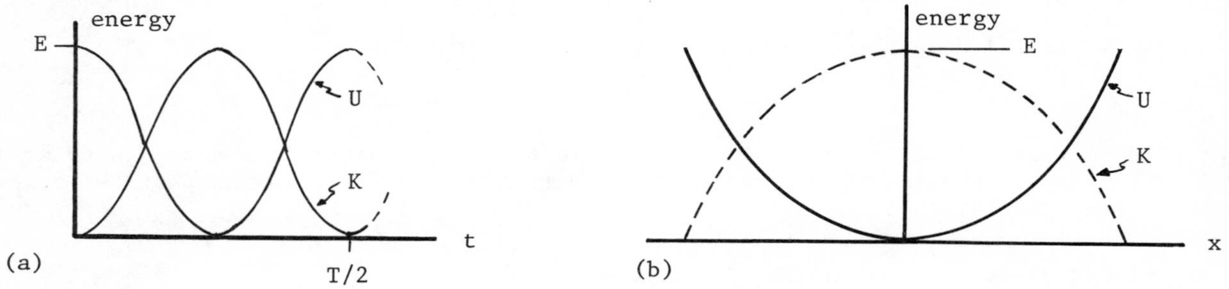

Figure 13-3. Energies of a simple harmonic oscillator: Kinetic energy (K), potential energy (U) and total energy (E) as functions of time (a) and as functions of position (b).

>>> Example 2. A particle executes linear SHM about the origin. At t = 0 it is at the origin with velocity v = -5 m/s. It returns to the origin 1 second later. Find the amplitude, frequency, angular frequency and displacement at any time.

Since the motion is SHM the displacement as a function of time has the form

$$x = A \cos(\omega t + \delta)$$

and the velocity is

$$v_x = dx/dt = -\omega A \sin(\omega t + \delta) \quad .$$

At $t = 0$ we have $x = 0$, so

$$x(0) = A \cos \delta = 0$$

which implies $\delta = \pm \pi/2$. (Note that the phase constant can have the range $-\pi$ to π.) At $t = 0$ the velocity is -5 m/s, so

$$-5 \text{ m/s} = -\omega A \sin \delta \quad .$$

Since ω and A are intrinsically positive we must have $\delta = \pi/2$. Now, a particle in SHM returns to a given position twice each period, once going one way and once in the other. Thus since it returns in 1 second we have that $T/2 = 1$ second so that $T = 2$ s. The frequency is

$$\nu = 1/T = 0.5 \text{ Hz}$$

and the angular frequency is

$$\omega = 2\pi\nu = \pi \text{ radians per second} \quad .$$

Finally since $\omega A = \pi A = 5$ m/s, the amplitude is

$$A = 5/\pi \text{ meters} \quad .$$ <<<

>>> Example 3. A block whose mass is 0.4 kg rests on a horizontal frictionless table and is attached to a spring. It is pulled + 0.08 m from equilibrium and given a velocity of + 0.16 m/s. Its maximum displacement is 0.12 m. Find the spring constant, angular frequency and displacement for arbitrary t.

Initially the kinetic energy is

$$K = \tfrac{1}{2} m v_i^2 = \tfrac{1}{2} (0.4)(0.16)^2 \text{kg}\cdot\text{m}^2/\text{s}^2 \quad .$$

The amplitude is 0.12 cm, so from Eq. (13-12b) we have

$$\tfrac{1}{2}(0.4)(0.16)^2 = \tfrac{1}{2}k[(0.12)^2 - (0.08)^2] \quad .$$

The spring constant is

$$k = 1.28 \text{ N/m} \quad .$$

The angular frequency is then given by

$$\omega = \sqrt{k/m} = \sqrt{1.28/0.4} = 4\sqrt{5}/5 \text{ radian/second} \quad .$$

Now, the displacement has the form

$$x = A \cos (\omega t + \delta)$$

and the velocity is

$$v_x = -\omega A \sin (\omega t + \delta) \quad .$$

At $t = 0$, $x = A \cos \delta$ and $v_x = -\omega A \sin \delta$. But v_x is positive at $t = 0$ so δ must be negative.

Then $$\cos \delta = x(0)/A = 0.08/0.12 = 2/3$$

and $$\delta = -48.2^o \quad .$$ <<<

13-5 Combination of Two SHM at Right Angles; Relationship of SHM to Uniform Circular Motion

If two SHM's having the same frequency are combined at right angles the resultant motion traces out a path which is in general an ellipse. For simplicity take

$$x = A_x \cos \omega t \tag{13-15a}$$

$$y = A_y \cos (\omega t + \alpha) \quad . \tag{13-15b}$$

To find the path one needs to eliminate t between the equations. To do so, expand Eq. (13-15b) to

$$y = A_y \cos \omega t \cos \alpha - A_y \sin \omega t \sin \alpha$$

or

$$\frac{y}{A_y} - \cos \omega t \cos \alpha = \sin \omega t \sin \alpha \quad . \tag{13-16}$$

From Eq. (13-15a)

$$\cos \omega t = x/A_x$$

and if we square both sides of Eq. (13-16) we have

$$\left(\frac{y}{A_y} - \frac{x}{A_x} \cos \alpha\right)^2 = \sin^2 \omega t \sin^2 \alpha = (1 - \cos^2 \omega t) \sin^2 \alpha = \left(1 - \frac{x^2}{A_x^2}\right) \sin^2 \alpha \quad .$$

Therefore the path is

$$\frac{x^2}{A_x^2} + \frac{y^2}{A_y^2} - \frac{2xy \cos \alpha}{A_x A_y} = \sin^2 \alpha \tag{13-17}$$

which is that of an ellipse. If $\alpha = 0$ the ellipse degenerates into a straight line through the origin and having slope $\tan^{-1} (A_y/A_x)$ as shown in Fig. 13-4a.

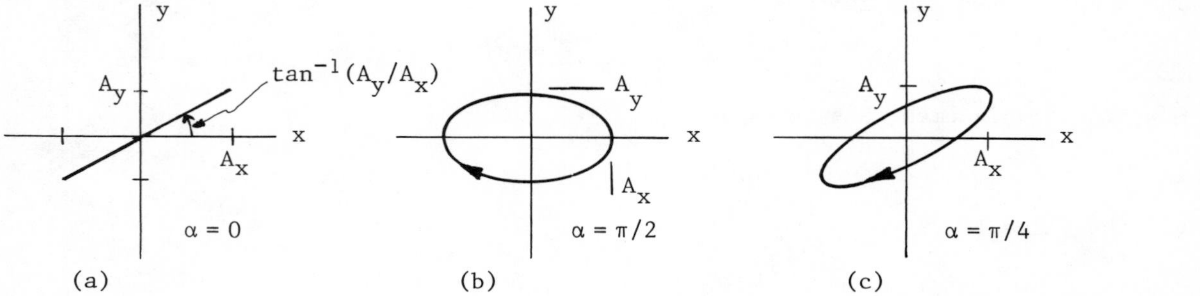

Figure 13-4. Combinations of SHM's at right angles. In (a) the phase difference is 0, in (b) it is $\pi/2$, and in (c) it is $\pi/4$. The arrow indicates the direction of motion of a particle.

In other cases the resultant is an ellipse whose orientation relative to the x,y axes depends upon the phase difference α. In Fig. 13-4b we show the case $\alpha = \pi/2$ and in Fig. 13-4c the case $\alpha = \pi/4$.

In the special case $\alpha = -\pi/2$ with $A_x = A_y = A$, the resultant is a circle

$$x = A \cos \omega t \tag{13-18a}$$

$$y = A \cos (\omega t - \pi/2) = A \sin \omega t \quad . \tag{13-18b}$$

If a particle has these as components of its position vector, then with

$$\theta = \omega t$$

we see from Eq. (10-6) that the particle moves in a circle of radius A with constant angular velocity ω.

Thus, the combination along perpendicular lines of two SHM's having the same amplitude and frequency and a phase difference of $\pi/2$ is equivalent to uniform circular motion. Conversely, the projected motion onto any diameter of uniform circular motion is SHM. The circle for the equivalent uniform circular motion in a SHM problem is often called the reference circle.

13-6 Programmed Problems

1. Now consider the motion of a mass connected to a spring as shown. The surface is frictionless.

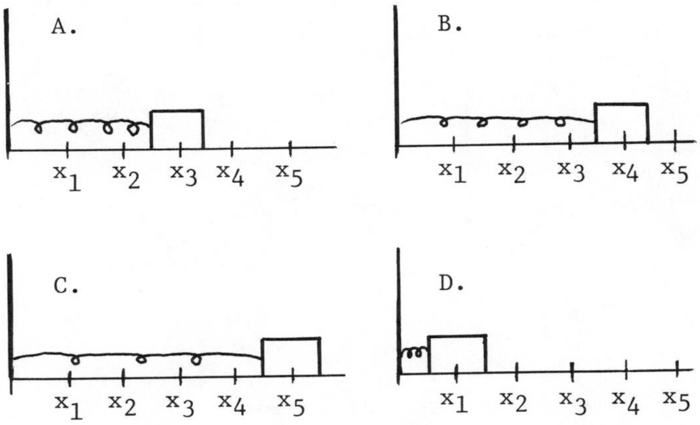

On the diagrams above draw vectors showing the force exerted by the spring on the mass m. Show specifically the directions and relative magnitudes. In A the spring is relaxed.

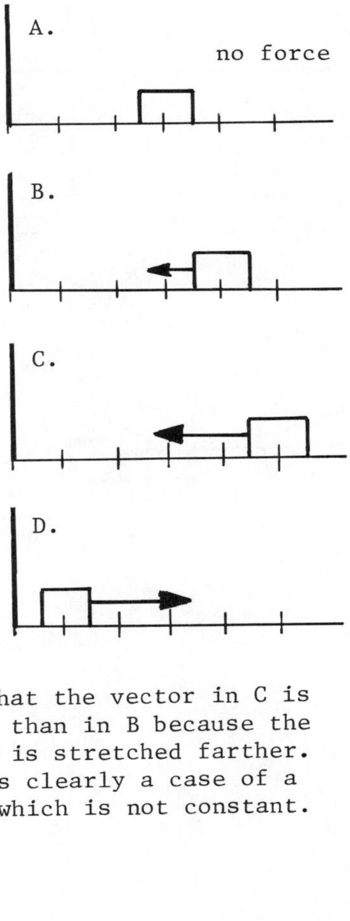

Note that the vector in C is larger than in B because the spring is stretched farther. This is clearly a case of a force which is not constant.

2\.

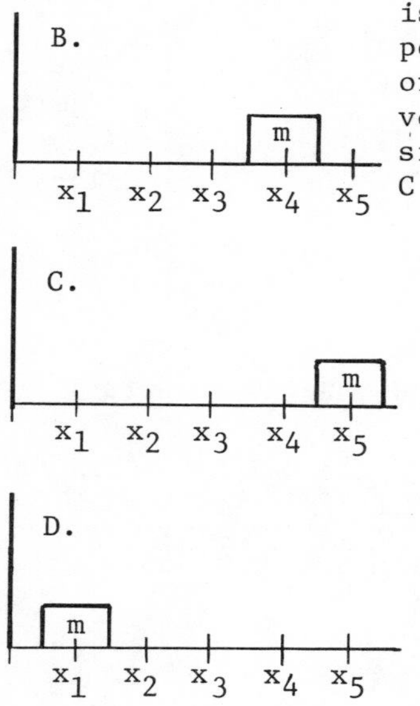

The position x_3 at which the spring exerts no force is called the equilibrium position. Using x_3 as the origin draw displacement vectors indicating the position of m in diagrams B, C and D.

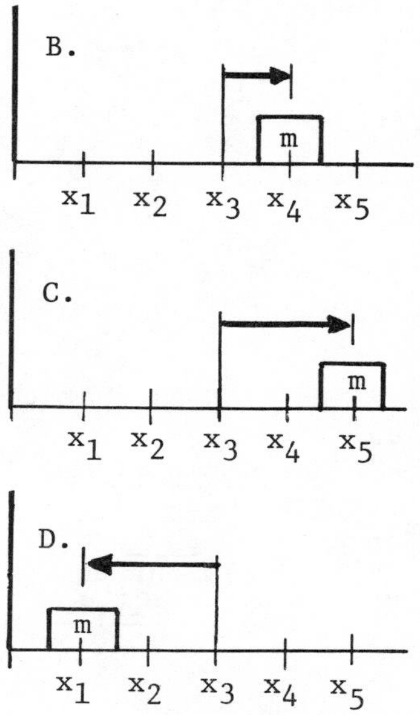

3\. What is the relationship between the force vector and the displacement vector in the answers of the two preceding frames.

$\underline{\underline{F}} = -k\underline{\underline{x}}$, i.e. oppositely directed <u>and</u> proportional.

4\. In this diagram note the relabeled axis. Answer the following questions:

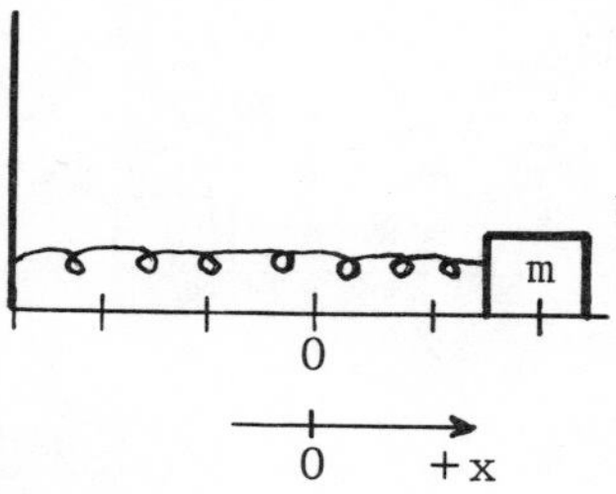

A. If each displacement is 0.1 meter and the spring constant k is 20 N/m the magnitude of the force is ______ .

B. What is the sign of the force?

C. If the object's mass is 0.5 kg, what is the magnitude and direction of its acceleration at this instant?

A. 4.0 N. $|\underline{\underline{F}}| = k|\underline{\underline{x}}|$

B. Minus, negative (opposite to the displacement which is positive).

C. $a_x = F_x/m$

$= (-\ 4.0\text{N})/(0.5\ \text{kg})$

$= -\ 8.0\ \text{m/s}^2$.

The acceleration is in the direction of the force; the negative sign indicates that direction here.

5. Having looked at the characteristics of the force acting on m in this example we can try to solve Newton's 2nd law $F_x = ma_x$. Writing a_x as d^2x/dt^2 we have

$$\_\_\_\_\_\_ = m\ d^2x/dt^2$$

Before writing the answer look again at the answer of frame 3.

Answer: $- kx = m\ d^2x/dt^2$

---

6. Try $x = x_o \cos \omega t$ as a solution of the equation

$$m\ d^2x/dt^2 = - kx\ .$$

Here x_o and ω are constants.

A. $dx/dt =$ ______ .
B. $d^2x/dt^2 =$ ________ .
C. Sketch x as a function of t.

Answer:

A. $- \omega x_o \sin \omega t$

B. $- \omega^2 x_o \cos \omega t$

C.

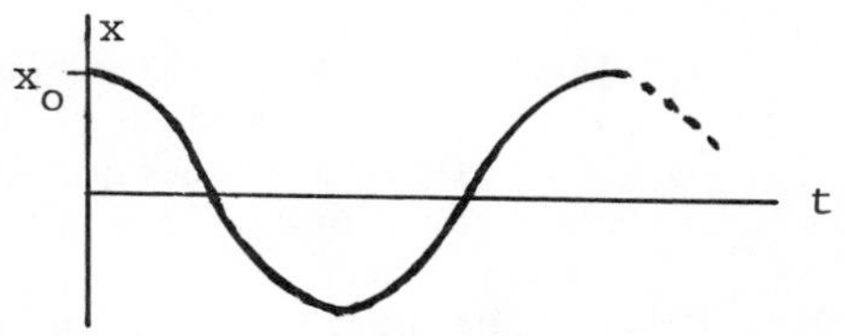

---

7. Substitute the assumed solution and its second time derivative in

$$m\ d^2x/dt^2 = - kx\ .$$

m ________ = - k _______ .

Answer: $- m\omega^2 x_o \cos \omega t = - kx_o \cos \omega t$

---

8. The equation in the answer of frame 7 will be true provided

$m\omega^2 =$ _______ .

Answer: k
If you missed this just look at the equation above.

---

9. We have then $\omega = \sqrt{k/m}$ as a condition that $x = x_o \cos \omega t$ be a solution of the equation of motion for this problem.

A. k has the units _______ .
B. m has the units _______ .
C. So ω has the units _______ .

Answer:

A. $N/m = kg/s^2$

B. kg

C. $\sqrt{(kg/s^2)/(kg)} = s^{-1}$

Our job of describing the motion of the mass is complete. We have a sketch of the position as a function of time as well as knowing the form of the function, i.e. $x = x_o \cos \omega t$. Of course it could be made more complicated by starting the motion differently or by starting the clock when the mass is at some position other than equilibrium or maximum excursion. We will look at a further example to elucidate these complications.

10. Study the graph below. It displays the position x of an oscillator as a function of time t.

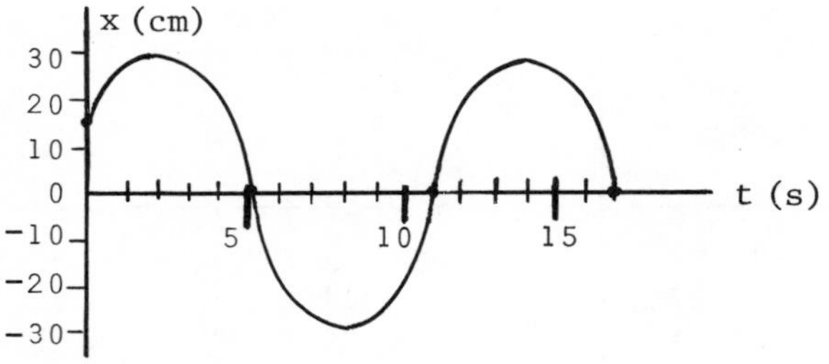

A. What is the amplitude of this oscillator?
B. What is the period of this oscillator?
C. What is the frequency of this oscillator?
D. What is the angular frequency of this oscillator?
E. Is the position of this object at time $t = 0$ described by either $x = x_o \sin \omega t$ or $x = x_o \cos \omega t$ where x_o is the amplitude of oscillation? Explain.

A. 30 cm (maximum displacement from equilibrium).

B. 12 s

C. $\nu = \frac{1}{T} = \frac{1}{12} \frac{\text{cycles}}{\text{s}} = 0.0833$ Hz

D. $\omega = 2\pi\nu = \frac{2\pi}{12} \frac{\text{rad}}{\text{s}}$

E. Neither will work. At $t = 0$, $x_o \sin \omega t = 0$ and $x_o \cos \omega t = 30$ cm. Neither is what the graph shows at $t = 0$.

---

11. Still referring to the diagram of the previous frame

A. What is the initial position, i.e., what is x at $t = 0$?
B. If we write the oscillating function as

$$x = x_o \cos (\omega t + \delta)$$

a. What is x_o?
b. What is ω?
c. In view of answer A above, what must δ be?

A. 15 cm

B. a. 30 cm

b. $\frac{2\pi}{12} \frac{\text{rad}}{\text{s}}$

c. At $t = 0$, $x = x_o \cos \delta$.

$\cos \delta = 15/30 = 0.5$

$\delta = \pm 60^o = \pm \pi/3$ (radian).

Since the initial velocity $v_o = -\omega x_o \sin \delta$ is positive we must have $\delta = -\pi/3$.

---

12. δ is called the phase constant. Let's calculate x at some arbitrary time and compare the answer to the graph. Using

$$x = x_o \cos (\omega t + \delta)$$

i.e.

$$x = 30 \cos (\frac{2\pi}{12} t - \frac{\pi}{3})$$

calculate x at $t = 5.0$ sec.
DON'T LOOK AT THE GRAPH YET.

$x = 30 \text{ cm} \cos [\frac{2\pi}{12} \times 5.0 - \frac{\pi}{3}]$

$x = 30 \text{ cm} \cos (\pi/2)$

$x = 0$

Check the curve at $t = 5.0$ s.

An object of mass 5 kg is hung gently from a vertical spring. The spring is observed to stretch 1.0 m at which point the mass hangs in equilibrium. The object is then given an upward velocity of 1.6 m/s. We want to consider some details of the ensuing motion.

13. What is the period? To determine this we need to know the spring constant k. Write the conditions which allow us to find k. Find k.	If the initial stretch is L then, at equilibrium $kL = mg$ $k = (5\ \text{kg})(9.8\ \text{m/s}^2)/(1\ \text{m})$ $= 49\ \text{N/m}$
14. Now that we know k we may write the equation $\omega =$ ______ and find the numerical value and unit $\omega =$ ______ .	$\omega = \sqrt{k/m}$ $= \sqrt{(49\ \text{N/m})/(5\ \text{kg})}$ $= 3.13\ \text{s}^{-1}$
15. To find the period then we write the equation T = ______ and substitute to find T = ______ s .	$T = 2\pi/\omega$ $= 2\pi/(3.13\ \text{s}^{-1})$ $= 2.0\ \text{s}$
16. Write the general equation for the displacement, y, from equilibrium. Call the amplitude A. y = ______	$y = A \cos(\omega t + \delta)$ For later use let us specify that the +y axis is upward.
17. Now we must insert the initial conditions. At t = 0 we know y = ______. Does this uniquely determine δ?	At $t = 0$, $y = 0$ so $\cos \delta = 0$. Thus $\delta = \pm \pi/2$. This condition does not therefore completely determine δ.
18. What else do we know about the motion of the object at t = 0?	We know $v = +1.6$ m/s. The plus sign occurs because of our choice of the +y axis upward.

19. To find δ then we must use the definition $v =$ ______ and then find $\delta =$ ______ .	$v = dy/dt = -\ \omega A \sin\ (\omega t + \delta)$ At $t = 0$ $v = -\ \omega A \sin \delta$ so of the two possibilities $\delta = \pm\ \pi/2$ we must choose $\delta = -\ \pi/2$
20. There are two ways we may find the amplitude A. List these and find A. i) ii) $A =$ ______	i) From $v = -\ \omega A \sin\ (\omega t - \pi/2)$ and $v(0) = 1.6$ m/s we have $\omega A = v(0) = 1.6$ m/s so $A = 0.51$ m ii) At $y = 0$ the total energy is entirely kinetic. But, the total energy may be written as $\frac{1}{2} kA^2$ because at $y = \pm$ A the total energy is entirely potential energy. Thus $\frac{1}{2} kA^2 = \frac{1}{2} m\ v^2(0)$ so $A^2 = \frac{m}{k} v^2(0) = \frac{v^2(0)}{\omega^2}$.
21. What is the maximum downward acceleration and at what value of y does it occur?	$a = dv/dt = -\ \omega^2 A \cos(\omega t - \pi/2)$ $= -\ \omega^2 y$. Thus, the maximum magnitude of a is $\omega^2 A = 5.0\ m/s^2$. The maximum downward (negative) acceleration is $-\omega^2 A$ which occurs at $y = +\ A = 0.51$ m.
22. What force does the spring exert on the mass at the point found in Frame 21? $F_{spring} =$ ______ in the direction ______.	When the object is at the top of its motion the spring is stretched downward by 1.0 - 0.51 = 0.49 m. So, the spring exerts an upward force of 24 N. If you answered 25N downward you computed the net force on the mass. This net force is the vector sum of the spring force and that of gravity. Gravity exerts a force of 49 N downward so the net force is indeed 25 N downward.

23. What is the maximum upward force on the object by the spring? At what value of y does this occur?

F_{spring} = ______

occurs at

y = ______

What is the net upward force on the object at this point?

F_{net} = ______

The spring exerts a maximum upward force at maximum downward stretch: the bottom of the motion. Thus at y = - A = - 0.51 m the spring is stretched by 1.51 m and exerts an upward force of 74 N. Gravity pulls downward with a force of 25 N so the net upward force is 49 N.

Note that at each point in the motion the spring is stretched by L - y. This is just enough so that

$$F_{net} = \underset{\text{upward}}{F_{spring}} - \underset{\text{downward}}{F_{gravity}}$$

$$= - ky$$

Therefore, the force law for SHM is satisfied.

Chapter 14: GRAVITATION

REVIEW AND PREVIEW

Previously you have studied the acceleration of gravity only in the near vicinity of the surface of the earth in which case you learned that $\underline{F} = m\underline{g}$ where $\underline{g}$ was a constant acceleration. This law was a deduction from experiment in the very restricted circumstances. Now, you will meet gravitation and the theory of the origin of the gravitational force in more general terms. The universal law of gravitation is a theory due to Newton and is also experimentally well founded. You will learn to apply conservation of energy to gravitational problems and will be able to solve many interesting problems such as those of satellite motion.

GOALS AND GUIDELINES

You have four major goals in this chapter:

1. Learning the definition of the law of universal gravitation and the restriction that it is formulated for point masses.
2. Learning how to apply the basic law to deduce the gravitation force for extended objects, principally uniform spherical objects. You should try to understand the basic concept even for non-spherical objects.
3. Understanding the connection between the form of gravity you met earlier and this more general theory. This is closely related with goal number 2.
4. Learning how to apply conservation of mechanical energy to problems in which the force is gravitational; this means, of course, that you will need to learn the form of the gravitational potential energy.

14-1 The Law of Universal Gravitation

The law is usually stated as:

> The force between any two particles of masses m_1 and m_2 separated by a distance r is an attraction acting along the line joining the two particles and having magnitude
>
> $$F = G\, m_1 m_2/r^2 \tag{14-1}$$
>
> where G is a universal constant which has the same value for all pairs of particles.

This is the magnitude of an action-reaction pair of forces. If as shown in Fig. 14-1,

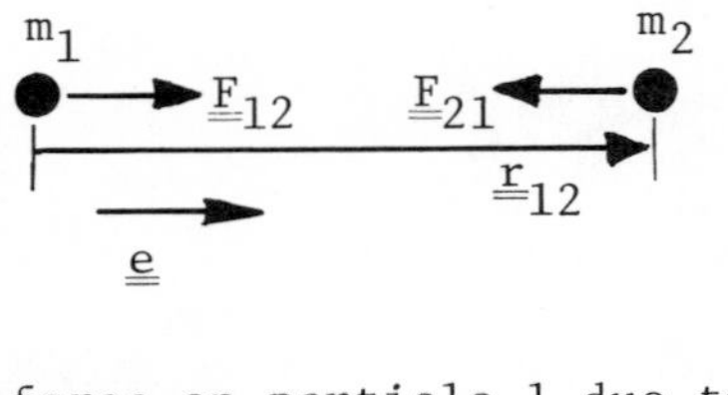

Figure 14-1. Illustration of the gravitational force between particles m_1 and m_2; the choice of the direction of the unit vector $\underline{e}$ is arbitrary.

$\underline{F}_{12}$ is the force on particle 1 due to particle 2 and $\underline{F}_{21}$ is the force on particle 2 due to particle 1, then

$$\underline{F}_{12} = G\, m_1 m_2\, \underline{e}/|\underline{r}_{12}|^2 = -\underline{F}_{21} \tag{14-2}$$

where r_{12} is the distance between the two particles, and $\underline{\underline{e}}$ is a unit vector along the direction from 1 to 2. The essential point is that the force on 1 is directed toward 2 and the (equal in magnitude) force on 2 is directed toward 1.

Note that G is a universal constant and should not be confused with the local acceleration of gravity g. The dimensions of G are

$$[G] = [F]L^2/M^2 = L^3M^{-1}T^{-2} \tag{14-3}$$

while those of g are acceleration

$$[g] = LT^{-2} \tag{14-4}$$

The force law of Eq. (14-1) has been verified by experiment many times and in the process the value of G is found. The currently accepted value is

$$G = 6.673 \times 10^{-11}\ \mathrm{N{\cdot}m^2/kg^2} = 3.436 \times 10^{-8}\ \mathrm{lb{\cdot}ft^2/slug^2} \quad .$$

14-2 Extended Masses

It is important to note that Eq. (14-1) applies only to point masses (particles). One must deduce the force between extended bodies by treating them as a composite of point masses. Solving such problems is often complicated and mathematically difficult. In special cases however, due to symmetry properties it may be relatively easy. Such a special case is that of a uniform spherical shell and a point mass M. The result is that if M is outside of the shell at a distance R from the center of mass of the shell, then the force on M is

$$\underline{\underline{F}}_{\substack{\text{due to}\\ \text{spherical}\\ \text{shell}}} = -\,GM_sM\underline{\underline{e}}/R^2 \quad , \quad R \geq R_s \tag{14-5}$$

where $\underline{\underline{e}}$ is a unit vector from the center of mass of the shell toward P, M_s is the shell mass, and R_s the shell radius. This result means that

> a uniform spherical shell attracts an external point mass as if all its mass were concentrated at its center of mass.

If $R < R_s$, then the net force is zero:

$$\underline{\underline{F}}_{\substack{\text{on M due to}\\ \text{spherical}\\ \text{shell}}} = 0 \quad , \quad R < R_s \quad . \tag{14-6}$$

If the density of a sphere is a function only of the magnitude of position (measured from the center) it can be made up of uniform spherical shells even though the density of a given shell need not be the same as those interior or exterior to it. Thus Eq. (14-5) applies to such spheres

$$F_{\text{sphere}} = -\,GM_sMe/R^2 \quad , \quad R \geq R_s; \quad \rho = \rho(|\underline{\underline{r}}|) \quad . \tag{14-7}$$

Here M_s is the mass of the sphere and R_s its radius. The second condition means that the density is a function only of the magnitude of position.

14-3 Variations in Acceleration of Gravity

To the extent that the earth can be treated as a sphere in the sense of Eq. (14-7), the force on a point mass m outside or on the earth's surface is

$$F = GMm/r^2$$

where M is the mass of the earth and r is the distance of the mass from the earth's center. This force may also be written as

$$F = mg$$

where g is the accelration of m due to gravity. From these two equations we have

$$g = GM/r^2 \qquad (14\text{-}8)$$

so in general,

i) g is not constant;
ii) $\underline{\underline{g}}$ is directed toward the center of the earth.

Eq. (14-8) can be used to deduce the mass of the earth. At the earth's surface g_s is known and

$$g_s = GM/R^2 \qquad (14\text{-}9)$$

where R is the radius of the earth which is also known. Eq. (14-9) can then be solved for M.

Eq. (14-8) can also be used to compute the variation of g over distances which are small compared to the earth's radius. In general

$$dg/dr = -\ 2\ GMr^{-3} = -\ 2\ g/r\ .$$

Thus

$$dg/g = -\ 2\ dr/r\ . \qquad (14\text{-}10a)$$

Eq. (14-10a) may be used as an approximation for finite changes in g due to finite changes in r provided $|\Delta r| \ll r$. In this form

$$\Delta g/g = -\ 2\ \Delta r/r\ . \qquad (14\text{-}10b)$$

>>> Example 1. Estimate the mass of the earth.
The mean radius of the earth is 6.37×10^6 m and the standard value of g is $9.807\ m/s^2$. Thus

$$M = g\,R^2/G = (9.807\ m/s^2)\ (6.37 \times 10^6\ m)^2/(6.673 \times 10^{-11}\ N{\cdot}m^2/kg^2)$$

$$M = 5.96 \times 10^{24}\ kg\ .$$

<<<

>>> Example 2. Determine approximately the relative change in the acceleration of gravity from the value at the earth's surface to the value at an elevation of 2,500 meters.

From Eq. (14-10b)

$$\Delta g/g = -2\ (2.5 \times 10^3\ \text{m})/(6.37 \times 10^6\ \text{m}) = -7.8 \times 10^{-4} \quad .$$ <<<

14-4 Planet and Satellite Motion

Kepler's three laws of planetary motion (which also apply to satellite motion about any planet -- for such motion read satellite for planet and planet for sun) are:

1. All planets move in elliptical orbits with the sun at one focus.
2. A line joining the planet to the sun sweeps out equal areas in equal times.
3. The square of the period of a planet's motion is proportional to the cube of its mean distance from the sun.

The first and third are properties of the particular force law; they depend for their validity upon the $1/r^2$ dependence. The second Kepler law is a statement of the conservation of angular momentum which would be valid for any gravitation law provided $\underline{\underline{F}}$ is directed along $\underline{\underline{r}}$. Then, the torque of the force is $\underline{\underline{\tau}} = \underline{\underline{r}} \times \underline{\underline{F}} = 0$ so the angular momentum is constant.

The explicit relationship for the third law is

$$T^2 = 4\pi^2 r^3/(GM) \tag{14-11}$$

where M is the mass of the sun and r is the distance between the center of the sun and that of the planet. Eq. (14-11) is valid for the special case of circular orbits; it is even then an approximation in that the mass of the sun is taken to be much greater than the mass of the planet. This is a good approximation. The equation is also valid for satellites of a planet. For earth it is particularly useful to know the period at which a satellite would orbit just above the earth's surface. Then $r = R_e$ so

$$T_s{}^2 = 4\pi^2 R_e{}^3/(GM_e) = 4\pi^2 R_e/g_s$$

where g_s is the acceleration of gravity at the surface. Thus

$$T_s = 2\pi\sqrt{R_e/g_s} = 84.4 \text{ minutes} \quad . \tag{14-12}$$

Then, the period for a satellite at any distance r is easily found by the ratio

$$T/T_s = (r/R_e)^{3/2} \tag{14-13a}$$

Often, one wants to use $r = R_e + h$ where h is the altitude of the orbit above the surface of the earth. If h is small in comparison with R_e we may then approximate

$$T/T_s = (1 + h/R_e)^{3/2} \simeq 1 + (3/2)h/R_e \quad . \tag{14-13b}$$

Equation (14-13b) also follows from the small change in T due to a small change in r, that is

$$dT/T_s = (3/2)(r/R_e)^{\frac{1}{2}}\ dr/R_e \quad .$$

Then we approximate $\Delta T \simeq dT$ and $\Delta h \simeq dr$ while $r/R_e \simeq 1$ so that

$$\Delta T/T_s = (3/2)h/R_e \quad .$$

With $T = T_s + \Delta T$, Eq. (14-13b) follows.

14-5 Gravitational Field, Potential Energy, and Potential

The gravitational force is an example of action at a distance which means that two particles interact although they do not come in contact necessarily. Often this concept is replaced by the notion of a field. The idea is that one particle is regarded as modifying the space or environment around it by setting up a field which itself acts on any other particle placed in it. The field plays an intermediate role and the problem has two separate parts:

i) Find the field from the sources;
ii) Find the effect of the field on a particle placed in it.

The field approach is most useful when the sources are very massive compared to the particle so that the particle motion does not essentially disturb the sources. Then the field of a single source is constant in time (in the rest frame of the source) and essentially independent of the motion of the particle.

The gravitational field strength $\underline{\underline{g}}$ of a particle of mass M is defined to be the force per unit mass on a small test particle of mass m placed at a position $\underline{\underline{r}}$ relative to the source. Thus

$$\underline{\underline{g}} = \underline{\underline{F}}/M = - GM\underline{\underline{e}}_r/r^2 \tag{14-14}$$

where $\underline{\underline{e}}_r$ is directed radially outward from M. Since $\underline{\underline{a}}$ is $\underline{\underline{F}}/m$ from Newton's law we see that $\underline{\underline{g}}$ is the local acceleration of gravity.

The gravitational potential energy difference between points a and b is the negative of the work by the force of gravity as the system moves from configuration a to configuration b. Thus

$$\Delta U_{ab} = U_b - U_a = - W_{ab} \quad .$$

To define the gravitational potential energy at a point we must specify the reference point where U has some arbitrarily agreed upon value; usually this value is chosen to be zero. In general then

$$U_b = - W_{\substack{\text{by } \underline{\underline{F}} \text{ from} \\ \text{reference} \\ \text{point to b}}} + U_{\substack{\text{reference} \\ \text{point}}} \quad . \tag{14-15}$$

By convention the reference point is generally taken to be that point where $\underline{\underline{F}}$ is zero and the reference potential energy is taken as zero. Since $\underline{\underline{F}}_{gravity}$ vanishes as $r \to \infty$, this convention yields

$$U(r) = - W_{\infty r} \quad . \tag{14-16}$$

Because $\underline{\underline{F}}_{grav}$ is conservative the path chosen to evaluate this result is immaterial; i.e. the result is path independent (see Section 7-2). Thus we choose a radial path ($\underline{\underline{e}}_r \cdot d\underline{\underline{r}} = dr$) to find

$$U(r) = - \int_\infty^r (- GMm\underline{\underline{e}}_r/r^2) \cdot d\underline{\underline{r}} = - GMm/r \Big|_\infty^r = - GMm/r \quad . \tag{14-17}$$

Eq. (14-16) applies to point masses at a separation r and to that case where M is a sphere in the sense of Eq. (14-7).

>>> Example 3. Find the gravitational potential energy at any point r for a uniform sphere of radius R and mass M and a point mass m.

For $r \geq R$ the force is

$$F = -GMm/r^2 \quad ,$$

so

$$U(r) = -GMm/r \quad , \quad r \geq R \quad .$$

For $r < R$, only that portion of the sphere interior to the point mass contributes. This result follows if one mentally divides the sphere into spherical shells and uses the result of Eq. (14-6). Thus for $r < R$

$$\underline{\underline{F}}(r) = -G\, M_{interior}\, m\underline{\underline{e}}_r/r^2$$

where $M_{interior}$ is the mass of the sphere interior to the point mass position r. Since the sphere is presumed uniform,

$$M_{interior} = \rho\, 4\pi r^3/3$$

where ρ is the density which is given by

$$\rho = M/(4\pi R^3/3) \quad .$$

Thus

$$M_{interior} = Mr^3/R^3$$

and

$$\underline{\underline{F}}(r) = -G\, Mm\underline{\underline{r}}/R^3 \quad .$$

Then for $r \leq R$ (note $\underline{\underline{r}}\cdot d\underline{\underline{r}} = rdr$)

$$U(r) = -\int_R^r \left(-G\frac{Mm}{R^3}\underline{\underline{r}}\right)\cdot d\underline{\underline{r}} + U(R) = G\frac{Mm}{2R^3}(r^2 - R^2) + \left(-\frac{GMm}{R}\right) = -\frac{3}{2}\frac{GMm}{R} + \frac{GMmr^2}{2R^3} \quad .$$

A graph of the potential energy is shown in Fig. 14-2.

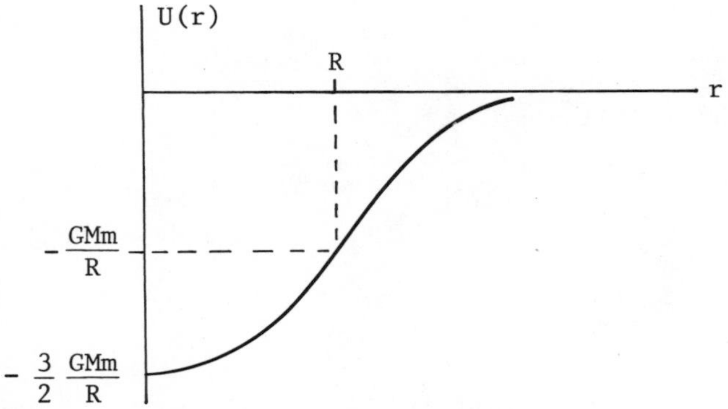

Figure 14-2. The potential energy of the system consisting of a point mass m and a uniform sphere of mass M and radius R as a function of the separation distance between m and the center of mass of the sphere.

<<<

It is important to remember that U is a property of both M and m, i.e. is characteristic of the system. In the case where M >> m, however, one often speaks of the potential energy of m in the field of M. This looseness of language is justified in this case by the fact that when this potential energy is converted into kinetic energy the less massive m obtains most of it.

The gravitational force F can be obtained from U(r) by (see Section 7-2)

$$\underline{\underline{F}} = - \underline{\underline{\nabla}}U(r) = - \underline{\underline{\nabla}}(-GMm/r) = - GMm\underline{\underline{e}}_r/r^2 \tag{14-18}$$

where $\underline{\underline{e}}_r$ is a unit radial vector. The gravitational field strength $\underline{\underline{g}}$ is a vector field associated with the source. One can also associate a scalar field, the gravitational potential, which is the gravitational potential energy per unit mass. Thus

$$V = U/m \quad . \tag{14-19}$$

For a point source

$$V = - GM/r \quad , \quad \text{point source} \quad . \tag{14-20}$$

The gravitational field strength $\underline{\underline{g}}$ is obtained from V in the same way as $\underline{\underline{F}}$ follows from U, i.e.

$$\underline{\underline{g}} = - \underline{\underline{\nabla}}V(r) \tag{14-21}$$

and for a point source, M

$$\underline{\underline{g}} = - \underline{\underline{\nabla}}(- GM/r) = - GM\underline{\underline{e}}_r/r^2 \quad , \quad \text{point source} \quad . \tag{14-22}$$

>>> Example 4. Show that in the near vicinity of the earth's surface the use of the expression U = mgh for the gravitational potential energy is justified.

To do so we shall use the result of Eq. (14-17) applied to a sphere. Then the change in potential energy as we go from the earth's surface (r = R) to a height, h, (r = R + h) where h << R is approximately given by

$$\Delta U/\Delta r \simeq dU/dr = GMm/r^2 \Big|_{r = R} \quad .$$

Now Δr = h, and from Eq. (14-9)

$$GM/R^2 = g$$

where g is the surface value of the acceleration of gravity. Thus

$$\Delta U = U(h) - U(0) = mgh \quad .$$

It is convenient then to shift the zero of potential energy to the surface so that

$$U(h) - U(0) = U(h) = mgh \quad .$$ <<<

14-6 Potential Energy of Many Particle Systems

If masses M_1, M_2, M_3 .... are brought together from infinity the total gravitational potential energy of the system is

$$- GM_1M_2/r_{12} - GM_1M_3/r_{13} - GM_1M_4/r_{14} - \cdot\cdot\cdot\cdot\cdot$$
$$- GM_2M_3/r_{23} - GM_2M_4/r_{24} - \cdot\cdot\cdot\cdot\cdot\cdot\cdot$$
$$- GM_3M_4/r_{34} - \cdot\cdot\cdot\cdot\cdot\cdot\cdot \text{ etc.} \qquad (14\text{-}23a)$$

If there are N particles we may write this as

$$U = -G \sum_{i=1}^{N} \sum_{j>i}^{N} M_iM_j/r_{ij} \qquad (14\text{-}23b)$$

where r_{ij} is the separation distance between particles i and j. Note that this is a scalar sum and no vector operations are required. If we wanted to remove particle number 4 from the system to infinity we would have to do work

$$W = GM_4 \sum_{\substack{i=1 \\ i\neq 4}}^{N} M_i/r_{i4}$$

which is positive.

>>> Example 5. How much work is required to transport a 30,000 kg mass M from the earth's surface to that of the moon? Use $R_m = .27\ R_e$ and $g_m = 0.17\ g_e$ where the subscripts refer to moon and earth. The distance, d, from the earth to the moon is about 3.8×10^8 meter and the earth radius is approximately 6.4×10^6 meter.

The exact answer to this problem depends upon where on the earth one starts and where on the moon the mass ends up. However since the earth-moon distance is quite large compared to either R_e or R_m the initial potential energy is approximately

$$U_i = -G\ M_eM/R_e - GM_mM/d - GM_mM_e/d$$

and

$$U_f = -G\ M_eM/d - GM_mM/R_m - GM_mM_e/d \quad .$$

The work done on the system by some outside agent is the change in potential energy, so

$$W = U_f - U_i = \left[GM_e/R_e - GM_m/R_m + GM_m/d - GM_e/d\right]M \quad .$$

But

$$GM_e/R_e = g_eR_e \qquad \text{and} \qquad GM_m/R_m = 4.6 \times 10^{-2}\ g_eR_e \quad .$$

Thus

$$W = g_eR_e\ M\{1 - 4.6 \times 10^{-2} + 4.6 \times 10^{-2}\ (1.7 \times 10^{-2}) - 1.7 \times 10^{-2}\}$$
$$= 1.8 \times 10^{12} \text{ joule} \quad .$$

<<<

14-7 Energy in Planet and Satellite Motion -- Circular Orbits (Also see Programmed Prob.)

If a body of mass m moves in a circular orbit of radius r about a body of mass M ($m << M$) the potential energy is

$$U = -GMm/r \quad .$$

The kinetic energy $K = \frac{1}{2} mv^2$ may also be written as $\frac{1}{2}$ GMm/r. This follows from the orbital velocity $v = \sqrt{GM/r}$. Hence the total energy is

$$E = K + U = \tfrac{1}{2} G\, Mm/r - GMm/r = -\tfrac{1}{2} GMm/r \quad .$$

Since the total energy is negative the system is bound. That is, energy $\frac{1}{2}$ GMm/r would have to be supplied to separate the two to infinity. For satellites in circular or elliptical motion both E and ℓ are constants of the motion. For circular orbits the constants are

$$\left.\begin{aligned} E &= -\tfrac{1}{2} G\, Mm/r \\ \ell &= m\omega r^2 \end{aligned}\right\}, \quad \text{satellite in circular orbit.} \qquad (14\text{-}24)$$

14-8 Programmed Problems

Find the linear velocity that must be imparted to an artificial satellite of the earth in order that it achieve a circular orbit of radius r measured from the center of the earth.

1. This is actually a problem of the type you have frequently done before. Begin by writing Newton's second law in the form appropriate to the circular motion of the satellite of mass m.

$F =$ ______ .

$F = mv^2/r$

where v is the linear velocity and r is the radius of the orbit.

---

2. 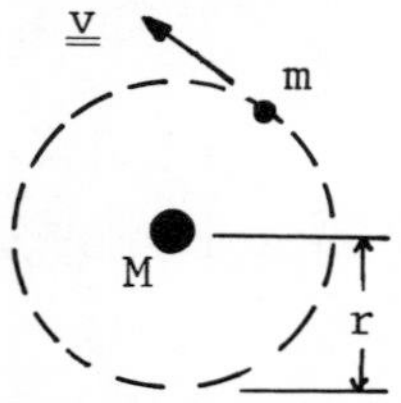

The motion as depicted to the left requires an attractive force acting on m due to its interaction with M. This force has the name ______ force.

The magnitude of this attractive force is given by

$F =$ ______ .

Consider the earth to be fixed and spherical.

Gravitational.

GMm/r^2

The distance r is measured from m to the center of mass of M. Homogeneous spherical objects can be considered as having their mass concentrated at their center.

---

3. The force acting on m is now explicit so it can be substituted into Newton's second law in the answer of frame 1. Do so and simplify the result obtaining an expression for v^2.

$v^2 =$ ______ .

$GMm/r^2 = mv^2/r$

simplifies to $v^2 = GM/r$.

Note that v does not depend upon m.

4. The problem is completed. We have v expressed in terms of quantities which are known. There are, however, one or two useful things that one can do to make this calculation less cumbersome.

Near the surface of the earth the force acting on a mass m is $F = GMm/R^2$ where R is the radius of the earth. We know that this force is equal to the mass m times the acceleration of m near the earth's surface. Using the symbol g_s to represent the acceleration of m near the earth's surface, express g_s in terms of G, M and R.

$F = GMm/R^2 = mg_s$

$g_s = GM/R^2$

---

5. Substitute algebraically g_sR^2 for GM in the answer to frame 3.

$$v^2 = \frac{g_sR^2}{r}$$

The quantity $v_s = \sqrt{g_sR} \simeq 7900$ m/s is the orbital speed that a satellite would have when orbiting just above the earth's surface. In other orbits of radius r the speed is $v = v_s \sqrt{R/r}$.

---

6. If $r = R + h$ where h is the altitude above the earth's surface and if $h/R \ll 1$, find an approximation for v correct to first order in h/R.

$v = v_s \sqrt{R/(R + h)} = v_s[1 + h/R]^{-\frac{1}{2}}$

or

$v = v_s\ [1 - \frac{1}{2}\ h/R]$ by expansion.

One can also derive this from $dv/v = -\ \frac{1}{2}\ dr/r$.

Then, approximate $dv \simeq \Delta v$, $dr \simeq \Delta r = h$, $v \simeq v_s$, $r \simeq R$ and then $\Delta v/v_s = -\ \frac{1}{2}\ h/R$

---

7. What is the orbital velocity of a satellite in orbit 100 miles above the earth's surface? Use 4000 miles as the radius of the earth.

From the approximation in the answer to Frame 6 one has

$v \simeq 7900$ m/s $[1 - \frac{1}{2}\ 100/4000]$

$= 7801$ m/s

The exact expression yields 7803 m/s .

---

8. In man's first voyage to the lunar surface the command ship "Columbia" orbited the moon at an altitude of about 10^5 meters above the surface. What orbital speed did Columbia have? The radius of the moon is about 1.7×10^6 m.

v = ______ m/s

To use the approximation formula we need to know v_s for the lunar surface. Since $g_{moon} \simeq 0.17\ g_{earth}$ and $R_m \simeq R_e$, we have for the lunar surface orbit

$$v_{\ell s} = \sqrt{0.17 \times 0.27} \times 7900 \text{ m/s}$$
$$\simeq 1700 \text{ m/s} .$$

Thus

$v = 1700\ \{1 - \frac{1}{2}\ 10^5/(1.7 \times 10^6)\}$m/s
$= 1650$ m/s .

Given the orbital speed v one can easily find the period of one orbit.

9. For a satellite of mass m moving in a circular orbit of radius r about a mass M we have $v^2 = GM/r$. What distance does the satellite travel in one orbital period T?	The circumference of the circular orbit, $2\pi r$.
10. Thus the orbital period T is given by the expression $T =$ ______ and then $T^2 =$ ______ .	$T = 2\pi r/v$ or $T = 2\pi r\sqrt{r/(GM)}$ so $T^2 = 4\pi^2 r^3/(GM)$ which is Eq. (14-11) .
11. All satellites in orbit about a given mass M have the same T^2/r^3 ratio. For satellites of the earth, $T^2/r^3 = 9.9 \times 10^{-14}\ s^2/m^3$. The moon is about 3.8×10^8 m from earth. What is the lunar period? $T_{moon} \simeq$ ______ days	$T^2_{moon} = (9.9 \times 10^{-14}\ s^2/m^3) \times (3.8 \times 10^8\ m)$ $T^2_{moon} = 5.4 \times 10^{12}\ s^2$ or $T_{moon} = 2.3 \times 10^6\ s \simeq 27$ days .
12. In frame 8 we saw that a satellite orbiting just above the lunar surface had a velocity of 1700 m/s. Find T^2/r^3 for satellites of the moon. $T^2/r^3 =$ ______ s^2/m^3 .	$2\pi R_{moon} = v_s T_s$ so $T_s = 6.3 \times 10^3\ s \simeq 105$ minutes and thus $T^2/r^3 = T_s^{\ 2}/R_{moon}^{\ 3}$ $= 8.1 \times 10^{-12}\ s^2/m^3$.
13. We associate both kinetic and potential energy with the motion of an object in the earth's gravitational field. For example as one tosses a ball vertically the kinetic energy is maximum to begin with, but is eventually zero as the ball stops rising. This change in kinetic energy is accompanied by a change in potential energy such that $\Delta K + \Delta U = 0$. This statement implies that the sum K + U is ______ .	Constant. Energy conservation means that the quantity K + U is a constant during the motion of some object.

14. Imagine tossing the ball repeatedly with an ever increasing initial speed until eventually on a given toss it is infinitely separated from you. What will be the kinetic energy of the ball at this infinite separation?	It may well be zero, but if you are strong enough it could be greater than zero. Certainly we will agree that it is less than it was when you tossed the ball.
15. Suppose we say that K = 0 at infinite separation at which time it starts falling toward you. Ignoring air friction, what change will occur in K as the ball approaches you?	It will increase.
16. The value for U at infinite separation is defined to be zero. Assuming as we have that K = 0 at infinity, what is the total mechanical energy E at that point?	Zero. $E = K + U = 0 + 0$.
17. Since we've already stated that K increases as the ball comes toward you, what must happen to U in order that the total energy E be conserved?	It must decrease.
18. Measuring the separation h from you to the ball, the separation (decreases/increases) as the ball returns to you.	Decreases.
19. Given the last answer, does the quantity GMm/r increase or decrease as the ball returns to you? M is the mass of the earth, m is the mass of the ball, and r = R + h is the distance from the center of the earth.	Increases since r is decreasing.
20. Does the quantity [- GMm/r] increase or decrease as the ball returns to you?	Decreases, i.e. it becomes more negative. Similarly as r increases this quantity increases, i.e. becomes less negative. This is the kind of quantity that we want to add to the increasing kinetic energy so that the total is a constant.
21. The quantity $U = - GMm/r$ is of course the gravitational potential energy. Thus the total mechanical energy $E = K + U$ is $E = \frac{1}{2} mv^2 - GMm/r$. The total energy of the ball as it rests in your hand on the surface of the earth is (positive/negative/zero)? We shall ignore the rotation of the earth.	Negative. $E = \frac{1}{2} mv^2 - GMm/r$. At rest, v = 0, so $E = - GMm/R$, where R is the radius of the earth.

22. Objects having negative total mechanical energy are said to be "bound", i.e., they can't escape.

How could you make the ball "unbound" in terms of its total mechanical energy

$$E = \tfrac{1}{2} mv^2 - GMm/R?$$

Give it an initial speed such that $E \geq 0$ or

$\tfrac{1}{2} mv^2 \geq GMm/R.$

The escape velocity is that value of v which makes this an equality. Thus

$$v_e = \sqrt{2GM/R} = \sqrt{2g_s R}$$

---

23. We have used the expression

$$v^2 = GM/r \quad \text{or} \quad v = \sqrt{GM/r}$$

for the speed of objects in circular orbits. Substitute this into the equation for total mechanical energy and simplify the result.

$$E = \underline{\qquad} .$$

$E = \tfrac{1}{2} mv^2 - GMm/r$

$E = \tfrac{1}{2} m\, GM/r - GMm/r$

$E = - \tfrac{1}{2} GMm/r$.

Note two points:
1. The total energy is negative resulting in bound orbits.
2. The kinetic energy is equal in magnitude to one-half the potential energy.

24.

Firing small thrust rockets in the manner shown would (increase/decrease) the energy E of the space vehicle in orbit.

Rocket aids motion.

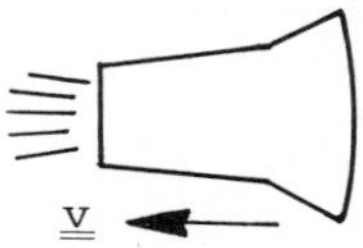

Firing small thrust rockets in the manner shown would (increase/decrease) the energy E of the space vehicle in orbit.

Rocket opposes motion.

Increase

In this case the energy E becomes less negative.

Decrease

This is called retro-firing.

In this case the energy E becomes more negative.

---

25. Since the total mechanical energy is

$$E = - \tfrac{1}{2} GMm/r \quad ,$$

what happens to r for the case of retro-firing?

E decreases, i.e., becomes more negative so r must be smaller.

$E = - \tfrac{1}{2} GMm/r$

---

26. Using the expression for v from frame 23, what happens to the speed of the satellite when the retro-rockets are fired?

$v = \sqrt{GM/r}$.

Since r becomes smaller according to the previous answer, v must become larger.

27. To review then, firing the retro-rockets results in:

1. The orbit becoming smaller;
2. The speed increasing.

What happens if you fire the "aiding" rockets?

1. The orbit becomes larger.
2. The speed decreases.

---

28. What procedure should the astronaut in A employ to attempt docking with B? Assume that B does not fire any rockets.

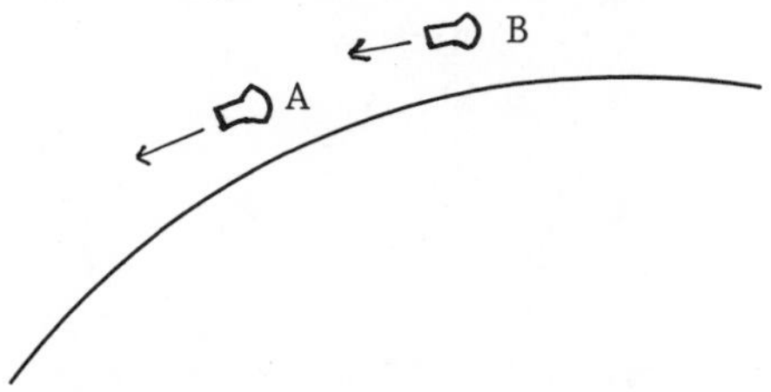

Since A wants to increase his orbit and slow down (to allow B to catch up) he fires thrust rockets as shown.

rocket aids motion

This seems to be intuitively incorrect, but that's the way it is. As A's kinetic energy starts to increase, its velocity is "too large" for its orbit. Since the total energy of A must increase, its orbital radius increases. The net effect is that its kinetic energy actually decreases while its potential energy increases (becomes less negative) by a larger amount than the kinetic energy decrease.

Chapter 15: FLUIDS

REVIEW AND PREVIEW

Previously you have dealt with particles or collections of particles which were rigidly related in position (rigid bodies). You have seen that the description of the motion of such objects requires at most six coordinates; three for the center of mass position and three for rotation orientation. You have even dealt with the case of two particles in the case of collisions. It would seem that the case of fluids with enormous numbers of particles would be hopelessly complicated. Indeed it would be if one were asked to follow each particle, that is to describe its position, velocity, and acceleration. So, one studies the property of fluids by looking at these properties averaged over a very large number of particles. Basically, one describes what happens in a small neighborhood of some point in space. The basic mechanical laws of particle physics apply but in special formulation. You will need to learn a bit of new vocabulary in order to understand the physics.

GOALS AND GUIDELINES

In the chapter you have two major goals:

1. Learning the vocabulary which applies to fluids. It will pay you to think of the corresponding particle quantity where there is a clear relationship. For example, in fluids one uses not kinetic energy but rather kinetic energy density which is kinetic energy per unit volume.
2. Learning the basic equations of a special type of fluid flow called steady, irrotational, incompressible, non-viscous flow (important vocabulary words by the way). You will study fluid statics first, but note that as you study fluid dynamics, fluid statics is a special case.

15-1 Fluids, Pressure and Density

A fluid is a substance that can flow; this definition includes liquids and gases. Fluid mechanics is concerned with those properties of fluids connected with their ability to flow. The basic mechanical laws of particle physics apply, but special formulations are needed.

The force which any surface in contact with a fluid exerts on the fluid must be at a right angle (normal) to the surface. Therefore the force which a fluid exerts on an element of boundary surface is perpendicular to that surface. This force is specified by the pressure, denoted p, which is the magnitude of the normal force per unit area. That is,

$$p = \Delta F/\Delta A \qquad (15\text{-}1a)$$

where ΔA is a unit of surface area. In order that p be independent of the size of the area element, more properly one defines

$$p = \lim_{\Delta A \to 0} \Delta F/\Delta A = dF/dA \quad . \qquad (15\text{-}1b)$$

The pressure may vary from point to point on a liquid boundary.

To describe fluids, one also needs the density

$$\rho = \text{mass/volume} = dM/dV \quad . \qquad (15\text{-}2)$$

The density, ρ, depends upon many factors such as temperature and pressure. For liquids ρ is approximately constant; for gases it is not.

15-2 Fluid Statics

If a fluid is in equilibrium, every portion of it is in equilibrium. Then, if y is the vertical position of a point in the fluid above some reference level, the variation in pressure with position is given by

$$dp/dy = -\rho g \tag{15-3}$$

where ρ is the fluid density. Eq. (15-3) is basic to fluid statics. If

i) ρ is independent of pressure (liquid) then Eq. (15-3) may be integrated. Let p_1 denote the pressure at y_1 and p_2 that at y_2. Then

$$p_2 - p_1 = \int_{p_1}^{p_2} dp = -\int_{y_1}^{y_2} \rho g \, dy \quad . \tag{15-4a}$$

If in addition to (i),

ii) ρ is independent of position (homogeneous liquid)

and

iii) $y_2 - y_1$ is sufficiently small so that g is approximately constant then

$$p_2 - p_1 = -\rho g(y_2 - y_1) \quad . \tag{15-4b}$$

For a free surface it is natural to take y_2 at the surface and denote the pressure at the surface by p_o. Then $y_2 - y_1 = h$ where h is the depth below the free surface. Then the pressure at this depth is from Eq. (15-4b)

$$p = p_o + \rho g h \quad . \tag{15-4c}$$

It is to be noted that the pressure is the same at a depth h independent of the shape of the container. This fact is used very often in fluid static problems.

>>> Example 1. A liquid of density ρ is at rest in a U-tube as shown in Fig. 15-1. On the left side the pressure at the surface is p_L while on the right it is p_R; the surface of the right side is at a height h above that on the left. Show that p_L is given by

$$p_L = p_R + \rho g h \quad .$$

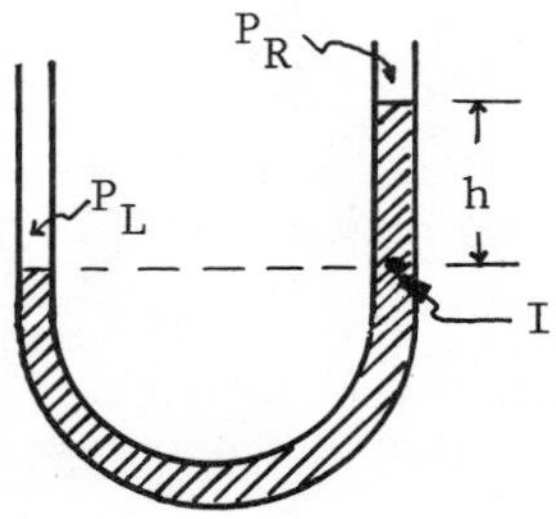

Figure 15-1. Example 1. Pressure difference in a U-tube.

At point I on the right the depth below the right surface is the same as that of the surface on the left. Therefore the pressure at I is p_L. From Eq. (15-4c)

$$p_L = p_R + \rho g h \quad .$$ <<<

Pressure gauges measure either absolute pressure or gauge pressure; the latter is the difference between the absolute pressure and the atmospheric pressure. If p_0 is the atmospheric pressure, $p - p_0$ is the gauge pressure. Tire pressure, for example, is gauge pressure.

>>> Example 2. Consider the U-tube arrangement shown in Fig. 15-2. The left side is connected to a system whose pressure is to be measured and the right side is open to the atmosphere whose pressure is p_0. Show that the gauge pressure is ρgh where h is the height of the right side above the left and ρ is the density of the liquid in the tube.

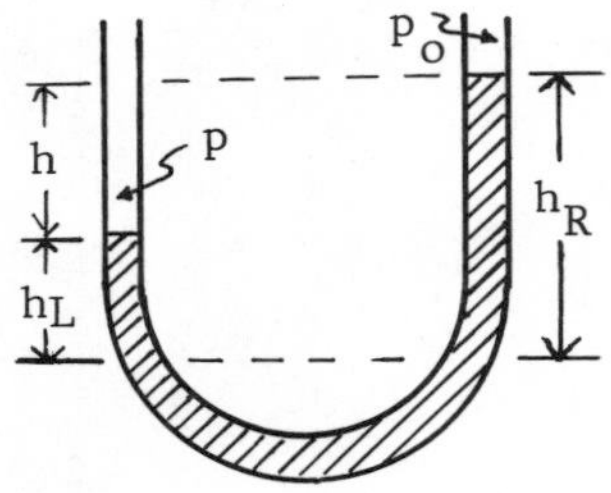

Figure 15-2. Example 2.
Measurement of gauge pressure.

At some point on the left a distance h_L below the surface the pressure is

$$p_L = p + \rho g h_L \quad .$$

At a point on the same elevation on the right, the pressure is

$$p_R = p_0 + \rho g h_R \quad .$$

But these are at the same elevation, so the two pressures are the same and hence

$$p + \rho g h_L = p_0 + \rho g h_R$$

or

$$p - p_0 = \rho g (h_R - h_L) = \rho g h \quad .$$ <<<

15-3 Pascal's Law; Archimedes Principle

Both of these are consequences of fluid mechanics rather than independent principles but they are so useful they are usually stated separately.

> Pascal's law: Pressure applied to an enclosed fluid is transmitted undiminished to every portion of the fluid and to the walls of its container.

For compressible fluids Pascal's law applies after equilibrium is re-established.

Archimedes principle: A body wholly or partially immersed in a fluid is buoyed up with a force equal to the weight of the fluid displaced. This force acts vertically upward through that point which was the center of gravity of the displaced fluid before displacement. This point is called the <u>center of buoyancy</u>.

>>> Example 3. Show that a boat is "self-righting" provided that the center of buoyancy is to the (right/left) of the center of gravity when the boat is heeled to the (right/left).

In Fig. 15-3 we show the case of the boat heeled to the right; if the center of buoyancy is to the right of the center of gravity of the boat, there is a counterclockwise torque about this center of gravity tending to rotate the boat to the left or to "right" it.

Figure 15-3. Example 3. A self-righting boat.

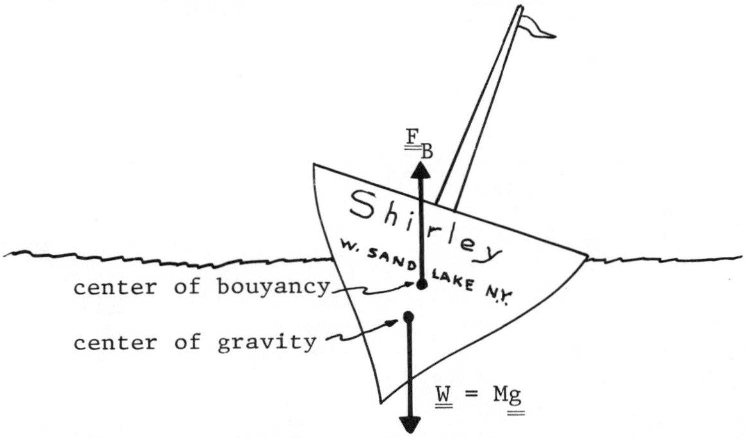

This torque has magnitude F_BD where D is the perpendicular distance between the lines of action of Mg and F_B. Because of equilibrium $F_B = Mg$. If the center of buoyancy were on the left side of the center of gravity of the boat, the torque would be clockwise and the boat would continue on over and capsize. Such considerations are obviously important in the design of the shape of the hull of a sailboat. <<<

>>> Example 4. A uniform cylinder of wood is floating in water with one fifth of its height above the water surface. If it is now floated in a certain oil, only one tenth of its height is above the oil surface. Find the relative density of the oil (the ratio of the density of the oil to that of the water).

As the cylinder floats in the water in equilibrium the upward force is the weight of water displaced and this is equal to the weight of the wood. This is nothing more or less than a statement that the sum of the forces on a body in equilibrium must be zero. Thus we have

$$g\rho_{H_2O}V_{displaced} = \rho_{wood}V_{wood}g \ .$$

Since $V = \pi r^2 h$ for a cylinder and the wood floats with 0.2 h above the water, the volume of the submerged wood and hence the volume of the water displaced is 0.8 V_{wood}. Thus

$$\rho_w = \rho_{H_2O}\ 0.8\ V/V = 0.8\ \rho_{H_2O} \ .$$

Similarly in the oil we find

$$\rho_w = 0.9\ \rho_{oil}$$

By equating the two we find

$$\rho_{oil}/\rho_{H_2O} = 8/9 \quad .$$ <<<

15-4 Fluid Dynamics

Again it is to be emphasized that fluids obey the same mechanical laws as formulated for particles. However, just as special formulations of these laws were useful for rigid bodies, so special formulations are useful and necessary for fluids. The appropriate quantities for the description of fluid dynamics are:

1. The density of the fluid -- $\rho(\underline{r},t)$. In general this may vary with position and time as indicated. This is the fluid analogy of the mass of a particle and is the mass per unit volume.
2. The fluid velocity -- $\underline{v}(\underline{r},t)$. This is the velocity of a small element of the fluid at the position r at time t.
3. The pressure -- $p(\underline{r},t)$.
4. The momentum density -- $\underline{j}(\underline{r},t)$. This is the fluid analogue of momentum and is related to the density and velocity by

$$\underline{j}(\underline{r},t) = \rho(\underline{r},t)\ \underline{v}(\underline{r},t) \quad . \tag{15-5}$$

It is sometimes also called the mass flux density because $\underline{j}\cdot d\underline{A}$, where $d\underline{A}$ is an infinitesimal area element, gives the mass of fluid transported past this area element per unit time. The vector $d\underline{A}$ is directed normal to the surface element and has magnitude equal to the infinitesimal surface area.

5. The kinetic energy density -- $\frac{1}{2}\rho\underline{v}\cdot\underline{v} = \frac{1}{2}\rho v^2$. This is the kinetic energy per unit volume; it may vary with position and time.
6. The potential energy density -- no particular form. This is the potential energy of an infinitesimal volume element of fluid due to a conservative force. For example, near the earth's surface if y is the height of a volume element of fluid above some reference point at which the gravitational potential energy is taken as zero, then the potential energy density is ρgy.

The fluid flow is classified in many ways; it may be

i) Steady or non-steady. If $\underline{v}(\underline{r},t)$ is independent of time t, then the fluid flow is steady. In other words, every fluid element passing a point $\underline{r}$ will have the same velocity as all others which passed this point earlier and the same velocity as all others to come. In steady flow, if a fluid element arrives at $\underline{r}$ at some time t_o, we know where it is going and can tell where it will be at all times later than t_o and indeed we know where it has been.

ii) Compressible or incompressible. If the density is a constant (independent of position and time) the fluid is incompressible. Liquids can usually be considered incompressible and for many applications the changes in density of a gas may be unimportant; in that case the gas flows incompressibly.

iii) Viscous or non-viscous. Viscosity is the fluid analogue of friction; it introduces tangential forces between fluid layers in relative motion and dissipates mechanical energy.

iv) Rotational or irrotational. If a small element of a fluid at some point has no rotational motion about its center of mass the fluid flow at that point is locally irrotational. If this is true for all elements of the fluid, its flow is irrotational. Conceptually one could imagine placing small paddle wheels in the fluid at various points. If these translate without rotating the fluid is irrotational.

In the beginning, one usually studies Steady, Irrotational, Incompressible, Non-viscous flow (SIIN). This results in great simplifications and still leaves many valid physical applications.

15-5 SIIN

For steady flow $\underline{\underline{v}}$ is independent of time although it may depend upon position. Every small element of fluid which arrives at $\underline{\underline{r}}$ follows a prescribed path which is called a streamline. Two streamlines never cross; a bundle of streamlines constitutes a tube of flow which acts like a pipe in that no fluid may pass in or out of the tube of flow except at its ends.

The equation of continuity is a statement of the conservation of mass. For steady flow each mass element of fluid that enters "one end" of a tube of flow results in an equal mass leaving at "the other end". Thus, the mass flux is constant throughout a tube of flow. Hence

$$\underline{\underline{j}} \cdot \underline{\underline{A}} = \text{constant} \quad , \tag{15-6}$$

where $\underline{\underline{A}}$ is the cross sectional area of the tube of flow perpendicular to the flow direction at some point and $\underline{\underline{j}}$ is the mass flux density at the same point. Then $\underline{\underline{A}}$ is parallel to $\underline{\underline{j}}$ and the equation of continuity is

$$\rho v A = \text{constant} \quad , \quad \text{rate of flow.} \tag{15-7a}$$

If in addition to being steady the flow is incompressible ρ may be taken into the constant on the right side of Eq. (15-7a) which becomes

$$vA = \text{constant} \quad . \tag{15-7b}$$

Bernoulli's equation is a statement of the conservation of energy. For steady, incompressible, non-viscous flow it is

$$p + \tfrac{1}{2}\rho v^2 + \rho g y = \text{constant along a streamline} \tag{15-8}$$

where p is the pressure at some point on the streamline, v is the fluid speed at that point and y the elevation of that point above some reference level at which the gravitational potential energy density is zero. If the flow is SIIN, (i.e. irrotational as well) then the constant is the same for all streamlines. Thus for SIIN flow the basic equations are

$$\left.\begin{array}{ll} p + \tfrac{1}{2}\rho v^2 + \rho g y = \text{constant} & (15\text{-}9a) \\ vA = \text{constant} & (15\text{-}9b) \end{array}\right\} \text{SIIN flow.}$$

There are several features of Eqs. (15-9) which you should notice. Eq. (15-9a) contains Eq. (15-4c) as a special case when v = 0. To see this, the constant is written as $p_o + \rho g y_o$. Also, for constant y, if the fluid speeds up then the pressure must fall and vice-versa. The volume rate of flow, vA, is constant. Hence, in a level pipe, the fluid velocity increases at constrictions and decreases whereever the pipe expands in cross section.

15-6 Applications

>>> Example 5. Water pours from a ½ inch diameter house pipe at the rate of 7×10^{-3} ft³/s. (a) Find the velocity with which it emerges from the pipe. (b) If the supply pipe to this ½ inch pipe is 2 inches in diameter what is the velocity in the supply pipe? (c) What is the gauge pressure in the supply pipe at the same height as the opening?

(a) The volume rate of flow is

$$vA = 7 \times 10^{-3} \text{ ft}^3/\text{s}$$

where the cross sectional area is

$$A = \pi \ (1/4)^2(1/12)^2 \text{ ft}^2 = 1.4 \times 10^{-3} \text{ ft}^2 \quad .$$

Thus

$$v = 5 \text{ ft/s} \quad .$$

(b) From Eq. (15-7b)

$$v_s A_s = vA \quad .$$

Thus, in the supply pipe the velocity is

$$v_s = v \ A/A_s = v \ \pi r^2/\pi r_s^2 = vd^2/d_s^2$$

so

$$v_s = v(½/2)^2 = (5 \text{ ft/s})/16 = 0.3 \text{ ft/s} \quad .$$

(c) Assuming we may treat this as SIIN flow, Bernoulli's equation applies. If p_s is the pressure in the supply pipe, then since the pressure at the ½ inch pipe opening is the atmospheric pressure p_o and both are at the same height

$$p_s + ½ \rho v_s^2 = p_o + ½ \rho v^2 \quad .$$

Thus

$$p_s = p_o + ½ \rho(v^2 - v_s^2) \quad .$$

The density of water is 1.94 slugs/ft³, so the gauge pressure is

$$p_s - p_o = ½ \ (1.94 \text{ slug/ft}^3)(5^2 - 0.3^2) \text{ ft}^2/\text{s}^2 = 24.2 \text{ lb/ft}^2 \quad .$$ <<<

>>> Example 6. Most light aircraft wings develop 75% of their lift from a decrease in pressure at the top of the wing and 25% from an increase in pressure at the bottom. The wing loading is the aircraft weight per unit wing area and a typical value is 15 lbs/ft². If the aircraft speed in level flight is 180 miles per hour what must be the airspeeds at the wing top and bottom? (Take the density of air to be 1.94×10^{-3} slug/ft³.)

The lift generated by the top surface is $(p - p_T)A$ where p is the normal air pressure; the lift generated by the bottom is $(p_b - p)A$. Thus since the top lift is 3 times that of the bottom

$$(p - p_T)\,A = 3\,(p_b - p)\,A \quad .$$

The total lift must equal the total weight in level flight, so

$$(p - p_T)\,A + (p_b - p)\,A = W \quad .$$

From these equations we have

$$p_b = p + \frac{1}{4}\frac{W}{A} \quad , \quad p_T = p - \frac{3}{4}\frac{W}{A} \quad .$$

From Bernoulli's equation

$$\tfrac{1}{2}\,\rho v_T^{\,2} + p_T = \tfrac{1}{2}\,\rho v^2 + p$$

and

$$\tfrac{1}{2}\,\rho v_b^{\,2} + p_b = \tfrac{1}{2}\,\rho v^2 + p$$

where v is the velocity of the air <u>relative to the airplane</u>. Thus

$$v_T^{\,2} = v^2 + \frac{2}{\rho}\,(p - p_T) = v^2 + \frac{2}{\rho}\,\frac{3}{4}\,\frac{W}{A}$$

and

$$v_b^{\,2} = v^2 - \frac{2}{\rho}\,(p_b - p) = v^2 - \frac{2}{\rho}\,\frac{W}{4A} \quad .$$

Now, $W/2\rho A$ is much smaller than v^2, so

$$v_T = v\left[1 + \frac{3W}{2\rho A v^2}\right]^{1/2} \simeq v\left[1 + \frac{3}{4}\,\frac{W}{\rho A v^2}\right]$$

and

$$v_b \simeq v\left[1 - \frac{W}{4\rho A v^2}\right] \quad .$$

From the given values (180 mph = 264 ft/sec)

$$\frac{W}{4\rho A v^2} = \frac{15}{4(1.94 \times 10^{-3})(264)^2} = 2.8 \times 10^{-2}$$

so

$$v_T \simeq 195 \text{ mph}$$

and

$$v_b \simeq 175 \text{ mph} \quad .$$

<<<

Chapter 16: WAVES IN ELASTIC MEDIA

REVIEW AND PREVIEW

Previously you studied oscillatory motion in which a particle oscillated back and forth (or circularly) retracing its path at regular intervals. There was a single frequency possible depending upon the inertia and elastic properties of the system. If a system which permits such elastic vibrations is part of a larger medium all of which has this property, a disturbance at one point in the medium shows up sooner or later as a disturbance elsewhere. There is a transmittal of energy without net movement of the system as a whole; such a phenomenon is called a wave. In this chapter you will study mechanical waves which require an elastic medium. Both the elastic and inertia properties are distributed throughout the medium. In studying oscillations you concentrated on single harmonic oscillations; here you will concentrate on simple harmonic waves in which the parts of the medium execute simple harmonic motion.

GOALS AND GUIDELINES

In this chapter you have four major goals:

1. Learning the vocabulary and definitions. As with any new area there are certain terms defined and some new vocabulary to be learned.
2. Understanding the form of simple harmonic waves. You should pay particular attention to being able to identify the physical terms in the expression given in Fig. 16-2. Notice the parallel with the corresponding form of simple harmonic motion.
3. Learning how to relate wave velocity to the inertial and elastic properties of a simple medium: an elastic string.
4. Learning and understanding the phenomenon of interference of waves. The forms of such interference are given in general form which does not depend upon the exact nature of the waves. A little extra time spent here is well invested as you will study this phenomenon again in optics. The form of the interference conditions will be the same in both cases.

16-1 Wave Classification

Waves in elastic or deformable media are mechanical waves originating in disturbances of the media from its normal equilibrium position. Waves of this sort transmit energy by transferring it from one portion of the medium to another; the medium as a whole does not move, but various portions execute oscillatory motion about their normal equilibrium positions. The speed with which a disturbance moves through a medium is determined by the elasticity of the medium which provides a restoring force against the disturbance, and by an inertia property of the medium which dictates how the medium responds to this restoring force.

Elastic or mechanical waves are classified by their spatial and temporal properties. The spatial classification depends upon the direction in which the particles of the medium move relative to the wave direction. There are two extremes:

1. Transverse waves: particles of the medium move perpendicular to the wave direction -- example, waves in a string (see Fig. 16-1a);
2. Longitudinal waves: particles of the medium move parallel to the wave direction -- example, waves in a coiled spring (see Fig. 16-1b).

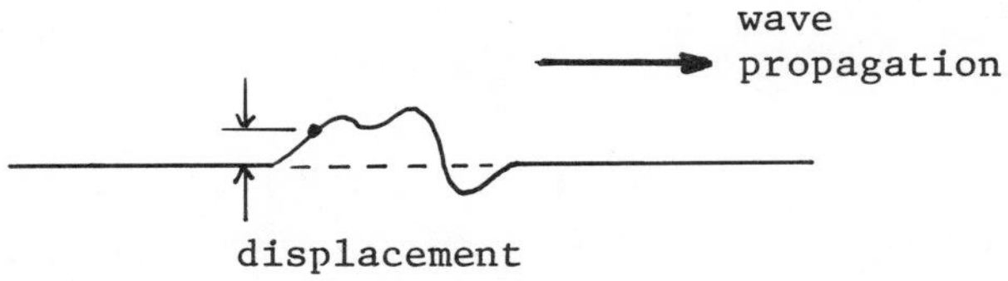

(a)

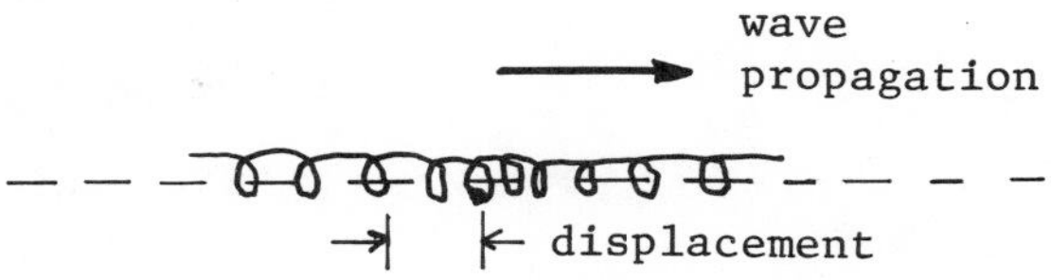

(b)

Figure 16-1. (a) Illustration of a transverse wave in a string; (b) Illustration of a longitudinal wave in a spring.

The spatial classification also includes whether the waves are in one, two or three dimensions. The temporal or time property depends upon how the particles in the medium behave in time as the wave moves through the medium. Waves may be:

1. A single pulse or single wave -- each particle of the medium remains at rest until the pulse arrives, is disturbed, and then returns to its equilibrium position;
2. A wavetrain -- more than a single pulse, usually a continuous disturbance of the particles;
3. A periodic train -- each particle of the medium undergoes periodic motion; the simplest case of this would be simple periodic motion associated with a simple harmonic wavetrain.

For waves in three dimensions there exists the concept of wavefronts which are found by constructing the surface which passes through all points undergoing a similar disturbance at a given instant. In time, the surface moves and perhaps distorts, showing how the wave propagates. For periodic waves there exist families of surfaces (they have a common normal -- i.e. a common direction perpendicular to the wavefronts). For a homogeneous, isotropic medium the direction of propagation is perpendicular to the wavefront and the normal direction is called a ray. An isotropic medium is one which "looks" the same in all directions.

16-2 Traveling Waves -- One Dimension

If y is the disturbance of a wave (e.g., the transverse displacement of a stretched string, the displacement from equilibrium of a coiled spring) and if

$$y = f\ (x \pm vt + \psi) \tag{16-1}$$

where f denotes some functional dependence, then Eq. (16-1) represents a wave traveling to the right (minus sign) or left (plus sign). In Eq. (16-1), ψ is an arbitrary constant. The actual shape of the wave is dictated by the functional form symbolized by f. The argument of the function $(x \pm vt + \psi)$ is called the phase. The phase velocity is the velocity with which a point of constant phase of the wave moves. Thus

$$x \pm vt + \psi = \text{constant}$$

so the phase velocity is

$$dx/dt = \mp v \quad . \tag{16-2}$$

The displacement of a simple harmonic traveling wave has the form

$$y = y_M \sin[2\pi(x \mp vt + \psi)/\lambda] \quad . \tag{16-3}$$

The functional form, the f of Eq. (16-1), is a sine or cosine function. The amplitude of the wave is y_M; it is intrinsically positive and gives the maximum positive value of y. The phase is

$$2\pi(x \mp vt + \psi)/\lambda$$

and λ is the wavelength, which is the distance between two adjacent points of the wave-train having the same phase at the same time. Alternatively the wavelength is defined by

$$y(x + n\lambda, t) = y(x,t) \quad , \quad n = 0,1,2, \ldots . \tag{16-4}$$

The period, T, is the time required for the wave to travel one wavelength, so

$$\lambda = vT \quad . \tag{16-5}$$

An alternative definition is

$$y(x, t + NT) = y(x,t) \quad , \quad N = 0,1,2, \ldots . \tag{16-6}$$

The wave number k is defined by

$$k = 2\pi/\lambda \quad . \tag{16-7}$$

The frequency, ν, of the wave is

$$\nu = 1/T \tag{16-8a}$$

and the angular frequency, ω, is

$$\omega = 2\pi\nu = 2\pi/T \quad . \tag{16-8b}$$

The phase velocity may thus be written as

$$v = \omega/k = \lambda/T \tag{16-9}$$

so the simple harmonic traveling wave displacement may be written as

$$y = y_M \sin(kx \mp \omega t + \phi) \tag{16-10}$$

where the phase constant has been written as $\phi = 2\pi\psi/\lambda$. In Fig. 16-2 we show Eq. (16-10) together with the names of the various parts.

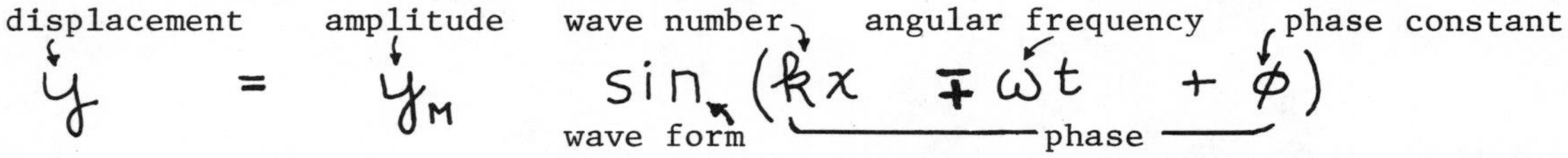

Figure 16-2. The parts of a standard description of a simple harmonic traveling wave in one dimension; the minus sign in front of ω is for wave propagation in the positive x direction while the positive sign is for propagation in the negative x direction.

>>> Example 1. A transverse wave in a string has the form

$$y = 5 \cos[\pi(0.02x + 3.00t)]$$

where y and x are expressed in centimeters and t is in seconds. (a) Write the equation in "standard" form and find the (b) amplitude (c) frequency (d) period (e) wavelength (f) phase constant (g) wave propagation direction and (h) maximum transverse speed of a particle in the string.

(a) To reduce the equation to "standard" form we need to write $y = y_M \sin(kx + \omega t + \phi)$. We identify, $k = .02\pi$, $\omega = 3.0\pi$ and write

$$y = 5 \sin(.02\pi x + 3.0\pi t + \phi)$$

$$= 5 \sin[\pi(0.02x + 3.00t)]\cos\phi + 5 \cos[\pi(0.02x + 3.00t)]\sin\phi \quad .$$

Thus $\sin\phi = 1$ and $\cos\phi = 0$, so (f) the phase constant is $\phi = \pi/2$, and the "standard" form is

$$y = 5 \sin[\pi(0.02x + 3.00t + 0.5)] \quad .$$

By comparing this with Eq. (16-10) we see that

(b) the amplitude = 5 cm.

(c) $\omega = 2\pi\nu = 3.00\pi \quad , \quad \nu = 1.50\ s^{-1}$.

(d) $T = 1/\nu = 2/3$ s .

(e) $k = 2\pi/\lambda = .02\pi \quad , \quad \lambda = 100$ cm .

(g) The wave propagation direction is along the negative x direction which is evident since x and t have the same sign.

(h) The transverse velocity is

$$dy/dt = (3\pi)\, 5 \cos[\pi(0.02x + 3.00t + 0.5)]$$

which has a maximum value of 15π cm/s . <<<

16-3 Superposition Principle -- One Dimensional Wave Equation -- Interference

The propagation of mechanical waves (indeed all waves) is governed by a differential equation which relates certain space and time derivatives (see Example 2). This differential equation is called the wave equation. The wave equation is said to be linear if it has the property that if y_1 is a solution and y_2 another then $A_1y_1 + A_2y_2$ is also a solution where A_1 and A_2 are arbitrary constants. In particular, if y_1 is the displacement of a portion of the medium due to the wave numbered one and y_2 that of wave two, then the total displacement,

$$y = y_1 + y_2$$

is also a solution of the wave equation. This is known as the superposition principle. Physically this means that two or more waves can traverse the medium independent of one another.

One important consequence of the superposition principle is that when it holds, a complicated wave can be expressed and analyzed as a linear combination of simple harmonic waves with varying wave numbers, amplitudes, and phase constants.

> For elastic media, the superposition principle holds when the amplitude of the disturbance is sufficiently small so that the restoring force is proportional to displacement.

>>> Example 2. Show that the wave equation for transverse waves in a string of tension F and mass per unit length μ is

$$\partial^2 y/\partial x^2 - (\mu/F)\partial^2 y/\partial t^2 = 0 \qquad (16\text{-}11)$$

for small displacements.

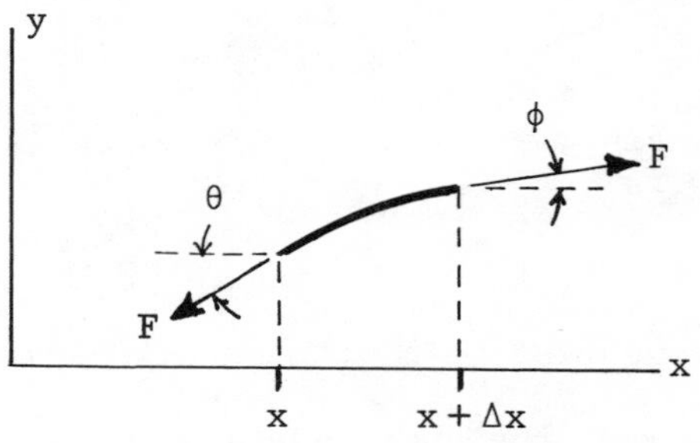

Figure 16-3. Example 2. A small segment of a transverse wave in a string with tension F.

First of all, the amplitude must be sufficiently small so that the tension does not change. Then, in Fig. 16-3 we show a small segment of the string. The tension is directed tangentially to the string and so the x component of this force at x + Δx is F cos ϕ while at x it is - F cos θ. The amplitude must be sufficiently small so that cos $\theta \simeq$ cos $\phi \simeq 1$ (i.e., θ and ϕ small) in order that the string segment be accelerated only in the y direction. In this case

$$\sin\theta \simeq \tan\theta$$

and

$$\sin\phi \simeq \tan\phi \quad .$$

But tan θ is the slope at x and tan ϕ that at x + Δx, so

$$\sin\theta \simeq (\partial y/\partial x)_x$$

and

$$\sin\phi \simeq (\partial y/\partial x)_{x+\Delta x}$$

where $(\ )_x$ means evaluated at x and similarly for $(\ )_{x+\Delta x}$. The total force in the y direction is

$$F\sin\phi - F\sin\theta = F\,[(\partial y/\partial x)_{x+\Delta x} - (\partial y/\partial x)_x] \ .$$

Now we process the term on the right side of the equation. If G is a function of x, the definition of a derivative is

$$G(x + \Delta x) - G(x) = (\partial G/\partial x)_x\ \Delta x$$

for small Δx. Here $G(x)$ is $\partial y/\partial x$ so we have

$$(\partial y/\partial x)_{x+\Delta x} - (\partial y/\partial x)_x = (\partial^2 y/\partial x^2)_x \, \Delta x \quad .$$

This is the bracketed term in our original equation and therefore the net force in the y direction is

$$F(\partial^2 y/\partial x^2)_x \, \Delta x \quad .$$

From Newton's second law this is equal to the mass of the segment times the acceleration which is

$$\underbrace{\mu\Delta x}_{\text{mass}} \times \underbrace{\partial^2 y/\partial t^2}_{\text{acceleration}} \quad .$$

Thus Newton's second law yields

$$F(\partial^2 y/\partial x^2)\,\Delta x = \mu\Delta x\,\partial^2 y/\partial t^2$$

so the wave equation is

$$\partial^2 y/\partial x^2 - (\mu/F)\partial^2 y/\partial t^2 = 0 \quad .$$

<<<

>>> Example 3. For the one dimensional wave equation above show that the wave speed is

$$v = \sqrt{F/\mu} \quad . \tag{16-12}$$

The velocity does not depend upon the wave form so let us use a simple harmonic wave

$$y = y_M \sin\,(kx - \omega t + \phi) \quad .$$

Then

$$\partial y/\partial x = k y_M \cos\,(kx - \omega t + \phi)$$

and

$$\partial^2 y/\partial x^2 = -\,k^2 y_M \sin\,(kx - \omega t + \phi) = -\,k^2 y \quad .$$

Similarly

$$\partial^2 y/\partial t^2 = -\,\omega^2 y \quad .$$

Substituting these results into the wave equation we have

$$-\,k^2 y + (\mu/F)\,\omega^2 y = 0 \quad .$$

Therefore

$$\omega^2/k^2 = v^2 = F/\mu$$

and since v is intrinsically a positive number

$$v = \sqrt{F/\mu} \ . \qquad <<<$$

>>> Example 4. One end of a rope of length 4 m and mass 0.12 kg is moved up and down a distance of 1 cm four times per second. If the tension in the rope is 1000 N what is the speed of propagation characteristic of this rope?

This problem is typical of what you'll often see -- it contains a lot of extraneous information! To find v we need *only* the tension and mass per unit length. The mass per unit length is

$$\mu = (0.12 \text{ kg})/(4 \text{ m}) = 0.03 \text{ kg/m}$$

and the tension is given as 1000 N. Thus

$$v = \sqrt{F/\mu} = \sqrt{(1000 \text{ N})/(0.03 \text{ kg/m})} = 182 \text{ m/s} \ . \qquad <<<$$

Interference is a physical effect of superposition. Of particular importance is the case of two traveling waves having the same frequency but different amplitudes and a phase difference. If

$$y_1 = A_1 \sin (kx - \omega t - \phi)$$

and

$$y_2 = A_2 \sin (kx - \omega t)$$

then the resultant is $y = y_1 + y_2$. And we write this as

$$y = y_1 + y_2 = A \sin (kx - \omega t - \alpha)$$

and find that the resultant amplitude is given by

$$A^2 = A_1{}^2 + A_2{}^2 + 2 A_1A_2 \cos \phi \ . \qquad (16\text{-}13)$$

This resultant amplitude is a maximum of $A_1 + A_2$ when $\phi = 0$ or an even multiple of π; the waves are said to interfere *constructively*. If $\phi = \pi$ or any odd multiple of π, the resultant is a minimum $A = |A_1 - A_2|$ and the waves interfere *destructively*.

Often, two wavetrains originate in a common source but a phase difference occurs because they follow different paths to the point of interference. A path difference Δx results in a phase difference $k\Delta x$ since the phase is $kx - \omega t + \phi$. *If this phase difference is an odd multiple of π the waves interfere destructively and if it is an even multiple of π they interfere constructively.* If there are no other sources of phase difference then this condition may be expressed in terms of the path difference:

$$\left.\begin{aligned} k\Delta_{path} &= (2\pi/\lambda)\ \Delta_{path} = (2n + 1)\ \pi \\ \text{or} \qquad \Delta_{path} &= \frac{2n + 1}{2}\ \lambda \end{aligned}\right\} \text{destructive interference} \qquad (16\text{-}14a)$$

$$\left.\begin{aligned} k\Delta_{path} &= (2\pi/\lambda)\ \Delta_{path} = (2n)\ \pi \\ \text{or} \qquad \Delta_{path} &= n\lambda \end{aligned}\right\} \text{constructive interference} \qquad (16\text{-}14b)$$

where $n = 0,1,2, \ldots.$ and there are no other sources of phase difference.

16-4 Power and Intensity in Wave Motion

The power transmitted by a wave is the rate at which the wave transports energy through the medium. Consider a string in which there is a transverse wave. We imagine that the string is divided into two parts, and then ask what is the rate at which energy is transmitted from the left part of the string to the right part. At this (mental) division of the string the transverse force exerted on the right portion of the string by the left portion is (see Example 2)

$$- F\, \partial y/\partial x$$

where F is the string tension. The transverse velocity is $\partial y/\partial t$. Since power may be expressed as $\underline{\underline{F}}\cdot\underline{\underline{v}}$ we have that the rate at which the left portion transmits energy to the right portion is

$$P = - F\, (\partial y/\partial x)(\partial y/\partial t) \quad .$$

For a simple harmonic wavetrain moving in the ± x direction,

$$y = y_M \sin\, (kx \mp \omega t + \phi)$$

and

$$P = \pm\, Fk\omega y_M^2 \cos^2\, (kx \mp \omega t + \phi) \quad .$$

This oscillates with time and one often wants the average over one period T which is

$$\bar{P}_T = \frac{1}{T} \int_t^{t+T} P\, dt$$

or

$$\bar{P}_T = \pm\, \tfrac{1}{2}\, y_M^2 \omega^2 \mu v \quad .$$

The + sign occurs when the wave moves to the right (positive x direction) and therefore the left portion is doing work <u>on</u> the right portion. If the wave moves to the left (minus x direction) the right portion does work on the <u>left</u> portion. In other words, <u>power is transmitted in the direction of wave propagation</u>.

In three dimensions, the <u>wave intensity</u> is the average power transmitted per unit area perpendicular to the wave propagation direction.

16-5 Standing Waves, One Dimension -- Resonance

Consider two simple harmonic wavetrains of the same amplitude and frequency traveling in opposite directions. Suppose they have the forms

$$y_1 = A \sin\, (kx - \omega t)$$

$$y_2 = A \sin\, (kx + \omega t) \quad .$$

The resultant displacement is

$$y = y_1 + y_2 = 2\ A \sin kx \cos \omega t \tag{16-15}$$

which is not a traveling wave since the x and t dependences are separated. This is a standing wave. All portions of the medium simultaneously execute simple harmonic motion with an amplitude that depends upon position. The amplitude of the SHM at x is

$$A_x = 2\ A \sin kx \tag{16-16}$$

which is a maximum at the antinodes given by

$$\left.\begin{aligned} kx &= [(2n + 1)/2]\ \pi \\ \text{or} \quad & \\ x &= [(2n + 1)/4]\ \lambda \end{aligned}\right\} \text{positions of antinodes } n = 0,1,2, \ldots . \tag{16-17a}$$

Also, A_x is zero at the nodes given by

$$\left.\begin{aligned} kx &= n\pi \\ \text{or} \quad & \\ x &= n\lambda/2 \end{aligned}\right\} \text{positions of nodes } n = 0,1,2, \ldots . \tag{16-17b}$$

There is no transport of energy in a standing wave because of the nodes; the energy remains standing, being sometimes kinetic and other times potential.

For standing waves in a string, a fixed end must be a node and a free end an antinode. If the standing wave is regarded as an incident traveling wave and a reflected one (such as we started with above),

i) at a fixed end the incident and reflected waves must differ in phase by π,
ii) at a free end the incident and reflected waves must be in phase.

If a system, which is capable of oscillating, is driven by a periodic force it oscillates at the frequency of the driver. In general, no matter how hard it is driven it doesn't oscillate with much amplitude. If, however, the frequency of the driver approaches a natural frequency of the system, it becomes easier for the system to oscillate and the amplitude becomes relatively larger for a given driving force amplitude. When the driver frequency approaches a natural frequency of the system, we speak of resonance.

The natural, proper, or eigen-frequencies of a string fixed at both ends are those frequencies such that the corresponding standing wave has nodes at each end. Since two successive nodes are separated by $\lambda/2$, (see Eq. (16-17b)) the string length ℓ must be an integral multiple of $\lambda/2$

$$\ell = n\ \lambda/2 \quad , \quad \text{fixed ends} \tag{16-18}$$

where n is the number of nodes less one (i.e., the node at one end is not counted). Thus the natural or eigen-wavelengths are

$$\lambda = 2\ell/n \quad , \quad \text{fixed ends,} \quad n = 1,2,3,\ldots \tag{16-19}$$

The corresponding eigen-frequencies are determined from λ by the tension and mass per unit length since $\lambda = v/\nu = (1/\nu)(\sqrt{F/\mu})$. Therefore the eigen-frequencies are

$$\nu = (n/2\ell)\sqrt{F/\mu} \quad , \quad \text{fixed ends} \; , \quad n = 1,2,3,\ldots \qquad (16\text{-}20)$$

>>> Example 5. What are the natural wavelengths of a string fixed at one end and effectively free at the other?

A node and an antinode are separated by 1/4 wavelength (see Eqs. (16-17)). Since the fixed end must be a node and the free end an antinode, the length of the string is related to λ by

$$\ell = n\,\lambda/2 + \lambda/4 = [(2n+1)/4]\lambda \quad .$$

Thus

$$\lambda = 4\ell/(2n+1) \quad , \quad \text{one free end} \; , \quad n = 0,1,2,\ldots \qquad (16\text{-}21)$$

<<<

16-6 Programmed Problems

As usual we wish to describe physical phenomena in mathematical terms. The next few frames will attempt to convince you that a particular functional form, i.e. $f(x - vt)$, is correct for traveling waves. We can think of the familiar "bump" that propagates along a rope when you flip one end.

1. We will begin by choosing the function

$$y = e^{-x^2}$$

to describe the y coordinate of portions of a rope located at various x's. In other words, at t = 0 this equation tells us what the shape of the rope is over all its length. We need some data to sketch the rope. Complete the table below. Included at the left is a table to help you.

n	e^{-n}
0	1
1	0.37
2	0.14
3	0.05
4	0.02
5	0.007
6	0.002
7	0.001
8	0.0003
9	0.0001

x	x^2	e^{-x^2}
0		
1,-1		
2,-2		
3,-3		

x	x^2	$y = e^{-x^2}$
0	0	1
1,-1	1	0.37
2,-2	4	0.02
3,-3	9	0.0001

This table relates the y coordinate of a portion of the rope at various positions x along the rope.

2. Plot $y = e^{-x^2}$ from the data of the previous frame. What does the rope look like?

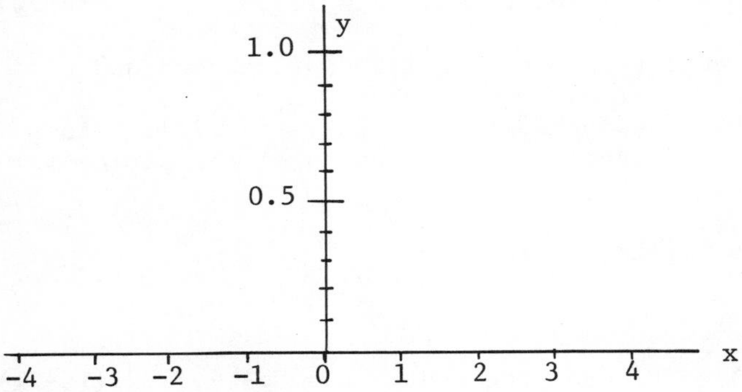

This is the "bump" on the rope at $t = 0$. Note that y is practically zero for all values of x larger than ± 3.

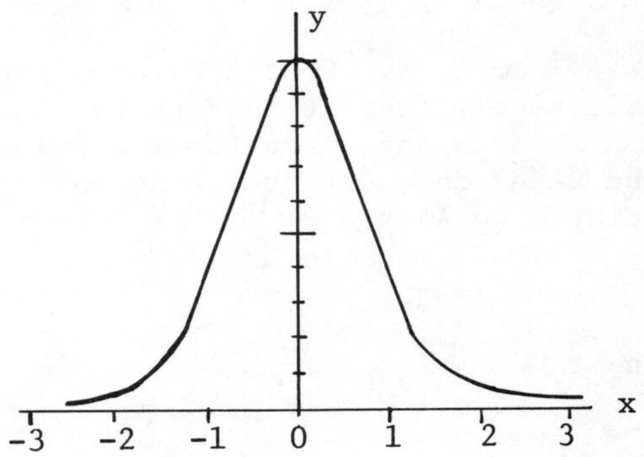

3. The rope has a bump. We seek an expression which will cause the bump to move (propagate).

If we write

$$y = e^{-(x-vt)^2}$$

this will at least be correct at $t = 0$, i.e. $y = e^{-x^2}$. Experience indicates that the speed v is constant for a given rope configuration. Complete the table below to determine the shape of the rope at $t = 2$ sec. Assume $v = 2$ m/s. Refer to the "help" table of frame 1 as required.

x meters	x-vt	$(x-vt)^2$	$y = e^{-(x-vt)^2}$
1			
2			
3			
4			
5			
6			
7			

We include only the values of x as shown because all other x's result in a y of essentially zero.

x	x-vt	$(x-vt)^2$	$y=e^{-(x-vt)^2}$
1	-3	9	0.0001
2	-2	4	0.02
3	-1	1	0.37
4	0	0	1.0
5	1	1	0.37
6	2	4	0.02
7	3	9	0.0001

4. Plot the data of the previous answer on the graph below. Included here is the shape of the rope at t = 0.

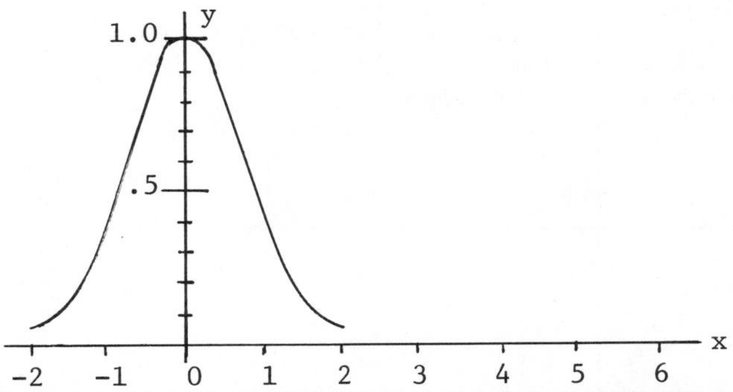

The bump has moved to the right!! Also the peak "phase" of the bump has moved a distance (2 m/s)(2 s) = 4 m.

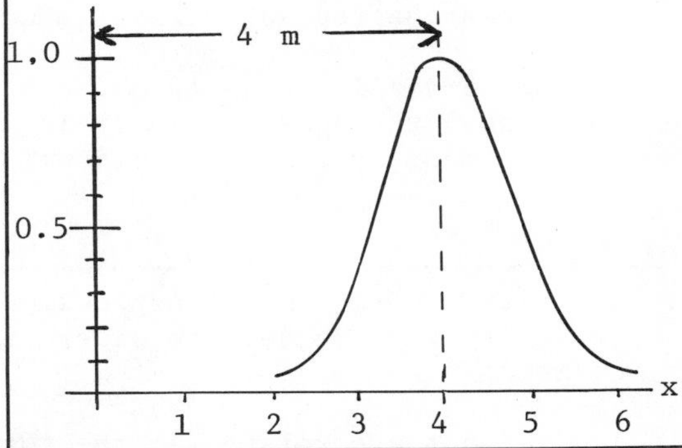

5. We could continue the above process for t = 3, 4, 5 s etc., but let us presume that our equation is correct for a "bump" traveling to the right. Suppose we started with the slightly different function

$$y = e^{-(x+vt)^2} \quad .$$

Describe the behavior of this disturbance.

Same as the previous one, but traveling to the left. You can check this yourself by sketching the curve again as in frames 2 and 4.

6. Perhaps you are now slightly convinced that the variables x and t must occur only in the combination $(x \pm vt)$ where v is constant. One more example is in order.

Above we show a very long rope attached at one end to a spring. Having attached the rope to m at x = 0 makes the description of y at that point particularly easy. If we lift the spring up and then release it

$$y(0,t) = \underline{\qquad} .$$

Here we mean the motion of that part of the rope at x = 0.

The motion of a mass on the end of a spring is

$y = y_m \cos \omega t$

or

$y = y_m \cos 2\pi\nu t$

since $\omega = 2\pi\nu$; ν is a constant depending upon the mass and the spring. Of course, the rope is attached to the mass at x = 0 so the end of the rope receives a transverse harmonic displacement

$$y(0,t) = y_m \cos 2\pi\nu t \quad .$$

7. Wiggling the end of the rope in this harmonic fashion will cause a disturbance to propagate along the rope. We want to generalize our expression in the previous answer so that it describes the rope at all x including $x = 0$.

For any traveling wave then (including an harmonic one) the equation involving the variables x and t must be of the functional form

$$y = f(\_\_\_\_\_\_) .$$

Answer: $y = f(x - vt)$

We can surely guess that our equation will be the cosine of something involving x and t. The behavior at $x = 0$ provides this clue.

---

8. To write a more general expression for the traveling wave we need to adhere to two conditions:

1. x and t are related in the form $(x - vt)$ where v is the constant wave speed.
2. The general expression must reduce to $y = y_m \cos 2\pi\nu t$ at $x = 0$.

Another way of stating condition one is:

"coefficient of t = − v times the coefficient of x".

From 2 above what do we require as the coefficient of t?

Answer: The coefficient of t is $2\pi\nu$.

---

9. One more step remains. We require that the coefficient of t = − v times coefficient of x, or $2\pi\nu = -v$ times coefficient of x.

The coefficient of x must be ______ .

Answer: $2\pi\nu = -v\ (-2\pi\nu/v)$.

The coefficient of x must be $-2\pi\nu/v$.

---

10. So our expression to satisfy the functional form is

$$(2\pi\nu t - 2\pi\nu x/v)$$

or

$$2\pi\nu\ (t - x/v) .$$

Can you now write the complete expression for the harmonic traveling wave?

$$y(x,t)\ \_\_\_\_\_\_\_ .$$

Answer: $y = y_m \cos [2\pi\nu\ (t - x/v)]$

Note that at $x = 0$, $y = y_m \cos 2\pi\nu t$ as required. Also, x and t are related correctly.

11. There are a variety of ways to rewrite

$$y = y_m \cos [2\pi\nu\ (t - x/v)]\ .$$

For example, we can factor v and write

$$y = y_m \cos [2\pi\ (\nu/v)\ (vt - x)]\ .$$

Also since $\cos(-\theta) = \cos\theta$ we can reverse the order of the terms in parentheses:

$$y = y_m \cos [2\pi\ (\nu/v)\ (x - vt)]\ .$$

Or we can put this in "standard" form:

$$y = y_m \sin [2\pi\ (\nu/v)\ (x - vt) + \pi/2]$$

since $\cos\alpha = \sin(\alpha + \pi/2)$.

What are the units of the constant ν/v in SI units?

$[\nu]$ = cycles/s = s^{-1}

$[v]$ = m/s

$$\left[\frac{\nu}{v}\right] = \frac{s^{-1}}{m/s} = m^{-1}\ .$$

We write $\nu/v = 1/\lambda$ where λ is the wave length or distance between two points having the same phase. Lambda (λ) has the units of length and is a constant for specific rope configurations.

---

12. Using the constant λ we have

$$y = y_m \sin [2\pi(x - vt)/\lambda + \pi/2]\ .$$

The period T is the time required for the wave to travel one wavelength, thus

$$\lambda/T = \_\_\_\_\_\_\ .$$

$\lambda/T = v$,

the speed of the wave.

---

13. Substitute for v from the previous answer into the traveling wave expression

$$y = y_m \sin [2\pi(x - vt)/\lambda + \pi/2]$$

$$y = y_m \sin [2\pi(\_\_\_\_) + \pi/2]\ .$$

$$y = y_m \sin [2\pi(x/\lambda - t/T) + \pi/2]$$

---

14. In frame 11 we had

$$\nu/v = 1/\lambda\ .$$

The frequency ν was that of the mass spring oscillator which drives the end of the rope. In frame 12 we had

$$\lambda/T = v\ .$$

From these two equations write an equation relating ν and T.

$v = \lambda\nu$; $v = \lambda/T$

so $\nu = 1/T$.

ν is the number of cycles per second while T is the number of seconds per cycle. As a check we note that

$$\frac{\text{cycles}}{\text{second}} = \frac{1}{\text{second/cycle}}$$

is dimensionally correct.

15. Using the answers of frames 13 and 14 write the specific equation for a traveling wave moving in the + x direction if the mass on the end of the spring vibrates with an amplitude 0.03 m at 60 Hz. The wave speed is 30 m/s.

$y =$ ________________ .

$y_m = 0.03$ m

$T = 1/\nu = (1/60)$ s

$\lambda = vT = 30 \text{ m/s} \times (1/60)$

$= 0.5$ m .

$y = y_m \sin [2\pi(x/\lambda - t/T) + \pi/2]$

$y = (0.03) \sin [4\pi x - 120\pi t + \pi/2]$

Here x,y are in meters and t is in seconds.

---

16. For traveling waves on ropes (or strings, etc.) we seek some way to determine the wave speed. Experience shows that the speed v depends upon the following:

1. The tension F in the rope, i.e. force;
2. The linear density μ of the rope, i.e. mass per unit length.

The unit of force in SI units is the newton which we can write as $kg \cdot m/s^2$. The linear density μ has the units kg/m. Show by dimensional analysis that the equation

$$v = \sqrt{F/\mu}$$

has the right units for the wave speed.

$v = \sqrt{F/\mu} = \sqrt{(kg \cdot m/s^2)/(kg/m)}$

$= \sqrt{m^2/s^2} = m/s$

which results in the proper units for speed. This dimensional analysis does not prove the equation, but it is correct.

Consider our original rope to be 10 meters long with a mass of 0.050 kg. One end of the rope is given a transverse motion with a frequency of 10 s^{-1}. If the tension in the rope is 200 N, what is the wave length of the resultant waves? Assume the rope to be long enough to ignore what happens at the far end.

17. Begin by calculating the wave velocity.

$v =$ ______ .

Obtain a numerical result.

$v = \sqrt{F/\mu}$

$\mu = 0.05$ kg/10 m

$\mu = 0.005$ kg/m

$F = 200$ N

$v = \sqrt{(200 \text{ N})/(0.005 \text{ kg/m})}$

$v = 200$ m/s .

18. We already have seen that the wave, say the peak of one bump, will move a distance λ as the driving force oscillates through one period, T.

$$\lambda = vT$$

$$\lambda = \_\_\_\_\_\_\_ \text{ m}$$

Obtain a numerical result.

$\lambda = vT$

$\lambda = 200 \text{ m/s} \times 0.1 \text{ s}$

$\lambda = 20 \text{ m}$.

---

19. We have been writing our wave in the form

$$y = y_m \cos[2\pi(x/\lambda - t/T)] \ .$$

By defining two quantities,

$$k = 2\pi/\lambda \quad [k = \text{wave number}]$$

and

$$\omega = 2\pi/T \quad [\omega = \text{angular frequency}]$$

we can write our wave in the form

$$y = y_m \cos(kx - \omega t) \ .$$

In the next sequence of frames we shall combine the following two waves:

$$y_R = y_m \cos(kx - \omega t)$$

and

$$y_L = y_m \cos(kx + \omega t) \ .$$

What is different about these two waves?

The wave y_R travels to the right, i.e. in the +x direction. The wave y_L travels to the left, i.e. in the −x direction. In all other respects, the two waves are identical.

20. The superposition principle states that the resultant displacement of the rope will be the sum of the displacements due to the separate waves.

$$y = y_R + y_L$$

$$y = y_m\ [\cos\ (kx - \omega t) + \cos\ (kx + \omega t)]\ .$$

Use the trigonometric identity

$$\cos\ (\alpha \pm \beta) = \cos\ \alpha\ \cos\ \beta \mp \sin\ \alpha\ \sin\ \beta$$

to expand the bracket term. Do not simplify, merely expand.

Answer:

$$y = y_m\ [\cos\ kx\ \cos\ \omega t + \sin\ kx\ \sin\ \omega t + \cos\ kx\ \cos\ \omega t - \sin\ kx\ \sin\ \omega t]\ .$$

---

21. Now simplify the previous answer.

y = ______ .

Answer: $y = 2\ y_m\ \cos\ kx\ \cos\ \omega t$.

---

22. Does the previous answer represent a traveling wave? Why?

Answer: No. It is not of the form $f(x - vt)$.

---

23. We can interpret this non-traveling wave as an amplitude times an oscillating term.

$$\underbrace{y = 2\ y_m\ \cos\ kx}_{\text{Amplitude}}\qquad \underbrace{\cos\ \omega t}_{\text{Oscillating term}}$$

Recalling that $k = 2\pi/\lambda$, what is the amplitude at the fixed values of x listed in the table below?

x	$\lambda/4$	$3\lambda/4$	$7\lambda/4$
$2\pi x/\lambda$			
$\cos(2\pi x/\lambda)$			
y			

Answer:

x	$\lambda/4$	$3\lambda/4$	$7\lambda/4$
$2\pi x/\lambda$	$\pi/2$	$3\pi/2$	$7\pi/2$
$\cos(2\pi x/\lambda)$	0	0	0
y	0	0	0

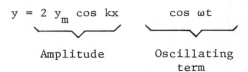

24. The previous answer implies that $y = 0$ at particular x's no matter what the value of the oscillating term. These points are permanently at rest and are called nodes. All other points on the rope have displacements which depend both upon their position x and the time t.

Will all non-node points oscillate with the same amplitude?

No.

The amplitude depends upon the the particular value of x.

$$y = \underbrace{2\,y_m \cos kx}_{\text{Amplitude}} \quad \underbrace{\cos \omega t}_{\text{Oscillating term}}$$

---

25. The amplitude term can have a maximum value of $\pm\, 2\, y_m$ wherever $\cos kx$ is $\pm\, 1$.

Using $k = 2\pi/\lambda$, which of the following will result in a maximum amplitude?

$$x = 0,\ \frac{\lambda}{4},\ \frac{\lambda}{2},\ \frac{3\lambda}{4},\ \lambda,\ \frac{5\lambda}{4},\ \frac{3\lambda}{2},\ \frac{7\lambda}{4},\ 2\lambda\ .$$

Encircle those that result in $y = \pm\, 2\, y_m$.

$$x = 0,\ \frac{\lambda}{2},\ \lambda,\ \frac{3\lambda}{2},\ 2\lambda\ .$$

---

26. These points of maximum displacement are called antinodes. At a node the string is permanently at rest. Is the string permanently at maximum displacement at an antinode?

No.

The displacement depends upon time. For example at $x = 0$,

$$y = 2\, y_m \cos \omega t\ .$$

At an antinode the string executes simple harmonic motion about the equilibrium position.

---

27. The waves on the rope described by the equation

$$y = 2\, y_m \cos kx \cos \omega t$$

are called <u>standing waves</u>.

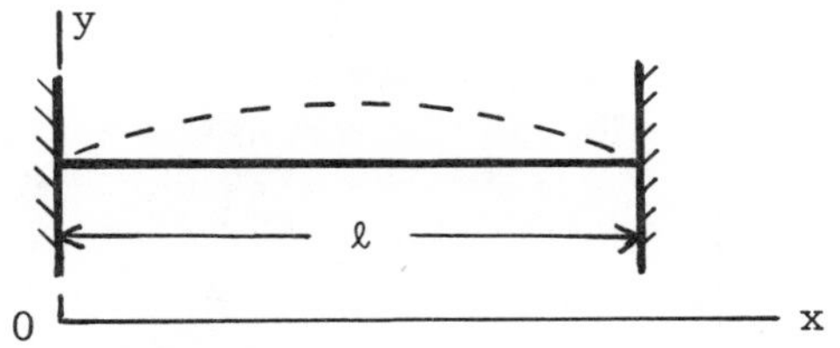

We could produce standing waves in the rope stretched between two fixed end points as shown above by plucking the string at the center. However, according to the equation above, the amplitude at the left end of the of the rope at $t = 0$ is

$y =$ ____________ .

Since $\cos \omega t = 1$ at $t = 0$, and $\cos kx = 1$ at $x = 0$, $y = 2\, y_m$ at $x = 0$, $t = 0$. This is by definition an antinode, but physically the fixed ends of the rope must be nodes.

28. Obviously our equation does not apply to this new situation. We could use instead

$$y = 2\ y_m \sin kx \cos \omega t$$

which would be correct provided the length of the rope satisfied the equation

$$\ell = n\lambda/2 \quad , \quad n = 1,2,3, \ldots .$$

Show that both ends will be nodes for a rope of length $\ell = 3\lambda/2$. Use the equation at the top of the frame.

We require $y = 0$ at $x = 0$ and $x = \ell$. In general

$$y = 2\ y_m \sin kx \cos \omega t \quad .$$

Then, since $\sin kx = 0$ at $x = 0$ and at $x = \ell = 3\lambda/2$

$$\sin kx = \sin[(2\pi/\lambda)(3\lambda/2)] = 0$$

29.

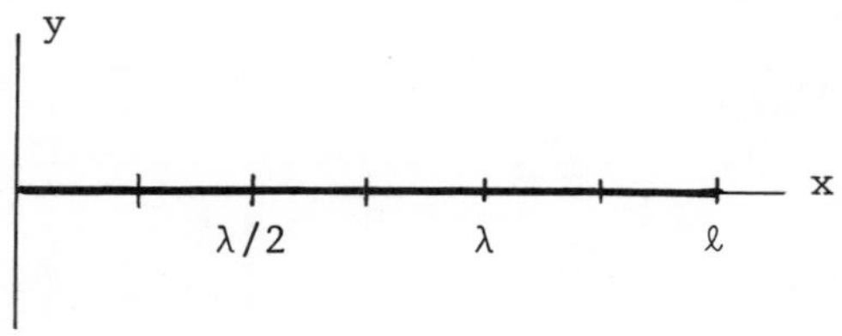

For the rope above our equation requires that both ends be nodes. Now, λ is the distance between points of a wave having the same phase. Sketch on the diagram above what you think the rope might look like at some instant t.

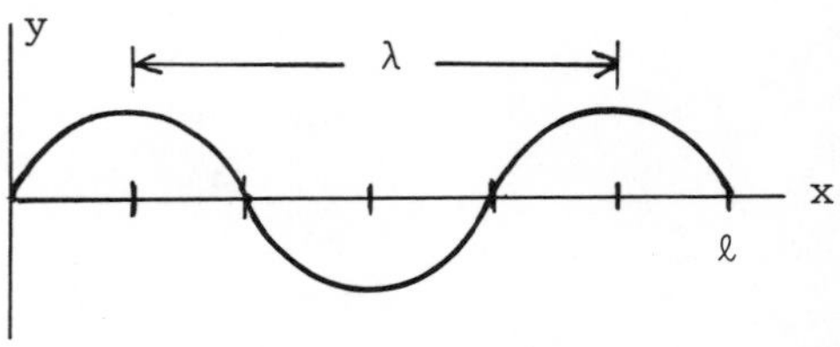

Both ends are nodes. Points of equal phase are one wave length apart. Also, note that adjacent nodes (as well as antinodes) are $\lambda/2$ apart. Your diagram might have a different amplitude or even be upside down.

30.

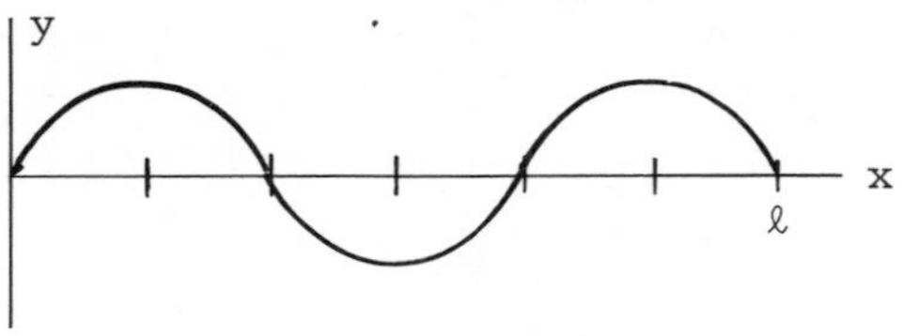

Superimpose on the diagram what you think the rope might look like at some later time t. Make it a very short time later.

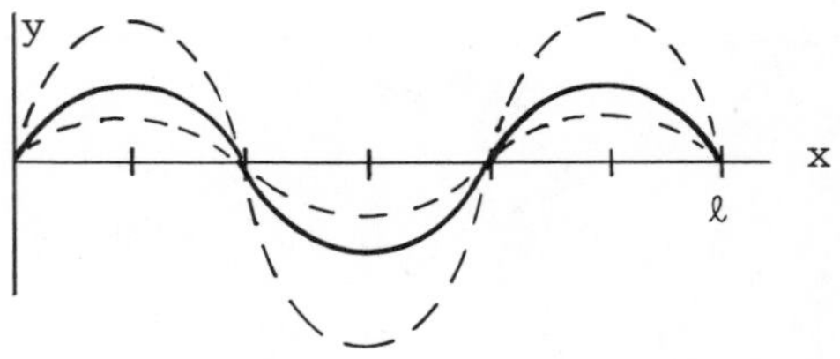

Either dashed curve will do. Note the nodes have not moved. Other parts of the rope are executing SHM at the frequency ω.

Chapter 17: SOUND WAVES

REVIEW AND PREVIEW

In the previous chapter you studied mechanical waves; most examples and illustrations dealt with the transverse waves found on a vibrating string. In this chapter you will deal with an example of longitudinal waves, namely sound waves. Only a little new vocabulary is introduced and two new phenomena, that of the Doppler shift and that of beats.

GOALS AND GUIDELINES

In this chapter you have three major goals:

1. Learning the special properties of sound waves. Basically this means learning to relate position variations to pressure variations.
2. Understanding the phenomena of standing sound waves and the temporal interference of sound waves which produce beats.
3. Learning and understanding the Doppler shift. Pay particular attention to the difference in the cause of the effect depending upon whether it is source or observer which moves.

17-1 Frequency Range of Longitudinal Waves

Sound waves are longitudinal mechanical waves which may be propagated in solids, liquids, and gases. What one usually calls sound waves are audible sound waves and these comprise only a small fraction of the frequency range possible for longitudinal mechanical waves. This range together with the appropriate names and some typical sources is shown in Fig. 17-1. The upper limit of 6×10^8 Hz should not be taken as fundamental, but merely reflects the present upper limit of man's efforts in this area. In all cases the generator of the wave compresses and rarefies the medium in (and opposite to) the direction of the wave propagation.

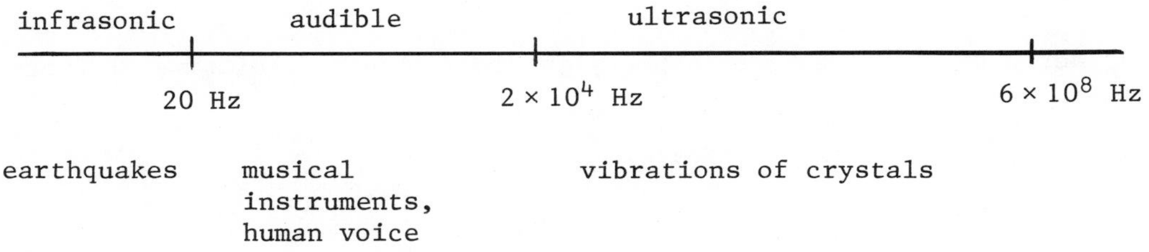

Figure 17-1. Present frequency range of longitudinal mechanical waves together with the names applied to certain ranges and typical sources.

17-2 Longitudinal Traveling Waves -- Wave Speed

The speed of a longitudinal wave is related to an inertial property of the medium, its density, and to a restoring force property of the medium which is called the bulk modulus of elasticity, B. This relationship is

$$v = \sqrt{B/\rho_0} \qquad (17\text{-}1)$$

where ρ_o is the normal density of the medium. One must realize that the longitudinal wave alternately compresses and rarefies the medium so that its density fluctuates about the normal value, ρ_o, but it is this normal value that determines the wave speed.

The bulk modulus of elasticity relates the change in pressure (due to the wave) to the corresponding change in volume by

$$B = -(\Delta p/\Delta V)\,V \quad . \tag{17-2}$$

For a gas one can show that

$$B = \gamma p_o \tag{17-3}$$

where p_o is the normal pressure (undisturbed by the wave) and γ is a constant typical of the gas (see Chapter 20).

If y represents the displacement from equilibrium (along the direction of wave propagation), then a simple harmonic traveling wave can have the form

$$y = y_M \sin\,(kx - \omega t + \phi) \quad . \tag{17-4}$$

For these waves one usually wants the pressure variations rather than displacements. From Eq. (17-2) $\Delta p = -\,B\,\Delta V/V$ is the change in pressure. It is usual to denote this change in pressure from the equilibrium value by p rather than Δp. Thus we use

$$p = -\,B\Delta V/V \quad , \quad p = \text{pressure variation.}$$

Now, if a layer of the medium has pressure p_o, thickness Δx and cross section A, its volume is $A\Delta x$. As the pressure changes, the volume changes by $A\Delta y$, so

$$\Delta V/V = (A\Delta y)/(A\Delta x)$$

or in the limit

$$\Delta V/V = \partial y/\partial x \quad .$$

(The partial derivative must be used since y is also a function of time.) Therefore one has

$$p = -\,B\partial y/\partial x \quad .$$

Thus we may say that if the position displacement is

$$y = y_M \sin\,(kx - \omega t + \phi)$$

the pressure displacement p is

$$p = -\,By_M k \cos\,(kx - \omega t + \phi)$$

$$= p_M \sin\,(kx - \omega t + \phi - \pi/2) \tag{17-5}$$

where the pressure amplitude p_M is

$$p_M = Bky_M = v^2\rho_o ky_M \quad . \tag{17-6}$$

Note:

i) the total pressure is $p + p_o$;
ii) the pressure displacement is out of phase with the displacement by 90° so that when $y = \pm y_M$, $p = 0$ and when $y = 0$, $p = \pm p_M$.

>>> Example 1. Show that the average intensity transmitted by a simple harmonic longitudinal wave is given by

$$\bar{I} = p_M{}^2/(2v\rho_o) = \tfrac{1}{2}\rho_o v\omega^2 y_M{}^2 \quad .$$

Consider an element of the medium of cross sectional area A perpendicular to the wave propagation direction. The net force on that segment is

$$F = pA$$

where p is the pressure displacement. Therefore, the power transmitted is the force times the velocity or

$$P = pA\ \partial y/\partial t$$

and since intensity is power per unit area

$$I = p\ \partial y/\partial t \quad .$$

From Eq. (17-4) which is an adequate form for a simple harmonic longitudinal wave

$$\partial y/\partial t = -\ \omega y_M \cos\ (kx - \omega t + \phi)$$

so from Eq. (17-5) we have

$$I = \omega y_M B y_M k \cos^2\ (kx - \omega t + \phi)$$

or from Eq. (17-1)

$$I = \omega y_M{}^2 k v^2 \rho_o \cos^2\ (kx - \omega t + \phi) \quad .$$

But $k = \omega/v$, so

$$I = \omega^2 y_M{}^2 v\rho_o \cos^2\ (kx - \omega t + \phi) \quad .$$

This is to be averaged over one period, since the average of $\cos^2$ is one-half,

$$\bar{I}_T = \tfrac{1}{2}\rho_o v\omega^2 y_M{}^2 \quad .$$

But from Eqs. (17-6) and (17-1) together with $k = \omega/v$, $y_M = p_M/(v\rho_o\omega)$.

Thus

$$\bar{I}_T = p_M{}^2/(2v\rho_o) \quad .$$ <<<

>>> Example 2. If a simple harmonic sound wave in air has an average intensity of 5×10^{-3} joule per second per square meter, what is the (a) pressure amplitude and (b) displacement amplitude at 20 cps, (c) pressure amplitude and (d) displacement amplitude at 20,000 Hz?

The velocity of sound in air is about 330 m/s and normal air density is about 1.2 kg/m^3. Therefore

$$p_M^{\ 2} = 2(330)(1.2)(5 \times 10^{-3})\ N^2/m^4$$

or

$$p_M = 1.99\ N/m^2 \quad .$$

This is independent of frequency, so it is the answer to both (a) and (c). The displacement amplitude follows from Eq. (17-6) as

$$y_M = p_M/(2\pi v \rho_o \nu)$$

or

$$y_M = (8 \times 10^{-4}/\nu)\ m$$

where ν is in Hz. For (b) ν = 20 Hz and

$$y_M = 4 \times 10^{-5}\ m$$

while for (d) ν = 20,000 Hz and therefore

$$y_M = 4 \times 10^{-8}\ m \quad .$$

<<<

17-3 Standing Longitudinal Waves -- Sources of Sound

All sources of sound consist of some object which vibrates. These vibrations cause pressure variations in the surrounding medium (usually air) resulting in sound waves at the same frequency. In other words, the frequency of the sound is determined by the source and the corresponding wavelength of the sound wave is determined by the velocity of sound in the medium and the frequency. The wavelength of the associated wave in the source is determined by the velocity of waves in the source and need not be the same as that of the sound wave. This is illustrated in Fig. 17-2 in which ν_s denotes the frequency of the waves in the source, λ_s their corresponding wavelength and v_s the velocity of waves in the source. Similarly, ν, λ and v denote the frequency, wavelength and velocity respectively of sound waves in the medium.

Figure 17-2. Schematic indication of a sound source in which the waves of frequency ν_s produce sound. They have velocity v_s which is characteristic of the source; the corresponding wavelength in the source is $\lambda = v_s/\nu_s$. In the medium, the sound waves have frequency $\nu = \nu_s$ and travel with velocity v which is characteristic of the medium; their wavelength is $\lambda = v/\nu_s$.

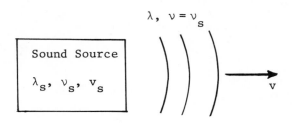

As far as the source is concerned, the waves in it are approximately standing waves -- the eigenfrequencies. Apart from percussion instruments (drums and the like) musical instruments are broadly classified as wind instruments or stringed instruments. Wind instruments are basically a vibrating membrane coupled to a vibrating air column and as such are, in the simplest sense, distorted organ pipes. The human vocal system is such an instrument. Stringed instruments consist of a vibrating string coupled to a resonant air chamber, but in the simplest sense one can consider the string only. At the introductory level then one considers as sources of sound

1) standing longitudinal waves -- the vibrating air column -- i.e., the organ pipe

2) standing transverse waves in a string.

If one has standing longitudinal waves in a tube there will be nodes and antinodes for both displacement and pressure. We tabulate these in Table 1.

TABLE 17-1

End of Tube	Pressure	Displacement
Closed	antinode	node
Open	node	antinode

Table 17-1. Tabulation of pressure and displacement nodes and antinodes for a longitudinal wave in a pipe.

Organ pipes are examples of such tubes for longitudinal waves. The waves in such a pipe are approximately standing. If they were exactly so, no sound would emanate. Organ pipes resonate at their natural frequencies. If such a pipe is open at both ends then there must be displacement antinodes or pressure nodes at the ends. Since successive nodes or antinodes are separated by one-half wavelength (see Eqs. (16-17)), the pipe length must be an integral multiple of a half wavelength. Thus if L is the pipe length

$$L = n\lambda/2$$

or, the proper wavelengths and frequencies are

$$\lambda_n = 2L/n \qquad n = 1,2,3 \ldots. \quad \text{pipe open at both ends} \tag{17-7a}$$

$$\nu_n = vn/(2L) \tag{17-7b}$$

where v is the speed of sound in air. The lowest value of ν_n is called the fundamental frequency and the others are called overtones. Overtones whose frequencies are integral multiples of a fundamental together with the fundamental are said to form a harmonic series. The quality of sound produced by a musical instrument is dictated by the number of overtones and their relative amplitudes. The fundamental and first three overtones are illustrated in Fig. 17-3a for an organ pipe open at both ends.

If the organ pipe is closed at one end, the closed end must be a displacement node or pressure antinode.

Since a node and the next antinode are separated by 1/4 wavelength (see Eqs. (16-17)) and successive nodes are separated by 1/2 wavelength, the length L of an organ pipe closed at one end is related to the wavelength by

$$L = n\lambda/2 + \lambda/4$$

so the eigen-wavelengths and eigenfrequencies are

$$\lambda_n = 4L/(2n + 1) \qquad (17\text{-}8a)$$

n = 0,1,2 ....
pipe closed at one end.

$$\nu_n = (2n + 1)v/(4L) \qquad (17\text{-}8b)$$

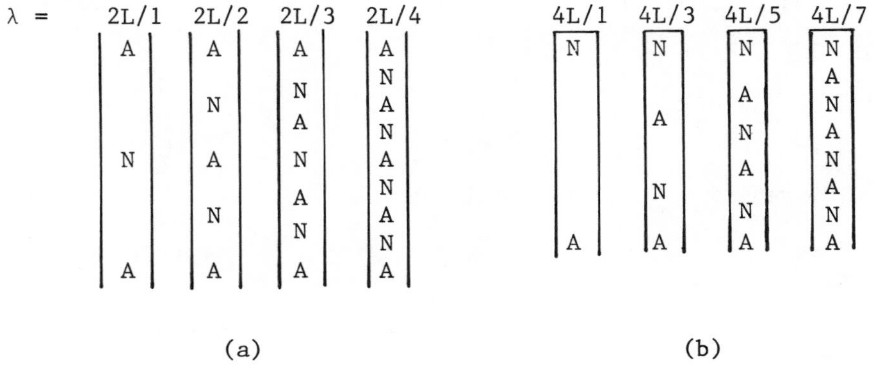

Figure 17-3. (a) Fundamental and first three overtones for an organ pipe open at both ends. (b) The first four modes of vibration of an organ pipe closed at the upper end. The letters A and N indicate the positions of the displacement antinodes and nodes respectively of the standing longitudinal wave. Pressure antinodes occur at displacement nodes and vice versa.

The fundamental and first three overtones are illustrated in Fig. 17-3b for an organ pipe closed at one end. It is important to notice that an organ pipe open at both ends has all harmonics whereas a pipe open at only one end has only odd harmonics. Closed end organ pipes therefore have a different quality or character to their sound from open-ended ones.

For standing waves in a string the eigenfrequencies are (see Eq. (16-20))

$$\nu_n = (n/2\ell)\sqrt{F/\mu} \quad , \qquad (17\text{-}9)$$

n = 1,2,3 ....
standing waves in a string

where ℓ is the string length, μ is mass per unit length, and F the tension in the string. The corresponding wavelengths of the transverse waves in the string are

$$\lambda_{\text{in string}} = 2\ell/n \qquad (17\text{-}10)$$

whereas in the medium adjacent to the string the longitudinal sound waves have wave-

$$\lambda_{\text{in medium}} = v_{\text{in medium}}/\nu_n \quad . \qquad (17\text{-}11)$$

The fundamental and first three overtones for waves in a string are illustrated in Fig. 17-4.

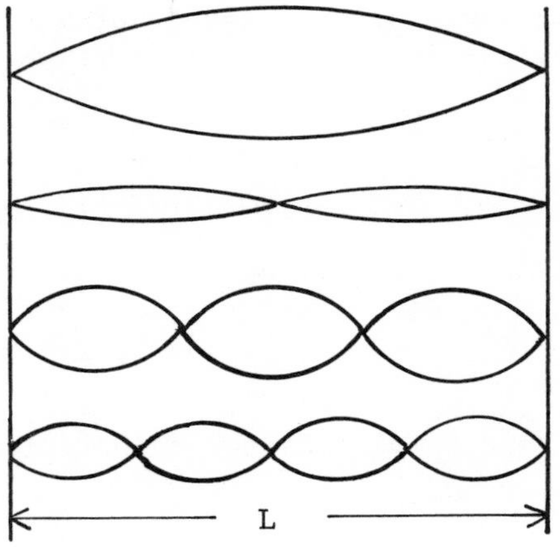

Figure 17-4. First four modes of vibration of a string fixed at both ends.

>>> Example 3. (a) What length organ pipe will produce sound in air of frequency 440 Hz as a fundamental frequency if the pipe is open at both ends? (b) What length organ pipe closed at one end will produce this same tone as its first overtone? (c) If a violin string is 50 cm long and has mass 2×10^{-3} kg, what must be its tension to produce this note as its fundamental? (d) What is the wavelength in the string which produces this note; compare it to the wavelength of the sound in air.

(a) Since the organ pipe is open at both ends Eq. (17-7b) applies with n = 1 (fundamental) and thus taking the velocity of sound in air to be approximately 330 m/sec we have

$$L_o = v/2\nu = (330 \text{ m/s})/(2 \times 440 \text{ s}^{-1}) = 37.5 \text{ cm} \quad .$$

(b) For a pipe closed at one end Eq. (17-8b) applies. Since we want the first overtone, n = 1, and we have

$$L_c = 3v/4\nu = (3/2)L_o = 56.2 \text{ cm} \quad .$$

(c) For a violin string Eq. (17-9) applies and since we want the fundamental frequency, n = 1. Thus

$$F = 4\mu\nu^2\ell^2 = 4(2 \times 10^{-3} \text{ kg}/0.5 \text{ m})(440 \text{ s}^{-1})^2(0.5 \text{ m})^2 = 774 \text{ N} \ .$$

(d) The wavelength of the transverse wave in the string depends only upon the <u>mode</u> (i.e., on n) and the length via Eq. (17-10). Thus since n = 1

$$\lambda_{\text{in string}} = 1 \text{ m}$$

whereas in air the compressional sound wave has wavelength given by Eq. (17-11)

$$\lambda_{\text{in air}} = 0.75 \text{ m} \quad .$$

<<<

17-4 Beats

Standing waves are the result of interference in space. The superposition principle also allows interference in time which occurs when two waves of slightly different frequency travel through the same region at the same time.

Let the displacement at some point x due to traveling wave number one be

$$y_1 = A_1 \sin (k_1x - \omega_1 t + \phi_1)$$

and let that due to traveling wave number two be

$$y_2 = A_2 \sin (k_2x - \omega_2 t + \phi_2) \quad .$$

Take the special case $x = 0$, $\phi_1 = \phi_2 = \pi$, and $A_1 = A_2 = A$ (this choice does not alter the physics, but simplifies the mathematics). Then the resultant displacement is

$$y = y_1 + y_2 = A [\sin \omega_1 t + \sin \omega_2 t]$$

or

$$y = 2A \cos [\tfrac{1}{2}(\omega_1 - \omega_2)t] \sin [\tfrac{1}{2}(\omega_1 + \omega_2)t]$$

which for $\omega_1 \simeq \omega_2$ (but not exactly equal) is a vibration with frequency $(\omega_1 + \omega_2)/2$ whose amplitude varies with a frequency $(\omega_1 - \omega_2)/2$. A beat is a maximum of amplitude without regard to sign and this occurs twice in each cycle of the amplitude. The beat frequency is

$$\omega_{beat} = \omega_1 - \omega_2 \qquad (17\text{-}12a)$$

or

$$\nu_{beat} = \nu_1 - \nu_2 \quad . \qquad (17\text{-}12b)$$

Note that if ω_2 is greater than ω_1, then $\omega_{beat} = \omega_2 - \omega_1$.

17-5 Doppler Effect

Whenever a source of sound and an observer are in relative motion the frequency at the observer is different from that emitted at the source. This is known as the Doppler effect. Let

ν = frequency at the source

ν' = frequency at the observer

v = speed of sound in medium (usually air)

v_o = speed of the observer relative to the medium

v_s = speed of the source relative to the medium

u_o = speed of the observer relative to the sound wave

u_s = speed of the source relative to the sound wave.

Then, for motion along a line joining source and observer we have

$$\nu' = \nu\, u_o/u_s \quad . \qquad (17\text{-}13)$$

If the observer moves toward the sound wavefronts

$$u_o = v + v_o \quad , \quad \text{observer toward wavefronts;} \qquad (17\text{-}14a)$$

the observer intercepts more wavefronts per unit time than he would if he were stationary with respect to the medium. The frequency at the observer increases. If the observer moves away from the sound wavefronts,

$$u_o = v - v_o \quad , \quad \text{observer away from wavefronts;} \qquad (17\text{-}14b)$$

the observer intercepts fewer wavefronts per unit time and the frequency decreases.

If the source moves into the sound wavefronts,

$$u_s = v - v_s \quad , \quad \text{source into wavefronts;} \qquad (17\text{-}15a)$$

the wavelength is decreased and the frequency increases above what it would be if the source were stationary with respect to the medium. If the source moves away from the wavefronts,

$$u_s = v + v_s \quad , \quad \text{source away from wavefronts;} \qquad (17\text{-}15b)$$

the wavelength increases and the frequency decreases.

Although the cause is different for source motion than for observer motion, the qualitative effect is the same: If source and observer have relative motion (toward/away from) one another, the frequency at the observer (increases/decreases).

>>> Example 4. An observer is to the right of a sound source whose frequency is 1000 Hz. Find the frequency detected by the observer if (a) the source moves to the right at 100 ft/s and (b) the source remains stationary while the observer moves toward the source at 100 ft/s. Assume that the air is still.

(a) Since the source moves into the wavefronts, $u_s = v - v_s = (1100 - 100)$ ft/s. Since the observer is not moving, $u_o = v = 1100$ ft/s. Therefore,

$$\nu' = 1000\ \text{Hz}\ \frac{1100}{1000} = 1100\ \text{Hz} \ .$$

(b) Now the observer moves toward the wavefronts so that $u_o = v + v_o = (1100 + 100)$ ft/s. Because the source is at rest, $u_s = 1100$ ft/s. Therefore,

$$\nu' = 1000\ \text{Hz}\ \frac{1200}{1100} = 1091\ \text{Hz}.$$

Although the frequency increases in both cases, the amounts are different since the causes are different. Also, note that if both the source and the observer move in the same direction with the same speed, there will still be a Doppler shift. <<<

>>> Example 5. If a source of sound moves to the right at 100 ft/sec and an observer moves also to the right at 50 ft/s, what is the frequency detected by the observer if the source frequency is 1000 Hz? Assume the observer is to the right of the source initially. The air is still.

Since the observer is on the right of the source, the source is moving into the wavefronts which reach the observer and thus

$$u_s = v - v_s = (1100 - 100)\ \text{ft/s} \quad .$$

The observer is moving away from the wavefronts which reach him, so

$$u_o = v - v_o = (1100 - 50)\ \text{ft/s} \quad .$$

Thus

$$\nu' = 1000\ \text{Hz}\ \frac{1050}{1000} = 1050\ \text{Hz} \quad . \qquad <<<$$

>>> Example 6. A source of sound of frequency 20,000 Hz is mounted on a vehicle (A) which moves at a speed of 44 ft/s (30 mph) toward a second vehicle (B) which is moving toward the first. Also mounted on (A) is a detector and the beat frequency between the signal reflected from (B) and that direct from the source is 4,400 Hz. What is the speed of (B)?

We consider first of all the source on (A). It is moving into the wavefronts which reach (B) which is also moving into the wavefronts it receives.
Thus the frequency received at (B) is

$$\nu' = \nu\ (v + v_B)/(v - v_A) \quad .$$

Now (B) acts as a "passive" source of sound waves of frequency ν' by reflection. Also, (B) moves into the wavefronts it passively emits and (A) moves into the wavefronts sent to it from (B), so the frequency received back at (A) is

$$\nu'' = \nu'\ (v + v_A)/(v - v_B) \quad .$$

The beat frequency is $\nu'' - \nu$ or

$$\Delta \equiv \nu'' - \nu = \nu \left[\left(\frac{v + v_A}{v - v_B}\right) \cdot \left(\frac{v + v_B}{v - v_A}\right) - 1\right] \quad .$$

We let $\Delta/\nu = \delta$, solve for v_B and find

$$v_B = \frac{\delta v\ (v - v_A) - 2vv_A}{v\ (2 + \delta) - \delta v_A}$$

$$= \frac{(0.22)(1100)(1056) - (2.0)(1100)(44)}{(1100)(2.22) - (0.22)(44)}\ \frac{\text{ft}}{\text{s}}$$

$$\simeq 65\ \text{ft/s} \simeq 45\ \text{mph} \quad . \qquad <<<$$

17-6 Programmed Problems

1. The phenomena of beats resulting from the superposition of two simple harmonic sound waves has as its displacement the form

$$y = 2y_m \cos\left[2\pi\tfrac{1}{2}(\nu_1 - \nu_2)t\right] \cos\left[2\pi\tfrac{1}{2}(\nu_1 + \nu_2)t\right].$$

As was the case with standing waves in the previous chapter we can interpret this equation as an amplitude times an oscillating term.

The amplitude varies in time at a frequency ______. The oscillating term varies with a frequency ______ .

Amplitude variation

$$\tfrac{1}{2}(\nu_1 - \nu_2) = \nu_{amp} \equiv \nu_a \quad .$$

Oscillation frequency

$$\bar{\nu} = \tfrac{1}{2}(\nu_1 + \nu_2) \quad .$$

2. From the equation of the previous frame what will be the amplitude of the sound when the beats are eliminated, i.e. $(\nu_1 - \nu_2) = 0$?

$2y_m \cos [2\pi \frac{1}{2}(\nu_1 - \nu_2)t] = 2y_m$ when $(\nu_1 - \nu_2) = 0$. This is true no matter what t is.

3. When we say that we eliminate the beats between two sources ν_1 and ν_2 we mean that the sound at frequency $(\nu_1 + \nu_2)/2$ has a constant amplitude. When $\nu_1 \neq \nu_2$ the amplitude will vary. The sound frequency $(\nu_1 + \nu_2)/2$ will get loud then soft then loud, etc. We know that the variation in amplitude goes as

$$\cos 2\pi\nu_a t \quad .$$

Also the period of the amplitude variation is

$$T = 1/\nu_a \quad .$$

Fill in the table below and sketch the variation in amplitude.

$\frac{t}{T}$	$2\pi\nu_a t$	$\cos 2\pi\nu_a t$
0		
$\frac{1}{4}$		
$\frac{1}{2}$		
$\frac{3}{4}$		
1		
$\frac{5}{4}$		

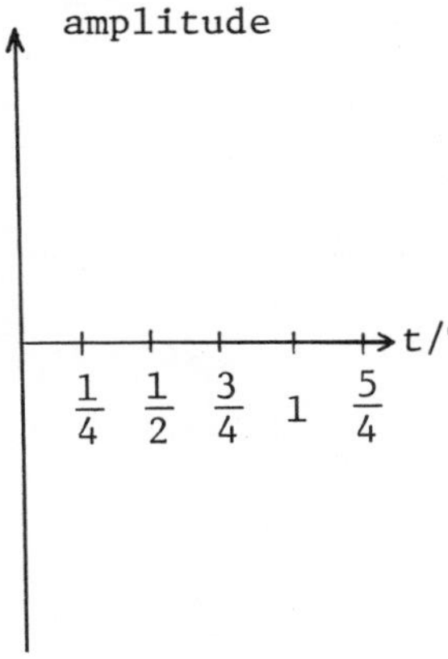

The units of time are given in fractions of a period, e.g. $t/T = 1/4$ means $t = T/4 = 1/4\nu_a$.

$\frac{t}{T}$	$2\pi\nu_a t$	$\cos 2\pi\nu_a t$
0	0	1
$\frac{1}{4}$	$\frac{\pi}{2}$	0
$\frac{1}{2}$	π	-1
$\frac{3}{4}$	$\frac{3\pi}{2}$	0
1	2π	1
$\frac{5}{4}$	$\frac{5\pi}{4}$	0

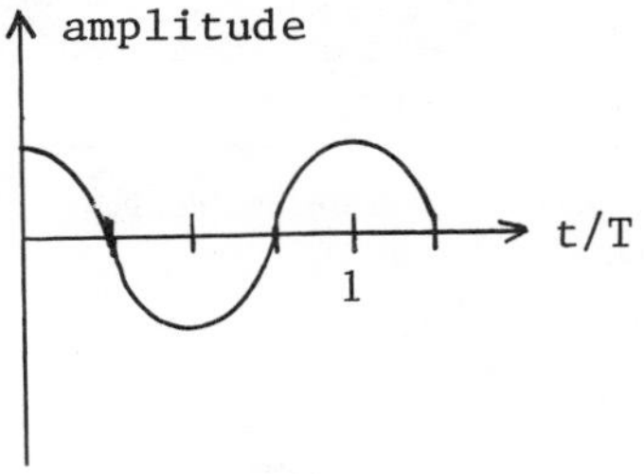

4. Over one cycle, say from $t = T/4$ to $t = 5T/4$ in the sketch of the last answer, how many times is the amplitude a maximum without regard to sign?

Twice. At $t = 1/2$ T and $t = T$.

5. One hears sound intensity which is proportional to the amplitude squared. Over one cycle of the amplitude variation, how many times will the sound be loud? Each intensity maximum is called a "beat".

Twice, corresponding to the two amplitude extrema each of which corresponds to an intensity maximum. Thus there are two beats per one cycle of ν_a.

6. Two sound sources $\nu_1 = 450$ Hz and $\nu_2 = 440$ Hz produce beats.
   1. The frequency of sound you will hear is ______ Hz.
   2. The number of beats per second is ______.

1. $\frac{1}{2}(\nu_1 + \nu_2) = 445$ Hz .

2. $\nu_a = \frac{1}{2}(\nu_1 - \nu_2) = 5$ Hz .

There will be two beats per cycle of ν_a so there will be 10 beats per second.

A tug boat is moving perpendicular to a cliff. The skipper wishes to use his fog horn and knowledge of the Doppler effect to assure the safety of his crew. For simplicity we assume the air is still. Obtaining a physical interpretation of the Doppler effect is a good way to begin this problem. Assume a fog horn of frequency 60 Hz, a tug speed of 22 ft/s (15 mph) and the speed of sound 1100 ft/s.

7.

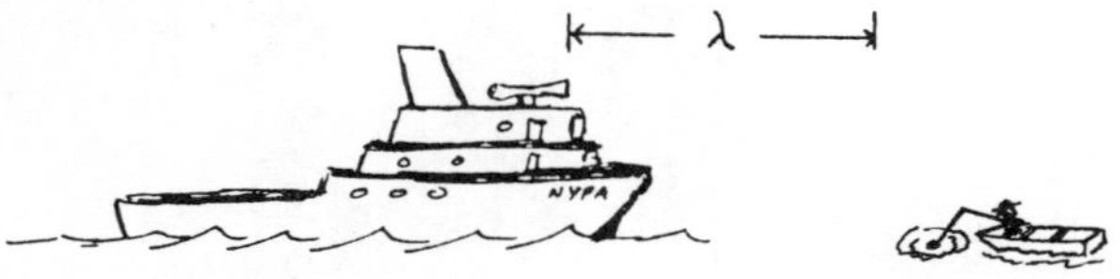

We represent the wave length λ of the fog horn for a boat at rest. The vertical lines represent wave fronts of similar pressure variations. What is the particular name given to the time between the emission of these wave fronts?

The period T of the horn.

---

8. How far will the boat move during one period?

$d =$ ______ .

Express the answer algebraically.

$d = v_sT$

where v_s is the speed of the boat (source).

9.

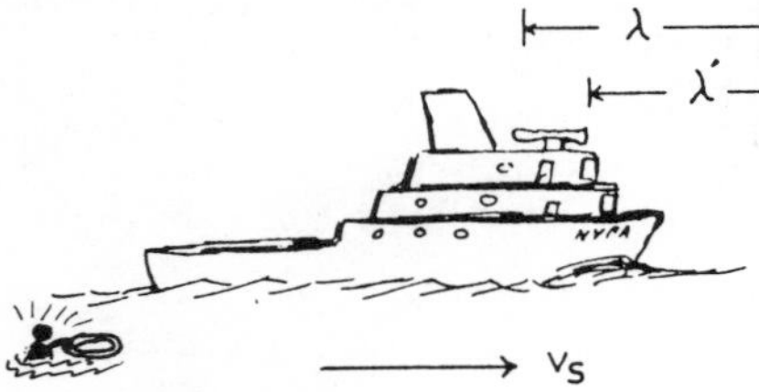

With the tug moving at speed v_s in the direction of propagation, the wave length will be shortened by the distance the tug moves during the period of the horn.

$\lambda' = \lambda$ ______ .

$\lambda' = \lambda - v_sT$

10. Although the wavelength of the emitted sound will appear shorter to a fixed observer, the velocity of propagation, v, is independent of source or observer motion. The wavelength, velocity of propagation and frequency are related by $\lambda = v/\nu$. For the boat in motion the wavelength is λ'. Use the answer of frame 9 to find an expression for ν'. ν' = ______ .	$\lambda' = \lambda - v_s T$ $v/\nu' = v/\nu - v_s/\nu$ $\nu' = [v/(v - v_s)]\nu$
11. Obtain the numbers from frame 7 and calculate the frequency that would be heard by a fixed observer as the tug moves toward that observer.	$\nu' = [v/(v - v_s)]\nu$ $\nu' = [1100/(1100 - 22)]$ 60 Hz . ν' = 61.2 Hz . Notice that the shorter wavelength means a higher frequency.
12. By the way, what will the horn on the tug sound like to the captain, i.e., what will the frequency be?	60 Hz all the time.
13. Back to the problem. The sound at 61.2 Hz strikes the cliff and is reflected back toward the tug. The captain is now an observer moving toward a fixed source. Because the captain is moving toward the source his ear will intercept (fewer/more) waves per second than if he were at rest.	More. Thus the reflected sound will appear to have a higher frequency than 61.2 Hz.
14. The frequency for the case of the captain (observer) moving toward the cliff (source) is given by $\nu'' = [(v + v_o)/v]\nu'$ with v_o = speed of observer toward source, and ν' = frequency of the reflected wave from cliff. What will be the frequency of the reflected sound that the captain will hear? ν'' = ______ . Obtain a number.	$\nu'' = [(1100 + 22)/1100]$ 61.2 Hz ν'' = 62.4 Hz .
15. As the captain blows his horn he will hear both the 60 Hz and the 62.4 Hz. The superposition of these two sound waves will result in his hearing 2.4 ______ per second.	Beats. $2 \times \frac{1}{2}(\nu'' - \nu) = 2.4$.

16. In practice it might be hard for the captain to detect the slightly higher frequency of 62.4 Hz which tells him that he is moving toward the cliff. He should be able to hear the beats which say that his fog horn is being Doppler shifted. Suppose he detects the 2.4 beats/s and decides to reverse his direction. We know that his fog horn sound will now be Doppler shifted to a longer wave length as it is reflected by the cliff. We use

$$\nu' = [v/(v + v_s)]\nu$$

to calculate this lower frequency.

$\nu' =$ ______ Hz .

ν' is the frequency of the sound wave which will strike the cliff.

58.8 Hz .

---

17. This sound wave will be reflected from the cliff, but its wave length will be lengthened since now the captain (observer) is moving away from the cliff (source).

$$\nu'' = [(v - v_o/v]\nu' \quad .$$

$\nu'' =$ ______ .

$\nu'' = 57.6$ Hz .

---

18. How many beats will the captain hear per second?

2.4 .

Well, it is too bad that the number of beats per second is the same as before. He cannot tell by listening to the beats whether he is going toward or away from the cliff. This is not fundamental, but occurs here because the speed of the boat is small in comparison with that of sound. In either case he had better turn 90° left or right just to be safe and eliminate those beats. Incidentally one could turn this problem around and by knowing the horn frequency and the number of beats determine the speed of the boat.

19.

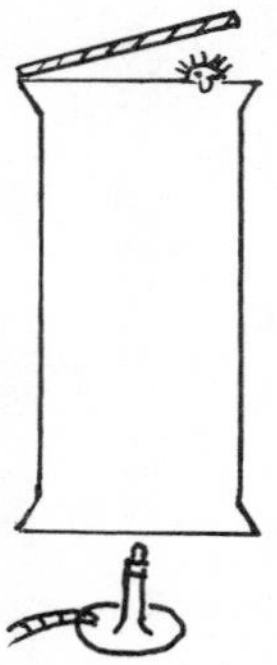

A favorite demonstration concerning sound resonance is depicted to the left. A long hollow tube is covered at the top with a flap and a Bunsen burner is placed at the bottom. After heating the air in the column, lifting the flap will cause the column to resonate as the best fog horn you ever heard.

For an open ended pipe both ends must be (nodes/antinodes)?

Antinodes.

The air is free to be displaced at both ends.

20.

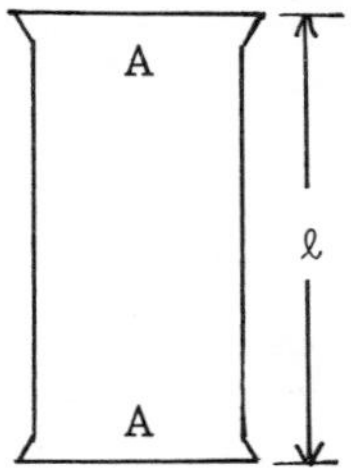

As in the case of standing waves with vibrating strings antinodes are spaced one-half wavelength apart. What is the minimum number of half wavelengths that one can have in this pipe? A denotes a displacement antinode.

One.

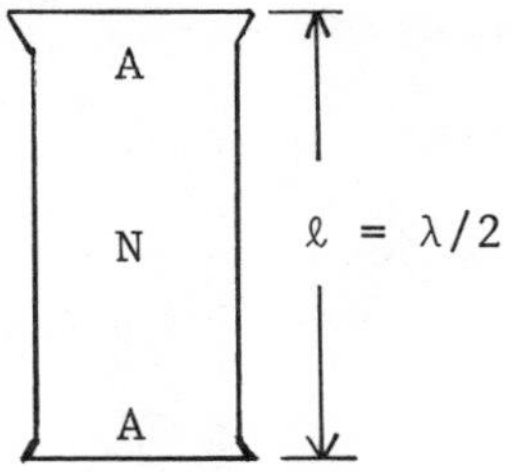

Nodes (N) must exist adjacent to antinodes (A).

21.

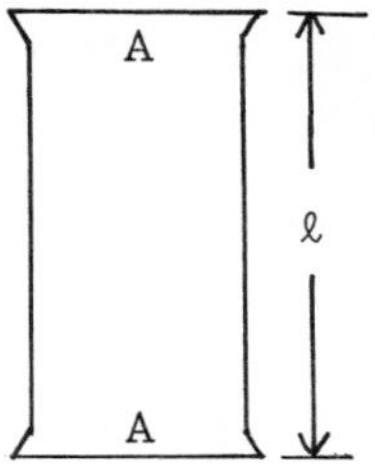

Indicate by the addition of appropriate node (N) and antinode (A) points a situation in which one would have three half wavelengths in the pipe.

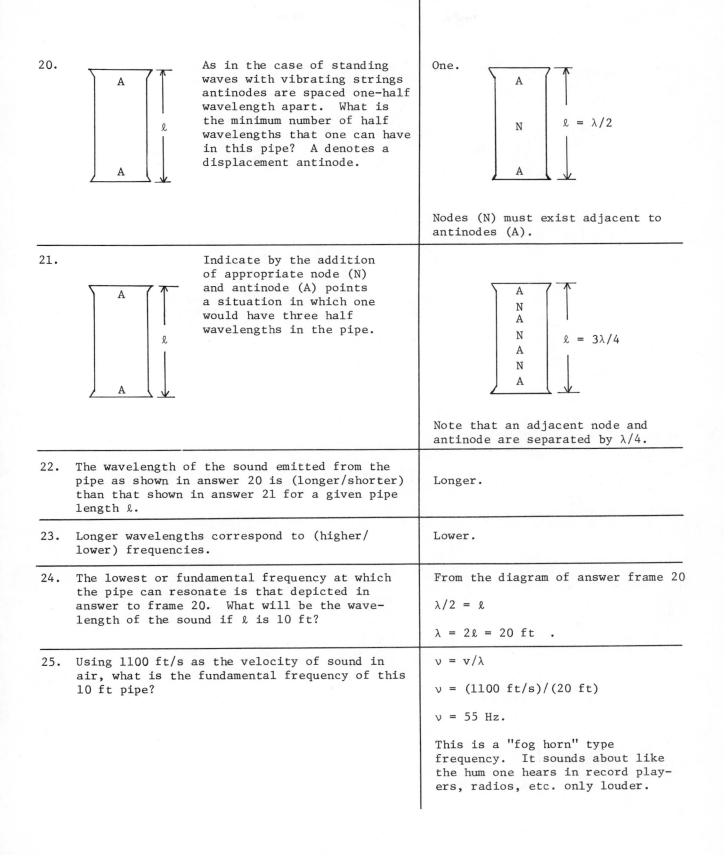

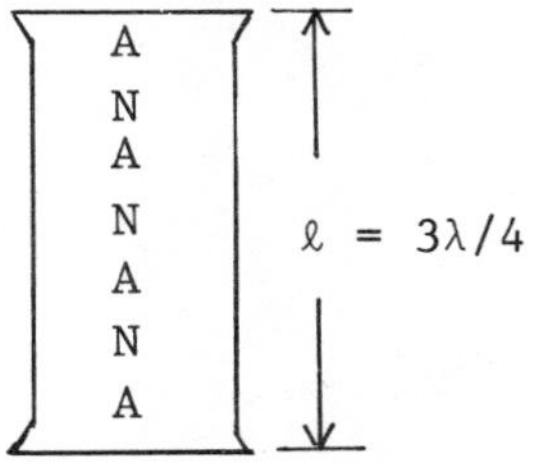

Note that an adjacent node and antinode are separated by $\lambda/4$.

---

22. The wavelength of the sound emitted from the pipe as shown in answer 20 is (longer/shorter) than that shown in answer 21 for a given pipe length ℓ.

Longer.

---

23. Longer wavelengths correspond to (higher/lower) frequencies.

Lower.

---

24. The lowest or fundamental frequency at which the pipe can resonate is that depicted in answer to frame 20. What will be the wavelength of the sound if ℓ is 10 ft?

From the diagram of answer frame 20

$\lambda/2 = \ell$

$\lambda = 2\ell = 20$ ft .

---

25. Using 1100 ft/s as the velocity of sound in air, what is the fundamental frequency of this 10 ft pipe?

$\nu = v/\lambda$

$\nu = (1100 \text{ ft/s})/(20 \text{ ft})$

$\nu = 55$ Hz.

This is a "fog horn" type frequency. It sounds about like the hum one hears in record players, radios, etc. only louder.

26. 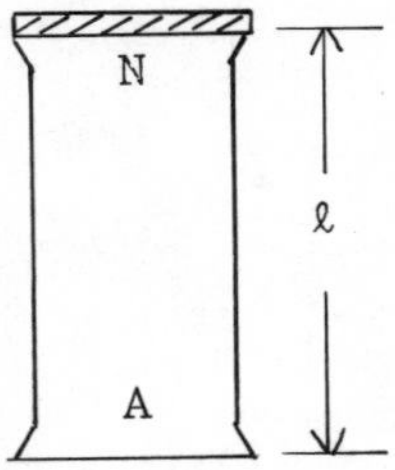

Suppose we had not lifted the flap after heating the air column. The bottom would be an antinode still, but the top is now a node. How is the wavelength here related to the pipe length at the fundamental frequency?

Distance between adjacent nodes and antinodes is $\lambda/4$.

$\ell = \lambda/4$.

27. What is the resonant frequency of this closed pipe if it is 10 ft long?

$\nu = v/\lambda = v/4\ell$

$\nu = (1100 \text{ ft/s})/(40 \text{ ft})$

$\nu = 27.5$ Hz .

Most people cannot hear this, so that's why the flap is used.

Chapter 18: TEMPERATURE

REVIEW AND PREVIEW

In previous chapters you have dealt with mechanical systems. You discovered that the concept of energy played a vital role. In mechanics, energy was in the form of kinetic energy (motion energy) or potential energy (position energy); it could be converted from one form to the other. Energy "losses" in mechanical systems were due to dissipative forces (friction) and were treated as a loss of mechanical energy. The manner in which this mechanical energy was lost was treated as rather incidental for this is the domain of thermodynamics which you now begin to study. Of primary importance in this area is the concept of temperature which is the topic of this chapter.

GOALS AND GUIDELINES

You have only two major goals in this chapter:

1. Learning and understanding the definition of temperature (from a more precise point of view than your intuitive ideas of this concept).
2. Learning the basic laws of thermal expansion and how to deal with linear, area, and volume expansions or contractions.

18-1 Microscopic and Macroscopic Descriptions of Matter

When one deals with a system containing a large number of particles two points of view are possible. One is the microscopic description which is concerned with the speeds, energies, masses, etc. of the particles. For a very large number of particles, averages of these particle properties are considered; this branch of science is called statistical mechanics. The second point of view is concerned with gross features of the system and forms the basis of a branch of science called thermodynamics. Examples of gross features are pressure, volume and temperature. These are macroscopic properties of the system.

18-2 Zeroth Law of Thermodynamics -- Temperature Measurement

The zeroth law of thermodynamics is the concept of temperature. If system A is hotter than system B and they are brought into contact, some time later they will both have the same temperature, or more formally, they will be in thermal equilibrium. If C is a thermometer, then the zeroth law says

> If A and B are in thermal equilibrium with C then A and B are in thermal equilibrium with each other.

A thermometer is made by:

1. choosing a thermometric substance and a particular thermometric property of the substance -- i.e., by choosing something which does something when its temperature changes,
2. defining the temperature scale by

$$T(X) \equiv aX \quad , \qquad (18\text{-}1)$$

where X is the thermometric property and a is a constant, and by
3. selecting a unique physical situation and defining the temperature which fixes a in Eq. (18-1).

This unique point is taken to be the triple point of water (ice, water and vapor coexist in equilibrium) and the temperature arbitrarily chosen to be 273.16 K where K stands for the kelvin, a unit step on a thermometric scale called the Kelvin scale.

Then, the temperature associated with any value X of the thermometric property of a given thermometric substance is

$$T(X) = 273.16\ K)X/X_{TP} \qquad (18\text{-}2)$$

where X_{TP} is the value at the triple point of water.

TABLE 18-1

Thermometric Substance	Thermometric Property
gas	pressure at constant volume
gas	volume at constant pressure
wire	electrical resistance
metal strip	length
liquid	volume as measured by height in a small tube

Table 18-1. Examples of thermometric substances and properties used to make thermometers.

Examples of thermometric substances and properties are given in Table 18-1. Such thermometers always agree at one point (the triple point) but need not agree at other points (say the boiling point of water at some atmospheric pressure). The least variation is among various gas thermometers at constant volume and this difference disappears as the amount of gas used and hence the pressure at the standard triple point decreases. The standard thermometer with which all others can be compared is the constant volume gas thermometer in the limit as the pressure at the triple point P_{TP} approaches zero. Then

$$T(P) = 273.16\ K \lim_{P_{TP}\to 0} (P/P_{TP}) \quad , \quad V \text{ constant} \quad . \qquad (18\text{-}3)$$

This is called the ideal gas thermometer.

18-3 Celsius and Fahrenheit Scales

A unit on the Celsius (formerly centigrade) scale is the same size as a kelvin and is called a Celsius degree (C°). Temperature is then given in degrees Celsius (°C). The triple point is assigned the value 0.01° C. Thus if T_C is the Celsius temperature and T_K the Kelvin temperature

$$T_C = T_K - 273.15 \quad . \qquad (18\text{-}4)$$

The ice point is the temperature at which ice and water are in equilibrium at one atmosphere pressure, and the steam point is that point at which steam and water are in equilibrium at one atmosphere pressure. With the choice of Celsius scale (Eq. (18-4))

$$T_C \text{ (ice point)} = 0.00^\circ \text{ C}$$

$$T_C \text{ (steam point)} = 100.00^\circ \text{ C} \quad .$$

On the Fahrenheit scale this ice point is 32° F (degrees Fahrenheit) and the steam point 212° F. Thus, a Fahrenheit degree is 5/9 of a Celsius degree. The Fahrenheit temperature scale is related to that of Celsius by

$$T_F = 32 \quad + 9/5\ T_C \quad . \qquad (18\text{-}5)$$

>>> Example 1. At what temperature are the Celsius and Fahrenheit readings numerically equal?

We want that temperature such that

$$T_F = 32 + 9/5\ T_C = T_C$$

or

$$T = -\ 40^\circ \text{ C} = -\ 40^\circ \text{ F} \quad .$$

This leads to a useful conversion from Fahrenheit to Celsius and vice versa. From Eq. (18-5) we have

$$T_C = 5/9\ (T_F - 32)$$

which is not very symmetric when compared to Eq. (18-5). However, one may also write more symmetrically

$$T_F = 9/5\ (T_C + 40) - 40$$

and

$$T_C = 5/9\ (T_F + 40) - 40 \quad .$$

To remember whether to use 9/5 or 5/9 it is only necessary to remember that the Fahrenheit degree is smaller than the Celsius degree. <<<

18-4 Thermal Expansion

As shown in Table 18-1, length can be a thermometric property. When the temperature of a substance changes so in general does the average spacing between the molecules of the substance. A change in a linear dimension is called linear expansion even though the dimension may decrease; the latter is negative linear expansion.

The coefficient of linear expansion, α, is defined by

$$\alpha = (1/\ell)d\ell/dT \qquad (18\text{-}6a)$$

where ℓ is the length at the temperature T. In general α is dependent upon the temperature but over fairly wide ranges of temperature a constant, average value may be used with negligible error.

Then, for finite changes of ℓ provided ΔT is not too large

$$\alpha = (1/\ell)\Delta\ell/\Delta T \quad . \qquad (18\text{-}6b)$$

For isotropic solids, every linear dimension changes by the same amount so we may express the change in area, ΔA, by

$$\Delta A = 2\alpha A\Delta T \qquad (18\text{-}7)$$

and the change in volume, ΔV, by

$$\Delta V = 3\alpha V\Delta T \quad . \qquad (18\text{-}8)$$

For liquids only the volume change is significant and a coefficient of volume expansion, β, is defined by

$$\beta = (1/V)dV/dT \simeq (1/V)\Delta V/\Delta T \qquad (18\text{-}9)$$

where β is treated as roughly constant.

>>> Example 2. By how much does the height of a steel television tower increase if its height is 1000 ft at 5° C and the temperature is 30° C? (Treat the tower as a single rod.)

From Eq. (18-6b),

$$\Delta\ell = \alpha\ell\Delta T$$

and since the tower is steel, $\alpha = 11 \times 10^{-6}$ per C^o. Thus

$$\Delta\ell = (11 \times 10^{-6}/C^o)(10^3 \text{ ft})(30 - 5)C^o$$

$$= 0.275 \text{ ft} = 3.3 \text{ inches} \quad .$$

This may seem a small amount, but such considerations must be taken into account in tower design. <<<

>>> Example 3. A steel rod and a brass rod have the same length at 0° C. What is their fractional difference in length at 100° C if $\alpha_{steel} = 11 \times 10^{-6}$ per C^o and $\alpha_{brass} = 19 \times 10^{-6}$ per C^o?

Let their common length at 0° C be ℓ. Then, for some temperature change ΔT their lengths will be

$$\ell_{brass} = \ell + \ell\alpha_b\Delta T$$

and

$$\ell_{steel} = \ell + \ell\alpha_s\Delta T \quad .$$

Their difference in lengths is

$$\ell\,(\alpha_b - \alpha_s)\,\Delta T = \Delta\ell_b - \Delta\ell_s \quad ,$$

so the fractional difference is

$$(\Delta\ell_b - \Delta\ell_s)/\ell = (\alpha_b - \alpha_s)\,\Delta T = 8 \times 10^{-6} \times 10^2 = 8 \times 10^{-4} \quad .$$ <<<

Chapter 19: THERMODYNAMICS

REVIEW AND PREVIEW

In mechanics energy is associated with position or motion and is directly related to particle or object properties. In thermodynamics (as with fluids) any description of individual particles is lost; one describes gross features of systems which involve many particles. Also, one is very interested in the transference of energy from one system to another by non-mechanical means; this is the concept of heat. Some old familiar friends from mechanics will remain, however, namely the conservation of total energy which will have a new name of the first law of thermodynamics. Also the concept of work done by a system will be used. You will first study the capacity of a system to accept or reject heat (energy) and then the flow of energy due to temperature difference which is called heat conduction. This is an area which has very many practical ramifications such as heating and cooling your house. You will be introduced to a model thermodynamic system, the ideal gas, and finally will study the second law of thermodynamics which is a statement of the natural direction in which physical processes proceed.

GOALS AND GUIDELINES

You have several major goals in this chapter:

1. Learning the definition of heat capacity and learning to solve calorimetry problems. Notice that these are nothing more than energy balancing equations and thought of that way they become quite simple.
2. Learning to solve the basic problem of linear heat conduction. There will be a few new vocabulary words here, but with a little practice heat conduction problems are quite easy to solve. They are worth your time from a practical standpoint as well as educational for they apply to the flow of energy from your house in winter and into it in summer.
3. Learning the basic vocabulary with definitions of thermodynamic processes (Section 19-3). You should read this carefully and then after you finish the chapter re-read it again when you will be able to think of concrete examples of the terms.
4. Learning and understanding how to use the first law of thermodynamics; this is your old friend conservation of total energy.
5. Learning the terms associated with and the basic properties of the model thermodynamics system you will study, the ideal gas. You will need to have this section well in hand for the next chapter.
6. Learning and understanding the second law of thermodynamics which will be stated in two equivalent forms. Mechanical paradoxes such as perpetual motion machines and the like can be laid to rest by arguments based upon this law. You will first meet the law in the form of engines and refrigerators (very practical energy devices) and a study of their efficiencies. You will study the law again in the form of the concept of entropy. This concept may seem a bit difficult because it is rather abstract, but it is also very important in learning the natural direction in which physical processes proceed. At this level you should concentrate on learning the definition and how to solve entropy problems for the model system, the ideal gas. If you pursue studies in this area further, entropy like energy will become an old friend (but never quite so familiar as energy!).
7. Learning and understanding the concept of a thermodynamic cycle by studying the very important Carnot cycle applied to the ideal gas. Try to understand each step of the cycle separately and then the cycle as a whole.

19-1 Heat

Heat is energy which flows from one system to another because of a temperature difference between them. The unit of heat is a kilocalorie. The modern definition of the kilocalorie (kcal) is 4186 J which recognizes that heat is energy. For all practical purposes this new kilocalorie, which is 1000 thermal calories, is the same as the original quantity which was defined as the amount of heat that must be supplied to raise the temperature of one kilogram of water from 14.5 $^{\circ}$C to 15.5 $^{\circ}$C. In the BES the unit of heat is the British thermal unit (Btu); this is the amount of heat required to raise the temperature of one pound of water from 63 $^{\circ}$F to 64 $^{\circ}$F. The conversion is

$$1 \text{ Btu} = 252 \text{ cal} = 0.252 \text{ kcal} \quad .$$

The amount of heat required to change the temperature of a substance by a given amount depends upon the substance, its mass, and the temperature. The ratio of the heat ΔQ supplied to the change in temperature ΔT is called the <u>heat capacity</u> C of a body. Thus

$$C = \Delta Q/\Delta T \quad .$$

You should be careful of the idea here; <u>a system has no capacity to hold heat</u>. Rather, it has the capacity to <u>accept</u> a given amount of heat, ΔQ, (or give it up) for a given change in temperature, ΔT, according to $\Delta Q = C\Delta T$. The heat capacity per unit mass is called the <u>specific heat</u>, c, and it is defined as

$$c = (1/m)(\Delta Q/\Delta T) \quad . \tag{19-1a}$$

As indicated before, c depends upon the temperature about which the interval ΔT takes place. Therefore one defines the specific heat <u>at any</u> temperature by

$$c = (1/m)(dQ/dT) \tag{19-1b}$$

and the heat Q which must be added to (or taken away from) a system to change its temperature from $T_{initial}$ to T_{final} is

$$Q = m \int_{T_i}^{T_f} c \, dT \quad . \tag{19-2}$$

For many substances, over a wide range of temperatures an average value of c may be used with the finite form of the heat absorption equation (Eq. 19-1a)).

The amount of heat required to change a substance's temperature depends also on <u>how</u> the heat is added or taken away. In general each different possible process leads to a different value of c. For example, for gases one has the <u>specific heat at constant</u> pressure, c_p, and that at <u>constant volume</u>, c_v. For solids one usually speaks of c_p.

Another useful form of specific heat is the <u>specific heat per mole or molar heat capacity</u>. A <u>mole</u> of a substance contains Avogadro's number, 6.02252×10^{23}, of molecules. The <u>molecular weight</u> is a dimensionless quantity expressing the number of grams per mole. The mole is therefore a sort of variable unit of mass whose value depends upon the chemical substance. The molar heat capacity is related to the specific heat by

$$\begin{matrix}\text{molar heat} \\ \text{capacity}\end{matrix} = \begin{matrix}\text{molecular} \\ \text{weight}\end{matrix} \times \begin{matrix}\text{specific} \\ \text{heat}\end{matrix}$$

in dimensions

$$(\text{cal/mol}\cdot\text{C}^{\circ}) = (\text{g/mol})(\text{cal/g}\cdot\text{C}^{\circ}) \quad .$$

>>> Example 1. A body of mass m_b and specific heat c_b is initially at a temperature T_b. It is immersed in a liquid of mass m_ℓ and specific heat c_ℓ which is in a container of mass m_c and specific heat c_c. The initial temperature of the liquid (and container) is T_ℓ. What is the final temperature of the system (i.e., body + liquid + container)?

This is a case of two systems initially at different temperatures reaching thermal equilibrium by transfer of heat from one to the other after they are combined into a single system. The heat lost by the body equals the heat gained by the liquid and container. This is just conservation of energy. Let T_f be the final equilibrium temperature. We assume here $T_b > T_f > T_\ell$. Then, we treat the specific heats as constants and use Eq. (19-1a). We have

$$m_b c_b (T_b - T_f) = \text{heat lost by body}$$

$$(m_\ell c_\ell + m_c c_c)(T_f - T_\ell) = \text{heat gained by liquid and container.}$$

Since these are equal we can solve for T_f and find

$$T_f = \frac{m_b c_b T_b + (m_\ell c_\ell + m_c c_c)\, T_\ell}{m_\ell c_\ell + m_c c_c + m_b c_b} \quad .$$

Notice that this same sort of problem can be phrased so that any one of the quantities (m_b, c_b, m_ℓ, c_ℓ, m_c, c_c, T_b, T_ℓ or T_f) is the unknown. The others must be specified in one way or another. <<<

Substances can absorb or reject heat without changing temperature but rather by changing state. For example, to convert one gram of ice at 0° C to one gram of water at 0° C requires that the ice absorb 80 cal of heat. This is called the heat of fusion. We write this as

$$Q = mL_f \tag{19-3a}$$

where L_f is the heat of fusion which is 80 cal/g or 80 kcal/kg for water. At 100° C and one standard atmosphere, one kg of water becomes one kg of steam, which is also at 100° C after the water has absorbed 539 kcal. In general we denote the heat of vaporization by L_v; it is the amount of energy required to convert one kilogram of a substance from liquid to vapor. Thus

$$Q = mL_v \tag{19-3b}$$

expresses the amount of heat for a mass m (where m is in kilograms).

>>> Example 2. An insulating cup (i.e., non-heat absorbing) contains 0.25 kg of coffee at 80° C. (Assume that the specific heat of coffee is the same as water.) How much ice initially at -10° C must be added if the final temperature is to be 60° C? The specific heat of ice is approximately 0.5 kcal/kg$\cdot$C$^\circ$.

This type of problem involves a change of state for the ice and must be done in three parts. Let m be the mass of the ice in kilograms. As the ice temperature goes from -10° C to 0° C the ice absorbs

$$10\ m\ (0.5)\text{kcal} = 5\ m\ \text{kcal} \quad .$$

This m kg of ice at 0° C is now converted to water at 0° C and absorbs a further

$$80\ m\ \text{kcal} \quad .$$

We now have m kg of water at 0° C and as its temperature is raised to 60° C it absorbs

$$60 \text{ m kcal}$$

since the specific heat of water is about 1.0 kcal/kg·C^o. Therefore the total heat absorbed by the ice is

$$145 \text{ m kcal} = \text{heat absorbed by ice}$$

and this must be the heat lost by the coffee which is

$$.250 \text{ kg } (1.0 \text{ kcal/kg}\cdot C^o)(20\ C^o) = 5.0 \text{ kcal} = 145 \text{ m kcal} \quad .$$

Hence,

$$m \simeq 0.035 \text{ kg} = 35 \text{ gram} \quad .$$ <<<

19-2 Heat Conduction

The transfer of energy from one part of a system to another by virtue of temperature difference is called heat conduction. If one face of an infinitesimally thin slab of material is at temperature T and the other face at T + dT, and if the area of the face of the slab is A, the time rate of heat transfer H across the slab face is given by

$$H = -\ kA\ dT/dx \tag{19-4}$$

where dx is the thickness and k is called the thermal conductivity of the substance. Eq. (19-4) is the fundamental equation of heat conduction; the situation described by this equation is indicated schematically in Fig. 19-1. Heat flows from higher to lower temperatures and the direction here is chosen to be that of increasing x. The minus sign in Eq. (19-4) insures that when the temperature gradient, dT/dx, is negative the rate of heat transfer, N, is positive. The direction of heat flow is perpendicular to the area.

Figure 19-1. Schematic indication of the heat conduction equation, Eq. (19-4). The heat flows at a rate H from the higher to lower temperature.

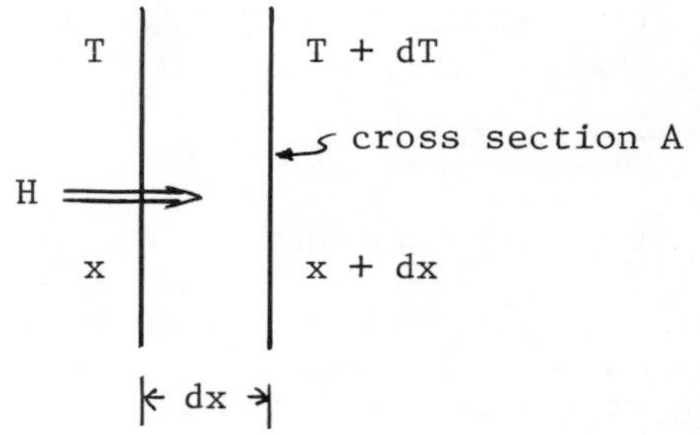

The usual units for H are kcal/s. Then A has the unit m^2 and dT/dx is expressed in C^o/m. Thus, thermal conductivity must have the unit kcal/s·m·C^o. The thermal conductivity in general depends not only upon the substance but also upon the temperature, but for moderate temperature differences a constant average value may be used for a given substance. If k is large, a substance is a good heat conductor whereas if k is small, the substance is a poor heat conductor.

Most thermal conductivity problems which you will be asked to solve involve steady state conditions. This means that both H and T are constant in time (they need not be constant spatially however).

When steady state is attained for a slab of material with heat flow perpendicular to its surfaces, H must also be independent of position. If it were not, then some small element of the substance would absorb more heat in some interval of time than it would reject and hence its temperature would change; steady state would not yet be attained. For a slab then, under steady state conditions one can integrate Eq. (19-4) and find

$$H = kA(T_L - T_R)/L \quad , \quad \text{steady state} \qquad (19\text{-}5)$$

where T_L is the temperature of the left face of the slab and T_R that of the right face. The slab thickness is L and we have assumed $T_L > T_R$.

>>> Example 3. Consider a hollow cylinder of inner radius R_1 and outer radius R_2 and length L (see Fig. 19-2). If the inner surface is maintained at T_1 and the outer surface at T_2 where $T_1 > T_2$ find the radial rate of heat transfer and the temperature at any value of r such that $R_1 \le r \le R_2$ for steady state conditions.

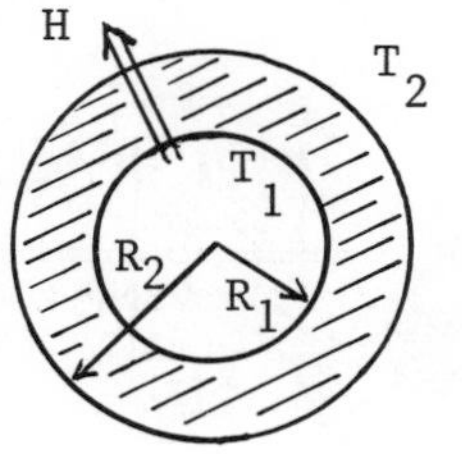

Figure 19-2. Example 3. Steady state radial heat flow through a cylinder of length L. A cross section is shown. The inner surface is maintained at T_1 and the outer surface at T_2; $T_1 > T_2$.

Since this is steady state flow, and the flow is radial, the rate of heat transfer into a cylindrical area $2\pi rL$ is constant. Because the heat flow is perpendicular to the area Eq. (19-4) applies and

$$H = \text{constant} = -\, k(2\pi rL)\, dT/dr \quad .$$

This can be integrated from $r = R_1$ to some value of r where the temperature is T(r). Thus

$$\int_{T_1}^{T(r)} dT = -\frac{H}{2\pi kL}\int_{R_1}^{r} \frac{dr}{r}$$

or

$$T(r) = T_1 - (H/2\,kL)\,\ell n(r/R_1) \quad .$$

Now when $r = R_2$, $T(r) = T_2$ and so we have

$$H = 2\pi kL(T_1 - T_2)/\ell n(R_2/R_1)$$

which gives

$$T(r) = T_1 - (T_1 - T_2)\,\ell n(r/R_1)/\ell n(R_2/R_1) \quad .$$ <<<

A compound slab consists of successive layers of different materials (hence different thermal conductivities) having (perhaps) different thicknesses, all have a common cross-sectional area A. If the higher temperature is on the left side and the lower temperature on the right, Eq. (19-5) applies to each layer. The hot temperature for each layer is the cold temperature for the layer to its left, and its cold face is in turn the hot face of the layer on its right.

The crucial point is that the rate of heat transfer H is the same for each part of the slab. Thus

$$\frac{L_1}{k_1 A} H = T_L - T_1$$

$$\frac{L_2}{k_2 A} H = T_1 - T_2$$

$$\frac{L_3}{k_3 A} H = T_2 - T_R$$

Hence the general relationship for such slabs is

$$H = \frac{A\ (T_L - T_R)}{\sum (L_i/k_i)}$$

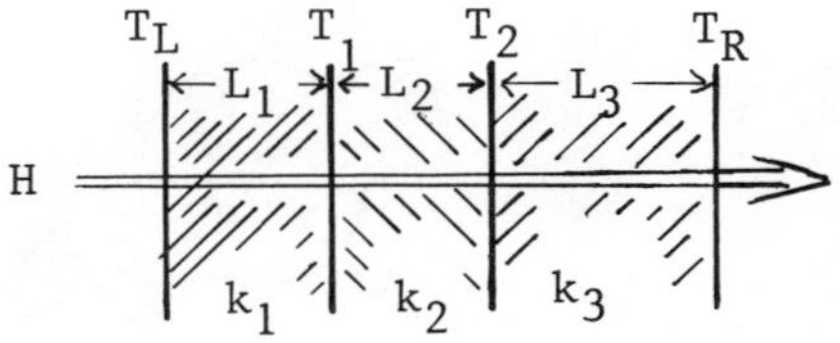

Figure 19-3. Heat transfer across a compound slab made of layers of thickness L_i and having thermal conductivity k_i.

19-3 Thermodynamic Systems, Coordinates, States, and Equilibrium

A thermodynamic coordinate is a macroscopic variable (examples are pressure, temperature, and volume) whose values have some bearing on the internal state of a physical system. A system whose properties of interest can be described in terms of such coordinates is called a thermodynamic system. Thermodynamic systems exist in various states. If such a system is in a state of mechanical equilibrium, there is no net force on the interior parts of the system nor between it and its relevant environment. In thermal equilibrium all parts of the system are at the same temperature as that of the relevant environment. Finally, if a system is in a state of chemical equilibrium, it does not undergo chemical changes. If all three forms of equilibrium are satisfied, a system is said to be in thermodynamic equilibrium which is characterized by the fact that a single value of each thermodynamic coordinate can be assigned to the system as a whole.

When a thermodynamic system interacts with its environment, it may do so by absorbing heat, rejecting heat, by performing work, or having work done on it. In doing so, it passes through non-equilibrium states which cannot be described by thermodynamic coordinates alone. For example, if the temperature of its environment is suddenly changed it takes time for this change to propagate throughout the system (see Section 19-2). All is not lost, however, because one can idealize the interaction by approximating it by a quasistatic process in which the system is infinitesimally near a thermodynamic equilibrium state at every step in the process. For example, instead of changing the temperature of the environment suddenly, change it very slowly over the same range.

A schematic representation of a system in an initial equilibrium state making a transition to a final equilibrium state is shown in Fig. 19-4.

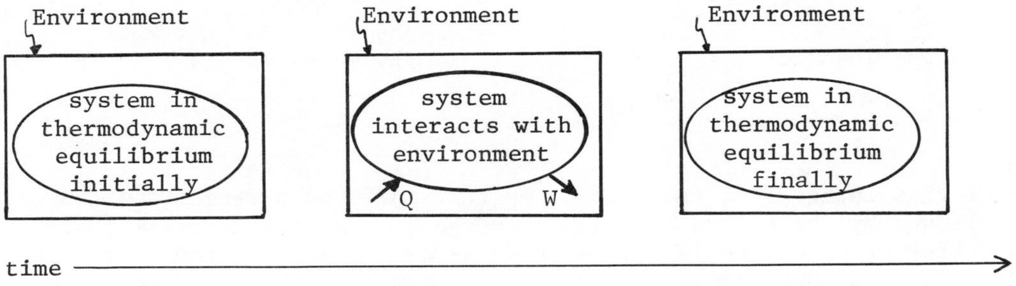

Figure 19-4. Schematic representation of a system in an initial equilibrium state making a transition to a final equilibrium state by interacting with its environment. Heat Q is absorbed by the system which performs work W; if Q is negative the arrow would be reversed and similarly for W.

19-4 Heat, Work, and the First Law of Thermodynamics

The amount of heat added to or taken from a system depends upon how this is done. (See Section 19-1.) Thus, Q is not a function of the thermodynamic coordinates alone. If the system goes from an initial state to a final state, Q depends not only upon the states but also upon the path taken between the two.* To indicate this state of affairs, a differential change in Q is often written as $đQ$ which means that

i) $đQ$ is infinitesimal in comparison to Q, but
ii) Q is not a function of the thermodynamic coordinates alone.

The work done on (or by) a system also depends in general upon the path between initial and final states. This is most easily seen for a gas confined to a cylinder as shown in Fig. 19-5. The thermodynamic coordinates are the pressure p, volume V, and temperature T of the gas.

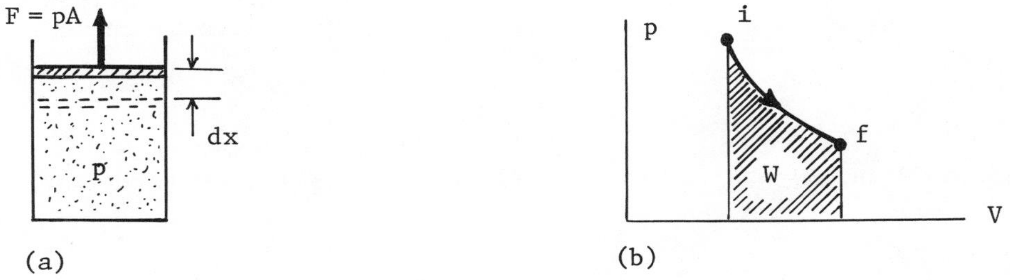

Figure 19-5. (a) A gas confined to a cylinder exerts a force F = pA on the piston where p is the pressure and A the piston area. When the gas volume expands by dV = A dx, the gas does work $đW = pA\,dx = p\,dV$. (b) The work done by the gas in going from an initial state i to a final state f is the area under the curve p versus V and depends upon the path.

For quasistatic processes the force exerted on the piston by the gas is pA where A is the piston area. If the piston moves by an amount dx, the work done by the gas is

$$đW = pA\,dx = p\,dV$$

*One is imagining a quasistatic approximation to the process so that a path in terms of a sequence of values of the thermodynamic coordinates has meaning.

and if the volume changes from $V_{initial}$ to V_{final} the work done is

$$W = \int_{V_i}^{V_f} p\, dV \quad .$$

The value of this integral is the area under that curve of p versus V which represents the path followed by the system in going from the initial to the final state. Note that again the differential change in W is written as $đW$. Neither W nor Q is a function only of the thermodynamic coordinates of a system. Indeed, there is no meaning to the phrase "work in a system" nor to "heat in a system". In short, they are not "of the system" and their values cannot depend upon its state.

The convention is adopted that Q indicates the heat added to the system and W the work done by it. Thus,

Q is + when heat enters the system;
Q is - when the system gives up heat;
W is + when the system does work;
W is - when work is done on the system.

If a system absorbs heat Q and performs work W in going from some initial state to some final state via some path, the difference between the two is found to be path independent and hence dependent only upon the end points, i.e., the initial and final states. From conservation of energy Q - W is the net energy left in (or taken from) the system in going from the initial to final state; it is called the internal energy, and since it depends upon the state of the system it is a function of the thermodynamic coordinates. The first law of thermodynamics is a statement of the conservation of energy and a statement of the existence of an internal energy function U. Mathematically, the first law is

$$U_f - U_i = Q_{if} - W_{if} \tag{19-6a}$$

where Q_{if} is the heat added to the system in going from the initial to final state, W_{if} is the work done by the system, and $U_f - U_i$ is the change in internal energy. In differential form

$$dU = đQ - đW \quad . \tag{19-6b}$$

Note that although $đQ$ and $đW$ are not exact differentials of the thermodynamic coordinates, dU is.

19-5 Ideal Gas Equation of State

The equation of state of a thermodynamic system is a relationship among the thermodynamic coordinates which describes the equilibrium states of the system. These equations are not deductions from thermodynamics but rather are additions to it. One of the simplest thermodynamic systems is that of an ideal gas which approximates real gases at low pressures. The equation of state is

$$pV = \mu RT \tag{19-7a}$$

where p is the pressure, V the volume, μ the number of moles of the gas, T the Kelvin temperature, and R the universal gas constant which has the SI value of

$$R = 8.314 \text{ J/mol}\cdot\text{K} = 1.986 \text{ cal/mol}\cdot\text{K} \quad .$$

There is a second condition that defines an ideal gas macroscopically and that is

$$U = f(T) \quad \text{only} \tag{19-7b}$$

or, the internal energy is a function of the temperature only.

>>> Example 4. Use the definition of an ideal gas and the first law to find a relationship between the specific heat per mole at constant pressure and that at constant volume.

From Section 19-4 we have that the work done by a gas is given by

$$đW = p\ dV \quad .$$

Consider a constant volume process where $đQ$ heat is added. From the first law

$$đQ = dU$$

since at constant volume $đW = 0$. Thus

$$(đQ/dT)_{\text{constant } V} = dU/dT = c_V$$

for one mole of the ideal gas. Now consider a constant pressure process. From Eq. (19-7a) we have

$$đW = p\ dV = R\ dT$$

for $\mu = 1$. The first law then yields

$$(đQ/dT)_{\text{constant } p} = (dU/dT)_{\text{constant } p} + (đW/dT)_{\text{constant } p} \quad .$$

But, since $U = f(T)$ only, dU/dT is always the same value, namely c_V. The left side of the above is c_p, and since $đW/dT = R$ we have

$$c_p = c_V + R \quad . \tag{19-8}$$

<<<

>>> Example 5. An <u>adiabatic process</u> is one in which dQ is zero. Show that for an adiabatic process with an ideal gas pV^γ = a constant where $\gamma = c_p/c_V$.

For μ moles of an ideal gas undergoing an adiabatic process the first law yields

$$đQ = 0 = \mu c_V\ dT + p\ dV \quad .$$

The first term is just dU. We rewrite this as

$$dT = -\ pdV/\mu c_V \quad .$$

The ideal gas equation, Eq. (19-7a), yields

$$p\ dV + V\ dp = \mu R\ dT$$

or

$$(pdV + Vdp)/\mu R = dT \quad .$$

We equate these two expressions for dT, rearrange, and obtain

$$dp/p = - [(c_v + R)/c_v]dV/V = - (c_p/c_v)dV/V = - \gamma dV/V \quad .$$

For an ideal gas c_p and c_v are assumed to be constants and this may be integrated to

$$pV^\gamma = \text{constant} \quad , \quad \text{adiabatic process.} \tag{19-9}$$

Thus if an ideal gas starts at some p_i, V_i, T_i and goes to p_f, V_f, T_f adaibatically, then $p_i V_i{}^\gamma = p_f V_f{}^\gamma$. This leads to another interesting relationship. Since $pV = \mu RT$,

$$TV^{\gamma-1} = \text{constant for an adiabatic process.} \tag{19-10}$$

<<<

19-6 Second Law of Thermodynamics

Any series of quasistatic processes which takes a thermodynamic system from some initial state and returns the system to that state via some path is called a cycle. For a cycle, the change in internal energy is zero. Let

Q_1 = heat absorbed by the system during the cycle
Q_2 = heat rejected by the system during the cycle
W = net work done by the system during the cycle.
} convention for heat engine

If $W > 0$ for a cycle, the mechanical device which causes the system to undergo the cycle is called a heat engine. The efficiency e of the heat engine is the ratio of the work output to the heat input, i.e.

$$e = W/Q_1 \quad . \tag{19-11a}$$

Since the system returns to its initial state in a cycle, $\Delta U = 0$ and from the first law

$$W = Q_1 - Q_2 \quad , \quad \text{heat engine} \tag{19-12}$$

so that

$$e = 1 - Q_2/Q_1 \quad . \tag{19-11b}$$

The heat engine is illustrated graphically in Fig. 19-6a. A reservoir is a system which maintains its temperature even when it gives up or absorbs a finite amount of heat. By convention the hotter reservoir is called T_1 and the colder one T_2. The engine absorbs Q_1 from T_1, performs work W and rejects Q_2 to T_2.

A refrigerator, as is illustrated in Fig. 19-6b, takes heat from a colder reservoir and by having work done on it rejects heat to the hotter reservoir. For a refrigerator

Q_1 = heat rejected to hot reservoir during cycle
Q_2 = heat absorbed from cold reservoir during cycle
W = work done on the system.
} convention for refrigerator

Here we expect Q_1 to be greater than Q_2 and W to be a positive number. From the first law

$$Q_2 - Q_1 = - W$$

where the minus sign is inserted because - W is done <u>by</u> the system. Thus

$$Q_1 = Q_2 + W \quad , \quad \text{refrigerator.} \tag{19-13}$$

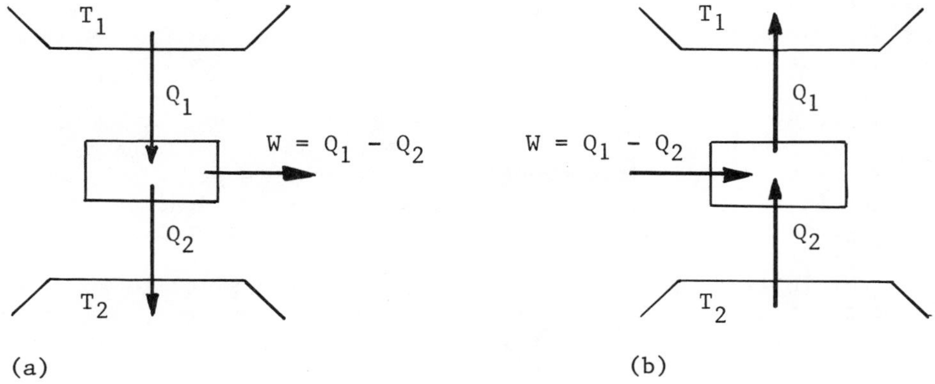

Figure 19-6. (a) Schematic indication of a heat engine which absorbs heat Q_1 from the reservoir at temperature T_1, rejects heat Q_2 to the reservoir T_2 and performs work $W = Q_1 - Q_2$. (b) Schematic representation of a refrigerator which absorbs heat Q_2 from T_2 and rejects heat Q_1 to T_1; work W is done <u>on</u> the system and the system does work - W.

Nothing in the first law prohibits Q_2 being zero for the heat engine nor W from being zero for the refrigerator; however both are <u>contrary</u> to experience. This is the spirit of the <u>second law of thermodynamics</u> which is a statement of experience. The Kelvin-Planck statement of the second law is

> It is impossible to construct an engine that absorbs heat energy Q from a reservoir and performs work equal to Q.

The Clausius statement is

> It is impossible to construct a cyclic device which transfers heat from a colder to a hotter body without some work input.

The two statements can be shown to be equivalent. The second law is a statement that certain processes are irreversible. For example, heat does not flow from a colder to a hotter body. Or to put it another way, if A is initially hotter than B and they are put into contact for a short time, A cools down while B warms up. If we leave them in contact, B does not spontaneously cool back to its original temperature, nor does A warm back up to its original temperature. To effect this, we must perform some work.

19-7 Reversible Processes, Carnot Cycle, Carnot's Theorem

Reversibility of a process in thermodynamics is defined in a very restricted way. To do so, the environment of a system is divided into two parts. The first is its local or relevant environment with which it interacts directly. For example the Bunsen burner that heats a gas is part of its local environment. The rest of its environment is called the rest of the universe (no cosmological implications are intended). The "gas works" that supplies gas to the Bunsen burner is part of the rest of the universe. A reversible process is then defined:

> A reversible thermodynamic process is one which is performed in such a way that at the end, both the system and its local environment may be restored to their initial conditions without change in the rest of the universe.

An example of an irreversible process is that of a gas confined to a cylinder fitted with a piston where friction is present between the piston and the cylinder walls.

All natural processes are irreversible; the reversible process is an idealization of a real process in the same sense that an ideal gas is an idealization of a real gas. A quasistatic process may or may not be reversible, but a reversible process is quasistatic.

The Carnot cycle is a reversible cycle which can be applied to any thermodynamic system. In addition to the system one imagines a hot reservoir T_1 and a cold reservoir T_2. The system is initially in equilibrium with the cold reservoir at temperature T_2. The following four reversible steps are then carried out:*

1. A reversible adiabatic process such that the temperature rises to that of the hot reservoir, T_1,
2. A reversible isothermal absorption of heat Q_1 at temperature T_1,
3. A reversible adiabatic process such that the temperature drops to that of the cold reservoir, T_2,
4. A reversible isothermal rejection of heat Q_2 at temperature T_2 until the initial state is reached.

The Carnot cycle for an ideal gas is shown in Fig. 19-7.

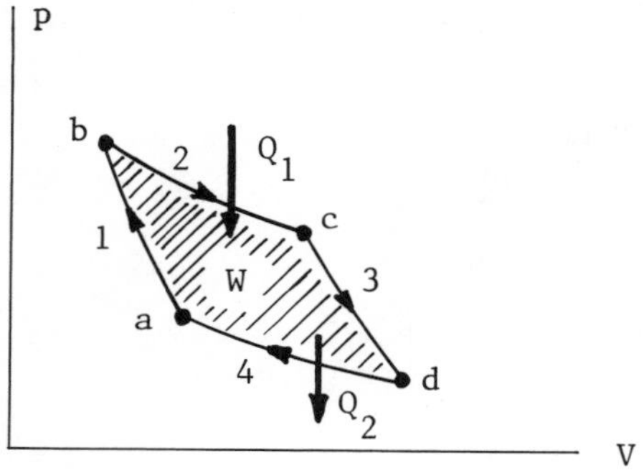

Figure 19-7. Carnot cycle for an ideal gas, abcd. Steps 1 to 4 are discussed in the text.

*A Carnot cycle can be defined more succinctly by insisting that i) there are two fixed reservoirs and ii) the thermodynamic system operating in a cycle absorbs or rejects heat only to these two fixed reservoirs. This fixes two parts of the cycle to being isotherms where heat may be absorbed or rejected and the other two can only be adiabatic because no heat may be absorbed or rejected. Hence between the two reservoirs with these conditions, all reversible cycles are Carnot cycles.

Step 1 starts at a and ends at b. Since this is an adiabatic process it is governed by (see Example 5)

$$pV^{\gamma} = \text{constant}$$

or

$$p_a V_a^{\gamma} = p_b V_b^{\gamma} .$$

Work is done <u>on</u> the gas.

Step 2 starts at b and ends at c. This is an isothermal process; hence (see Programmed Problems)

$$p_b V_b = p_c V_c$$

and the heat absorbed is given by

$$Q_1 = \mu R T_1 \ln (V_c/V_b) .$$

Work is done <u>by</u> the gas.

Step 3 starts at c and ends at d. Again this is adiabatic and the gas does work. Also

$$p_c V_c^{\gamma} = p_d V_d^{\gamma} .$$

Step 4 starts at d and ends back at a. Again this is isothermal so

$$p_d V_d = p_a V_a$$

and the heat <u>rejected</u> by the gas (which is the negative of heat absorbed) is a positive number Q_2 given by

$$Q_2 = \mu R T_2 \ln (V_d/V_a) .$$

The efficiency e of this Carnot engine is given by Eq. (19-11b) as

$$e = 1 - Q_2/Q_1 = 1 - [T_2 \ln (V_d/V_a)]/[T_1 \ln (V_c/V_b)] .$$

But

$$(p_a V_a^{\gamma})(p_b V_b)(p_c V_c^{\gamma})(p_d V_d) = (p_b V_b^{\gamma})(p_c V_c)(p_d V_d^{\gamma})(p_a V_a) ,$$

or

$$V_d/V_a = V_c/V_b$$

and

$$e = 1 - Q_2/Q_1 = 1 - T_2/T_1 , \quad \text{Carnot cycle, ideal gas.}$$

The equation on the preceding page is valid for a Carnot cycle applied to <u>any</u> thermodynamic system although it has only been derived for an ideal gas. Hence Carnot's theorem

> All reversible engines operating between the same two fixed reservoirs have the same efficiency,* namely

$$e = 1 - T_2/T_1 \quad . \tag{19-14}$$

A second theorem or sometimes the second part of the same theorem is

> The efficiency of an irreversible engine operating between two fixed reservoirs cannot be greater than that of a reversible engine operating between the same two fixed reservoirs.

Thus

$$e_{\text{irreversible}} \leq e_{\text{Carnot}} \quad , \quad \text{between same two reservoirs.} \tag{19-15}$$

>>> Example 6. A Carnot cycle run backwards is an ideal refrigerator. The coefficient of performance, K, is the heat extracted from the cold reservoir divided by the work necessary to do this. Show for an ideal gas that K is given by

$$K = T_2/(T_1 - T_2) \quad .$$

From the first law, Eq. (19-13), the heat rejected to the hot reservoir Q_1 is related to the heat absorbed from the cold reservoir Q_2 by

$$Q_1 = Q_2 + W$$

where W is the work done on the system. Thus

$$K = Q_2/W = Q_2/(Q_1 - Q_2) \quad .$$

Because this is a Carnot cycle the ratio Q_2/Q_1 is equal to the ratio T_2/T_1. (This is true for a refrigerator also because the sign convention for both Q_1 and Q_2 is changed.) Thus

$$K = T_2/(T_1 - T_2) \quad .$$

As with e, this result is independent of the thermodynamic system used in the Carnot cycle. <<<

>>> Example 7. If an air conditioner is driven by a motor which has an output of 3/4 hp and the room temperature is 72° F while the outside temperature is 100° F, what is the maximum amount of heat that can be extracted from the room per hour expressed in Btu? While these are the room and outside temperatures we must use the coil temperatures. Typical values might be 50° F and 130° F.

We treat such devices by idealizing them as a Carnot refrigerator in which case the heat absorbed from the room Q_2 is related to the work input by

$$Q_2 = KW = [T_2/(T_1 - T_2)]\, W \quad .$$

---

* See previous footnote.

We must remember that in such problems the temperature is the Kelvin or absolute temperature. Therefore we must convert the room and outside temperatures and find

$$T_2 \simeq 283 \text{ K}$$

$$T_1 \simeq 327 \text{ K} \quad .$$

Thus

$$Q_2 = (283/44)W = 6.4 \text{ W} \quad .$$

Now W is 3/4 hp·hr which is

$$0.75 \text{ hp·hr} = \frac{(0.75 \text{ hp·hr})(550 \text{ ft·lb/s·hp}(3600 \text{ s/hr})}{778 \text{ ft·lb/Btu}} = 1900 \text{ Btu} \quad .$$

Thus $Q_2 \simeq 12{,}200$ Btu, or the maximum possible rate of heat extraction under these conditions is 12,200 Btu per hour. (An actual air conditioner of this size might have a K of 5 in which case it would actually extract heat at about 9,500 Btu/hr.) <<<

19-8 Entropy, Mathematical Formulation of the Second Law

When dealing with heat engines or refrigerators it is customary to express the Q's and W as positive numbers. Whether an amount of heat Q is absorbed or rejected is a subsidiary condition. These are the conventions of the Carnot engine in contradistinction to that normally used in the first law of thermodynamics. With the normal convention for a Carnot cycle one has

$$Q_1/T_1 + Q_2/T_2 = 0 \quad , \quad \text{Carnot cycle, normal convention.}$$

Any reversible path can be approximated as closely as one likes by a series of Carnot cycles with a common isotherm between adjacent cycles. Since the above equation is valid for each Carnot cycle one concludes that,

$$\sum (Q/T) = 0 \quad .$$

The number of Carnot cycles required to approximate an arbitrary closed cycle becomes infinite and in this limit as the number of cycles necessary becomes infinite we write

$$\oint_{\substack{\text{arbitrary}\\\text{reversible}\\\text{cycle}}} đQ/T = 0 \quad . \tag{19-16}$$

This result is called Clausius' theorem. If i is some initial state and f some final state, then if 1 represents some reversible path connecting them and 2 is another, Eq. (19-16) asserts

$$\underset{①}{\int_i^f} đQ/T + \underset{②}{\int_f^i} đQ/T = 0$$

or

$$\underset{①}{\int_i^f} đQ/T = -\underset{②}{\int_f^i} đQ/T = \underset{②}{\int_i^f} đQ/T$$

where the <u>last exchange is only possible because the paths</u> are reversible. Since ① and ② were arbitrary one has that

$$\int_{i}^{f} đQ/dT \quad \text{is path independent.}$$
$$\text{arbitrary reversible path}$$

It follows that there is a function of the thermodynamic variables whose difference is given by this integral (compare with potential energy, Section 7-2). This function is called entropy and is defined by

$$dS = đQ/T \quad , \quad \text{heat transferred reversibly} \tag{19-17}$$

or

$$\int_{i}^{f} đQ/T = S_f - S_i \quad . \tag{19-18}$$
$$\text{arbitrary reversible path}$$

<u>For an irreversible process between equilibrium states i and f, the entropy change is evaluated using Eq. (19-17) and any convenient reversible path which connects i to f;</u> the result is of course path independent.

In a thermodynamic process the change in the entropy of the system plus that of its environment is called the change in entropy of the universe. That is

$$\Delta S_{system} + \Delta S_{environment} = \Delta S_{universe} \quad .$$

The second law of thermodynamics is formulated as the principle of increase of entropy. This is

$$\Delta S_{universe} \geq 0 \tag{19-19}$$

where the equality applies to reversible processes only. For natural (hence irreversible) processes one can say,

> natural processes starting in an equilibrium state and ending in another proceed in such a direction that the entropy of the universe increases.

>>> Example 8. An ideal gas starts at some p_i, V_i, T_i and expands isothermally to some p_f, V_f, T_f. Now consider that it reaches the same final state by expanding adiabatically to V_f and then absorbs heat at this constant volume until p_f, V_f, T_i is reached. Treat these as reversible paths and use entropy to show that $TV^{\gamma-1}$ = constant for adiabatic processes.

During the isothermal expansion the system absorbs heat

$$Q = \mu R T_i \; \ell n(V_f/V_i)$$

so the entropy change of the gas is

$$S_f - S_i = Q/T_i = \mu R \; \ell n(V_f/V_i) \quad .$$

Along the adiabatic path $đQ = 0$, so $\Delta S = 0$. Along the constant volume path

$$đQ = \mu c_v \, dT \quad .$$

Thus along these paths the entropy change of the gas is

$$S_f - S_i = \int_{\substack{\text{constant}\\ \text{volume}\\ \text{path}}} đQ/T = \int_{T_a}^{T_i} \mu c_v \, dT/T$$

where T_a is the temperature at $V_f = V_a$ on the adiabatic path. Thus this path yields

$$S_f - S_i = \mu c_v \, \ell n(T_i/T_a) \quad .$$

Since these are reversible paths the entropy changes are the same, so

$$\mu c_v \, \ell n(T_i/T_a) = \mu R \, \ell n(V_a/V_i)$$

Thus

$$\ell n(T_i/T_a) = \ell n[(V_a/V_i)^{R/c_v}]$$

or

$$\ell n[T_i V_i^{R/c_v}] = \ell n[T_a V_a^{R/c_v}] \quad .$$

since a is any point on the adiabatic path

$$TV^{R/c_v} = TV^{\gamma - 1} = \text{constant} \quad .$$

<<<

19-9 Programmed Problems

The first problem will be of the so-called "calorimeter" type. The emphasis here will be on understanding the main theme of such problems rather than all possible variations.

1.

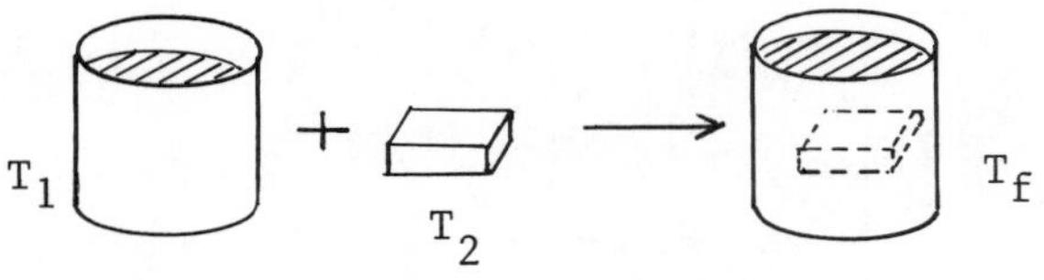

A copper block of mass m_C at temperature T_2 is placed in a calorimeter of mass m_A at T_1. The calorimeter contains water of mass m_W also at T_1. In general the problem is to calculate quantities such as T_f, m_C, etc. for a given set of conditions. For all problems of this type we consider the heat transferred from one part of the system to another. For the case $T_2 > T_1$ which element(s) of this system will absorb heat? Why?

The calorimeter plus water will absorb heat. Heat flows "natually" from hot objects to cooler ones.

2. Given the notion that heat is energy, the previous answer implies that the calorimeter plus water gains energy, i.e. absorbs heat.

Invoking the principle of energy conservation requires that the copper block ______ energy so that the energy change of the system is zero.

loses, gives up.

As is usually the case in problems of this type we are assuming that no heat is lost to or absorbed from the environment.

---

3. Combining the two previous answers we have the central idea of calorimeter problems:

$$\text{Heat Gained} = \text{Heat Lost} \quad .$$

This is equivalent to stating that energy is conserved.

Write an equation appropriate to the following statement: The quantity of heat Q necessary to change the temperature of a body is proportional to the temperature change and the mass of the body.

$Q = mc\Delta T$

The proportionality constant is written as c and is called the specific heat. We consider c as a constant in most problems. Its value depends upon the material of the body.

---

4. If we write this as

$$Q = mc(T_f - T_i)$$

it will be positive (heat gained) or negative (heat lost) depending upon whether T_f is greater than or less than T_i.

Use the algebraic variables given below to write the energy conservation statement for our problem (heat gained = heat lost).

m_A = mass of aluminum calorimeter

c_A = specific heat of calorimeter

m_W = mass of water

c_W = specific heat of water

m_C = mass of copper

c_C = specific heat of copper

T_1 = initial temperature of water

T_2 = initial temperature of copper

T_f = final temperature of combined system.

Heat gained = Heat lost

$$m_Ac_A(T_f - T_1) + m_Wc_W(T_f - T_1) = m_Cc_C(T_2 - T_f) \quad .$$

The water and calorimeter gain heat thus raising their temperature. The copper loses heat resulting in a lower temperature.

$T_1 < T_f < T_2$

The next few frames involve heat conduction through a compound slab. Consider the slab as shown. The cross-sectional area is A. The higher temperature is T_h and the colder side is at T_c; T_i is the temperature at the interface between the materials whose thermal conductivities are k_1 and k_2.

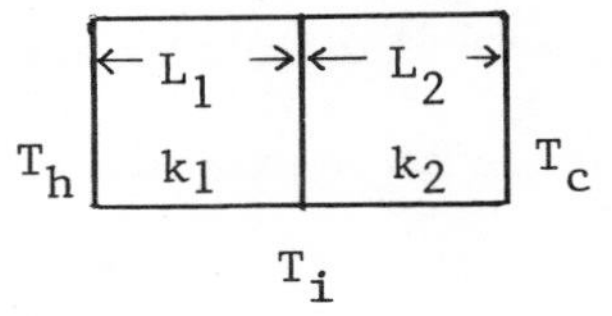

5. Write the equation which governs the rate of heat transfer, H_1, across slab 1.

$H_1 = k_1 A(T_h - T_i)/L_1$

6. What are the SI units for each quantity in this expression?

$[H_1] = \text{kcal/s}, [T_h - T_i] = C^o$,

$[A] = m^2$, $[L_1] = m$,

$[k] = \text{kcal}/(m \cdot s \cdot C^o)$

7. Write the expression for H_2, the rate of heat transfer across slab 2.

$H_2 = k_2 A(T_i - T_c)/L_2$

8. At steady state how are H_1 and H_2 related?

They must be equal, $H_1 = H_2$. If this were not so, then somewhere in the slab more heat would be absorbed than rejected and the temperature there would change. This would not be steady state.

9. Eliminate T_i from the heat transfer equation and find

$$H = H_1 = H_2 \quad .$$

One could solve one equation for T_i; substitute this in the other and find

$$H = \frac{A(T_h - T_c)}{L_1/k_1 + L_2/k_2}$$

10. The quantity L/k is often called the thermal resistance, R. What SI units must R have? Express H in terms of R_1 and R_2.

$[R] = m^2 \cdot s \cdot C^o/\text{kcal}$

$$H = A(T_h - T_c)/(R_1 + R_2)$$

For a compound slab the total thermal resistance is the sum of the individual resistances.

11. Express T_i in terms of T_h, T_c and the R_i.

T_i, T_h, T_c, R_i

From

$$H = A(T_h - T_i)/R_1 = A(T_i - T_c)/R_2$$

One finds

$$T_i = \frac{R_2}{R_1 + R_2} T_h + \frac{R_1}{R_1 + R_2} T_c$$

12. Now, suppose $R_1 >> R_2$.
(a) Which slab is the better insulator?

(b) Write an appropriate approximation for H.

(c) Write an appropriate approximation for T_i.

(a) Slab 1 is the better insulator; it has higher thermal resistance. It is the poorer conductor.

(b) $H \simeq A(T_h - T_c)/R_1$ since $R_1 + R_2 \simeq R_1$. The rate of heat transfer is limited by the ability of the poorest conductor (heat insulator) to transmit this energy.

(c) $T_i \simeq T_c$. The poorer conductor needs more temperature difference than the better conductor to transmit heat at the same rate.

---

13. Now consider a triple slab whose area is $1 m^2$ and whose thermal resistances are $R_1 = 25$, $R_2 = 10$ and $R_3 = 50$ $m^2 \cdot s \cdot C^o/kcal$. Suppose $T_h = 100^o$ C and $T_c = 0^o$ C.

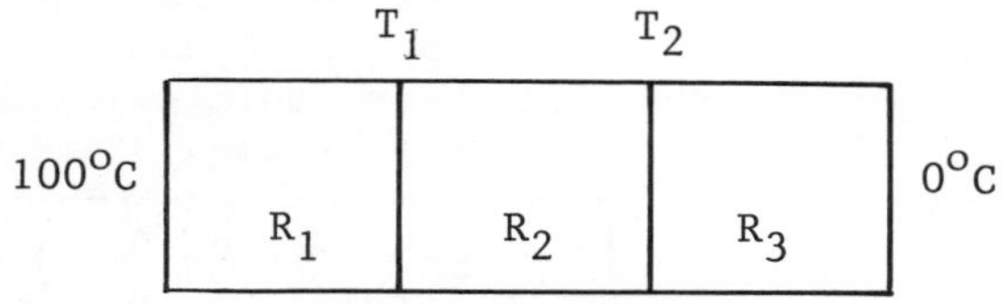

What is the rate of heat transfer across the slab?

$$\text{Since } R_{Total} = R_1 + R_2 + R_3$$
$$= 85\ m^2 \cdot s \cdot C^o/kcal$$

$$H = (1\ m^2)(100\ C^o)/(85\ m^2 \cdot s \cdot C^o/kcal)$$
$$= 1.18\ kcal/s\ .$$

---

14. Test your intuition. Guess the interface temperatures T_1 and T_2 and then compute them.

Clearly the temperature difference is least across R_2 and greatest across R_3. One might guess then $T_2 \simeq 60^oC$ because $R_1 + R_2 = 35$ compared with $R_3 = 50$. Then T_1 ought to be about 2/3 of the way down from 100° C to 60° C or $T_1 \simeq 75^o$ C. One easily finds T_1 from say

$$H = 1.18\ kcal/s$$
$$= (1\ m^2)(100 - T_i C^o/(25\ m^2 \cdot s \cdot C^o/kcal)$$

so $T_1 = 71^o$ C. Also $T_2 = 59^o$ C.

The remaining frames will constitute a review of the ideas of thermodynamics. The basic aim is to help you understand the Carnot cycle, but we will utilize that to cover a variety of topics.

15.

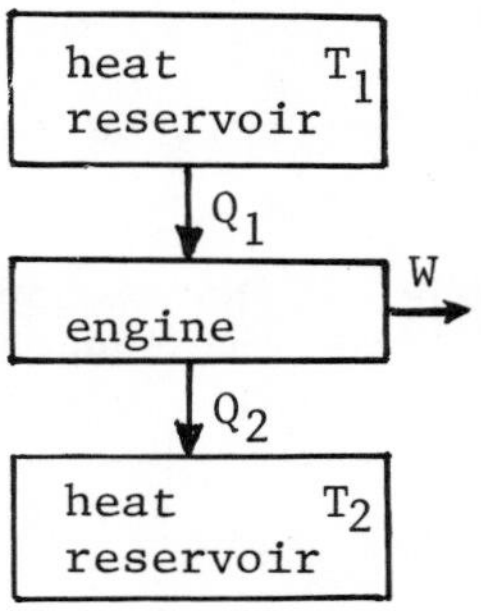

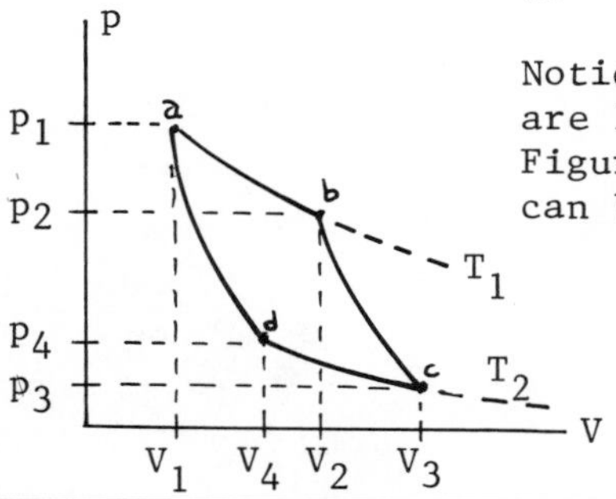

We begin by reviewing the basic ideas of a cyclic heat engine. To the left we show schematically a heat engine as well as a p-V diagram for a Carnot cycle.

With the engine in the condition p_1, V_1, T_1 at a, what is its condition at b as it moves isothermally from a to b?

Notice that the points a,b,c,d are labeled differently from Figure 19-7; a Carnot cycle can be started anywhere.

p_2, V_2, T_1

Isothermally means that the temperature does not change.

---

16. If we think of the engine as containing a piston, the change from V_1 to V_2 during the isothermal expansion will result in the gas, i.e. the engine, doing work on the piston. This non-adiabatic process requires the addition of heat Q_1 to the engine.

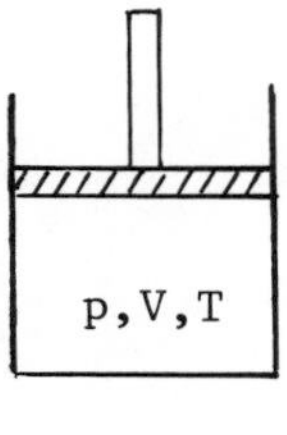

For the adiabatic expansion from b to c:

1. Work is done <u>by</u> the piston. (True/False)
2. The temperature of the gas remains constant. (True/False)
3. Heat is added to the system. (True/False)

1. True, the volume V_3 at c is greater than V_2 at b.

2. False. See p-V diagram of frame 15.

3. False. Adiabatic means no heat enters or leaves the system.

---

17. Thus far in the cycle from a to b to c the gas has performed work and heat has been added. Now we continue from c to d which is an isothermal compression. The piston now does work on the gas. Ordinarily compression would raise the temperature of the gas. What aspect of the engine schematic in frame 15 indicates that the temperature can remain constant?

The arrow marked Q_2 means that an amount of heat Q_2 is removed from the gas during the isothermal compression from c to d. Similarly Q_1 on the diagram represents heat added from a to b.

18. Finally the system proceeds adiabatically from d to a. Describe what happens during this path in terms of work, heat, temperature, etc. Look at the p-V diagram and be specific.

1. Heat neither enters nor leaves, i.e. adiabatic.

2. The temperature returns to the higher temperature T_1.

3. Work is done by the piston on the gas since it is compressed, i.e. work is done <u>on</u> the system.

---

19. To review aspects of the complete cycle:

1. Work is done by the system.
2. Heat is absorbed by the system.
3. Heat is rejected by the system.
4. Work is done on the system.
5. The system returns to its original configuration of p_1, V_1 and T_1.

The first law of thermodynamics relates the change in the internal energy ($U_f - U_i$) of a system to the heat added to the system and the work done by the system. The internal energy of a system is considered to be only a function of the temperature.

What is the change in the internal energy U of this Carnot engine over one cycle?

Zero.

In one cycle, the Carnot engine returns to its original state, so $U_f = U_i$.

---

20. Mathematically the first law of thermodynamics is

$$U_f - U_i = Q_{if} - W_{if}$$

where Q_{if} is the <u>net heat added</u> to the system in going from the initial to final state and W_{if} is the work done by the system.

Looking again at the diagrams of frame 15, what is the specific form of the first law for a cycle of our heat engine?

$U_f - U_i = Q_{if} - W_{if}$

$0 = Q_1 - Q_2 - W$

usually written $Q_1 - Q_2 = W$.

Q_1 = heat absorbed

Q_2 = heat rejected

$Q_1 - Q_2$ = net heat added

W = work done by system

$U_f - U_i = 0$ over one cycle.

21.

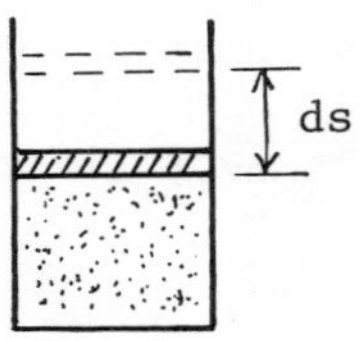

Let us take a closer look at the isothermal expansion from a to b. Recalling that the gas does work on the piston during this part of the path, we show the piston displaced a distance ds. Obviously the volume of the gas will change by a small amount dV. How is this change related to ds and the piston area A?

dV = ______ .

dV = A ds

The gas must expand into the volume vacated by the piston which is A ds. A is the area of the piston.

22. Multiplying the equation in the previous answer by the pressure p exerted by the gas on the piston we have

$$p\,dV = pA\,ds \quad .$$

Since pressure is defined as force per unit area, pA is ______.

Force.

---

23. We have then

$$p\,dV = F\,ds \quad .$$

Verbally, what is meant by the right side of this equation?

F ds is the work $\bar{d}W$ done <u>by</u> the gas on the piston in displacing the piston a distance ds.

24.

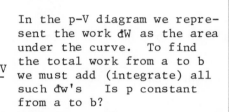

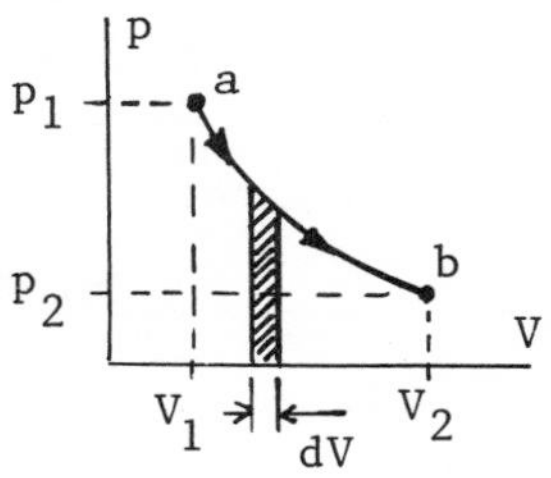

We have then that

$$\bar{d}W = p\,dV \quad .$$

In the p-V diagram we represent the work $\bar{d}W$ as the area under the curve. To find the total work from a to b we must add (integrate) all such $\bar{d}W$'s Is p constant from a to b?

No.

---

25. To calculate the work

$$W = \int_a^b p\,dV$$

we need to know how the pressure varies. If we assume that our gas is ideal, this is given by the equation of state

$$pV = \mu RT$$

where μ is the number of moles of gas and R is a constant.

For the path a to b as shown in the previous frame:
1. p (is/is not) constant.
2. V (is/is not) constant.
3. T (is/is not) constant.
4. μ (is/is not) constant.

1. Is not.
2. Is not.
3. Is (isothermal).
4. Is (no gas escapes from the piston).

26. In view of the last answer we can write the equation of state during the isothermal expansion as

$$pV = \mu RT = \text{constant}.$$

By the way this is known as Boyle's law. Solve the above equation for p and substitute that in the integral of the previous frame.

$$W = \int_a^b \_\_\_\_\_\_ .$$

$p = \mu RT/V$

where μRT = constant.

$W = \mu RT \int_a^b dV/V$

where at a, $V = V_1$ and at b, $V = V_2$. $V_2 > V_1$.

27. The total work then is

$$W = \mu RT \int_{V_1}^{V_2} dV/V \quad .$$

Do this integral.

$$W = \_\_\_\_\_\_ .$$

$W = \mu RT_1 \ \ell n(V_2/V_1)$

where $V_2 > V_1$ and we have added the appropriate subscript to T.

28. The ratio V_2/V_1 is (greater than/less than) one.

Greater than.

29. $\ell n\ (V_2/V_1)$ for $V_2 > V_1$ is (positive/negative).

Positive.

Thus $W = \mu RT_1 \ \ell n\ (V_2/V_1)$ for $V_2 > V_1$ is positive. We say the system <u>does</u> work.

30.

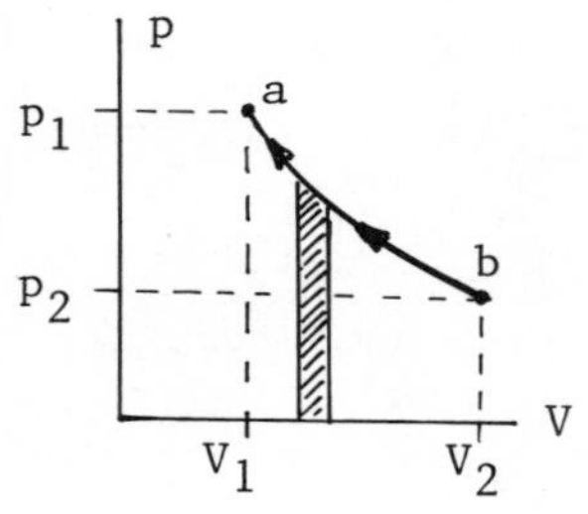

Suppose we began again at frame 24 but this time considered the reverse situation as shown to the left. For this isothermal compression the answer would be

$$W = \mu RT_1 \ \ell n\ (V_1/V_2) \quad .$$

(a) The ratio (V_1/V_2) is (greater than/less than) one.

(b) $\ell n\ (V_1/V_2)$ for $V_1 < V_2$ is (positive/negative).

(a) Less than, $V_1 < V_2$.

(b) Negative. Thus for $V_1 < V_2$

$W = \mu RT_1 \ \ell n\ (V_1/V_2)$ is negative. We say work is done <u>on</u> the system.

31. In summary then

$$W = \mu RT \ \ln (V_f/V_i)$$

where W is positive for isothermal expansion and W is negative for isothermal compression. We say that expanding gases <u>do</u> work while work must be <u>done</u> <u>on</u> gases to compress them.

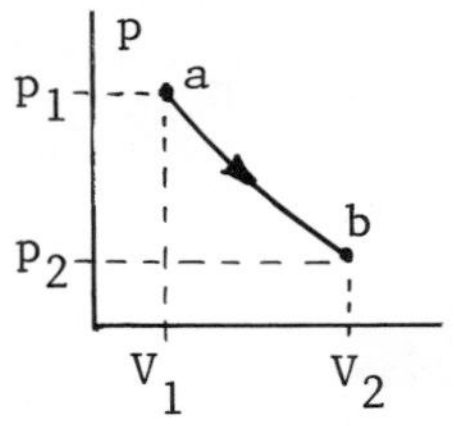

Going back to the isothermal expansion we can now use our result for the work done to compute the heat added.

The first law of thermodynamics relates the change in internal energy U to the work done and the heat added in going from some initial state to some final state.

What is the change in U for our path as shown in the p-V diagram?

Zero.

This is an isothermal expansion; for an ideal gas, U depends only upon the temperature.

---

32. Write the first law of thermodynamics for the isothermal expansion now under discussion.

$U_f - U_i = Q_{if} - W_{if}$.

$0 = Q_{if} - W_{if}$

where by Q_{if} we mean the heat absorbed and by W_{if} the work done by the system.

---

33. Using the result of our calculation for the work we have

$$Q_{if} = W_{if} = \mu RT_1 \ \ln(V_f/V_i) \quad .$$

To what aspect of the schematic representation of the heat engine of frame 15 does this correspond?

Q_1

This is the heat added in going from a to b.

$Q_1 = \mu RT_1 \ \ln(V_2/V_1)$

---

34. Now to determine the efficiency of this heat engine. What we get out of an engine in terms of work must in a sense be "paid" for by the heat we put into the engine. For this engine of frame 15

efficiency e = ______ .

$e = W/Q_1$

Of course one would like this to be 1 (100%).

---

35. In frame 20 we obtained the result that

$$Q_1 - Q_2 = W$$

for one cycle of the engine. Use this equation to eliminate W from

$$e = W/Q_1 \quad .$$

Thus

e = ______ .

$e = W/Q_1$

$W = Q_1 - Q_2$

$$e = \frac{Q_1 - Q_2}{Q_1} = 1 - Q_2/Q_1$$

Chapter 20: KINETIC THEORY OF GASES

REVIEW AND PREVIEW

In the previous chapter you studied the model thermodynamic system, the ideal gas, from a purely thermodynamic part of view. No attempt was made to use the mechanical nature of the particles which make up the gas. In this chapter the microscopic description of an ideal gas will relate average mechanical properties of the gas particles (molecules) to the macroscopic thermodynamic properties which you have struggled to learn. The concepts of this chapter will lead to a deeper understanding of the thermodynamic properties from a more fundamental point of view. For example, previously you met specific heats as a property of materials which was measurable. Now for gases you will learn to relate these to properties of the molecules which make up the gas.

GOALS AND GUIDELINES

You have three major goals in this chapter:

1. Learning the basic microscopic properties of an ideal gas and how these are applied to relate the pressure of the gas to the average of the square of the speed of the gas molecules. The equation (20-7) that results is the basic connection with the thermodynamics description. From this connection the concept of temperature is placed on a microscopic basis.
2. From goal number 1 will follow then the description of the theory of the specific heats of ideal gases. You should learn the basic theory and how to apply it to various types of gases.
3. Learning the concept of mean free path which will lead to a deeper understanding of the concept of density and averages of microscopic properties of gases to relate to the macroscopic properties familiar from thermodynamics.

20-1 Basic Ideas

The individual particles of a system containing an enormous number of particles (one mole of a gas contains about 6×10^{23} molecules) obey the laws of ordinary mechanics which may be classical or quantum. One cannot use these laws as such, however, because no possible experiment could obtain such information as the positions and velocities of all the particles as functions of time. In other words, there is too much information to be useful. One therefore uses certain average quantities to describe such systems. The ordinary mechanical laws are applied statistically and this branch of physics is called statistical mechanics. Using the laws of statistical mechanics, one can derive the laws of thermodynamics. The coordinates and momenta of the individual particles are called microscopic quantities or variables and their appropriate statistical averages are the macroscopic variables of thermodynamics.

Kinetic theory is a subbranch of statistical mechanics in which no attempt is made to derive the thermodynamic laws, but rather their meaning in terms of averages of microscopic variables is made clearer. For mathematical simplicity only the kinetic theory of gases is considered in this section.

20-2 Ideal Gas -- Microscopic Description

An ideal gas microscopically is described by six conditions:

1. The gas consists of identical particles called molecules.
2. The molecules obey Newton's laws of motion and move randomly; i.e. they move in all directions at various speeds.
3. The number of molecules is very large. This justifies the use of statistics -- i.e. the use of averages.

4. The total volume of the molecules is a negligible fraction of the volume occupied by the gas.
5. No appreciable forces act on the molecules except during collisions.
6. All collisions are elastic and of negligible duration; conservation of momentum and kinetic energy are assumed. The change from kinetic energy to potential energy and back again during a collision is ignored.

20-3 Kinetic Theory Interpretation of Pressure and Temperature

In any textbook a relationship is derived which relates the gas pressure p to the density ρ and the average of the square of the molecular speed, $\bar{v}^2$. Such derivations are made in varying degrees of mathematical and physical rigor. All, however, are based upon the microscopic description of an ideal gas and rest upon the idea that the macroscopic quantity, pressure p, is some sort of average of the microscopic collisions (presumed elastic) between the molecules and the walls of the container. The wall must exert a force on an individual molecule to reverse the momentum of the molecule as it collides with the wall. Conversely then, the molecules exert forces on the walls. The statistically appropriate average of these forces per unit area is the pressure.

The result is

$$p = 1/3 \; \rho \; \bar{v}^2$$

in which p is the pressure (a macroscopic quantity), ρ is the density (a macroscopic quantity) and $\bar{v}^2$ is the average of a microscopic quantity, v^2, which is the square of the velocity of a gas molecule.

At your stage of development the actual derivation is not so important. We give one which is reasonably rigorous and which you may want to follow through to develop some insights. It is not necessary that you do so and if you want to skip this then go to Eq. (20-1) in this section.

An ideal gas consists of molecules each of which has mass m. Since the number of molecules is very large and their motion assumed random one cannot actually specify the velocity of any one particle. Rather we specify the number per unit volume having velocity between $\underline{v}$ and $\underline{v} + d\underline{v}$. Let n denote the number of molecules per unit volume and let dn denote the number per unit volume with speed between v and v + dv and velocity direction into a direction between θ and $\theta + d\theta$ and between ϕ and $\phi + d\phi$. We use spherical coordinates here. The direction is said to be into a solid angle $d\Omega$. A solid angle is illustrated in Fig. 20-1. It is the area subtended at a distance r divided by r^2.

The differential solid angle $d\Omega$ is given by $d\Omega = \sin\theta \; d\theta \; d\phi$

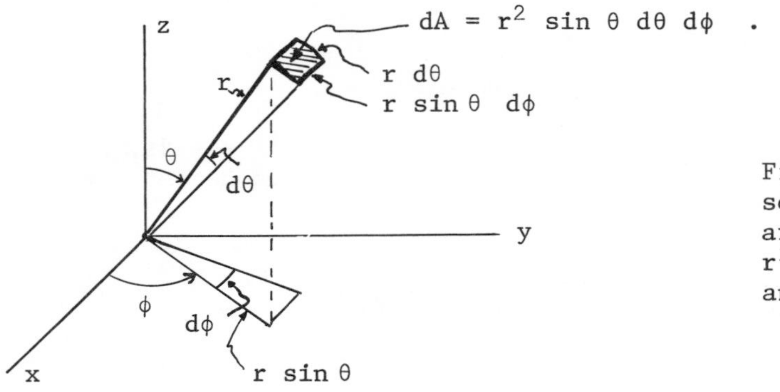

Figure 20-1. Illustration of a solid angle $d\Omega = \sin\theta \; d\theta d\phi$; the area at a distance r is $r^2 d\Omega = r^2 \sin\theta \; d\theta d\phi$. The solid angle of any sphere is 4π.

According to condition 2 for an ideal gas, all directions are equally likely so the number of molecules whose velocity direction is into $d\Omega$ is proportional to the size of the angle. Thus we write

$$dn = [\tilde{n}(v)/4\pi] \sin\theta \, d\theta \, d\phi \, dv \qquad \text{(i)}$$

where $\tilde{n}(v)\, dv$ is the number whose speed lies between v and v + dv and $\tilde{n}(v)$ is a function whose dependence upon v is as yet undetermined. Indeed, it is not needed here. The number per unit volume, n, is however given by

$$n = (1/4\pi) \int_0^\infty \tilde{n}(v)\, dv \int_0^\pi \sin\theta \, d\theta \int_0^{2\pi} d\phi = \int_0^\infty \tilde{n}(v)\, dv \quad .$$

Now, suppose a molecule of mass m collides with a wall of the container at an angle θ relative to the wall normal and with speed v. Its change in momentum is

$$2\, mv \cos\theta \quad . \qquad \text{(ii)}$$

Consider a very small area A of the wall. If all the molecules in a cylinder of cross sectional area $A \cos\theta$ have speed v they collide with this area of the wall. Let this wall be perpendicular to the z axis as shown in Fig. 20-2 and let the cylinder be of length v dt. The number of molecules in this cylinder is by (i)

$$dnA \cos\theta \, v \, dt = A[\tilde{n}(v)/4\pi]\, v \, dv \sin\theta \cos\theta \, d\theta \, d\phi \, dt \quad . \qquad \text{(iii)}$$

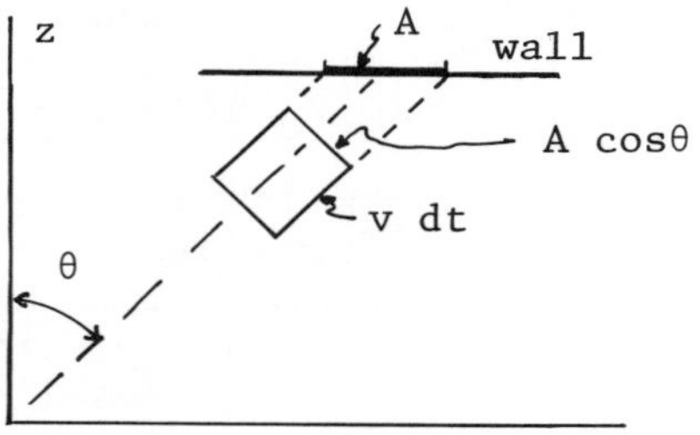

Figure 20-2. All the molecules in a cylinder of cross sectional area A cos θ and length v dt strike an area A of the wall which is perpendicular to the z axis.

In the limit of very small area all the molecules have approximately the speed v and so the rate of change of momentum per unit area (the differential of pressure, dp) due to the molecules in $d\Omega$ is given by multiplying Eqs. (ii) and (iii) and dividing by A and dt. Thus, the contribution to the pressure is

$$dp = 2m\, [\tilde{n}(v)/4\pi]\, v^2 \, dv \cos^2\theta \sin\theta \, d\theta \, d\phi \quad .$$

The total pressure p is obtained by integrating this result over all speeds, all angles ϕ, and over θ from 0 to $\pi/2$. Those in solid angles with $\theta > \pi/2$ are headed toward the opposite wall and are thus not counted. Therefore,

$$p = (2m/4\pi) \int_0^\infty \tilde{n}(v) v^2 \, dv \int_0^{\pi/2} \cos^2\theta \sin\theta \, d\theta \int_0^{2\pi} d\phi$$

$$= (m/3) \int_0^\infty \tilde{n}(v) v^2 \, dv \quad . \qquad \text{(iv)}$$

If f(v) is any function of v its average $\bar{f}$ is

$$\bar{f} = \left[\int_0^\infty f(v)\tilde{n}(v)\,dv \int_0^\pi \sin\theta\,d\theta \int_0^{2\pi} d\phi\right]/4\pi n$$

$$\bar{f} = (1/n)\int_0^\infty f(v)\tilde{n}(v)\,dv \quad .$$

Thus Eq. (iv) may be written as

$$p = (m/3)\,\overline{v^2}n = (1/3)\,\rho\overline{v^2} \qquad (20\text{-}1)$$

where ρ is the density (mass per unit volume) given by

$$\rho = nm \quad . \qquad (20\text{-}2)$$

Eq. (20-1) is a fundamental equation of the kinetic theory of gases because it establishes a relationship between the average of a microscopic quantity (v^2) and a macroscopic quantity (p).

The positive square root of $\overline{v^2}$ is called the root-mean-square speed and denoted v_{rms}. In equation form

$$v_{rms} = \sqrt{\overline{v^2}} = \sqrt{3p/\rho} \quad . \qquad (20\text{-}3)$$

In the following we will define the terms

m = mass of a molecule

$\underline{M}$ = total mass of the gas

M = molecular weight

μ = number of moles

N = total number of molecules

N_o = Avogadro's number, the number of molecules per mole.

These are related by

$$N = \mu N_o \qquad (20\text{-}4)$$

and

$$\underline{M} = M\mu = mN = m\mu N_o \quad . \qquad (20\text{-}5)$$

Also in terms of the density ρ and volume V of the gas the total mass is

$$\underline{M} = \rho V \quad . \qquad (20\text{-}6)$$

The kinetic theory interpretation of temperature is obtained by equating the kinetic theory expression for pressure, Eq. (20-1), to the corresponding thermodynamic expression, Eq. (19-7a). Thus

$$p = \mu RT/V = (1/3)\rho\overline{v^2}$$

so

$$(3/2)\mu RT = (1/2)\rho V\overline{v^2} = (1/2)\underline{M}\overline{v^2} \quad . \tag{20-7}$$

The interpretation of Eq. (20-7) is that

> The total translational kinetic energy of the molecules of an ideal gas is proportional to the gas temperature.

Several alternative forms of Eq. (20-7) are possible. Using $\underline{M} = M\mu$ one finds

$$(3/2)RT = (1/2)M\overline{v^2} = \text{total translational kinetic energy per mole.} \tag{20-8a}$$

Finally the average translational kinetic energy per molecule is

$$(1/2)m\overline{v^2} = (1/2)M/N_o\ \overline{v^2} = (3/2)(R/N_o)\ T = (3/2)\ kT \tag{20-8b}$$

where k is the Boltzmann constant given by

$$k = R/N_o = 1.38 \times 10^{-23}\ \text{joule/(molecule·K)} \quad ; \tag{20-9}$$

k plays the role of the gas constant per molecule.

If a gas consists of a mixture of two types of molecules of masses m_1 and m_2 the kinetic theory of temperature says

$$m_1\overline{v_1{}^2} = 3\ kT = m_2\overline{v_2{}^2}$$

or

$$\sqrt{m_1}\ v_{1,rms} = \sqrt{m_2}\ v_{2,rms} \quad . \tag{20-10}$$

Eq. (20-10) has been amply confirmed by experiment.

>>> Example 1. Compute the root-mean-square speed of He at room temperature (72° F). The molecular weight of He is 4 gram/mole. At what temperature would the root-mean-square speed be 10 times this value?

From Eq. (20-8a)

$$v_{rms} = \sqrt{3RT/M} \quad .$$

Now 72° F ≃ 295 K (only the Kelvin or absolute temperature is used in these ideal gas relationships). Thus

$$v_{rms} = \sqrt{(8.137\ \text{J/mol·K})(295\ \text{K})(3)/[4 \times 10^{-3}\ \text{kg/mol}]}$$

$$= 1356\ \text{m/s} \quad .$$

Since v_{rms} is proportional to the square root of the absolute temperature

$$v_{rms}(T_1)/v_{rms}(T_2) = \sqrt{T_1/T_2} \quad .$$

To increase v_{rms} by a factor of 10, the temperature must increase by a factor of 100.
<<<

20-4 Specific Heats of an Ideal Gas

For an ideal gas the internal energy U is translational only* and is given by

$$U = (3/2)NkT = (3/2)\mu RT \quad . \tag{20-11}$$

From Example 4 of Section 19-5 the molar (μ = 1 mole) specific heat at constant volume is

$$c_v = dU/dT = d[(3/2)RT]/dT = (3/2)\ R \quad . \tag{20-12a}$$

Then, from Eq. (19-8) the molar specific heat at constant pressure is

$$c_p = c_v + R = (5/2)\ R \quad . \tag{20-12b}$$

>>> Example 2. If the volume of an ideal gas is halved adiabatically, find the ratio of initial to final pressures.

From Eq. (19-9) for an adiabatic process

$$p_i V_i^{\gamma} = p_f V_f^{\gamma} \quad .$$

For an ideal gas, $\gamma = c_p/c_v = 5/3 = 1.67$. Thus, if $V_f = \frac{1}{2} V_i$ we have

$$p_i/p_f = 1/2^{\gamma} = 1/2^{1.67} \simeq 0.31\overline{5} \ .$$ <<<

20-5 Equipartition of Energy

Simple kinetic theory which is based upon point molecules fails to properly account for the specific heats of other than monatomic gases, i.e. one atom per molecule. This is because polyatomic molecules can have energy in forms other than translational. The total energy of a molecule of finite extent may be written as

$$E = \underbrace{\tfrac{1}{2}\, m \sum_{j=1}^{3} v_j^2}_{\text{translation}} + \underbrace{\tfrac{1}{2} \sum_{j=1}^{3} I_j \omega_j^2}_{\text{rotation}} + \underbrace{E_{vib}}_{\text{vibration}}$$

where I_j is a moment of inertia. The vibrational energy will have the form

$$\underbrace{\tfrac{1}{2}\, B\, (d\delta/dt)^2}_{\substack{\text{vibrational}\\ \text{kinetic}\\ \text{energy}}} + \underbrace{\tfrac{1}{2}\, C\delta^2}_{\substack{\text{vibrational}\\ \text{potential}\\ \text{energy}}}$$

*See Section 20-5, however.

for each mode of vibration where δ is the appropriate vibrational coordinate. Each mode of motion by which a molecule possesses energy is called a degree of freedom. Note that each term in the total energy has the form $\frac{1}{2} A\psi^2$ where ψ is either a coordinate or velocity component and A is the appropriate constant. The law of equipartition of energy which may be proved by statistical mechanics states that the average of each of these is the same and depends only on the temperature. Thus

$$\frac{1}{2} A\overline{\psi^2} = \frac{1}{2} kT \quad . \tag{20-13}$$

If f is the number of degrees of freedom, (one must always remember that a vibrational mode yields two, one for kinetic and one for potential energy), the total internal energy per mole is

$$u = \frac{1}{2}f\, N_o kT = \frac{1}{2}\, fRT \quad . \tag{20-14}$$

Then the molar specific heat at constant volume is

$$c_v = \frac{1}{2}\, fR \tag{20-15a}$$

and that at constant pressure is

$$c_p = \frac{1}{2}\,(f + 2)\, R \tag{20-15b}$$

so the ratio of specific heats γ is

$$\gamma = (f + 2)/f \quad . \tag{20-16}$$

In Table 20-1 are shown the values of c_p, c_v and γ for several types of gases.

Table 20-1

Type	f	Motion	c_v	c_p	γ
monatomic	3	translation	3R/2	5R/2	5/3
diatomic	5	3 translation 2 rotation	5R/2	7R/2	7/5
diatomic	7	3 translation 2 rotation 1 vibration	7R/2	9R/2	9/7
polyatomic	6	3 translation 3 rotation	3 R	4 R	4/3
polyatomic	8	3 translation 3 rotation 1 vibration	4 R	5 R	5/4
polyatomic	10	3 translation 3 rotation 2 vibration	5 R	6 R	6/5

20-6 Mean Free Path

The mean free path, $\bar{\ell}$, is the average distance traveled by a molecule of a gas between collisions with other molecules. Its value depends upon the particle density (number of particles per unit volume), n, and the molecular size. Let d be the molecular diameter, then the expression for the mean free path is

$$\bar{\ell} = 1/(\sqrt{2}\ \pi d^2 n) \quad . \qquad (20\text{-}17)$$

The collision frequency Z, i.e. the number of collisions per unit time, is obtained by dividing the mean or average speed* by $\bar{\ell}$. Thus

$$Z = \bar{v}/\bar{\ell} \quad . \qquad (20\text{-}18)$$

>>> Example 3. Find the number of molecules per unit volume at room temperature if the pressure is 10^{-4} mm-Hg. If the molecular diameter is 10^{-8} cm, what is the mean free path?

First of all, the pressure at 1.0 mm-Hg is $1.33 \times 10^2\ N/m^2$ so at 10^{-4} mm-Hg

$$p = 1.33 \times 10^{-2}\ N/m \quad .$$

Now

$$pV = \mu RT = \mu N_o kT = NkT \quad ,$$

so

$$n = N/V = p/kT \quad .$$

At room temperature (22° C)

$$kT = 4.07 \times 10^{-21}\ J/molecule$$

so

$$n = (1.33 \times 10^{-2}\ N/m^2)/(4.07 \times 10^{-21}\ J/molecule)$$

or

$$n = 3.3 \times 10^{18}\ molecule/m^3 \quad .$$

The mean free path then is

$$\bar{\ell} = 1/\{\sqrt{2}\ \pi(10^{-10}\ m)^2(3.3 \times 10^{18}\ m^{-3})\} \simeq 7.0\ m \quad .$$

If the average molecular speed is of the order of 10^3 m/s the collision frequency is

$$Z = \bar{v}/\bar{\ell} = (10^3\ m/s)/(7.0\ m) \simeq 140 \text{ collisions per second.}$$

<<<

20-7 Distribution of Molecular Speeds

The distribution of molecular speeds was first found by C. Maxwell. His expression, known as a Maxwell distribution, gives the most probable distributions of speeds for a gas

*See Section 20-7.

sample containing a very large number of molecules. With n the number of molecules of mass m per unit volume and $\tilde{n}(v)$ dv the number per unit volume with speed between v and v + dv the Maxwell distribution function is

$$n(v) = 4\pi n \ (m/2\pi kT)^{3/2} \ v^2 e^{-(mv^2/2kT)} \quad . \tag{20-19}$$

The number of molecules per unit volume with speed between v and v + dv irrespective of direction is n(v) dv and

$$n = \int_0^\infty \tilde{n}(v) \ dv \quad . \tag{20-20}$$

If $\tilde{n}(v)$ is integrated over the gas volume then $\tilde{N}(v)$ dv is the number of molecules <u>in the sample</u> with speed between v and v + dv, and

$$\tilde{N}(v) = \int \tilde{n}(v) \ dV \quad .$$

The total number of molecules in the sample is

$$N = \int_0^\infty \tilde{N}(v) \ dv \quad .$$

Three sorts of speeds are usually considered. They are the <u>average</u> or <u>mean</u> speed $\bar{v}$ defined by

$$\bar{v} = (1/n) \int_0^\infty v\tilde{n}(v) \ dv = \sqrt{8kT/\pi m} \quad , \tag{20-21a}$$

the <u>mean-square</u> speed (which appears for example in Eqs. (20-1) and (20-8)) defined by

$$\overline{v^2} = (1/n) \int_0^\infty v \ \tilde{n}(v) \ dv = 3kT/m$$

from which the <u>root-mean-square</u> speed v_{rms} follows

$$v_{rms} = \sqrt{3kT/m} \tag{20-21b}$$

and the most probable speed which is the speed at which $\tilde{n}(v)$ has its maximum. Thus v_p is determined by the condition

$$d\tilde{n}(v)/dv = 0$$

and

$$v_p = \sqrt{2kT/m} \quad . \tag{20-21c}$$

So long as the Maxwell distribution law applies, the ordering for any temperature is

$$v_p < \bar{v} < v_{rms} \quad .$$

PART II

PRELIMINARY CONCEPTS

Vector Algebra

In Part II we assume that the student has a familiarity with vectors and certain operations involving vectors (addition, subtraction, dot and cross products). A summary of these concepts is given in Chapter 2. However, if these are not well understood, we suggest that the student review these topics in detail.

Fields

Suppose a fluid is flowing as indicated in the Figure. At some point P in the fluid there is a certain temperature T and a certain velocity $\underline{\underline{v}}$. These quantities will in general vary from point to point; their values at P might differ from their values at another point Q. Thus to be more explicit we should write T(x,y,z) and $\underline{\underline{v}}$(x,y,z) to indicate that T and $\underline{\underline{v}}$ are functions of the location (x,y,z). This example illustrates what is meant by a <u>field</u>: a field is a quantity which is a function of location.

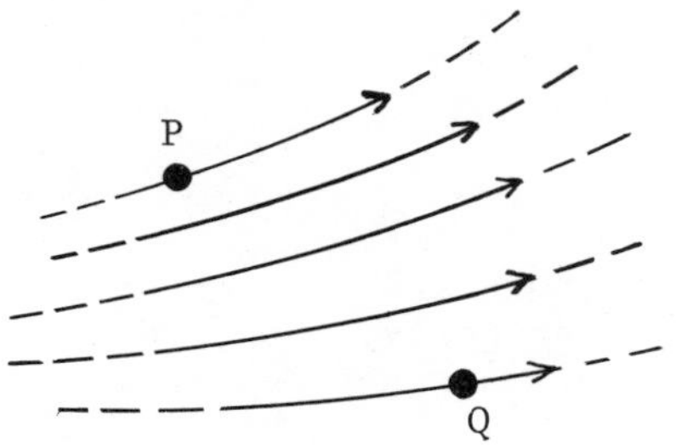

The pattern illustrates fluid flow. The temperatures at P and Q in general will differ; similarly, the velocities at these two points will be different (both in magnitude and direction). The temperature is a scalar field; the velocity is a vector field.

When specifying the value of a field, one should always state the location with which that value is associated. To say that <u>the</u> temperature is 300 K is insufficient; rather we should say that the temperature <u>at point P</u> is 300 K. Sometimes this location is understood instead of being stated explicitly. For example, when we say that the magnitude of the acceleration due to gravity is 980 cm/s^2 we really mean that it is 980 cm/s^2 at (or near) the surface of the earth.

The fields which we will encounter in physics may be classified as either <u>scalar fields</u> or <u>vector fields</u>. Some examples may clarify these concepts:

a) The mass of an atom, the length of a stick are both scalars but not fields.
b) The temperature of a moving fluid, the pressure in a vibrating column of air are both scalar fields.
c) The velocity of a projectile, the angular momentum of a spinning top are both vectors but not fields.
d) The velocity of a moving fluid, the acceleration due to the earth's gravitational attraction are both vector fields.

Question

Classify (in the above manner) all the quantities you can think of pertaining to a vibrating string (e.g. mass, transverse velocity, etc.).

Chapter 21: CHARGE AND MATTER

REVIEW AND PREVIEW

You have previously encountered various force laws. For example, you have studied the force due to a stretched spring ($F = -kx$) and the gravitational force exerted between point masses ($F = G\,m_1 m_2/r^2$). In this chapter you will deal with another force: the electrostatic force. Like the gravitational force, the electrostatic force exists between a pair of bodies which need not be in contact with each other. The electrostatic force is related to a property of the bodies called charge.

GOALS AND GUIDELINES

In this chapter there are two major goals:

1. Learning the concepts of
   a. charge - a property of a body
   b. current - describes a motion of charges.
2. Learning and understanding how to use Coulomb's law. This is the force law which gives the electrostatic force between point charges.

Since force is a vector quantity, the problem of calculating the total electrostatic force exerted on a single charged body due to several other charged bodies leads to a vector addition of forces. You may wish to review vector addition (Chapter 2).

21-1 Charge

Atoms may be considered to consist of electrons, protons and neutrons; these particles possess certain definite properties. One such property, mass, relates to the inertia of the particle and has been studied in mechanics. Another intrinsic property of these particles is their electrical charge, q, the unit of charge being the coulomb, C. This property is related to the particles being able to exert electrical forces on each other. Unlike mass, charge can be of either algebraic sign (plus or minus). The following table lists some charges and masses which occur frequently in problems.

Particle	Charge	Mass
electron	$-e$	9.11×10^{-31} kg
proton	e	1.67×10^{-27} kg
neutron	0	1.67×10^{-27} kg
deuteron (proton + neutron)	e	3.34×10^{-27} kg
alpha particle (Helium nucleus)	$2e$	6.68×10^{-27} kg

The charge $e = 1.60 \times 10^{-19}$ C is called the electronic charge (note that the charge of the electron is $-e$, not e).

21-2 Conductors and Insulators

Most materials fall into one of the following two categories:

1. Conductors (such as metals, salt water solution) allow the motion of charges (usually electrons for metals and ions for solutions) under the application of external forces.

2. Insulators (such as glass, rubber) do not allow the motion of charges through them or along their surfaces.

For example, a charge placed on the tip of a glass rod will remain there; a charge placed on the tip of a copper rod is free to redistribute itself throughout the rod.

21-3 Current

Current, i, describes a flow of charge. If a charge q flows past a given point (say in a metal wire) in a time interval t, then the (average) current is given by

$$i = q/t \quad \text{(average current).} \tag{21-1}$$

To obtain the instantaneous current, i(t), we must take the charge dq which flows past the given point in the infinitesimal time interval dt and form the quotient

$$i(t) = dq/dt \quad \text{(instantaneous current).} \tag{21-2}$$

The unit of current is the ampere, A (=coulomb/second). Note that the charge "q" appearing in the definition of current is not the charge in the wire (the net charge in the wire is usually zero) but rather the charge which has been transported past a given point.

Although current is a scalar, we may assign a sense to it as shown in Figure 21-1. By convention the sense of the current is that of the motion of the positive charges (if these constitute the current) or opposite to the motion of the negative charges (if these constitute the current).

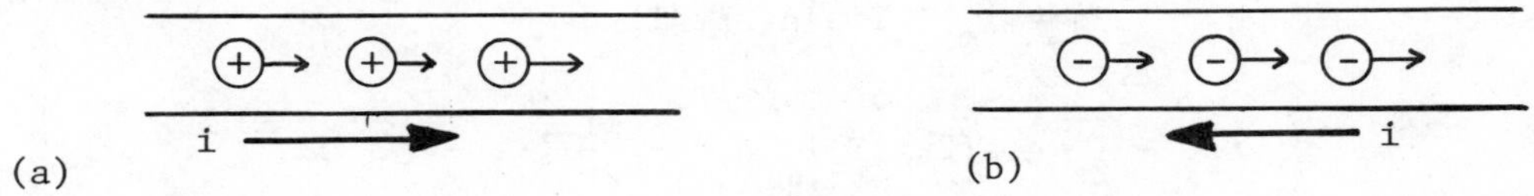

Figure 21-1. Illustrating the relationship between the flow of charge and the sense of the current. (a) Positive charges moving to the right. By convention the current is to the right. (b) Negative charges moving to the right. By convention the current is to the left.

>>> Example 1. An electron beam constitutes a current of 5 μA. After one hour of time, a) What is the magnitude of the charge which has been transported by this beam? b) How many electrons have been transported? c) What is the total mass of these electrons?

a) Since the current is constant, i = q/t or

$$q = it = (5 \times 10^{-6}\ \text{A})(1\ \text{h})(3.6 \times 10^{3}\ \text{s/h})$$

$$= 1.80 \times 10^{-2}\ \text{A·s} = 1.80 \times 10^{-2}\ \text{C} \ .$$

Since the electrons have a negative charge, -1.80×10^{-2} coulombs have actually been transported.

b) Let N = number of electrons transported. Then $q = Ne$,

$$N = q/e = (1.80 \times 10^{-2}\ \text{C})/(1.60 \times 10^{-19}\ \text{C/electron})$$
$$= 1.13 \times 10^{17}\ \text{electrons}\quad .$$

"Electron" is of course not a unit; however, its use can be helpful in problems of this type.

c) Let M = total mass of these electrons. Then

$$M = Nm_e = (1.13 \times 10^{17}\ \text{electron})(9.11 \times 10^{-31}\ \text{kg/electron})$$
$$= 1.03 \times 10^{-13}\ \text{kg}\quad .$$ <<<

>>> Example 2. The current, $i(t)$, produced by a "half-wave rectifier" is given by

$$i(t) = \begin{cases} I \sin(2\pi t/T), & 0 \leq t \leq T/2; \\ 0, & T/2 \leq t \leq T; \end{cases}$$

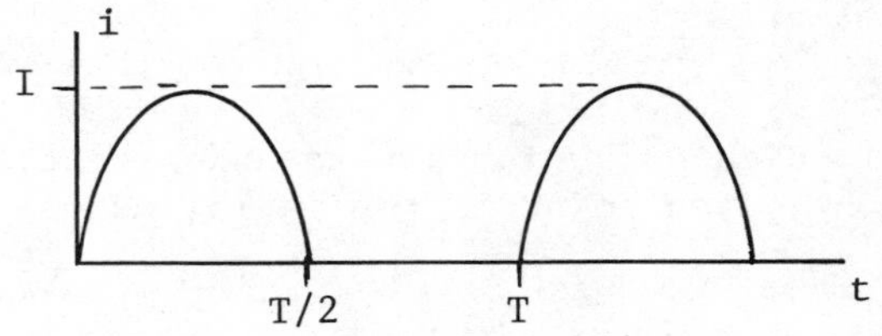

etc. as shown. The constant I is the maximum current. a) What is the total charge which has been transported during one complete cycle (say from $t = 0$ to $t = T$)? b) What constant current would transport the above charge in a time interval T?

a) In this problem the current is not constant; therefore, "$q = it$" is incorrect. Since $i = dq/dt$, $dq = i\,dt$. The total charge q is then the sum (integral) of all the dq's, i.e.

$$q = \int dq = \int_0^{T/2} I \sin(2\pi t/T)\,dt + \int_{T/2}^{T} 0\,dt$$
$$= -\frac{IT}{2\pi}\cos(2\pi t/T)\Big|_{t=0}^{T/2} = -\frac{IT}{2\pi}[-1\ -1] = IT/\pi\quad .$$

Note that it was necessary to divide the time interval into two pieces (0 to T/2, T/2 to T) in order to do the integration.

b) Let i_{av} be that constant current which will transport the above charge q in a time interval T. Then

$$i_{av} = q/T = I/\pi\quad .$$

Remark: The current i_{av} (the "average current") is in general given by

$$i_{av} = \frac{1}{T}\int_0^T i(t)\,dt\quad .$$ <<<

21-4 Coulomb's Law

A point charge refers to a charged body whose size is negligible in comparison with the distance between such bodies. Two point charges will exert electrostatic forces on each other as shown in Figure 21-2. In the figure the notation $\mathbf{F}_{12}$ means the force exerted on charge #1 by charge #2.

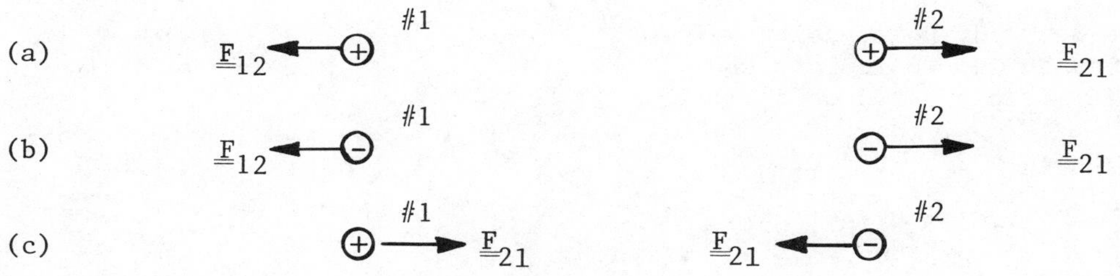

Figure 21-2. The electrostatic force between point charges. (a) Two positive charges repel each other. (b) Two negative charges repel each other. (c) Two opposite charges attract each other.

These forces obey the following rules:

1. The forces lie along the line connecting the two charges. (This would be meaningless if we were not dealing with point charges.)
2. The forces form an equal and opposite pair: $\mathbf{F}_{12} = -\mathbf{F}_{21}$. This is true regardless of the magnitudes and signs of the two charges.
3. The magnitude of the force $|\mathbf{F}_{12}| = |\mathbf{F}_{21}|$ is directly proportional to the product of the magnitudes of the two charges and inversely proportional to the square of the distance between the charges. (Again, the distance between the charges has meaning only for point charges.)
4. The sense of the pair of forces is repulsive for charges of the same sign and attractive for charges of opposite sign.

The above facts are summarized in Coulomb's law which gives the electrostatic force between a pair of point charges (q_1, q_2) which are separated by a distance r:

$$F = \frac{1}{4\pi\varepsilon_o}\frac{q_1 q_2}{r^2} \qquad (21\text{-}3)$$

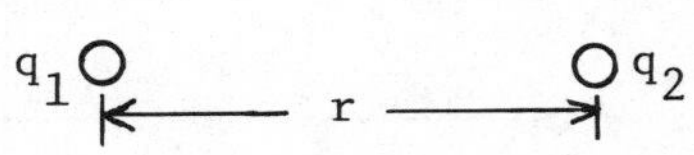

The constant of proportionality, $1/4\pi\varepsilon_o$, has the value $1/4\pi\varepsilon_o = 9.0 \times 10^9\ \mathrm{N{\cdot}m^2/C^2}$.* In applying Coulomb's law to problems one must keep in mind not only the equation (21-3), but also the four properties listed above. It is particularly important to remember that:

a) Coulomb's law holds only for point charges.
b) The force $\mathbf{F}_{12}$ is a vector. If there are several point charges present the total electrostatic force, $\mathbf{F}_1$, exerted on charge #1 by all the other charges is given by the vector sum:

$$\mathbf{F}_1 = \mathbf{F}_{12} + \mathbf{F}_{13} + \mathbf{F}_{14} + \ldots .$$

---

*The constant $\varepsilon_o = 8.85 \times 10^{-12}\ \mathrm{C^2/N{\cdot}m^2}$ is called the permittivity constant.

Students frequently have trouble with <u>signs</u> in applying Coulomb's law to problems. There are two methods of consistently handling this difficulty:

Method #1. If the signs of all the charges are known, then the actual directions (and senses) of all the electrostatic forces are known. A force diagram (showing <u>all</u> the forces acting <u>on</u> the body in question) can be drawn consistent with these known directions. The magnitudes of these forces can then be computed using the <u>magnitudes</u>, $|q|$, of the various charges in the Coulomb's law formula (21-3). This method is illustrated below.

>>> Example 3. Four point charges form a rectangle as shown. Calculate the electrostatic force exerted on q_1 if

$$q_1 = q_2 = 2 \times 10^{-9} \text{ C},$$

$$q_3 = q_4 = -3 \times 10^{-9} \text{ C}.$$

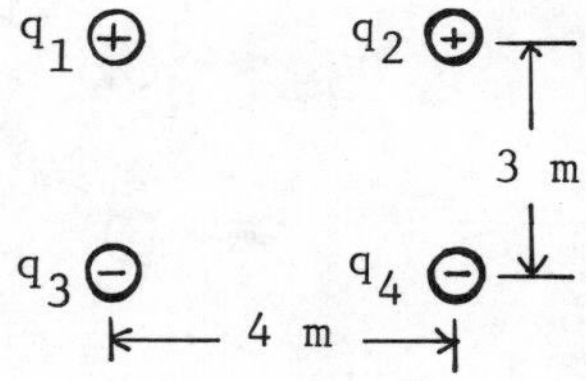

We first draw the force diagram showing all the electrostatic forces acting on q_1.

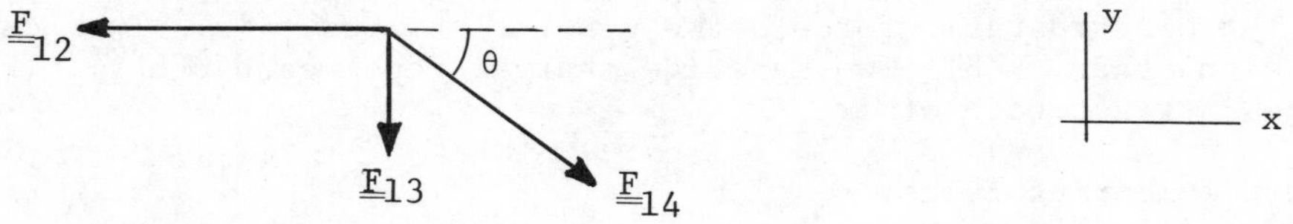

Note that $\mathbf{F}_{12}$ is repulsive while $\mathbf{F}_{13}$ and $\mathbf{F}_{14}$ are attractive. We then resolve these forces into their x and y components. The work is best done in tabular form:

	x component	y component
$\mathbf{F}_{12}$	$-\lvert\mathbf{F}_{12}\rvert = -\dfrac{1}{4\pi\varepsilon_o}\dfrac{\lvert q_1\rvert\lvert q_2\rvert}{r_{12}^2}$	0
$\mathbf{F}_{13}$	0	$-\lvert\mathbf{F}_{13}\rvert = -\dfrac{1}{4\pi\varepsilon_o}\dfrac{\lvert q_1\rvert\lvert q_3\rvert}{r_{13}^2}$
$\mathbf{F}_{14}$	$+\lvert\mathbf{F}_{14}\rvert \cos\theta = \dfrac{1}{4\pi\varepsilon_o}\dfrac{\lvert q_1\rvert\lvert q_4\rvert}{r_{14}^2}\dfrac{4}{5}$	$-\lvert\mathbf{F}_{14}\rvert \sin\theta = -\dfrac{1}{4\pi\varepsilon_o}\dfrac{\lvert q_1\rvert\lvert q_4\rvert}{r_{14}^2}\dfrac{3}{5}$

Adding these columns we have the components, F_{1x} and F_{1y}, of the resultant electrostatic force $\mathbf{F}_1 = \mathbf{F}_{12} + \mathbf{F}_{13} + \mathbf{F}_{14}$.

$$F_{1x} = \frac{|q_1|}{4\pi\varepsilon_o}\left(-\frac{|q_2|}{r_{12}^2} + \frac{|q_4|}{r_{14}^2}\frac{4}{5}\right)$$

$$= (2 \times 10^{-9}\ \text{C})(9 \times 10^{9}\ \text{N}\cdot\text{m}^2/\text{C}^2)\left(-\frac{2 \times 10^{-9}\ \text{C}}{4^2\ \text{m}^2} + \frac{3 \times 10^{-9}\ \text{C}}{5^2\ \text{m}^2}\frac{4}{5}\right)$$

$$= -\ 5.2 \times 10^{-10}\ \text{N}.$$

$$F_{1y} = \frac{|q_1|}{4\pi\varepsilon_o}\left(-\frac{|q_3|}{r_{13}^2} - \frac{|q_4|}{r_{14}^2}\frac{3}{5}\right)$$

$$=(2 \times 10^{-9}\ \text{C})(9 \times 10^{9}\ \text{N}\cdot\text{m}^2/\text{C}^2)\left(-\frac{3 \times 10^{-9}\ \text{C}}{3^2\ \text{m}^2} - \frac{3 \times 10^{-9}\ \text{C}}{5^2\ \text{m}^2}\frac{3}{5}\right)$$

$$= -\ 7.3 \times 10^{-9}\ \text{N}.$$

These components completely specify the resultant force $\mathbf{F}_1$.

Remarks:

1. Only the magnitudes, $|q|$, of the various charges were used in the calculation; their signs were taken into account in drawing the force diagram.

2. By adding the entries in the table algebraically instead of numerically some computational labor was saved: $|q_1|/4\pi\varepsilon_o$ was a common factor. <<<

Method #2. A force diagram is drawn as if all the electrostatic forces were repulsive. Then the algebraic charges, q, including sign are substituted into the Coulomb's law formula (21-3). This method is particularly useful when the sign of some charge is not known. The following example illustrates this method.

>>> Example 4. Three point charges are arranged in a line as shown. What must q_3 be in order that the total electrostatic force exerted on q_2 be 1.0×10^{-8} N toward the right? Take $q_1 = -\ 4 \times 10^{-9}$ C, $q_2 = 2 \times 10^{-9}$ C.

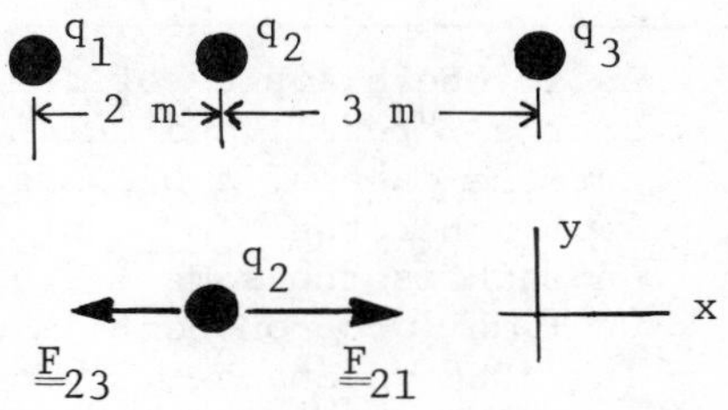

We first draw the force diagram for q_2 as if all the electrostatic forces were repulsive.

The resultant force $\mathbf{F}_2 = \mathbf{F}_{21} + \mathbf{F}_{23}$ has the components

$$F_{2x} = F_{21} - F_{23} = \frac{1}{4\pi\varepsilon_o}\frac{q_2 q_1}{r_{21}^2} - \frac{1}{4\pi\varepsilon_o}\frac{q_2 q_3}{r_{23}^2}\ ,$$

$$F_{2y} = 0.$$

Here F_{21} and F_{23} are the electrostatic forces as given by (21-3). It is given that $F_{2x} = 1.0 \times 10^{-8}$ N and $F_{2y} = 0$. Solving for q_3,

$$q_3 = -4\pi\varepsilon_o \frac{r_{23}^2}{q_2} F_{2x} + q_1 \left(\frac{r_{23}}{r_{21}}\right)^2 .$$

Substituting the numerical values and suppressing the units for brevity,

$$q_3 = -\frac{1}{9 \times 10^9} \frac{3^2}{(+2 \times 10^{-9})} (1.0 \times 10^{-8}) + (-4 \times 10^{-9})\left(\frac{3}{2}\right)^2$$

$$q_3 = -1.4 \times 10^{-8} \text{ C} .$$

Remark: The required charge q_3 turned out to be negative. Thus both $\underline{F}_{21}$ and $\underline{F}_{23}$ are really attractive forces; by drawing the force diagram as if they were repulsive and using the algebraic values of the charges throughout the calculation we obtained the correct answer. This may be checked by computing the actual forces exerted on q_2 using method #1. <<<

21-5 Programmed Problems

1. Let us begin with a few general questions about Coulomb's law to insure that you comprehend the basic elements of this law. Below are shown three different cases in which two point charges interact with each other. In each case indicate by a vector the Coulomb force acting on each charge.

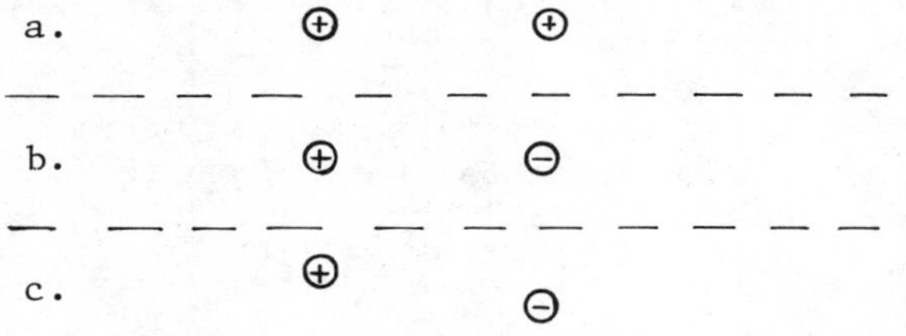

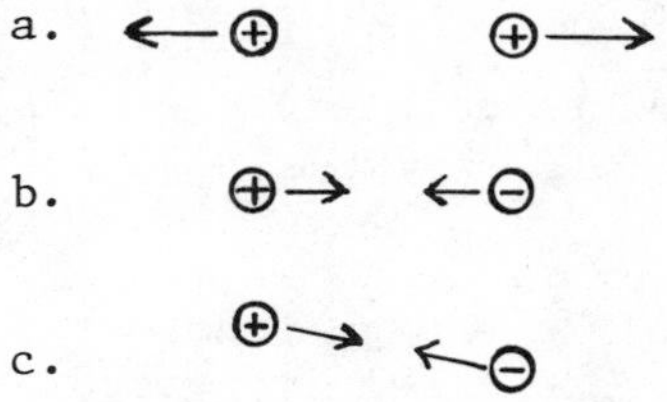

In c note that the forces are still along the line joining the charges.

---

2. As in the examples of the first frame, like charges "repel" and unlike charges "attract". For the case of two charges interacting with each other the ______ of the Coulomb force on each is the same. The ______ of the Coulomb force on each is along the line joining the two charges.

Magnitude.
Direction.

---

3. For the configuration of charges shown, indicate by vectors the electrostatic forces exerted on q_1 by q_2 and by q_3. Indicate only the forces on q_1. All charges have the same magnitude, the signs of the charges are as shown.

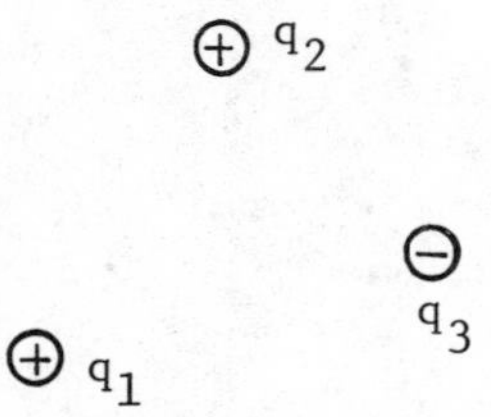

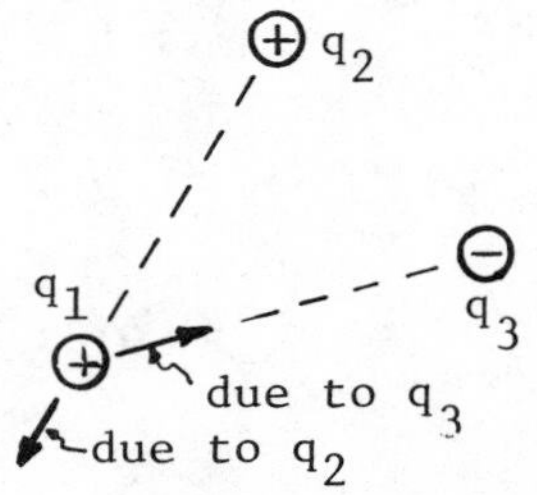

4. Here we have labeled as $\underline{\underline{F}}_{12}$ the electrostatic force acting on q_1 by its interaction with q_2. $\underline{\underline{F}}_{13}$ is similarly the force on q_1 by q_3.

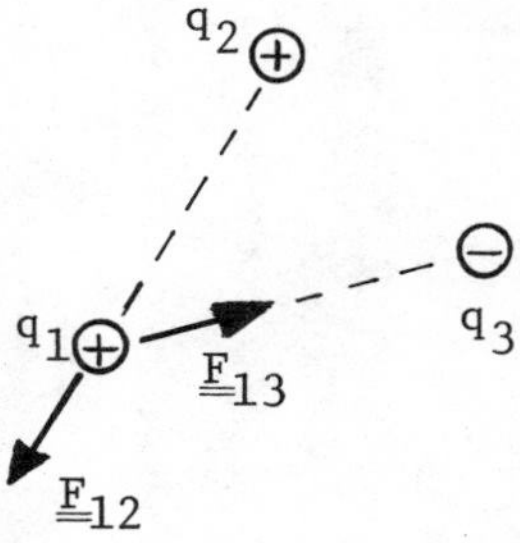

The electrostatic forces $\underline{\underline{F}}_{12}$ and $\underline{\underline{F}}_{13}$ are vector forces acting on q_1. To completely specify these vectors we need to know the magnitude and direction of each.

Will the directions (senses) of the forces in the diagram be as shown even if the magnitudes of the charges were altered?

Yes. The directions of the forces are determined by the sign (like or unlike) of the paired charges and their configuration, i.e., the line joining the pair. The directions are not determined by the magnitudes of the charges.

---

5. Although we have not selected a reference direction, i.e., coordinate system, we do know the directions of the electrostatic forces between q_1 and q_2 as well as q_1 and q_3. In your own words what is the direction of the electrostatic force exerted on one charge by a second charge?

Along the line joining the charges, being either repulsive or attractive depending upon whether the charges are of the same or opposite sign.

---

6. The quantity to be determined now is the magnitude of the electrostatic force. Write the scalar equation which gives the force on a charge q_1 due to another charge q_2. Assume the charges to be separated by a distance r_{12}.

$$|\underline{\underline{F}}_{12}| = \_\_\_\_\_\_$$

Coulomb's law

$$|\underline{\underline{F}}_{12}| = \frac{1}{4\pi\varepsilon_o}\left|\frac{q_1 q_2}{r_{12}^2}\right| .$$

r_{12} is the distance from q_1 to q_2.

$$1/4\pi\varepsilon_o = 9.0 \times 10^9 \text{ N}\cdot\text{m}^2/\text{C}^2 .$$

---

7. Two point charges $q_1 = 4.0 \times 10^{-6}$ C and $q_2 = -8.0 \times 10^{-6}$ C are separated by 4 meters as shown.

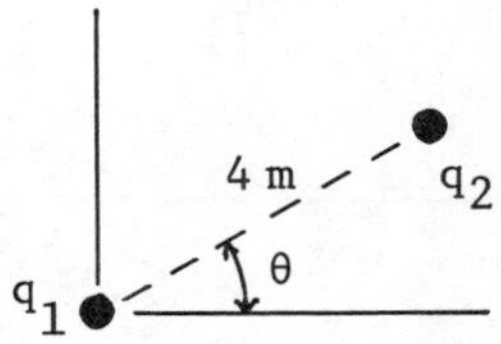

a. What is the direction of the electrostatic force on q_1?

b. What is the magnitude of the force on q_1? Obtain a numerical result.

a. Toward q_2 along the line connecting q_1 and q_2.

b. $$|\underline{\underline{F}}_{12}| = \frac{1}{4\pi\varepsilon_o}\left|\frac{q_1 q_2}{r_{12}^2}\right|$$

$$|\underline{\underline{F}}_{12}| = 18.0 \times 10^{-3} \text{ N} .$$

8. Let θ in the problem be 30°. What are the x and y components of $\underline{\underline{F}}_{12}$?

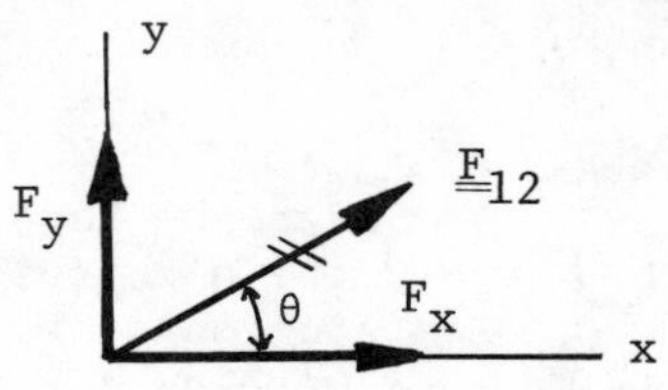

$F_x = |\underline{\underline{F}}_{12}| \cos\theta$

$F_x = 15.6 \times 10^{-3}$ N .

$F_y = |\underline{\underline{F}}_{12}| \sin\theta$

$F_y = 9.0 \times 10^{-3}$ N .

---

9. Describe fully the electrostatic force acting on q_2.

It is oppositely directed to $\underline{\underline{F}}_{12}$ and of the same magnitude.

10.

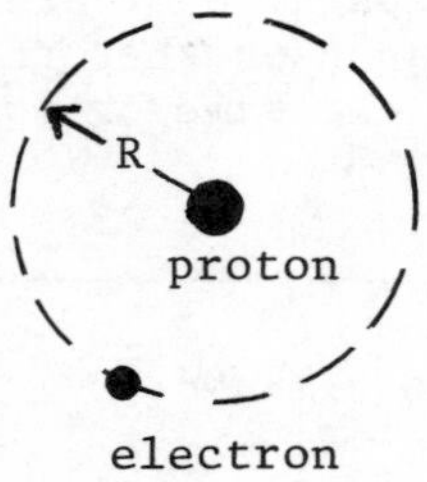

The electron in the hydrogen atom moves in a circular orbit of radius 5.3×10^{-11} meter about the proton. If the Coulomb force acting on the electron is 8.2×10^{-8} N, what is the speed of the electron?

First write "F = ma" which is appropriate to uniform circular motion.

$F = mv^2/R$

where v is the speed of the electron in its orbit.

---

11. The Coulomb force is due to the interaction between the proton and the electron. In the answer to the previous frame m is the mass of the (electron/proton).

Electron. You wrote Newton's second law for the <u>electron</u>.

---

12. Using $m = 9.1 \times 10^{-31}$ kg calculate the speed of the electron in its orbit.

$F = mv^2/R$

$v = \sqrt{FR/m}$

$v = 2.2 \times 10^6$ m/s .

13.

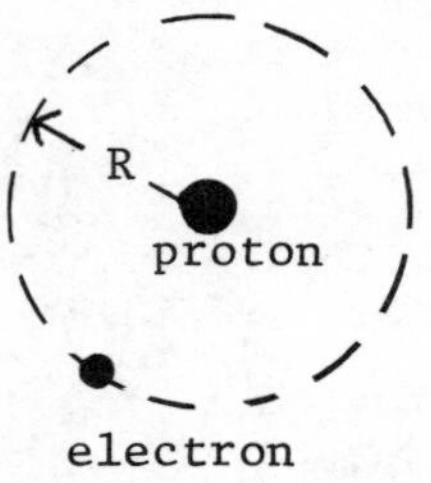

How long does it take the electron to make one revolution? Let T be the period (time) of one revolution.

T = _____ sec

$T = 2\pi R/v$

$T = 15.1 \times 10^{-17}$ s .

14. If you were able to "stand" at one position in the orbit and observe the number of electrons passing by, you would see one (the same one of course) every 15.1×10^{-17} second. The amount of charge passing a given point is called current; one ampere is a current of one coulomb per second. What constant current (in amperes) is equivalent to the motion of the electron in the hydrogen atom? The magnitude of the charge of the electron is 1.6×10^{-19} C.

$I = q/t$

$$I = \frac{1.6 \times 10^{-19}\ \text{C}}{15.1 \times 10^{-17}\ \text{s}}$$

$I = 1.06 \times 10^{-3}$ A .

15.

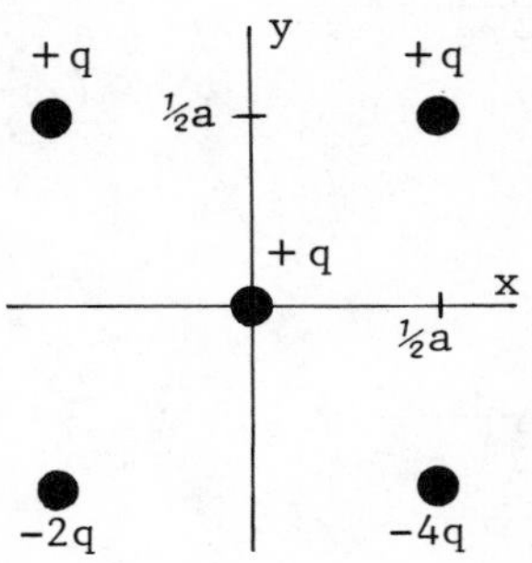

Four charges are arranged symmetrically about the origin of a coordinate system as shown. What is the resultant electrostatic force on the charge at the origin?

The approach to this problem invokes the principle of superposition. Superposition means that we can calculate the total force on the charge at the center by first pairing the central charge q with each of the four surrounding charges and then determining the vector sum of the four forces.

Draw vectors on the diagram representing the forces acting on +q at the center due to its separate interactions with the four other charges. Label the forces as UL (due to upper left charge), LR (due to lower right charge), etc.

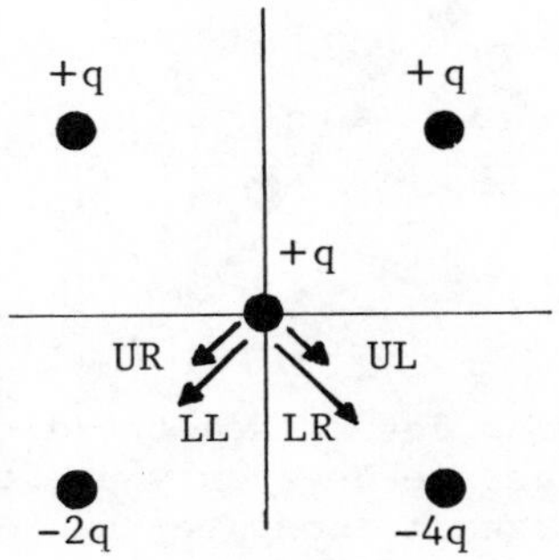

Coulomb's law can be used to calculate (for example) the force UL by ignoring the other three charges.

---

16. Now you know the directions of the four separate forces. Which forces have equal magnitudes? Identify as UL, UR, etc.

UL = UR

17.

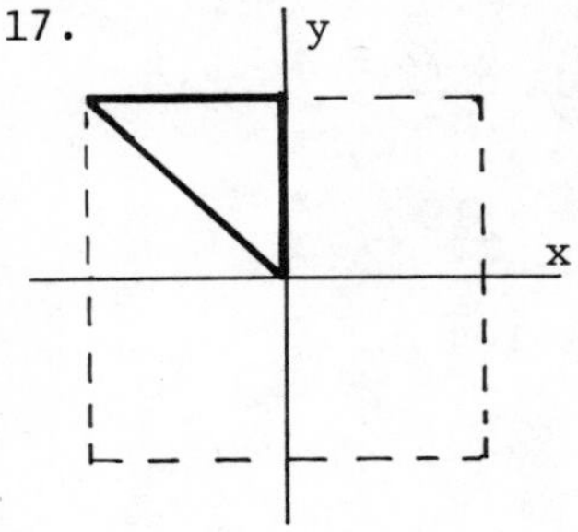

Prior to calculating the magnitude we need to do a bit of geometry. Here is a piece of the diagram from frame 15. Identify all the angles of this triangle as well as the length of the horizontal and vertical sides in terms of the given length a of the problem.

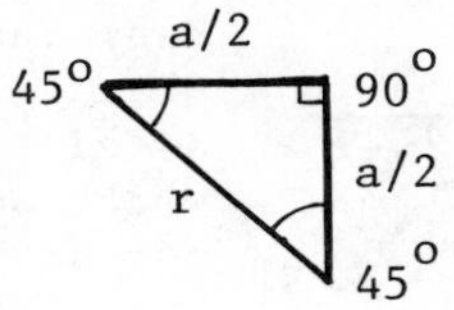

18\. From the theorem of Pythagoras, what is the hypotenuse r of the triangle?

Those of you who are conversant with elementary trigonometry may choose to answer this question by a different method. Please do so.

$r = \sqrt{a^2/4 + a^2/4}$

$r = a/\sqrt{2}$.

19\.

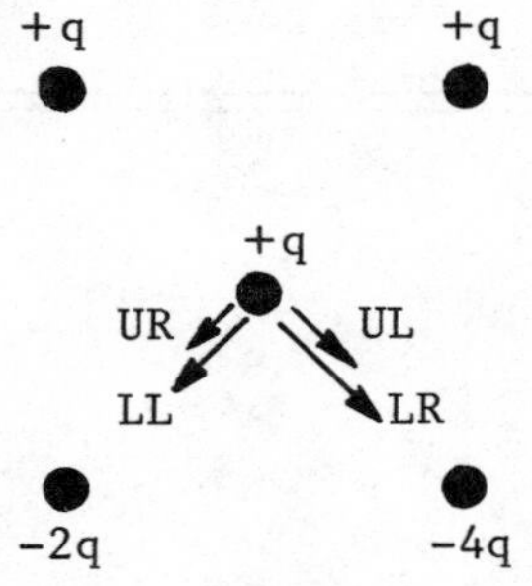

The separation between +q at the center and each of the four charges is thus $a/\sqrt{2}$.

From Coulomb's law then

a. $|\mathbf{F}_{UL}|$ = _____

b. $|\mathbf{F}_{LL}|$ = _____

c. $|\mathbf{F}_{UR}|$ = _____

d. $|\mathbf{F}_{LR}|$ = _____

Having determined the directions of all the forces we need only use the magnitudes of the charges to compute the forces from Coulomb's law.

a. $|\mathbf{F}_{UL}| = \dfrac{1}{4\pi\varepsilon_o}\dfrac{q^2}{a^2/2}$

b. $|\mathbf{F}_{LL}| = \dfrac{1}{4\pi\varepsilon_o}\dfrac{2q^2}{a^2/2}$

c. Same as a.

d. $|\mathbf{F}_{LR}| = \dfrac{1}{4\pi\varepsilon_o}\dfrac{4q^2}{a^2/2}$

20\. From the previous frame

a. $|\mathbf{F}_{UL}| + |\mathbf{F}_{LR}|$ = __________ .

b. $|\mathbf{F}_{UR}| + |\mathbf{F}_{LL}|$ = __________ .

a. $\dfrac{1}{4\pi\varepsilon_o}\left[\dfrac{q^2}{a^2/2} + \dfrac{4q^2}{a^2/2}\right]$

$= \dfrac{1}{4\pi\varepsilon_o}\dfrac{10q^2}{a^2}$

b. $\dfrac{1}{4\pi\varepsilon_o}\dfrac{6q^2}{a^2}$.

21\. Why was it appropriate to add the magnitudes $|\mathbf{F}_{UL}|$ and $|\mathbf{F}_{LR}|$ in the previous frame?

Because $\mathbf{F}_{UL}$ and $\mathbf{F}_{LR}$ were in the same direction, i.e. they are colinear.

22\.

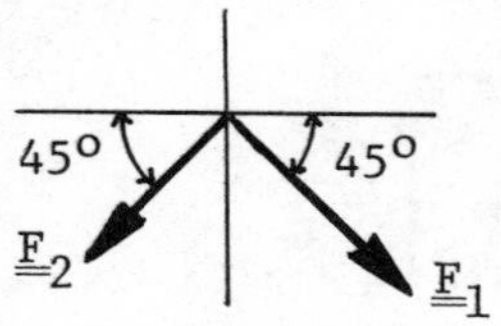

Here the diagram is somewhat simpler as we now have added those pairs of forces which were colinear. This is not drawn to scale.

$\mathbf{F}_1 = \mathbf{F}_{UL} + \mathbf{F}_{LR}$

$\mathbf{F}_2 = \mathbf{F}_{UR} + \mathbf{F}_{LL}$

Data: $q = 2 \times 10^{-4}$ C, $a = 1$ m.

Calculate the magnitude of $\mathbf{F}_1$ and of $\mathbf{F}_2$. The formulas are in frame 20.

$|\mathbf{F}_1| = 9 \times 10^9 \dfrac{\text{N}\cdot\text{m}^2}{\text{C}^2} \times \dfrac{10 \times 4 \times 10^{-8}\ \text{C}^2}{1\ \text{m}^2}$

$|\mathbf{F}_1| = 3.6 \times 10^3$ N.

$|\mathbf{F}_2| = 2.16 \times 10^3$ N.

23.

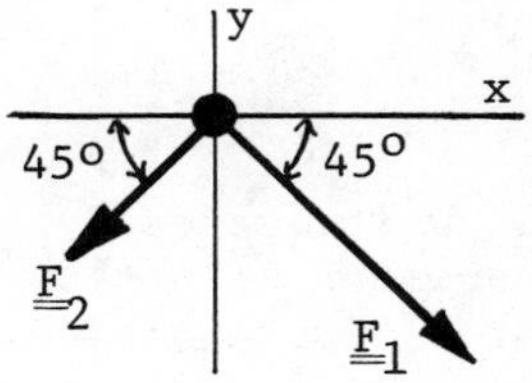

Vectors as now drawn are roughly to scale. Diagrammatically resolve $\underline{\underline{F}}_1$ and $\underline{\underline{F}}_2$ into their respective vertical and horizontal components. We need to do this in order to find the final resultant force.

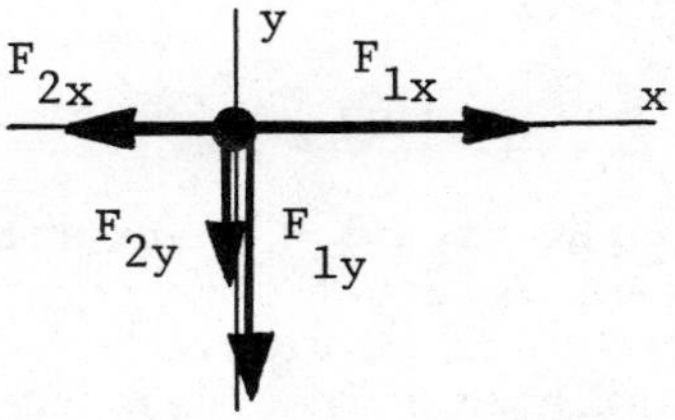

---

24. Utilizing the standard sign conventions calculate

a. F_{1x} = ______

b. F_{1y} = ______

c. F_{2x} = ______

d. F_{2y} = ______

a. $F_{1x} = |\underline{\underline{F}}_1| \cos 45^\circ$

$= 3.6 \times 10^3\ N \times \sqrt{2}/2$

$= 2.55 \times 10^3\ N$.

b. $F_{1y} = -\ 2.55 \times 10^3\ N$.

c. $F_{2x} = -\ 1.53 \times 10^3\ N$.

d. $F_{2y} = -\ 1.53 \times 10^3\ N$.

---

25. Now we are almost finished. The components of the total force F_{total} are

a. $\sum F_x$ = ______ .

b. $\sum F_y$ = ______ .

This can be done algebraically of course. That is the utility of resolving forces into components.

a. $1.02 \times 10^3\ N$

b. $-\ 4.08 \times 10^3\ N$

26.

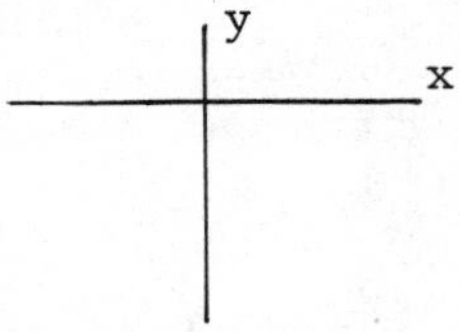

Draw vectors representing the answers to the last frame and show their resultant diagrammatically. Use an approximate scale.

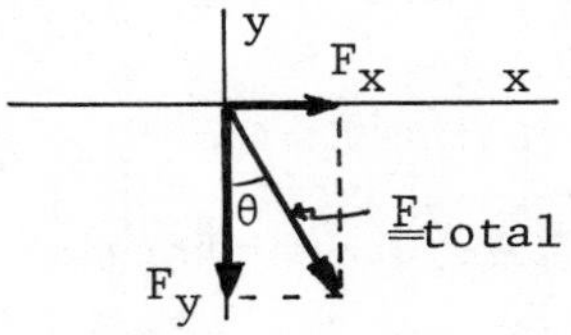

---

27. Finally then what is:

a. $\underline{\underline{F}}_{total}$ in magnitude?

b. Direction of $\underline{\underline{F}}_{total}$ with respect to the -y axis?

a. $|\underline{\underline{F}}_{total}| = \sqrt{F_x^2 + F_y^2}$

$= 4.21 \times 10^3\ N$.

b.

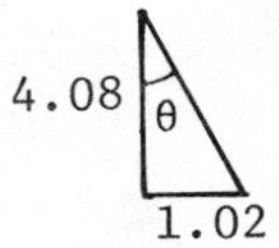

The total electrostatic force makes an angle θ to the right of the -y axis. This angle is given by $\tan\theta = 1.02/4.08 = 0.25$: $\theta = 14^\circ$.

Chapter 22: THE ELECTRIC FIELD

REVIEW AND PREVIEW

In the previous chapter the electrostatic force between charges was discussed. In this chapter rather than considering the effect of one charge upon another directly, we take the point of view that each charge creates an electric field $\underline{E}$; it is this electric field which in turn causes the force to be exerted on another charge.

GOALS AND GUIDELINES

In this chapter there are three major goals:

1. Learning the definition of the electric field $\underline{E}$. This involves the force exerted on a "test charge" q_0. You might want to review the general concept of a "field" (as discussed in the Preliminary Concepts for Part II immediately preceding Chapter 21).

2. Learning the formula (Eq. 22-2) for the electric field due to a single point charge. You should be able to quickly derive this using Coulomb's law and the definition of $\underline{E}$. This simple but important case (in which the source of $\underline{E}$ is a single point charge) is the key to doing many more complicated problems.

3. Learning how to calculate $\underline{E}$ due to any given distribution of charge. This involves dividing the distribution of charge into point charges and then applying Eq. 22-2 to each such point charge.

In this and several succeeding chapters the concept of the definite integral as a sum is used. You might want to review the definite integral (see appendix).

22-1 Electric Field

If a point charge q_0 is placed at a point P in space, we may find that an electrical force $\underline{F}$ is exerted on q_0 (by other charges). The electric field*, $\underline{E}$, at the point P (due to these other charges) is then defined as

$$\underline{E} = \underline{F}/q_0 \ . \tag{22-1}$$

We see that the units of electric field are those of force/charge, i.e. N/C. The charge q_0, used to measure $\underline{E}$, is called a test charge. The force $\underline{F}$ will always be proportional to q_0, thus the electric field $\underline{E} = \underline{F}/q_0$ is really independent of both the magnitude and sign of the test charge q_0.

[Not only does an electric field exert a force $\underline{F} = q\underline{E}$ on a point charge, it also exerts a torque $\underline{\tau} = \underline{p} \times \underline{E}$ on an electric dipole $\underline{p}$. An electric dipole consists of two equal but opposite charges (q, -q) separated by a distance 2a. The electric dipole moment, $\underline{p}$, is of magnitude $|\underline{p}| = 2aq$ and is directed from -q to q. See section 32-3.]

---

*Sometimes called "electric field strength", "electric intensity", etc.

>>> Example 1. Derive an expression for the electric field at a point P, a distance r away from a point charge q. (Note that q is a charge, while P is a point in space.)

We imagine placing a test charge q_o at the point P. According to Coulomb's law, the electrostatic force exerted on q_o by q is given by $F = (1/4\pi\varepsilon_o)(qq_o/r^2)$. Thus

$$E_{\text{at P}} = F/q_o = \frac{1}{4\pi\varepsilon_o}\frac{q}{r^2}$$

The student should be able to verify from the vector definition, $\mathbf{E} = \mathbf{F}/q_o$, that the direction of $\mathbf{E}$ is radially <u>away</u> from q if q is positive and radially <u>toward</u> q if q is negative. <<<

The electric field $\mathbf{E}$ depends (both in magnitude and direction) upon the location of the point P; that is, $\mathbf{E}$ is a <u>vector field</u>. When specifying the value of $\mathbf{E}$, one must always give the point at which that value applies. Thus in the above example, $E = (1/4\pi\varepsilon_o)(q/r^2)$ gives the electric field at a distance r from the point charge q.

The force $\mathbf{F}$ in $\mathbf{E} = \mathbf{F}/q_o$ is only the electrical force. Sometimes we encounter forces of more than one type as in the following example.

>>> Example 2. What electric field must be present so that a proton is in equilibrium under the combined influence of this electric field and gravity?

We first draw the force diagram. Clearly the electrostatic force must be "up". For equilibrium we must then have Eq = mg. Thus

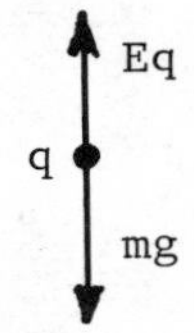

$$E = mg/q = (1.67 \times 10^{-27}\text{ kg})(9.80\text{ m/s}^2)/(1.60 \times 10^{-19}\text{ C})$$

$$= 1.02 \times 10^{-7}\text{ N/C.}$$

The direction of this electric field is "up". If q were negative (electron instead of proton, for example), then $\mathbf{E}$ would be "down" in order that the electrostatic force $\mathbf{F} = \mathbf{E}q$ be "up".

22-2 Calculating the Electric Field

As shown in Example 1, the electric field at a point P a distance r away from a point charge q is given by

$$E_{\text{at P}} = \frac{1}{4\pi\varepsilon_o}\frac{q}{r^2} \qquad (22\text{-}2)$$

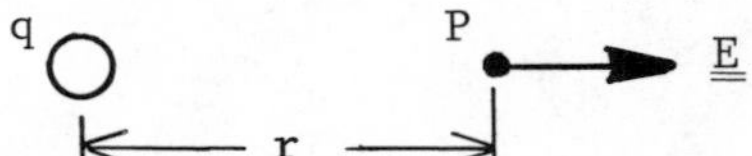

Again, P is <u>not</u> a charge; it is merely a point in space. <u>If</u> a test charge q_o <u>were</u> placed at P, then there would be an electrostatic force $\mathbf{F} = \mathbf{E}q_o$ exerted on it.

Equation (22-2) permits the calculation of $\mathbf{E}$ at a point P due to any finite number of point charges q_1, q_2, q_3, ... We need only calculate the electric fields $\mathbf{E}_1$, $\mathbf{E}_2$, $\mathbf{E}_3$, ... at the point P due to each charge <u>separately</u> and then take the <u>vector</u> sum $\mathbf{E} = \mathbf{E}_1 + \mathbf{E}_2 + \mathbf{E}_3 + \ldots$. As with Coulomb's law, there may be some difficulty with signs.

The two methods suggested there are also useful here:

Method #1. If we know the signs of all the point charges q_1, q_2, q_3, ... causing the electric field, we may draw a vector diagram showing $\mathbf{E}_1$, $\mathbf{E}_2$, $\mathbf{E}_3$, ... with their correct directions and senses. Then in (22-2) we use the magnitudes, $|q_i|$, of the charges. This method is illustrated in the text and in the programmed problems.

Method #2. A vector diagram is drawn showing $\mathbf{E}_1$, $\mathbf{E}_2$, $\mathbf{E}_3$, ... at the point P as if all the point charges q_1, q_2, q_3, ... causing these fields were positive (that is, each field $\mathbf{E}_i$ is directed radially away from its source q_i). Then the algebraic charges, q_i, including sign are used in (22-2). This method is illustrated in the following example.

>>> Example 3. Two equal and opposite point charges (q, -q) separated by a distance 2a form a dipole. Point P lies on the perpendicular bisector of the line connecting the two charges, the perpendicular distance of P from this line being r. Calculate the electric field at point P.

We draw $\mathbf{E}_1$ and $\mathbf{E}_2$ as if the point charges causing them were positive. Note that the distance "r" here is not the same as the "r" in equation (22-2). The components of the total electric field, $\mathbf{E} = \mathbf{E}_1 + \mathbf{E}_2$, are

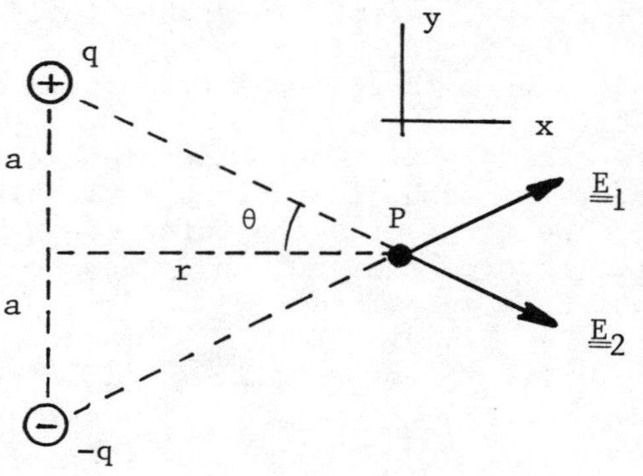

$$E_x = E_1 \cos\theta + E_2 \cos\theta$$

$$= (E_1 + E_2)\cos\theta$$

$$E_y = E_1 \sin\theta - E_2 \sin\theta$$

$$= (E_1 - E_2)\sin\theta$$

where from the diagram $\cos\theta = r/(r^2 + a^2)^{1/2}$ and $\sin\theta = a/(r^2 + a^2)^{1/2}$. Using equation (22-2),

$$E_1 = \frac{1}{4\pi\varepsilon_0}\frac{(-q)}{(r^2 + a^2)}$$

$$E_2 = \frac{1}{4\pi\varepsilon_0}\frac{(q)}{(r^2 + a^2)}$$

Therefore

$$E_x = 0$$

$$E_y = -\frac{2qa}{4\pi\varepsilon_0 (r^2 + a^2)^{3/2}}$$

Assuming that q is positive the direction of $\mathbf{E}$ is along the negative y axis. <<<

We now know how to calculate the electric field due to any finite number of point charges. Another class of problems concerns the case in which the charge is continuously distributed (charged rod, plate, etc.). Basically we follow the same procedure as before, with integration replacing summation. First the given charge distribution is divided into infinitesimal point charges dq. Equation (22-2) applies to each such dq (since it is a point charge). At a point P a distance r away from dq there will be an infinitesimal electric field given by

$$dE_{\text{at P}} = \frac{1}{4\pi\varepsilon_0}\frac{dq}{r^2} \qquad (22\text{-}3)$$

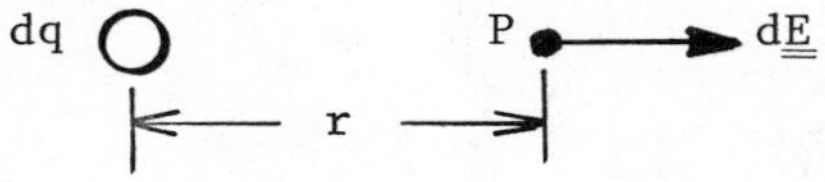

This is of the same form as (22-2) except that dq is used instead of q, and dE instead of E. Although one may think of (22-3) as being obtained from (22-2) by suitably taking differentials, it is perhaps better to regard (22-3) as being the same as (22-2) except that the source (dq) of the field is infinitesimal. (In particular, note that there is no "dr" in (22-3). Rather, we have the small field "$d\mathbf{E}$" at the finite distance "r" from the small point charge "dq".) The various $d\mathbf{E}$'s must then be added (integrated) vectorially to obtain the resultant electric field, $\mathbf{E} = \int d\mathbf{E}$, at P.

>>> Example 4. A thin glass rod of length L carries a constant charge per unit length λ (units of λ are C/m). Calculate the electric field at point P, a perpendicular distance r from one end of the rod as shown.

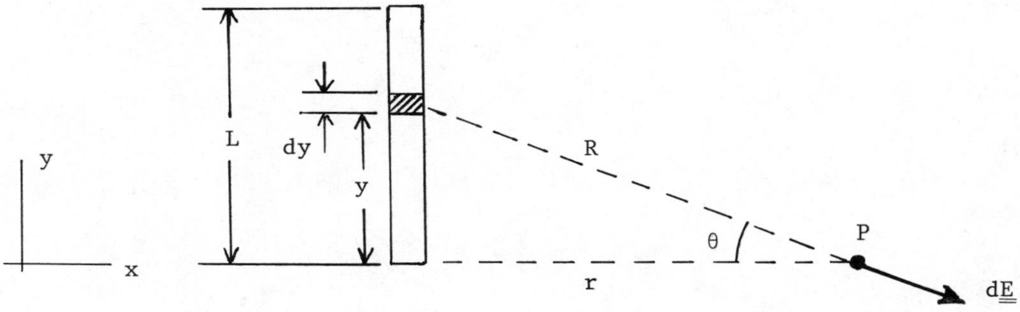

We divide the rod into infinitesimal pieces dy. Each such piece carries a charge $dq = \lambda\,dy$. The electric field at point P due to this point charge dq is given by (22-3),

$$dE = \frac{1}{4\pi\varepsilon_0}\frac{dq}{R^2}$$

We cannot add (integrate) these to obtain $E = \int dE = \int \frac{1}{4\pi\varepsilon_0}\frac{dq}{R^2}$. This is because $d\mathbf{E}$ is a vector and we must add (integrate) these infinitesimal vectors properly. We resolve $d\mathbf{E}$ into its components:

$$dE_x = \frac{1}{4\pi\varepsilon_o}\frac{dq}{R^2}\cos\theta = \frac{1}{4\pi\varepsilon_o}\frac{\lambda\, dy}{R^2}\cos\theta$$

$$dE_y = -\frac{1}{4\pi\varepsilon_o}\frac{dq}{R^2}\sin\theta = -\frac{1}{4\pi\varepsilon_o}\frac{\lambda\, dy}{R^2}\sin\theta$$

The x component, E_x of the total electric field, $\underline{\underline{E}} = \int d\underline{\underline{E}}$, is the sum (integral) of the x components of the $d\underline{\underline{E}}$'s:

$$E_x = \int dE_x = \frac{\lambda}{4\pi\varepsilon_o}\int \frac{\cos\theta}{R^2}\, dy$$

We are now faced with more than one variable in the integrand (y, R, θ are all variables). Usually, in problems of this type, the simplest integration arises if we express all the variables in terms of the angle θ. Thus we must find y and R in terms of θ. From the right triangle formed by y, r, R we have

$$y = r\tan\theta \quad \text{therefore} \quad dy = r\sec^2\theta\, d\theta\ ,$$

$$R = r\sec\theta$$

Substituting into the integral for E_x,

$$E_x = \frac{\lambda}{4\pi\varepsilon_o}\int \frac{\cos\theta}{(r\sec\theta)^2}(r\sec^2\theta\, d\theta)$$

$$= \frac{\lambda}{4\pi\varepsilon_o r}\int_{\theta_1}^{\theta_2}\cos\theta\, d\theta = \frac{\lambda}{4\pi\varepsilon_o r}(\sin\theta_2 - \sin\theta_1).$$

From the diagram $\sin\theta_2$ (θ_2 is the largest value of θ) is given by

$$\sin\theta_2 = \frac{L}{\sqrt{r^2 + L^2}}$$

while $\sin\theta_1 = 0$. Therefore

$$E_x = \frac{\lambda}{4\pi\varepsilon_o r}\frac{L}{\sqrt{r^2 + L^2}}$$

You should be able to perform a similar calculation for E_y to obtain

$$E_y = -\frac{\lambda}{4\pi\varepsilon_0 r}\left[1 - \frac{r}{\sqrt{r^2 + L^2}}\right]$$

Remarks:

1. To find e.g. R in terms of θ, we wrote $R = r \sec\theta$. This expresses R in terms of θ and the <u>constant</u> r. r is a constant (for purposes of the integration) since all the various dq's have the same value of r. Suppose we had instead written $R = y \csc\theta$ (a correct equation). This does not result in the desired simplification since it expresses R in terms of θ and the <u>variable</u> y. Usually we have that one side of the right triangle is fixed, in this case the side "r". The idea is to find a trigonometric relation which expresses the undesired variable (R) in terms of the desired variable (θ) and constants (r). Thus $R = r \sec\theta$ is the useful relation.

2. In the original integrand, "y" did not appear explicitly, only "dy" was present. Nonetheless it is much easier to first find y ($y = r\tan\theta$) and then obtain dy by taking differentials. The reason for this is that the length dy is not the side of any simple right triangle, whereas y itself is. (It is especially important here that we first find y in terms of only θ and constants. Had we written e.g. $y = R\sin\theta$ instead, we would then have $dy = R\cos\theta\, d\theta + (dR)\sin\theta$ which involves several differentials.)

3. The integrations produced terms such as $\sin\theta_2$, $\cos\theta_2$, etc. Instead of finding θ_2 itself (e.g. $\theta_2 = \tan^{-1}(L/r)$ and substituting this, it is by far <u>easier to find</u> the desired <u>trigonometric</u> functions <u>directly</u> from the diagram: $\sin\theta_2 = L/\sqrt{r^2 + L^2}$, $\cos\theta_2 = r/\sqrt{r^2 + L^2}$.

4. Students sometimes have trouble with "missing" differentials in this type of problem. <u>All</u> integrals of course must contain a differential (e.g. $\int f(x)$ is meaningless, $\int f(x)dx$ is meaningful). If the problem involves an integration, as this one does, then there <u>must</u> be a differential present. This differential should always come into the calculation in a natural manner. For example $dE = (1/4\pi\varepsilon_0)(dq/R^2)$ [not $(1/4\pi\varepsilon_0)(q/R^2)$], $dq = \lambda\, dy$ (not λy), etc. If the differential seems to be "missing", the student should <u>not</u> multiply by a differential just in order to have one present (among other things this will usually make the equation dimensionally incorrect). Instead, he should review his work to find the mistake. As a guide to finding such errors, note that if one side of an equation is infinitesimal (i.e. contains a differential factor) then the other side of the equation must also be infinitesimal. <<<

22-3 Programmed Problems

1. In this chapter you will be doing exercises which in many ways are very similar to the exercises of the previous chapter. The major difference is that now we will think in terms of the concept of the <u>electric field</u>. The first few frames will be used to develop this concept for the case of a point charge.

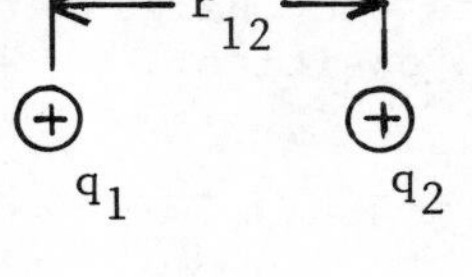

a. What is the magnitude of the Coulomb force acting on q_2 as shown to the left?

b. What is the direction of the force on q_2? Show this by an appropriate vector.

a. $F_{21} = \frac{1}{4\pi\varepsilon_0}\frac{q_1 q_2}{r_{12}^2}$

The subscript 21 means the force on q_2 by its interaction with q_1.

b.

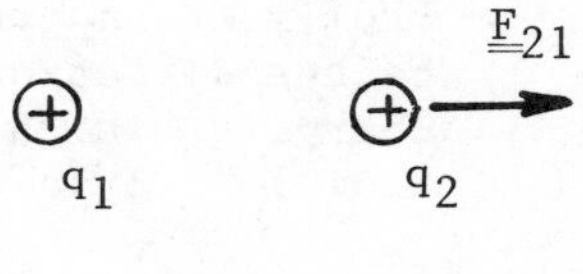

2. If we increase the magnitude of the charge q_2, the force of F_{21} on q_2 will of course increase. We can rewrite F_{21} as

$$\frac{F_{21}}{q_2} = \frac{1}{4\pi\varepsilon_o} \frac{q_1}{r_{12}^{\,2}}$$

The left side of this equation is the force per unit charge of q_2. For given r_{12} and q_1 the right side of this equation is a constant. Under these conditions as q_2 increases, F_{21} (increases/decreases). Remember that the spatial configuration remains fixed.

Increases.

If the right side of the equation is a constant then the left side must remain a constant.

---

3. Looking again at the equation

$$\frac{F_{21}}{q_2} = \frac{1}{4\pi\varepsilon_o} \frac{q_1}{r_{12}^{\,2}}$$

we introduce the concept of an electric field to interpret the left side of the equation.

We define the term F_{21}/q_2 to be:

a. The electric field strength due to (q_1, q_2),

b. at the distance r_{12} as measured from (q_1, q_2).

a. q_1

b. q_1

We are talking about the electric field due to q_1 at a distance r_{12} away from q_1.

---

4. This is the new concept introduced in this chapter. We associate with the space around a point charge q the vector field E. The magnitude of this field due to the charge q at a distance r away from q is

E = __________.

$$E = \frac{1}{4\pi\varepsilon_o} \frac{q}{r^2}$$

---

5. Since the electric field is a vector field we need to specify its direction as well as its magnitude.

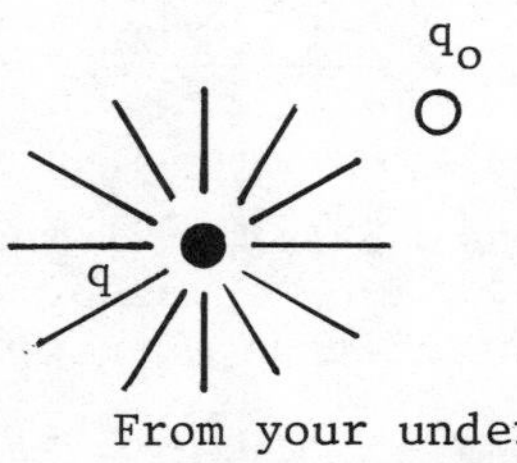

To the left we represent by radial lines the electric field in the space around a positive charge q.

From your understanding of Coulomb's law, what would be the direction of the force exerted on the positive charge q_o which is shown in the electric field of q? Show this by a vector representing the force F on q_o.

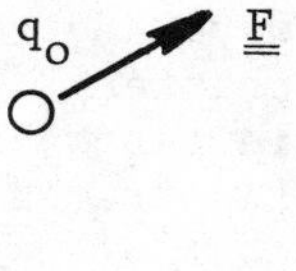

Directed away from q since the charges are both positive.

6. The direction of $\underline{\underline{F}}$ on the positive charge q_o in the previous frame is to be taken as the direction of $\underline{\underline{E}}$. The charge q_o is called a test charge.

Now completely specify the electric field of q at the position of q_o as shown in frame 5.

a. $|\underline{\underline{E}}|$ = __________
b. Direction of $\underline{\underline{E}}$: __________

a. $|\underline{\underline{E}}| = \frac{1}{4\pi\varepsilon_o} \frac{q}{r^2}$

b. Directed radially away from q.

---

7. We can summarize the discussion so far with the equation

$$\underline{\underline{E}} = \underline{\underline{F}}/q_o$$

The idea here is that the test charge q_o can be used to investigate a region of space for the existence of electric fields. If a force $\underline{\underline{F}}$ (other than gravitational, magnetic,etc.) is observed to act on q_o then both the magnitude and direction of $\underline{\underline{E}}$ can be determined. Referring to the equation above:

a. $\underline{\underline{E}}$ and $\underline{\underline{F}}$ are in the same direction. (True/False)

b. $\underline{\underline{E}}$ is the field due to q_o. (True/False)

c. $\underline{\underline{E}}$ and $\underline{\underline{F}}$ have the same magnitude. (True/False)

d. $\underline{\underline{F}}$ is the force acting on q_o. (True/False)

a. True, provided q_o is positive.
b. False. $\underline{\underline{E}}$ is the field due to charges <u>other</u> than q_o.
c. False (look at the defining equation).
d. True.

---

8. From the defining equation for $\underline{\underline{E}}$ as given in frame 7, the units of $\underline{\underline{E}}$ are __________ in the SI system.

newton/coulomb.
$\underline{\underline{E}} = \underline{\underline{F}}/q_o$
The units of $\underline{\underline{F}}$ are newtons.
The units of q_o are coulombs.
The units of $\underline{\underline{E}}$ are therefore N/C.

---

9. In problems involving a distribution of 3 point charges we can find the total electric field at some point P by using the following:

$$\underline{\underline{E}}_{at\ P} = \sum_{i=1}^{3} \underline{\underline{E}}_i$$

In your own words what does this equation mean?

The total electric field $\underline{\underline{E}}$ is the <u>vector sum</u> of the electric field due to <u>each individual</u> charge. For example, $\underline{\underline{E}}_1$ at P is determined as though q_1 were the only charge present.

10. 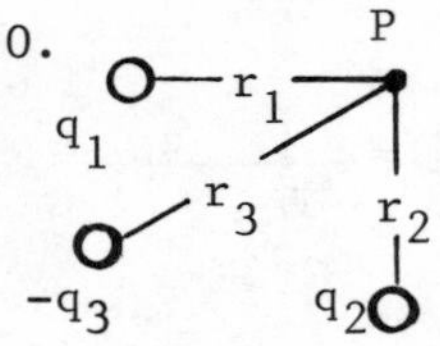

a. Draw vectors representing the direction of the individual electric fields due to q_1, q_2 and $-q_3$ at the point P. Label them $\mathbf{E}_1, \mathbf{E}_2$ and $\mathbf{E}_3$. Here the symbols q_1, q_2, q_3 are positive.

b. $|\mathbf{E}_1|$ = ________ .

$|\mathbf{E}_2|$ = ________ .

$|\mathbf{E}_3|$ = ________ .

a. 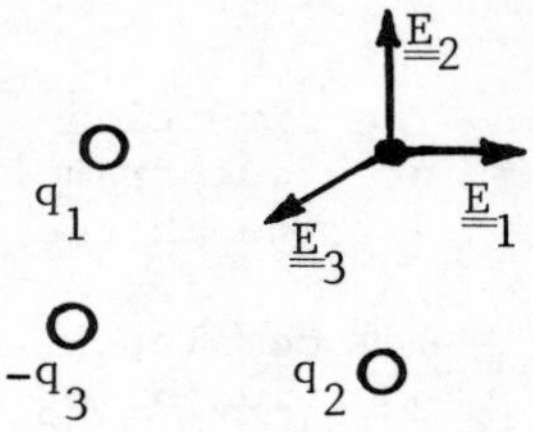

b. $|\mathbf{E}_1| = \dfrac{1}{4\pi\varepsilon_o} \dfrac{q_1}{r_1^{\,2}}$

$|\mathbf{E}_2| = \dfrac{1}{4\pi\varepsilon_o} \dfrac{q_2}{r_2^{\,2}}$

$|\mathbf{E}_3| = \dfrac{1}{4\pi\varepsilon_o} \dfrac{q_3}{r_3^{\,2}}$

---

11. Presumably you have been using this study guide for some time now and you have had many, many opportunities to find vector sums. You know that this involves coordinate systems, vector components, geometry, trigonometry,etc. All these techniques are still appropriate to finding the vector sum of electric fields.

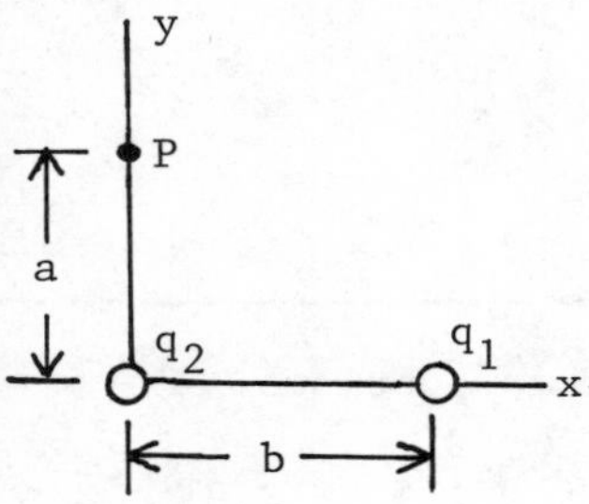

Two positive point charges q_1 and q_2 are located in a coordinate system as shown. We seek the electric field at the point P.

We know that the magnitude of the electric field a distance r away from a point charge is

$$E = \frac{1}{4\pi\varepsilon_o} \frac{q}{r^2}$$

a. What is r^2 for q_2 in this problem?

b. What is r^2 for q_1 in this problem?

a. a^2

b. $a^2 + b^2$

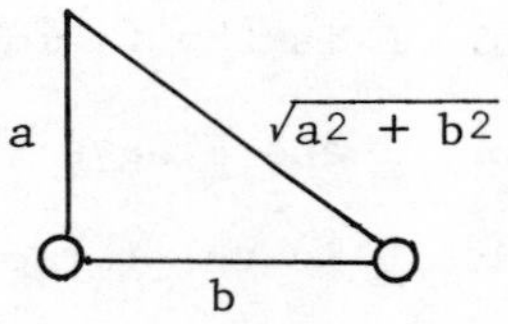

Note that the answer in part b is <u>not</u> $\sqrt{a^2 + b^2}$, it is $(\sqrt{a^2 + b^2})^2 = a^2 + b^2$.

---

12. What are the expressions for the electric field magnitudes at the point P?

$|\mathbf{E}_1|$ = ________ .

$|\mathbf{E}_2|$ = ________ .

Subscript 1 means the electric field due to q_1.

$|\mathbf{E}_1| = \dfrac{1}{4\pi\varepsilon_o} \dfrac{q_1}{a^2 + b^2}$

$|\mathbf{E}_2| = \dfrac{1}{4\pi\varepsilon_o} \dfrac{q_2}{a^2}$

13. For the case of $q_1 = 2.0 \times 10^{-8}$ C, $q_2 = 0.3 \times 10^{-8}$ C, a = 0.5 m and b = 1.0 m

$|\underline{\underline{E}}_1|$ = __________ .

$|\underline{\underline{E}}_2|$ = __________ .

$1/4\pi\varepsilon_o = 9.0 \times 10^9$ N·m²/C². Obtain numerical answers.

$|\underline{\underline{E}}_1|$ = 144 N/C

$|\underline{\underline{E}}_2|$ = 108 N/C

14.

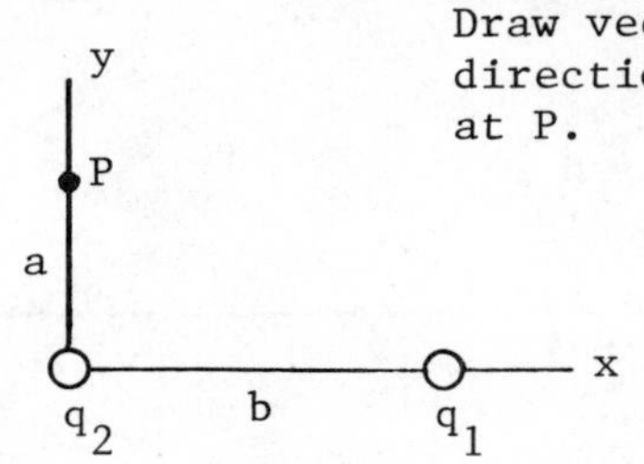

Draw vectors showing the directions of $\underline{\underline{E}}_1$ and $\underline{\underline{E}}_2$ at P.

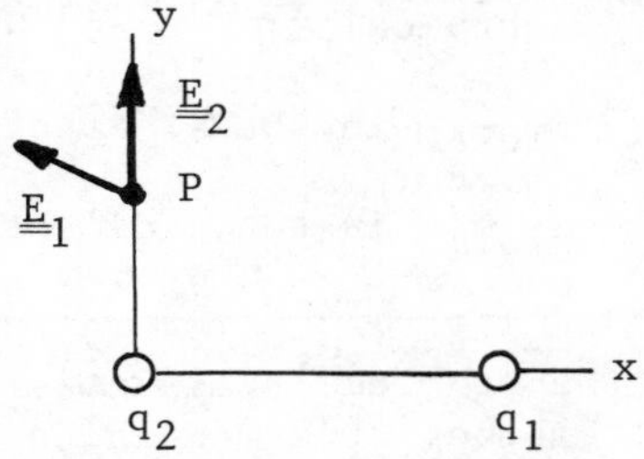

15.

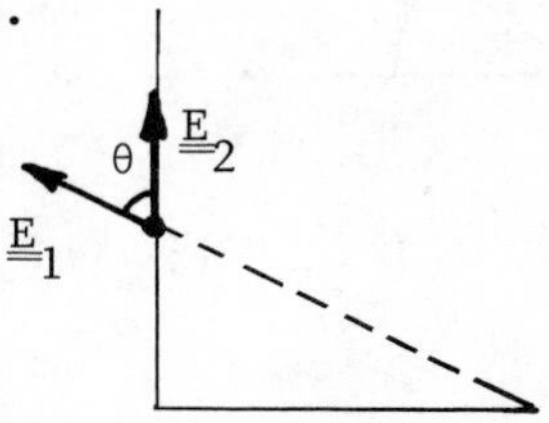

The problem is how to find the vector sum of $\underline{\underline{E}}_1$ and $\underline{\underline{E}}_2$. This is most conveniently done by resolving $\underline{\underline{E}}_1$ into x and y components.

In terms of the angle θ:

a. Draw vectors on the diagram representing E_{1x} and E_{1y}.

b. E_{1x} = __________ .

c. E_{1y} = __________ .

a.

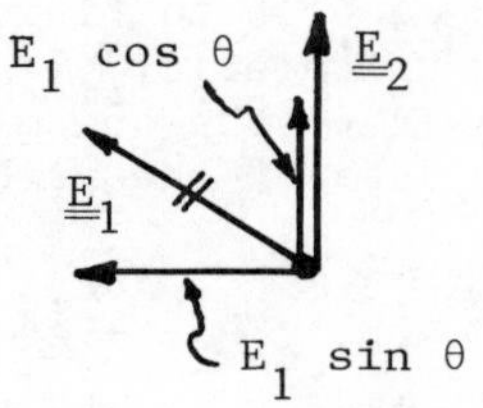

b. $E_{1x} = - E_1 \sin\theta$

c. $E_{1y} = E_1 \cos\theta$

---

16. Because θ is not explicitly given in the problem we need to express the sin θ and cos θ terms using the variables a and b which are given.

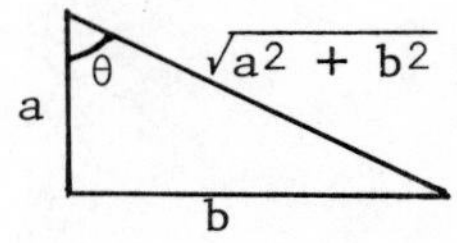

In terms of the lengths a and b,

a. cos θ = __________ .

b. sin θ = __________ .

a. $\cos\theta = \dfrac{a}{\sqrt{a^2 + b^2}}$

b. $\sin\theta = \dfrac{b}{\sqrt{a^2 + b^2}}$

17. Substituting the answers of frame 16 into the answers of frame 15 we have

$$E_{1x} = -E_1 \frac{b}{\sqrt{a^2 + b^2}}$$

$$E_{1y} = E_1 \frac{a}{\sqrt{a^2 + b^2}}$$

Having resolved $\underline{\underline{E}}_1$ we can now find the components of the vector $(\underline{\underline{E}}_1 + \underline{\underline{E}}_2)$. Algebraically,

E_{TOTAL} in the x direction = __________

E_{TOTAL} in the y direction = __________ .

$$E_{Tx} = -\frac{1}{4\pi\varepsilon_o}\left[\frac{q_1}{a^2 + b^2}\frac{b}{\sqrt{a^2 + b^2}}\right]$$

$$E_{Ty} = \frac{1}{4\pi\varepsilon_o}\left[\frac{q_1}{a^2 + b^2}\frac{a}{\sqrt{a^2 + b^2}} + \frac{q_2}{a^2}\right]$$

---

18. The problem is essentially finished. You will remember from mechanics that if you know the components of a vector then you know everything about it.

If you wish to finish the problem by doing the arithmetic, please do so. Compare your answer to the one given in the answer frame. The data of the problem is in frame 13.

E_{Tx} = __________ ,

E_{Ty} = __________ ,

α = __________ .

Here α is the angle (measured counter-clockwise from the positive x axis) which describes the direction of $\underline{\underline{E}}_T$.

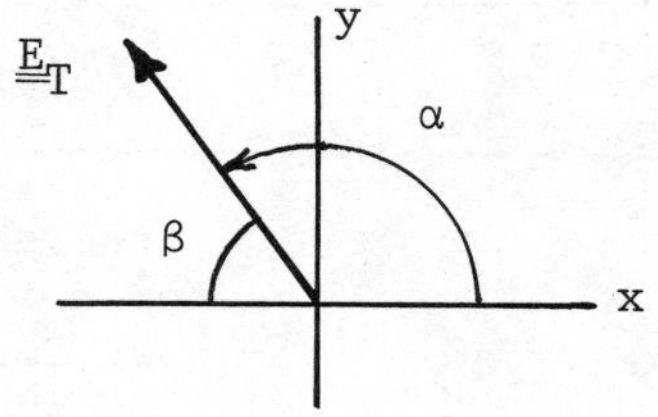

$E_{Tx} < 0$

$E_{Ty} > 0$

$E_{Tx} = -128$ N/C,

$E_{Ty} = 172$ N/C,

$\alpha = 127^o$.

Note that $\tan\beta = \frac{172}{128}$

$\beta = 53^o$

$\alpha = 180^o - \beta = 127^o$

19. A crucial element in determining the motion of a particle was to "find the force". The force $\underline{F} = q\underline{E}$ acting on the charge q in an electrostatic field $\underline{E}$ is an example of "finding a force". We can thus discuss the motion of charged particles in electric fields.

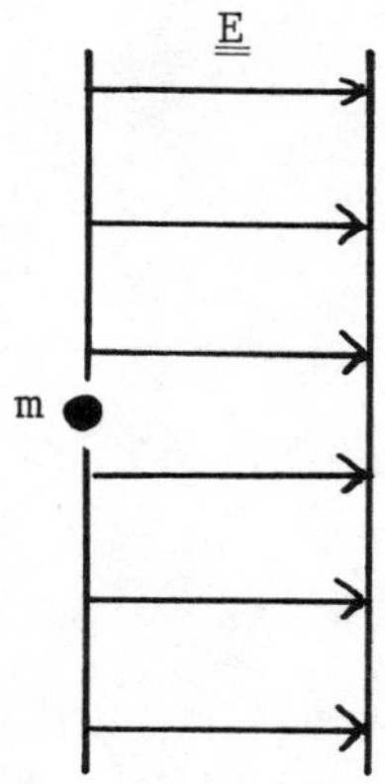

Consider a proton (mass m and charge q) which enters a region of uniform electric field $\underline{E}$. The proton is initially at rest. For convenience we think of the proton as entering through a hole.

a. Show by a vector the force $\underline{F}$ acting on the proton while it is in the field $\underline{E}$.

b. What will be the magnitude of this force?

c. What does a "uniform electric field" mean?

a.

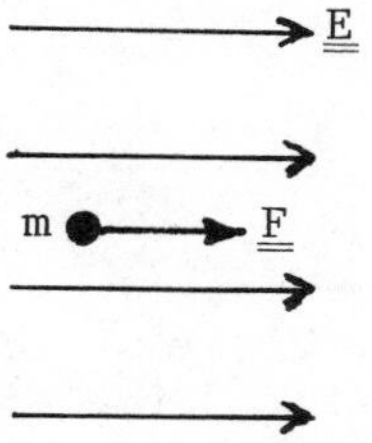

The electric force is in the direction of $\underline{E}$ for positive charges.

b. $F = qE$ where q is the proton charge and E is the magnitude of the electric field.

c. The electric field $\underline{E}$ has the same magnitude and direction everywhere. Hence here $\underline{F}$ is constant.

20.

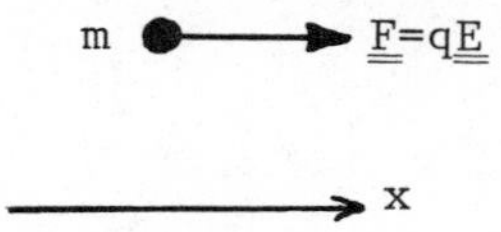

Considering this as a one dimensional problem and ignoring gravitational effects, write Newton's second law of motion for the proton.

$qE = ma_x$

---

21. Since q, E, and m are constant in this problem the acceleration a_x is a constant. Algebraically what will be the velocity of the proton after a time t assuming it to start from rest?

$v_x = a_x t$

Note: The initial velocity was given to be zero.

---

22. For uniformly accelerated motion of this type the average speed is

$$\bar{v} = 1/2\ v_x$$

where v_x is given in the previous frame answer. From this fact the distance ℓ that the proton goes in a time t is

$$\ell = \underline{\hspace{2cm}}\ t\ .$$

$\ell = 1/2\ v_x t$

23. Eliminate t between the answers of frame 21 and 22.	$v_x = a_x t$ $\ell = \frac{1}{2} v_x t$ $v_x^2 = 2\, a_x \ell$
24. Surely you recognize this expression as a kinematic formula from long ago. Rewrite $v_x^2 = 2\, a_x \ell$ using a_x from the answer of frame 20.	$v_x^2 = 2 \frac{qE}{m} \ell$
25. Rewriting the previous answer we have $\frac{1}{2} m v_x^2 = qE\ell$. a. What do we call the left side? b. What do we call the right side?	a. Kinetic energy. b. Work done by the electric field (force qE times distance ℓ). See, all that stuff you learned is still useful. Remember the work-energy theorem?
26. We say that the proton has acquired kinetic energy. Devices which do this job are called accelerators. If a proton (mass = 1.67×10^{-27} kg, charge = 1.6×10^{-19} C) is accelerated from rest through an electric field of strength 2.5×10^5 N/C for a distance 0.4 m, a. Proton's final kinetic energy = __________. b. Proton's final speed = __________. Consider the case as shown in frame 19.	a. 1.6×10^{-14} J. b. 4.4×10^6 m/s.

Chapter 23: GAUSS'S LAW

REVIEW AND PREVIEW

In the previous chapter you learned how to calculate the electric field due to a given distribution of charge. If the charge distribution has a certain symmetry then, in many cases, the resulting electric field also possesses a symmetry. For example, the electric field due to a uniformly charged sphere is spherically symmetric.

In this chapter you will study Gauss's law. This law relates an integral of the electric field over a closed surface to the charge contained within that surface. Although Gauss's law is perfectly general (applying to all charge distribution and all electric fields), for purposes of calculating $\underline{\underline{E}}$ it is useful only for problems having sufficient symmetry. In these cases you will learn how to take advantage of the symmetry to calculate $\underline{\underline{E}}$ in a relatively simple manner using Gauss's law.

GOALS AND GUIDELINES

In this chapter there are three major goals:

1. Understanding the definitions
   a. vectorial surface area element;
   b. flux of a vector field; (the flux of fields other than $\underline{\underline{E}}$ will be involved in later chapters.)
   c. open and closed surfaces.
2. Learning the formula (Eq. 23-2) for Gauss's law and very clearly understanding the meaning of the symbols in it.
3. Learning how to apply Gauss's law to problems. This is probably more difficult than "applying" most of the previous physical laws to problems. You will find that the problems are usually one of the following two types.
   a. Calculating $\underline{\underline{E}}$ due to a given distribution of charge. The charge distribution must have sufficient symmetry.
   b. Calculating something about the distribution of charge on a conductor. Here use is made of the fact that $\underline{\underline{E}} = 0$ within a conductor in electrostatic equilibrium.

You should try to develop a feeling for which problems can successfully be solved using Gauss's law and which can not.

23-1 Vectorial Surface Area Element

Consider a small (infinitesimal) piece of surface area. We are going to associate with this area an infinitesimal vector, $d\underline{\underline{S}}$, called the vectorial surface area element. $d\underline{\underline{S}}$ is that vector

(a) whose magnitude is equal to the amount of the area (the units of $d\underline{\underline{S}}$ are therefore m^2), and
(b) whose direction is normal to the surface area. (Of course there are really two such normals, directed oppositely to each other. In any actual problem we must know which of these two is meant.)

23-2 Flux of a Vector Field

Now suppose we have

(a) a surface S (this is not infinitesimal), and

(b) a vector field, for example the electric field $\underline{\underline{E}}$.

We imagine dividing the area S into vectorial surface area elements $d\underline{\underline{S}}$. Choose one such element $d\underline{\underline{S}}$ and evaluate $\underline{\underline{E}}$ at that location. Then form the dot product of these two vectors, $\underline{\underline{E}} \cdot d\underline{\underline{S}}$. Some properties of this product are:

(a) It is an infinitesimal scalar.
(b) It is positive if the angle between $\underline{\underline{E}}$ and $d\underline{\underline{S}}$ is less that 90°, negative if this angle is greater than 90°.
(c) It is zero if $\underline{\underline{E}}$ is perpendicular to $d\underline{\underline{S}}$, i.e. if $\underline{\underline{E}}$ is parallel to the area represented by $d\underline{\underline{S}}$.

We now imagine doing this for all the various $d\underline{\underline{S}}$'s and then adding (integrating) these quantities. The result

$$\Phi_E = \int \underline{\underline{E}} \cdot d\underline{\underline{S}} \qquad (23\text{-}1)$$

is called the flux of $\underline{\underline{E}}$ through the surface S. See Figure 23-1.

The flux, a scalar, depends upon both the surface S (its size, location, orientation, etc.) and the vector field $\underline{\underline{E}}$. If we had taken the other choice for the normal direction then each vector $d\underline{\underline{S}}$ would be reversed by 180°, thus changing merely the sign (but not the magnitude) of the flux Φ_E.

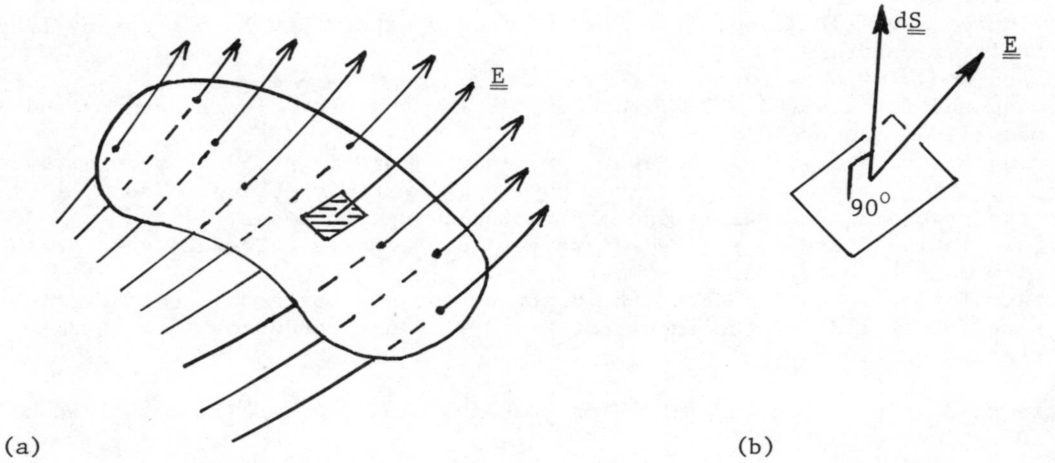

Figure 23-1. (a) A surface S and a vector field $\underline{\underline{E}}$. The area S is divided into vectorial surface area elements $d\underline{\underline{S}}$. To calculate the flux of $\underline{\underline{E}}$ through S we must add (integrate) all the dot products $\underline{\underline{E}} \cdot d\underline{\underline{S}}$, $\Phi_E = \int \underline{\underline{E}} \cdot d\underline{\underline{S}}$. (b) Detail view of one vectorial surface area element $d\underline{\underline{S}}$. $\underline{\underline{E}}$ is the vector field evaluated at the location of this $d\underline{\underline{S}}$. Note that $d\underline{\underline{S}}$ is normal to the area.

23-3 Open and Closed Surfaces

Surfaces may be classified as either:

(a) Open surfaces. These are surfaces which have an edge (or rim, or boundary). The surface of a sheet of paper is an example. Open surfaces need not be plane; the curved surface of a hemisphere is an open surface.

or

(b) Closed surfaces. These are surfaces with no edge (or rim, or boundary). The surface of a sphere (or a cube) is an example.

A good way to tell open surfaces from closed surfaces is the following. A closed surface separates all space into two distinct regions: an "inside" volume and an "outside" volume. It is impossible to go from the "inside" to the "outside" without actually crossing the closed surface. Open surfaces do not have this property.

The flux of a vector field may pertain to either open or closed surfaces. For the special case of a closed surface, we will always choose the vector $d\underline{\underline{S}}$ to be directed along the outward-pointing normal. To indicate that we are dealing with a closed surface, a small circle is put through the integral sign: $\oint$. In summary, $\oint \mathrm{E} \cdot d\mathrm{S}$ denotes the flux of $\underline{\underline{E}}$ through a closed surface, the direction of $d\underline{\underline{S}}$ being along the outward-pointing normal.

23-4 Gauss's Law

Lines of electric field seem to originate from positive charges and terminate at negative charges. The precise statement of this intuitive idea is called Gauss's law:

$$\oint \underline{\underline{E}} \cdot d\underline{\underline{S}} = q/\varepsilon_0 \ . \qquad (23\text{-}2)$$

The meaning of the left hand side of this equation was discussed above; it is the flux of $\underline{\underline{E}}$ through a closed surface S. On the right hand side, "q" means the net charge contained within this surface (i.e. in the volume bounded by this surface). For example in Figure 23-2 the flux of $\underline{\underline{E}}$ through the closed surface S would equal $(+3\ \mu C - 5\ \mu C)/\varepsilon_0$. The $+10\ \mu C$ charge lies outside the closed surface and therefore is not to be included as part of the charge "q". (This does not mean that the $+10\ \mu C$ charge has no effect on $\underline{\underline{E}}$ along the surface S. It does affect $\underline{\underline{E}}$, but in such a manner as to not change the flux $\oint \underline{\underline{E}} \cdot d\underline{\underline{S}}$ through S.) Although the surface S will be chosen to be especially simple and symmetric in problem applications, it must be emphasized that Gauss's law holds for all closed surfaces.

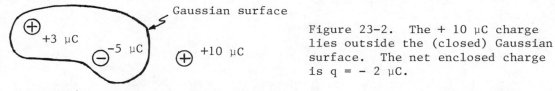

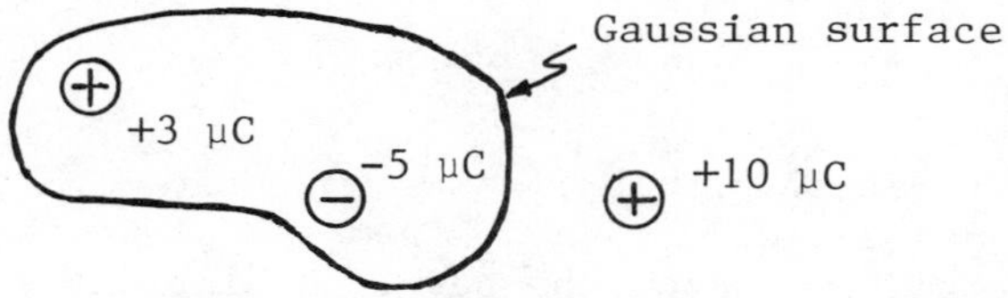

Figure 23-2. The + 10 μC charge lies outside the (closed) Gaussian surface. The net enclosed charge is $q = -\ 2\ \mu C$.

The formula $\oint \underline{\underline{E}} \cdot d\underline{\underline{S}} = q/\varepsilon_0$ is very concise. When applying it to problems the student must also keep in mind:

(a) It applies only to closed surfaces.
(b) $d\underline{\underline{S}}$ denotes the outward-pointing normal.
(c) In general $\underline{\underline{E}}$ varies from point to point along the surface S and hence cannot be "factored out" of the integral.
(d) $\underline{\underline{E}}$ and $d\underline{\underline{S}}$ are vectors; we must first form their dot product $\underline{\underline{E}} \cdot d\underline{\underline{S}}$ and then add (integrate) these to find the flux through the surface S.
(e) q means the net charge contained within the surface S.

Usually we are given a distribution of charge (e.g. a point charge, a charged cylinder, a charged plane, etc.) and we are asked to find $\underline{\underline{E}}$ at a given point P. But Gauss's law is an integral equation; it cannot in general be solved for $\underline{\underline{E}}$. Recall that an integral is a sum. Gauss's law then says that the sum (integral) of many $\underline{\underline{E}} \cdot d\underline{\underline{S}}$'s equals something (q/ε_0). We are interested in finding one of these $\underline{\underline{E}}$'s. In order to successfully obtain this from Gauss's law, we must know some more information concerning $\underline{\underline{E}}$; this additional information usually lies in the symmetry of the problem.

An analogy might make this more clear. Consider the integral equation $\int_0^1 f(x)\,dx = 1$. What is f(1/4)? Now we do not have enough information to answer this; there are many functions f(x) whose integral $\int_0^1 f(x)\,dx$ is unity. We need to know something about the <u>form</u> of the function f(x). For example if we know that f(x) is a constant, then it must be that f(x) = 1 so that f(1/4) = 1. Or, if we knew that f(x) was proportional to x, then it must be that f(x) = 2x so that f(1/4) = 1/2.

23-5 Application of Gauss's Law to Problems

A typical Gauss's law problem involves finding the electric field at some point P due to a given charge distribution. The essential steps to be followed are:

(1) Make certain assumptions concerning $\underline{\underline{E}}$ based on the symmetry of the problem. These are
   (a) an assumption about the direction of $\underline{\underline{E}}$, and
   (b) an assumption about what variables the magnitude of $\underline{\underline{E}}$ may depend upon.

(2) Choose an appropriate closed surface (called a "Gaussian surface") to take advantage of the symmetry of the particular problem. Gauss's law will be applied to this surface.

(3) Using the assumptions (1), somehow remove "E" from under the integral sign in the left hand side of Gauss's law. This is the crucial step in the procedure.

(4) Evaluate the net enclosed charge q, substitute this into the right hand side of Gauss's law and solve for E.

These steps will become more clear in the following two examples.

>>> Example 1. An infinite plane carries a uniform charge per unit area of σ (units of σ are C/m^2). Calculate the electric field $\underline{\underline{E}}$ at a point P, a perpendicular distance r away from the plane.

We follow the four steps listed above.

(1) Based on the symmetry of this particular problem, we assume that
   (a) the direction of $\underline{\underline{E}}$ is perpendicularly away from the plane, and
   (b) the magnitude of $\underline{\underline{E}}$ depends only upon the perpendicular distance r from the plane.

(2) Choose the cylinder shown in Figure 23-3 to be the Gaussian surface. This cylinder has "end caps" to make it a closed surface. The right cap contains the given point P, the left cap contains a similar point P´ which is also at a distance r away from the plane.

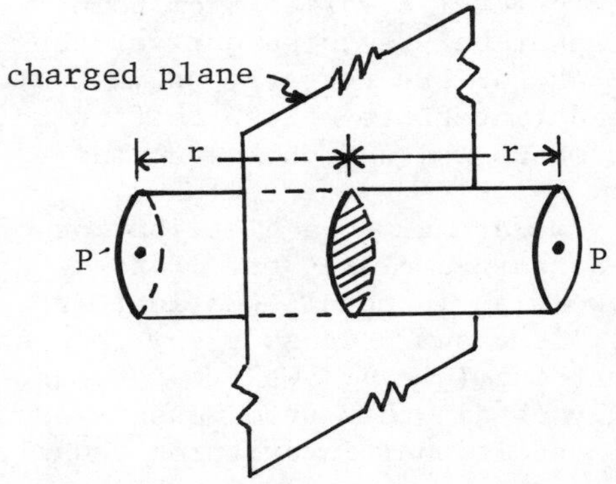

Figure 23-3. Gaussian surface for Example 1.

(3) Recall that an integral is a sum. We can divide the integral for the flux of $\underline{\underline{E}}$ through the Gaussian surface into three parts,

$$\oint \underline{\underline{E}}\cdot d\underline{\underline{S}} = \int_{\substack{\text{round}\\ \text{part}}} \underline{\underline{E}}\cdot d\underline{\underline{S}} + \int_{\substack{\text{right}\\ \text{cap}}} \underline{\underline{E}}\cdot d\underline{\underline{S}} + \int_{\substack{\text{left}\\ \text{cap}}} \underline{\underline{E}}\cdot d\underline{\underline{S}} \; .$$

We will now consider these three parts separately.

Along the round part of the Gaussian surface, $\underline{\underline{E}}$ is perpendicular to $d\underline{\underline{S}}$ (by assumption (a)), i.e. $\underline{\underline{E}}\cdot d\underline{\underline{S}} = 0$. Therefore

$$\int_{\substack{\text{round}\\ \text{part}}} \underline{\underline{E}}\cdot d\underline{\underline{S}} = 0 \; .$$

Along the right cap, $\underline{\underline{E}}$ is parallel to $d\underline{\underline{S}}$ (by (a)), i.e. $\underline{\underline{E}}\cdot d\underline{\underline{S}} = E\, dS$. Also, E along the right cap equals E at point P (by (b)). This (constant) value of E may be factored out of the integral:

$$\int_{\substack{\text{right}\\ \text{cap}}} \underline{\underline{E}}\cdot d\underline{\underline{S}} = \int_{\substack{\text{right}\\ \text{cap}}} E\, dS = E_{\text{at P}} \int_{\substack{\text{right}\\ \text{cap}}} dS \; .$$

The last integral, $\int dS$, is merely the sum of the little areas which make up the right cap. Thus

$$\int_{\substack{\text{right}\\ \text{cap}}} \underline{\underline{E}}\cdot d\underline{\underline{S}} = (E_{\text{at P}})A \; ,$$

where A is the area of the right cap.

Similarly, along the left cap,

$$\int_{\substack{\text{left}\\ \text{cap}}} \underline{\underline{E}}\cdot d\underline{\underline{S}} = (E_{\text{at P}'})A = (E_{\text{at P}})A \; .$$

This is because the area of the left cap is the same as the area A of the right cap and also because $E_{\text{at P}'} = E_{\text{at P}}$(by (b)).

The flux of $\underline{\underline{E}}$ through the entire Gaussian surface is the sum of the above three integrals,

$$\oint \underline{\underline{E}}\cdot d\underline{\underline{S}} = 0 + (E_{\text{at P}})A + (E_{\text{at P}})A = 2(E_{\text{at P}})A$$

(4) The charge "q" consists of that charge which is located on the cross-hatched area of the charged plane as shown in the figure. To compute this charge we multiply the charge per unit area σ by this cross-hatched area. Since this area is the same as the area A of the right cap we have $q = \sigma A$.

Finally, substituting into Gauss's law ($\oint \underline{\underline{E}}\cdot d\underline{\underline{S}} = q/\varepsilon_0$),

$$2(E_{\text{at P}})A = \sigma A/\varepsilon_0$$

$$E_{\text{at P}} = \sigma/2\varepsilon_0 \; .$$

The direction of $\underline{\underline{E}}$ is then perpendicularly away from the charged plane if σ is positive and perpendicularly toward the plane if σ is negative. <<<

In the example above, note that the area "A" cancelled out. This is to be expected, since the electric field at point P should not depend upon an irrelevant dimension of the Gaussian surface. The answer also happens to be independent of the distance "r" of the point P from the charged plane. This is not an assumption but is rather a result (note that assumption (a) allowed E to depend upon r).

In this sample problem all the reasoning was presented in complete detail. Once the student has mastered these ideas, he can do Gauss's law problems in a more abbreviated manner as shown in the next example.

>>> Example 2. An infinitely long solid cylinder of radius R carries a uniform charge per unit volume of ρ (units of ρ are C/m^3). Calculate the electric field $\underline{\underline{E}}$ at point P, a distance $r \le R$ away from the axis.

(1) Assume: (a) direction of $\underline{\underline{E}}$ is radially outward, and
(b) magnitude of $\underline{\underline{E}}$ depends only upon r.

(2) Gaussian surface: a cylinder (with end caps) of radius r, length L, coaxial with the given charged cylinder as shown in Figure 23-4.

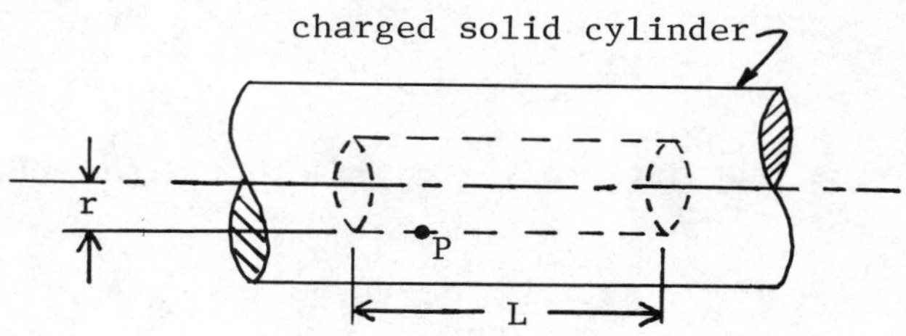

Figure 23-4. Gaussian surface for Example 2.

(3) $$\oint \underline{\underline{E}}\cdot d\underline{\underline{S}} = \int_{\substack{\text{round}\\\text{part}}} \underline{\underline{E}}\cdot d\underline{\underline{S}} + \int_{\substack{\text{right}\\\text{cap}}} \underline{\underline{E}}\cdot d\underline{\underline{S}} + \int_{\substack{\text{left}\\\text{cap}}} \underline{\underline{E}}\cdot d\underline{\underline{S}}$$

$$= \int_{\substack{\text{round}\\\text{part}}} E\, dS + 0 + 0 \qquad \text{(because of (a))}$$

$$= E_{\text{at P}} \int_{\substack{\text{round}\\\text{part}}} dS \qquad \text{(because of (b))}$$

$$= E_{\text{at P}}\,(2\pi r L)$$

(4) $$q = \rho(\pi r^2 L)$$

Note that since $r \le R$, the entire volume within the Gaussian surface carries a charge per unit volume of ρ. Substituting into Gauss's law,

$$\oint \underline{\underline{E}}\cdot d\underline{\underline{S}} = q/\varepsilon_0$$

$$E_{\text{at P}}(2\pi r L) = \rho\pi r^2 L/\varepsilon_0$$

$$E_{\text{at P}} = \rho r/2\varepsilon_0$$

The direction of $\mathbf{E}$ is then radially outward if ρ is positive and radially inward if ρ is negative. Note that the dimension "L" cancelled out. <<<

In each of the above examples the same shape Gaussian surface (a cylinder) was used, but for entirely different reasons: in the first example all the flux of $\mathbf{E}$ was through the end caps, in the second example the flux of $\mathbf{E}$ was through the round part.

23-6 Gauss's Law and Conductors

If a conductor is in electrostatic equilibrium, the force on the free electrons in the interior of the conductor must vanish. The consequences of this are:

1. In the interior of the conductor, $\mathbf{E} = 0$.
2. Immediately outside the conductor, the electric field is normal to the surface of the conductor.

Using these and Gauss's law, it is shown in the text that there is no net charge in the interior of the conductor and that the (outwardly) normal electric field immediately outside the conductor is given by

$$E = \sigma/\varepsilon_0 \, . \tag{23-3}$$

Here σ is the surface charge density (charge per unit area) on the adjacent surface of the conductor. An important thing to remember concerning (23-3) is that it gives the <u>total</u> (resultant) electric field due to all the charges in the problem (<u>not</u> just those on the nearby surface of the conductor).

>>> Example 3. A thin plate of metal (1 m × 1 m × 1 cm) carries a charge of 10 μC. Point P lies just outside the plate, somewhere near the center of one of the 1 m^2 faces. Calculate the electric field at point P.

Since this is a conductor, the 10 μC charge will redistribute itself over the surface of the metal. By symmetry we assume there will be 5 μC uniformly distributed on each of the two large faces (the smaller faces are being ignored). Thus we have <u>two</u> planes of charge, each with a surface charge density $\sigma = 5\ \mu C/m^2$. An edge view of this situation is shown in Figure 23-5.

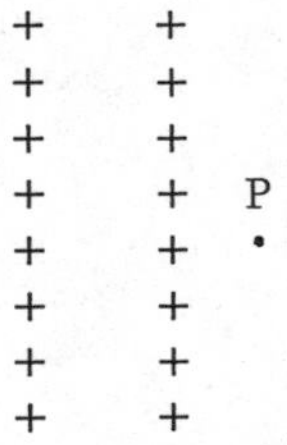

Figure 23-5. The two planes each carry a surface charge density $\sigma = 5\ \mu C/m^2$. Each plane produces an electric field $E = \sigma/2\varepsilon_0$ (directed toward the right) at point P.

(a) Using $E = \sigma/\varepsilon_0$ where $\sigma = 5\ \mu C/m^2$ is the surface charge density of the <u>single</u> adjacent conducting surface, we have

$$E_{at\ P} = (5 \times 10^{-6}\ C/m^2)/(8.85 \times 10^{-12}\ C^2/N{\cdot}m^2)$$

$$= 5.6 \times 10^5\ N/C.$$

This is the total electric field at point P.

(b) Each of the two planes of charge (regarding them as infinite planes) produces an electric field given by $E = \sigma/2\varepsilon_0$. At point P both these fields are directed toward the right; the total electric field at P is then

$$E_{\text{at P}} = \sigma/2\varepsilon_0 + \sigma/2\varepsilon_0 = \sigma/\varepsilon_0 = 5.6 \times 10^5 \text{ N/C}.$$

(In the interior of the conductor these two electric fields are equal in magnitude and oppositely directed. The total electric field in the interior is then zero, as it must be.)

We thus obtain the same answer upon correct application of either the "σ/ε_0" or the "$\sigma/2\varepsilon_0$" formula. <<<

In the above example, we were able to guess the charge distribution. Another type of Gauss's law problem involves finding the distribution of charge over the surface(s) of a conductor. In order to do this we make use of the fact that $\underline{\underline{E}} = 0$ in the interior of a conductor.

>>> Example 4. A conductor carries a net charge of 10 μC. Inside the conductor there is a hollow cavity. A point charge Q = 3 μC is located within this cavity. Calculate the charge q_1 on the inner surface of the conductor (i.e. on the cavity wall), and the charge q_2 on the outer surface of the conductor.

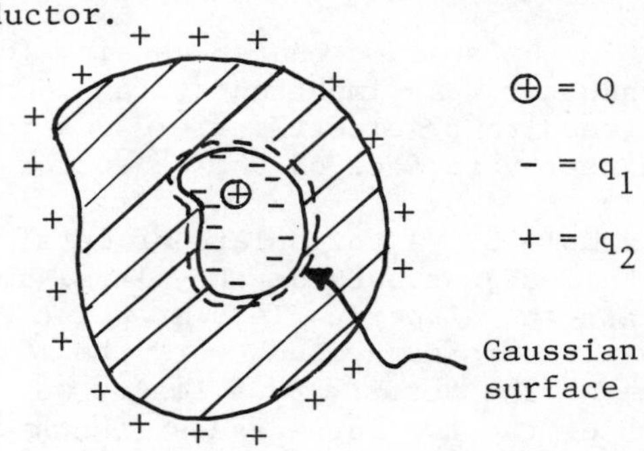

Figure 23-6. Gaussian surface for Example 4. The shaded region is the conductor. Charge Q is in the cavity, charge q_1 is on the walls of the cavity, charge q_2 is on the outer surface of the conductor.

Choose a Gaussian surface to lie in the interior of the conductor surrounding the cavity wall as shown in Figure 23-6. Since the Gaussian surface lies entirely within the conductor, $\underline{\underline{E}} = 0$ everywhere along this surface. Therefore the flux of $\underline{\underline{E}}$ through the Gaussian surface vanishes, $\oint \underline{\underline{E}}\cdot d\underline{\underline{S}} = 0$. Now the net charge within the Gaussian surface is $q = q_1 + Q$. Thus

$$\oint \underline{\underline{E}}\cdot d\underline{\underline{S}} = q/\varepsilon_o$$

$$0 = (q_1 + Q)/\varepsilon_0$$

$$q_1 = -Q = -3\ \mu\text{C} .$$

Since $q_1 + q_2 = 10\ \mu$C (given), $q_2 = 10\ \mu\text{C} - q_1 = 13\ \mu$C. Thus the 10 μC charge on the conductor redistributes itself as follows:

$$q_1 = -3\ \mu\text{C} \quad \text{(on the inner surface)} \ ,$$

$$q_2 = +13\ \mu\text{C} \quad \text{(on the outer surface)} \ . \qquad <<<$$

23-7 Programmed Problems

This chapter has to do with things like flux, surface integrals, enclosed charges, surface elements, normal components, etc. A few general questions may help to untangle these ideas.

1. 

The balloon to the left is a closed surface. An element of area of the balloon is shown and is labeled dS. Draw on the sketch a vector representing this element of area.

By convention the vector $d\mathbf{S}$ is perpendicular to the surface element and points <u>outward</u>.

---

2. In using Gauss's law we want to calculate the flux of the electric field through a closed surface. Let us first consider the general notion of flux. To make things even simpler we won't consider the entire balloon. Rather we will take only the area element $d\mathbf{S}$.

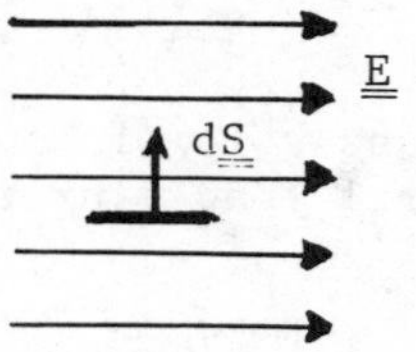

To the left we represent a constant electric field by evenly spaced lines. What is the flux of $\mathbf{E}$ through the area element $d\mathbf{S}$, i.e. how many lines of $\mathbf{E}$ cross the surface in the direction of $d\mathbf{S}$?

Zero.

We say that the flux of $\mathbf{E}$ through the surface element $d\mathbf{S}$ is zero.

---

3. 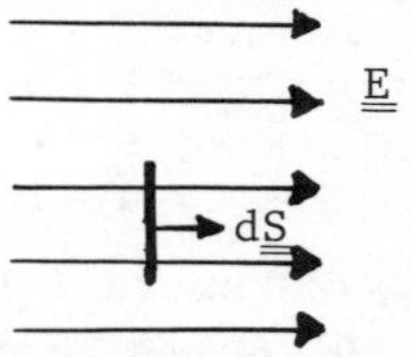

If we rotate the element of area 90° the flux is no longer zero. In this case what two things determine the flux of $\mathbf{E}$ through the the area?

1. The magnitude of $\mathbf{E}$.
2. The magnitude of the area element $d\mathbf{S}$.

---

4. 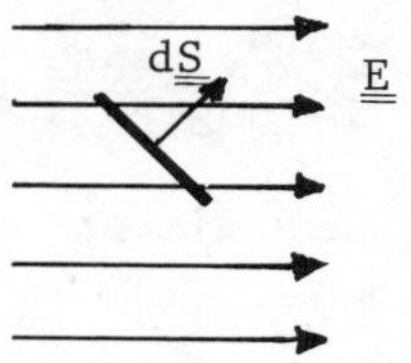

Here we have the intermediate case. Now we must take into account the relative orientations of $\mathbf{E}$ and $d\mathbf{S}$. The shorthand way to write this is:

Flux of $\mathbf{E}$ = ________.

$\mathbf{E}\cdot d\mathbf{S}$, which can be written

$$\underbrace{|\mathbf{E}|}_{\text{Strength of } \mathbf{E}} \quad \underbrace{|d\mathbf{S}|}_{\text{Magnitude of the area}} \quad \underbrace{\cos\theta}_{\text{Relative orientation}}$$

---

5. Gauss's law is written

$$\oint \mathbf{E}\cdot d\mathbf{S} = q/\varepsilon_0 \; .$$

a. The left hand side means the flux of the electric field through a ________ surface.
b. The meaning of q is ________.

a. Closed.

b. The q in the right hand side of Gauss's law is the net charge enclosed by the surface.

6. 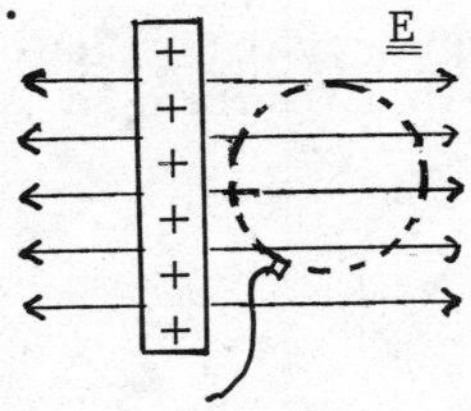

To the left is shown a charged plate which gives rise to a field $\mathbf{E}$ as represented by the field lines. Consider the closed surface of the imaginary balloon. There are charges only on the plate.

a. If you had a device to detect electric fields, would you detect a field if you were "inside" the imaginary balloon? (Yes/No)
b. The "q" on the right hand side of the Gauss's law is the net charge enclosed by the imaginary balloon surface. What net charge is enclosed by the surface?

a. Yes. There are field lines inside the surface.

b. Zero. There are charges only on the plate.

---

7. Looking at Gauss's law again

$$\oint \mathbf{E} \cdot d\mathbf{S} = q/\varepsilon_0 \ .$$

a. The left side is the ________ of $\mathbf{E}$ through the closed surface.
b. If q is zero, what is the flux of $\mathbf{E}$?

a. Flux, Φ_E.

b. Zero. Gauss's law is an equation relating flux to charge.

---

8. The crucial point illustrated by the last two frames is that the flux of $\mathbf{E}$ may be zero, but this does not mean that $\mathbf{E}$ is zero. In the diagram of frame 6 the flux of $\mathbf{E}$ through the closed surface is zero, but $\mathbf{E}$ is not zero inside the surface. The distinction is so important that we must look at our example a little closer.

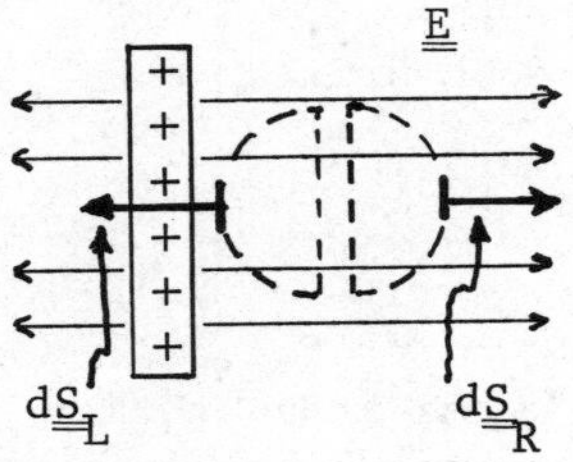

To the left we have divided the Gaussian surface into a left hand part and a right hand part.

Consider the flux contributions.

a. $\mathbf{E} \cdot d\mathbf{S}_L$ = ________.

b. $\mathbf{E} \cdot d\mathbf{S}_R$ = ________.

Considering that $|\mathbf{E}|$ and $|d\mathbf{S}|$ are the same respectively in both cases, in what way do these two terms differ?

a. $\mathbf{E} \cdot d\mathbf{S}_L = -\ E\ dS$ because the cosine of the angle is $-\ 1$.

b. $\mathbf{E} \cdot d\mathbf{S}_R = +\ E\ dS$ because the cosine of the angle is $+\ 1$.

The two terms thus differ only in sign.

9. Gauss's law requires that we add (integrate) all flux contributions of the type discussed in the preceding frame.

Can you now say why the total flux through our balloon Gaussian surface will be zero?

The left half will have negative terms which when added to the positive terms for the right half will result in zero.

---

10. Consider now a situation in which the net electric flux through a closed surface is not zero.

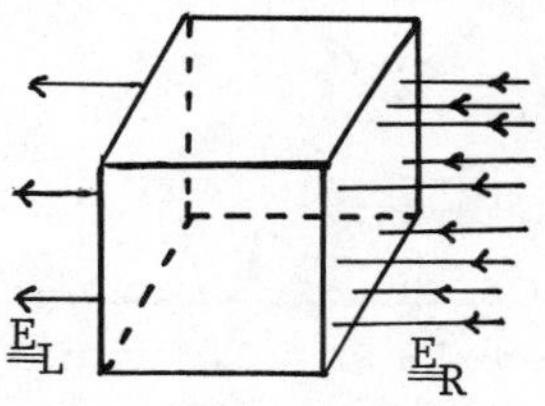

At the left is shown a cubical closed surface. The electric field is perpendicular to the right and left side of the cube. Is the electric field the same on both the right and left hand faces of the cube?

For the moment do not concern yourself about the situation within the cube.

No.

In representing electric fields by field lines, a greater density of lines means a stronger field. The field is stronger at the right face than at the left.

---

11. Draw on the sketch above two vectors to represent an element of area $d\mathbf{S}$ on both the left and right hand faces of the cubical.

Note that both vectors $d\mathbf{S}$ point "out" from the closed surface.

---

12. a. The algebraic sign of the electric flux through the right face is ________.

b. The algebraic sign of the electric flux through the left face is ________.

a. Negative.

$$\mathbf{E}\cdot d\mathbf{S} = |\mathbf{E}|\,|d\mathbf{S}|\underbrace{\cos 180^\circ}_{=\,-\,1}$$

b. Positive

$$\mathbf{E}\cdot d\mathbf{S} = |\mathbf{E}|\,|d\mathbf{S}|\underbrace{\cos 0^\circ}_{=\,+\,1}$$

---

13. The magnitude of the electric flux through the right hand side is (larger/smaller) than the electric flux through the left hand side.

Larger.
Because $|\mathbf{E}|$ is larger on the right hand side (and because the two areas are equal).

14. Thus we see that the negative flux through the right side is larger in magnitude than the positive flux through the left side. The flux through the entire cube is

$$\Phi_E = \oint_{cube} \underline{\underline{E}} \cdot d\underline{\underline{S}}$$

$$= \int_{\substack{Left \\ side}} \underline{\underline{E}} \cdot d\underline{\underline{S}} + \int_{\substack{Right \\ side}} \underline{\underline{E}} \cdot d\underline{\underline{S}} = \_\_\_\_\_\_\_\_.$$

Qualitatively describe the answer (positive, negative or zero). Note that all other sides of the closed surface are left out because their contribution to the flux is zero.

Negative,

since $|\Phi_L| < |\Phi_R|$.

---

15. Gauss's law states that the electric flux through a closed surface is proportional to the charge enclosed by that surface.

a. Is there a net charge inside the cubical surface?
b. If so, can you tell what the sign of the net charge is?

a. Yes, there must be. For if there were no charge, then the electric flux would have been zero.
b. Since the flux is negative, the net charge is also negative. There may be some positive as well as negative charges, but the net charge is negative.

16.

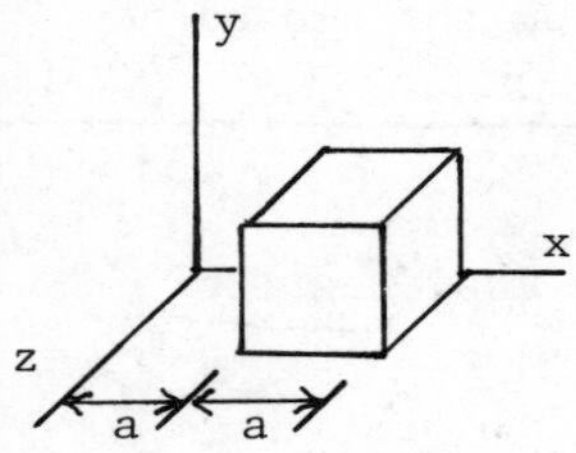

The left face of the cube shown is a distance "a" away from the origin. Each side of the cube is also "a" in length. An electric field given by

$$E_x = bx^{1/2}, \quad E_y = 0, \quad E_z = 0$$

is present. We wish to calculate the flux of $\underline{\underline{E}}$ through the cube and use this result to obtain the charge enclosed by the cube.

Qualitatively describe the given electric field.

The field is in the x direction only. The strength of the field becomes larger as you go away from the origin in the direction of x.

---

17. Will electric field lines pass through all faces of the cubical closed surface? Explain.

No.
Since the electric field is only in the x direction, field lines will be parallel to all faces except the left and right faces.

18. Will the electric field have the same intensity at all surface elements $d\underline{\underline{S}}$ of the left face of the cube?

Yes.
All surface elements of the left side are at the same value of x and the electric field has a fixed intensity at any particular x, i.e.

$E_x = bx^{1/2}$.

---

19. Is the electric field intensity the same for every surface element of the right face of the cube?

Yes.
Same reason as the previous answer.

---

20. Is the electric field intensity the same at the left face as it is at the right face?

No.
$E = bx^{1/2}$
At the left face x = a , while at the right face x = 2a .

21.

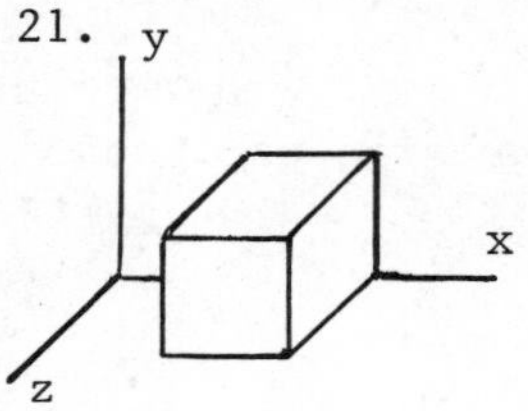

For $b = 800\ \mathrm{N/(C \cdot m^{1/2})}$ and $a = 0.1$ m,

E_L = _______ N/C at left face.

E_R = _______ N/C at right face.

$E_L = bx^{1/2}$

$E_L = 800 \dfrac{\mathrm{N}}{\mathrm{C \cdot m^{1/2}}} \times (0.1\ \mathrm{m})^{1/2}$

$E_L = 250$ N/C.

Similarly for x = 2a = 0.2 m,
$E_R = 360$ N/C

---

22. We have already established that the total flux through the closed cube involves only the left and right sides.

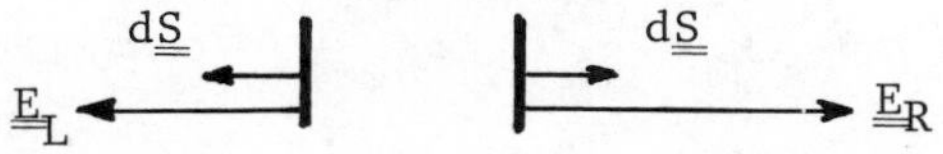

$$\Phi_E = \oint \underline{\underline{E}} \cdot d\underline{\underline{S}} = \int_L \underline{\underline{E_L}} \cdot d\underline{\underline{S}} + \int_R \underline{\underline{E_R}} \cdot d\underline{\underline{S}} \; .$$

Rewrite this equation in the correct scalar form, i.e. consider the orientation of the relevant vectors.

Φ_E = _______ .

$$\Phi_E = - \int_L E_L \, dS + \int_R E_R \, dS.$$

The minus sign because $\cos 180^o = -1$ and the plus sign because $\cos 0^o = +1$.

---

23. In view of frames 18 and 19 concerning the value of E_L and E_R we can write

$$\Phi_E = - E_L \int_L dS + E_R \int_R dS$$

where

$\int_L dS$ = ________ .

$\int_R dS$ = ________ .

$a^2 = 0.01\ \mathrm{m}^2$

$a^2 = 0.01\ \mathrm{m}^2$.

The area of each face.

24. From frame 21 we have $a = 0.1$ m, $E_L = 250$ N/C and $E_R = 360$ N/C so

$$\Phi_E = \_\_\_\_\_\_\_\_\_\_.$$

Obtain a number.

$$\Phi_E = -E_L \int_L dS + E_R \int_R dS$$

$$\Phi_E = -E_L a^2 + E_R a^2$$

$$\Phi_E = a^2[-E_L + E_R]$$

$$\Phi_E = 0.01 \text{ m}^2[-250 \text{ N/C} + 360 \text{ N/C}]$$

$$\Phi_E = 1.1 \text{ N}\cdot\text{m}^2/\text{C}.$$

---

25. From Gauss's law we can now determine the charge enclosed.

$$\oint_{\text{cube}} \underline{\underline{E}}\cdot d\underline{\underline{S}} = q/\varepsilon_0 .$$

$$q = \_\_\_\_\_\_\_\_\_\_.$$

$$\varepsilon_o = 8.85 \times 10^{-12} \text{ C}^2/\text{N}\cdot\text{m}^2.$$

$$q = \varepsilon_0 \oint \underline{\underline{E}}\cdot d\underline{\underline{S}}$$

$$q = 1.1 \text{ N}\cdot\text{m}^2/\text{C} \times 8.85 \times 10^{-12} \text{ C}^2/\text{N}\cdot\text{m}^2$$

$$q = 9.7 \times 10^{-12} \text{ C}.$$

Note that the net charge enclosed is positive since the flux of $\underline{\underline{E}}$ was positive.

Chapter 24: ELECTRIC POTENTIAL

REVIEW AND PREVIEW

In Part 1 you studied the concept of the potential energy U associated with a conservative force. For example, $U = mgy$ is associated with the gravitational force near the surface of the earth and $U = \frac{1}{2}kx^2$ is associated with the force of a stretched spring. In this chapter you will study a somewhat similar quantity associated with the electrostatic force. This quantity is called the electric potential V.

GOALS AND GUIDELINES

In this chapter there are four major goals:

1. Learning the definition of potential difference $V_B - V_A$ and its relation to the potential energy difference $U_B - U_A$.
2. Learning the formula (Eq. 24-6) for the potential due to a single point charge.
3. Learning how to apply the above formula in order to calculate V due to any distribution of charge (Examples 4 and 5).
4. Learning the relation between power, current and potential difference (Eq. 24-9). This will be used in later applications to electric circuits.

Both $\mathbf{E}$ and V are fields. Each can be calculated from a knowledge of the sources (i.e., the distribution of charges causing the field). You should also be able to calculate V directly from $\mathbf{E}$ (Eq. 24-4) and $\mathbf{E}$ directly from V (Eq. 24-8). Furthermore you should have a feeling for the relation between the lines of $\mathbf{E}$ and the equipotential surfaces (Section 24-6).

24-1 Work Done Against an Electric Field

Suppose an electrostatic field $\mathbf{E}$ acts on a test charge q_o. The force exerted on q_o by this field is $\mathbf{F}_{field} = \mathbf{E}q_o$. If another force (say supplied by a man) holds this charge in equilibrium, then this force must be $\mathbf{F}_{man} = -\mathbf{F}_{field} = -\mathbf{E}q_o$. Now let this charge move from a point A to a point B along some path L. We divide this path into infinitesimal segments $d\boldsymbol{\ell}$ as shown in Figure 24-1. The direction of each vector $d\boldsymbol{\ell}$ is tangent to the path, the sense is such that these vectors go from A to B along the path. The work $W_{A\to B}$ done by the man (against the electric field) in moving the charge q_o from A to B is then $W_{A\to B} = \int \mathbf{F}_{man}\cdot d\boldsymbol{\ell}$. Since $\mathbf{F}_{man} = -\mathbf{E}q_o$,

$$W_{A\to B} = -q_o \int_A^B \mathbf{E}\cdot d\boldsymbol{\ell} . \qquad (24\text{-}1)$$

The meaning of the integral $\int \mathbf{E}\cdot d\boldsymbol{\ell}$ is the following: Choose a segment $d\boldsymbol{\ell}$ and evaluate $\mathbf{E}$ at that location. Form the dot product $\mathbf{E}\cdot d\boldsymbol{\ell}$ and then add (integrate) these for all the various $d\boldsymbol{\ell}$'s along the path L.

Figure 24-1. The path L from point A to point B is divided into segments $d\boldsymbol{\ell}$.

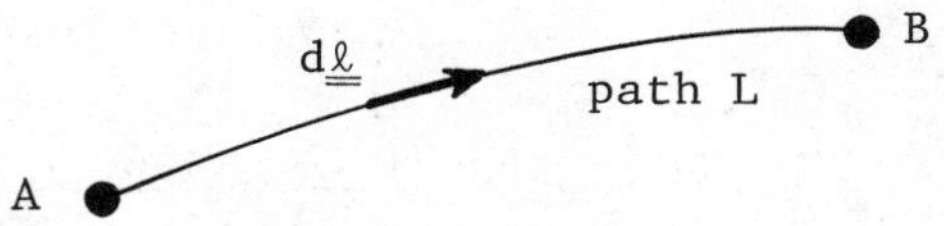

24-2 Electric Potential

The work $W_{A\to B}$ is proportional to the test charge q_o. If we divide this work by q_o we obtain the work per unit charge. This quantity is called the potential difference, $V_B - V_A$, between B and A. That is

$$V_B - V_A = W_{A\to B}/q_o . \qquad (24\text{-}2)$$

From equation 24-1 we then have

$$V_B - V_A = -\int_A^B \underline{\underline{E}}\cdot d\underline{\underline{\ell}} \ . \tag{24-3}$$

The unit of potential is the <u>volt</u>, V (= joule/coulomb). A convenient unit for electric field is then the V/m (= N/C).

The choice of the path L between points A and B is arbitrary; $V_B - V_A$ is the same for all paths starting at A and ending at B. This is because the electrostatic force is <u>conservative</u>.

Since only potential differences are defined by (24-2), we may define the potential at a particular point to be any convenient value. Usually, the potential at infinity is defined to be zero. Then the potential at a point P is simply

$$V_P = -\int_\infty^P \underline{\underline{E}}\cdot d\underline{\underline{\ell}} \tag{24-4}$$

with the understanding that $V_\infty = 0$. For each point P there is some value of the potential V_P; that is, the potential V is a <u>scalar</u> <u>field</u>.

24-3 Potential Energy

Recall from mechanics that the potential energy U associated with a given force is defined as the work done against that force (i.e. the negative of the work done by that force). From equation (24-2) we have the potential energy difference associated with the electric force,

$$U_B - U_A = q(V_B - V_A) . \tag{24-5}$$

Here $U_B - U_A$ means the change in (electric) potential energy when a charge q moves <u>from</u> point A <u>to</u> point B. For example, suppose that B is at a higher potential than A (i.e. V_B is greater than V_A); then as q moves from A to B, U will increase if q is positive and decrease if q is negative.

Now that we know the change in potential energy, we can apply the principle of conservation of energy to problems involving electrostatic forces.

>>> Example 1. As it passes a point A, a proton is moving with a speed of 10^5 m/s. It travels to a point B which is at a potential 100 volts lower than A. What is the speed of the proton at point B?

We use conservation of energy; the total energy (kinetic plus potential) must be constant.

$$K_B + U_B = K_A + U_A$$

$$K_B = K_A + (U_A - U_B)$$

$$\tfrac{1}{2}mv_B^2 = \tfrac{1}{2}mv_A^2 + q(V_A - V_B)$$

$$v_B = [v_A^2 + (2q/m)(V_A - V_B)]^{1/2}$$

$$v_B = \left[(10^5 \text{ m/s})^2 + \frac{(2)\times(1.60\times10^{-19}\text{ C})}{1.67\times10^{-27}\text{ kg}}(+\,10^2\text{ V})\right]^{1/2}$$

$$v_B = 1.71\times10^5 \text{ m/s} .$$

Note that $(V_A - V_B) = +100$ V (because B is 100 volts <u>lower</u> than A). Since the (positively charged) proton went through a potential <u>decrease</u>, its speed <u>increased</u> (i.e. U decreased causing K to increase). <<<

24-4 Electron Volt

The unit of energy is, of course, the joule. It is convenient sometimes to use another unit of energy called the electron volt (eV). As its name suggests, it is the electronic charge multiplied by one volt:

$$1 \text{ eV} = e \times 1 \text{ volt} = (1.60 \times 10^{-19} \text{ C}) \times (1 \text{ J/C}) = 1.60 \times 10^{-19} \text{ J}.$$

The conversion factor between electron volts and joules is then (numerically) the value of the electronic charge. The physical significance of one electron volt is that it is the change in potential energy of an electron when it moves through a potential difference of one volt. If a proton were to similarly go through a one volt potential difference then its potential energy would also change by one electron volt; for an alpha particle the corresponding change would be 2 eV (since its charge is 2e).

Sometimes we speak of e.g. "a 3 MeV proton". This refers to a proton whose kinetic energy is three million electron volts. This could be obtained by accelerating a proton (which was initially at rest) through a potential difference of 3×10^6 volt.

>>> Example 2. A 1 MeV alpha particle is accelerated through a potential difference of two million volts. What is its final kinetic energy?

In accelerating through the two million volt potential difference the alpha particle will gain an additional 4 MeV in kinetic energy (since its charge is 2e). Its kinetic energy after this acceleration will then be 1 MeV + 4 MeV = 5 MeV. <<<

24-5 Calculating the Potential

If we know the electric field $\mathbf{E}$ then equation (24-3) tells us how to calculate the potential difference $V_B - V_A$. We merely choose any convenient path between points A and B and then evaluate the integral (24-3). The potential V at a point P can similarly be obtained using equation (24-4).

>>> Example 3. Given a uniform* electric field field $\mathbf{E}$ calculate the potential difference, $V_B - V_A$, between the two points shown.

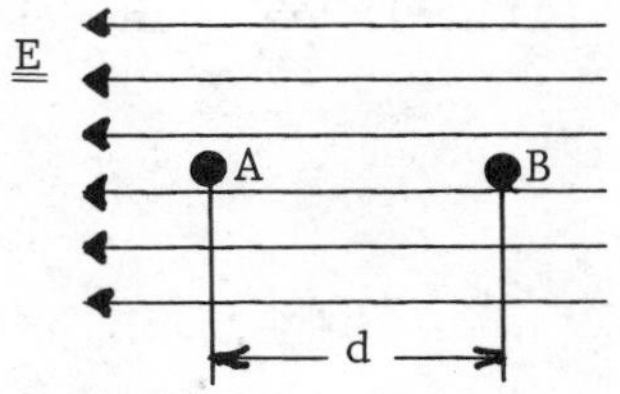

We will apply equation (24-3) to this situation because $\mathbf{E}$ is known. Choose a straight line as the path L from A to B. Then

$$V_B - V_A = -\int_A^B \mathbf{E}\cdot d\boldsymbol{\ell}$$

$$= +\int_A^B E\, d\ell \quad \text{(since the angle between } \mathbf{E} \text{ and } d\boldsymbol{\ell} \text{ is } 180^\circ)$$

$$= E\int_A^B d\ell \quad \text{(since E is constant)}$$

$$V_B - V_A = E\, d \quad \text{(since the sum of all the segments } d\ell \text{ is the length d between the two points).}$$

*"Uniform" means that the field has the same magnitude and direction at all points in space.

Remarks:

1. It turned out that B is at a higher potential than A. Thus the lines of electric field tend to run from regions of higher potential (B) toward regions of lower potential (A).

2. In many problems involving uniform electric fields, the question of sign is unimportant. The above result is then written simply as "V = Ed" where "V" is the magnitude of the potential <u>difference</u> between two points which are separated by a distance "d", this distance being along the $\underline{\underline{E}}$ field lines. We emphasize that "V = Ed" holds only for uniform <u>electric</u> fields. <<<

As mentioned above, if we know the electric field $\underline{\underline{E}}$ we can calculate the potential V at a point P by means of (24-4). But $\underline{\underline{E}}$ itself depends upon the distribution of the various charges which cause this electric field. It seems reasonable then that we should be able to calculate V directly from a given charge distribution, without first having to find $\underline{\underline{E}}$. The simplest case is when the charge distribution consists of merely one point charge q. The electric field due to this charge is given by equation (22-2). Substituting this expression for $\underline{\underline{E}}$ into (24-4) gives the potential. (See text for the details of the integration.) The result for the potential at a point P, a distance r away from the point charge q is

$$V_P = \frac{1}{4\pi\varepsilon_o}\frac{q}{r} \qquad (24\text{-}6)$$

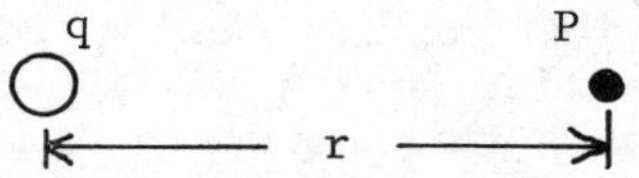

Again, P is not a charge; it is merely a point in space. If a test charge q_o were brought from infinity to P, an amount of work $W_{\infty\to P} = q_oV_P$ would be required.

Now that the potential due to a single point charge is known, we can calculate V due to any finite number of point charges. We treat then one at a time, using equation (24-6) to find V due to each charge. The total potential is then the sum of these, this being of course a <u>scalar</u> addition.

>>> Example 4. Find the potential at point P due to the three point charges q_1, q_2, q_3 as shown. Take

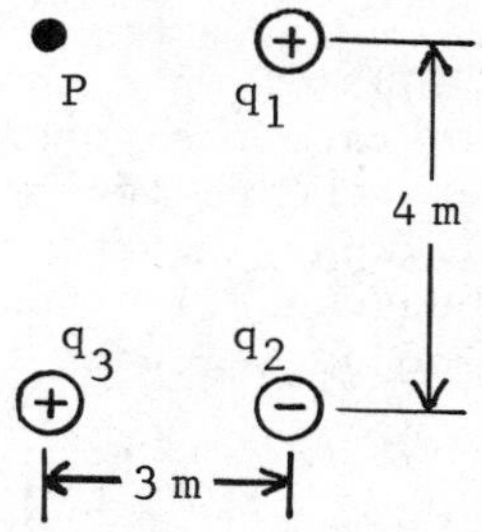

$$q_1 = 3\ \mu C\ ,$$

$$q_2 = -\ 10\ \mu C\ ,$$

$$q_3 = 2\ \mu C\ .$$

Let r_1 be the distance from q_1 to P, etc. Using equation (24-6),

$$V_P = \frac{1}{4\pi\varepsilon_o}\frac{q_1}{r_1} + \frac{1}{4\pi\varepsilon_o}\frac{q_2}{r_2} + \frac{1}{4\pi\varepsilon_o}\frac{q_3}{r_3} = \frac{1}{4\pi\varepsilon_o}\left[\frac{q_1}{r_1} + \frac{q_2}{r_2} + \frac{q_3}{r_3}\right]$$

$$= (9.0 \times 10^9\ N\cdot m^2/C^2)\left[\frac{3 \times 10^{-6}\ C}{3\ m} + \frac{-10 \times 10^{-6}\ C}{5\ m} + \frac{2 \times 10^{-6}\ C}{4\ m}\right]$$

$$V_P = -\,4.5 \times 10^3 \text{ V} .$$

Remarks:

1. Note that we used scalar addition, only the distances (r_1, r_2, r_3) counted. There is no such thing as the "direction" of any of these potentials.

2. The result for V_P happened to be negative. This means that if a positive test charge were brought from infinity to P, it would do work (say on the man who was carrying the charge from infinity to P). <<<

Another class of problems concerns the case in which the charge is continuously distributed. To treat this case, we divide the charge distribution into elementary point charges dq. At a point P a distance r away from dq there will be an infinitesimal potential given by

$$dV_P = \frac{1}{4\pi\varepsilon_o}\frac{dq}{r} \qquad (24\text{-}7)$$

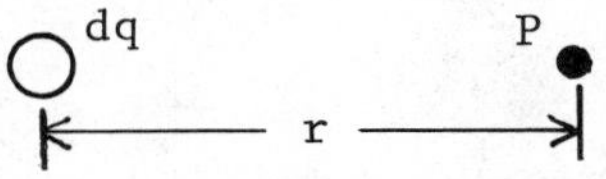

This is of course just equation (24-6) again, with dq instead of q and dV instead of V. The total potential is the sum (integral) of these:

$$V_P = \int dV_P = \frac{1}{4\pi\varepsilon_o}\int \frac{dq}{r} .$$

>>> Example 5. A thin glass rod of length L carries a charge Q uniformly distributed along its length. Find the potential at point P on the axis of the rod, a distance r away from one end as shown.

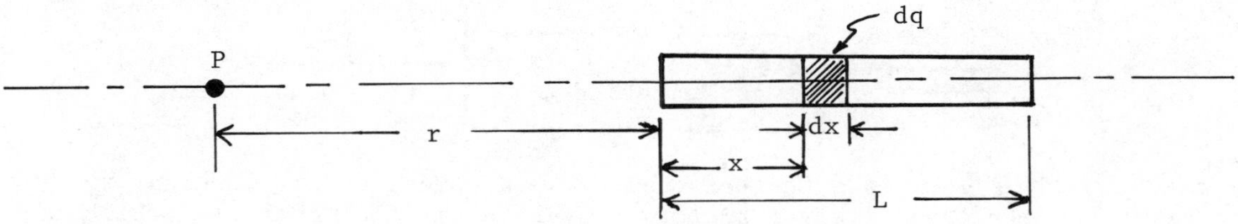

The charge per unit length is $\lambda = Q/L$; λ is a constant since the charge is uniformly distributed along the length of the rod. We divide the rod into segments dx, the charge on each segment is $dq = \lambda\, dx$. Since the distance from dq to P is $r + x$, we have from (24-7),

$$dV_P = \frac{1}{4\pi\varepsilon_o}\frac{dq}{r+x} = \frac{1}{4\pi\varepsilon_o}\frac{dx}{r+x}$$

$$V_P = \int dV_P = \frac{\lambda}{4\pi\varepsilon_o}\int_0^L \frac{\lambda\, dx}{r+x} \quad \frac{\lambda}{4\pi\varepsilon_o}\,\ell n(r+x)\Bigg|_{x=0}^{x=L}$$

$$V_P = \frac{Q}{4\pi\varepsilon_o L}\,\ell n(1 + L/r) .$$

Here we have used the fact that $\ell n(r + L) - \ell n(r) = \ell n(1 + L/r)$. <<<

24-6 Relation Between $\underline{\underline{E}}$ and V

The potential V is an integral of $\underline{\underline{E}}$, $V_P = -\int \underline{\underline{E}}\cdot d\underline{\underline{\ell}}$. The electric field $\underline{\underline{E}}$ can be obtained from V by differentiation, the components of $\underline{\underline{E}}$ being given by

$$E_x = -\partial V/\partial x \ , \qquad (24\text{-}8a)$$

$$E_y = -\partial V/\partial y \ , \qquad (24\text{-}8b)$$

$$E_z = -\partial V/\partial z \ . \qquad (24\text{-}8c)$$

If V is a function only of r (where r is the distance from the origin for problems having spherical symmetry, or the distance from the axis for problems having cylindrical symmetry), then $\underline{\underline{E}}$ has only an "r" component,

$$E_r = -\partial V/\partial r \ . \qquad (24\text{-}8d)$$

Points of constant V form an equipotential surface. The surface of any conductor in electrostatic equilibrium is such an equipotential surface. Lines of electric field always cross equipotential surfaces at right angles.

24-7 Power Supplied to an Electrical Circuit

In electrical circuits we sometimes need to know the power supplied to a circuit element. Resistors, capacitors and batteries are examples of circuit elements; these will be considered in later chapters. Consider the following circuit which has two terminals ("x" and "y").

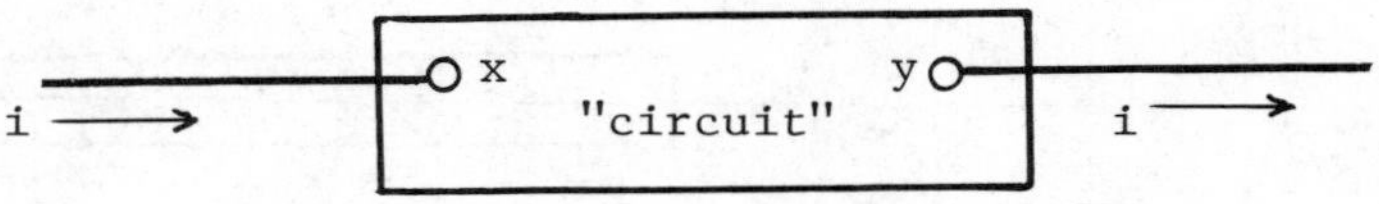

A current $i = dq/dt$ goes into the terminal "x" and comes out the terminal "y". The contents of the box is entirely arbitrary. Every element of charge dq which flows through the circuit will gain a potential energy $dq(V_y - V_x)$, i.e. it will lose a potential energy $dq(V_x - V_y)$. The rate of potential energy loss by these charges is then

$$\frac{dq}{dt}(V_x - V_y) = i(V_x - V_y) \ .$$

This rate of energy* loss by the charges as they pass through the circuit represents the power supplied to the circuit.

$$P_{\text{to circuit}} = i(V_x - V_y) \ . \qquad (24\text{-}9)$$

Note that in this formula, i is the current going through the circuit from x to y. The units of P are of course joule/second = watt. Equation (24-9) is completely general, independent of what is actually inside the box.

---

*The kinetic energy of the charges as they enter and leave the circuit is negligible. Thus a loss in potential energy is equivalent to a loss in total energy.

24-8 Programmed Problems

In actual practice electric fields are seldom measured directly. In the television picture tube for example, the brightness (florescence) on the screen is the result of energetic electrons colliding with phosphor sprayed onto the glass screen. These energetic electrons are produced by acceleration in electric fields. However, advertisements always refer to the electron beams as "25,000 volt electron beams" rather than stating the electric field and the distance through which the electrons traveled.

1.

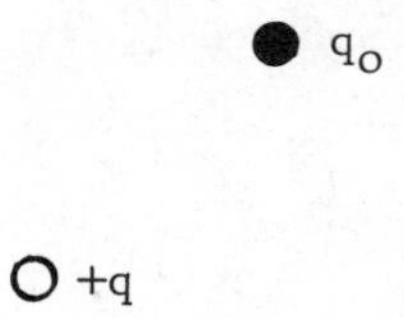

Describe qualitatively what would happen to a positive test charge q_o if it were placed, but not held, in the electric field of a positive charge q which is fixed in space.

The test charge would be accelerated in the direction of the electric field of +q. In this case the test charge would accelerate away from q.

---

2. As the previous answer indicates the acceleration of q_o is "away" from +q. What is the direction of the resultant force on the so-called "free" test charge q_o.

As always the acceleration and resultant force are in the same direction.

3.

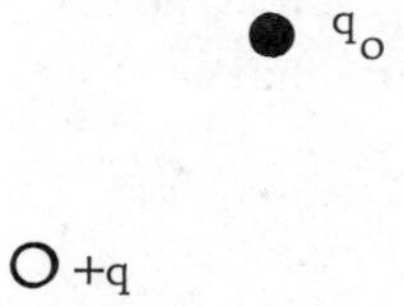

Draw two forces on q_o to represent

1. the electrical force
2. an external force (say of a man)

necessary to keep q_o in equilibrium.

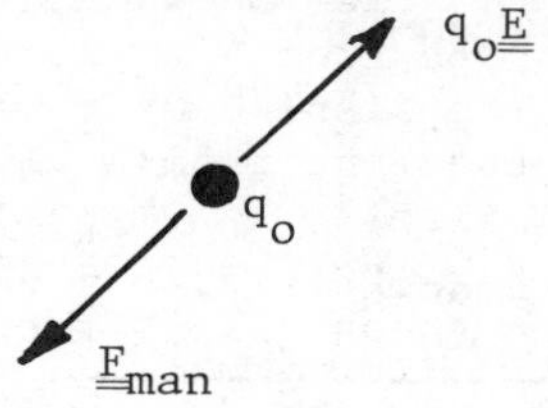

---

4. In the previous answer the electrical force is labeled $q_o\underline{\underline{E}}$. What is the source of $\underline{\underline{E}}$?

The charge +q.

---

5. How are the force vectors $\underline{\underline{F}}_{man}$ and $q_o\underline{\underline{E}}$ related in the answer of frame 3?

Equal magnitude and oppositely directed.

$\underline{\underline{F}}_{man} = - q_o\underline{\underline{E}}$ or

$\underline{\underline{F}}_{man} + q_o\underline{\underline{E}} = 0$

since q_o is in equilibrium.

6. The force $\mathbf{F}_{man}$ has been introduced so that the test charge can be used to explore the electric field of +q. We have already seen that q_o would accelerate away if it were "free". In this sequence of frames we shall use a positive q_o.

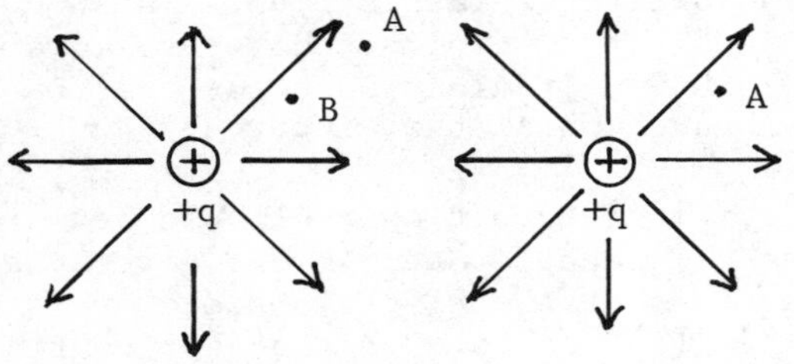

Above are shown two identical charges. In both cases we wish to calculate the work done by the external force $\mathbf{F}_{man}$ in going from A to B under the condition that at all times $\mathbf{F}_{man} = - q_o\mathbf{E}$. Let us take the simple radial path.

a. $\mathbf{F}_{man}$ is the same along all points of the path from A to B. (True/False)
b. The displacement vector $d\boldsymbol{\ell}$ is directed along the line from A to B in both cases. (True/False)
c. $\mathbf{F}_{man}$ points in a direction from A to B in both cases. (True/False)

a. False. Since $|\mathbf{F}_{man}| = q_o|\mathbf{E}|$, $|\mathbf{F}|$ is not constant because $|\mathbf{E}|$ is not constant.

$$|\mathbf{E}| = \frac{1}{4\pi\varepsilon_o}\frac{q}{r^2}$$

b. True.

c. False. In both cases $\mathbf{F}_{man}$ points "toward" the +q charge. (If the test charge q_o were negative then in both cases $\mathbf{F}_{man}$ would point "away" from the +q charge.)

---

7. In view of the answers b and c of the previous frame what chief difference would there be in the integral

$$\int_A^B \mathbf{F}_{man} \cdot d\boldsymbol{\ell}$$

for the two cases shown.

One would be positive and one would be negative.

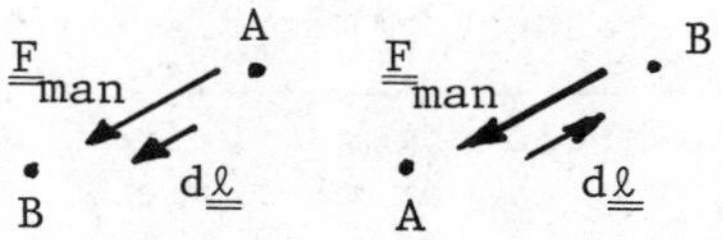

Positive, $\cos\theta = 1$

Negative, $\cos\theta = -1$

---

8. The above integral is called _________.

Work. This is the work done by the man in moving q_o from A to B.

---

9. From our example of frame 6 then
a. (Positive/Negative) work is done by the man in moving a test charge against electric field lines.
b. (Positive/Negative) work is done by the man in moving a test charge in the direction of electric field lines.

a. Positive.

b. Negative.

The result noted in the previous frame is summarized as follows: For a positive charged particle in an electric field,

1. If an external agent does positive work in going from A to B then we say that B is at a higher electric potential than A (or that A is at a lower electric potential than B).
2. If an external agent does negative work in going from A to B then we say that B is at a lower electric potential than A (or that A is at a higher electric potential than B).

Just the opposite of the above two statements is true for a negative charged particle.

10.

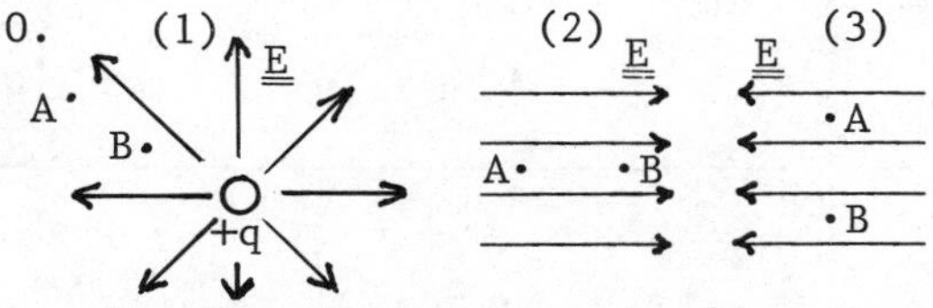

Imagine that a man supplies an external force such that $\mathbf{F}_{man} = -q_o\mathbf{E}$ in the three examples above. By considering the sign of

$$\int_A^B \mathbf{F}_{man} \cdot d\boldsymbol{\ell} \ ,$$

determine whether A is at a (higher/lower) electric potential than B. Consider all three of the above cases.

1. Lower.
2. Higher.
3. Both at the same potential! In this case the work is zero since

$$\mathbf{F}_{man} \cdot d\boldsymbol{\ell} = 0.$$

When the work is zero in going from A to B then A and B are at the same electric potential.

11. We can now write down formally that

$$W_{AB} = \int_A^B \mathbf{F}_{man} \cdot d\boldsymbol{\ell}$$

as the work done by an outside agent in moving a test charge q_o from point A to point B. Since $\mathbf{F}_{man} = -q_o\mathbf{E}$ we can write

$$W_{AB} = -\, q_o \int_A^B \mathbf{E} \cdot d\boldsymbol{\ell} \ .$$

Dividing by q_o we have

$$W_{AB}/q_o = -\int_A^B \mathbf{E} \cdot d\boldsymbol{\ell} \ .$$

The quantity $-\int \mathbf{E} \cdot d\boldsymbol{\ell}$ will be

1. (Positive/negative/zero) when $d\boldsymbol{\ell}$ is in the direction of $\mathbf{E}$.
2. (Positive/negative/zero) when $d\boldsymbol{\ell}$ is in the opposite direction of $\mathbf{E}$.
3. (Positive/negative/zero) when $d\boldsymbol{\ell}$ is perpendicular to $\mathbf{E}$.

1. Negative: $[-|\mathbf{E}|\,|d\boldsymbol{\ell}|\times(+1)]$
2. Positive: $[-|\mathbf{E}|\,|d\boldsymbol{\ell}|\times(-1)]$
3. Zero: $[-|\mathbf{E}|\,|d\boldsymbol{\ell}|\times(0)]$

12. The left side of the equation

$$W_{AB}/q_o = -\int_A^B \underline{\underline{E}}\cdot d\underline{\underline{\ell}}$$

can thus be positive, negative or zero.

1. If positive work is done in going from A to B then B is at a (higher/lower) electric potential than A.
2. If negative work is done in going from A to B then B is at a (higher/lower) electric potential than A.

1. Higher.
2. Lower.

---

13. The work per unit charge in going from A to B is called the electric potential difference ($V_B - V_A$) so the equation becomes

$$V_B - V_A = -\int_A^B \underline{\underline{E}}\cdot d\underline{\underline{\ell}} .$$

If the above integral is independent of the actual path from A to B then $\underline{\underline{E}}$ is called a ____________ field.

Conservative.

All electrostatic fields have this property.

---

14. What is the formula for the electric potential V due to a charge q at a distance r from that charge?

$$V = \frac{1}{4\pi\varepsilon_o}\frac{q}{r}$$

---

15. Is there any convention associated with the above answer?

Yes. $V \equiv 0$ at large distances (i.e. for $r\to\infty$).

---

16. Calculate V at the midpoint P of the line connecting the two charges shown below. Do this by considering V due to each charge separately.

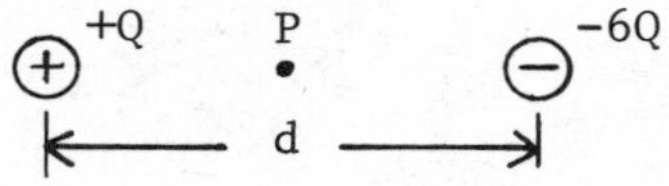

a. $V_{\text{due to } +Q} =$ ____________

b. $V_{\text{due to } -6Q} =$ ____________

c. $V_{\text{total}} =$ ____________

a. $\frac{1}{4\pi\varepsilon_o}\frac{Q}{d/2}$

since "q"=Q and "r"=d/2.

b. $\frac{1}{4\pi\varepsilon_o}\frac{-6Q}{d/2}$

since "q"=-6Q and "r"=d/2.

c. $\frac{-5Q}{2\pi\varepsilon_o d}$

(The total potential is <u>scalar</u> sum of the potentials due to the individual charges.)

Chapter 25: CAPACITORS AND DIELECTRICS

REVIEW AND PREVIEW

In this chapter you will study an electrical circuit device called a capacitor. This consists of two conducting plates carrying equal but opposite charges $\pm Q$. Unlike e.g. the charged rods and spheres considered in previous chapters, capacitors permit the storage of relatively large amounts of charge Q with only relatively modest potential differences V involved. Capacitors have many applications to electrical circuits (e.g. energy storage, frequency filtering, etc.).

GOALS AND GUIDELINES

In this chapter there are four major goals:

1. Learning and understanding the definition of the capacitance C of a capacitor (Section 25-1). Using this definition you should be able to calculate C for a given capacitor (Example 1).
2. Learning the formulas for the case of several capacitors connected in series (Eq. 25-2) and for the case of several capacitors connected in parallel (Eq. 25-3) you should be able to apply these formulas to find the equivalent capacitance of more complicated circuits consisting of capacitors (Example 2).
3. Learning the formulas for the energy U stored in a capacitor (Eq. 25-4) and for the energy density u associated with the electric field in the capacitor (Eq. 25-5). You should be able to calculate C for a given capacitor using energy considerations (Example 4).
4. Understanding the effect of inserting a dielectric material between the plates of a capacitor.

Some students have difficulty in being able to see which capacitors are actually connected in series or parallel in a given circuit. This requires some practice. Frequently, redrawing the circuit diagram will help. Study Example 2 very carefully as an illustration of how to treat capacitor circuits.

25-1 Capacitance

Consider a system consisting of two isolated conductors of any shape, not connected to each other. This forms what is known as a capacitor, each conductor (regardless of its shape) being called a plate. Now suppose that we put equal and opposite charges of magnitude Q on the plates (Q on one plate, -Q on the other). This process is called "charging the capacitor", the amount Q is called the "charge on the capacitor" (although the net charge is actually zero). Each plate, since it is a conductor, is an equipotential surface. There is then a certain potential difference between the plates; let V be the magnitude of this potential difference*. The capacitor has a property called capacitance, C , defined by

$$C = Q/V \quad . \tag{25-1}$$

The unit of capacitance is the farad, F(=coulomb/volt).

The potential difference V will automatically be proportional to the charge Q. $C = Q/V$ is then independent of Q, the capacitor having the same capacitance whether or not it is actually charged. What does the capacitance depend upon?

---

*Sometimes we say that the capacitor has been "charged to a potential V". Of course this really means that the potential difference between the plates is V.

It depends only upon the <u>geometry</u> of the capacitor (i.e. the sizes, shapes, separation, etc. of the two plates). As an example, two parallel plates of metal each of area A separated by a small distance d form a "parallel plate capacitor" whose capacitance is $C = \varepsilon_o A/d$. (From this we have a useful identity: farad equals units of ε_o times meter.)

Sometimes we deal with the capacitance of a single conductor instead of a pair. The same defining formula is used: $C = Q/V$. Here "Q" is the charge on the conductor, "V" is the potential of the conductor (assuming that the potential at infinity has been taken to be zero). The single conductor capacitor may be regarded as a two conductor one, the second conductor being a sphere of infinite radius.

In electrical circuits, capacitors are represented by the following symbol: 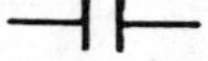 . Although this symbol looks like a parallel plate capacitor, it is used for all capacitors whatever their actual shape. The question of <u>signs</u> can be important in electrical circuits. Consider the following capacitor.

The algebraically correct equation for this is $V_x - V_y = q/C$ where "q" is the charge on that plate which is connected to the terminal "x" (-q is then the charge on the other plate). This equation is true regardless of whether q itself is positive or negative.

25-2 Calculation of Capacitance (Method #1)

To calculate the capacitance associated with two given conductors we can follow the definition directly. The steps are:

1. Imagine the conductors are charged (Q and -Q).
2. Find the electric field $\underline{\underline{E}}$ caused by this charge (perhaps by means of Gauss's law).
3. Find the magnitude of V of the potential difference between the conductors (using $V_B - V_A = -\int \underline{\underline{E}}\cdot d\underline{\underline{\ell}}$). This will be proportional to Q.
4. Using this expression for V, apply the definition of capacitance: $C = Q/V$.

Usually, calculation of the electric field (step 2) is the difficult part. This procedure for calculating capacitance is illustrated in the following example.

>>> Example 1. Find the capacitance of two long hollow coaxial metal cylinders. The radii of the cylinders are a < b; their length is L.

We follow the steps outlined above.

1. Imagine charging (say) the inner cylinder with a charge Q, the outer one with -Q.

2. Neglecting fringing, in the region $a < r < b$ the electric field is directed radially outward. Using Gauss's law it can be shown that this field is given by $E = Q/(2\pi\varepsilon_o Lr)$.

3. To find $V_b - V_a$, choose a radial path from the inner to the outer cylinder as the path of integration.

$$V_b - V_a = -\int_a^b \underline{\underline{E}}\cdot d\underline{\underline{\ell}} = -\int_a^b E\, d\ell \qquad \text{(since } \underline{\underline{E}} \text{ is parallel to } d\underline{\underline{\ell}}\text{)}$$

$$= -\int_a^b E\, dr \qquad \text{(since } d\ell = dr\text{)}$$

$$= \frac{-Q}{2\pi\varepsilon_o L}\int_a^b \frac{dr}{r} = \frac{-Q}{2\pi\varepsilon_o L}\ \ell n(b/a) \quad .$$

Since ℓn (b/a) is positive, the magnitude of this potential difference is

$$V = \frac{Q}{2\pi\varepsilon_o L}\ \ell n(b/a) \quad .$$

4.
$$C = Q/V = \frac{2\pi\varepsilon_o L}{\ell n\ (b/a)} \quad .$$

Remarks:

1. As a check, note that the units of C are those of ε_o times meters.
2. In step 3, we need only the magnitude V. Once this is understood the student should just simply ignore
   (a) the minus sign in $-\int \underline{\underline{E}}\cdot d\underline{\underline{\ell}}$,
   (b) whether the angle between $\underline{\underline{E}}$ and $d\underline{\underline{\ell}}$ is 0° or 180°,
   (c) the order of the limits of integration.

All of these merely affect the overall sign of the potential difference. <<<

25-3 Equivalent Capacitance

Consider any circuit consisting only of capacitors and let "x" and "y" be the only two terminals which emerge from the circuit. An example of this is shown in Figure 25-1a. Suppose we charge this circuit by putting a charge q in through terminal x and withdrawing a similar charge from terminal y. There will then be a certain potential difference $V = V_x - V_y$ across the circuit. As far as the terminals x and y are concerned, the entire circuit behaves as if it were a single capacitor (Figure 25-1b). This equivalent capacitance, C, is defined by C = q/V. It is important to note that the only external connections to the circuit are x and y (e.g. there is no external connection to z).

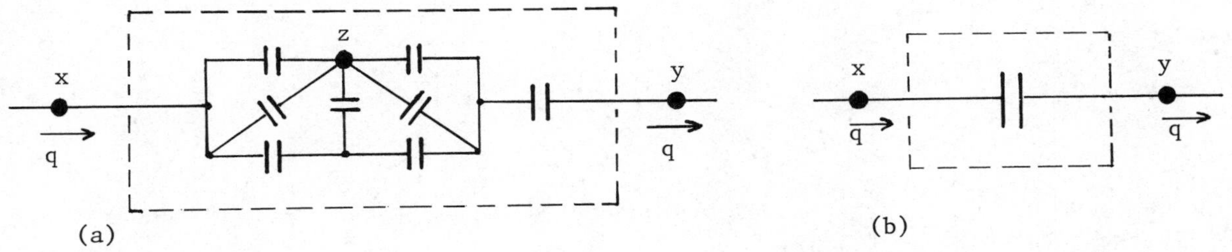

Figure 25-1. (a) A circuit consisting of capacitors. The only external connections are x and y. (b) The equivalent capacitor.

If the (initially uncharged) capacitors are as shown in Figure 25-2a, they are said to be connected in series.

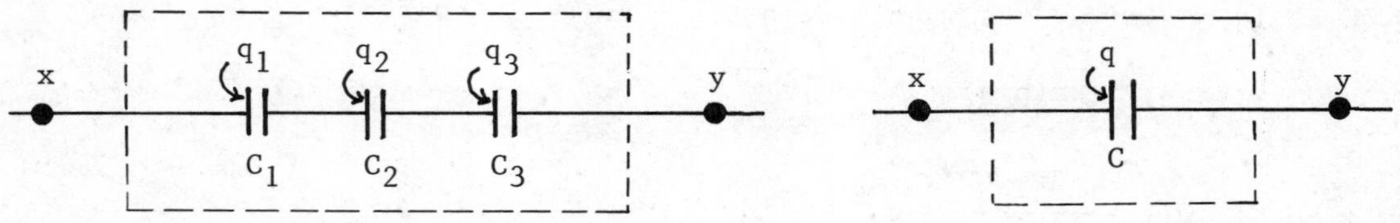

(a) (b)

Figure 25-2. (a) Three capacitors connected in series. (b) The equivalent capacitor: $1/C = 1/C_1 + 1/C_2 + 1/C_3$.

In the series case

$$1/C = 1/C_1 + 1/C_2 + 1/C_3 \tag{25-2a}$$

$$q = q_1 = q_2 = q_3 \tag{25-2b}$$

$$V = V_1 + V_2 + V_3 \quad . \tag{25-2c}$$

If the capacitors are as shown in Figure 25-3a, they are said to be connected in parallel.

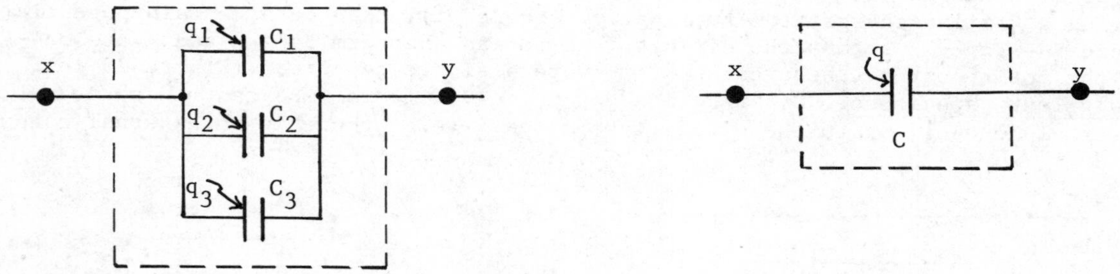

(a) (b)

Figure 25-3. (a) Three capacitors connected in parallel. (b) The equivalent capacitor: $C = C_1 + C_2 + C_3$.

In the parallel case

$$C = C_1 + C_2 + C_3 \tag{25-3a}$$

$$q = q_1 + q_2 + q_3 \tag{25-3b}$$

$$V = V_1 = V_2 = V_3 \quad . \tag{25-3c}$$

In the series case all the charges are the same, while the voltages obey a sum rule. Just the opposite is true in the parallel case: the voltages are all the same, the charges obey a sum rule. The equivalent capacitance for a series circuit is always <u>less</u> than the <u>smallest</u> individual capacitance; the equivalent capacitance for a parallel circuit is always <u>more</u> than the <u>largest</u> capacitance.

Sometimes it is convenient to use other units in applying formulas (25-2) and (25-3). For example, (25-2a) is true if all the C's are measured, say, in μF. As long as the same unit is used throughout, these formulas remain correct.

Of course not every circuit is a series or a parallel one. Many capacitor circuits however can be simplified with successive applications of the series and parallel formulas. This is illustrated below.

>>> Example 2. In the circuit (a) shown below, what is the equivalent capacitance between points x and y? If an external battery supplies 100 volts between these two points (i.e. if $V_x - V_y = 100$ volt), what is the voltage across the 6 μF capacitor?

We apply formulas (25-2a) and (25-3a) to simplify the circuit. The units for capacitance in these will be taken to be μF.

(a)

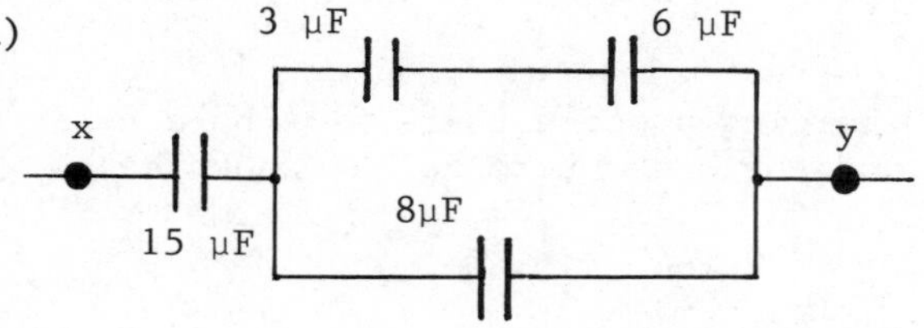

$1/C = 1/3 + 1/6$, $C = 2$ μF

(b)

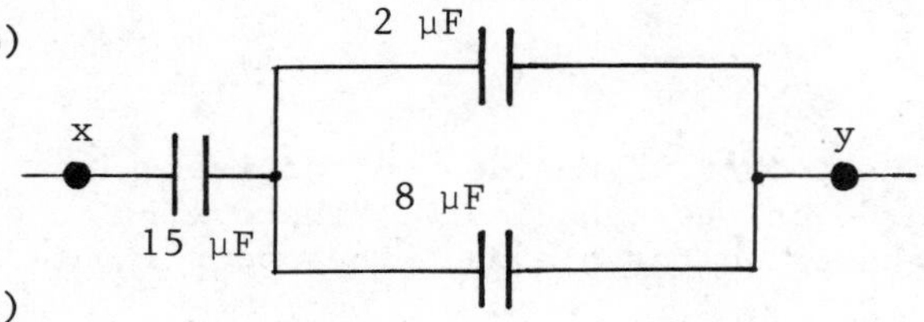

$C = 2 + 8$, $C = 10$ μF

(c)

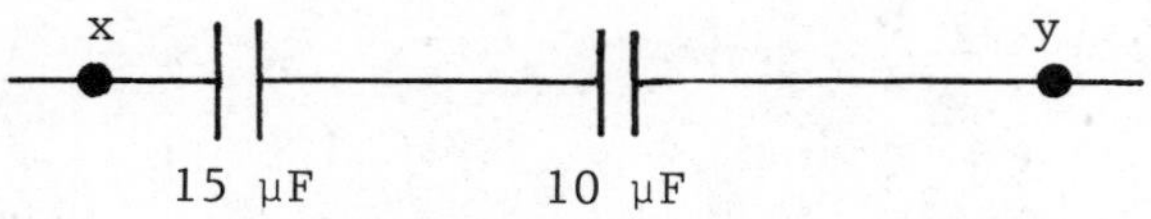

$1/C = 1/15 + 1/10$, $C = 6$ μF

(d)

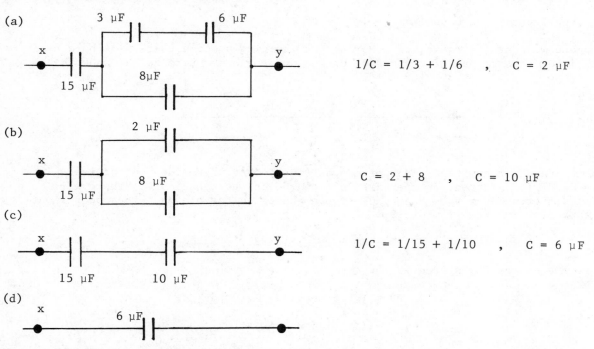

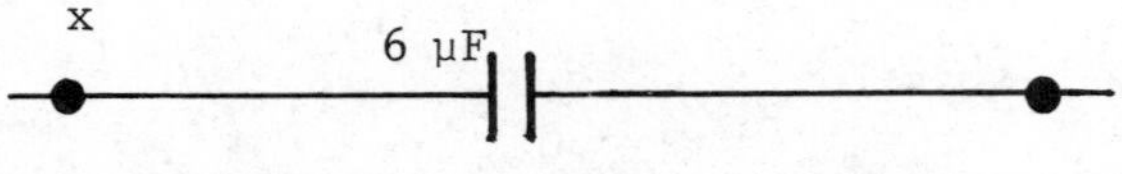

The equivalent capacitance is then 6 μF.

To do the remaining part, we use the fact that if (for a particular capacitor) any two of C, q, V are known then the third can be found using $C = q/V$.

1. In (d), $V = 100$ volt (given), $C = 6$ μF. Then for this capacitor,

$$q = CV = (6\ \mu F) \cdot (100\ V) = 600\ \mu C \quad .$$

2. In (c), both the 15 μF and the 10 μF capacitors must also have this same <u>charge</u> (600 μC). Then for the 10 μF capacitor,

$$V = q/C = (600\ \mu C)/(10\ \mu F) = 60\ V \quad .$$

3. In (b), both the 2 μF and the 8 μF capacitors must also have this same voltage (60 volt). Then for the 2 μF capacitor,

$$q = CV = (2\ \mu F)\cdot(60\ V) = 120\ \mu C \quad .$$

4. In (a), both the 3 μF and the 6 μF capacitors must also have this same charge (120 μC). Then for the 6 μF capacitor,

$$V = q/C = (120\ \mu C)/(6\ \mu F) = 20\ V \quad .$$

Remarks:

1. In finding the equivalent capacitance, the circuit was sequentially simplified form (a) through (d). To avoid mistakes, we drew a new diagram after each step.

2. In the second part of this problem we started with the equivalent (d) and worked back toward the original circuit (a). At each such step either (25-2b) or (25-3c) was used. For example, in (c) the two capacitors are in series and hence must have the same charge as their equivalent capacitor in (d). <<<

25-4 Energy Stored in a Capacitor

Suppose we charge a capacitor. During the charging process there will be a current, i, into one terminal of the capacitor and out the other terminal as shown below.

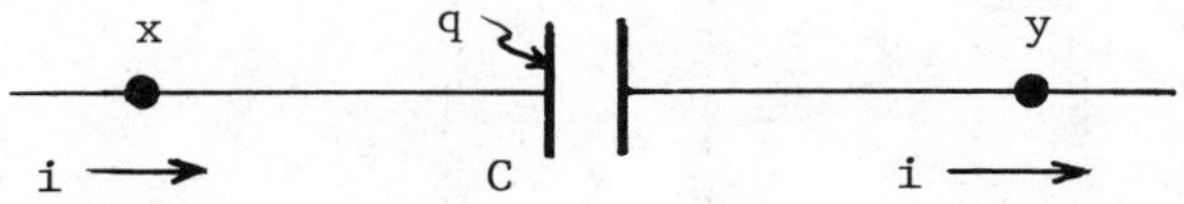

All the charge which has been transported past the point "x" resides on the left plate. Thus $i = dq/dt$ (i = current in the wire, q = charge on the capacitor). There will be a potential difference $V = V_x - V_y = q/C$ between the capacitor plates. The power input to the capacitor is then from (24-9),

$$P = Vi = (q/C)\cdot(dq/dt) \quad .$$

Therefore the energy, U, stored in the capacitor is

$$U = \int P\ dt = (1/C)\int q(dq/dt)\ dt$$

$$= (1/C)\int q\ dq = \tfrac{1}{2}\ q^2/C \quad .$$

Since $q = CV$, there are three forms of expressing this stored energy:

$$U = \tfrac{1}{2}\ q^2/C = \tfrac{1}{2}\ qV = \tfrac{1}{2}\ CV^2 \quad . \qquad (25\text{-}4)$$

The student need remember only one of these, the others follow easily using $q = CV$. Of the three forms, $U = \frac{1}{2}\ CV^2$ is the most useful in practice since we usually know C and V.

>>> Example 3. An 8.0 μF parallel plate capacitor is connected to a 100 volt battery. (a) Calculate the stored energy. The battery is then disconnected. A man then pulls the plates of the capacitor apart, doubling their separation. (b) What is the final value of the stored energy?

(a) $$U = \tfrac{1}{2}\, CV^2 = \tfrac{1}{2}\, (8.0 \times 10^{-6}\ F) \cdot (10^2\ V)^2$$

$$= 4.0 \times 10^{-2}\ J \quad .$$

(b) Let the subscript "1" refer to the original situation (battery connected), and let the subscript "2" refer to the final situation (plate separation doubled). Once the battery is disconnected, the charge on the capacitor must remain constant (there is no way for it to leave the plates); therefore, $q_1 = q_2$. Doubling the plate separation will halve the capacitance ($C = \varepsilon_o A/d$), i.e. $C_2 = \tfrac{1}{2}\, C_1$. Using $U = \tfrac{1}{2}\, q^2/C$ we have

$$U_1 = \tfrac{1}{2}\, q_1{}^2/C_1 \quad ,$$

$$U_2 = \tfrac{1}{2}\, q_2{}^2/C_2 \quad .$$

Dividing,

$$U_1/U_2 = C_2/C_1 = 1/2$$

since $q_1 = q_2$ and $C_2 = \tfrac{1}{2}\, C_1$, Thus

$$U_2 = 2\, U_1 = (2) \cdot (4.0 \times 10^{-2}\ J) = 8.0 \times 10^{-2}\ J \quad .$$

Remarks:

1. In part (b) we chose to use the $\tfrac{1}{2}\, q^2/C$ formula (rather than $\tfrac{1}{2}\, CV^2$) because we knew something about how q and C changed when the plate separation increased.
2. By working with the ratio U_1/U_2, the charges cancelled out.
3. The increase in the stored energy is a result of the man doing mechanical work in separating the plates of the capacitor. <<<

25-5 Electric Energy Density

Capacitors can store energy. This energy is regarded as being associated with the electric field rather than the plates where the charges are. There is an (electric) energy density, u,* which gives the amount of energy stored per unit volume; i.e. in a differential volume element dv there is an energy dU = u dv. The units of u are those of energy per volume (joule/meter^3).

The energy density is proportional to the square of the electric field; it is given by

$$u = \tfrac{1}{2}\, \varepsilon_o E^2 \quad . \qquad (25\text{-}5)$$

Since $\underline{\underline{E}}$ is a function of location, u must also be a function of location. The energy density is then a scalar field.

---

*Sometimes written as u_E to emphasize that it is an electric energy density.

25-6 Calculation of Capacitance (Method #2)

An alternative method of calculating the capacitance associated with two given conductors uses the concept of energy density. The steps are:

1. Imagine the conductors are charged (Q and -Q).
2. Find the electric field $\underline{E}$ caused by this charge. Using this, find the energy density $u = \frac{1}{2}\,\varepsilon_o E^2$.
3. Integrate the energy density to find the stored energy, $U = \int u\,dv$ (dv = differential volume element). This will be proportional to Q^2.
4. Using this expression for U, solve the equation $U = \frac{1}{2}\,Q^2/C$ for C.

This procedure is illustrated below.

>>> Example 4. Find the capacitance of two long hollow coaxial metal cylinders. The radii of the cylinders are $a < b$; their length is L.

We follow the steps outlined above.

1. Imagine charging (say) the inner cylinder with a charge Q, the outer one with -Q.

2. Neglecting fringing, in the region $a < r < b$ the electric field is directed radially outward. Using Gauss's law it can be shown that this field is given by $E = Q/(2\pi\varepsilon_o Lr)$. Then

$$u = \tfrac{1}{2}\,\varepsilon_o E^2 = \tfrac{1}{2}\,\varepsilon_o[Q/(2\pi\varepsilon_o Lr)]^2$$

$$= Q^2/(8\pi^2\varepsilon_o L^2 r^2) \quad , \quad a < r < b \quad .$$

In the regions $r < a$ and $r > b$ we have $u = 0$ since $\underline{E} = 0$.

3. Take a cylindrical shell (radius r, length L, thickness dr) as a volume element. Its volume is $dv = 2\pi rL\,dr$. The stored energy is

$$U = \int u\,dv = \int_a^b [Q^2/(8\pi^2\varepsilon_o L^2 r^2)]\,2\pi rL\,dr$$

$$= [Q^2/(4\pi\varepsilon_o L)] \int_a^b dr/r$$

$$= [Q^2/(4\pi\varepsilon_o L)]\,\ln(b/a) \quad .$$

4. $\frac{1}{2}\,Q^2/C = [Q^2/(4\pi\varepsilon_o L)]\,\ln(b/a)$. Solving for C,

$$C = \frac{2\pi\varepsilon_o L}{\ln(b/a)} \quad .$$

The answer is in agreement with Example 1. <<<

25-7 Dielectrics

So far the capacitor plates were always located in empty space (or air; the effect of air is very small). If an insulating material is inserted between the plates the capacitance will generally increase. When used in this manner the insulating material is called a dielectric.

Dielectric materials are characterized by their dielectric constant κ (a dimensionless quantity). If the dielectric completely fills the space between the capacitor plates the capacitance becomes

$$C = \kappa C_o \qquad (25\text{-}6)$$

where C_o is the capacitance without the dielectric.

The increase in capacitance ($\kappa \geq 1$) is due to the appearance of induced charges on the surface of a dielectric when it is placed in an electric field; these induced charges result from the alignment of dipoles within the dielectric. Positive induced charge (q') appears on the dielectric surface near the negatively charged ($-q$) capacitor plate, negative induced charge ($-q'$) appears on the dielectric surface near the positively charged (q) capacitor plate. Within the dielectric the electric field due to the induced charge (q', $-q'$) therefore tends to oppose the electric field due to the charges (q, $-q$) on the capacitor plates. For a given charge q on the capacitor, this results in a decrease in the potential difference V between the capacitor plates. Thus the presence of the dielectric causes the capacitance $C = q/V$ to increase.

>>> Example 5. A parallel plate capacitor contains a paper dielectric ($\kappa = 3.5$). Its capacitance (with this dielectric) is 1.0 μF. The capacitor is connected to a 50 volt battery. (a) What is the charge q on the capacitor? (b) What is the induced charge q' on the surface of the dielectric?

(a) $$q = CV = (1.0\ \mu F)\cdot(50\ V) = 50\ \mu C \quad .$$

(b) We first draw an edge view of the capacitor, see Figure 25-4.

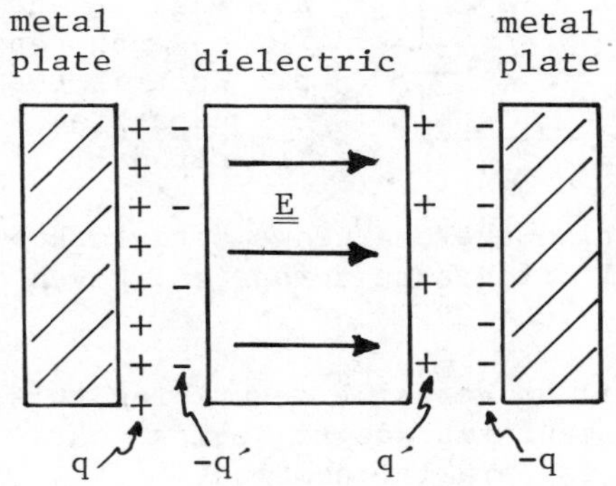

Figure 25-4. Parallel plate capacitor with dielectric. The charge on the capacitor plates is q, the induced charge on the surface of the dielectric is q'. Within the dielectric, the electric field due to q' opposes the electric field due to q.

The dielectric really fills the entire space between the plates; it is shown slightly separated for clarity. Note that the negative induced charge ($-q'$) appears on that dielectric surface adjacent to the positively charged capacitor plate (q).

In the interior of the dielectric the electric field is uniform and is given by $E = V/d$ (d = plate separation). What causes this field? It is caused by four (approximately) infinite planes of charge (q, $-q$, q', $-q'$). The field due to each of these planes of charge can be calculated using the "$\sigma/2\varepsilon_o$" formula. Adding these four electric fields vectorially gives

$$E = q/(2\varepsilon_o A) + q/(2\varepsilon_o A) - q'/(2\varepsilon_o A) - q'/(2\varepsilon_o A)$$

$$E = (q - q')/(\varepsilon_o A) \qquad \text{(where } A = \text{plate area)}$$

$$q' = q - \varepsilon_o AE$$
$$= q - \varepsilon_o AV/d \qquad \text{(since } E = V/d\text{)}$$
$$= q - CV/\kappa \qquad \text{(since } C = \kappa\varepsilon_o A/d\text{)}$$
$$= q - q/\kappa \qquad \text{(since } q = CV\text{)}$$
$$= q(1 - 1/\kappa) = (50\ \mu C) \cdot (1 - 1/3.5) = 36\ \mu C \quad . \qquad <<<$$

For a parallel plate capacitor $C = \kappa C_o$ becomes $C = (\kappa\varepsilon_o)A/d$. Many such formulas which apply only to free space (vacuum) become valid for dielectrics if we replace "ε_o" by "$\kappa\varepsilon_o$". In particular, the energy density (25-5) becomes

$$u = \tfrac{1}{2}\,\kappa\varepsilon_o E^2. \qquad (25\text{-}7)$$

25-8 Programmed Problems

1.

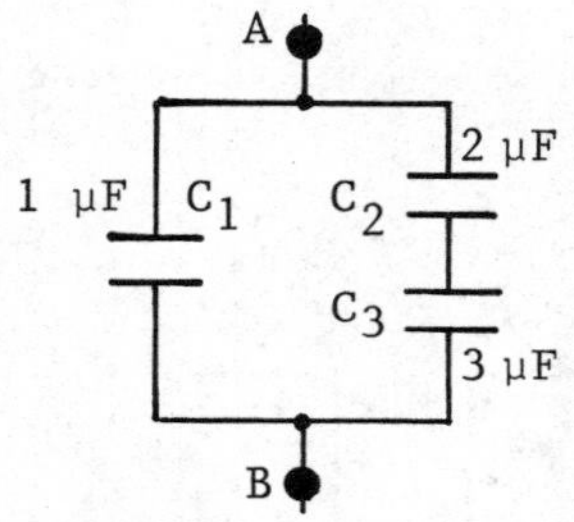

The potential difference between the points A and B is 30 volts.

a. What is the charge on each capacitor?
b. What is the equivalent capacitance of this network?

To find the charge we need to know the potential difference across any given capacitor since Q = CV.

Begin by verbally describing this circuit configuration as to series, parallel, or series parallel combination.

C_1 is in parallel with the series combination of C_2 and C_3.

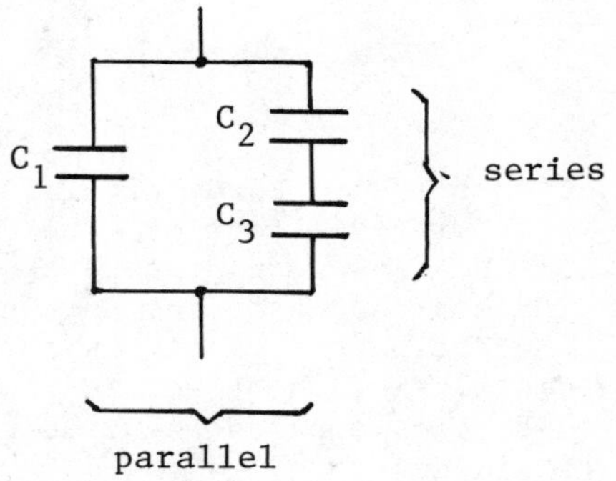

---

2. What is the potential difference across C_1? What is the potential difference across the C_2, C_3 series combination?

30 volts in both cases. C_1 is in parallel with the C_2, C_3 combination. Parallel configurations have the same potential difference.

---

3. What is the charge Q_1 on C_1? Express your answer in coulombs.

$Q_1 = C_1 V$

$Q_1 = 1.0 \times 10^{-6}$ F × 30 V

$Q_1 = 30 \times 10^{-6}$ coulomb.

4. 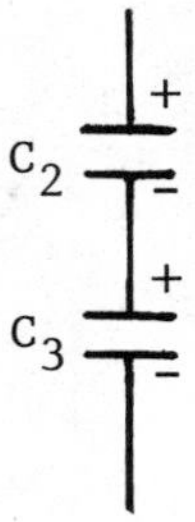

To the left are the charge conditions of the capacitors. Assuming no sparking, which charges are the result of some external agent (e.g. a battery) and which are the result of induction?

The positive charge on the upper plate of C_2 and the negative charge on the lower plate of C_3 are the result of some external agent. The inner charges (lower plate of C_2 and and upper plate of C_3) are the result of induction. Note that the net charge on these two inner plates is zero (they are connected to each other but are insulated from everything else).

---

5. For the series capacitors C_2 and C_3 we have $Q_2 = Q_3 = Q'$ where Q' is the charge delivered to this pair. Thus, since voltages add for a series configuration,

$$30 \text{ volts} = \frac{Q_2}{C_2} + \frac{Q_3}{C_3} = Q'(\frac{1}{C_2} + \frac{1}{C_3}) \quad .$$

Express $(1/C_2 + 1/C_3)$ in terms of a common denominator. (This will turn out to be a very useful computational device to remember.)

$$\frac{1}{C_2} + \frac{1}{C_3} = \frac{C_3}{C_2C_3} + \frac{C_2}{C_2C_3}$$

$$\frac{1}{C_2} + \frac{1}{C_3} = \frac{C_2 + C_3}{C_2C_3} \quad .$$

---

6. What charge has been delivered to the series combination?

$Q' =$ __________ coulomb.

From frame 5, 30 volts $= Q'(C_2 + C_3)/(C_2C_3)$

$$Q' = \frac{30 \text{ volt}}{\frac{C_2 + C_3}{C_2C_3}}$$

$$Q' = (30 \text{ volt})[\frac{C_2C_3}{C_2 + C_3}]$$

$$Q' = (30 \text{ volt})[\frac{6}{5} \ \mu\text{F}]$$

$$Q' = 36 \times 10^{-6} \text{ coulomb.}$$

---

7. 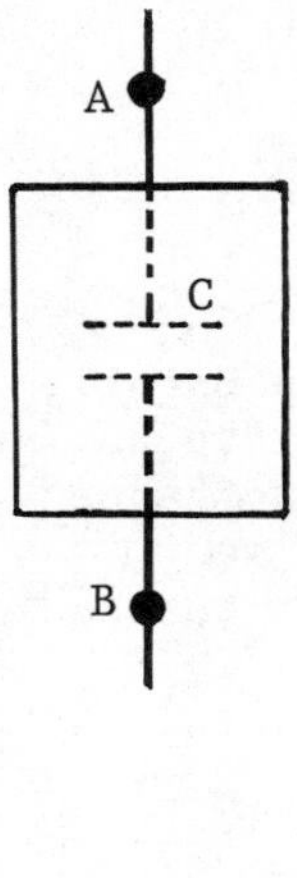

Now that we know the charge delivered to the circuit we can determine the equivalent capacitance. We can think of the situation as shown to the left. A total charge Q_1 ($= 30 \times 10^{-6}$ coulomb) plus $Q' = 36 \times 10^{-6}$ coulomb is delivered to a "capacitor" such that the potential difference is 30 volts between points A and B. Thus

$C = \frac{Q_T}{V} =$ __________ μfarad.

Obtain a numerical result.

$$C = \frac{66 \times 10^{-6} \text{ coulomb}}{30 \text{ volt}}$$

$$C = 2.2 \times 10^{-6} \text{ farad}$$

$$C = 2.2 \ \mu\text{farad} \quad .$$

8. Let us check the answer of the previous frame. First we look at the equivalent capacitance of the C_2, C_3 combination.

In frame 5 we had the equation

$$30 \text{ volts} = \left[\frac{1}{C_2} + \frac{1}{C_3}\right] Q'$$

which could be written as

$$30 \text{ volts} = \left[\frac{C_2 + C_3}{C_2C_3}\right] Q' \quad .$$

If we put this in the standard form

$$C = Q/V \quad ,$$

then

$$C_{23} = \underline{\qquad\qquad}?$$

(Express algebraically in terms of C_2 and C_3.)

$$C_{23} = \frac{C_2C_3}{C_2 + C_3}$$

C_{23} means the equivalent capacitance of the capacitors C_2 and C_3.

---

9. The result is that two capacitors in series, such as C_2 and C_3 of our problem, have an equivalent capacitance C_{23} equal to the "product over the sum" of their individual values. For $C_2 = 2.0\ \mu F$ and $C_3 = 3.0\ \mu F$

$$C_{23} = \underline{\qquad\qquad}\ \mu F.$$

Caution: This rule applies only to two capacitors.

$$C_{23} = \frac{C_2C_3}{C_2 + C_3} = \frac{\text{Product}}{\text{Sum}}$$

$$C_{23} = \frac{(2.0\ \mu F)\cdot(3.0\ \mu F)}{2.0\ \mu F + 3.0\ \mu F}$$

$$C_{23} = 1.2\ \mu F.$$

(Note how the units, μF, emerge.)

---

10. Incidentally the product over the sum rule is consistent of course with the formula

$$\frac{1}{C_{23}} = \frac{1}{C_2} + \frac{1}{C_3} \quad .$$

This formula holds for any number of capacitors in series.

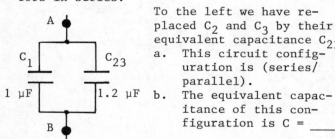

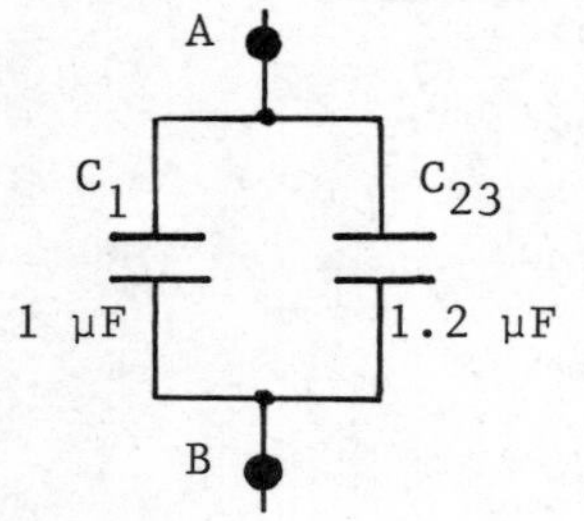

To the left we have replaced C_2 and C_3 by their equivalent capacitance C_{23}.

a. This circuit configuration is (series/parallel).

b. The equivalent capacitance of this configuration is $C = \underline{\qquad}$.

a. Parallel.

b. $C = C_1 + C_{23}$

$C = 2.2\ \mu F.$

Capacitors in a parallel configuration have an equivalent capacitance equal to their sum. Thus the answer checks with that of frame 7.

11.

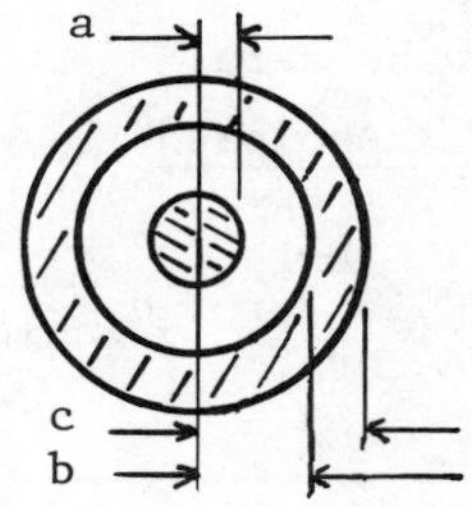

We now will calculate the capacitance of a conducting sphere surrounded by a thick spherical conducting shell. The radii are a, b and c as shown.

This problem will provide you with experience in calculating capacitance as well as a review of Gauss's law.

Gauss's law is written as

$$\oint \underline{\underline{E}} \cdot d\underline{\underline{S}} = q/\varepsilon_o \quad .$$

a. What is the "physical" meaning of $d\underline{\underline{S}}$?
b. What particular charge is meant by q?
c. What is the name given to the left side of this equation?

a. $d\underline{\underline{S}}$ is an element of area of a <u>closed</u> Gaussian surface.

b. The charge enclosed by a particular closed Gaussian surface.

c. The flux of the electric field $\underline{\underline{E}}$.

12.

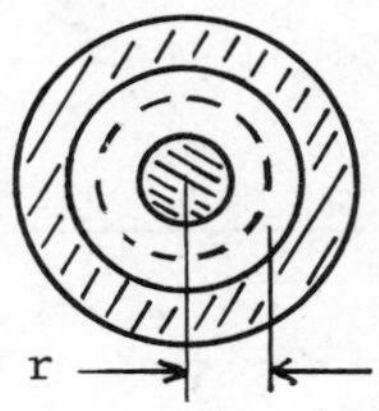

We start by imagining that the spheres have charge +Q on the inner sphere and -Q on the outer sphere. In the figure we show a Gaussian surface symmetrical with the inner sphere and located at a distance r. By symmetry $\underline{\underline{E}}$ is normal to this surface and constant at the surface.

Using Gauss's law, calculate $\underline{\underline{E}}$ at r.

$$\oint \underline{\underline{E}} \cdot d\underline{\underline{S}} = Q/\varepsilon_o \quad .$$

$|\underline{\underline{E}}|$ = __________.

Direction is __________.

$E \oint dS = Q/\varepsilon_o$, but

$\oint dS = 4\pi r^2$, so that

$E = \dfrac{1}{4\pi\varepsilon_o} \dfrac{Q}{r^2}$ and is

directed radially outward.

This result is not surprising. It is the same as that of a point charge. This is true for spherical charge distributions.

---

13. Looking again at the diagram in frame 11, what will be the field for $b < r < c$?

Zero. This is inside the spherical conducting shell. Electric fields are zero inside conductors for the static case.

14. 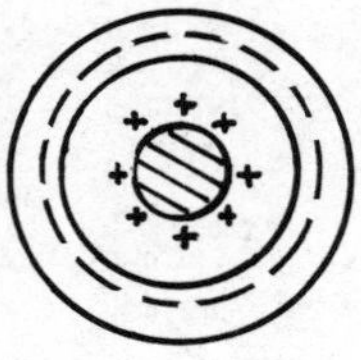

Here we show a Gaussian surface inside the spherical shell. In view of the previous answer, what is the net charge enclosed inside this Gaussian surface?

It must be zero.

$$\oint \underline{\underline{E}} \cdot d\underline{\underline{S}} = \frac{q_{enclosed}}{\varepsilon_o} .$$

Since $\underline{\underline{E}}$ is zero, $q_{enclosed}$ must be zero.

---

15. How would you doctor up the diagram of the previous frame so that it was consistent with the answer of that frame?

The charge on the inner surface (r = b) must be -Q. Since the <u>total</u> charge on the shell is -Q, the charge on the outer surface (r = c) must be zero.

---

16. We now wish to calculate

$$V_a - V_b = -\int_a^b \underline{\underline{E}} \cdot d\underline{\underline{\ell}} .$$

Since $E = (1/4\pi\varepsilon_o)(Q/r^2)$ the integration is exactly like that for a point charge (except for the limits of integration). Find $V_a - V_b$.

$$V_a - V_b = \frac{1}{4\pi\varepsilon_o} \frac{Q}{r} \Bigg|_{r=b}^{r=a}$$

$$= \frac{Q}{4\pi\varepsilon_o} \left[\frac{1}{a} - \frac{1}{b}\right] .$$

---

17. Then the capacitance is given by C = Q/V where $V = V_a - V_b$.

C = ___________.

$$C = \frac{Q}{\frac{Q}{4\pi\varepsilon_o}\left[\frac{1}{a} - \frac{1}{b}\right]}$$

$$C = 4\pi\varepsilon_o \frac{ab}{b - a} .$$

---

18. What does C depend upon?

C depends upon the geometry (the radii a, b). The answer happens to be independent of the radius c. Note that the charge Q cancelled out.

Chapter 26: CURRENT AND RESISTANCE

REVIEW AND PREVIEW

So far, Part 11 has dealt chiefly with electrostatics (in particular, fixed charges). Recall that current i is the rate of flow of charge through a conductor. In this chapter you will study the effects of currents in a conductor. There are two aspects to this:

a. The macroscopic point of view. Here one considers such "external" properties as the current i, the potential difference V between the ends of the conductor, and a property of the conductor called the resistance R.
b. The microscopic point of view. Here one considers such "internal" properties as the current density $\underline{j}$, the electric field $\underline{E}$ within the conductor, and a property of the material of which the conductor is made called the resistivity ρ.

The topics discussed in this chapter are essential to an understanding of all electrical circuits, from those as simple as a flashlight to those as complicated as a modern computer.

GOALS AND GUIDELINES

In this chapter there are four major goals:

1. Learning the definition of resistance R (Eq. 26-1) and Ohm's law (Section 26-2).
2. a. Learning the formulas for the equivalent resistance for resistors in series (Eq. 26-2) and for resistors in parallel (Eq. 26-3).
   b. Learning how to apply these to simplify more complicated circuits involving resistors.
3. Learning how to calculate the power dissipated by a resistor (Section 26-4).

The above three goals pertain to the macroscopic point of view. Most problems you encounter can probably be done with these concepts. However, you should also try to understand the phenomenon of a current in a conductor from a microscopic point of view:

4. You should learn the defnitions of, and relations between, the following quantities: the current density $\underline{j}$, the electric field $\underline{E}$ within the conductor, the resistivity ρ, and the drift velocity v_d of the electrons in the conductor.

26-1 Resistance

Consider a conductor with two terminals x and y. Suppose there is a current i into terminal x and out terminal y as shown below.

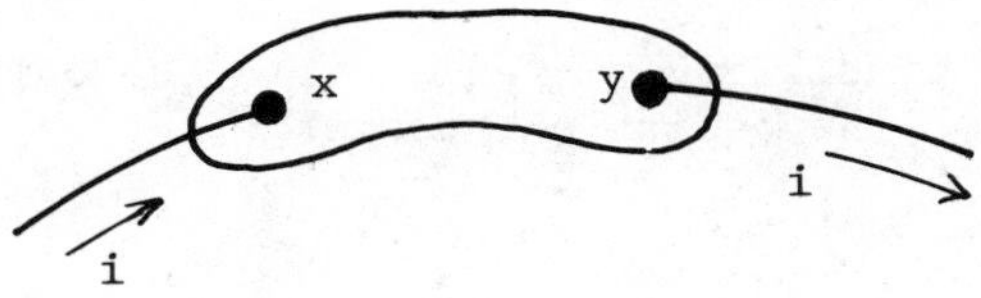

This is a <u>non</u>-electrostatic situation. An electric field is required to maintain the current (recall that for the electrostatic case, $\underline{\underline{E}} = 0$ in the interior of a conductor). There is then a certain potential difference $V = V_x - V_y = -\int \underline{\underline{E}}\cdot d\underline{\underline{\ell}}$ between the two terminals. The <u>resistance</u>, R, between the terminals x and y is defined as

$$R = V/i \quad . \tag{26-1}$$

Resistance is a macroscopic property. The unit of resistance is the ohm, Ω (= volt/ampere). When we are concerned with this property of resistance, the material is called a <u>resistor</u>.

In electrical circuits, resistors are represented by the following symbol: . The question of <u>signs</u> can be important in electrical circuits. Consider the following resistor.

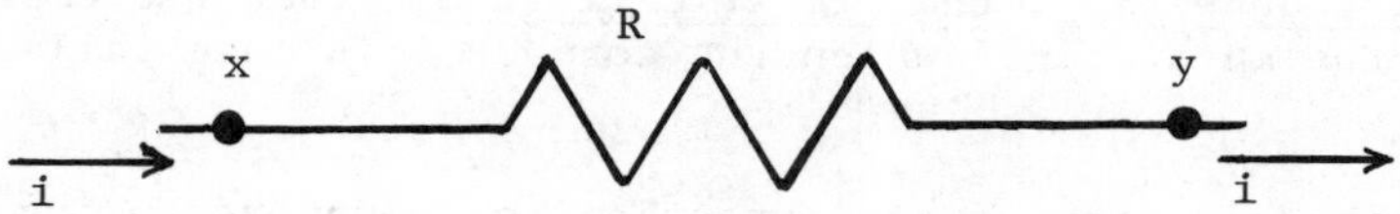

The algebraically correct equation for this is $V_x - V_y = iR$ where "i" is the current into the terminal "x" (and out terminal "y"). This equation is true regardless of whether i itself is positive or negative.

26-2 Ohm's Law

For many materials, the ratio V/i (= R) is essentially a constant independent of the current. In this case we say that the material obeys <u>Ohm's law</u>. Again, "R = V/i" is the <u>definition</u> of resistance while "R = constant" is the statement of Ohm's law.

26-3 Equivalent Resistance

Consider any circuit consisting only of resistors and let "x" and "y" be the only two terminals which emerge from the circuit. An example of this is shown in Figure 26-1a. Suppose there is a current i into terminal x and out terminal y. There will then be a certain potential difference $V = V_x - V_y$ across the circuit. As far as the terminals x and y are concerned, the entire circuit behaves as if it were a single resistor (Figure 26-1b). The <u>equivalent resistance</u>, R, is defined by R = V/i. It is important to note that the only external connections to the circuit are x and y (e.g. there is no external connection to z).

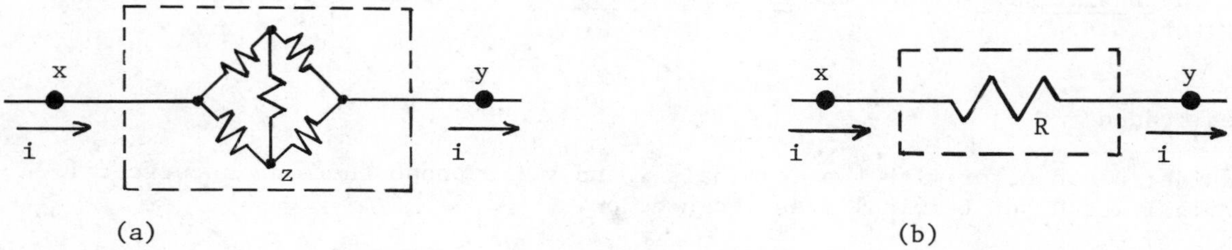

Figure 26-1. (a) A circuit consisting of resistors. The only external connections are x and y. (b) The equivalent resistor.

If the resistors are connected in series (Figure 26-2a) the following formulas apply:

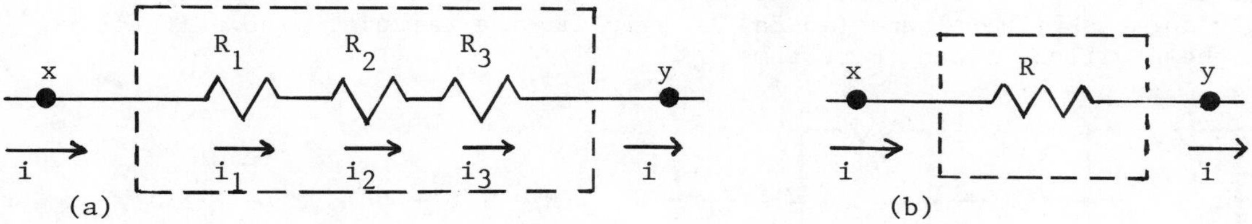

Figure 26-2. (a) Three resistors connected in series. (b) The equivalent resistor: $R = R_1 + R_2 + R_3$.

$$R = R_1 + R_2 + R_3 \tag{26-2a}$$

$$i = i_1 = i_2 = i_3 \tag{26-2b}$$

$$V = V_1 + V_2 + V_3 \quad . \tag{26-2c}$$

If the resistors are connected in parallel (Figure 26-3a) the following formulas apply:

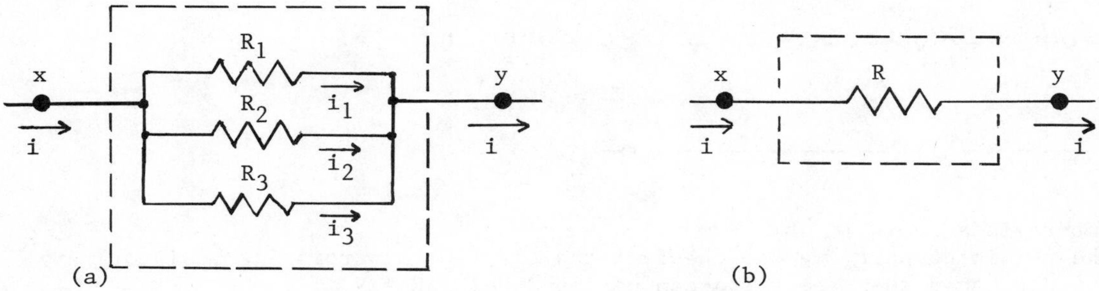

Figure 26-3. (a) Three resistors connected in parallel. (b) The equivalent resistor: $1/R = 1/R_1 + 1/R_2 + 1/R_3$.

$$1/R = 1/R_1 + 1/R_2 + 1/R_3 \tag{26-3a}$$

$$i = i_1 + i_2 + i_3 \tag{26-3b}$$

$$V = V_1 = V_2 = V_3 \quad . \tag{26-3c}$$

In the series case all the currents are the same, while the voltages obey a sum rule. Just the opposite is true in the parallel case: the voltages are all the same, the currents obey a sum rule. The equivalent resistance for a series circuit is always more than the largest individual resistance; the equivalent resistance for a parallel circuit is always less than the smallest individual resistance. Note that formulas (26-2a), (26-3a) for the series and parallel equivalent resistors are just the opposite in form as those for the series and parallel equivalent capacitors.

Of course not every circuit is a series or parallel one. Many resistor circuits however can be simplified with successive applications of the series and parallel formulas.

>>> Example 1. In the circuit (a) shown below, what is the equivalent resistance between points x and y? If an external battery supplies 100 volts between these two points (i.e. if $V_x - V_y = 100$ volt), what is the current in the 3 ohm resistor?

We apply formulas (26-2a) and (26-3a) to simplify the circuit. The unit for resistance in these will be taken to be ohms.

(a)

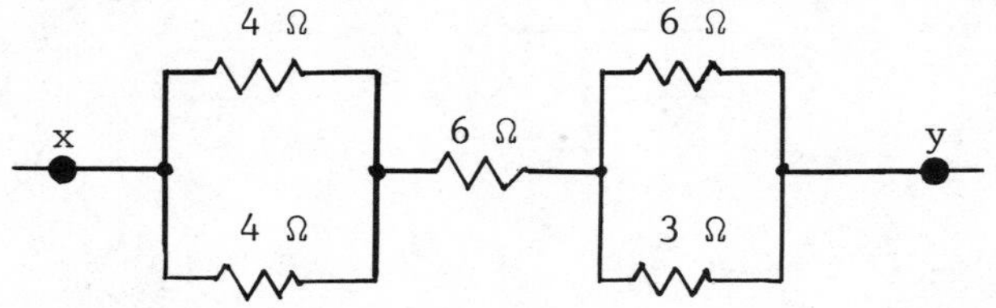

$1/R = 1/6 + 1/3 \quad , \quad R = 2\ \Omega$

(b)

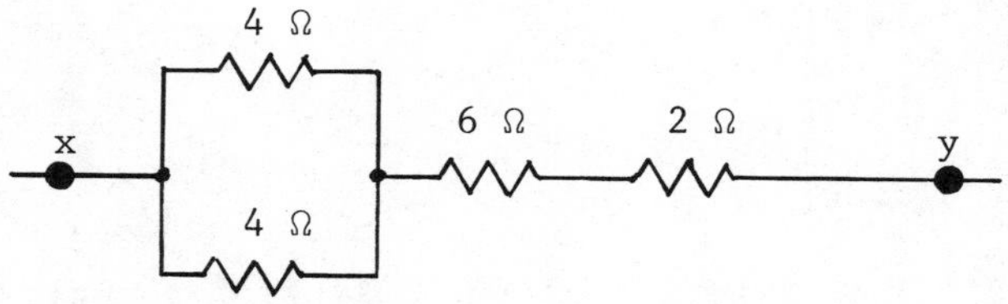

$1/R = 1/4 + 1/4 \quad , \quad R = 2\ \Omega$

(c)

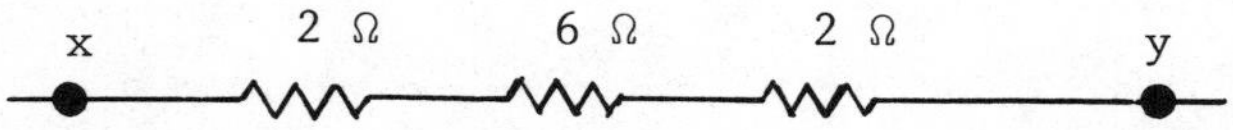

$R = 2 + 6 + 2 \quad , \quad R = 10\ \Omega$

(d)

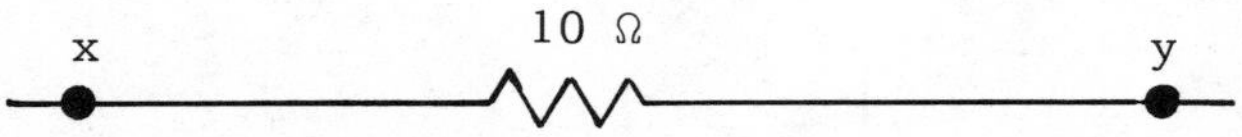

The equivalent resistance is 10 ohm.

To do the remaining part, we use the fact that if (for a particular resistor) any two of R, V, i are known then the third can be found using $R = V/i$.

1. In (d), V = 100 volt (given), R = 10 ohm. Then for this resistor,

$$i = V/R = (100 \text{ volt})/(10 \text{ ohm}) = 10 \text{ ampere} \quad .$$

2. In (c) all three resistors must also have this same current (10 amp). Then for the 2 ohm resistor in (b).

$$V = iR = (10 \text{ ampere})\cdot(2 \text{ ohm}) = 20 \text{ volt} \quad .$$

3. In (a) both the 6 ohm and 3 ohm parallel resistors must have this same voltage (20 volt). Then for the 3 ohm resistor,

$$i = V/R = (20 \text{ volt})/(3 \text{ ohm}) = 6.67 \text{ ampere} \quad .$$

Remarks:

1. In finding the equivalent resistance, the circuit was sequentially simplified from (a) through (d). To avoid mistakes, a new diagram was drawn after each step.

2. In the second part of this problem we started with the equivalent (d) and worked back toward the original circuit (a).

At each such step either (26-2b) or (26-3c) was used. For example, in (c) the three resistors are in series and hence have the same current as their equivalent in (d). <<<

26-4 Power Dissipated by a Resistor

The general formula (24-9) for the electrical power supplied to any circuit element is $P = Vi$. For the case of a resistor there are three ways of expressing this power (using $V = iR$):

$$P = i^2R = Vi = V^2/R \quad . \qquad (26\text{-}4)$$

Note that this is always positive (or zero). It is impossible then to get any (electrical) power from a resistor; this power always flows into the resistor. What happens to the energy $W = \int P\,dt$? It is not stored in the resistor; rather it is dissipated in the form of heat. In other words there is really zero total power flowing into a resistor: an amount i^2R of electrical power flows in, an equal amount of power in the form of heat flows out.

>>> Example 2. An immersion heater is connected to a 120 volt power supply. It requires four minutes to bring an 8 oz cup of water from 15°C to the boiling point. Assume that there is no loss of heat from the water to its surroundings. (a) What is the (electrical) power consumed by the heater? (b) What is the resistance of the heater?

(a) The mass of the water is approximately 237 gm. The power required to heat this water is

$$P = Q/t = mc\Delta T/t \qquad (Q = \text{heat, } m = \text{mass, } c = \text{specific heat})$$

$$= \frac{(237\text{ g})\cdot(1\text{ cal/g}\cdot\text{C}^{\text{o}})\cdot(100 - 15)\text{C}^{\text{o}}\cdot(4.18\text{ J/cal})}{(4\text{ min})\cdot(60\text{ s/min})}$$

$$= 350\text{ watt} \quad .$$

(b)

$$P = V^2/R$$

$$R = V^2/P = (120\text{ volt})^2/(350\text{ watt})$$

$$= 41\text{ ohm} \quad .$$

<<<

26-5 Current Density

Consider the interior of a conductor in which there is a current. Let $d\underline{S}$ be a vectorial element of area within the conductor. There is then a certain differential current, di, through this area. This current may be written as

$$di = \underline{\underline{j}}\cdot d\underline{S} \quad . \qquad (26\text{-}5)$$

Equation (26-5) defines the current density (current per unit area), $\underline{\underline{j}}$. The dot product takes care of any "tilt" which the area $d\underline{S}$ may have relative to $\underline{\underline{j}}$. The units of current density are A/m^2. $\underline{\underline{j}}$ is a vector, its direction at any point is that of the current at that point (i.e. opposite to the electrons' velocity, assuming that the current consists of a motion of electrons). In general $\underline{\underline{j}}$ is a function of location; thus $\underline{\underline{j}}$ is a vector field.

To obtain the current through a macroscopic area "S" we integrate equation (26-5):

$$i = \int \mathbf{j} \cdot d\mathbf{S} \quad , \tag{26-6}$$

the region of integration being over the area S.

26-6 Drift Velocity

A current is a motion of charge. The (average) velocity of these charges is called the drift velocity, v_d. Generally, the drift velocity is very small and is superimposed upon the much larger random thermal velocities of the charges.

>>> Example 3. Derive a relation between the current density j and the drift velocity v_d in a conductor. Assume that the current consists of a motion of electrons.

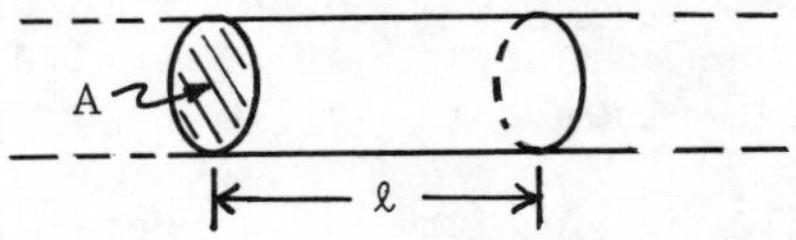

Consider a segment (length ℓ, cross sectional area A) of the conductor. Let n be the number of conduction electrons per unit volume. (This may be calculated using the mass density of the conductor, Avogadro's number, and the number of conduction electrons per atom.) The corresponding charge per unit volume is then ne. The charge in the segment is

$$q = (\text{charge/unit volume})\cdot(\text{volume}) = (ne)\cdot(A\ell) \quad .$$

The time t required for this charge to move past a given cross section is

$$t = \text{distance/speed} = \ell/v_d \quad .$$

The current i is then

$$i = q/t = (neA\ell)/(\ell/v_d) = neAv_d \quad .$$

Dividing by A,

$$j = i/A = nev_d \quad .$$ <<<

26-7 Resistivity

An electric field is required to maintain a current. We regard this electric field $\mathbf{E}$ as the cause and the current density $\mathbf{j}$ as the effect. In isotropic* materials $\mathbf{j}$ is parallel to $\mathbf{E}$. We then write

$$\mathbf{j} = \left(\frac{1}{\rho}\right) \mathbf{E} \quad . \tag{26-7}$$

The scalar quantity ρ (a property of the material) is called the resistivity, its units are $\Omega\cdot m$. For many materials ρ is a constant, independent of $\mathbf{E}$. These are the materials which obey Ohm's law. We say that "R = constant" is a macroscopic statement of Ohm's law, while "ρ = constant" is a microscopic statement of this law.

The resistance R of a sample of material may be calculated using (26-7). The result depends upon the material (through its resistivity) and the geometry of the particular resistor.

---

*Isotropic means that the material behaves equally in all orientations. A single non-cubic crystal is a good example of a non-isotropic material.

In general it is difficult to calculate R since both $\underline{j}$ and $\underline{E}$ are unknown. For resistors with simple enough geometry, the form of $\underline{j}$ as a function of location can be determined by symmetry. This is illustrated in the following example.

>>> Example 4. What is the resistance (between the ends) of a wire of length ℓ, cross sectional area A, and resistivity ρ? The current may be taken to be uniformly distributed over the cross section.

Since the current is uniformly distributed over the cross section of the wire, $j = i/A$. Since A is constant (e.g. there is no taper), j is the same everywhere within the wire. From (26-7), $E = \rho j = \rho i/A$. Thus the electric field is uniform within the wire. The potential difference between the ends of the wire is then simply

$$V = E\ell = (\rho i/A)\,\ell \quad .$$

Therefore

$$R = V/i = \rho\ell/A \quad .$$

Remarks:

1. Note that "$V = E\ell$" holds only because the electric field is uniform within the wire.
2. From the answer it is easy to see why the units of resistivity are $\Omega\cdot$m. <<<

26-8 Effects of a Temperature Change

In general the resistivity of a material will vary with temperature. For many materials, over reasonable ranges of temperature, the resistivity varies linearly with temperature. In this case the temperature dependence of resistivity is given by

$$\rho = \rho_0[1 + \overline{\alpha}(T - T_0)] \qquad (26\text{-}8)$$

where ρ_0 is the resistivity at the temperature T_0. The quantity $\overline{\alpha}$ is called the mean temperature coefficient of resistivity* of the material. The units of $\overline{\alpha}$ are $(\text{deg})^{-1}$.

A change in resistivity causes a change in the resistance of a body. From equation (26-8) the temperature dependence of resistance is given by

$$R = R_0[1 + \overline{\alpha}(T - T_0)] \qquad (26\text{-}9)$$

where R_0 is the resistance at the temperature T_0. In obtaining (26-9) from (26-8) the effects of a temperature change upon the geometry of the resistor (linear expansion) have been ignored; these effects are small compared with that due to the change in the resistivity.

---

*The temperature coefficient of resistivity, α, is defined by $\alpha = (1/\rho)(d\rho/dT)$. The constant quantity $\overline{\alpha}$ may be regarded as a kind of effective value of α over some temperature range.

26-9 Programmed Problems

1.

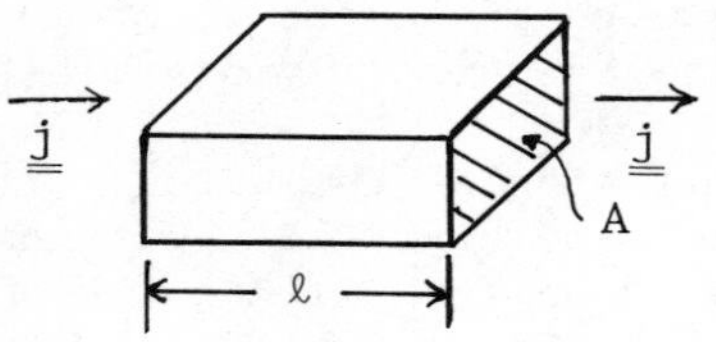

A uniform current density j ($j = i/A$) exists in a copper conductor of length ℓ and cross sectional area A as shown. If the number of conduction electrons per unit volume is n, how many conduction electrons are present in the wire at any instant? Let N = number of conduction electrons.

$N =$ ________.

$N = nA\ell$.

$N = \frac{\text{number}}{\text{unit volume}} \times \text{volume}$,

$\text{volume} = A\ell$.

---

2. What is the total charge constituted by these conduction electrons at any instant?

$q_{total} =$ __________.

$q_{total} = Ne$.

The number of charges times the charge of each.

$q = nA\ell e$.

---

3. Imagine that all electrons are moving with the same drift speed v_d. At a time when all those charges at the right hand end leave, a similar group is entering the left hand end. This new group of charges will thus leave the right hand end in a time t such that

$v_d t =$ __________.

ℓ

The distance that the electrons must travel to leave the right hand end.

---

4. In a time $t = \ell/v_d$ all conduction electrons in the wire at the beginning of that time will have drifted out of the wire. (Of course they will be replaced by others.) This net transfer of charge per unit time is called a current. Write the equation defining current.

$i =$ __________.

$i = q/t$.

[In general, $i = dq/dt$; for the case of a steady current we can just use charge/time.]

---

5. Using the first of the equations in the previous answer and the information of frames 2, 3 and 4 write i in terms of n, A, e and v_d.

$i =$ _________.

From frame 4, $i = q/t$.

From frame 3, $t = \ell/v_d$.

From frame 2, $q = nA\ell e$.

Thus $i = nAev_d$.

6. The drift velocity can be written then as

$$v_d = \frac{i}{nAe} \quad \text{or} \quad \frac{j}{ne}$$

where $j = i/A$.

For a given conductor

a. What determines n?
b. What determines A?

a. The metal of which the conductor is made. The metal is characterized by its n value.
b. The geometry of the wire, i.e. the size.

Let us obtain some feel for the n in the equation of the previous frame.

$$n = \frac{\text{number of conduction electrons}}{\text{unit volume}} \ .$$

In general one can roughly estimate that the number of conduction electrons is about one (or a few) per atom. We need to know something about the number of atoms per unit volume for various conductors which will thus give a rough estimate of the number of conduction electrons per unit volume.

7. From chemistry we have the term mole. A mole of any substance is that mass of the substance which contains 6.02252×10^{23} molecules (or atoms in our case). This number is called ___________ number.

Avogadro's

---

8. Let us arrive at a way to calculate the number of atoms in a certain mass of a substance by using dimensional analysis.

Let M = atomic weight of a substance in gm/mole,

N_0 = Avogadro's number in atom/mole,

m = mass of a substance in g,

N = number of atoms.

Write an equation combining these four terms such that

N = ___________.

Your answer should have the units "atom".

$$N = \frac{N_0 m}{M} \ .$$

Note that

$$\frac{(\text{atom/mole}) \times g}{g/\text{mole}} = \text{atom} \ .$$

9. Assuming that there is one (conduction) electron per atom, use dimensional analysis to write an equation involving

M = atomic weight in g/mole,
N_o = Avogadro's number in atom/mole,
d = density (i.e. mass per unit volume) in g/cm^3,

to calculate n. Your answer should have the units "electron/cm^3".

n = ___________.

$$n = \frac{N_o d}{M} \times \frac{1 \text{ electron}}{\text{atom}} .$$

Note that

$$\frac{(\text{atom/mole}) \times (\text{g/cm}^3)}{(\text{g/mole})} \times \frac{\text{electron}}{\text{atom}} = \frac{\text{electron}}{\text{cm}^3} .$$

Here "electron" means conduction electron.

---

10. Using the result of the previous frame, calculate the number of conduction electrons per unit volume (n) for copper. You may assume that there is one conduction electron per atom.

Let $N_o = 6.0 \times 10^{23}$ atom/mole,

$M = 64$ g/mole,

$d = 9.0$ g/cm^3.

Then n = ___________.

$$n = \frac{N_o d}{M} \times \frac{1 \text{ electron}}{\text{atom}}$$

$$= \frac{(6.0 \times 10^{23} \text{ atom/mole})}{(64 \text{ g/mole})} \times 9.0 \frac{\text{g}}{\text{cm}^3} \times \frac{1 \text{ electron}}{\text{atom}}$$

$$= 8.4 \times 10^{22} \text{ electron/cm}^3 .$$

---

11. We now return to our equation for drift velocity ity,

$$v_d = \frac{i}{nAe} .$$

Suppose we were to use a straight copper wire from the earth to the moon in order that telephone communications might be possible between earth people and astronauts. Let us say that a current of one microampere (10^{-6} A) is sufficient and that the cross sectional area of the wire is 0.00005 cm^2.

a. Calculate the drift velocity. Use the value of n as calculated in the previous frame.

v_d = ___________.

b. Assuming that the distance to the moon is 3.8×10^{10} cm, how long would it take a given electron to make the trip from earth to moon along the wire?

t = ___________.

a.

$$v_d = \frac{i}{nAe}$$

$$= \frac{10^{-6} \text{ A}}{(8.4 \times 10^{22} \text{ electron/cm}^3} \times \frac{1}{(5\times10^{-5}\text{cm}^2)(1.6\times10^{-19}\text{C/elec.})}$$

$$= 15 \times 10^{-7} \text{ cm/s}.$$

To make the units come out correctly, we must use $e = 1.6 \times 10^{-19}$C/electron (instead of simply "coulomb").

b.

$$t = \frac{\text{distance}}{\text{velocity}} = \frac{3.8 \times 10^{10}\text{cm}}{15 \times 10^{-7} \text{ cm/s}}$$

$$= 25 \times 10^{15} \text{ s}.$$

(A radio signal takes slightly more than one second to reach the moon. Fortunately, fast communication does not depend upon drift velocities.)

Chapter 27: ELECTROMOTIVE FORCE AND CIRCUITS

REVIEW AND PREVIEW

In the previous chapter you studied properties of a conductor carrying a current; in particular you learned that a potential difference $V = iR$ is required to maintain the current. In this chapter you will study the sources of such potential differences. These sources are called seats of electromotive force (emf). You will learn how to solve more complicated circuits which contain several seats of emf and several resistors. Finally you will also study some simple circuits containing a resistor and a capacitor.

GOALS AND GUIDELINES

In this chapter there are three major goals:

1. Understanding what is meant by a seat of electromotive force (emf) $\mathcal{E}$. Learning how to include the effect of an internal resistance of a seat of emf. (Sections 27-1, 27-2).
2. Learning how to handle more complicated circuits consisting of seats of emf and resistors. (Section 27-5).
3. Understanding the properties of simple circuits containing a resistor and a capacitor (Sections 27-6, 27-7, 27-8). Here you will deal with some simple differential equations.

In general, when you encounter an electrical circuit there are two aspects you should keep in mind:

(i) You must understand the properties of the individual circuit elements. Table 27-1 can be very helpful in this regard.

(ii) You must be able to apply these properties of the individual circuit elements to the entire circuit. This involves Kirchhoff's laws (Eq. 27-1, 27-2). The question of algebraic signs is very important here (see Example 2).

27-1 Electromotive Force

A circuit device which can supply a constant potential difference $\mathcal{E}$ between two points (independent of the current) is called a seat of electromotive force (emf). The unit of $\mathcal{E}$ is of course the volt.

In electrical circuits a seat of emf is represented by the following symbol: —| |—. Consider the following seat of emf.

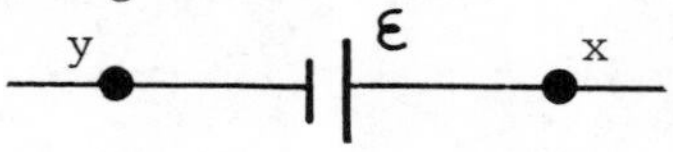

The algebraically correct equation for this is $V_x - V_y = \mathcal{E}$. (Note that the "long line" part of the symbol is at the higher potential.) This equation is true regardless of the amount or direction of any current through the seat.

27-2 Internal Resistance of a Seat of EMF

Many practical seats of an emf $\mathcal{E}$ (such as chemical batteries) also have an internal resistance r. Such a battery behaves as if it were an ideal (i.e. resistanceless) seat of the same emf $\mathcal{E}$ in series with a resistance r as shown in Figure 27-1.

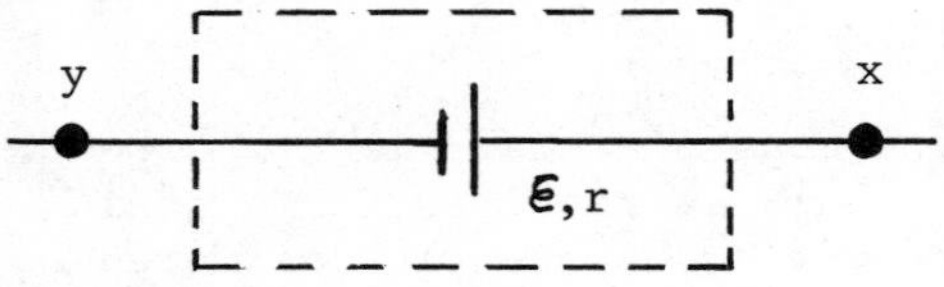

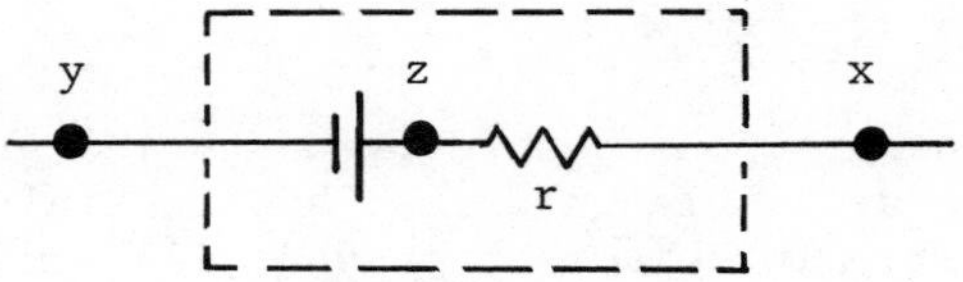

Figure 27-1. (a) A battery having an emf $\mathcal{E}$ and an internal resistance r. (b) The equivalent circuit. The seat of emf and the resistor are in series.

In the figure the point "z" cannot be reached experimentally; it is for "bookkeeping" purposes only. The external terminals "x" and "y" of course can be reached experimentally; $V_x - V_y$ is called the "terminal potential difference"* of the battery.

>>> Example 1. A battery ($\mathcal{E}$ = 10 volts, r = 0.5 ohm) carries a current of 2.0 amperes. The sense of this current is such that the current comes out the positive terminal of the battery. Calculate the terminal potential difference.

The given information is shown in the following diagram.

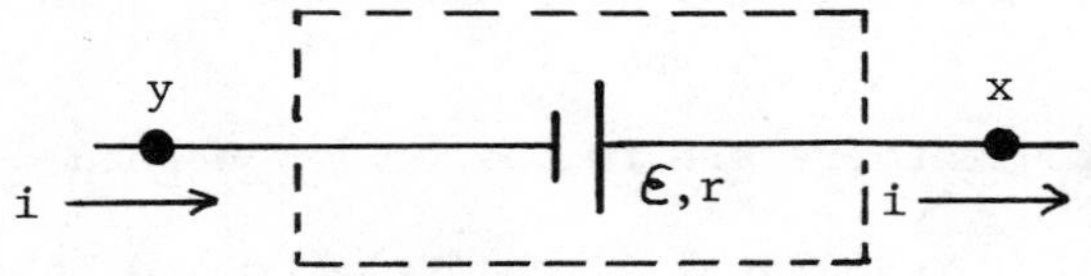

$\mathcal{E}$ = 10 volt

r = 0.5 ohm

i = 2.0 ampere

We first replace the battery by its equivalent circuit (as in Figure 27-1).

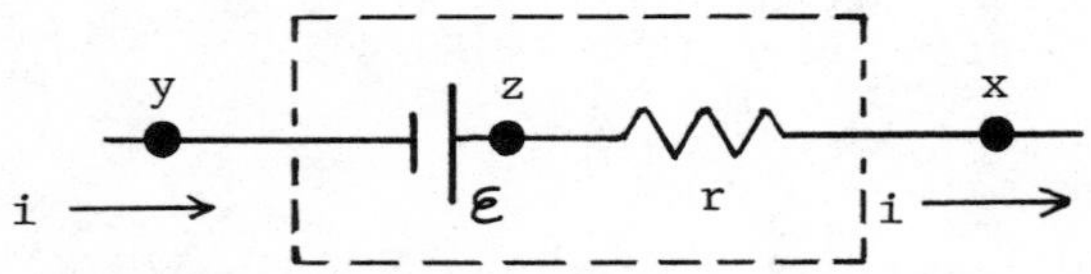

We want $V_x - V_y$. Noting that the current through r is from z to x we have

$$\begin{aligned} V_x - V_y &= (V_x - V_z) + (V_z - V_y) \\ &= -(V_z - V_x) + (V_z - V_y) \\ &= -(ir) + (\mathcal{E}) \\ &= -(2.0\ \text{ampere})(0.5\ \text{ohm}) + (10\ \text{volt}) \\ &= 9\ \text{volt}\ . \end{aligned}$$

Remark: In this case the battery is being discharged; this causes the terminal potential difference to be less than the emf. If the battery were being charged (i.e. if the current were reversed), the terminal potential difference would be more than the emf. If there were no current, the terminal potential difference would equal the emf (this is called the "open circuit voltage"). <<<

*Sometimes simply called the "terminal voltage".

27-3 Power Supplied by a Seat of EMF

The general formula for the power supplied to any circuit is (24-9). Usually (i.e. when a battery is being discharged), a seat of emf supplies power. Reversing the sign in (24-9) gives for the power supplied by a seat of emf

$$P = \mathcal{E} i$$

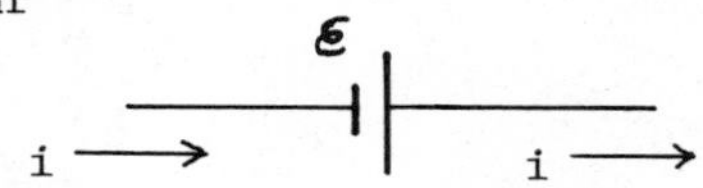

where i is the current through the seat of emf in the direction shown (i.e. coming out the positive terminal).

27-4 Circuit Elements

We have discussed seats of emf, resistors, and capacitors. These circuit elements can be connected in various ways to form an electrical circuit. Table 27-1 summarizes the relevant properties of circuit elements.

Table 27-1

Circuit Element	Property (units)	Circuit Symbol	Equations
seat of emf	emf (volt, V)	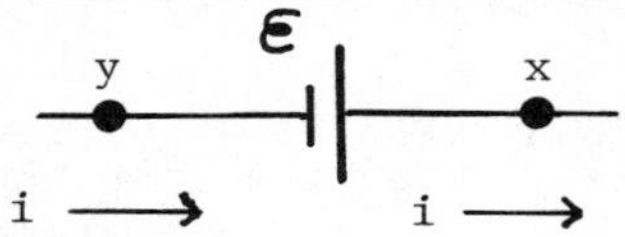	$V = V_x - V_y = \mathcal{E}$ Power supplied by seat: $P = \mathcal{E} i$
capacitor	capacitance (farad, F)	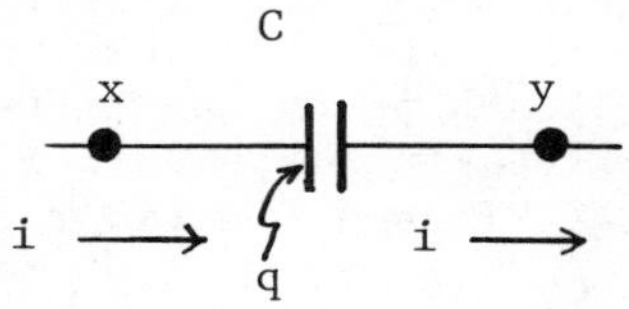	$V = V_x - V_y = q/C$, $i = dq/dt$ Stored energy: $U_C = \frac{1}{2}q^2/C = \frac{1}{2}qV = \frac{1}{2}CV^2$
resistor	resistance (ohm, Ω)	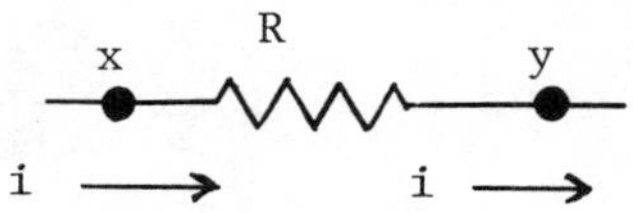	$V = V_x - V_y = iR$ Power dissipated: $P = i^2R = iV = V^2/R$
inductor*	inductance (henry, H)	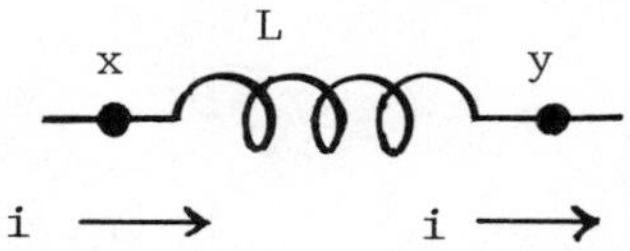	$V = V_x - V_y = L\, di/dt$ Stored energy: $U_L = \frac{1}{2}Li^2$

Table 27-1. Properties of various circuit elements. The equations are algebraically correct for all cases provided that the sign conventions shown under "circuit symbol" are followed.

*Inductors are treated in Chapter 31. They are included in this table for completeness.

27-5 Kirchhoff's Rules

As an example, consider the following form of an electrical circuit.

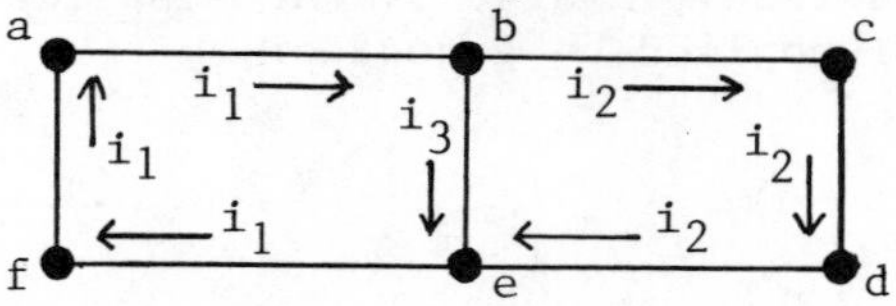

The circuit elements have been suppressed (e.g. there might be a resistor between a and b, a seat of emf between b and c, etc.) The unknown currents i_1, i_2, i_3 have been labelled in an arbitrary manner. Note that e.g. the <u>same</u> current (i_1) into a (from f) must also leave a (toward b). This is because there can be no accumulation (or depletion) of charge at the junction a. To apply this idea at the junction b, we must have that the total current into b (i_1) must equal the total current leaving b ($i_2 + i_3$). That is, $i_1 - i_2 - i_3 = 0$. This illustrates <u>Kirchhoff's first rule</u> (also called the <u>junction theorem</u>):

The algebraic sum of all the currents into any junction must vanish. (27-1)

Now consider a loop in this circuit, say a-b-e-f-a. Then

$$(V_a - V_b) + (V_b - V_e) + (V_e - V_f) + (V_f - V_a) = 0 \quad . \tag{27-2a}$$

This equation is an <u>identity</u>, requiring no derivation or explanation. It is an illustration of <u>Kirchhoff's second rule</u> (also called the <u>loop theorem</u>):

The algebraic sum of the potential differences around any loop must vanish. (27-2b)

Since the question of signs is so important here, it is better to remember this rule in equation form (27-2a) rather than in verbal form (27-2b). When applying (27-2a) to circuit problems, one need only replace each potential difference appearing in it by the appropriate formula from Table 27-1 (taking care to use the proper signs). The following example illustrates the use of Kirchhoff's two rules as applied to a resistive circuit.

>>> Example 2. For the circuit shown below, find the current through each resistor.

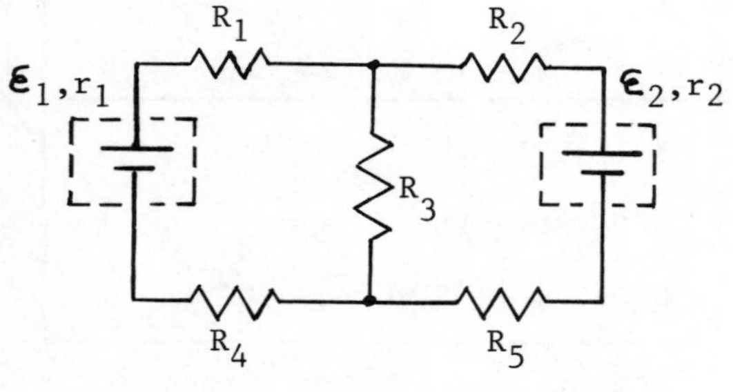

$R_1 = 5$ ohm , $R_2 = 1$ ohm,

$R_3 = 10$ ohm , $R_4 = 4$ ohm,

$R_5 = 3$ ohm , $\mathcal{E}_1 = 10$ volt,

$\mathcal{E}_2 = 2$ volt , $r_1 = r_2 = 1$ ohm.

Each battery is replaced by its equivalent circuit. The unknown currents are labelled; their directions having been arbitrarily assigned. For brevity, all units will be omitted.

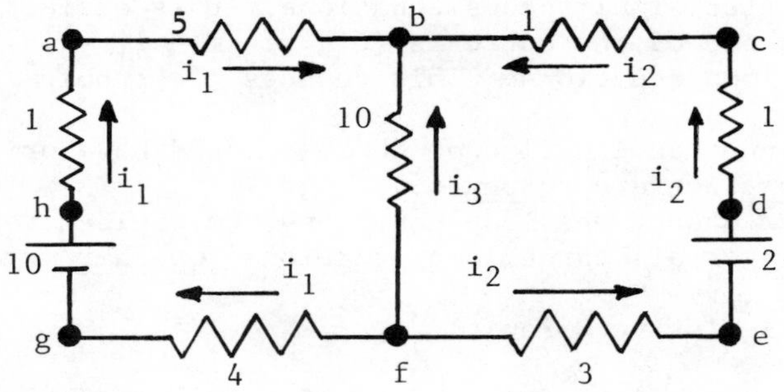

Junction theorem at "b":

$$i_1 + i_2 + i_3 = 0 \quad . \tag{a}$$

Loop theorem for "a-b-f-g-h-a":

$$(V_a - V_b) + (V_b - V_f) + (V_f - V_g) + (V_g - V_h) + (V_h - V_a) = 0$$

$$(5\ i_1) + (-\ 10\ i_3) + (4\ i_1) + (-\ 10) + (1\ i_1) = 0$$

$$10\ i_1 - 10\ i_3 - 10 = 0 \tag{b}$$

Loop theorem for "b-c-d-e-f-b":

$$(V_b - V_c) + (V_c - V_d) + (V_d - V_e) + (V_e - V_f) + (V_f - V_b) = 0$$

$$(-\ 1\ i_2) + (-\ 1\ i_2) + (2) + (-\ 3\ i_2) + (10\ i_3) = 0$$

$$-\ 5\ i_2 + 10\ i_3 + 2 = 0 \tag{c}$$

Equations (a), (b), and (c) are three linear equations in the three unknowns: i_1, i_2, i_3. The solution of these equations is:

$$i_1 = 13/20 \text{ A} \quad , \quad i_2 = -\ 3/10 \text{ A} \quad , \quad i_3 = -\ 7/20 \text{ A} \quad .$$

The currents through the various resistors are:

R_1: 13/20 A (right) , R_2: 3/10 A (right) , R_3: 7/20 A (down) ,

R_4: 13/20 A (left) , R_5: 3/10 A (left) .

Remarks:

1. To process e.g. the term $(V_b - V_f)$ in the first loop equation we write

$$(V_b - V_f) = -\ (V_f - V_b) = -\ (10\ i_3) \quad .$$

Here Table 27-1 has been used with f playing the role of "x", and b playing the role of "y".

2. The current through e.g. R_2 is $i_2 = - 3/10$ A to the left; therefore, + 3/10 A to the right.

3. In solving the three simultaneous equations, it is easiest to first solve the junction equation (a) for one of the currents (e.g. $i_3 = - i_1 - i_2$) and then substitute this expression into the loop equations. This reduces the problem to two equations in two unknowns.

4. Instead of the junction equation at "b", we could have written one at "f". However both equations together are redundant.

5. Instead of either one of the loop equations, we could have written one for "a-b-c-d-e-f-g-h-a". However all three loop equations together are redundant. <<<

27-6 Power Balance in a Resistive Circuit

Resistors dissipate power (i^2R) while seats of emf supply power ($\mathcal{E}i$). The total power dissipated by all the resistors (including internal resistances of the seats of emf) must equal the total power supplied by all the seats of emf. Note that this represents one equation for the entire circuit. It does not say that the power supplied by one seat of emf equals the power dissipated by any particular resistor (or resistors).

>>> Example 3. Check the answer to the previous example by accounting for the various powers.

The power dissipated by those resistors carrying the current i_1 is:

$$i_1^2R_1 + i_1^2R_4 + i_1^2r_1 = i_1^2(R_1 + R_4 + r_1)$$
$$= (13/20)^2(5 + 4 + 1) = 1690/400 \text{ watt} \quad .$$

The power dissipated by those resistors carrying the current i_2 is:

$$i_2^2R_2 + i_2^2R_5 + i_2^2r_2 = i_2^2(R_2 + R_5 + r_2)$$
$$= (3/10)^2(1 + 3 + 1) = 45/100 \text{ watt} \quad .$$

The power dissipated by those resistors carrrying the current i_3 is:

$$i_3^2R_3 = (7/20)^2(10) = 490/400 \text{ watt} \quad .$$

The total power dissipated by all the resistors (including internal resistances) is:

$$(1690/400) + (45/100) + (490/400) = 2360/400 = 59/10 \text{ watt} \quad .$$

The power supplied by emf $\mathcal{E}_1$ (it is being discharged) is:

$$(10 \text{ volt})\cdot(13/20 \text{ amp}) = 13/2 \text{ watt} \quad .$$

The power supplied by emf $\mathcal{E}_2$ (it is being charged) is:

$$- (2 \text{ volt})\cdot(3/10 \text{ amp}) = - 3/5 \text{ watt} \quad .$$

The total power supplied by all the seats of emf is:

$$(13/2) + (- 3/5) = 59/10 \text{ watt} \quad .$$

Thus the total power supplied by all the seats of emf (59/10 watt) equals the total power dissipated by all the resistors (59/10 watt). <<<

27-7 Single Loop RC Circuits

In the single loop RC circuit shown below the switch S is closed at time $t = 0$. The capacitor C (assumed to be initially uncharged) will begin to acquire a charge. The charge q on the capacitor is then a function of time.

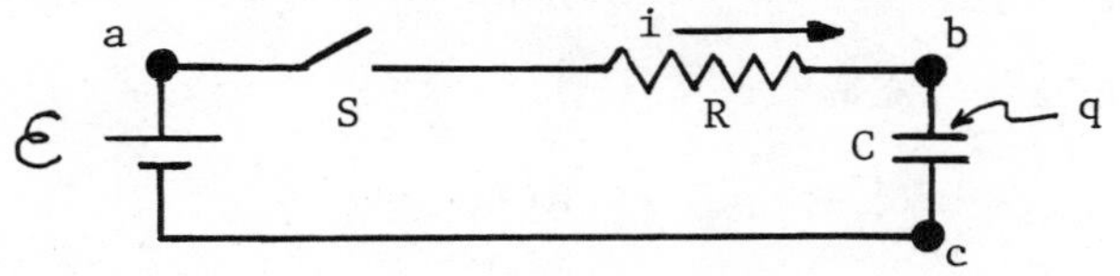

Since this is a single loop circuit, we need write only one loop equation. After the switch is closed ($t \geq 0$) we have

$$(V_a - V_b) + (V_b - V_c) + (V_c - V_a) = 0$$

$$(iR) + (q/C) + (-\mathcal{E}) = 0$$

Here Table 27-1 has been used in substituting for the various potential differences. From this table we also have $i = dq/dt$ (note that the signs are correct: i is the current into that capacitor plate which has the charge q). Then

$$R\, dq/dt + q/C = \mathcal{E} \quad . \tag{27-3}$$

This is a differential equation for q. Unlike an algebraic equation (whose solution is a number), a differential equation has a function (in this case, of time) as its solution. The solution of (27-3), subject to the given initial condition (at $t = 0$, $q = 0$), is

$$q = C\mathcal{E}(1 - e^{-t/RC}) \quad . \tag{27-4}$$

Here "e" is the base of natural logarithms ($e = 2.718...$), not the electronic charge.

\>>> Example 4. Solve the differential equation

$$R\, dq/dt + q/C = \mathcal{E} \tag{27-3}$$

subject to the initial condition that when $t = 0$, $q = 0$.

We try to "separate the variables". This means that we seek to have all the q's and dq's on one side of the equation, all t's and dt's on the other side.

$$R\, dq/dt + q/C = \mathcal{E}$$

$$R\, dq/dt = (\mathcal{E} - q/C)$$

$$\frac{R\, dq}{\mathcal{E} - q/C} = dt \quad .$$

The variables have been separated. Integrating both sides (this is possible because the variables have been separated) gives

$$- RC \ln(\mathcal{E} - q/C) = t + A$$

where "A" is the constant of integration. Now we use the initial condition: when $t = 0$, $q = 0$. Substituting this gives

$$- RC \ln(\mathcal{E}) = A \quad .$$

Using this expression for A we have

$$- RC[\ln(\mathcal{E} - q/C) - \ln(\mathcal{E})] = t$$

$$\ln(1 - q/(C\mathcal{E})) = - t/RC \quad .$$

Here we have used the fact that the difference of the two logarithms is the logarithm of the quotient: $(\mathcal{E} - q/C)/\mathcal{E} = 1 - q/(C\mathcal{E})$. Finally, taking exponentials of both sides,

$$1 - q/(C\mathcal{E}) = e^{-t/RC}$$

$$q = C\mathcal{E}(1 - e^{-t/RC}) \quad . \qquad (27\text{-}4)$$

<<<

Equation (27-4) has certain properties which can easily be checked. For example,

(a) The current at any time t is $i = dq/dt = (\mathcal{E}/R)e^{-t/RC}$. At $t = 0$ the current is $\mathcal{E}/R$; this is the current which would exist in a simple $\mathcal{E}R$ circuit. Since the capacitor is initially uncharged, there is no potential difference across it at $t = 0$. Therefore iR must initially "balance" $\mathcal{E}$, i.e. $i = \mathcal{E}/R$.

(b) For $t \to \infty$, $q = C\mathcal{E}$. After a long time the current approaches zero. Then q/C must balance $\mathcal{E}$, i.e. $q = C\mathcal{E}$.

Another example of a single loop RC circuit is shown below. The capacitor has an initial charge $q = q_o$. At time $t = 0$ switch S is closed.

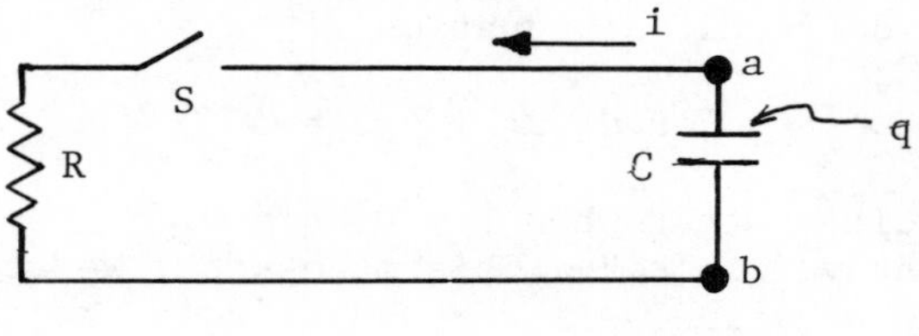

For $t \geq 0$ the loop equation for this circuit is

$$(V_a - V_b) + (V_b - V_a) = 0$$

$$(q/C) + (- iR) = 0 \quad .$$

In this case (note the assumed direction of the current in the diagram), $i = - dq/dt$. Thus

$$q/C + R\, dq/dt = 0 \quad . \qquad (27\text{-}5)$$

This may be solved using the same technique as in Example 4. The solution (subject to the initial condition that when $t = 0$, $q = q_o$) is

$$q = q_o e^{-t/RC} \quad . \tag{27-6}$$

27-8 Time Constant

The two solutions above (equations (27-4,6)) involve the term $e^{-t/RC}$. Since the exponent must be dimensionless, the product RC must have the dimensions of time. It is called the (capacitive) time constant, τ:*

$$\tau = RC \quad . \tag{27-7}$$

Regardless of whether the capacitor is charging or discharging, q changes from some initial value (at $t = 0$) to some final value (at $t = \infty$). After one time constant (i.e. when $t = \tau$), 63% of this change has been accomplished. After the next time constant (i.e. when $t = 2\tau$), 63% of the remaining 37% will be overcome, etc. (The number 63% comes from $1 - e^{-1} = 0.632...$.)

27-9 Power Balance in an RC Circuit

In the $\mathcal{E}$RC circuit considered above, some of the power supplied by the seat of emf is dissipated by the resistor. The remaining power is used to change the energy stored in the capacitor. Thus (power supplied by seat of emf) = (power dissipated by resistor) + (rate of change of energy stored in capacitor). That is,

$$\mathcal{E}i = i^2R + \frac{d}{dt}(U) = i^2R + \frac{d}{dt}(\tfrac{1}{2}q^2/C)$$

$$\mathcal{E}i = i^2R + (1/C)(q\,dq/dt) \quad .$$

Cancelling the common factor $i = dq/dt$,

$$\mathcal{E} = iR + q/C$$

which is the loop equation (27-3). For a single loop RC circuit we can thus derive the loop equation by considering the power balance.

*Sometimes written τ_C to emphasize that it is the capacitive time constant.

27-10 Programmed Problems

1.

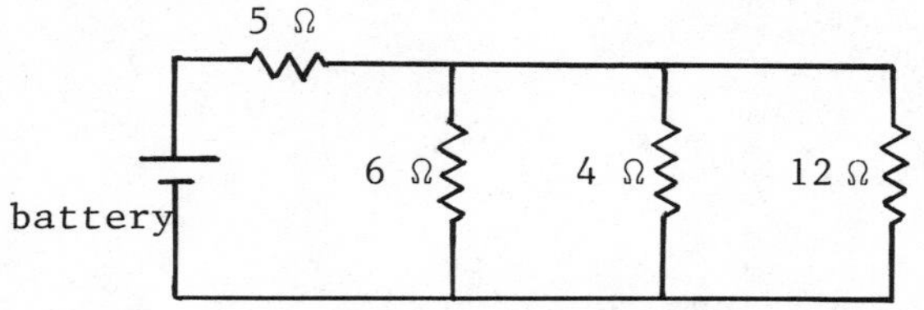

In the above circuit the battery has an emf of 15 V and an internal resistance of 0.5 Ω. We seek to calculate the current through the 12 Ω resistor.

To solve this problem, first redraw the diagram showing the internal resistance of the battery.

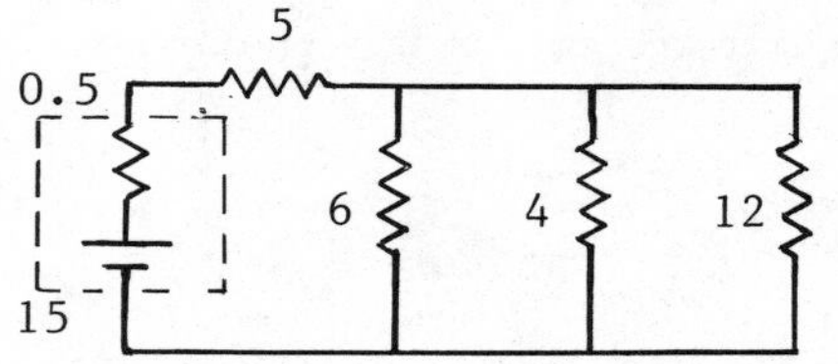

The physical battery has been replaced by an ideal seat of emf and a series resistor (as shown in the dotted box). For brevity all units have been omitted (resistances are in ohms, emfs in volts).

---

2. Is there anything in this circuit which can be simplified?

Yes. The 6, 4 and 12 ohm resistors are all in parallel.

---

3. What is the formula for finding the equivalent resistance of several parallel resistors?

$1/R = 1/R_1 + 1/R_2 + 1/R_3 + \ldots$.

There are as many terms on the right hand side as there are resistors in parallel.

---

4. Apply this formula to find the equivalent resistance in this case.

R = ___________.

2 ohm.

$1/R = 1/6 + 1/4 + 1/12$

$1/R = 1/2$

$R = 2$.

---

5. Redraw the circuit using the above equivalent resistance.

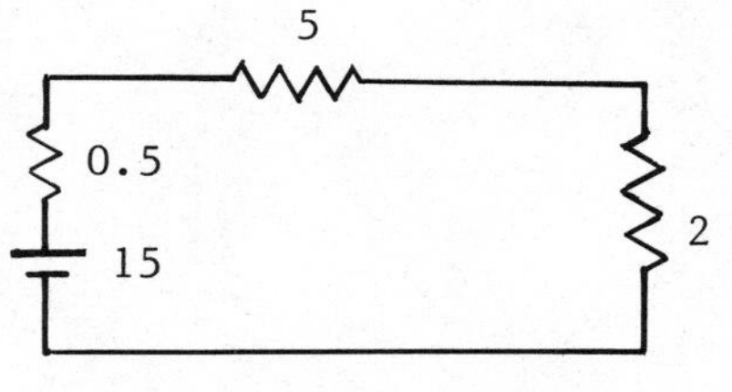

---

6. Is there anything in this circuit which can be simplified?

Yes. The 0.5, 5 and 2 ohm resistors are all in series.

7. What is the formula for finding the equivalent resistance of several series resistors?

$R = R_1 + R_2 + R_3 + \ldots$.

There are as many terms on the right hand side as there are resistors in series.

---

8. Apply this formula to find the equivalent resistance in this case.

R = _______________.

7.5 ohm.

$R = 0.5 + 5 + 2$

$R = 7.5$.

---

9. Redraw the circuit using the above equivalent resistance.

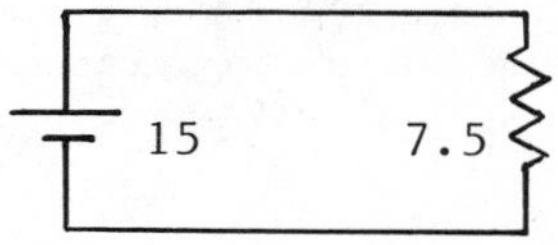

---

10. What is the potential difference across the 7.5 Ω resistor?

15 volt.

It is in parallel with the 15 volt seat of emf.

---

11. State Ohm's law and give the meaning of each of the three quantities appearing in it.

$i = V/R$ (where R is constant).

i = current in the resistor.

V = potential difference across the resistor.

R = resistance of the resistor.

---

12. Apply Ohm's law to find the current through the resistor shown in the answer to frame 9.

2 ampere.

$i = V/R = 15/7.5 = 2$.

---

13. The 7.5 ohm resistor was the equivalent to the 0.5, 5 and 2 ohm series resistors:

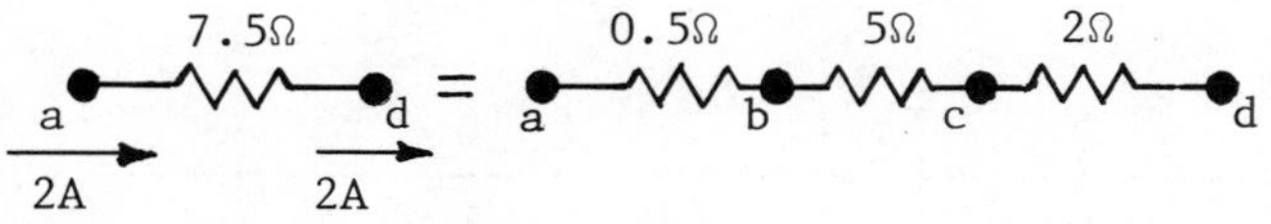

What is the current through each of the 0.5, 5 and 2 ohm resistors?

2 ampere.

The current through series resistors is the same.

14. What is the potential difference $V = V_c - V_d$ across the 2 ohm resistor above?

4 volt.

$V = iR = (2)\cdot(2) = 4$.

---

15. Therefore in frame 13, $V_c - V_d = 4$ volt. The 2 ohm resistor was equivalent to the 6, 4 and 12 ohm parallel resistors:

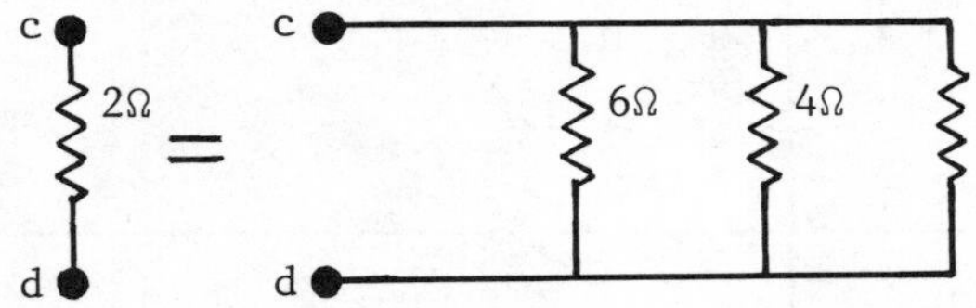

What is the potential difference across the 6, 4 and 12 ohm resistors?

V = ____________.

4 volt.

The potential difference across parallel resistors is the same.

---

16. Finally, what is the current through the 12 ohm resistor?

i = ____________.

0.333 ampere.

$i = V/R = 4/12 = 0.333$.

---

17. Capacitor C initially has a charge Q. When t = 0, switch S is closed. Write down (by inspection and recollection of the general form) the charge as a function of time.

q(t) = ________________.

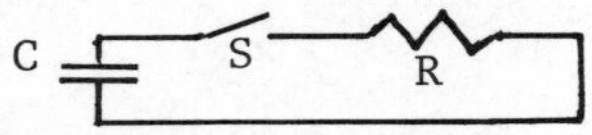

$$q = Q\,e^{-t/RC}$$

This is obtained by realizing the process is exponential with time constant $\tau = RC$ and that $q(0) = Q$.

---

18. By differentiation,

i = dq/dt = ________________.

$$i = -(Q/RC)\,e^{-t/RC}$$

---

19. The power dissipated by the resistor is

$P = i^2R =$ ________________

$$P = (Q^2/RC^2)\,e^{-2t/RC}$$

---

20. The total energy dissipated by the resistor from t = 0 to t = ∞ is

$U = \int_0^\infty P\,dt =$ ________________.

$$U = (Q^2/RC^2)(-RC/2)\,e^{-2t/RC}\Big|_0^\infty$$

$$= \tfrac{1}{2}\,Q^2/C$$

---

21. Can you interpret the previous answer?

The energy dissipated by the resistor turned out to equal the energy initially stored in the capacitor. The capacitor's lost energy appears as heat in R.

Chapter 28: THE MAGNETIC FIELD

REVIEW AND PREVIEW

So far, Part II has been concerned with electric forces on charged particles. This led to the introduction of the $\mathbf{E}$ field ($\mathbf{E} = \mathbf{F}/q$). There is another type of force which can be exerted on a charged particle, namely a magnetic force. This force is associated with another field ($\mathbf{B}$ field). The magnetic force not only depends on q and $\mathbf{B}$ but also upon the velocity $\mathbf{v}$ of the particle. In this chapter you will study some properties and consequences of the magnetic force exerted on a moving charged particle.

GOALS AND GUIDELINES

In this chapter there are three major goals:

1. Understanding the nature of the magnetic force and learning the formula for this force in terms of q, $\mathbf{v}$, $\mathbf{B}$ (Eq. 28-1). You should be able to calculate the resultant force due to simultaneous $\mathbf{E}$ and $\mathbf{B}$ fields.
2. Understanding, as an important application of the magnetic force, the motion of a charged particle in a uniform $\mathbf{B}$ field (Section 28-3).
3. A very important example of moving charged particles occurs when there is a current in a wire. You should be able to calculate the force exerted on a current carrying wire in a $\mathbf{B}$ field (Section 28-4 and especially Eq. 28-7).

The formula for the magnetic force (as well as other formulas occurring in magnetism) involves the vector (or "cross") product of two vectors. You should review this vector product if you do not fully understand it (Section 2-6).

28-1 Magnetic Field

An electric field $\mathbf{E}$ exerts an electric force $\mathbf{F}_E$ on a charge q. This force, given by $\mathbf{F}_E = q\mathbf{E}$, is independent of the motion of the particle. Its direction is parallel to $\mathbf{E}$.

There exists another kind of field, $\mathbf{B}$, which can also exert a force on a charge. This magnetic force depends upon the velocity $\mathbf{v}$ of the particle. Its direction is perpendicular to both $\mathbf{v}$ and $\mathbf{B}$. The magnetic force, $\mathbf{F}_B$, exerted on a (point) charge q moving with velocity $\mathbf{v}$ is

$$\mathbf{F}_B = q\mathbf{v} \times \mathbf{B} \quad . \tag{28-1}$$

The field $\mathbf{B}$ is called the magnetic field*, its unit is tesla, T (= (N/C)/(m/s)).**

>>> Example 1. An electron moves in the x-y plane with a speed of 10^6 meter/second. Its velocity vector makes an angle of 60° with the x-axis as shown. There is a $\mathbf{B}$ field of magnitude 10^{-2} T directed along the y-axis. Calculate the magnetic force exerted on the electron.

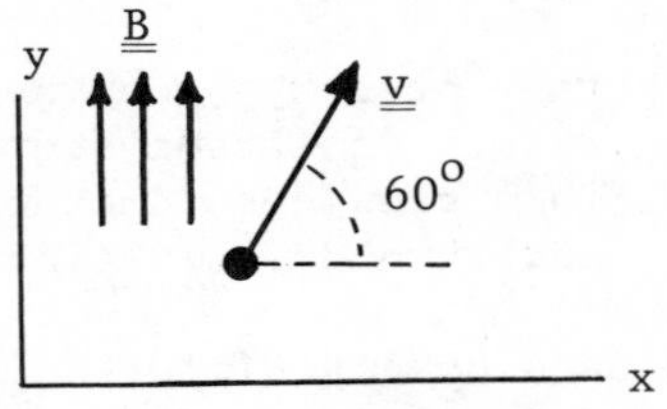

*The magnetic field $\mathbf{B}$ is also known simply as the "$\mathbf{B}$ field". An electric field is similarly called an "$\mathbf{E}$ field".

**The weber (Wb) is defined as the unit for the flux of $\mathbf{B}$ ($\Phi_B = \int \mathbf{B}\cdot d\mathbf{S}$). The unit tesla is also written as Wb/m^2.

$$\underline{\underline{F}} = q\underline{\underline{v}} \times \underline{\underline{B}} = - e\underline{\underline{v}} \times \underline{\underline{B}}$$

$$|\underline{\underline{F}}| = e|\underline{\underline{v}}|\ |\underline{\underline{B}}| \sin\theta$$

where $\theta = 30^o$ is the angle between $\underline{\underline{v}}$ and $\underline{\underline{B}}$.

$$|\underline{\underline{F}}| = (1.60 \times 10^{-19}\ \text{C})(10^6\ \text{m/s})(10^{-2}\ \text{T})(0.500)$$

$$|\underline{\underline{F}}| = 8.0 \times 10^{-16}\ \text{N} \quad .$$

The vector $\underline{\underline{v}} \times \underline{\underline{B}}$ points out of the page. Since $q = - e$ is negative, $q\underline{\underline{v}} \times \underline{\underline{B}}$ has the opposite direction. The direction of $\underline{\underline{F}} = q\underline{\underline{v}} \times \underline{\underline{B}}$ is therefore into the page. <<<

If both $\underline{\underline{E}}$ and $\underline{\underline{B}}$ fields are present, their combined force exerted on a charge q moving with velocity $\underline{\underline{v}}$ is given by the vector sum

$$\underline{\underline{F}} = q(\underline{\underline{E}} + \underline{\underline{v}} \times \underline{\underline{B}}) \quad . \qquad (28\text{-}2)$$

This is called the "Lorentz force".

>>> Example 2. An electron moves in the +x direction. A $\underline{\underline{B}}$ field is into the page as shown. Is it possible to apply an electric field in such a manner as to make the resultant force on the electron vanish? If so, find the magnitude and direction of this electric field.

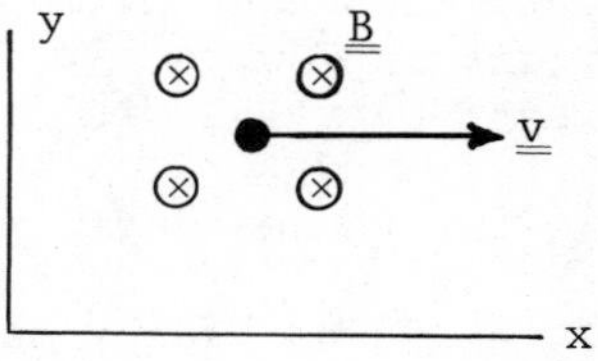

We want the Lorentz force on the electron to vanish.

$$0 = \underline{\underline{F}} = q(\underline{\underline{E}} + \underline{\underline{v}} \times \underline{\underline{B}})$$

$$\underline{\underline{E}} = - \underline{\underline{v}} \times \underline{\underline{B}} \quad .$$

This is the desired electric field. Its magnitude is vB (since $\underline{\underline{v}}$ is perpendicular to $\underline{\underline{B}}$) and it is in the -y direction.

Remarks:

1. Note that the answer is independent of both the magnitude and sign of the charge q.
2. With the Lorentz force vanishing, the electron is undeflected. If the transverse field E were different from vB the electron would be deflected from its straight line path. This provides a method for measuring B: E is adjusted until the electron is undeflected, then B = E/v (Thompson's experiment). <<<

28-2 Work Done by a $\underline{\underline{B}}$ Field

The work done on a charge q by a $\underline{\underline{B}}$ field is

$$W_B = \int \underline{\underline{F}}_B \cdot d\underline{\underline{\ell}} = q \int (\underline{\underline{v}} \times \underline{\underline{B}}) \cdot d\underline{\underline{\ell}} \quad .$$

Now $(\underline{\underline{v}} \times \underline{\underline{B}})$ is perpendicular to $\underline{\underline{v}}$ while $d\underline{\underline{\ell}}$ is parallel to $\underline{\underline{v}}$ ($\underline{\underline{v}} = d\underline{\underline{\ell}}/dt$). Since $(\underline{\underline{v}} \times \underline{\underline{B}})$ and $d\underline{\underline{\ell}}$ are then perpendicular, their dot product vanishes. No matter what the $\underline{\underline{B}}$ field is, we must have

$$W_B = 0. \tag{28-3}$$

The work done on a charge by a B field is always zero. Therefore a B field cannot change the kinetic energy ($\frac{1}{2}mv^2$) of a particle. That is, a B field cannot change the speed of a particle; it can only change the direction of its velocity.

The work done by a B field is always zero. This is one reason why we cannot define a potential for the B field in the same manner that we did for the E field.

28-3 Motion in a Uniform B Field

Consider a particle (mass m, charge q) moving in a uniform B field. For simplicity, suppose that the velocity v is in a plane perpendicular to B. The force, $\underline{F} = q\underline{v} \times \underline{B}$, exerted on the particle will change the direction (but not the magnitude) of the velocity. The particle will follow a circular path of some radius r as shown in Figure 28-1. We have

$$F = ma$$

$$qvB = mv^2/r \quad . \tag{28-4}$$

The angular velocity*, $\omega = v/r$, is then

$$\omega = (q/m)\,B \quad . \tag{28-5}$$

This is called the cyclotron frequency. For a given value of B, it depends only upon the charge to mass ratio (q/m) of the particle.

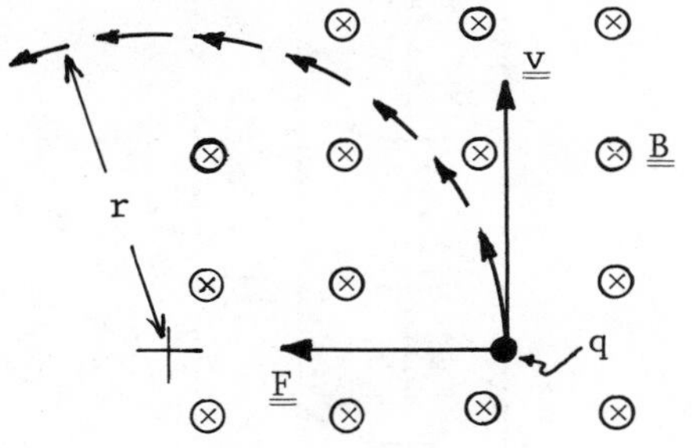

Figure 28-1. A charged particle moving in a uniform B field. The B field is into the plane of the page. The velocity vector v lies in the plane of the page, perpendicular to B. The force $\underline{F} = q\underline{v} \times \underline{B}$ is perpendicular to both v and B, and assuming that q is positive is directed as shown. As a result of this force, the particle will follow a circular path of some radius r.

>>> Example 3. In a certain cyclotron a proton moves in a circle of radius r = 0.5 meter. The magnitude of the B field is 1.2 T. (a) What is the cyclotron frequency? (b) What is the kinetic energy (in Mev) of the proton?

(a)
$$qvB = mv^2/r$$

$$\omega = v/r = Bq/m$$

$$= (1.2\text{ T})(1.60 \times 10^{-19}\text{ C})/(1.67 \times 10^{-27}\text{ kg})$$

$$= 1.15 \times 10^{8}\text{ s}^{-1} \quad \text{(i.e. radian/second)} \quad .$$

The corresponding frequency ν (in cycles/second) is

$$\nu = \omega/2\pi = 1.83 \times 10^{7}\text{ Hz} \quad \text{(i.e. cycle/second)} \quad .$$

---

*Also called the angular frequency. Its units are "radian"/second.

(b) $$qvB = mv^2/r$$

$$v = qBr/m$$

$$K = \tfrac{1}{2}mv^2 = \tfrac{1}{2}m(qBr/m)^2 = \tfrac{1}{2}q^2B^2r^2/m$$

$$= [\tfrac{1}{2}(1.60 \times 10^{-19}\ \mathrm{C})^2(1.2\ \mathrm{T})^2(0.5\ \mathrm{m})^2/(1.67 \times 10^{-27}\ \mathrm{kg})]\cdot[(1\ \mathrm{eV})/(1.60 \times 10^{-19}\ \mathrm{J})]$$

$$= 1.7 \times 10^7\ \mathrm{eV} = 17\ \mathrm{MeV} \quad .$$

Remark: Note that we always start with "F = ma" in the form (28-4): $qvB = mv^2/r$. Most cyclotron type problems can be approached from this single concept. <<<

So far we have assumed that the particle's velocity was perpendicular to $\mathbf{B}$; this led to circular motion. In the general case we resolve the velocity into two components: $\mathbf{v} = \mathbf{v}_{\parallel} + \mathbf{v}_{\perp}$, where $\mathbf{v}_{\parallel}$ is parallel to the $\mathbf{B}$ field and $\mathbf{v}_{\perp}$ is perpendicular to the $\mathbf{B}$ field. The projection of the motion onto a plane perpendicular to $\mathbf{B}$ remains a circle; equations (28-4) and (28-5) also apply here provided that $v_{\perp}$ is used for the speed in (28-4). The component $\mathbf{v}_{\parallel}$ remains constant (because there is no force parallel to $\mathbf{B}$). The path that the particle follows is then a <u>helix</u> as shown in Figure 28-2.

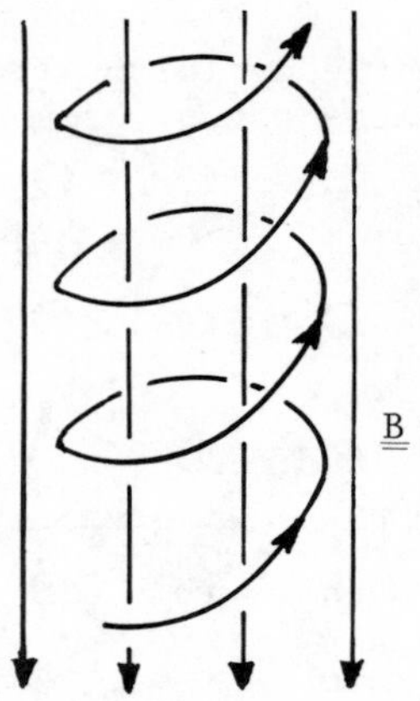

Figure 28-2. A charged particle moving in a uniform $\mathbf{B}$ field. The velocity vector in this case is not perpendicular to $\mathbf{B}$. The resulting motion is a helix, the component $\mathbf{v}_{\parallel}$ of velocity parallel to $\mathbf{B}$ remaining constant. The figure is drawn for the case of a positive charge.

28-4 Force on a Current Element

Equation (28-1) gives the magnetic force exerted on a moving charge. If this charge is differential (dq), then the magnetic force exerted on it is also differential ($d\mathbf{F}$). In this case (28-1) becomes

$$d\mathbf{F} = dq\ \mathbf{v} \times \mathbf{B} \quad . \tag{28-6}$$

Consider a wire carrying a current i. Let $d\boldsymbol{\ell}$ be a segment of this wire, the sense of $d\boldsymbol{\ell}$ is that of the current (left to right in the diagram below); $i\,d\boldsymbol{\ell}$ is called a <u>current element</u>.*

*Note that $i\,d\boldsymbol{\ell}$ is not a differential current. Rather $d\boldsymbol{\ell}$ is a differential length of wire in which there is a finite current i.

The conduction charge dq in the segment $d\boldsymbol{\ell}$ will move through the segment in some time dt. Assuming that dq is positive, its (drift) velocity is $\mathbf{v} = d\boldsymbol{\ell}/dt$. The term $dq\ \mathbf{v}$ in (28-6) becomes

$$dq\ \mathbf{v} = dq\ (d\boldsymbol{\ell}/dt) = (dq/dt)\ d\boldsymbol{\ell} = i\ d\boldsymbol{\ell}$$

where $i = dq/dt$ is the current in the wire.* Substitution of this into (28-6) gives the magnetic force exerted on the segment $d\boldsymbol{\ell}$:

$$d\mathbf{F} = i\ d\boldsymbol{\ell} \times \mathbf{B} \quad . \tag{28-7}$$

Equation (28-7) is really equivalent to (28-6); it is merely expressed in terms of the more convenient quantities i and $d\boldsymbol{\ell}$ rather than dq and $\mathbf{v}$. Since a current is a motion of charge, it is reasonable that there is a magnetic force exerted on a current-carrying wire.

In the special case of a uniform $\mathbf{B}$ field, equation (28-7) may be integrated to give the magnetic force exerted on the entire wire. The result is

$$\mathbf{F} = i\boldsymbol{\ell} \times \mathbf{B} \qquad \text{(uniform } \mathbf{B} \text{ field only).} \tag{28-8}$$

Here $\boldsymbol{\ell} = \int d\boldsymbol{\ell}$ is the vector from one end of the wire (the "current input" end) to the other (the "current output" end).

>>> Example 4. A 5.0 cm length of wire carries a current of 3.0 A. There is a uniform $\mathbf{B}$ field of magnitude 10^{-3} T whose direction is shown in the diagram. Calculate the magnetic force exerted on the wire.

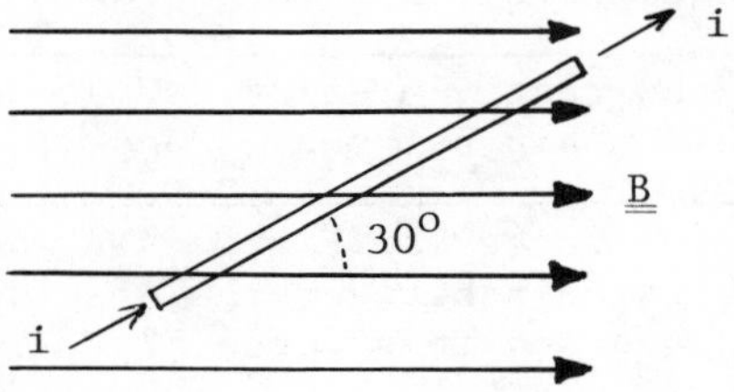

$$d\mathbf{F} = i\ d\boldsymbol{\ell} \times \mathbf{B}$$

Since $\mathbf{B}$ is uniform this may be integrated to give

$$\mathbf{F} = i\boldsymbol{\ell} \times \mathbf{B}$$

$$|\mathbf{F}| = i|\boldsymbol{\ell}|\,|\mathbf{B}| \sin\theta$$

where θ (= 30°) is the angle between $\boldsymbol{\ell}$ and $\mathbf{B}$.

$$|\mathbf{F}| = (3.0\ \text{A})\cdot(5.0 \times 10^{-2}\ \text{m})\cdot(10^{-3}\ \text{T})\cdot(0.5)$$

$$|\mathbf{F}| = 7.5 \times 10^{-5}\ \text{N} \quad .$$

The direction of this force is into the page. <<<

---

*The result, $dq\ \mathbf{v} = i\ d\boldsymbol{\ell}$, is true regardless of the sign of dq. If dq were negative, then $\mathbf{v} = -\ d\boldsymbol{\ell}/dt$ (since $d\boldsymbol{\ell}$ is in the direction of the current) and $i = -\ dq/dt$ (since dq is negative). These two sign changes cancel each other out.

[Not only does a magnetic field exert a force $\underline{F} = i\, d\underline{\ell} \times \underline{B}$ on a current element, it also exerts a torque $\underline{\tau} = \underline{\mu} \times \underline{B}$ on a magnetic dipole $\underline{\mu}$. A magnetic dipole consists of a current loop I of area $\bar{A}$. The magnetic dipole moment, $\underline{\mu}$, is of magnitude $|\underline{\mu}| = IA$ and is directed perpendicularly to the plane of the loop. See section 32-3]

28-5 Programmed Problems

1. As in the case of forces such as $G(m_1m_2/R^2)$, $-kx$, qE, $(1/4\pi\varepsilon_o)(q_1q_2/r^2)$, etc. each force frequently implies certain consequences. For instance $-kx$ resulted in oscillatory motion. $G(m_1m_2/R^2)$ resulted in conservation of angular momentum, etc. Now we have in this chapter still another force: $$\underline{F} = q\underline{v} \times \underline{B}\ .$$ Let us discuss some consequences of this force. From what you know about cross-products describe the spatial relationship between $\underline{F}$ and $\underline{v}$ required by this force law.	$\underline{F}$ and $\underline{v}$ are always perpendicular.
2. We know that $\underline{v} = d\underline{s}/dt$. What is the spatial relationship between $d\underline{s}$ and $\underline{v}$?	Parallel.
3. For the force law $\underline{F} = q\underline{v} \times \underline{B}$ what is the spatial relationship between the displacement $d\underline{s}$ of a charge q and this force $\underline{F}$?	$\underline{F} \perp \underline{v}$, frame 1. $\underline{v} \parallel d\underline{s}$, frame 2. Therefore $\underline{F} \perp d\underline{s}$.
4. Now if we take this last result and ask how much work is done by $\underline{F}$ on a charge q as q is displaced an amount $d\underline{s}$ we would say $$dW = \underline{F}\cdot d\underline{s}\ .$$ In view of the previous answer, what is the value of dW?	$dW = 0$. $\underline{F}\cdot d\underline{s} = \lvert\underline{F}\rvert\,\lvert d\underline{s}\rvert \cos\theta$ but $\cos\theta = 0$ for $\theta = 90^o$ and this θ is <u>always</u> 90° since $\underline{F} \perp d\underline{s}$ for this magnetic force.
5. Now there is a theorem which states that Work = Change in Something . What is that something?	Kinetic energy $\frac{1}{2}\ mv^2$
6. But we just said that the force $\underline{F} = q\underline{v} \times \underline{B}$ does no work on a charge q. What does the theorem say will happen to the speed of q?	It will not change. $W = 0,\quad \Delta K = 0.$ $\frac{1}{2}\ mv^2$ = constant, v = constant.

7\. Consider the following:
1. A force acts on a particle of mass m and charge q. The force is $\mathbf{F} = q\mathbf{v} \times \mathbf{B}$.
2. The force $\mathbf{F}$ does no work on the particle so its speed is constant.
3. The force is perpendicular to the displacement.

Will the particle be accelerated?

Yes. If you missed this question you shouldn't. When a force acts on a particle it is accelerated. Newton's second law!!

8\. How is it possible that the particle can be accelerated even though its speed remains constant?

The direction of its velocity will change while its speed remains constant.

9\. To summarize: When a force $\mathbf{F} = q\mathbf{v} \times \mathbf{B}$ on a charged particle,
1. The force is perpendicular to the displacement.
2. The particle is accelerated.
3. The particle's speed is constant.

If we further stipulate that $\mathbf{v}$ has no component in the direction of $\mathbf{B}$, i.e., $\mathbf{v} \perp \mathbf{B}$, then these facts characterize a particular kind of motion. What is that motion?

Circular motion.

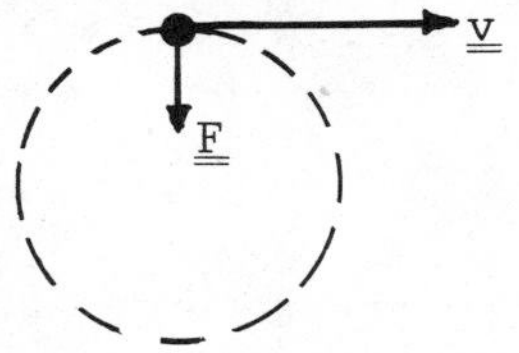

$\mathbf{B}$ is considered to be coming out of the page for a positive charge q.

10\. This will be an example of a cyclotron problem. We will assume that the B field has a value equal to the magnitude of the earth's field (0.4×10^{-4} T) and that the cyclotron radius R is 10 km.

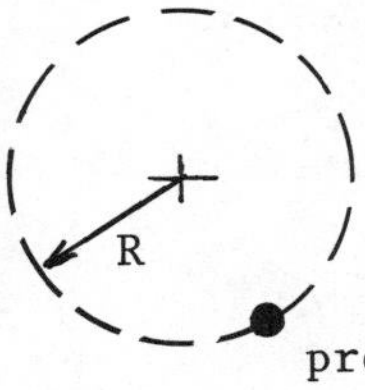

proton

The figure shows a proton in the cyclotron orbit. $\mathbf{B}$ is directed out of the plane of the paper.

On the diagram, show the direction of the force ($\mathbf{F} = q\mathbf{v} \times \mathbf{B}$) necessary to make the proton go in the circle of radius R. Describe the velocity direction.

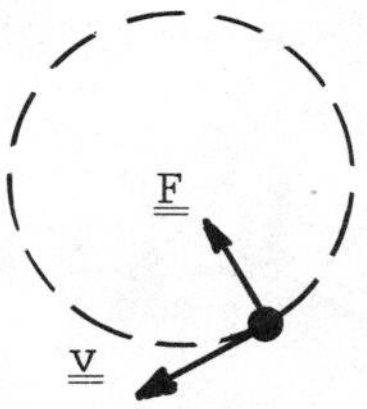

$\mathbf{F} = q\mathbf{v} \times \mathbf{B}$.

$\mathbf{v}$ must be directed as shown so that $q\mathbf{v} \times \mathbf{B}$ is directed toward the center of the circle.

Note that an electron (negative charge) would have to circulate in the opposite sense.

11\. The magnitude of the force on the proton is

$|\mathbf{F}| =$ ________.

$|\mathbf{F}| = qvB$

because the angle between $\mathbf{v}$ and $\mathbf{B}$ is 90°, and $\sin 90^\circ = 1$.

12. We have uniform circular motion. The special form of Newton's law for this case is

$|\underline{\underline{F}}|$ = __________ .

$|\underline{\underline{F}}| = mv^2/R$

where m is the proton mass. Recall that for circular motion the acceleration is v^2/R, directed toward the center of the circle.

---

13. From frames 11 and 12 we have

$$qvB = mv^2/R$$

so

$$v = qBR/m \quad .$$

The proton charge q is 1.60×10^{-19} C, and its mass m is 1.67×10^{-27} kg. What proton speed is required for this orbit?

v = __________ .

Also, what is the kinetic energy of the proton in MeV?

K = _______ MeV .

$v = 3.84 \times 10^7$ m/s .

$K = \frac{1}{2} mv^2$

$= 1.23 \times 10^{-12}$ J .

$1 \text{ J} = 6.24 \times 10^{12}$ MeV .

K = (1.23)(6.24) MeV

= 7.7 MeV .

This is not much energy for such a big radius accelerator, but we've used a very small B field.

---

14. Show that a magnetic field can be used to focus a beam of charged particles. This is in fact the method used in focusing an electron beam in electron microscopes.

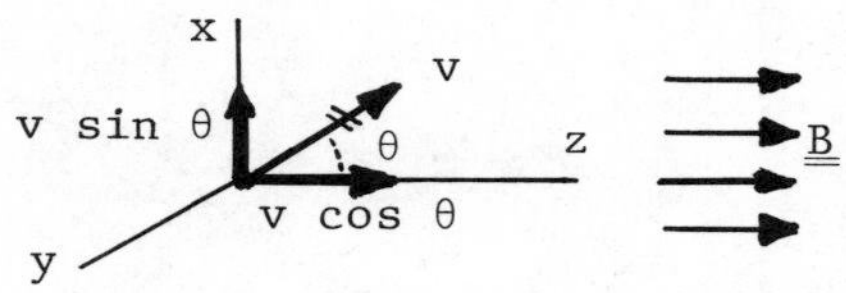

A source of charged particles is situated at the origin. The $\underline{\underline{B}}$ field is along the z-axis. We show a charged particle emerging at an angle θ; its velocity components are $v_{\parallel} = v_z = v \cos\theta$ and $v_{\perp} = v_x = v \sin\theta$ as shown. We seek to find the position at which the particle will again cross the z-axis.

Describe qualitatively the effect of $\underline{\underline{v}}$ due to the interaction of the charge with $\underline{\underline{B}}$. Remember Newton's second law.

$q\underline{\underline{v}}_{\parallel} \times \underline{\underline{B}} = 0$.

Therefore $\underline{\underline{v}}_{\parallel}$ = constant. Also,

$q\underline{\underline{v}}_{\perp} \times \underline{\underline{B}} = qv_{\perp}B$.

This will rotate $\underline{\underline{v}}_{\perp}$ in a circle. The magnitude of $\underline{\underline{v}}_{\perp}$ will not change.

---

15. The motion of the particle can be considered to consist of two components:
1. Motion in the _________ direction with a constant speed of _________, and
2. ___________motion in the _____ plane with a constant speed of __________.

1. z, $v_{\parallel} = v \cos\theta$.

2. uniform circular, x-y, $v_{\perp} = v \cos\theta$.

16. This combination results in uniform helical motion.

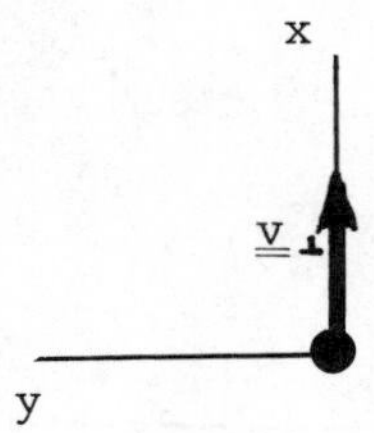

Imagine looking directly into the beam with the $\underline{\underline{B}}$ field pointing toward you and $\underline{\underline{v}}_\perp$ initially as shown. For this positive charge, draw $\underline{\underline{F}}$ to represent $\underline{\underline{F}} = q\underline{\underline{v}}_\perp \times \underline{\underline{B}}$. ($|\underline{\underline{F}}| = qv_\perp B$).

Note that $\underline{\underline{v}}_\perp$ is initially in the x direction as shown in frame 14 and has constant magnitude v sin θ.

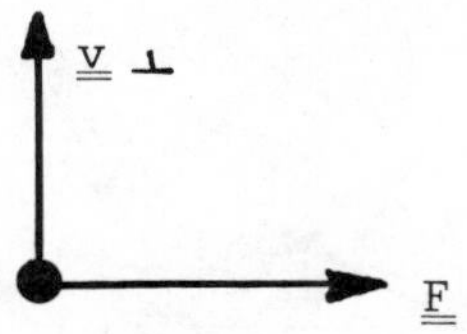

Right hand rule. $\underline{\underline{F}}$ is initially in the -y direction.

17.

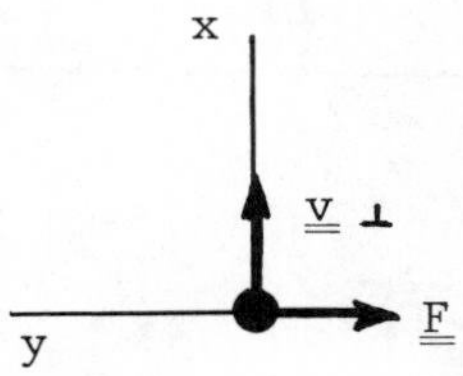

Now we are looking at the motion in the x-y plane. Using a dotted line draw the motion as a result of the force $\underline{\underline{F}}$.

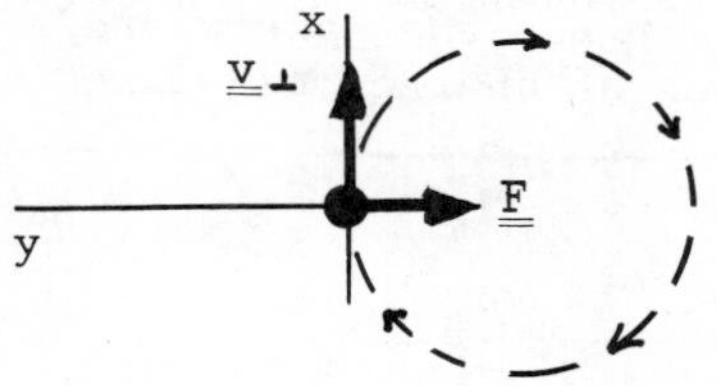

A circle.

---

18. From frame 16 we have that $F = qv_\perp B$. What will be the radius of this circular aspect of the motion?

$$F = m\frac{v_\perp^2}{R}.$$

$$R = \_\_\_\_\_\_\_\_ .$$

Express in terms of m, $v_\perp$, q and B.

$$F = \frac{mv_\perp^2}{R}.$$

Note that the acceleration is $v_\perp^2/R$. Only the component $v_\perp$ of the velocity in the plane of the circular motion is used in this formula.

$$qv_\perp B = \frac{mv_\perp^2}{R}$$

$$R = \frac{mv_\perp}{qB}.$$

---

19. Looking at the answer to frame 17, we note that the particle will cross the z axis again after one revolution. The circular orbit distance is $2\pi R$. How long will this single revolution take?

$$T = \_\_\_\_\_\_\_\_ .$$

Express in terms of m, $v_\perp$, q and B.

$$2\pi R = v_\perp T$$

Using the answer to frame 17 for R,

$$T = \frac{2\pi m}{qB}.$$

Note that this time T happens to be independent of $v_\perp$.

20.

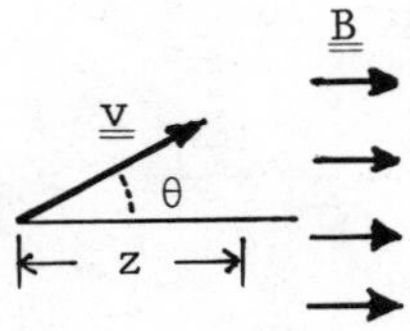

Remembering that while the particle has been rotating through one revolution, it has also been moving along the z axis; calculate z.

z = __________ .

Express in terms of v, θ, m, q and B.

$z = v_{\parallel} T$

$z = v_{\parallel} \frac{2\pi m}{qB}$

$z = (v \cos \theta) \frac{2\pi m}{qB}$.

Our answer $z = (v \cos \theta) \frac{2\pi m}{qB}$ for small angles ($\cos \theta \cong 1$) is $z = \frac{v2\pi m}{qB}$. This is the position at which the charged particles would be focused. That is, all particles leaving the origin at different but small angles, would be focused at $z = v2\pi m/qB$. This is important for the case where z is on a television screen and the clarity depends upon the number of charged particles striking the screen over a small area.

Chapter 29: AMPERE'S LAW AND BIOT-SAVART LAW

REVIEW AND PREVIEW

It turns out that the sources (causes) of the $\mathbf{B}$ field are moving charges, the most important example of this being a current. In this chapter you will study two ways of calculating the $\mathbf{B}$ field due to a given distribution of currents. The first method, called Ampere's law involves an integral of $\mathbf{B}$ along a closed path. Although perfectly general, like Gauss's law, it proves to be useful only for situations which possess enough symmetry. The second method you will study is called the Biot-Savart law. This law permits you to calculate $\mathbf{B}$ directly from its sources.

GOALS AND GUIDELINES

In this chapter there are five major goals:

1. Before learning Ampere's law it is very important that you understand a certain sign convention (Section 29-1). This sign convention relates an integral around a closed path to an integral over the surface bounded by the path. This sign convention will also be used in later chapters.
2. Learning the formula (Eq. 29-3) for Ampere's law and clearly understanding the meaning of the symbols in it.
3. Learning how to apply Ampere's law to problems (Section 29-4, Example 1).
4. Learning the formula (Eq. 29-4) for the Biot-Savart law and clearly understanding the meaning of the symbols in it.
5. Learning how to apply the Biot-Savart law to problems (Example 2).

You will find it useful to compare and contrast the technique of using Ampere's law for calculating $\mathbf{B}$ with the Gauss's law calculations for $\mathbf{E}$ (Chapter 23, Examples 1 and 2). It is also useful to compare the technique of using the Biot-Savart law for calculating $\mathbf{B}$ with the analogous $\mathbf{E}$ field calculation (Chapter 22, Example 4).

29-1 Sign Convention for Integral Laws

Before proceeding any further, it will be necessary to discuss a certain sign convention. Consider a closed path,* L. As shown in Figure 29-1, this path can be divided into infinitesimal segments $d\boldsymbol{\ell}$. Now consider an area, S, bounded by this path. Since it has a boundary, S is an open surface. This surface can be divided into vectorial surface area elements $d\mathbf{S}$, as shown in the figure. We will be concerned with certain "integral laws" of the form

$$\oint \mathbf{A}_1 \cdot d\boldsymbol{\ell} = \int \mathbf{A}_2 \cdot d\mathbf{S} \quad . \tag{29-1}$$

Here $\mathbf{A}_1$, $\mathbf{A}_2$ are two vector fields. The integral $\oint \mathbf{A}_1 \cdot d\boldsymbol{\ell}$ is taken around the path L (the small circle on the integral sign reminds us that we are dealing with a closed path), the integral $\int \mathbf{A}_2 \cdot d\mathbf{S}$ is taken over the area S bounded by this path.

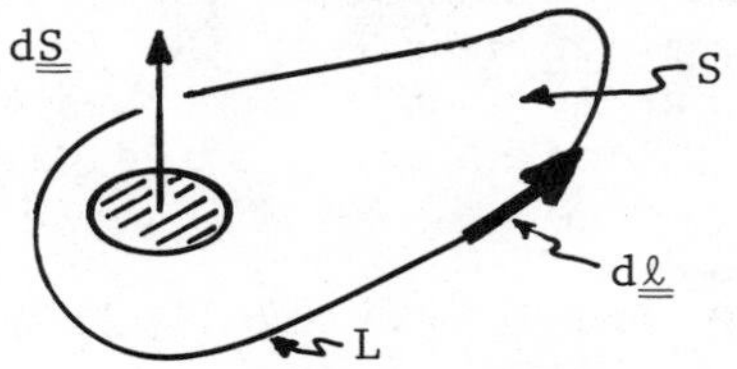

Figure 29-1. A closed path L is divided into infinitesimal segments $d\boldsymbol{\ell}$. Area S, bounded by the path L, is divided into vectorial surface area elements $d\mathbf{S}$.

*A closed path is one whose two endpoints coincide.

There are two things which are ambiguous regarding a law of the form (29-1):

(a) What sense do we choose for the $d\ell$ vector? $d\ell$ is of course tangent to the path L; however, there are two possible senses for this tangent. Reversing this choice will change the sign of $\oint A_1 \cdot d\ell$.

(b) What sense do we choose for the dS vector? dS is of course normal to the area S; however there are two possible senses for this normal.* Reversing this choice will change the sign of $\oint A_2 \cdot dS$.

Referring to Figure 29-1, we now state the sign convention:

> If the $d\ell$'s go (generally) counterclockwise around the path L, we choose for dS that normal which is (generally) out of the page. If the $d\ell$'s go (generally) clockwise around the path L, we choose for dS that normal which is (generally) into the page. (29-2a)

This convention can be summarized by a "right hand rule":

> If the fingers of the right hand go around the path L in the (general) direction of the $d\ell$'s, then the thumb indicates the (general) direction of dS. (29-2b)

The sign convention (29-2) does not tell us which $d\ell$ (or dS) to choose. Rather it relates the choice of $d\ell$ to the choice of dS.

29-2 Cause of the B Field

A moving charge experiences a force due to a B field. It turns out that a moving charge is also a cause of a B field. In practice, the most important example of this is a current. Thus we want to consider the problem: What is the B field due to a given current?

29-3 Ampere's Law

Lines of a B field tend to form loops about currents. The precise statement of this idea is called Ampere's law:

$$\oint B \cdot d\ell = \mu_o i \quad . \tag{29-3a}$$

On the left hand side, the integral $\oint B \cdot d\ell$ is taken over any closed path L. On the right hand side, the term "i" is the net current crossing an area S which is bounded by the path L. The sign convention for i follows the right hand rule (29-2b). For example, in Figure 29-2 we arbitrarily choose the $d\ell$'s to go around the path L in a counterclockwise direction. Therefore currents which come out of the page are considered positive, those going into the page are considered negative. In this example, $i = 5\text{ A} - 2\text{ A} = +3\text{ A}$. The 7 A current does not cross the surface S and therefore is not to be included as part of the current "i". (This does not mean that the 7 A current has no effect on B along the path L. It does affect B, but in such a manner as to not change the integral $\oint B \cdot d\ell$ along L.) The proportionality constant μ_o has the value $\mu_o = 4\pi \times 10^{-7}$ Wb/A·m.**

---

*Note that S is an open surface; hence there is no "inward" or "outward" normal.

**The constant μ_o is called the permeability constant.

Since the current crossing an area S is given by $i = \int \mathbf{j} \cdot d\mathbf{S}$ where $\mathbf{j}$ is the current density, Ampere's law may be written as

$$\oint \mathbf{B} \cdot d\boldsymbol{\ell} = \int \mathbf{j} \cdot d\mathbf{S} \quad . \qquad (29\text{-}3b)$$

This is now exactly in the form of equation (29-1). The sign convention (29-2) applies regarding the senses for the $d\boldsymbol{\ell}$ and $d\mathbf{S}$ vectors. The form (29-3a) of Ampere's law is more useful when the thicknesses of the wires carrying the currents are negligible. When these thicknesses are not negligible (e.g. in the interior of a wire) the form (29-3b) is more useful.

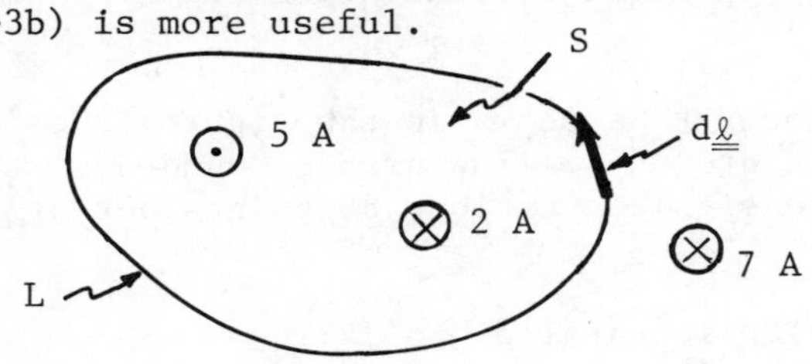

Figure 29-2. The 5 ampere current comes out of the page; the 2 and 7 ampere currents go into the page. The net current crossing the area S is $i = +3$ ampere.

In Chapter 35 we shall have to modify Ampere's law. This modification involves certain effects of time varying electric fields. In "magneto<u>static</u>" problems (steady currents), this modification is unimportant.

29-4 Application of Ampere's Law to Problems

A typical Ampere's law problem involves finding the $\mathbf{B}$ field at some point P due to a given distribution of current. The essential steps to be followed are:

(1) Make certain assumptions concerning $\mathbf{B}$ based on the symmetry of the problem. These are
    (a) an assumption about the direction of $\mathbf{B}$, and
    (b) an assumption about what variables the magnitude of $\mathbf{B}$ may depend upon.

(2) Choose an appropriate closed path L to take advantage of the symmetry of the particular problem. Ampere's law will be applied to this path.

(3) Using the assumptions (1), somehow remove "B" from under the integral sign in the left hand side of Ampere's law. This is the crucial step in the procedure.

(4) Evaluate the net current i which crosses an area S bounded by the path L (using the sign convention to determine the correct sign for this current). Substitute this into the right hand side of Ampere's law and solve for B.

These steps will become clearer in the following example.

>>> Example 1. A hollow cylindrical wire (inner radius "a", outer radius "b") carries a current I, uniformly distributed over its cross section. Calculate $\mathbf{B}$ at some point P a distance R ($a < R < b$) away from the axis.

Referring to Figure 29-3, we follow the steps outlined above.

(1) Assume (because of symmetry) that

(a) the direction of $\mathbf{B}$ is circular around the axis, the sense of these circles is arbitrarily taken to be counterclockwise in Figure 29-3a; and

(b) the magnitude of $\mathbf{B}$ depends only upon the distance from the axis.

(2) Path L: a circle of radius R passing through P as shown in the figure. We arbitrarily choose the $d\boldsymbol{\ell}$'s to go counterclockwise around L. The area S bounded by L is taken to be a circle of radius R; following the sign convention, $d\mathbf{S}$ points out of the page in Figure 29-3a.

(3)
$$\oint \mathbf{B}\cdot d\boldsymbol{\ell} = \oint B\, d\ell \qquad \text{(by (a) } \mathbf{B} \text{ is parallel to } d\boldsymbol{\ell})$$
$$= B_{\text{at P}} \oint d\ell \qquad \text{(by (b))}$$
$$= B_{\text{at P}}(2\pi R) \quad .$$

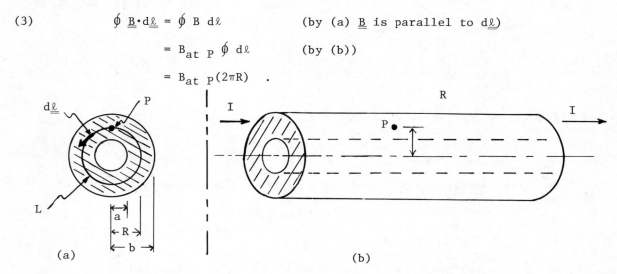

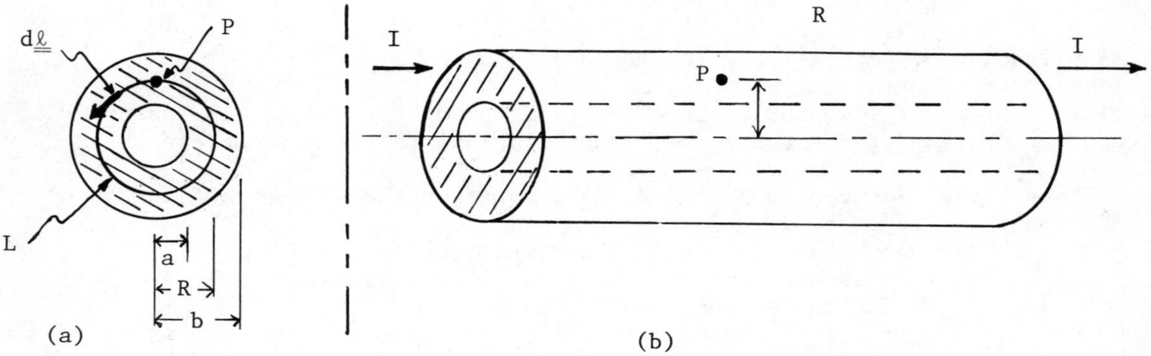

Figure 29-3. A hollow wire carrying a current I. Point P lies within the conductor a distance R ($a < R < b$) from the axis. Path L is a counterclockwise circle as shown in the left view (a).

(4) The magnitude of the current density ($a < r < b$) is $j = I/[\pi(b^2 - a^2)]$, the direction of $\mathbf{j}$ being into the page in Figure 29-3a. For $r < a$, $\mathbf{j} = 0$. The current "i" crossing the area S is then

$$i = \int_{r=0}^{r=R} \mathbf{j}\cdot d\mathbf{S} = \int_{a}^{R} \mathbf{j}\cdot d\mathbf{S} \qquad \text{(since } \mathbf{j} = 0 \text{ for } r < a)$$

$$= -\int_{r}^{R} j\, dS \qquad \text{(since the angle between } \mathbf{j} \text{ and } d\mathbf{S} \text{ is } 180^{\circ})$$

$$= - j \int_a^R dS \qquad \text{(since j is constant for } a < r < R)$$

$$= - j[\pi(R^2 - a^2)] = - I(R^2 - a^2)/(b^2 - a^2) \quad .$$

Substituting into Ampere's law,

$$\oint \underline{\underline{B}}\cdot d\underline{\underline{\ell}} = \mu_o i$$

$$B_{at\ P}(2\pi R) = - \mu_o I(R^2 - a^2)/(b^2 - a^2) \quad .$$

This yields a negative value for $B_{at\ P}$. The magnitude of $\underline{\underline{B}}$ at point P is therefore

$$|\underline{\underline{B}}|_{at\ P} = \frac{\mu_o I}{2\pi R}\frac{(R^2 - a^2)}{(b^2 - a^2)} \quad .$$

The negative value for $B_{at\ P}$ means that the actual direction of $\underline{\underline{B}}$ is opposite to that assumed in (a), i.e. $\underline{\underline{B}}$ is clockwise in Figure 29-3a.

Remarks:

1. In (a) we assumed the "wrong" sense (counterclockwise) for $\underline{\underline{B}}$. We do need to know that $\underline{\underline{B}}$ consists of circles around the axis of the wire, but we do not need to know the correct sense for $\underline{\underline{B}}$. It is obtained automatically as part of the result.
2. Step 4 could have been simplified by realizing that the fraction of the current I which crosses S is simply the ratio of the two annular areas, $(R^2 - a^2)/(b^2 - a^2)$.

<<<

29-5 Biot-Savart Law

Ampere's law is useful for problems having sufficient symmetry. For the more general case we need to know the magnetic field, $d\underline{\underline{B}}$, due to an element of current.

In the diagram below (Equation (29-4)), the wire carries a current i. The infinitesimal quantity $i\, d\underline{\underline{\ell}}$ is a current element ($d\underline{\underline{\ell}}$ is tangent to the wire; its sense is that of the current). Point P lies at a distance r from the current element. The Biot-Savart law gives the (static) magnetic field, $d\underline{\underline{B}}$, at point P due to this current element:

$$d\underline{\underline{B}}_{at\ P} = \frac{\mu_o}{4\pi}\frac{i\, d\underline{\underline{\ell}} \times \underline{\underline{e}}_r}{r^2} \qquad (29\text{-}4)$$

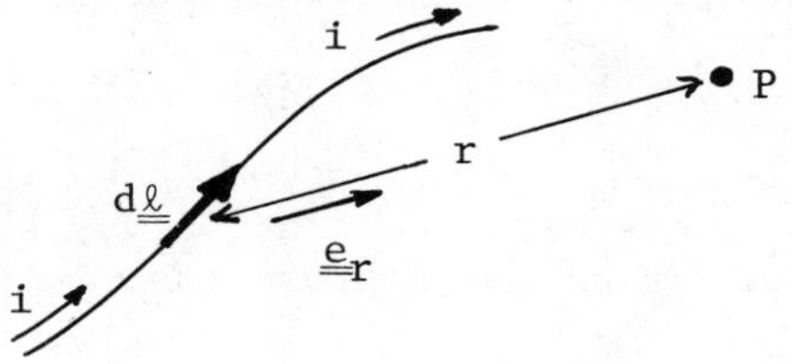

Here $\underline{\underline{e}}_r$ is a unit vector* directed along the line from the current element to point P. By way of analogy, the electric field $d\underline{\underline{E}}$ due to a charge element dq can be written as:

$$d\underline{\underline{E}}_{at\ P} = \frac{1}{4\pi\varepsilon_o}\frac{dq\ \underline{\underline{e}}_r}{r^2} \qquad (29\text{-}5)$$

*Recall that a unit vector has magnitude one. Unit vectors are dimensionless.

The following comparisons can be made between (29-4) and (29-5).

1. The proportionality constant: Both have the $1/4\pi$ factor, in the electric case ε_o is in the denominator while in the magnetic case μ_o is in the numerator.*
2. Dependence upon the distance r: Both are $1/r^2$ laws.
3. Dependence of the field upon its cause: Both are proportional to the quantity causing the field (dq in the electric case, i $d\underline{\underline{\ell}}$ in the magnetic case).
4. Direction of the field: In the electric case there is only one vector ($\underline{\underline{e}}_r$) involved, the direction of $d\underline{\underline{E}}$ is along this vector. In the magnetic case there are two vectors ($d\underline{\underline{\ell}}$ and $\underline{\underline{e}}_r$) involved, the direction of $d\underline{\underline{B}}$ is along the cross product of these two vectors. In both cases the unit vector $\underline{\underline{e}}_r$ is directed along the line <u>from</u> the cause of the field (either dq or i $d\underline{\underline{\ell}}$) <u>to</u> the point P.

Now that we know the field $d\underline{\underline{B}}$ due to a current element we can find the $\underline{\underline{B}}$ field due to a wire of any length. The total $\underline{\underline{B}}$ field is the vector sum (integral) of the various $d\underline{\underline{B}}$'s,

$$\underline{\underline{B}} = \int d\underline{\underline{B}} = \frac{\mu_o}{4\pi} \int \frac{i\, d\underline{\underline{\ell}} \times \underline{\underline{e}}_r}{r^2} \quad .$$

>>> Example 2. A thin infinitely long straight wire carries a current I. Find the $\underline{\underline{B}}$ field at some point P which is a distance R from the wire.

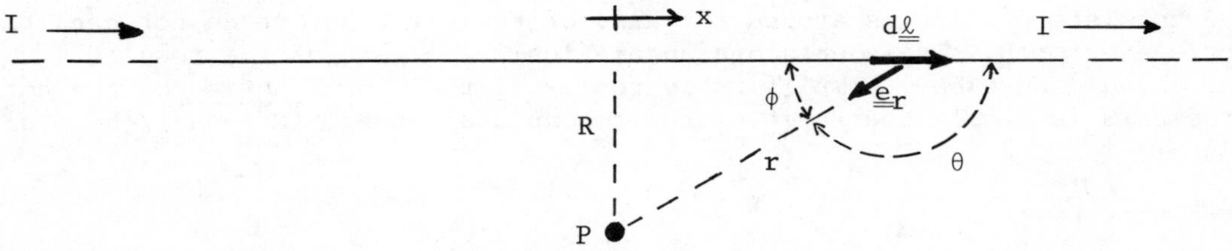

Figure 29-4. Segment of a long straight wire.

We divide the wire into infinitesimal segments $d\underline{\underline{\ell}}$ as shown in Figure 29-4. The direction of $d\underline{\underline{\ell}} \times \underline{\underline{e}}_r$ (and hence of $d\underline{\underline{B}}$) is into the page. Thus all the various $d\underline{\underline{B}}$'s have the <u>same</u> direction. Therefore in this case the magnitude of $\underline{\underline{B}}$ is simply the scalar sum (integral) of the magnitudes of the $d\underline{\underline{B}}$'s.

$$|\underline{\underline{B}}| = \int |d\underline{\underline{B}}| = \frac{\mu_o I}{4\pi} \int \frac{|d\underline{\underline{\ell}} \times \underline{\underline{e}}_r|}{r^2} = \frac{\mu_o I}{4\pi} \int \frac{d\ell \sin\theta}{r^2}$$

where θ is the angle between $d\underline{\underline{\ell}}$ and $\underline{\underline{e}}_r$ as shown. We are now faced with mixed variables ($d\ell$, r, θ). Let us try to express them in terms of the angle ϕ shown in the figure. We have

$$\sin\theta = \sin\phi \qquad (\text{since } \theta = 180^o - \phi)$$

$$r = R \csc\phi$$

$$d\ell = dx = d(R\cot\phi) = -R\csc^2\phi\, d\phi \quad .$$

*Note that in the right hand side (q/ε_o) of Gauss's law ε_o is in the denominator while in the right hand side ($\mu_o i$) of Ampere's law μ_o is in the numerator.

Substituting these expressions into the integral,

$$|\mathbf{B}| = \frac{\mu_o I}{4\pi} \int \frac{(-\,R \csc^2 \phi \; d\phi) \sin \phi}{(R \csc \phi)^2} = \frac{-\mu_o I}{4\pi R} \int_{\phi=180^o}^{\phi=0^o} \sin \phi \; d\phi$$

$$|\mathbf{B}| = \left[\frac{\mu_o I}{4\pi R}\right] \cos \phi \Bigg|_{\phi=180^o}^{\phi=0^o} = \left[\frac{\mu_o I}{4\pi R}\right][1 - (-\,1)] = \frac{\mu_o I}{2\pi R} \; .$$

The direction of $\mathbf{B}$ is into the page.

Remark: The correct limits for ϕ were obtained by referring to the figure: When the current element is located far to the left ($x \to -\infty$) then $\phi = 180^o$, when the current element is located far to the right ($x \to \infty$) then $\phi = 0^o$.

29-6 Programmed Problems

As in the case of Gauss's law in Chapter 23 the exercises here require the use of a variety of skills and concepts. Again it seems wise to consider these skills and concepts rather than specific problems.

Ampere's law

$$\oint \mathbf{B} \cdot d\boldsymbol{\ell} = \mu_o i$$

relates the line integral of $\mathbf{B}$ around a closed path to the current enclosed by that path.

1.

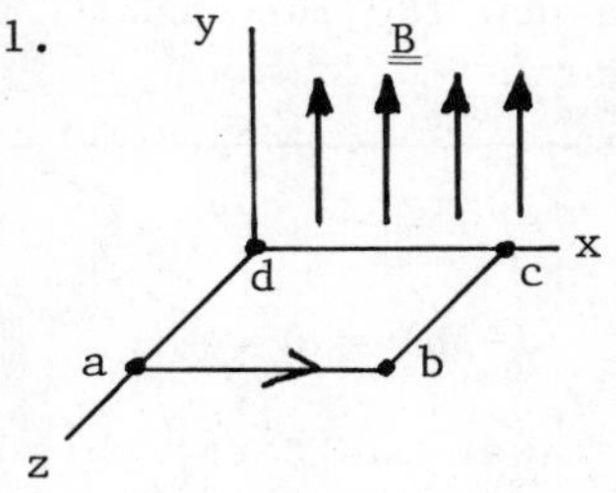

To the left is shown a closed path (here a square) labelled abcda. This closed path is in a plane perpendicular to a constant magnetic field $\mathbf{B}$.

Consider $\mathbf{B} \cdot d\boldsymbol{\ell}$ from a to b. What is the spatial relationship of $\mathbf{B}$ and $d\boldsymbol{\ell}$?

Perpendicular.

$\mathbf{B}$ is in the y direction and $d\boldsymbol{\ell}$ is in the x direction.

---

2. What is $\int \mathbf{B} \cdot d\boldsymbol{\ell}$ for each of the four segments ab, bc, cd, and da?

Zero for each. $\mathbf{B}$ is perpendicular to $d\boldsymbol{\ell}$.

---

3. We have then for the closed path abcda

$$\oint \mathbf{B} \cdot d\boldsymbol{\ell} = 0 \; .$$

The current i enclosed by the path abcd must be ____________.

Zero from Ampere's law:

$$\oint \mathbf{B} \cdot d\boldsymbol{\ell} = \mu_o i$$

Note that $\mathbf{B} \cdot d\boldsymbol{\ell}$ is zero even though $\mathbf{B}$ and $d\boldsymbol{\ell}$ are not zero.

4. 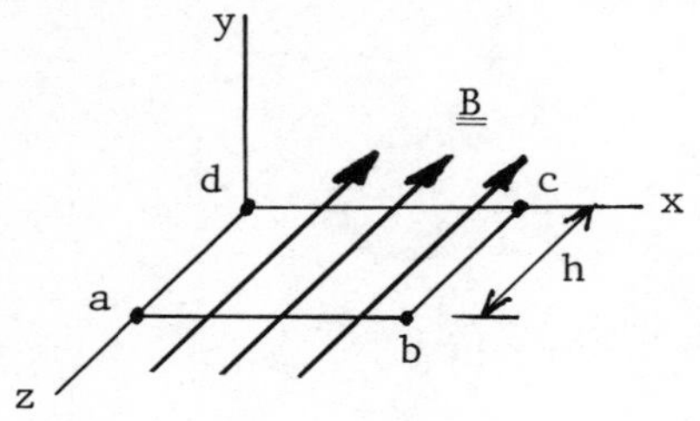

Here is a slightly different case. Again $\underline{\underline{B}}$ is constant but now $\underline{\underline{B}}$ is parallel to the plane of the path abcda.

$$\int_a^b \underline{\underline{B}} \cdot d\underline{\underline{\ell}} = \_\_\_\_\_\_\_\_\_\_ .$$

Zero since $\underline{\underline{B}}$ is perpendicular to $d\underline{\underline{\ell}}$.

---

5. For the segment bc

$$\int_b^c \underline{\underline{B}} \cdot d\underline{\underline{\ell}} \neq 0$$

but rather

$$\int_b^c \underline{\underline{B}} \cdot d\underline{\underline{\ell}} = Bh$$

Fill in the reasoning which gives this result.

$$\int_b^c \underline{\underline{B}} \cdot d\underline{\underline{\ell}} = \int B\, d\ell \cos(0)$$

$$= \int B\, d\ell \quad [\cos(0)=1]$$

$$= B \int d\ell \quad [B \text{ is const.}]$$

$$= Bh \quad [\int_b^c d\ell = h]$$

6.

$$\int_c^d \underline{\underline{B}} \cdot d\underline{\underline{\ell}} = \_\_\_\_\_\_\_\_\_\_ .$$

Zero since $\underline{\underline{B}}$ is perpendicular to $d\underline{\underline{\ell}}$.

7.

$$\int_d^a \underline{\underline{B}} \cdot d\underline{\underline{\ell}} = \_\_\_\_\_\_\_\_\_\_ .$$

-Bh. The calculation is similar to that in frame 5 except that the angle between $\underline{\underline{B}}$ and $d\underline{\underline{\ell}}$ is now 180° and that $\cos(180^o) = -1$.

---

8. Finally the complete line integral around the closed path abcda is

$$\int_a^b \underline{\underline{B}} \cdot d\underline{\underline{\ell}} + \int_b^c \underline{\underline{B}} \cdot d\underline{\underline{\ell}} + \int_c^d \underline{\underline{B}} \cdot d\underline{\underline{\ell}} + \int_d^a \underline{\underline{B}} \cdot d\underline{\underline{\ell}}$$

$$= \_\_\_\_\_\_\_\_\_\_ .$$

We have calculated all the pieces; what is the answer?

Zero.

$$0 + Bh + 0 + (-Bh) = 0 .$$

Again the answer was zero; this tells us that the rectangular area abcd is not crossed by any (net) current.

---

9. 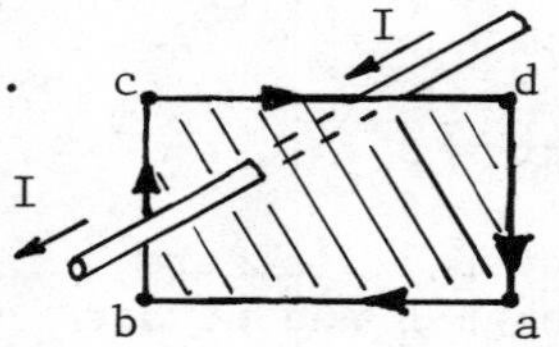

$i = \_\_\_\_\_\_\_\_$.

Here we have a closed path again but this time the area bounded by the path <u>is</u> pierced by a current. What is the value of the symbol "i" appearing in Ampere's law in this case? (The rectangular path is to be traversed in the sense abcda.)

-I. The negative sign comes from the right hand rule sign convention: the path is <u>clockwise</u>, therefore currents are considered positive if they go <u>into</u> the page.

10.

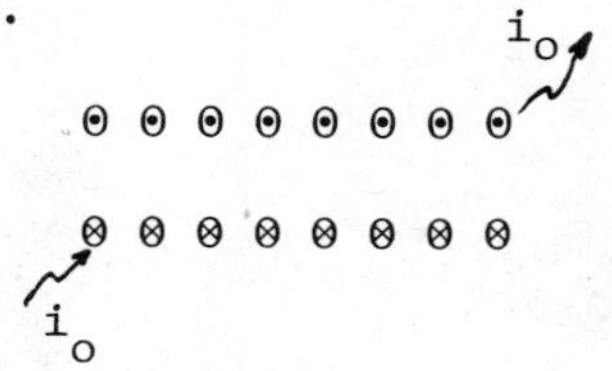

To the left is represented a cross-section of a solenoid. The current i_o enters at the lower left, goes through the solenoid, and emerges at the upper right as shown.

Sketch what you think the $\mathbf{B}$ field is inside this solenoid.

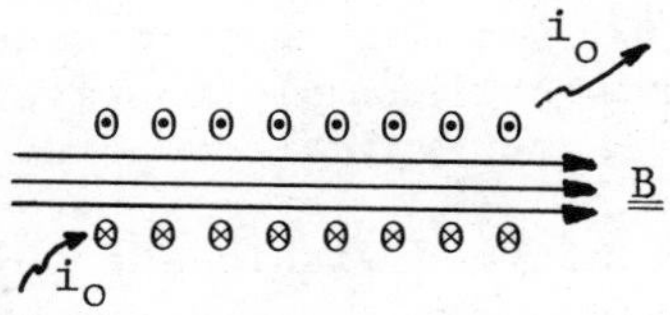

The $\mathbf{B}$ field within the solenoid is uniform (if the solenoid can be considered to be infinitely long). The correct sense of $\mathbf{B}$ as shown can be remembered by the rule: curl the fingers of your right hand around the solenoid following the current; the thumb then gives the correct direction of $\mathbf{B}$.

11.

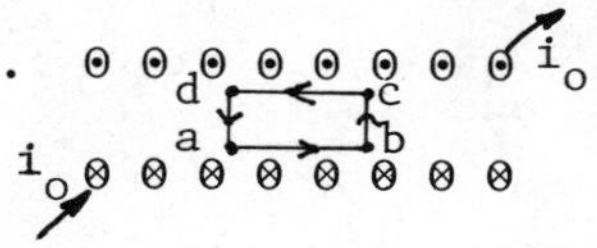

We will use the solenoid of the previous frame to correct a frequent error.

Student #1 claims that $\mathbf{B} = 0$ inside the solenoid because for the path shown

$$\oint \mathbf{B}\cdot d\boldsymbol{\ell} = \mu_o i = 0 \quad (\text{since } i = 0).$$

"Therefore" $\mathbf{B} = 0$. Student #2 disagrees with his reasoning since "$i = i_o \neq 0$". Student #3 (correctly) disagrees with both previous students. Why?

Her response to her classmates is as follows:

to #1: Just because $\oint \mathbf{B}\cdot d\boldsymbol{\ell} = 0$, one cannot conclude that $\mathbf{B} = 0$. Part of the line integral is positive and part is negative; the situation is similar to that of frame 4.

to #2: You are confusing the current i_o in the solenoid wire with the current (zero in this case) piercing the rectangular area bounded by the chosen path. It is this latter current which is denoted by the symbol "i" appearing in Ampere's law.

12.

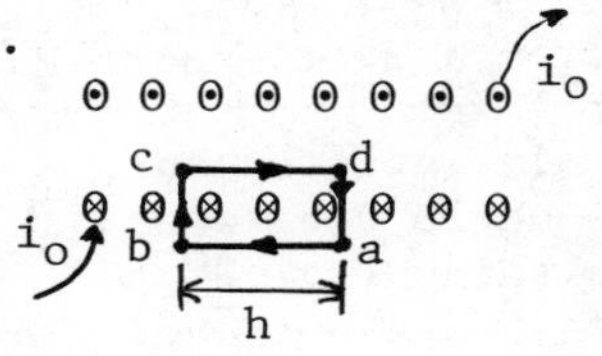

Suppose that there are n turns per unit length. Calculate B using Ampere's law. Use the rectangular path abcda shown.

$$\oint \mathbf{B}\cdot d\boldsymbol{\ell} = Bh$$

since only the segment cd contributes and since $\mathbf{B}$ is constant along this segment and parallel to $d\boldsymbol{\ell}$. The current i is

$$i = nhi_o$$

since nh wires cross the area.

Ampere's law, $\oint \mathbf{B}\cdot d\boldsymbol{\ell} = \mu_o i$, gives

$$B = \mu_o n i_o \quad .$$

Note that the length h cancelled out.

13. We will complete this sequence with a numerical example to give you a feeling for some typical numbers.

Calculate B within the solenoid if

$$n = 30{,}000 \text{ (turn)/meter}$$

$$i_0 = 0.1 \text{ ampere/(turn)}$$

Note: $\mu_0 = 4\pi \times 10^{-7}$ weber/ampere·meter.

$B = \mu_0 n i_0$

$= [4\pi \times 10^{-7} \frac{\text{Wb}}{\text{A}\cdot\text{m}}][3 \times 10^4 \frac{\text{turn}}{\text{m}}][0.1 \frac{\text{A}}{\text{turn}}]$

$= 3.8 \times 10^{-3}$ T .

Recall that T (tesla) = Wb/m^2. Note that "turn", although not strictly a unit, cancelled out.

Next we shall look at the Biot-Savart law. In Ampere's law (see above frames), the $d\mathbf{\ell}$'s had nothing to do with the wire which is carrying the current; rather the $d\mathbf{\ell}$'s constituted a fictitious path for calculating the line integral $\oint \mathbf{B}\cdot d\mathbf{\ell}$. In the Biot-Savart law, however, the $d\mathbf{\ell}$'s do have something to do with the wire which is carrying the current; in fact each $d\mathbf{\ell}$ is physically an element of that wire.

---

14. The Biot-Savart law is

$$d\mathbf{B} = \frac{\mu_0 i}{4\pi} \frac{d\mathbf{\ell} \times \mathbf{e_r}}{r^2}$$

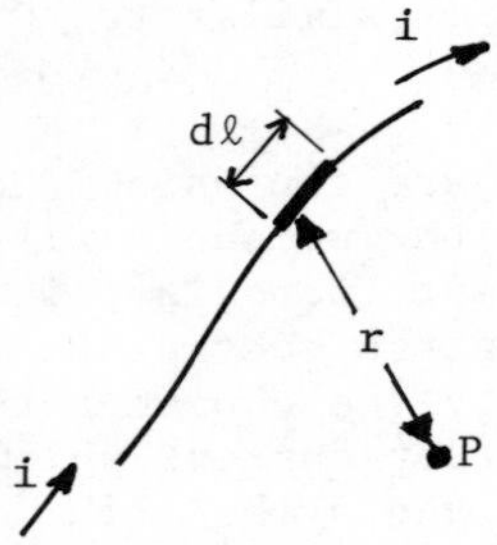

A wire carrying a current i is shown. We seek to calculate the magnetic field contribution $d\mathbf{B}$ at point P due to the current element i $d\mathbf{\ell}$ a distance r from P.

Indicate on the diagram the vectors $d\mathbf{B}$, $d\mathbf{\ell}$, and the unit vector $\mathbf{e_r}$.

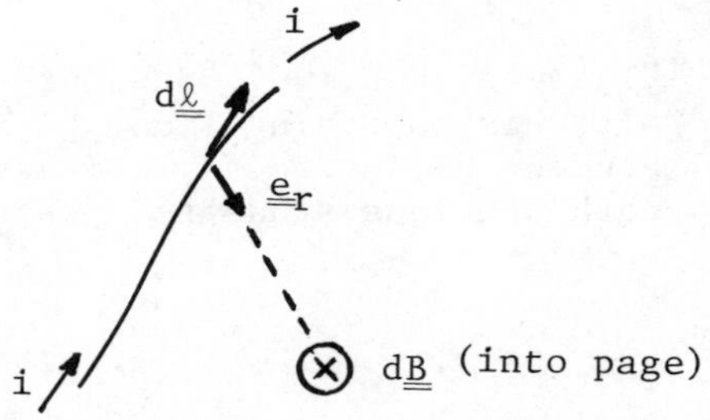

Note that $d\mathbf{\ell}$ is directed along the wire with the <u>same sense as the current</u> i; $\mathbf{e_r}$ is directed <u>from</u> i $d\mathbf{\ell}$ <u>to</u> point P.

---

15. The direction of $d\mathbf{B}$ is determined by the vector cross product $d\mathbf{\ell} \times \mathbf{e_r}$. Also the magnitude of $d\mathbf{B}$ is in part determined by the relative orientation of $d\mathbf{\ell}$ and $\mathbf{e_r}$:

$$|d\mathbf{B}| = \frac{\mu_0 i}{4\pi} \frac{|d\mathbf{\ell}||\mathbf{e_r}| \sin\theta}{r^2}$$

Is θ the same for <u>every</u> $d\mathbf{\ell}$ in frame 14?

No. This is a point of frequent difficulty. The angle θ between $d\mathbf{\ell}$ and $\mathbf{e_r}$ can be anything from 0° to 180°; nontheless $d\mathbf{B}$ is always perpendicular to both $d\mathbf{\ell}$ and $\mathbf{e_r}$. The angle θ involved in the above frame is shown below; its value is approximately 125°.

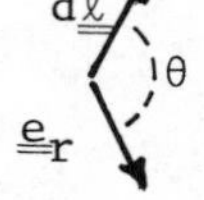

16. Using the Biot-Savart law to find B at point P, we must add (integrate) the dB's contributed by all $d\boldsymbol{\ell}$'s. We have shown three representative $d\boldsymbol{\ell}$'s.

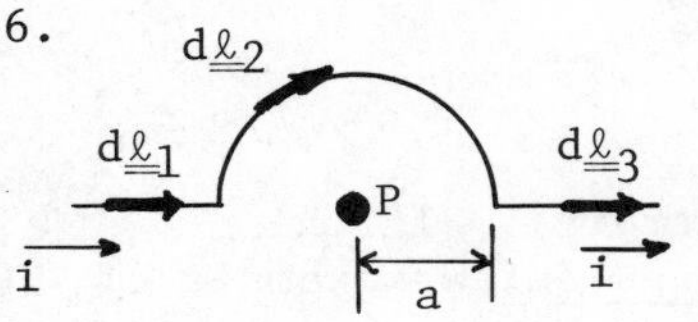

For each of the three $d\boldsymbol{\ell}$'s shown, sketch the corresponding unit vector $\mathbf{e}_r$.

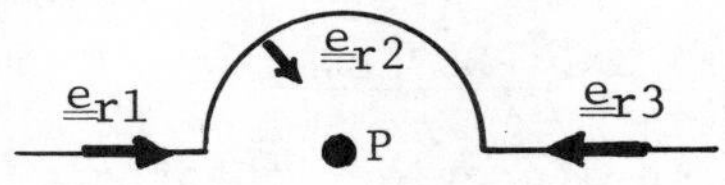

Note that $\mathbf{e}_r$ points from the current element to the point at which $d\mathbf{B}$ is to be evaluated.

---

17. What is the particular value of

a. $d\boldsymbol{\ell}_1 \times \mathbf{e}_r =$ ______

b. $d\boldsymbol{\ell}_2 \times \mathbf{e}_r =$ ______ .

Both are zero. In (a) the angle between $d\boldsymbol{\ell}$ and $\mathbf{e}_r$ is 0° while in (b) the angle is 180°. The cross product involves the sine of this angle.

---

18. In view of the above answer we need only consider the curved portion of the wire since the straight segments contribute zero to the $\mathbf{B}$ field at P.

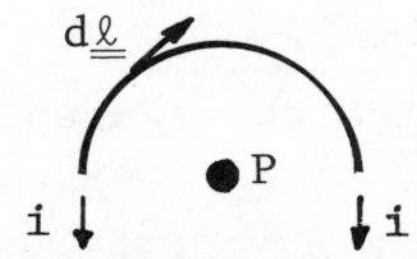

Indicate on the diagram $\mathbf{e}_r$, r and $d\mathbf{B}$ for the $d\boldsymbol{\ell}$ shown.

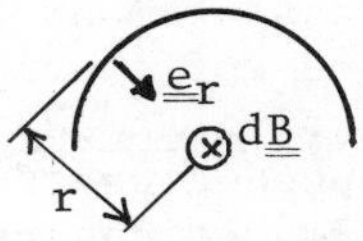

Note: $d\mathbf{B}$ is into the page at the point P.

---

19. The total $\mathbf{B}$ field will be the integral of all the above $d\mathbf{B}$'s. For each such $d\mathbf{B}$,

a. direction of $d\mathbf{B}$ is ________

b. angle between $d\boldsymbol{\ell}$ and $\mathbf{e}_r$ is ______

c. value of r is ________ .

a. into the page

b. 90°

c. a (the radius of the semicircular arc)

---

20. Justify each of the indicated steps below.

a. $B = \int |d\mathbf{B}|$

b. $B = \dfrac{\mu_0 i}{4\pi} \int \dfrac{|d\boldsymbol{\ell} \times \mathbf{e}_r|}{r^2}$

c. $B = \dfrac{\mu_0 i}{4\pi a^2} \int d\ell$

d. $B = \mu_0 i/4a$

a. all the $d\mathbf{B}$'s are parallel.

b. Biot-Savart law.

c. $|d\boldsymbol{\ell} \times \mathbf{e}_r| = |d\boldsymbol{\ell}||\mathbf{e}_r| \sin(90^\circ)$

$= d\ell$ (since $|\mathbf{e}_r| = 1$).

In the denominator, $r = a$.

d. $\int d\ell = \pi a$ (the arc length of the semi-circle).

Chapter 30: FARADAY'S LAW

REVIEW AND PREVIEW

In a previous chapter you studied seats of emf (e.g. a battery). In this chapter you will study a particular type of emf called an induced emf an example of which is the ordinary electric generator. The physical law you will study is called Faraday's law: this law relates the induced emf to a changing flux of $\underline{B}$.

GOALS AND GUIDELINES

In this chapter there are four major goals:

1. Understanding motional emf (Section 30-1). This arises when a conductor is moved through a $\underline{B}$ field. Although motional emf can be understood using previously studied concepts alone, it serves as an excellent introduction to induced emfs in general.
2. Learning the formula for induced emf (Eq. 30-4).
3. Learning and understanding Lenz's law (Section 30-3). This law is a qualitative statement from which you should be able to obtain the "sense" of any induced emf (here the "sense" is synonymous with the "polarity").
4. Learning Faraday's law in its integral form (Eq. 30-6). This law relates an induced electric field to a changing flux of $\underline{B}$. From the induced electric field one can calculate the induced emf.

The flux of $\underline{B}$, $\Phi_B = \int \underline{B}\cdot d\underline{S}$ is used in this chapter. You should review the concept of flux in general (Sections 23-1, 23-2).

30-1 Motional EMF

The concept of motional emf can best be understood by considering the following example. In Figure 30-1a, the metal rod can slide on the two metal tracks. The rod is pulled to the right (by some external agency) at a constant velocity $\underline{v}$. The entire apparatus is immersed in a uniform $\underline{B}$ field directed into the page.

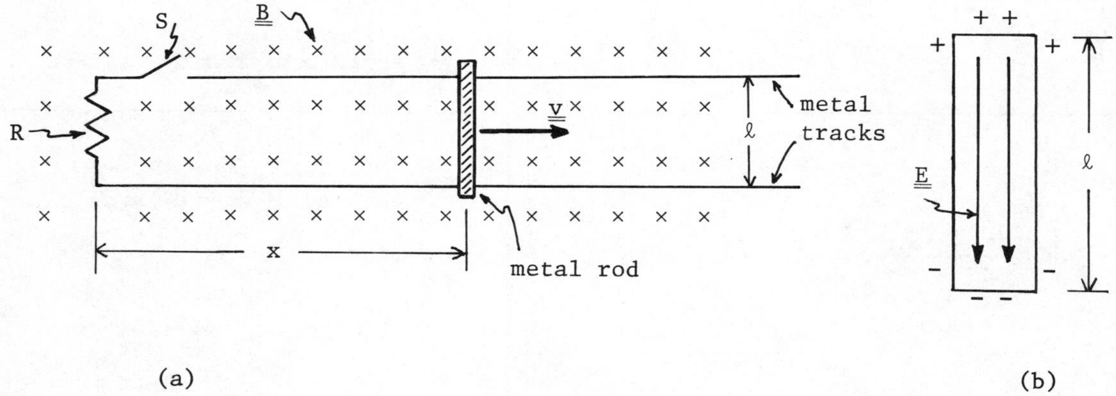

Figure 30-1. (a) The metal rod moves to the right with velocity $\underline{v}$. There is a uniform $\underline{B}$ field directed into the page. (b) The distribution of charge on the rod causes an electric field $\underline{E}$ to exist within the rod.

Because of the magnetic force ($\underline{\underline{F}}_{mag} = q\underline{\underline{v}} \times \underline{\underline{B}}$) exerted on them, the conduction electrons within the rod tend to move down toward the bottom of the rod. This results in a charge distribution as indicated in Figure 30-1b. The charge distribution causes an electric field to exist, the (general) direction of this $\underline{\underline{E}}$ field being from the top toward the bottom of the rod. This process continues until the Lorentz force exerted on the interior conduction electrons vanishes,* i.e. $q\underline{\underline{E}} + q\underline{\underline{v}} \times \underline{\underline{B}} = 0$. Then

$$\underline{\underline{E}} = - \underline{\underline{v}} \times \underline{\underline{B}} \qquad (30\text{-}1)$$

where $\underline{\underline{E}}$ is the electric field within the rod. Note that this electric field is uniform ($\underline{\underline{v}} \times \underline{\underline{B}}$ is a constant). There is then a potential difference $E\ell = vB\ell$ across the rod. As far as the rest of the circuit is concerned, the rod acts as a seat of emf $\mathcal{E} = vB\ell$. (For example, if switch S is closed a counterclockwise current $i = \mathcal{E}/R$ will exist.) This emf (since it arises because of the motion of the rod) is called a <u>motional emf</u>:

$$\mathcal{E} = B\ell v \quad . \qquad (30\text{-}2)$$

Equation (30-2) is a rather special case: the $\underline{\underline{B}}$ field must be uniform; the rod, velocity of the rod, and the $\underline{\underline{B}}$ field must all be mutually perpendicular. In the more general case we must deal with the emf $d\mathcal{E}$ generated across an infinitesimal segment $d\underline{\underline{\ell}}$ of a wire moving with velocity $\underline{\underline{v}}$. This is given by

$$d\mathcal{E} = (\underline{\underline{v}} \times \underline{\underline{B}}) \cdot d\underline{\underline{\ell}} \quad . \qquad (30\text{-}3)$$

To obtain the total emf one would integrate this: $\mathcal{E} = \int d\mathcal{E}$.

If the switch S in Figure 30-1a is closed, a current (called an <u>induced current</u>) will exist. The source of the resulting electrical power (i^2R) is the external agency which pulls the rod.

>>> Example 1. When switch S is closed in Figure 30-1a, the induced emf causes an induced current $i = \mathcal{E}/R$. Show that the (heat) power dissipated by the resistor equals the (mechanical) power supplied by the external agency which pulls the rod.

The induced current is $i = \mathcal{E}/R = B\ell v/R$.

(a) The power dissipated by the resistor is

$$P_{heat} = i^2R = (B\ell v/R)^2 R = B^2\ell^2v^2/R \quad .$$

(b) The magnetic force exerted on the rod is

$$\underline{\underline{F}}_{mag} = i\underline{\underline{\ell}} \times \underline{\underline{B}}$$

$$|\underline{\underline{F}}_{mag}| = i\ell B = (B\ell v/R)\ \ell B = B^2\ell^2 v/R \quad .$$

Since the bar moves at constant velocity, the external force $\underline{\underline{F}}_{ext}$ must balance the magnetic force $\underline{\underline{F}}_{mag}$. Thus

$$\underline{\underline{F}}_{ext} = - \underline{\underline{F}}_{mag}$$

$$|\underline{\underline{F}}_{ext}| = |\underline{\underline{F}}_{mag}| = B^2\ell^2 v/R \quad .$$

---

*The actual time that it takes to reach the equilibrium condition (30-1) is extremely small.

The power supplied by the external agency is

$$P_{ext} = \underline{\underline{F}}_{ext} \cdot \underline{\underline{v}} = |\underline{\underline{F}}_{ext}|\, v = (B^2 \ell^2 v/R)\, v$$

$$P_{ext} = B^2 \ell^2 v^2/R \quad .$$

The two powers are thus equal.

Remark: Note that the sense of the induced emf is such as to force the current to be directed <u>down</u> through the resistor and <u>up</u> through the rod. The magnetic force $\underline{\underline{F}}_{mag} = i\underline{\underline{\ell}} \times \underline{\underline{B}}$ is then directed to the <u>left</u> in Figure 30-1a. Thus $\underline{\underline{F}}_{mag}$ <u>opposes</u> $\underline{\underline{F}}_{ext}$ as it should. <<<

30-2 Induced EMF

Equation (30-2) may be rewritten in terms of the flux, Φ_B, of $\underline{\underline{B}}$ through the rectangular area bounded by the loop. Referring to Figure 30-1a, $\mathcal{E} = B\ell v = B\ell\, dx/dt = d(B\ell x)/dt$. Since $\underline{\underline{B}}$ is uniform and perpendicular to the plane of the loop, $\Phi_B = \int \underline{\underline{B}} \cdot d\underline{\underline{S}} = B(\ell x)$. Thus*

$$\mathcal{E} = d(\Phi_B)/dt \quad . \tag{30-4}$$

Equation (30-4) is completely general; it say that there is an emf $\mathcal{E}$ (called an <u>induced emf</u>) in a closed loop whenever there is a <u>changing</u> flux of $\underline{\underline{B}}$ through an area bounded by the loop. Motional emf is then merely a special case of an induced emf.

The flux of $\underline{\underline{B}}$ might change for many reasons. Some of the possibilities are:

(a) a change in the size of the loop (as in the above example of motional emf),
(b) a change in the magnitude of $\underline{\underline{B}}$,
(c) a change in the direction of $\underline{\underline{B}}$,
(d) a change in the orientation of the loop,
(e) a motion of the loop to a region of different $\underline{\underline{B}}$, etc.

The essential point is that no matter why the flux of $\underline{\underline{B}}$ is changing, the induced emf is given simply by $d(\Phi_B)/dt$. Case (d) above is illustrated in the following example.

>>> Example 2. A closed ten turn rectangular loop of wire (5 cm × 10 cm) rotates about an axis as shown with an angular velocity of 100 (radian)/s. The resistance of the wire is 5 Ω . There is a uniform $\underline{\underline{B}}$ field perpendicular to the axis of rotation; the magnitude of this field is 2×10^{-2} T . (a) What is the maximum magnitude of the induced current in the loop? (b) What is the orientation of the loop at the time of this maximum current?

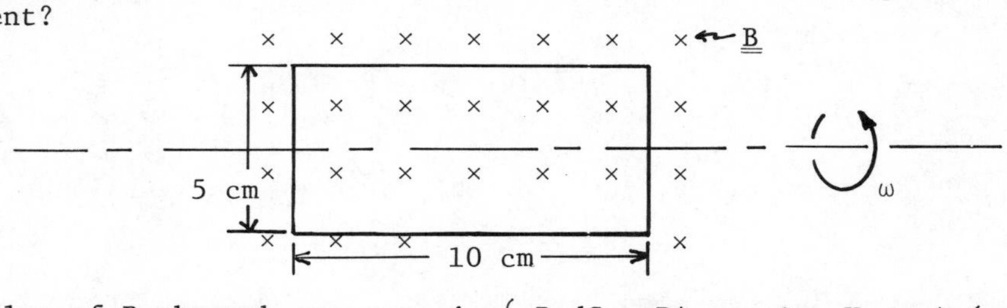

The flux of $\underline{\underline{B}}$ through <u>one</u> turn is $\int \underline{\underline{B}} \cdot d\underline{\underline{S}} = BA \cos\theta$. Here A (= 50 cm^2) is the area of one turn and θ is the angle between $\underline{\underline{B}}$ and the <u>normal</u> to the plane of the loop. The <u>flux of $\underline{\underline{B}}$ through</u> the N (= 10) turns is $\Phi_B = NBA \cos\theta$.

*The text inserts a minus sign in (30-4), the question of this sign is considered in the next section.

The induced emf is

$$\mathcal{E} = d(\Phi_B)/dt = - NBA \sin\theta \; d\theta/dt$$

$$= - NBA\omega \sin\theta$$

where $\omega = d\theta/dt$ is the angular velocity of the loop. The induced current is

$$i = \mathcal{E}/R = (- NBA\omega \sin\theta)/R \quad .$$

(a) The maximum magnitude of this current is therefore

$$|i|_{max} = NBA\omega/R$$

$$= (10)(2 \times 10^{-2}\ Wb/m^2)(50 \times 10^{-4}\ m^2)(10^2\ s^{-1})/(5\ \Omega)$$

$$= 2 \times 10^{-2}\ A \quad .$$

(b) From the result $i = -(NBA\omega \sin\theta)/R$, we see that the maximum magnitude of the current occurs when $\theta = \pm 90^o$, i.e. when the plane of the loop is parallel to $\underline{\underline{B}}$.

Remarks:

1. The minus sign in the expression obtained for "i" is of no significance here.
2. In problems involving coils of N turns, we sometimes write the induced emf as "$\mathcal{E} = N\, d(\Phi_B)/dt$". Here Φ_B is the flux of $\underline{\underline{B}}$ through <u>one</u> turn, $N\Phi_B$ is therefore the flux of $\underline{\underline{B}}$ through the entire N turns. Of course the geometry of the problem has to be such that the same flux of $\underline{\underline{B}}$ goes through each of the turns. <<<

30-3 Lenz's Law

We have so far disregarded the question of signs. For example, in computing the flux of $\underline{\underline{B}}$ ($\Phi_B = \int \underline{\underline{B}} \cdot d\underline{\underline{S}}$), we have not specified which <u>sense</u> to choose for the $d\underline{\underline{S}}$ vector. It is really unnecessary to do this here since there is no sign convention for the emf anyway (the fact that $\mathcal{E}$ might turn out positive or negative still does not tell us the sense of this emf). The correct sense of the induced emf can be obtained by <u>Lenz's law</u>. There are several ways to phrase this law, one of which is:

"The sense of the induced emf is such that the resulting induced current tends to oppose the cause of the emf." (30-5)

To illustrate Lenz's law, consider the example shown in Figure 30-1a. We first imagine that switch S is closed so that there will be an induced current. Now, what causes the induced emf? Answer: the fact that the flux of $\underline{\underline{B}}$ through the loop is <u>into</u> the page and is <u>increasing</u>. How can the induced current tend to oppose this? Answer: by creating its own $\underline{\underline{B}}$ field which is directed through the loop <u>out</u> of the page. A <u>counterclockwise</u> induced current will do this. Therefore the induced emf is such as to force the induced current counterclockwise around the loop, i.e. it is positive at the top of the rod and negative at the bottom.

As another example, suppose the rod in Figure 30-1a were moving to the left. What causes the induced emf? Answer: the fact that the flux of $\underline{\underline{B}}$ through the loop is <u>into</u> the page and is <u>decreasing</u>. How can the induced current tend to oppose this? Answer: by creating its own $\underline{\underline{B}}$ field which is directed through the loop <u>into</u> the page. A <u>clockwise</u> induced current will do this. Therefore the induced emf is such as to force the induced current clockwise around the loop, i.e. it is negative at the top of the rod and positive at the bottom.

Note that in the first example the induced current creates a $\underline{B}$ field which opposes the given flux while in the second example the induced current creates a $\underline{B}$ field which aids the given flux. The point is that the $\underline{B}$ field created by the induced current always tends to oppose the change in the given flux.

30-4 Faraday's Law

In the case of motional emf we can see the reason why the induced current exists: the magnetic force $\underline{F}_{mag} = q\underline{v} \times \underline{B}$ on the conduction electrons pushes them along the moving rod. In other cases, where there is no motion of the conductors (e.g. a time varying $\underline{B}$ field causing a change in the flux through the loop), another explanation is needed: an electric field is required to push the conduction electrons through the wire. This is the basis of Faraday's law: a time varying $\underline{B}$ field creates an electric field; this electric field tends to form loops around the lines of $\underline{B}$. The mathematical statement of Faraday's law is

$$\oint \underline{E}\cdot d\underline{\ell} = - d(\Phi_B)/dt \quad . \tag{30-6a}$$

Using the definition of Φ_B, another way of writing this law is

$$\oint \underline{E}\cdot d\underline{\ell} = - \frac{d}{dt} \left[\int \underline{B}\cdot d\underline{S} \right] \quad . \tag{30-6b}$$

This law is of the form discussed in Section 29-1 (under "sign convention"). The integral on the left hand side is taken around any (stationary) closed path, the integral on the right hand side is taken over a surface bounded by this path. The sign convention in Section 29-1 regarding the senses of $d\underline{\ell}$ and $d\underline{S}$ applies here. The minus sign in Faraday's law then has a definite meaning. This sign is really a mathematical way of invoking Lenz's law.

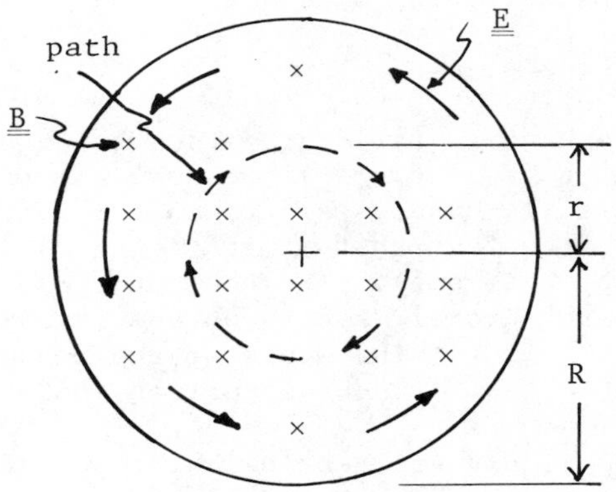

>>> Example 3. A uniform $\underline{B}$ field exists in a region $0 \leq r \leq R$ as shown. The field is into the page and is increasing with time. Derive an expression for the induced electric field at point P, a distance r ($\leq R$) from the center.

Based on the symmetry of this problem we assume that:

(a) The direction of the induced electric field $\underline{E}$ is that of circles, concentric with the center of the region of the $\underline{B}$ field. The sense of these circles is arbitrarily taken to be counterclockwise.

(b) The magnitude of $\underline{E}$ depends only upon r (and of course t), i.e. at a given time $|\underline{E}|$ is the same at all points located a distance r from the center.

We now choose to apply Faraday's law to a circular path of radius r; the sense of the $d\underline{\ell}$ vectors is arbitrarily taken to be clockwise around this path.

$$\oint \underline{E}\cdot d\underline{\ell} = - \oint E \, d\ell$$ (since the angle between $\underline{E}$ and $d\underline{\ell}$ is 180° by (a)).

$$= - E_{at\ P} \oint d\ell \qquad \text{(by (b))}$$

$$= - E_{at\ P}\, 2\pi R \quad .$$

This is the left hand side of Faraday's law. To evaluate the right hand side of this law we need $\int \mathbf{B}\cdot d\mathbf{S}$, the flux of $\mathbf{B}$ through the circular area of radius r. The sign convention tells us that the $d\mathbf{S}$ vectors point into the page (because the $d\boldsymbol{\ell}$'s are clockwise).

$$\oint \mathbf{B}\cdot d\mathbf{S} = \int B\ dS \qquad \text{(since } \mathbf{B} \text{ and } d\mathbf{S} \text{ are parallel)}$$

$$= B \int dS \qquad \text{(since } \mathbf{B} \text{ is uniform)}$$

$$= B\pi r^2 \quad .$$

Substituting into Faraday's law,

$$\oint \mathbf{E}\cdot d\boldsymbol{\ell} = - \frac{d}{dt}\left[\int \mathbf{B}\cdot d\mathbf{S}\right]$$

$$- E_{at\ P}\, 2\pi r = - \pi r^2\, dB/dt$$

$$E_{at\ P} = \tfrac{1}{2}\, r\, dB/dt \quad .$$

Since this is positive, the assumed sense of $\mathbf{E}$ is correct, i.e. $\mathbf{E}$ is counterclockwise.

<<<

30-5 Programmed Problems

1. 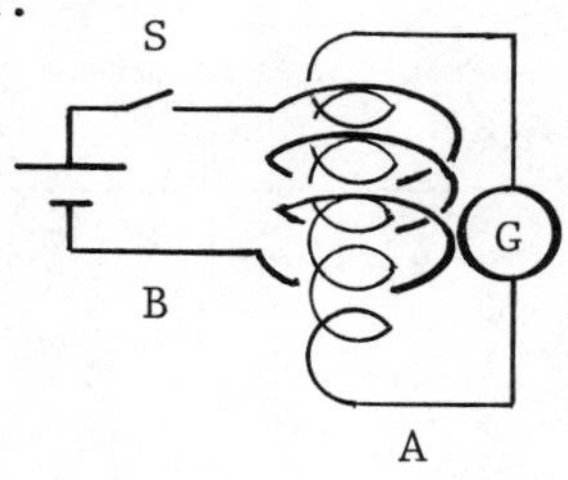 Faraday's famous experiments on induced currents consisted of one arrangement as shown to the left. Oersted had earlier shown that magnetic fields were associated with current-carrying wires. What was it that Faraday was trying to show with this experiment?	Faraday wanted to show just the opposite of what Oersted had done, i.e. he wanted to show that currents in wires could result from the interaction of those wires with magnetic fields.
2. The galvanometer G in the circuit above is the device Faraday used to detect an "induced" current in the wire A. What in the circuit is the source of the magnetic field with which he sought to induce the current in A?	The coil B with the battery supplying the current which produces the magnetic field in B.
3. Faraday found that the galvanometer deflected (indicated a current in A) only at the instant that the switch S was either opened or closed. How do we interpret that fact in terms of the ability of the magnetic field in B to induce a current in A?	It is only a <u>changing</u> magnetic field that can induce a current in A. When the switch is closed the field is increasing, when opened the magnetic field decreases.

4. 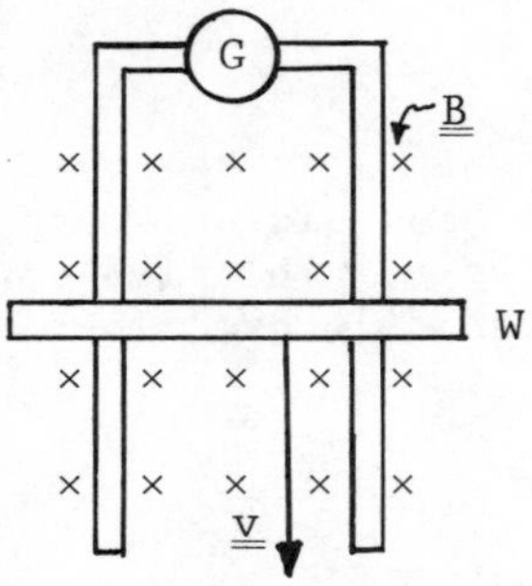

Here we show a different experiment in which we have a constant magnetic field into the paper. As you slide the wire W down, the galvanometer G deflects. This is slightly different from the above experiment. What was the condition of G in Faraday's experiment when the field of B was constant?

Undeflected.

His experiment required a changing magnetic field.

---

5. We can explain both phenomena if we talk not about the magnetic field itself, but rather talk about the flux of the magnetic field. Give the definition of the flux of B.

Φ_B = ______________.

$\Phi_B = \int \mathbf{B} \cdot d\mathbf{S}$.

---

6. We represent our two experiments below.

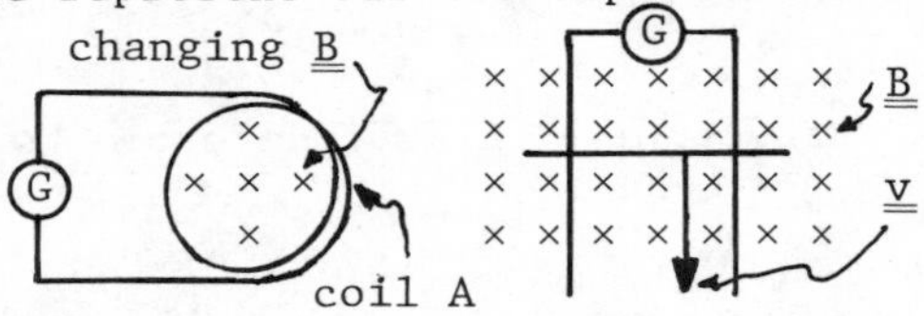

If we assert that both galvanometers deflect because of a changing flux of B, what in particular is causing the change in Φ_B in each case?

In the Faraday experiment B is changing while the area of the loop of coil A is not.

In the second experiment B is constant but the area ($\int dS$) is changing. Actually the area of the loop is increasing.

---

7. If we speak of induced emf's instead of currents the same idea about a changing flux of B still holds. Faraday's law is:

$$\mathcal{E} = \frac{d\Phi}{dt}$$

where $\Phi = \int \mathbf{B} \cdot d\mathbf{S}$. We can write the equation as:

$$\mathcal{E} = \frac{d}{dt} \int \mathbf{B} \cdot d\mathbf{S} \quad .$$

So far we have considered the following two cases:

1. Constant area and changing magnetic field.
2. Constant magnetic field and changing area.

From what you know about scalar products, how could we have a changing flux with a constant magnetic field and a constant area (say a loop of a certain size)?

Since the magnitude of $\mathbf{B} \cdot d\mathbf{S}$ depends upon the relative orientation of B and dS, we could obtain a change of flux by changing their relative orientation, e.g. rotating the loop.

8. The previous answer is of course the explanation for the basic generator.

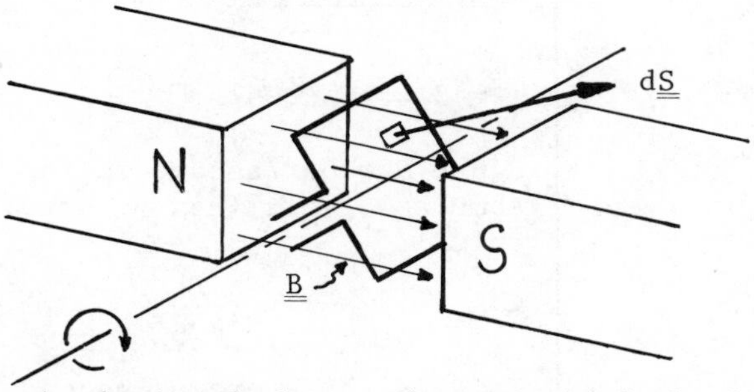

As depicted above, the area A of the loop is a constant as is the reasonably uniform $\underline{\underline{B}}$ field produced by the two permanent magnets. However their relative orientation is changed as we rotate the loop (armature) of this elementary generator.

Looking at a side view and considering only a small element of area we can imagine two situations as shown below.

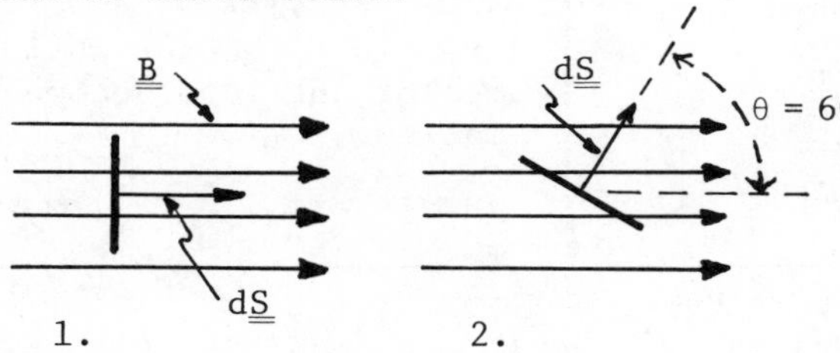

In both cases, use $|\underline{\underline{B}}| = 10^{-2}$ T (about what you could expect from an inexpensive bar magnet) and $S = 0.01$ m^2 to calculate the flux of Φ_B.

Case 1: Φ_B = ______. Case 2: Φ_B = ______.

Case 1.

$\Phi_B = \int \underline{\underline{B}} \cdot d\underline{\underline{S}}$

$\Phi_B = \int B \cos 0^o \, dS$

$\Phi_B = B \int dS \quad , \cos 0^o = 1$

$\Phi_B = BS = 10^{-4}$ Wb .

Case 2.

$\Phi_B = \int \underline{\underline{B}} \cdot d\underline{\underline{S}}$

$\Phi_B = \int B \cos 60^o \, dS$

$\Phi_B = 0.5 \, B \int dS \quad , \cos 60^o = \frac{1}{2}$

$\Phi_B = 0.5 \, BS$

$= 0.5 \times 10^{-4}$ Wb .

9. What is the magnitude of the change $\Delta\Phi_B$ in the flux of $\underline{\underline{B}}$ through S when the loop is rotated 60° as in the previous frame?

Change in flux =

final flux - initial flux.

$= 0.5 \times 10^{-4}$ Wb $- 10^{-4}$ Wb

$= - 0.5 \times 10^{-4}$ Wb .

The sign here is unimportant. The magnitude of the change in flux is then

$\Delta\Phi_B = 0.5 \times 10^{-4}$ Wb .

10. If the generator took 1/10 s to rotate through 60°, what (average) emf was induced in the loop?

$$\mathcal{E} = \frac{\text{change in flux}}{\text{change in time}}$$

$$= \frac{0.5 \times 10^{-4}\ \text{Wb}}{0.1\ \text{s}}$$

$$= 5 \times 10^{-4}\ \text{Wb/s} \quad .$$

Since a weber/second is the same as a volt (can you show this?),

$$\mathcal{E} = 5 \times 10^{-4}\ \text{volt} \quad .$$

This is the average induced emf during the 0.1 second time interval.

---

11. Consider now the effect of an increasing or decreasing flux.

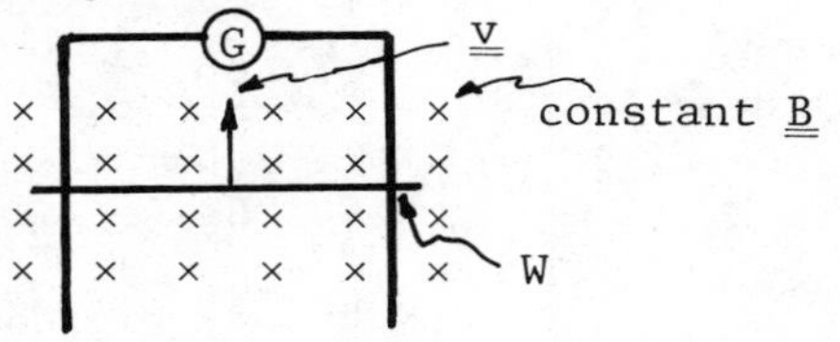

The flux of $\mathbf{B}$ (increases/decreases) as the slide wire W is moved up.

Decreases. [Of course here we are talking about the <u>magnitude</u> of the flux; the actual <u>sign</u> of Φ_B depends upon the sense of the $d\mathbf{S}$ vector.]

$\mathbf{B}$ remains constant while the area of the loop becomes smaller.

---

12. Lenz's law has something to say about this decreasing flux of $\mathbf{B}$. There are many ways to state this law and some of them are confusing to the beginner. Try this one:

Lenz's law states that "induced currents tend to oppose changes in flux".

Earlier in this program we stated three ways to change the flux of $\mathbf{B}$. What were they?

$\Phi_B = \int \mathbf{B} \cdot d\mathbf{S}$.

1. Change $\mathbf{B}$.
2. Change the area, $\int dS$.
3. Change the relative orientation of $\mathbf{B}$ and $d\mathbf{S}$.

---

13. Which of the above three ways is being employed in frame 11 to decrease the flux of $\mathbf{B}$?

Answer 2.
Change the area.

---

14. It seems clear that induced currents in the loop can only affect one of the three options available to "oppose" this decreasing change of flux. Which is it? (Consult the frame 12 answer again.)

Change $\mathbf{B}$.

For this somewhat rigid situation we can rule out the changing of the area or the relative orientation.

15. In what way should the induced current change $\underline{\underline{B}}$ in order that the decreasing flux be "opposed"?

The induced current must provide a stronger $\underline{\underline{B}}$ through the loop to "oppose" the decrease of the flux.

16. 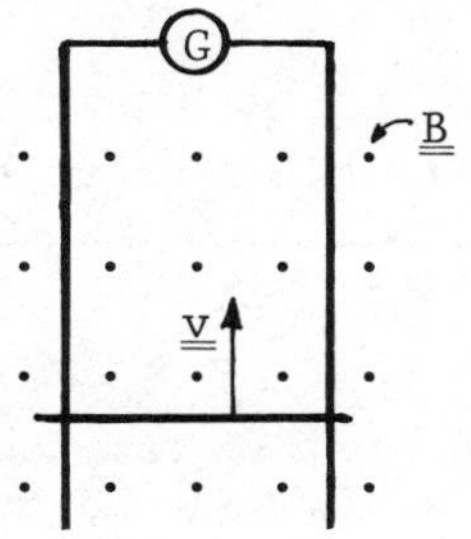

The constant field $\underline{\underline{B}}$ as shown is to be increased in this case by application of Lenz's law. What direction must the induced current have to accomplish this, i.e., what direction of the induced current will provide an additional magnetic field to aid the constant $\underline{\underline{B}}$?

Since we want to increase the magnetic field we want an induced current which will provide an additional magnetic field coming out of the paper.

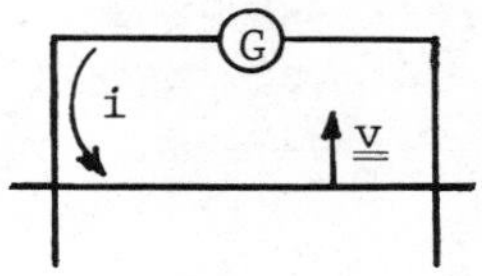

Curve your right hand fingers in the current direction and your thumb will point in the direction of $\underline{\underline{B}}$.

Try imagining other situations of changing flux and invoking Lenz's law. Mathematically we include Lenz's law in Faraday's law of induction by the use of a minus sign:

$$\mathcal{E} = -\frac{d\Phi_B}{dt} \quad .$$

This sign is of no real algebraic significance. It really serves as a reminder to invoke Lenz's law.

17. A device known as a "search coil" consists (for example) of N = 50 turns of wire with a cross-sectional area A = 4 cm^2 and a resistance R = 25 Ω. Such a device can be used to determine the value of a constant magnetic field.

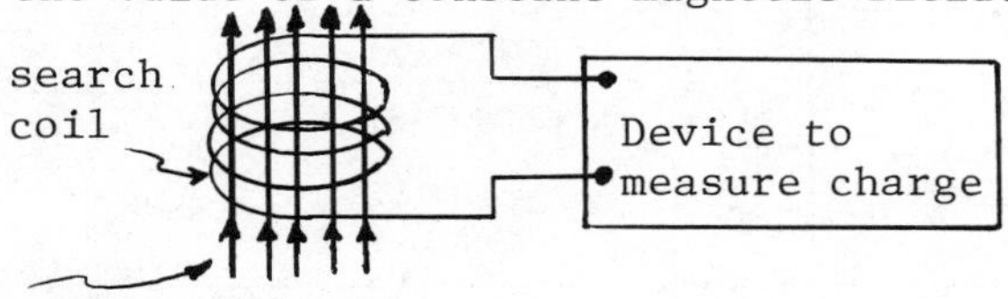

In the sketch we show the plane of the search coil perpendicular to the unknown field. What is the flux of $\underline{\underline{B}}$ at this time?

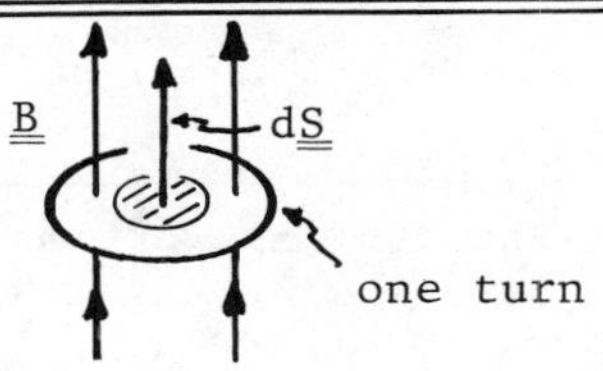

For one turn

$$\Phi_B = \int \underline{\underline{B}}\cdot d\underline{\underline{S}}$$

$$= \int B\,dS = B\int dS = BA$$

Here we have used the fact that the angle between $\underline{\underline{B}}$ and $d\underline{\underline{S}}$ is 0° and that B is a constant.

For N turns,

$$\Phi_B = NBA \quad .$$

18. If now we quickly flip the search coil 90°, what is the new flux of $\underline{\underline{B}}$?	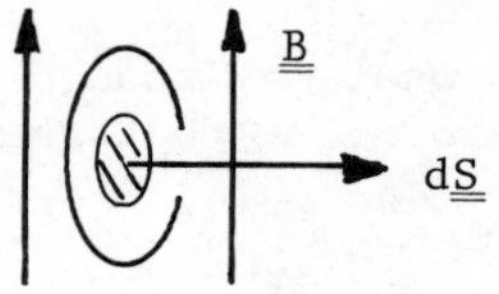 $\Phi_B = \int \underline{\underline{B}} \cdot d\underline{\underline{S}} = 0$, since $\underline{\underline{B}}$ is perpendicular to $d\underline{\underline{S}}$.
19. What is the change in the flux during the time Δt required to flip the coil?	$\Delta\Phi_B = 0 - NBA = - NBA$. Remember that the sign has to do with the direction of the induced current. We won't worry about that aspect, and shall use NBA as the change in flux.
20. What is the induced emf in the coil?	$\mathcal{E} = \frac{\Delta\Phi_B}{\Delta t} = \frac{NBA}{\Delta t}$.
21. Assuming the resistance to be only that of the search coil, what current is induced in the search coil during the time Δt? Express your answer in terms of B, A, R and Δt.	$i = \frac{\mathcal{E}}{R} = \frac{NBA}{R\Delta t}$.
22. Using the definition of current, rewrite the previous answer. $i = \Delta q/\Delta t$, $\Delta q =$ __________ .	$i = \frac{\Delta q}{\Delta t}$ $i = \frac{\Delta q}{\Delta t} = \frac{NBA}{R\Delta t}$ $\Delta q = \frac{NBA}{R}$.
23. The last answer implies that measuring Δq and knowing A and R allows one to determine B. Use $R = 25\ \Omega$, $A = 4 \times 10^{-4}\ m^2$ (note the change of units) and $\Delta q = 4 \times 10^{-5}$ C to determine B.	$B = \frac{(\Delta q)R}{NA}$ $B = \frac{4 \times 10^{-5}\ C \times 25\ \Omega}{4 \times 10^{-4}\ m^2 \times 50}$ $B = .05$ T .

Note that it turns out that the result is independent of time. That is, the search coil may be rotated in any manner (fast, slow, not at a constant rate, etc.). All that matters is that it was flipped through the 90° angle as shown in frame 18.

There are several ways of measuring the charge quantitatively so that this method of measuring $\underline{\underline{B}}$ fields is in fact used. One could, for example, measure Δq with a calibrated electroscope. Another instrument used to measure Δq is the so-called ballistic galvanometer.

24. What would the answer to frame 22 become if instead of 90° the angle were (a) 180°? $\Delta q =$ ______ . (b) 360°? $\Delta q =$ ______ .	(a) $\Delta q = 2NBA/R$ because $\Delta\Phi_B$ would be twice as large. (b) $\Delta q = 0$ because $\Delta\Phi_B = 0$.

Chapter 31: INDUCTANCE

REVIEW AND PREVIEW

In previous chapters you have studied several circuit elements: capacitors, resistors, seats of emf. In this chapter you will study another circuit element called an inductor. An example of an inductor is a coil of wire. Like capacitors, inductors can be used for energy storage, frequency filtering, etc. In addition to properties of a single inductor, you will learn how to deal with inductors connected in series and parallel. Finally you will study properties of a simple circuit containing a resistor and an inductor.

GOALS AND GUIDELINES

In this chapter there are five major goals:

1. Learning and understanding the definition of the inductance L of a given inductor (Section 31-1). Using this definition you should be able to calculate L for a given inductor (Example 2).
2. Learning the formulas for inductors connected in series (Eq. 31-4) and in parallel (Eq. 31-5).
3. Learning the formula for the energy U stored in an inductor (Eq. 31-6) and for the energy density u (Eq. 31-7) associated with the $\underline{\underline{B}}$ field in an inductor. You should be able to calculate L of a given inductor using energy considerations (Example 4).
4. Understanding the properties of a simple circuit containing both a resistor and an inductor (Section 31-7). You should compare this "RL" circuit with the "RC" circuit studied in Chapter 27.
5. Learning and understanding the definition of mutual inductance M (Section 31-9).

By the end of this chapter you will have studied four circuit elements. You may want to compare and contrast their properties by referring to Table 27-1.

31-1 Inductance

According to Faraday's law, a changing flux of $\underline{\underline{B}}$ through a loop of wire causes an induced emf $\mathcal{E} = d(\Phi_B)/dt$ to exist in the wire. We now want to consider the case in which this $\underline{\underline{B}}$ field is caused by the current in the wire itself, rather than by an external current.

Suppose we have a loop of wire which carries a current I. This current will cause a $\underline{\underline{B}}$ field to exist. The flux $\Phi_B = \int \underline{\underline{B}}\cdot d\underline{\underline{S}}$ of this $\underline{\underline{B}}$ field through the loop is proportional to $\underline{\underline{B}}$; $\underline{\underline{B}}$ itself is proportional to I (by the Biot-Savart law). Thus Φ_B is proportional to I; we write $\Phi_B = LI$ or

$$L = \Phi_B/I \quad . \qquad (31\text{-}1)$$

The positive quantity L is called the inductance; a system possessing this property is called an inductor. The unit of inductance is the henry, H (= weber/ampere). Inductance may be regarded as the magnetic analog of capacitance (an electrical quantity). Both inductance and capacitance depend only upon the geometry (sizes, shapes, etc.) of the particular device.

Sometimes we deal with N turns of wire arranged so that all the lines of $\underline{\underline{B}}$ which pass through any one turn also pass through all the others (e.g. a long solenoid). In this case equation (31-1) is usually written as

$$L = N\Phi_B/I \qquad (31\text{-}2)$$

where Φ_B now means the flux of $\underline{\underline{B}}$ through <u>one</u> turn.

In electrical circuits, inductors are represented by the following symbol:  . Since $\mathcal{E} = d(\Phi_B)/dt$, differentiation of (31-1) yields

$$\mathcal{E} = L\ di/dt \tag{31-3}$$

for the potential difference across the terminals of an inductor which carries a current i.* As for the question of <u>signs</u>, consider the following inductor:

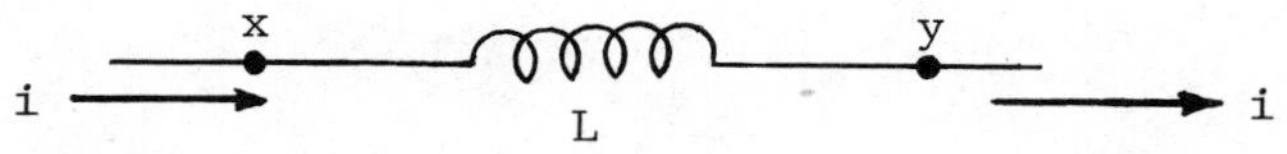

The algebraically correct equation for this is $V_x - V_y = L\ di/dt$ where i is the current through the inductor from x to y.

>>> Example 1. The current through a 2.0 henry inductor varies sinusoidally in time with an amplitude of 0.5 ampere and a frequency of 60 cycles per second. Calculate the potential difference across the terminals of the inductor.

From the given data, the current through the inductor is

$$i = I \sin(2\pi ft)$$

where I = 0.5 A and f = 60 Hz. The potential difference across the terminals of the inductor is

$$V = L\ di/dt = L \frac{d}{dt}(I \sin(2\pi ft))$$

$$= 2\pi fLI \cos(2\pi ft) \quad .$$

This is oscillatory with the same frequency f = 60 Hz as the current. The amplitude of this oscillating voltage is

$$2\pi fLI = 2\pi(60\ s^{-1})(2.0\ H)(0.5\ A)$$

$$= 380\ \text{volt} \quad .$$

Note that V is 90° out of phase with respect to i. <<<

31-2 Calculation of Inductance (Method #1)

To calculate the inductance associated with a given conductor (say a loop of wire) we can follow the definition directly. The steps are:

1. Imagine a current I exists in the conductor.
2. Find the $\underline{\underline{B}}$ field caused by this current (perhaps by means of Ampere's law).
3. Find the magnitude, Φ_B, of the flux of $\underline{\underline{B}}$ through the area bounded by the loop (using $\Phi_B = \int \underline{\underline{B}} \cdot d\underline{\underline{S}}$). This will be proportional to I.
4. Using this expression for Φ_B, apply the definition to inductance: $L = \Phi_B/I$.

*From this we have the useful dimensional identity: henry = volt·second/ampere.

Usually, calculation of the $\underline{\underline{B}}$ field (step 2) is the difficult part. This procedure is illustrated in the following example.

>>> Example 2. A long solenoid has length ℓ, cross sectional area A, and N turns. Calculate the inductance of this solenoid.

Imagine a current I exists in the solenoid. To find $\underline{\underline{B}}$, we apply Ampere's law to the rectangular path shown in Figure 31-1.

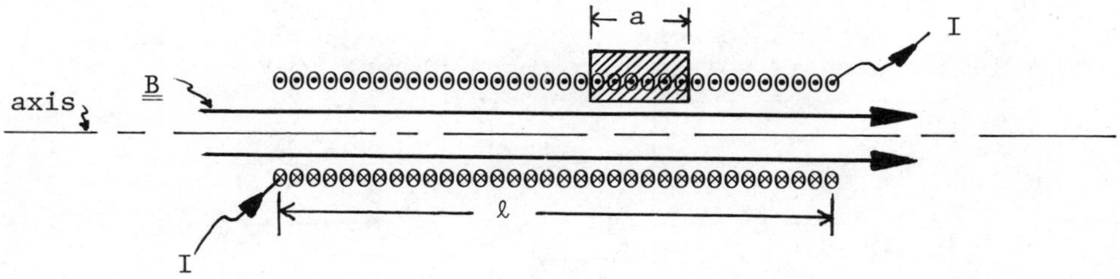

Figure 31-1. A solenoid of length ℓ. Ampere's law is applied to the rectangular path. The symbol ⊗ indicates a current I into the page while the symbol ⊙ indicates a current I out of the page.

Assuming that the $\underline{\underline{B}}$ field within the solenoid is uniform and parallel to the axis of the solenoid, and that the $\underline{\underline{B}}$ field outside the solenoid is negligible, we have

$$\oint \underline{\underline{B}}\cdot d\underline{\underline{\ell}} = Ba \quad .$$

Note that only one side of the rectangular path contributes to the integral. On the right hand side ($\mu_o i$) of Ampere's law, "i" is the net current crossing the shaded area shown in the figure. Since there are N/ℓ turns per unit length and each turn carries the current I,

$$i = (N/\ell)\, aI \quad .$$

Thus Ampere's law becomes

$$\oint \underline{\underline{B}}\cdot d\underline{\underline{\ell}} = \mu_o i$$

$$Ba = \mu_o (N/\ell)\, aI \quad .$$

$$B = \mu_o NI/\ell \quad .$$

Since $\underline{\underline{B}}$ is uniform, the total flux of $\underline{\underline{B}}$ (through all N turns) is simply

$$\Phi_B = NBA = \mu_o N^2 IA/\ell \quad .$$

Finally, the inductance L is given by

$$L = \Phi_B/I = \mu_o N^2 A/\ell \quad .$$ <<<

31-3 Equivalent Inductance

Consider any circuit consisting only of inductors and let "x" and "y" be the only two terminals which emerge from the circuit. An example of this is shown in Figure 31-2a. Suppose there is a (time varying) current i going into terminal x and out terminal y. There will then be a certain potential difference $V = V_x - V_y$ across the circuit. As far as the terminals x and y are concerned, the entire circuit behaves as if it were a single inductor (Figure 31-2b). The equivalent inductance, L, is defined by $L = V/(di/dt)$. It is important to note that the only external connections to the circuit are x and y (e.g. there is no external connection to z).

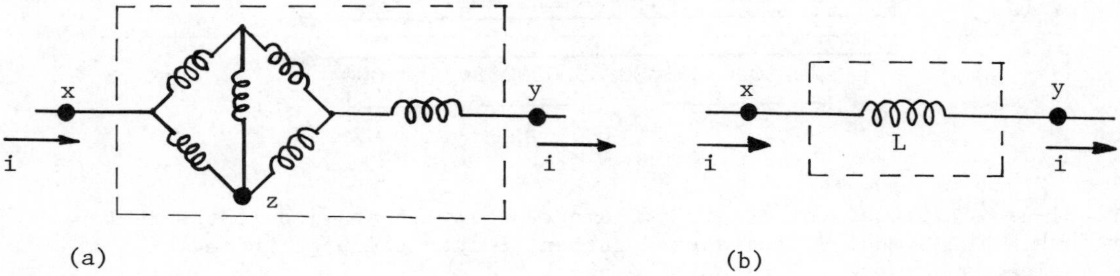

Figure 31-2. (a) A circuit consisting of inductors. The only external connections are x and y. (b) The equivalent inductor.

If the inductors are connected in series (Figure 31-3a) the following formulas apply:

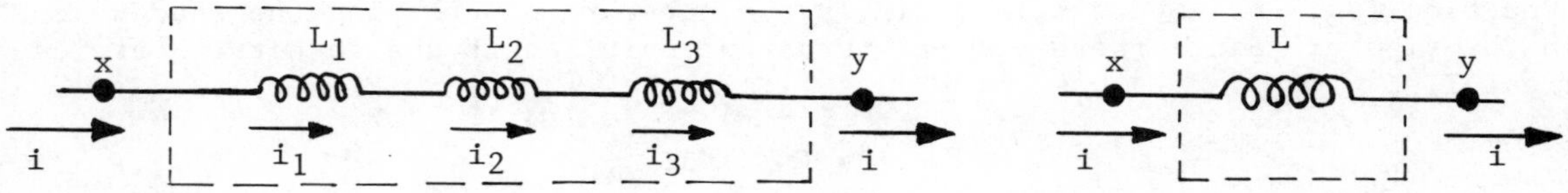

Figure 31-3. (a) Three inductors connected in series. (b) The equivalent inductor: $L = L_1 + L_2 + L_3$.

$$L = L_1 + L_2 + L_3 \tag{31-4a}$$

$$i = i_1 = i_2 = i_3 \tag{31-4b}$$

$$V = V_1 + V_2 + V_3 \quad . \tag{31-4c}$$

If the inductors are connected in parallel (Figure 31-4a) the following formulas apply:

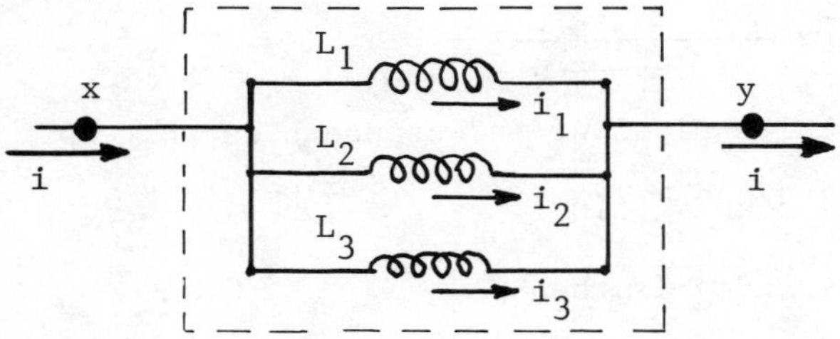

Figure 31-4. (a) Three inductors connected in parallel. (b) The equivalent inductor: $1/L = 1/L_1 + 1/L_2 + 1/L_3$.

$$1/L = 1/L_1 + 1/L_2 + 1/L_3 \tag{31-5a}$$

$$i = i_1 + i_2 + i_3 \tag{31-5b}$$

$$V = V_1 = V_2 = V_3 \quad . \tag{31-5c}$$

In the series case all the currents are the same, while the voltages obey a sum rule. Just the opposite is true in the parallel case: the voltages are all the same, the currents obey a sum rule. The equivalent inductance for a series circuit is always more than the largest individual inductance; the equivalent inductance for a parallel circuit is always less than the smallest individual inductance. Note that formulas (31-4a,5a) for the series and parallel equivalent inductors are the same in form as those for the series and parallel equivalent resistors.

Of course not every circuit is a series or parallel one. Many inductor circuits however can be simplified with successive applications of the series and parallel formulas.

>>> Example 3. In the circuit shown below L_1 = 4 mH, L_2 = 12 mH, L_3 = 5 mH (1 mH = 10^{-3} H). The current i is increasing steadily at the rate of 20 amperes per second. Calculate the potential difference between the terminals x and y.

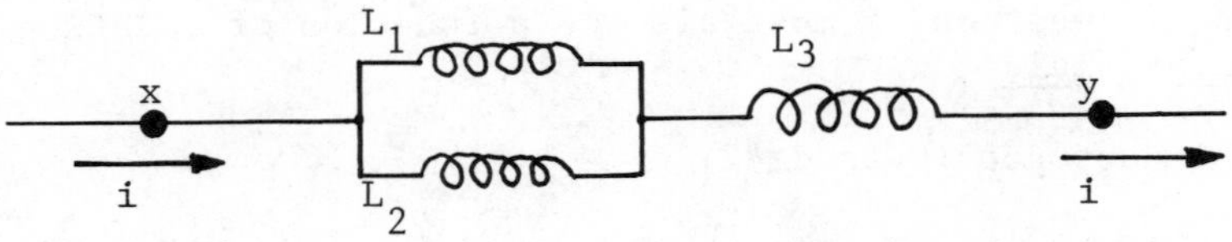

Inductors L_1 and L_2 are in parallel. Their equivalent inductance is given by

$$1/L = 1/L_1 + 1/L_2 = 1/(4 \text{ mH}) + 1/(12 \text{ mH})$$

$$L = 3 \text{ mH} \quad .$$

The circuit has now been simplified to the following:

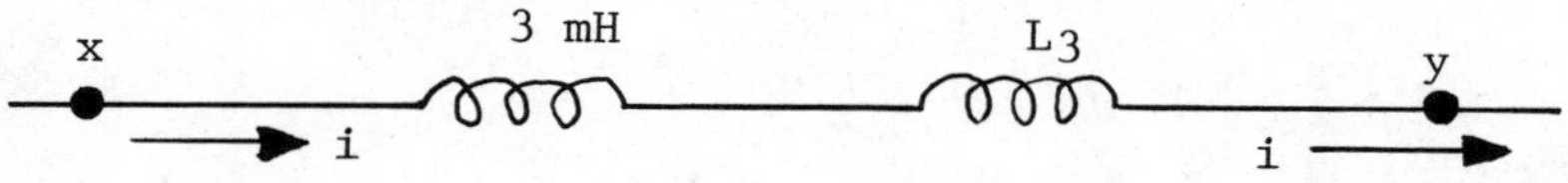

The 3 mH inductor is in series with L_3. The total equivalent inductance L is therefore

$$L = 3 \text{ mH} + L_3 = 3 \text{ mH} + 5 \text{ mH} = 8 \text{ mH} \quad .$$

The potential difference between the terminals x and y is then

$$V = V_x - V_y = L\, di/dt$$

$$= (8 \times 10^{-3} \text{ H})(20 \text{ A/s}) = 0.16 \text{ volt} \quad .$$ <<<

31-4 Energy Stored in an Inductor

Suppose the current in an inductor is increased from zero to some final value i. During this process there will be a potential difference $V = L\, di/dt$ across the inductor. The power input to the inductor is then

$$P = Vi = (L\, di/dt)\cdot(i) \quad .$$

Therefore the energy, U, stored in the inductor is

$$U = \int P\, dt = L \int i\, (di/dt)\, dt = L \int i\, di$$

$$U = \tfrac{1}{2} Li^2 \quad . \qquad (31\text{-}6)$$

31-5 Magnetic Energy Density

Inductors can store energy. This energy is regarded as being associated with the B field, rather than the wires where the currents are. There is a (magnetic) energy density, u,* which gives the amount of energy stored per unit volume; i.e. in a differential volume element dv there is an energy $dU = u\, dv$. The units of u are those of energy per volume (J/m^3).

The energy density is proportional to the square of the B field; it is given by

$$u = \frac{1}{2\,\mu_o} B^2 \quad . \qquad (31\text{-}7)$$

Since B is a function of location, u must also be a function of location. The energy density is then a scalar field.

31-6 Calculation of Inductance (Method #2)

An alternative method of calculating the inductance associated with a given conductor uses the concept of energy density. The steps are:

1. Imagine a current I exists in the conductor.
2. Find the B field caused by this current. Using this, find the energy density $u = B^2/(2\mu_o)$.
3. Integrate the energy density to find the stored energy, $U = \int u\, dv$ (dv = differential volume element). This will be proportional to I^2.

---

*Sometimes written as u_B to emphasize that it is a magnetic energy density.

4. Using this expression for U. solve the equation $U = \frac{1}{2} LI^2$ for L.

This procedure is illustrated below.

>>> Example 4. Calculate the inductance of the solenoid in Example 2 by considering the magnetic stored energy.

Imagine a current I exists in the solenoid. Using Ampere's law (as in Example 2), the B field within the solenoid is found to be

$$B = \mu_0 NI/\ell \quad .$$

The magnetic energy density is

$$u = \frac{1}{2\,\mu_0} B^2 = \frac{1}{2\,\mu_0} (\mu_0 NI/\ell)^2$$

$$u = \mu_0 N^2 I^2/(2\ell^2) \quad .$$

Since u is constant within the solenoid (and zero outside the solenoid) the magnetic stored energy is simply

$$U = \int u\,dv = u \int dv = u(A\ell)$$

$$U = \mu_0 N^2 I^2 A/(2\ell) \quad .$$

Here we have used the fact that $\int dv$ = volume of solenoid = $A\ell$. Since the magnetic stored energy is also given by $U = \frac{1}{2} LI^2$,

$$\tfrac{1}{2} LI^2 = \mu_0 N^2 I^2 A/(2\ell) \quad .$$

Solving this for the inductance L,

$$L = \mu_0 N^2 A/\ell \quad .$$

Note that the I^2 term cancelled out. The answer is in agreement with Example 2. <<<

31-7 Single Loop RL Circuits

In the single loop RL circuit shown below the switch S is closed at time t = 0. The current i (which is initially zero) will begin to increase; the voltage across the inductor (L di/dt) does not permit i to instantaneously go from zero to its final value. The current is thus a function of time.

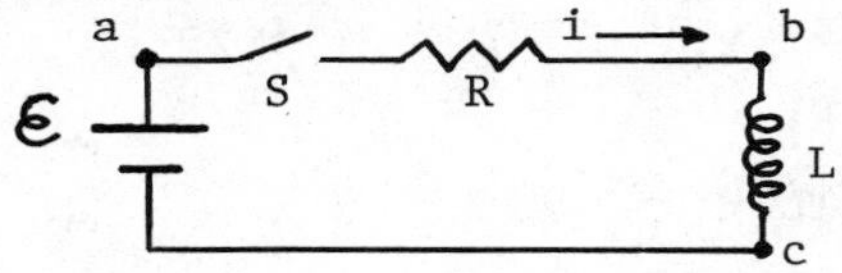

After the switch is closed ($t \geq 0$) we have

$$(V_a - V_b) + (V_b - V_c) + (V_c - V_a) = 0$$

$$(iR) + (L\,di/dt) + (-\mathcal{E}) = 0$$

$$L\,di/dt + Ri = \mathcal{E} \quad . \qquad (31\text{-}8)$$

This equation is similar in form to that for the RC circuit (although now i is the dependent variable, rather than q). This may be solved by the same technique that was used in the RC case. The solution (subject to the initial condition that when t = 0, i = 0) is

$$i = \frac{\mathcal{E}}{R}\,(1 - e^{-Rt/L}) \quad . \tag{31-9}$$

From the form of this equation we see that the current approaches its final value ($\mathcal{E}/R$) with an (inductive) time constant, τ,* given by

$$\tau = L/R \quad . \tag{31-10}$$

31-8 Power Balance in an RL Circuit

In the RL circuit considered above, some of the power supplied by the seat of emf is dissipated by the resistor. The remaining power is used to change the energy stored in the inductor. Thus (power supplied by seat of emf) = (power dissipated by resistor) + (rate of change of energy stored in inductor). That is,

$$\mathcal{E}\,i = i^2R + \frac{d}{dt}\,(U) = i^2R + \frac{d}{dt}\,(\tfrac{1}{2}\,Li^2)$$

$$\mathcal{E}\,i = i^2R + Li\;di/dt \quad .$$

Cancelling the common factor i,

$$\mathcal{E} = iR + L\;di/dt$$

which is the loop equation (31-8). For a single loop RL circuit we can thus derive the loop equation by considering the power balance.

31-9 Mutual Inductance

We have seen that a changing current in a coil causes an induced emf $\mathcal{E} = L\,di/dt$. If two coils exist in close proximity then a changing current in coil #2, di_2/dt, will result in a changing flux through coil #1; hence there will be an induced emf in coil #1. This can be written as

$$\mathcal{E}_1 = M_{12}\;di_2/dt \quad . \tag{31-11}$$

The quantity M_{12} is called the mutual inductance; its units are the same as those of L (i.e. henry). Similarly a changing current in coil #1 causes an induced emf in coil #2,

$$\mathcal{E}_2 = M_{21}\;di_1/dt \quad . \tag{3]-12}$$

It can be shown that the two M's are equal: $M_{12} = M_{21}$; this is analogous to Newton's third law (action-reaction). Thus we may delete the subscripts and simply write M (= $M_{12} = M_{21}$) which is a joint geometric property of the two coils and their relative locations and orientations.

*Sometimes written τ_L to emphasize that it is the inductive time constant.

>>> Example 5. Write down the two loop equations which apply to the circuit shown when switch S is closed

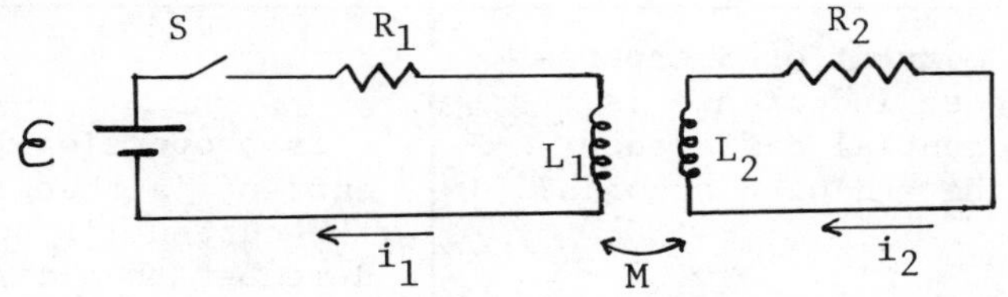

Solution: Currents i_1 and i_2 have been drawn with arbitrary senses. For the left loop we have

$$L_1\ di_1/dt + M\ di_2/dt + i_1R_1 - \mathcal{E} = 0 \quad .$$

Similarly for the right loop we have

$$L_2\ di_2/dt + M\ di_1/dt + i_2R_2 = 0 \quad .$$

Remarks: (a) The sign of the "M" term depends upon the relative orientations of the two coils as well as the assumed senses of the two currrents. If a positive current i_2 causes a flux in coil #1 which aids the flux due to a positive current i_1, then the L_1di_1/t term and the Mdi_2/dt term will have the same sign. [It is always true that the relative sign of the Ldi/dt term and the Mdi/dt term in one of the loop equation is the same as the corresponding relative sign in the other loop equation.]
(b) The quantity $[L_1di_1/dt + Mdi_2/dt]$ is the total potential difference across coil #1; the first part of this is due to a changing current in that coil and the second part is due to a changing current in another coil. <<<

31-10 Programmed Problems

1. 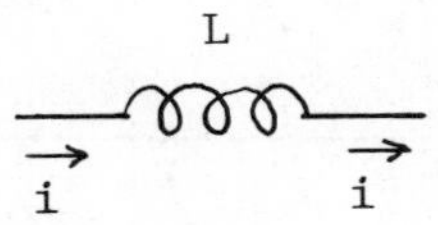 The diagram shows an inductor with a current i through it. This current causes a ______ to exist within the inductor.	magnetic field
2. If this current i changes, there is a potential difference $V = V_x - V_y$ across the terminals of the inductor. What causes this potential difference?	The $\underline{\underline{B}}$ field is caused by the current i. When i changes, $\underline{\underline{B}}$ will change. Thus the flux of $\underline{\underline{B}}$ through the coils of the inductor will change. By Faraday's law there will then be an induced emf equal to the rate of change of the magnetic flux.
3. The potential difference V is proportional to the rate of change of the current through the inductor. Write down the equation which expresses this. V = ______ .	$V = L\ di/dt$.

4. The quantity L is called the ______; its unit is the ______ .

inductance, henry

---

5. Suppose that a constant current of 3 amperes exists in an inductor whose inductance is 2 henries. What is the potential difference across the inductor at the end of 6 seconds?

Zero.

V is proportional to how fast the current is changing, in this case the current is not changing at all. Remember that di/dt is not the same thing as i/t.

6.

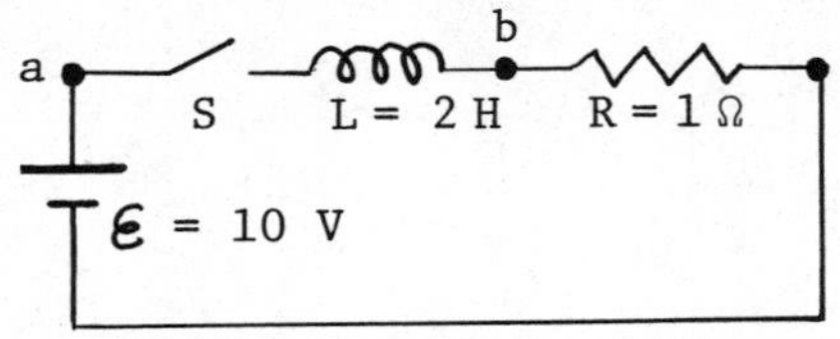

When t = 0 the switch S in the above circuit is closed. We seek to find the behavior of the current i as a function of time. Start by writing down the loop equation for this circuit.

$$(V_a - V_b) + (V_b - V_c) + (V_c - V_a) = 0 \quad .$$

$$(\qquad) + (\qquad) + (\qquad) = 0 \quad .$$

Substitute into the above parentheses the correct expressions in terms of resistance, inductance, current, etc.

$$(L\, di/dt) + (iR) + (-\mathcal{E}) = 0 \quad .$$

---

7. Solve the above equation for di/dt and substitute the numerical values for L, R, $\mathcal{E}$.

di/dt = ______ .

$$di/dt = \frac{\mathcal{E} - iR}{L}$$

$$= 5 - (0.5)\, i \ .$$

For brevity, the units have been suppressed. The current i is understood to be in amperes, the time t in seconds.

---

8. Initially (t = 0), what is the value of di/dt? Remember that the initial current is zero.

di/dt = ______ .

$$di/dt = 5 - (0.5)\, i$$

$$= 5 - (0.5)(0) = 5 \quad .$$

Of course the units are ampere/second.

9. We have that the current is initially zero and is increasing at the rate of 5 ampere/second. In a reasonably small time interval this rate of increase (5 ampere/second) will not appreciably change. Therefore a good estimate of the current at t = 0.5 seconds is

$i =$ ______ .

$\frac{\Delta i}{\Delta t} \simeq \frac{di}{dt}$ ("≃" means approximately)

$i \simeq (\frac{di}{dt})(\Delta t)$

$\simeq 5 \times 0.5 = 2.5$.

This gives the change in the current. Since the current at t = 0 is zero, the estimate for the current at t = 0.5 seconds is

$i \simeq 2.5$ A .

From now on we will delete the "≃" sign.

---

10. Returning to the answer to frame 7,

$di/dt = 5 - (0.5)\ i$,

what is the value of di/dt at t = 0.5 second?

$di/dt =$ ______ .

$di/dt = 5 - (0.5)\ i$

$= 5 - (0.5)(2.5)$

$= 3.75$ A/s .

---

11. Estimate the change in the current between t = 0.5 and t = 1.0 seconds.

$\Delta i =$ ______ .

$\Delta i = (di/dt)(\Delta t)$

$= (3.75)(0.5)$

$= 1.88$ A .

---

12. The current at t = 1.0 s is the sum of the current at t = 0.5 s plus the change in the current between t = 0.5 and t = 1.0 s, i.e. the current at t = 1.0 s is

$i =$ ______ + ______ = ______ .

$i = 2.5 + 1.88 = 4.38$ A .

13. Now let's get a little more organized. If we know the current at any time t we can find the rate of change of this current from

$$di/dt = 5 - (0.5)\ i \quad .$$

The change in i (during a small time interval Δt) can then be found from

$$\Delta i = (di/dt)(\Delta t) \quad .$$

Thus complete the following table (carry your work to the hundreth's place):

t (s)	i (A)	di/dt (A/s)
0.00	0.00	5.00
0.50	2.50	3.75
1.00	4.38	
1.50		
2.00		
2.50		
3.00		

t	i	di/dt
0.00	0.00	5.00
0.50	2.50	3.75
1.00	4.38	2.81
1.50	5.78	2.11
2.00	6.84	1.58
2.50	7.63	1.18
3.00	8.22	0.89

14. Return to the answer to frame 6,

$$L\ di/dt + iR - \mathcal{E} = 0 \quad .$$

You probably remember that the solution to this differential equation is exponential but perhaps you have trouble recalling the details of the solution. Let's try one of the form

$$i = A + Be^{-t/\tau}$$

where A, B, τ are constants which are somehow related to L, R, $\mathcal{E}$.

a. Referring to the circuit diagram, what is the current after a long time?

i = __________.

b. What does this tell us about A, B, τ?

a. After a long time the current will approach a constant value. The potential difference across the inductor is then zero ($L\ di/dt = 0$). Therefore the potential difference across the resistor must equal that across the seat of emf, i.e.

$$i = \mathcal{E}/R.$$

b. Noting that $e^{-\infty} = 0$,

$$\mathcal{E}/R = i = A + Be^{-\infty} = A$$

$$A = \mathcal{E}/R$$

We obtain no information from this concerning B, τ.

15. Now our solution becomes

$$i = \mathcal{E}/R + B\ e^{-t/\tau}$$

What can you say about the constants B, τ using the fact that the initial current is zero?

Noting that $e^0 = 1$,

$$0 = i = \mathcal{E}/R + Be^0 = \mathcal{E}/R + B$$

$$B = -\mathcal{E}/R.$$

We learn nothing about τ from this.

16. Now our solution becomes

$$i = \mathcal{E}/R - (\mathcal{E}/R)\ e^{-t/\tau}$$

which is usually factored so as to be written

$$i = (\mathcal{E}/R)(1 - e^{-t/\tau}) \quad .$$

What must be the units of the constant τ?

seconds

The exponent must be dimensionless, therefore τ must have the same units as t.

---

17. Express the units of L, R, $\mathcal{E}$ in terms of volt, ampere, second.

units of L: ______

units of R: ______

units of $\mathcal{E}$: ______

Hints: V = L di/dt, V = iR.

L: volt·second/ampere

R: volt/ampere

$\mathcal{E}$: volt

---

18. We know that the constant must have the units of seconds. What combination of L, R, $\mathcal{E}$ has the units of seconds?

L/R

$$\frac{\text{volt}\cdot\text{second/ampere}}{\text{volt/ampere}} = \text{second} \quad .$$

---

19. It turns out that this is the correct expression for τ:

$$\tau = L/R \quad .$$

a. τ is called the ________ .
b. The value of τ in this problem is τ = ____.

a. time constant (or inductive time constant).

b. 2 seconds.

Note that one may find that $\tau = L/R$ by substituting

$$i = (\mathcal{E}/R)(1 - e^{-t/\tau})$$

into

$$L\ di/dt + iR - \mathcal{E} = 0 \quad .$$

---

20. Express i as a function of time. Substitute the numerical values for L, R, .

i = ______ .

$$i = (\mathcal{E}/R)(1 - e^{-Rt/L})$$
$$= 10\ (1 - e^{-0.5t}) \quad .$$

21. Using this formula, complete the following table.

t (s)	0.5 t	$e^{-0.5t}$	i (A)
0.00	0.00	1.00	
0.50	0.25	0.78	
1.00	0.50	0.61	
1.50	0.75	0.47	
2.00	1.00	0.37	
2.50	1.25	0.29	
3.00	1.50	0.22	

t	i
0.00	0.0
0.50	2.2
1.00	3.9
1.50	5.3
2.00	6.3
2.50	7.1
3.00	7.8

22. These exact values for the current agree reasonably well with those obtained in frame 13.

a. Why isn't the agreement perfect?

b. Why is the agreement reasonably good?

a. The time intervals Δt are finite. Thus $\Delta i/\Delta t$ and di/dt are only approximately equal.

b. The time intervals $\Delta t = 0.5$ sec are reasonably smaller than the time constant $\tau = 2.0$ sec.

We see that the time constant τ serves as a measure of the duration of the process. Times may be considered to be "small" if they are much less than τ, times may be considered to be "large" if they are much greater than τ.

Chapter 32: MAGNETIC PROPERTIES OF MATTER

REVIEW AND PREVIEW

In Chapter 25 you studied some of the electric properties of an insulator (dielectric) when placed in an $\underline{\underline{E}}$ field; in Chapter 26 you studied some of the electric properties of a conductor. In this chapter you will study some of the magnetic properties of materials when placed in a $\underline{\underline{B}}$ field. The behavior of these materials will be discussed in terms of "magnetic dipoles" within the material.

GOALS AND GUIDELINES

In this chapter the chief goal is to gain an understanding of the three basic types of magnetic materials:

a. paramagnetic,
b. diamagnetic,
c. ferromagnetic.

The first four sections of this chapter are concerned with magnetic poles and magnetic dipoles. The discussion emphasizes the analogies between these and the corresponding electric charges and electric dipoles.

32-1 Magnetic Poles

Lines of an $\underline{\underline{E}}$ field diverge from positive charges and converge into negative charges. We say that a positive charge acts as a source for the $\underline{\underline{E}}$ field; similarly a negative charge is said to act as a sink for the $\underline{\underline{E}}$ field.

What are the analogous quantities for the $\underline{\underline{B}}$ field? Corresponding to the concept of an electric charge there is the concept of a magnetic pole. A source for the $\underline{\underline{B}}$ field is called a north pole, a sink for the $\underline{\underline{B}}$ field is called a south pole.

32-2 Gauss's Law for Magnetism

Experiment shows that there are no (isolated) magnetic poles. The mathematical formulation of this fact is called Gauss's law for magnetism:

$$\oint \underline{\underline{B}} \cdot d\underline{\underline{S}} = 0 \quad . \tag{32-1}$$

Note the similarity of this formula to Gauss's law for electricity: $\oint \underline{\underline{E}} \cdot d\underline{\underline{S}} = q/\varepsilon_o$. Since there are no magnetic poles in nature, the magnetic analog of "q" is zero. We may interpret (32-1) as saying that the lines of $\underline{\underline{B}}$ never terminate.

32-3 Magnetic (and Electric) Dipoles

An electric dipole consists of two equal and opposite charges (+q and -q) separated by a distance 2a as shown in Figure 32-1a.

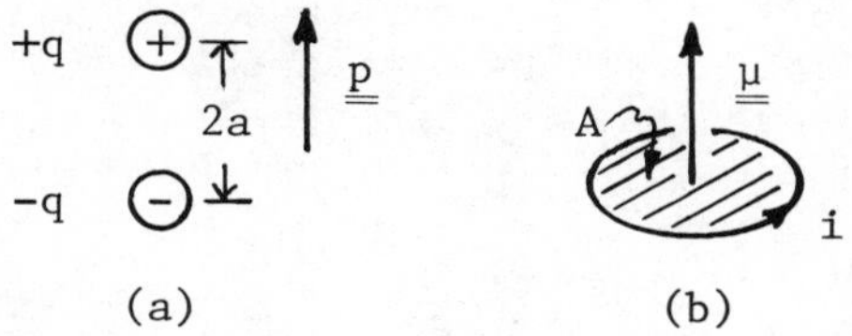

Figure 32-1. (a) An electric dipole consists of two equal and opposite charges. (b) A magnetic dipole consists of a current loop.

If this electric dipole is placed in a uniform (externally caused) $\mathbf{E}$ field there will be no net force acting on it. However there will be a torque acting on the dipole. This torque $\boldsymbol{\tau}$ is given by

$$\boldsymbol{\tau} = \mathbf{p} \times \mathbf{E} \quad , \quad |\mathbf{p}| = 2\,aq \quad . \tag{32-2}$$

Here $\mathbf{p}$ is the (electric) dipole moment; the direction of the vector $\mathbf{p}$ is from $-q$ to $+q$.

>>> Example 1. An electric dipole is placed in a uniform (externally caused) $\mathbf{E}$ field. Calculate the torque exerted on the dipole by this field.

The diagram shows an electric dipole (of dipole moment $\mathbf{p}$) in a uniform $\mathbf{E}$ field. Using $\mathbf{F} = q\mathbf{E}$ we see that there are two equal and opposite forces of magnitude $q|\mathbf{E}|$ present. These two forces form a couple; the magnitude of the associated torque is

$$|\boldsymbol{\tau}| = (2\,a)(q|\mathbf{E}|)\sin\theta$$

$$= |\mathbf{p}|\,|\mathbf{E}|\sin\theta = |\mathbf{p} \times \mathbf{E}| \quad .$$

Here θ is the angle between $\mathbf{p}$ and $\mathbf{E}$. The direction of this torque is into the page. Thus $\boldsymbol{\tau}$ has the same magnitude and direction as $\mathbf{p} \times \mathbf{E}$, i.e.

$$\boldsymbol{\tau} = \mathbf{p} \times \mathbf{E} \quad .$$ <<<

What is the magnetic analog of the electric dipole? As shown in Figure 32-1b, a <u>magnetic dipole</u> consists of a current loop (current i in a plane loop of area A). If this magnetic dipole is placed in a uniform (externally caused) $\mathbf{B}$ field there will be no net force acting on it. However there will be a torque acting on the dipole. This torque $\boldsymbol{\tau}$ is given by

$$\boldsymbol{\tau} = \boldsymbol{\mu} \times \mathbf{B} \quad , \quad |\boldsymbol{\mu}| = iA \quad . \tag{32-3}$$

Here $\boldsymbol{\mu}$ is the (<u>magnetic</u>) <u>dipole moment</u>; the direction of the vector $\boldsymbol{\mu}$ is perpendicular to the plane of the current loop, its sense is given by a "right hand rule"*.

>>> Example 2. A magnetic dipole is placed in a uniform (externally caused) $\mathbf{B}$ field. Calculate the torque exerted on the dipole by this field.

For simplicity choose the loop to be rectangular with sides of length a and b, oriented with respect to the $\mathbf{B}$ field as shown (the sides of length a are perpendicular to the page). Also shown in the diagram is the magnetic moment $\boldsymbol{\mu}$ (check that the sense of this vector is correctly shown). Using $F = i\boldsymbol{\ell} \times \mathbf{B}$ we have that the two equal and opposite forces on the sides of length a are each of magnitude $ia|\mathbf{B}|$ (note that the angle between $\boldsymbol{\ell}$ and $\mathbf{B}$ is 90°).

*Orient your right hand so that the fingers follow around the loop in the direction of the current. Then the thumb gives the correct sense for the vector $\boldsymbol{\mu}$.

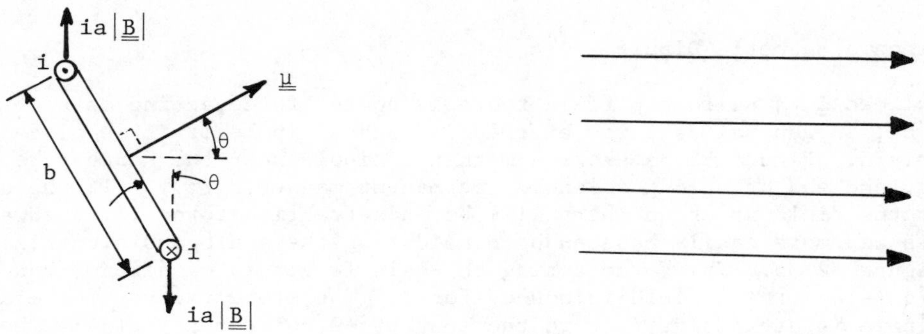

The magnitude of the associated torque is

$$
\begin{aligned}
|\underline{\underline{\tau}}| &= (b)(ia|\underline{\underline{B}}|) \sin\theta \\
&= (iA)|\underline{\underline{B}}| \sin\theta = |\underline{\underline{\mu}}||\underline{\underline{B}}| \sin\theta \\
&= |\underline{\underline{\mu}} \times \underline{\underline{B}}| \quad .
\end{aligned}
$$

Here A = ab is the area of the loop and θ is the angle between the side of length b and the force causing this torque. Note that this angle θ is <u>also</u> the angle between $\underline{\underline{\mu}}$ (which is normal to the loop) and $\underline{\underline{B}}$. Also, the direction of this torque is into the page. Thus $\underline{\underline{\tau}}$ has the same magnitude and direction as $\underline{\underline{\mu}} \times \underline{\underline{B}}$, i.e.

$$\underline{\underline{\tau}} = \underline{\underline{\mu}} \times \underline{\underline{B}} \quad .$$ <<<

In the above example, only the special case of a rectangular current loop was treated. Equation (32-3) however is true for a plane current loop of arbitrary shape.

Note that for both the electric and magnetic cases the direction of the torque exerted on the dipole is such as to tend to <u>align</u> the dipole with the external field. That is, the torque tends to make $\underline{\underline{p}}$ parallel to $\underline{\underline{E}}$ and to make $\underline{\underline{\mu}}$ parallel to $\underline{\underline{B}}$.

To further justify a current loop as the magnetic analog of an electric dipole, consider the fields shown in Figure 32-2. At large distances the $\underline{\underline{E}}$ field due to an electric dipole (Figure 32-2a) is exactly similar to the $\underline{\underline{B}}$ field due to a magnetic dipole (Figure 32-2b). At distances comparable with the dimensions of the dipole these fields are quite different; in particular the lines of $\underline{\underline{E}}$ terminate at the charges while the lines of $\underline{\underline{B}}$ never terminate. However if we consider the $\underline{\underline{B}}$ field at large distances only, it acts as if it were caused by the two fictitious poles N and S in Figure 32-2c.

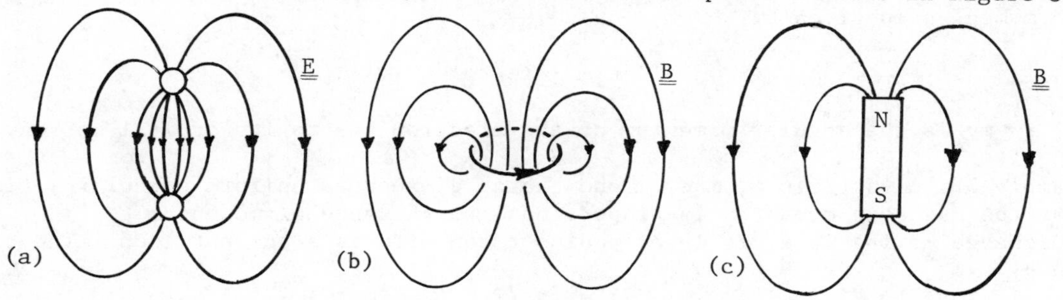

Figure 32-2. (a) The $\underline{\underline{E}}$ field due to an electric dipole. (b) The $\underline{\underline{B}}$ field due to a magnetic dipole. (c) The $\underline{\underline{B}}$ field due to two (fictitious) equal and opposite poles N, S.

32-4 Force on a Magnetic Dipole

When placed in a uniform $\mathbf{B}$ field there is no net force acting on a magnetic dipole. There will be, in general, a force exerted on such a dipole if it is placed in a non-uniform $\mathbf{B}$ field. Figure 32-3a shows a magnetic dipole near the north pole, N, of a magnet (e.g. one end of a long solenoid, permanent magnet, etc.). If the dipole aligns itself with the field as shown there will be an attractive force (i.e. toward N) acting on it. This can more easily be seen by considering the analogous electrical situation shown in Figure 32-3b. The force acting on -q is larger in magnitude than the force acting on +q (since the $\mathbf{E}$ field is non-uniform). Therefore the net force acting on the electric dipole is attractive (toward the source, +Q, of the $\mathbf{E}$ field). The student should verify that if the magnetic dipole in Figure 32-3a were placed near a south pole instead of a north pole, the force acting on it would still be attractive (note that the dipole will align itself the other way in this case).

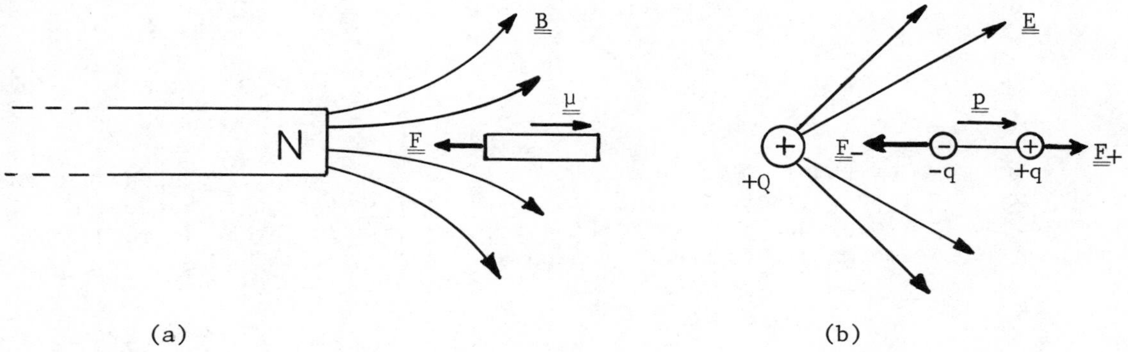

Figure 32-3. (a) A magnetic dipole in a non-uniform $\mathbf{B}$ field. If the dipole moment $\boldsymbol{\mu}$ aligns itself with $\mathbf{B}$ the dipole will experience an attractive force. (b) An electric dipole in a non-uniform $\mathbf{E}$ field. If the dipole moment $\mathbf{p}$ aligns itself with $\mathbf{E}$ the force $\mathbf{F}_-$ is larger in magnitude that the force $\mathbf{F}_+$. The dipole will then experience an attractive force.

32-5 Magnetic Moment of an Atom

An atom can be thought of as consisting of electrons orbiting about a nucleus. Each such electron orbit acts as a current loop and thus has a magnetic moment. This magnetic moment $\boldsymbol{\mu}$ is given by

$$\boldsymbol{\mu} = -\ (e/2m)\ \mathbf{L}\ . \tag{32-4}$$

Here $\mathbf{L} = \mathbf{r} \times m\mathbf{v}$ is the angular momentum of the electron due to its orbital motion.

>>> Example 3. A particle of mass m and charge q executes uniform circular motion. Show that the magnetic moment $\boldsymbol{\mu}$ is proportional to the angular momentum $\mathbf{L}$.

The charge moving in a circle of radius r constitutes a current loop. The current is given by

$$i = |q|/T$$

where $T = 2\pi r/v$ is the period (i.e. the time required for the particle to travel the distance $2\pi r$ at a speed v). The magnitude of the associated magnetic moment is

$$|\underline{\underline{\mu}}| = iA = (|q|/T)\ A = (|q|v/2\pi r)(\pi r^2)$$

$$|\underline{\underline{\mu}}| = |q|vr/2 \quad .$$

The magnitude of the angular momentum is

$$|\underline{\underline{L}}| = mvr \quad .$$

Eliminating v between these two equations gives

$$|\underline{\underline{\mu}}| = (|q|/2m)\ |\underline{\underline{L}}| \quad .$$

The student should verify that the direction of $\underline{\underline{\mu}}$ is the same as that of $\underline{\underline{L}}$ if q is positive and opposite to that of $\underline{\underline{L}}$ if q is negative. Therefore regardless of the actual sign of q,

$$\underline{\underline{\mu}} = (q/2m)\ \underline{\underline{L}} \quad .$$

In particular, for an electron ($q = -e$, $m = m_e$) we have

$$\underline{\underline{\mu}} = -\ (e/2m_e)\ \underline{\underline{L}} \quad .$$

For example, in the diagram below an electron circulates in a counterclockwise direction. The current i is then clockwise. The direction of $\underline{\underline{\mu}}$ is <u>into</u> the page while that of $\underline{\underline{L}}$ is <u>out</u> of the page.

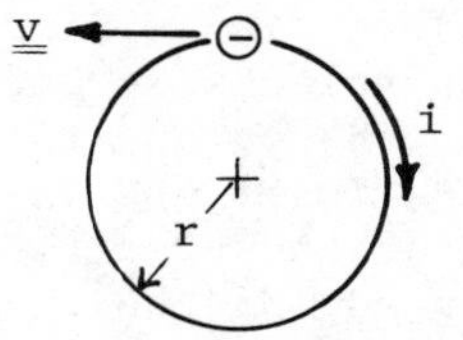

<<<

In addition to this "orbital" magnetic moment (related to its orbital angular momentum), an electron also has a "spin" magnetic moment (related to its spin angular momentum). The magnetic moment of the atom is the (vector) sum of the magnetic moments (both orbital and spin) of all the electrons in the atom.* In some atoms this sum is non-zero; these atoms then possess a <u>permanent magnetic moment</u>. In other atoms this sum vanishes; these atoms have no permanent magnetic moment.

32-6 Magnetic Properties of Matter

Suppose a sample of matter is placed in an externally caused field of magnetic induction $\underline{\underline{B}}_o$. The total $\underline{\underline{B}}$ field within the sample is given by

$$\underline{\underline{B}} = \underline{\underline{B}}_o + \underline{\underline{B}}_M \qquad (32\text{-}5)$$

where $\underline{\underline{B}}_M$ is the $\underline{\underline{B}}$ field due to the magnetic dipoles within the sample. If these dipoles align themselves with $\underline{\underline{B}}_o$ then $\underline{\underline{B}}_M$ tends to <u>aid</u> $\underline{\underline{B}}_o$. Such a material will be <u>attracted</u> by a magnetic pole (e.g. by either end of a solenoid). If, for some reason, the dipoles tend to align themselves opposite to $\underline{\underline{B}}_o$, then $\underline{\underline{B}}_M$ tends to <u>oppose</u> $\underline{\underline{B}}_o$.

---

*The nuclei (protons and neutrons) also have a spin magnetic moment. This is much smaller (by a factor of several hundred) than the spin magnetic moment of an electron.

Such a material will be repelled by a magnetic pole.

The behavior of most materials falls into one of the three categories: paramagnetic, diamagnetic, ferromagnetic. These three types of materials are described below.

32-7 Paramagnetism

In a paramagnetic substance the atoms have a permanent magnetic moment. The magnetic interaction between adjacent atoms is small; the effect of thermal agitation of the atoms is to cause them to be randomly oriented with respect to each other. Thus although each atom has a magnetic moment, a sample of a paramagnetic material exhibits no permanent magnetic moment. When placed in an externally caused $\underline{\underline{B}}$ field the magnetic moment of the atoms tend to align themselves with the field. The fraction of atoms that actually are aligned is usually small due to thermal agitation (i.e. the thermal kinetic energy of the atom, 3/2 kT, is usually much larger than the energy required to turn the dipole in the $\underline{\underline{B}}$ field). A paramagnetic material is therefore slightly attracted by a magnetic pole.

32-8 Diamagnetism

In a diamagnetic substance the atoms have no permanent magnetic moment. As a model of such a material, consider the atom as consisting of two electrons in orbits which are identical except for the sense of their motion as shown in Figure 32-4. Further, suppose that their spin magnetic moments cancel. Then since the two orbital angular moments are equal and opposite, the atom will have no permanent magnetic moment.

Suppose an external $\underline{\underline{B}}$ field is now applied (say into the page). This will affect the motion of the electrons. One way to analyze this is by Faraday's law: $\oint \underline{\underline{E}}\cdot d\underline{\underline{\ell}} = -d(\Phi_B)/dt$. Application of this law shows that a counterclockwise induced $\underline{\underline{E}}$ field will exist during the time that $\underline{\underline{B}}$ is increasing. Since the electron has a negative charge, the force $\underline{\underline{F}} = q\underline{\underline{E}}$ due to this counterclockwise $\underline{\underline{E}}$ field is clockwise, i.e. it is such as to slow down the electron in Figure 32-4a and to speed up the electron in Figure 32-4b. The magnetic moment of the electron in Figure 32-4a (which is into the page) is then decreased and the magnetic moment of the electron in Figure 32-4b (which is out of the page) is increased. The resulting sum of the two magnetic moments is therefore out of the page. Thus the atom, which had no permanent magnetic moment, now has an "induced" magnetic moment aligned so as to oppose the applied $\underline{\underline{B}}$ field. In practice this effect is quite small. A diamagnetic material is therefore slightly repelled by a magnetic pole.

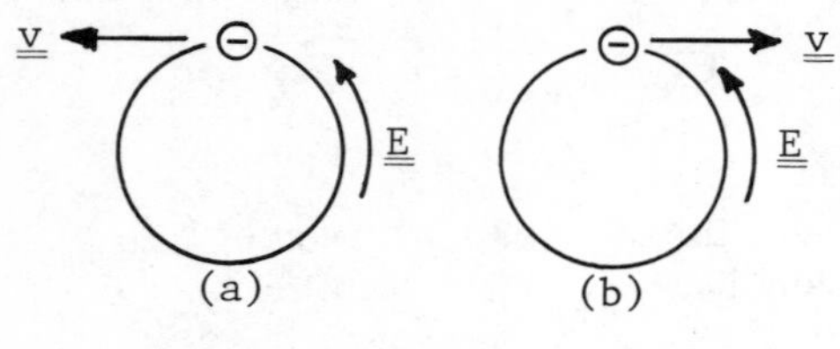

Figure 32-4. An increasing $\underline{\underline{B}}$ field into the page causes a counterclockwise induced $\underline{\underline{E}}$ field. The force $\underline{\underline{F}} = q\underline{\underline{E}} = -e\underline{\underline{E}}$ is then clockwise. (a) The electron's speed is decreased by this force. (b) The electron's speed is increased by this force.

>>> Example 4. Using Faraday's law, derive an expression for the change in the (orbital) magnetic moment of an electron in a diamagnetic material when an external $\underline{\underline{B}}$ field is applied.

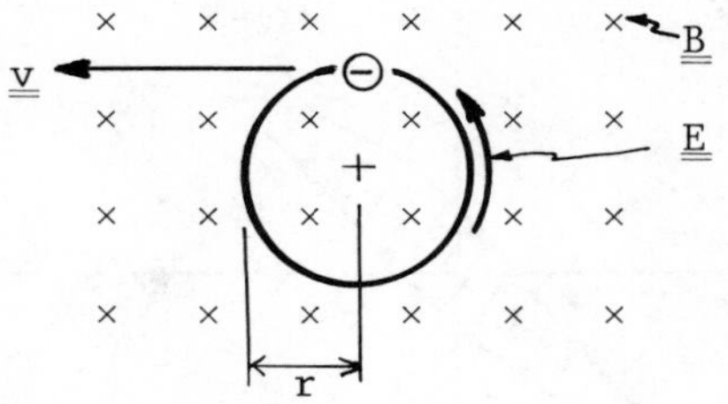

Assume that the electron is in a circular orbit of radius r, the plane of the orbit being perpendicular to $\underline{B}$ as shown. Application of Faraday's law (plus some symmetry assumptions) gives the induced $\underline{E}$ field due to the increasing $\underline{B}$ field:

$$\oint \underline{E}\cdot d\underline{\ell} = -\, d(\Phi_B)/dt$$

$$2\pi r|\underline{E}| = \pi r^2\, d(|\underline{B}|)/dt$$

$$|\underline{E}| = \tfrac{1}{2}\, r\, d(|\underline{B}|)/dt \quad .$$

The torque $\underline{\tau} = \underline{r} \times q\underline{E}$ associated with the force $\underline{F} = q\underline{E}$ is of magnitude

$$|\underline{\tau}| = r|q\underline{E}| = \tfrac{1}{2}\, er^2\, d(|\underline{B}|)/dt \quad .$$

The change in the orbital angular momentum is the angular impulse,

$$\Delta|\underline{L}| = \int |\underline{\tau}|\, dt = \int \tfrac{1}{2}\, er^2[d(|\underline{B}|)/dt]\, dt = \tfrac{1}{2}\, e \int r^2\, d(|\underline{B}|) \quad .$$

Assuming that r does not appreciably change,

$$\Delta|\underline{L}| = \tfrac{1}{2}\, er^2 \int_0^{|\underline{B}|} d(|\underline{B}|) = \tfrac{1}{2}\, er^2|\underline{B}| \quad .$$

Finally, since $\underline{\mu} = -\,(e/2m)\,\underline{L}$

$$\Delta|\underline{\mu}| = (e/2m)\,\Delta|\underline{L}|$$

$$\Delta|\underline{\mu}| = (e^2r^2/4m)\,|\underline{B}| \quad .$$

Remark: Some texts derive this result by considering the change ($= q\underline{v}\times\underline{B}$) in the central force assuming that the orbital radius remains the same. <<<

32-9 Ferromagnetism

In a ferromagnetic substance the atoms have a permanent magnetic moment. If an alternating external field $\underline{B}_o$ is applied to a sample of ferromagnetic material, the $\underline{B}$ field within the sample ($\underline{B} = \underline{B}_o + \underline{B}_M$) is large and follows the general pattern shown in Figure 32-5. The state of the system (as specified by a point P in the graph) follows along the curve in the direction of the arrow. Ferromagnetic materials exhibit hysteresis, i.e. the curve does not retrace itself. For example, the points P and P′ both correspond to the same external field $\underline{B}_o$. At the points where the curve crosses the "B" axis the sample has a magnetic moment with no applied external field (the sample is then a "permanent magnet"). The shaded area in the figure (a result of hysteresis) is proportional to the energy dissipated per cycle. This energy appears in the form of heat (note the analogy to the area enclosed in a P-V graph for a gas).

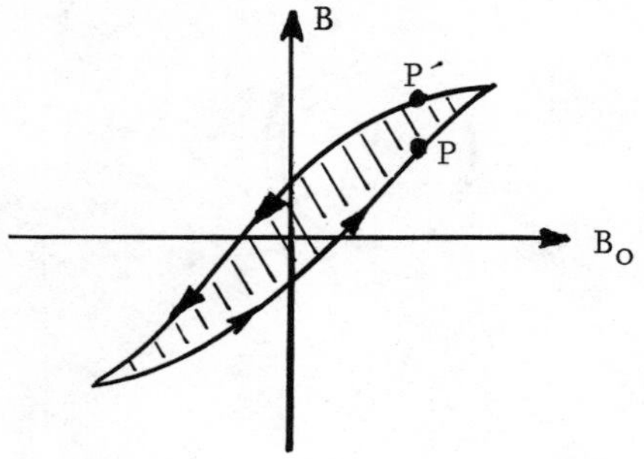

Figure 32-5. The B field within a ferromagnetic material as a function of the applied external field B_o. As B_o alternates, the state of the system (as specified by a point P) follows along the curve in the direction of the arrow. The axes are not drawn to scale; usually B is much larger than B_o.

The behavior of a ferromagnetic material is explained in terms of regions called domains. The interaction between adjacent atoms is large causing the atoms in any one domain to be essentially aligned with one another. Each domain then has a large magnetic moment. The various domains however might be randomly oriented with respect to each other, thus the sample may have no net magnetic moment. When an external $\underline{\underline{B}}$ field is applied the sample acquires a large magnetic moment, this magnetic moment being aligned with the external field. This is due to:

(a) Those domains which are aligned with $\underline{\underline{B}}$ may grow in size at the expense of some adjacent domains which are not aligned with $\underline{\underline{B}}$.
(b) The atoms in any given domain may become more aligned with $\underline{\underline{B}}$, the entire domain swinging around as a unit.

When the external field is removed, the above processes may not completely reverse themselves. This results in hysteresis.

In summary, ferromagnetic materials are generally strongly attracted by a magnetic pole. Such materials can acquire a permanent magnetic moment by being temporarily placed in an external $\underline{\underline{B}}$ field; the material is then said to be "permanently magnetized".

Chapter 33: ELECTROMAGNETIC OSCILLATIONS

REVIEW AND PREVIEW

In previous chapters you studied RC and RL circuits. These led to exponential functions of time (for the charge in the RC case and for the current in the RL case). In this chapter you will study LC and LCR circuits. You will see that these lead to oscillatory functions of time.

GOALS AND GUIDELINES

In this chapter the major goal consists of understanding some of the important properties of single loop circuits containing an inductor, a capacitor and a resistor.

a. Learning how to treat a simple LC circuit. From L and C you should be able to calculate the frequency of the resulting oscillation (Eq. 33-2). You might want to review simple harmonic motion in general (Chapter 13).
b. Learning how to treat a simple LCR circuit. Here you should not try to remember the detailed formula for the charge as a function of time; rather you should try to understand the qualitative effects caused by the addition of the resistance R.
c. Understanding the forced oscillation of an LCR circuit when connected to an alternating source of emf. In particular, you should try to gain an understanding of resonance.

33-1 Single Loop LC Circuits

In the single loop LC circuit shown below, q denotes the charge on the capacitor C and i the current in the inductor L. Both of these will be functions of time: $q = q(t)$, $i = i(t)$. We may obtain an equation for this circuit by summing the potential differences around the loop.

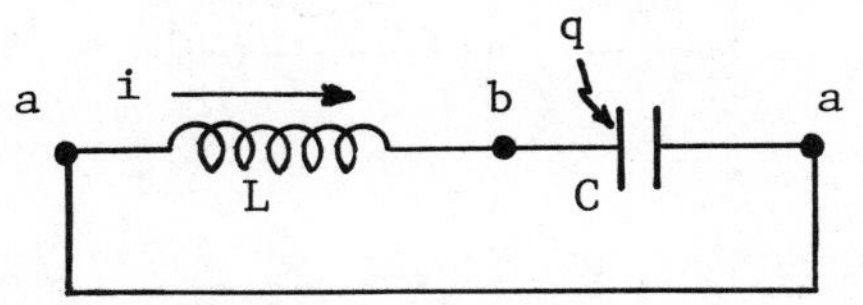

$$(V_a - V_b) + (V_b - V_a) = 0$$

$$(L\, di/dt + (q/C) = 0 \quad .$$

For q and i defined as shown, we have $i = + dq/dt$. Then $di/dt = d^2q/dt^2$ and

$$L\, d^2q/dt^2 + q/C = 0 \quad . \qquad (33\text{-}1)$$

>>> Example 1. Derive the differential equation for the LC circuit using energy considerations.

The energy stored in the circuit is

$$U = U_L + U_C$$

$$= \tfrac{1}{2} Li^2 + \tfrac{1}{2} q^2/C \quad .$$

Since for this circuit there is no loss of energy (the circuit contains no resistance), the stored energy U must be a constant. We have

$$dU/dt = 0$$

$$\frac{d}{dt}\left(\tfrac{1}{2}\, Li^2 + \tfrac{1}{2}\, q^2/C\right) = 0$$

$$Li\, di/dt + (q/C)\, dq/dt = 0 \ .$$

Using $i = dq/dt$ and cancelling the common factor of i we find

$$L\, d^2q/dt^2 + q/C = 0 \quad .$$

<<<

Equation (33-1) is a differential equation for q as a function of t. Since it is of second order, its general solution will contain two arbitrary constants. According to (33-1), q is proportional to the negative of d^2q/dt^2; we are therefore led to try a solution of the form $q(t) = A \cos(\omega t + \phi)$ where A, ω, ϕ are constants. Substitution shows that this satisfies (33-1) provided that $\omega = \sqrt{1/LC}$.

$$q(t) = A \cos(\omega t + \phi) \quad , \ \omega = \sqrt{1/LC} \ . \tag{33-2}$$

The two remaining constants, A and ϕ, are arbitrary. Their values may be determined for any particular problem if we know the initial conditions (q and dq/dt at t = 0). The behavior of q as a function of time is shown in Figure 33-1.

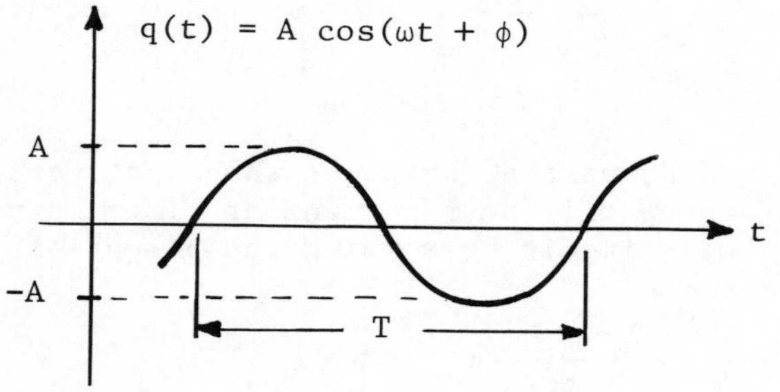

Figure 33-1. Behavior of q as a function of time for an LC circuit. The period $T = 2\pi/\omega = 2\pi\sqrt{LC}$.

Equation (33-2) says that the circuit (i.e. q and i = dq/dt) oscillates according to simple harmonic motion (SHM) with an angular frequency $\omega = \sqrt{1/LC}$. The student should now review the basic concepts concerning SHM. Of particular importance to oscillating circuits are:

ω, the angular frequency (radians/time);

ν, the frequency (cycles/time);

T, the period (time/cycle).

These are related by $\nu = \omega/2\pi$, $T = 1/\nu$. The argument (such as $\omega t + \phi$) of any trigonometric function which appears in SHM is always understood to be expressed in radians.*

---

*The familiar differentiation formulas such as $d(\sin\theta)/d\theta = \cos\theta$ are valid only if θ is expressed in radians.

>>> Example 2. In the LC circuit shown, $C = 1\ \mu F$. With C charged to 100 volts, switch S is suddenly closed at time $t = 0$. The circuit then oscillates at 10^3 cycles per second.
(a) Calculate ω, T.
(b) Express q as a function of time.
(c) Calculate the inductance L.
(d) Calculate the average current during the first quarter cycle.

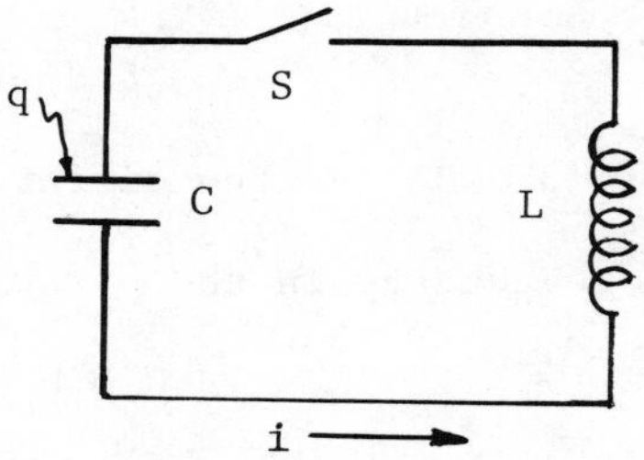

(a)

$$\omega = 2\pi\nu = 2\pi(10^3\ s^{-1}) = 6.28 \times 10^3\ \text{(radian)/second} \quad .$$

$$T = 1/\nu = 1/(10^3\ s^{-1}) = 10^{-3}\ \text{second} \quad .$$

(b) In general, $q(t) = A \cos(\omega t + \phi)$. In this problem $\phi = 0$ (since $i = dq/dt$ is zero initially), and A = initial charge = $CV = (10^{-6}\ F)\cdot(10^2\ V) = 10^{-4}$ coulomb. Thus

$$q(t) = (10^{-4}\ C) \cos((6.28 \times 10^3\ s^{-1})\ t) \quad .$$

(c)

$$\omega = \sqrt{1/LC}$$

$$L = \frac{1}{\omega^2 C} = \frac{1}{4\pi^2\nu^2 C} = \frac{1}{4\pi^2(10^3\ s^{-1})^2(10^{-6}\ F)}$$

$$= 2.53 \times 10^{-2}\ H \quad .$$

(d) During the first quarter cycle, q changes from 10^{-4} C to zero. Thus $\Delta q = -\ 10^{-4}$ C. Also, $\Delta t = T/4 = 0.25 \times 10^{-3}$ s. The average current during this time is then C

$$i_{av} = \Delta q/\Delta t = (-\ 10^{-4}\ C)/(0.25 \times 10^{-3}\ s)$$

$$i_{av} = -\ 0.400\ A \quad .$$

Remark: The instantaneous current at the end of the quarter cycle is

$$i = dq/dt = -\ \omega A \sin(\omega(T/4)) = -\ \omega A \sin(\pi/2)$$

$$= -\ \omega A = -\ (6.28 \times 10^3\ s^{-1})\cdot(10^{-4}\ C)$$

$$= 0.628\ A \quad .$$

<<<

33-2 Single Loop LCR Circuits

We may obtain an equation for the single loop LCR circuit shown below by summing the potential differences around the loop.

$$(V_a - V_b) + (V_b - V_c) + (V_c - V_a) = 0$$

$$(L\ di/dt) + (iR) + (q/C) = 0 \quad .$$

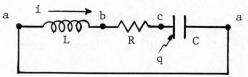

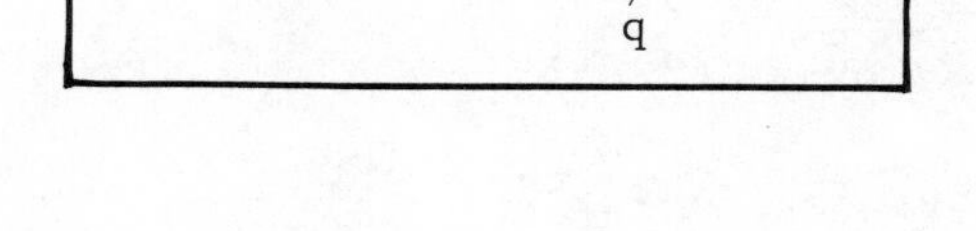

For q and i defined as shown, we have $i = + dq/dt$. Then

$$L\, d^2q/dt^2 + R\, dq/dt + q/C = 0 \quad . \qquad (33\text{-}3)$$

>>> Example 3. Derive the differential equation for the LCR circuit using energy considerations.

The stored energy in the circuit is

$$U = U_L + U_C = \tfrac{1}{2} Li^2 + \tfrac{1}{2} q^2/C \quad .$$

In this circuit there is an energy loss due to Joule heating in the resistor. The rate of this energy loss is the power dissipated by the resistor (i^2R). Thus the rate of change of the stored energy in the circuit is

$$du/dt = - i^2R$$ (the minus sign arises since i^2R is the rate of energy <u>loss</u>)

$$\frac{d}{dt} (\tfrac{1}{2} Li^2 + \tfrac{1}{2} q^2/C) = - i^2R$$

$$Li\, di/dt + (q/C)\, dq/dt = - i^2R \quad .$$

Using $i = dq/dt$ and cancelling the common factor of i yields

$$L\, d^2q/dt^2 + R\, dq/dt + q/C = 0 \quad .$$ <<<

The general solution to the second order differential equation (33-3) is

$$q(t) = Ae^{-Rt/2L} \cos(\omega' t + \phi) \quad , \quad \omega' = \sqrt{(1/LC) - (R/2L)^2} \quad . \qquad (33\text{-}4)$$

The LCR solution (33-4) has certain properties which should be compared with the LC solution (33-2):

(a) Both contain the two arbitrary constants A and ϕ. These may be determined for any particular problem if we know the initial conditions (q and dq/dt at t = 0).

(b) Both contain an <u>oscillatory term</u>. For the LCR circuit the angular frequency is $\omega' = \sqrt{(1/LC) - (R/2L)^2}$, for the LC circuit it is $\omega = \sqrt{1/LC}$. In many problems R is small enough so that these two expressions are essentially equal.

(c) The LCR solution contains a <u>damping</u> term $e^{-Rt/2L}$. The LC solution is of course undamped.

In summary, the presence of the resistance R has the following two effects:

(i) It lowers the angular frequency of oscillation slightly from $\omega = \sqrt{1/LC}$ to $\omega' = \sqrt{(1/LC) - (R/2L)^2}$.

(ii) It introduces a damping term $e^{-Rt/2L}$.

The behavior of q as a function of time is shown in Figure 33-2.

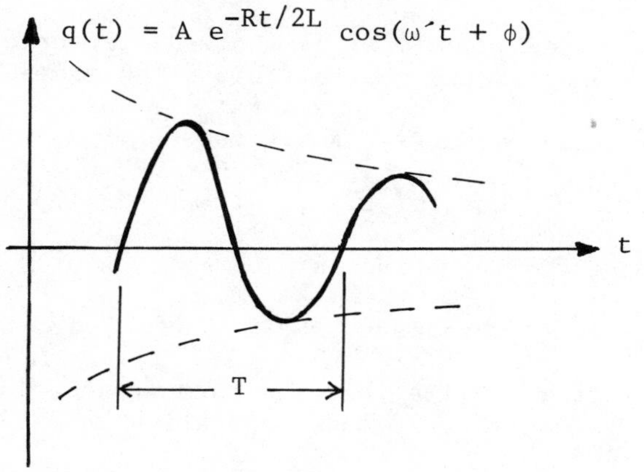

Figure 33-2. Behavior of q as a function of time for an LCR circuit. The period $T = 2\pi/\omega' = 2\pi/\sqrt{(1/LC) - (R/2L)^2}$. The envelope (dotted line) is given by the damping term $e^{-Rt/2L}$.

>>> Example 4. An oscillating LCR circuit has R = 1 Ω, C = 1 μF, L = 0.01 H. (a) Calculate the period. (b) After how many cycles will the stored energy be reduced to one half of its initial value?

(a)
$$\omega' = \sqrt{(1/LC) - (R/2L)^2}$$

$$= \sqrt{\frac{1}{(10^{-2}\text{ H})(10^{-6}\text{ F})} - \left[\frac{1\ \Omega}{2 \times 10^{-2}\text{ H}}\right]^2}$$

$$= \sqrt{10^8 - 2.5 \times 10^3}\ \text{s}^{-1} = 10^4\ \text{(radian)/second.}$$

Note that the "R/2L" term was negligible. The period is given by

$$T = 1/\nu = 2\pi/\omega' = 2\pi/(10^4\ \text{s}^{-1}) = 6.28 \times 10^{-4}\ \text{second.}$$

(b) Both q and i contain the damping term $e^{-Rt/L}$. Therefore both q^2 and i^2 will contain a damping term $(e^{-Rt/2L})^2 = e^{-Rt/L}$. The stored energy $U = \frac{1}{2}Li^2 + \frac{1}{2}q^2/C$ will then exhibit damping according to the term $e^{-Rt/L}$. To calculate the time at which this term is equal to one half of its initial value we write

$$e^{-Rt/L} = \tfrac{1}{2}e^0 = \tfrac{1}{2}\ .$$

Taking natural logarithms,

$$-\ Rt/L = \ln(\tfrac{1}{2}) = -\ \ln(2) = -\ 0.69$$

$$t = 0.69\ L/R = (0.69)(10^{-2}\ \text{H})(1\ \Omega) = 6.9 \times 10^{-3}\ \text{second.}$$

To find how many cycles this time represents,

$$\frac{t}{T} = \frac{6.9 \times 10^{-3}\ \text{s}}{6.28 \times 10^{-4}\ \text{s/(cycle)}} = 11\ \text{cycles.}$$

<<<

33-3 Forced Oscillations in LCR Circuits

Suppose that an alternating source of emf $\mathcal{E}(t) = \mathcal{E}_m \cos(\omega'' t)$ is applied (in series) to an LCR circuit. Here ω'' is the angular frequency of the source (this is an independent quantity, not related to the values of L, C, R). The constant $\mathcal{E}_m$ is the "strength" of the source. The differential equation for this circuit is

$$L\, d^2q/dt^2 + R\, dq/dt + q/C = \mathcal{E}_m \cos(\omega'' t) \quad . \tag{33-5}$$

The general solution to this second order differential equation is

$$q(t) = (\mathcal{E}_m/G) \sin(\omega'' t - \phi) + \text{"transient"} \quad . \tag{33-6}$$

The "transient" term, which becomes negligible after a while, has the form of equation (33-4); it contains two arbitrary constants. The constants G and ϕ appearing in (33-6) are not arbitrary, they are given by

$$G = \omega'' \sqrt{(\omega'' L - 1/\omega'' C)^2 + R^2} \quad , \tag{33-7a}$$

$$\phi = \cos^{-1}(R\omega''/G) \quad . \tag{33-7b}$$

The relations (33-7) can easily be remembered by referring to the geometric construction shown in Figure 33-3.

Ignoring the "transient" term, the important features of the solution (33-6) are:

(a) The circuit oscillates with the same angular frequency, ω'', as the source; this is known as a forced oscillation.

(b) These oscillations are undamped.

(c) For a given source strength $\mathcal{E}_m$, the amplitude of the oscillations depends upon the angular frequency, ω'', of the source. There is a resonance (large current amplitude) when ω'' approaches $\sqrt{1/LC}$. This resonance is "sharp" if R is small, "broad" if R is large.

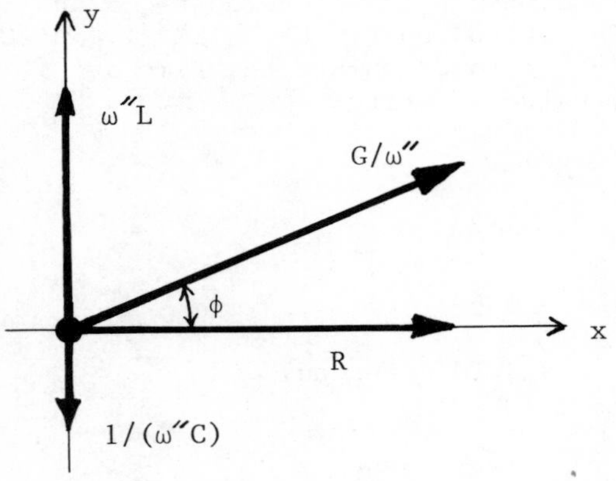

Figure 33-3. This geometric construction is a method of remembering equations (33-7). Consider three vectors: a vector of magnitude $\omega'' L$ in the +y direction, a vector of magnitude $1/\omega'' C$ in the -y direction, a vector of magnitude R in the +x direction. The magnitude of their vector sum is $\sqrt{(\omega'' L - 1/\omega'' C)^2 + R^2} = G/\omega''$. Also, the angle between this vector sum and the +x axis is $\phi = \cos^{-1}(R\omega''/G)$. Note that all the vectors in the figure have the units of "ohms".

33-4 Programmed Problems

1\. 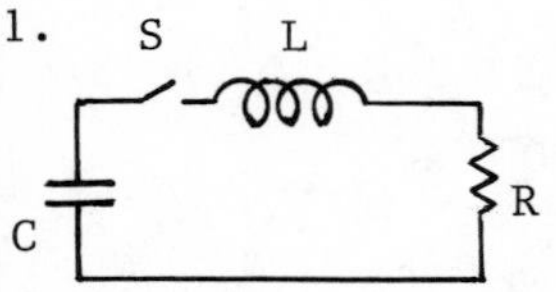

A capacitor initially charged as shown is connected in series with an inductor L and a resistor R through the switch S.

The voltage on the capacitor after the switch S is closed is given by

$C = 0.001\ \mu F$
$L = 25 \times 10^{-3}$ H
$R = 1000\ \Omega$

$$V = V_m e^{-Rt/2L} \cos \omega t \quad .$$

For the value of R given here the angular frequency of oscillation of this LCR circuit is given by

$$\omega = \sqrt{(1/LC) - (R/2L)^2}$$

ν = _______ cycles/second (i.e. Hz).

Obtain a numerical result.

$2\pi\nu = \omega$

$\omega = 2 \times 10^5\ s^{-1}$

$\nu = 32 \times 10^3$ Hz .

Note that $(R/2L)^2$ is here much less than 1/LC. [Because of this the initial condition involving the current (i = 0 when t = 0) can be approximately satisfied using only a cos ωt term. In general, one would need cos $(\omega t + \phi)$].

The symbol ω used here (i.e. the angular frequency of this undriven LCR series circuit) is denoted by ω' in section 33-2.

2\. What is the period of this oscillation?

T = _________ second/(cycle).

$T = 1/\nu$

$T = 1/(32 \times 10^3$ Hz)

$T = 30 \times 10^{-6}$ s/(cycle)

or 30 μs/(cycle) .

3\. This circuit is called a damped oscillator because the oscillations die out as time goes on, e.g., the amplitude of the voltage on C becomes progressively smaller.

We can see this from the equation

$$V = V_m e^{-Rt/2L} \cos \omega t$$

by looking at the amplitude term

$$V_m e^{-Rt/2L} \quad .$$

What is R/2L for this circuit? The values are given in frame 1.

R/2L = _________ .

(Ignore units for now.)

$$\frac{R}{2L} = \frac{1 \times 10^3}{50 \times 10^{-3}} = 2 \times 10^4 \quad .$$

4. The quantity 2L/R serves as a "time constant" τ for this problem. Compute the value of τ in this problem. (Note: henry/ohm = second.)

$$\tau = 2L/R = \_\_\_\_\_\_\_\_ \quad .$$

$$\frac{R}{2L} = 2 \times 10^4$$

$$\frac{2L}{R} = \tau = 50 \times 10^{-6} \text{ s}$$

$$\tau = 50 \ \mu s \quad .$$

5. Having defined this constant τ the equation for the voltage on the capacitor can be written as a ratio

$$\frac{V}{V_m} = e^{-t/\tau} \cos \omega t \quad .$$

What is the maximum value of V/V_m?

1

It occurs initially at t = 0 .

6. Fill in the table below with the aid of the table to the left. From frame 4, τ = 50 μs.

x	e^{-x}	t (μs)	t/τ	$e^{-t/\tau}$
0	1	0		
0.2	0.82	10		
0.4	0.67	20		
0.6	0.55	30		
0.8	0.45	40		
1.0	0.37	50		
1.2	0.30	60		
1.4	0.25	70		
1.6	0.20	80		
1.8	0.16	90		
2.0	0.14	100		

t (μs)	t/τ	$e^{-t/\tau}$
0	0	1
10	0.2	0.82
20	0.4	0.67
30	0.6	0.55
40	0.8	0.45
50	1.0	0.37
60	1.2	0.30
70	1.4	0.25
80	1.6	0.20
90	1.8	0.16
100	2.0	0.14

7. Plot the data from the previous answer above and below the time axis on the graph below.

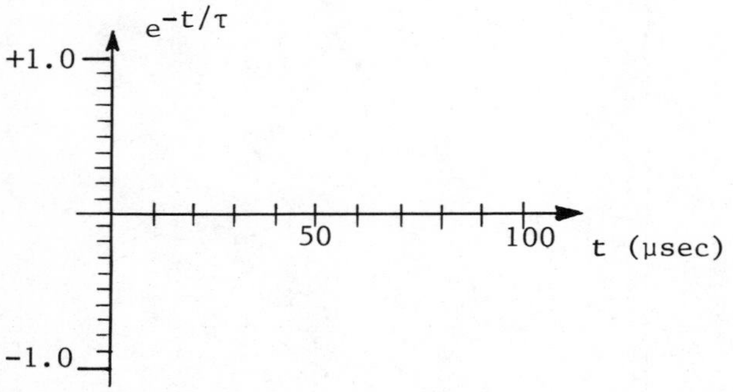

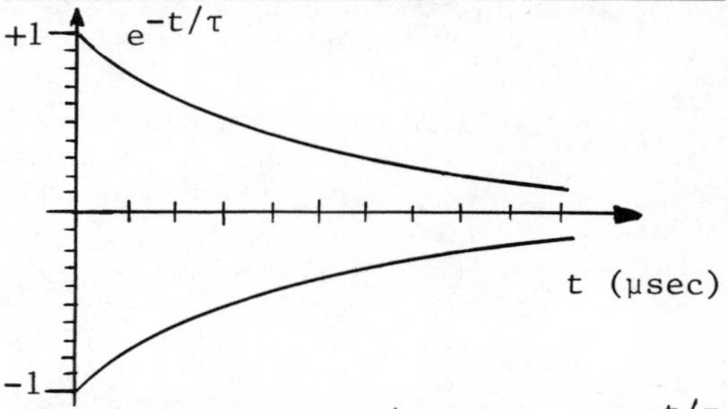

The amplitude of V/V_m goes as $e^{-t/\tau}$. Here we have sketched the envelope of the amplitude of oscillation. The oscillation amplitude at any time t cannot exceed this envelope. You were asked to include the bottom exponential since the oscillating term is sometimes negative.

8. 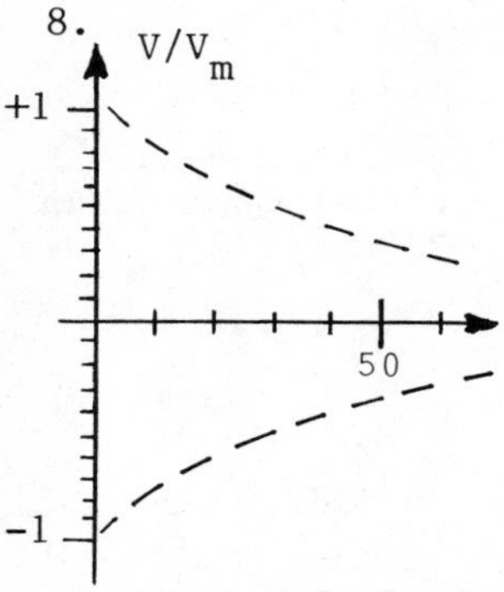

On the graph we have relabeled the ordinate. The equation to be plotted is

$$V/V_m = e^{-t/\tau} \cos \omega t.$$

The exponential lines are guides within which the oscillations must occur.

At t = 0, both the amplitude term, $e^{-t/\tau}$, and $\cos \omega t$ are 1. Look back at frame 2 and sketch V/V_m as a function of time.

From frame 2, T = 30 μs/cycle which is the duration of one cycle.

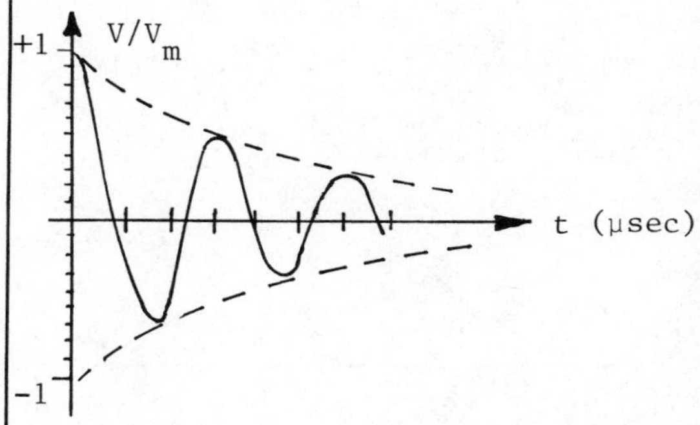

Note that one cycle covers 30 μs as required. This is how damped oscillations appear.

---

9. 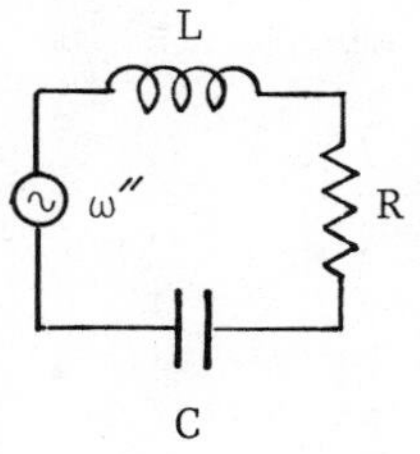

At what frequency of the driving source will this circuit resonate?

The source of emf has amplitude $\mathcal{E}_m$ and angular frequency ω''. If we consider only the steady state condition the charge on the capacitor as a function of time is

$$q(t) = \frac{\mathcal{E}_m}{G} \sin (\omega'' t - \phi)$$

where

$$G = \omega'' \sqrt{(\omega'' L - 1/\omega'' C)^2 + R^2}$$

and

$$\phi = \cos^{-1} \frac{R\omega''}{G} .$$

Differentiate q(t) with respect to time to obtain an expression for the current.

i(t) = __________ .

$$i(t) = \omega'' \frac{\mathcal{E}_m}{G} \cos (\omega'' t - \phi) .$$

Again this is an amplitude $\omega'' \mathcal{E}_m / G$ times an oscillating term $\cos (\omega'' t - \phi)$.

---

10. The oscillating term varies between the extremes of ______ and _______ .

+1, -1 .

11. At either one of the extremes of the oscillating term, i(t) would be the maximum possible current. At the extreme +1, the current as given by

$$i(t) = \omega'' \frac{\mathcal{E}_m}{G} \cos (\omega'' t - \phi)$$

is $i =$ ____________ .

$$\omega'' \frac{\mathcal{E}_m}{G}$$

This is the maximum possible current. Note that it depends upon $\mathcal{E}_m$, ω'', R, L and C.

$$i_m = \omega'' \frac{\mathcal{E}_m}{G} .$$

The subscript m indicates maximum.

12. Using the expression for G, the above answer can be written

$$i_m = \frac{\mathcal{E}_m}{\sqrt{(\omega'' L - 1/\omega'' C)^2 + R^2}}$$

For a given circuit R, C, and L are constants. If we imagine that the amplitude $\mathcal{E}_m$ of the source is fixed but that its frequency ω'' can be varied, then the actual value of i_m will change due to the term $(\omega'' L - (1/\omega'' C))^2$ in the denominator.

What value of this term will make the denominator a minimum?

Zero.

The denominator is the sum of two terms, one of which is a constant R^2. The whole thing is a minimum when $(\omega'' L - 1/\omega'' C))^2$ is zero.

13. With the stipulation that $(\omega'' L - (1/\omega'' C))^2$ is zero, the current amplitude will be a maximum, which for a series RLC circuit is called the resonant condition.

For maximum current amplitude

$$\omega'' L = \frac{1}{\omega'' C} .$$

Solve this equation for ω''.

$\omega'' =$ __________ .

$$\omega'' = \frac{1}{\sqrt{LC}} .$$

This is the resonant condition.

Chapter 34: ALTERNATING CURRENTS

IMPORTANT NOTICE

There are two versions of this Study Guide Chapter 34 on Alternating Currents.

If your text is titled "PHYSICS" (whose chapter on alternating currents is number 39) use Study Guide pages 391 P to 397 P.

If your text is titled "FUNDAMENTALS OF PHYSICS" (whose chapter of alternating currents is number 36) use Study Guide pages 391 F to 397 F.

The Table of Contents as well as the Index of this Study Guide remain valid provided one uses the correct ("P" or "F") set of pages.

SEE IMPORTANT NOTE ON p. 391 SEE IMPORTANT NOTE ON p. 391

Chapter 34: ALTERNATING CURRENTS

REVIEW AND PREVIEW

In previous chapters you studied circuits containing resistors, inductors, capacitors and sources of electromotive force. Both steady state and time-varying cases were considered; the time-varying case led to a differential equation.

In this chapter you will study the special but important case in which the time dependence of the source is a sinusoidal function. Special techniques will be introduced to handle this problem using only algebra rather than differential equations.

GOALS AND GUIDELINES

Basically the goal is to understand the algebraic technique for handling sinusoidally varying voltages and currents. This goal can be broken down into the following pieces:

(a) reactance--a property of an inductor or a capacitor.
(b) impedance--a property of a circuit containing R, L or C.
(c) phase angle--the amount by which the voltage and current are out of phase.
(d) power factor--used to calculate the power dissipated by a circuit.

Special formulas will be introduced which permit an algebraic treatment of circuits. You should keep in mind that these apply only to sinusoidally varying sources.

34-1 Sources of Alternating EMF

We shall deal with circuits of the following form:

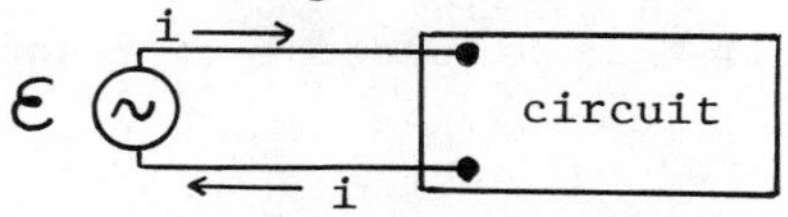

The symbol (∿) denotes a source of potential difference which is sinusoidal in time,

$$\mathcal{E}(t) = \mathcal{E}_m \sin \omega t \ . \qquad (34\text{-}1)$$

Here

$\mathcal{E}_m$ = amplitude (maximum value) of the emf [volt] ,

ω = angular frequency [(radian)/s] ,

t = time [s] .

34-2 General Response of the Circuit

It turns out that the response (i.e. the current) due to a source of the form (34-1) is

$$i(t) = i_m \sin(\omega t - \phi) + \text{"transient"} \ .$$

The "transient" term consists of decaying exponentials. After a short while it is not important; hence we shall deal only with the "steady state" term

$$i(t) = i_m \sin(\omega t - \phi) \ . \qquad (34\text{-}2)$$

SEE IMPORTANT NOTE ON p. 391

The important qualitative points concerning this response are:

(a) $i(t)$ is also sinusoidal with the <u>same</u> (angular) <u>frequency</u> as the source.

(b) There can be a <u>phase difference</u> ϕ between the voltage and the current. In particular if

(i) $\phi > 0$ then the "voltage <u>leads</u> the current by ϕ",

(ii) $\phi < 0$ then the "voltage <u>lags</u> the current by ϕ".

(c) The amplitude of the response, i_m, is proportional to the amplitude of the source, $\mathcal{E}_m$.

From (c) we may write

$$i_m = \mathcal{E}_m/Z \tag{34-3}$$

where the quantity Z (to be defined later) is called the impedance of the circuit [unit: ohm].

Equation (34-3) looks somewhat like Ohm's law. One must keep in mind though that

(i) it relates only the two amplitudes, not the two functions (in fact since voltage and current are generally out of phase they cannot possibly be related by proportionality);

(ii) in general Z is a function of frequency, $Z = Z(\omega)$.

34-3 Single Element Circuits

In this section we shall consider the three single element circuits shown in Figure 34-1. In each case the emf $\mathcal{E}(t)$ is given by (34-1). We shall see that the current $i(t)$ is indeed of the form (34-2).

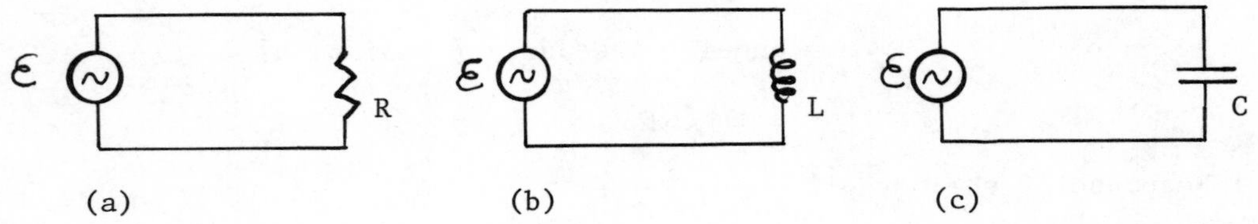

Figure 34-1. A source of sinusoidally varying emf is connected to a single element circuit: (a) resistor R, (b) inductor L, (c) capacitor C.

SEE IMPORTANT NOTE ON p. 391

For the single element circuit consisting of a resistor as shown in Figure 34-1a, the loop equation gives

$$i(t) = \mathcal{E}(t)/R$$
$$= (\mathcal{E}_m \sin \omega t)/R$$
$$= [\mathcal{E}_m/R] \sin(\omega t - 0) \quad .$$

By comparison with (34-2) and (34-3) we see that

$$Z = R \quad , \qquad (34\text{-}4a)$$

$$\phi = 0 \quad . \qquad (34\text{-}4b)$$

Thus for a resistor, Z is simply equal to R; also the voltage and the current are in phase.

For the single element circuit consisting of an inductor as shown in Figure 34-1b, the loop equation gives

$$L\, di(t)/dt = \mathcal{E}(t)$$
$$di/dt = \mathcal{E}/L$$
$$= (\mathcal{E}_m \sin \omega t)/L \quad .$$

This can be integrated to give

$$i(t) = -(\mathcal{E}_m \cos \omega t)/(\omega L)$$
$$= [\mathcal{E}_m/(\omega L)] \sin(\omega t - \pi/2) \quad .$$

By comparison with (34-2) and (34-3) we see that

$$Z = \omega L \quad ,$$

$$\phi = \pi/2 \quad .$$

The quantity ωL, a joint property of the inductor and the frequency, is called the inductive reactance X_L [unit: ohm]. Thus

$$Z = X_L \quad , \qquad (34\text{-}5a)$$

$$X_L = \omega L \quad , \qquad (34\text{-}5b)$$

$$\phi = \pi/2 \ . \qquad (34\text{-}5c)$$

Thus for an inductor, Z is equal to the inductive reactance; also the voltage leads the current by $\pi/2$ (= 90°).

SEE IMPORTANT NOTE ON p. 391

For the single element circuit consisting of a capacitor as shown in Figure 34-1c, the loop equation gives

$$q(t)/C = \mathcal{E}(t)$$

$$i(t) = dq/dt = C\,\mathcal{E}_m\,\omega \cos \omega t$$

$$= [\mathcal{E}_m/(1/\omega C)] \sin(\omega t - (-\pi/2)) \quad .$$

Again by comparison with (34-2) and (34-3) we see that

$$Z = 1/(\omega C) \quad ,$$

$$\phi = -\pi/2 \quad .$$

The quantity $1/(\omega C)$, a joint property of the capacitor and the frequency is called the capacitive reactance X_C [unit: ohm]. Thus

$$Z = X_C \quad , \qquad (34\text{-}6a)$$

$$X_C = 1/\omega C \quad , \qquad (34\text{-}6b)$$

$$\phi = -\pi/2 \quad . \qquad (34\text{-}6c)$$

Thus for a capacitor, Z is equal to the capacitive reactance; also the voltage lags the current by $\pi/2$ (= 90°).

>>> Example 1. Given an inductor L = 1 H and a capacitor C = 1 μF, calculate their reactances at frequencies of (a) 60 Hz, (b) 1000 Hz.

Solution: Note that the frequency, f, is given (Hz = cycle/second). We must use the angular frequency ω (radian/second). The conversion formula is $\omega = 2\pi f$.

(a)
$$X_L = \omega L = 2\pi f L = 2\pi(60)(1) = 377\ \Omega \quad ,$$

$$X_C = 1/\omega C = 1/2\pi f C = 1/[2\pi(60)(10^{-6})] = 2653\ \Omega \quad .$$

(b) Similarly with f = 1000 Hz,

$$X_L = 6283\ \Omega \ , \quad X_C = 159\ \Omega \quad .$$

Remark: Note that either circuit element (L or C) can have the larger reactance depending on the frequency. <<<

34-4 RLC Series Circuit

We now generalize to a series circuit consisting of all three circuit elements: resistance R, inductance L and capacitance C. The circuit diagram is shown in Figure 34-2.

SEE IMPORTANT NOTE ON p. 391

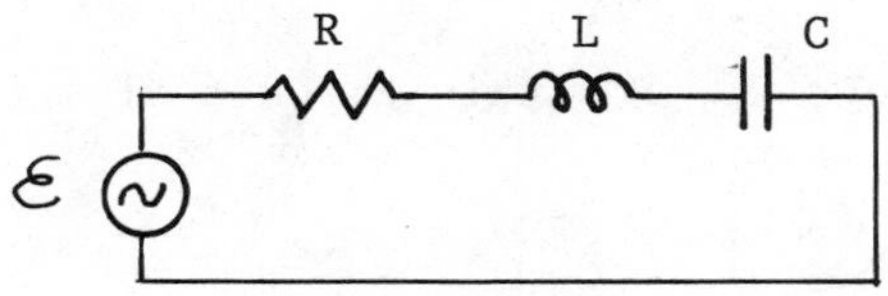

Figure 34-2. An RLC series circuit connected to a sinusoidally varying source of emf. Note that the current is the same for each of the circuit elements since they are connected in series.

We seek a generalization of our previous equations (34-4), (34-5) and (34-6) each of which applied to a single element circuit. This generalization must reduce to one of these three in the special case where there is only one circuit element. It can be shown that the correct generalization of these formulas is

$$Z = \sqrt{R^2 + (X_L - X_C)^2} \quad , \tag{34-7a}$$

$$\phi = \tan^{-1}[(X_L - X_C)/R] \quad . \tag{34-7b}$$

The quantity Z is called the <u>impedance</u> [unit: ohm]. It is the proportionality factor which relates voltage and current amplitudes. The angle ϕ tells us by how much the voltage leads the current. The term $(X_L - X_C)$ is called the <u>total reactance</u>, X. In terms of X the above equations become

$$Z = \sqrt{R^2 + X^2} \quad , \tag{34-8a}$$

$$\phi = \tan^{-1}(X/R) \quad . \tag{34-8b}$$

With regard to limiting cases of these formulas, note that if R or L is absent we simply set that quantity equal to zero; however if C is absent we must formally set $C = \infty$ [physically a "short circuit"] to obtain $X_c = 0$.

The concept of impedance can be applied to any part of the circuit as well as to the circuit as a whole. The current-voltage (amplitude) relation is $i_m = V_m/Z$ where Z is the impedence of that part of the circuit under consideration. The following example illustrates some of these ideas.

>>> Example 2. At a certain frequency the circuit shown has the following parameters.

$X_L = 5000\ \Omega$

$X_C = 1000\ \Omega$

$R = 3000\ \Omega$

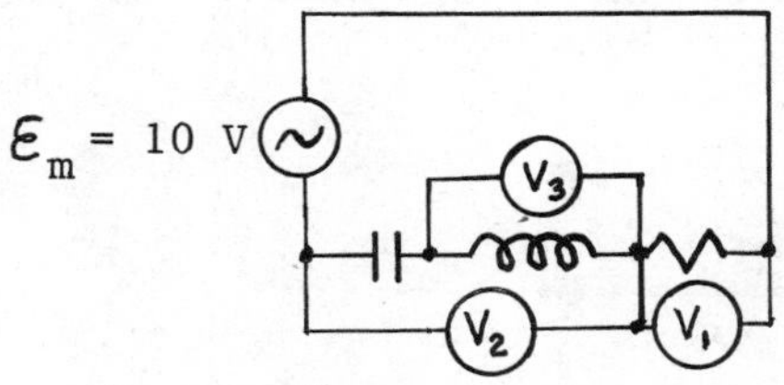

Calculate (a) Z and ϕ, (b) i_m, (c) the voltages read by the three voltmeters V_1, V_2 and V_3. [Note: assume, for purposes of this example, that these voltmeters read the amplitude of the voltage.]

Solution: (a) $Z = \sqrt{R^2 + (X_L - X_C)^2}$

$Z = \sqrt{(3000)^2 + (5000 - 1000)^2} = 5000\ \Omega$.

$\phi = \tan^{-1}[(X_L - X_C)/R] = \tan^{-1}(4000/3000) = 53^\circ$.

(b) $i_m = \mathcal{E}_m/Z = (10\ \text{V})/(5000\ \Omega) = 2 \times 10^{-3}\ \text{A}$.

SEE IMPORTANT NOTE ON p. 391

(c) To find V_1: V_1 is the voltage across R. For this circuit element

$$Z = R$$

$$V_1 = i_m Z = (2 \times 10^{-3}\ \text{A})(3000\ \Omega) = 6\ \text{V} \quad .$$

To find V_2: V_2 is the voltage across the LC part of the circuit.

$$Z = \sqrt{(X_L - X_C)^2} = 4000\ \Omega$$

$$V_2 = i_m Z = (2 \times 10^{-3}\ \text{A})(4000\ \Omega) = 8\ \text{V} \quad .$$

To find V_3: V_3 is the voltage across L. For this circuit element

$$Z = X_L$$

$$V_3 = i_m Z = (2 \times 10^{-3}\ \text{A})(5000\ \Omega) = 10\ \text{V} \quad .$$ <<<

Remarks: We have used the fact that the current is the same in every part of this series circuit. Note that the voltages do not "add up" as you might guess. For example $V_1 + V_2 \neq \mathcal{E}_m$. The reason is that we are dealing with voltage amplitudes; the voltages across various parts of the circuit attain their maximum values at different times during the cycle. It is true that the instantaneous voltages would obey the loop theorem.

34-5 Geometric Construction for Z and ϕ

There is a simple geometric construction for showing the impedence Z and the phase angle ϕ. Refer to Figure 34-3. One simply draws an arrow of magnitude X_L in the positive y direction, an arrow of magnitude X_C in the negative y direction and an arrow of magnitude R in the positive x direction. Then the "vector" sum of these arrows is of magnitude Z and the angle which this resultant makes with the x axis (measured counterclockwise from the positive x axis) is the angle ϕ. Note that all the arrows in the figure have the units of ohm.

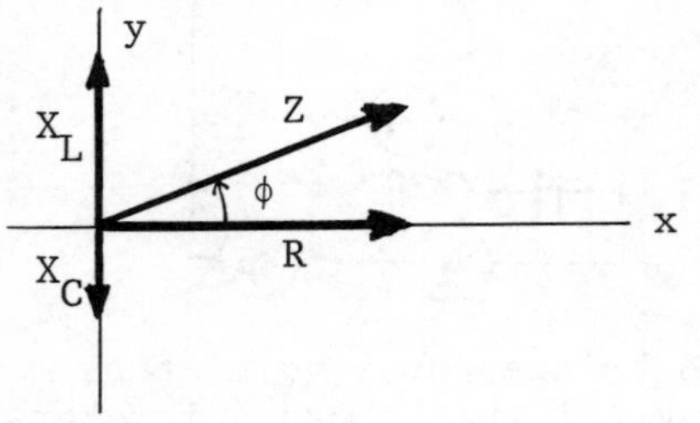

Figure 34-3. A geometric construction for obtaining the impedance Z and the phase angle ϕ. Z is the resultant of the three arrows, ϕ is the angle shown. Note that all arrows have the units of ohm.

34-6 Resonance

Refer to Figure 34-3. Noting that X_L is proportional to the frequency and that X_C is inversely proportional to the frequency and that R is independent of the frequency we can qualitatively understand what happens in a series RLC circuit as the frequency is varied. Starting at a very low frequency, X_C dominates; Z is very large and the voltage lags the current by almost 90°. As the frequency is increased X_C gets smaller and X_L gets larger. Eventually there is a certain frequency, called the resonance frequency, at which $X_L = X_C$. At this "resonance" the impedance has its minimum value (Z = R); furthermore the voltage and current are now in phase since $\phi = 0$. As the frequency is made still greater, the X_L term dominates and the voltage leads the current by almost 90°.

At the resonance frequency the current amplitude has its largest possible value for a given voltage amplitude $\mathcal{E}_m$. Thus as the frequency is "swept" through resonance, the current amplitude i_m will exhibit a peak. This peak is very "sharp" if the resistance R is small; the peak is "broad" if R is large.

34-7 Power in an RLC Circuit

We might expect that power = "voltage times current". Even for a resistor this is not true if we deal with voltage and current amplitudes. The reason for this is that (except at the peaks of the cycle) the instantaneous voltage and current are less than their maximum values. To take this into account we use the rms [root-mean-square] values rather than the amplitudes. For sinusoidally varying quantities:

$$V_{rms} = V_m/\sqrt{2} = 0.707\ V_m \quad , \qquad (34\text{-}9a)$$

$$i_{rms} = i_m/\sqrt{2} = 0.707\ i_m \quad . \qquad (34\text{-}9b)$$

Even with this, however, the answer for the case of anything other than a pure resistor needs further modification. In general the voltage and current are not in phase. Hence if we multiply them to obtain the power, for part of the cycle their product is positive but for part of the cycle their product is negative. Thus some power flows out of the circuit back to the source (even though the net power flow when averaged over a cycle is into the circuit). A term called the power factor = cos ϕ takes care of this effect. Note that this factor is one for a pure resistor and zero for a circuit with no resistance. Thus the formula for calculating the (average) power is

$$P = V_{rms}\, i_{rms} \cos\phi \quad . \qquad (34\text{-}10)$$

In actual practice, one uses the rms values all the time. The familiar "120 volt" source supplied by most electric companies is an rms number. Since the rms quantity differs from the maximum quantity by merely a factor (0.707) the relation "V = iZ" holds if V and i are both maximum values or are both rms values.

\>>> Example 3. Calculate the rms voltage and current for the case of the previous example. Compute the power factor and the power.

Solution: $V_{rms} = (0.707)\mathcal{E}_m = (0.707)(10\ V) = 7.07\ V$

$i_{rms} = (0.707)\ i_m = (0.707)(2 \times 10^{-3}\ A) = 1.41 \times 10^{-3}\ A$

Referring to Fig. 34-3 we see that $\cos\phi = R/Z = 3000/5000 = 0.6$; this is the power factor. Finally,

$$P = V_{rms}\, i_{rms} \cos\phi = (7.07\ V)(1.41 \times 10^{-3}\ A)(0.6) = 6 \times 10^{-3}\ W \ . \qquad <<<$$

SEE IMPORTANT NOTE ON p. 391 SEE IMPORTANT NOTE ON p. 391

Chapter 34: ALTERNATING CURRENTS

REVIEW AND PREVIEW

In previous chapters you studied circuits containing resistors, inductors, capacitors and sources of electromotive force. Both steady state and time-varying cases were considered; the time varying case led to a differential equation.

In this chapter you will study the special but important case in which the time dependence of the source is a sinusoidal function. Special techniques will be introduced to handle this problem using only algebra rather than differential equations.

GOALS AND GUIDELINES

Basically the goal is to understand the algebraic technique for handling sinusoidally varying voltages and currents. This goal can be broken down into the following pieces:

(a) reactance--a property of an inductor or a capacitor.
(b) impedance--a property of a circuit containing R, L or C.
(c) phase angle--the amount by which the voltage and current are out of phase.
(d) power factor--used to calculate the power dissipated by a circuit.

Special formulas will be introduced which permit an algebraic treatment of circuits. You should keep in mind that these apply only to sinusoidally varying sources.

34-1 Sources of Alternating EMF

We shall deal with circuits of the following form:

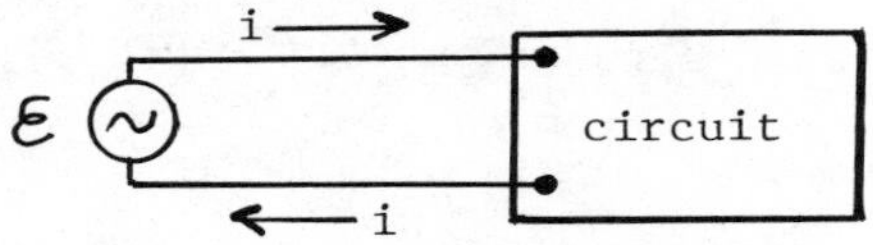

The symbol (~) denotes a source of potential difference which is sinusoidal in time,

$$\mathcal{E}(t) = \mathcal{E}_m \sin \omega t \ . \tag{34-1}$$

Here

$\mathcal{E}_m$ = amplitude (maximum value) of the emf [volt] ,

ω = angular frequency [(radian)/s] ,

t = time [s] .

34-2 General Response of the Circuit

It turns out that the response (i.e. the current) due to a source of the form (34-1) is

$$i(t) = i_m \sin(\omega t + \phi) + \text{"transient"} \ .$$

The "transient" term consists of decaying exponentials. After a short while it is not important; hence we shall deal only with the "steady state" term

$$i(t) = i_m \sin(\omega t + \phi) \ . \tag{34-2}$$

SEE IMPORTANT NOTE ON p. 391

The important qualitative points concerning this response are:

(a) i(t) is also sinusoidal with the same (angular) frequency as the source.

(b) There can be a phase difference ϕ between the voltage and the current. In particular if

(i) $\phi > 0$ then the "voltage lags the current by ϕ",

(ii) $\phi < 0$ then the "voltage leads the current by ϕ".

(c) The amplitude of the response, i_m, is proportional to the amplitude of the source, $\mathcal{E}_m$.

From (c) we may write

$$i_m = \mathcal{E}_m / Z \qquad (34\text{-}3)$$

where the quantity Z (to be defined later) is called the impedance of the circuit [unit: ohm].

Equation (34-3) looks somewhat like Ohm's law. One must keep in mind though that

(i) it relates only the two amplitudes, not the two functions (in fact since voltage and current are generally out of phase they cannot possibly be related by proportionality);

(ii) in general Z is a function of frequency, $Z = Z(\omega)$.

34-3 Single Element Circuits

In this section we shall consider the three single element circuits shown in Figure 34-1. In each case the emf $\mathcal{E}(t)$ is given by (34-1). We shall see that the current i(t) is indeed of the form (34-2).

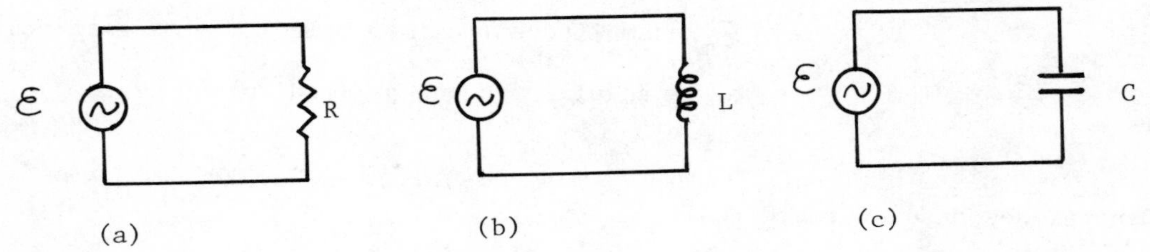

Figure 34-1. A source of sinusoidally varying emf is connected to a single element circuit: (a) resistor R, (b) inductor L, (c) capacitor C.

SEE IMPORTANT NOTE ON p. 391

For the single element circuit consisting of a resistor as shown in Figure 34-1a, the loop equation gives

$$\begin{aligned} i(t) &= \mathcal{E}(t)/R \\ &= (\mathcal{E}_m \sin \omega t)/R \\ &= [\mathcal{E}_m/R] \sin(\omega t + 0) \quad . \end{aligned}$$

By comparison with (34-2) and (34-3) we see that

$$Z = R \quad , \qquad (34\text{-}4a)$$

$$\phi = 0 \quad . \qquad (34\text{-}4b)$$

Thus for a resistor, Z is simply equal to R; also the voltage and the current are in phase.

For the single element circuit consisting of an inductor as shown in Figure 34-1b, the loop equation gives

$$\begin{aligned} L\, di(t)/dt &= \mathcal{E}(t) \\ di/dt &= \mathcal{E}/L \\ &= (\mathcal{E}_m \sin \omega t)/L \quad . \end{aligned}$$

This can be integrated to give

$$\begin{aligned} i(t) &= -\ (\mathcal{E}_m \cos \omega t)/(\omega L) \\ &= [\mathcal{E}_m/(\omega L)] \sin(\omega t + (-\pi/2)) \quad . \end{aligned}$$

By comparison with (34-2) and (34-3) we see that

$$Z = \omega L \quad ,$$

$$\phi = -\pi/2 \ .$$

The quantity ωL, a joint property of the inductor and the frequency, is called the inductive reactance X_L [unit: ohm]. Thus

$$Z = X_L \quad , \qquad (34\text{-}5a)$$

$$X_L = \omega L \quad , \qquad (34\text{-}5b)$$

$$\phi = -\pi/2 \ . \qquad (34\text{-}5c)$$

Thus for an inductor, Z is equal to the inductive reactance; also the voltage leads the current by $\pi/2$ (= 90°).

SEE IMPORTANT NOTE ON p. 391

For the single element circuit consisting of a capacitor as shown in Figure 34-1c, the loop equation gives

$$q(t)/C = \mathcal{E}(t)$$

$$i(t) = dq/dt = C\,\mathcal{E}_m\,\omega \cos \omega t$$

$$= [\mathcal{E}_m/(1/\omega C)] \sin(\omega t + \pi/2) \quad .$$

Again by comparison with (34-2) and (34-3) we see that

$$Z = 1/(\omega C) \ ,$$

$$\phi = \pi/2 \quad .$$

The quantity $1/(\omega C)$, a joint property of the capacitor and the frequency is called the <u>capacitative reactance</u> X_C [unit: ohm] . Thus

$$Z = X_C \ , \tag{34-6a}$$

$$X_C = 1/\omega C \ , \tag{34-6b}$$

$$\phi = \pi/2 \quad . \tag{34-6c}$$

Thus for a capacitor, Z is equal to the capacitative reactance; also the voltage <u>lags</u> the current by $\pi/2$ (= 90°).

>>> Example 1. Given an inductor L = 1 H and a capacitor C = 1 μF, calculate their reactances at frequencies of (a) 60 Hz, (b) 1000 Hz.

Solution: Note that the frequency, f, is given (Hz = cycle/second). We must use the angular frequency ω (radian/second). The conversion formula is $\omega = 2\pi f$.

(a) $$X_L = \omega L = 2\pi f L = 2\pi(60)(1) = 377\ \Omega \ ,$$

$$X_C = 1/\omega C = 1/2\pi f C = 1/[2\pi(60)(10^{-6})] = 2653\ \Omega \quad .$$

(b) Similarly with f = 1000 Hz,

$$X_L = 6283\ \Omega \ , \ X_C = 159\ \Omega \quad .$$

Remark: Note that either circuit element (L or C) can have the larger reactance depending on the frequency. <<<

34-4 RLC Series Circuit

We now generalize to a series circuit consisting of all three circuit elements: resistance R, inductance L and capacitance C. The circuit diagram is shown in Figure 34-2.

SEE IMPORTANT NOTE ON p. 391

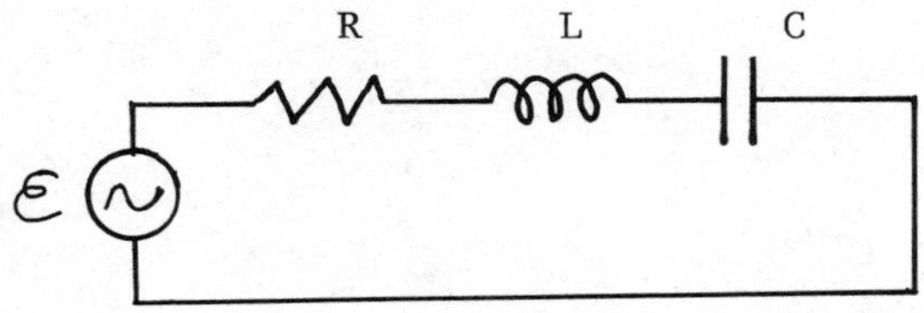

Figure 34-2. An RLC series circuit connected to a sinusoidally varying source of emf. Note that the current is the same for each of the circuit elements since they are connected in series.

We seek a generalization of our previous equations (34-4), (34-5) and (34-6) each of which applied to a single element circuit. This generalization must reduce to one of these three in the special case where there is only one circuit element. It can be shown that the correct generalization of these formulas is

$$Z = \sqrt{R^2 + (X_C - X_L)^2} \quad , \tag{34-7a}$$

$$\phi = \tan^{-1}[(X_C - X_L)/R] \quad . \tag{34-7b}$$

The quantity Z is called the impedance [unit: ohm]. It is the proportionality factor which relates voltage and current amplitudes. The angle ϕ tells us by how much the voltage lags the current. The term $(X_C - X_L)$ is called the total reactance, X. In terms of X the above equations become

$$Z = \sqrt{R^2 + X^2} \quad , \tag{34-8a}$$

$$\phi = \tan^{-1}(X/R) \quad . \tag{34-8b}$$

With regard to limiting cases of these formulas, note that if R or L is absent we simply set that quantity equal to zero; however if C is absent we must formally set $C = \infty$ [physically a "short circuit"] to obtain $X_C = 0$.

The concept of impedance can be applied to any part of the circuit as well as to the circuit as a whole. The current-voltage (amplitude) relation is $i_m = V_m/Z$ where Z is the impedance of that part of the circuit under consideration. The following example illustrates some of these ideas.

>>> Example 2. At a certain frequency the circuit shown has the following parameters.

$X_L = 5000\ \Omega$,

$X_C = 1000\ \Omega$,

$R = 3000\ \Omega$.

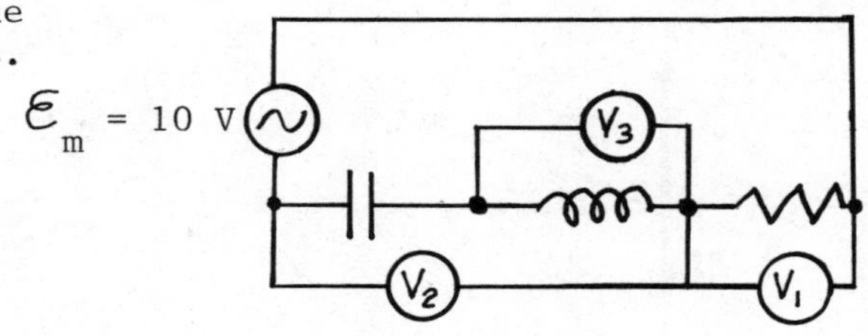

Calculate (a) Z and ϕ, (b) i_m, (c) the voltages read by the three voltmeters V_1, V_2 and V_3. [Note: assume, for purposes of this example, that these voltmeters read the amplitude of the voltage.]

Solution: (a) $Z = \sqrt{R^2 + (X_C - X_L)^2}$

$Z = \sqrt{(3000)^2 + (1000 - 5000)^2} = 5000\ \Omega$.

$\phi = \tan^{-1}[(X_C - X_L)/R] = \tan^{-1}[(1000 - 5000)/3000] = -53^\circ$.

(b) $i_m = \mathcal{E}_m/Z = (10\ \text{V})/(5000\ \Omega) = 2 \times 10^{-3}$ A .

SEE IMPORTANT NOTE ON p. 391

(c) To find V_1: V_1 is the voltage across R. For this circuit element

$$Z = R$$

$$V_1 = i_m Z = (2 \times 10^{-3}\ \text{A})(3000\ \Omega) = 6\ \text{V} \quad .$$

To find V_2: V_2 is the voltage across the LC part of the circuit.

$$Z = \sqrt{(X_C - X_L)^2} = 4000\ \Omega$$

$$V_2 = i_m Z = (2 \times 10^{-3}\ \text{A})(4000\ \Omega) = 8\ \text{V} \quad .$$

To find V_3: V_3 is the voltage across L. For this circuit element

$$Z = X_L$$

$$V_3 = i_m Z = (2 \times 10^{-3}\ \text{A})(5000\ \Omega) = 10\ \text{V} \quad .$$ <<<

Remarks: We have used the fact that the current is the same in every part of this series circuit. Note that the voltages do not "add up" as you might guess. For example, $V_1 + V_2 \neq \mathcal{E}_m$. The reason is that we are dealing with voltage amplitudes; the voltages across various parts of the circuit attain their maximum values at <u>different times</u> during the cycle. It is true that the <u>instantaneous</u> voltages would obey the loop theorem.

34-5 Geometric Construction for Z and ϕ

There is a simple geometric construction for showing the impedance Z and the phase angle ϕ. Refer to Figure 34-3. One simply draws an arrow of magnitude X_C in the positive y direction, an arrow of magnitude X_L in the negative y direction and an arrow of magnitude R in the positive x direction. Then the "vector" sum of these arrows is of magnitude Z and the angle which this resultant makes with the x axis (measured counterclockwise from the positive x axis) is the angle ϕ. Note that all the arrows in the figure have the units of ohm.

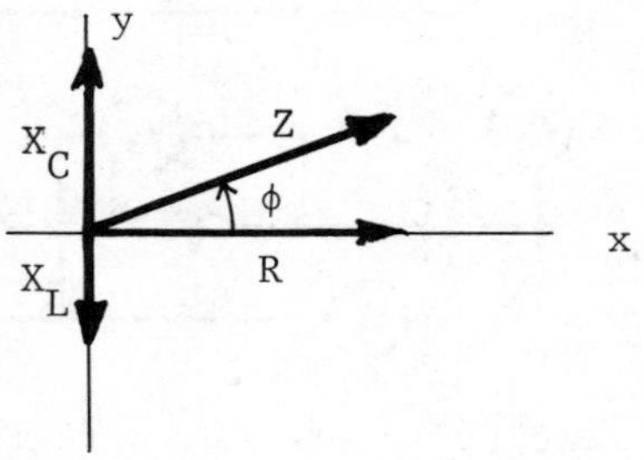

Figure 34-3. A geometric construction for obtaining the impedance Z and the phase angle ϕ. Z is the resultant of the three arrows, ϕ is the angle shown. Note that all arrows have the units of ohm.

34-6 Resonance

Refer to Figure 34-3. Noting that X_L is proportional to the frequency and that X_C is inversely proportional to the frequency and that R is independent of the frequency we can qualitatively understand what happens in a series RLC circuit as the frequency is varied. Starting at a very low frequency, X_C dominates; Z is very large and the voltage lags the current by almost 90°. As the frequency is increased X_C gets smaller and X_L gets larger. Eventually there is a certain frequency, called the <u>resonance frequency</u>, at which $X_L = X_C$. At this "resonance" the impedance has its minimum

SEE IMPORTANT NOTE ON p. 391

value ($Z = R$); furthermore the voltage and current are now in phase since $\phi = 0$. As the frequency is made still greater, the X_L term dominates and the voltage leads the current by almost 90°.

At the resonance frequency the current amplitude has its largest possible value for a given voltage amplitude $\mathcal{E}_m$. Thus as the frequency is "swept" through resonance, the current amplitude i_m will exhibit a peak. This peak is very "sharp" if the resistance R is small; the peak is "broad" if R is large.

34-7 Power in an RLC Circuit

We might expect that power = "voltage times current". Even for a resistor this is not true if we deal with voltage and current amplitudes. The reason for this is that (except at the peaks of the cycle) the instantaneous voltage and current are less than their maximum values. To take this into account we use the rms [root-mean-square] values rather than the amplitudes. For sinusoidally varying quantities:

$$V_{rms} = V_m/\sqrt{2} = 0.707\ V_m \ , \tag{34-9a}$$

$$i_{rms} = i_m/\sqrt{2} = 0.707\ i_m \ . \tag{34-9b}$$

Even with this, however, the answer for the case of anything other than a pure resistor needs further modification. In general the voltage and current are not in phase. Hence if we multiply them to obtain the power, for part of the cycle their product is positive but for part of the cycle their product is negative. Thus some power flows out of the circuit back to the source (even though the net power flow when averaged over a cycle is into the circuit). A term called the power factor = $\cos\phi$ takes care if this effect. Note that this factor is one for a pure resistor and zero for a circuit with no resistance. Thus the formula for calculating the (average) power is

$$P = V_{rms}\ i_{rms}\ \cos\phi \ . \tag{34-10}$$

In actual practice, one uses the rms values all the time. The familiar "120 folt" source supplied by most electric companies is an rms number. Since the rms quantity differs from the maximum quantity by merely a factor (0.707) the relation "$V = iZ$" holds if V and i are both maximum values or are both rms values.

>>> Example 3. Calculate the rms voltage and current for the case of the previous example. Compute the power factor and the power.

Solution: $V_{rms} = (0.707)\mathcal{E}_m = (0.707)(10\ V) = 7.07\ V$,

$i_{rms} = (0.707)\ i_m = (0.707)(2 \times 10^{-3}\ A) = 1.41 \times 10^{-3}\ A$.

Referring to Fig. 34-3 we see that $\cos\phi = R/Z = 3000/5000 = 0.6$; this is the power factor. Finally,

$$P = V_{rms}\ i_{rms}\ \cos\phi = (7.07\ V)(1.41 \times 10^{-3}\ A)(0.6) = 6 \times 10^{-3}\ W \ .$$ <<<

Chapter 35: MAXWELL'S EQUATIONS AND ELECTROMAGNETIC WAVES

REVIEW AND PREVIEW

So far in Part II you studied certain laws of electricity and magnetism (Gauss's law, Ampere's law, etc.). After introducing a new concept (displacement current), we will gather four of these laws together and present them as Maxwell's equations. Their importance, in this chapter, is that they can have wave-like solutions.

In Part I you studied various types of waves (e.g. transverse waves on a string, longitudinal sound waves in a gas, etc.). In this chapter you will study electromagnetic waves. Examples of electromagnetic waves are: radio and television waves, heat, light, X-ray.

GOALS AND GUIDELINES

In this chapter there are two major goals:

1. The first goal pertains to Maxwell's equations (Equations 35-3).
   a. Understanding why Ampere's law must be modified so as to include the displacement current.
   b. Learning Maxwell's equations. Except for the above modification these are all relations which you have studied in previous chapters. You should try to appreciate the symmetry of these equations and develop a feeling for the geometric relations between the various paths, surfaces and volumes involved.
2. The second goal pertains to electromagnetic waves. [Section 35-3 briefly reviews some of the important concepts involving waves. If you do not know these you should study Chapter 16 for a more detailed review of waves in general.]
   a. Understanding the derivation of electromagnetic waves as presented in the text. Here two of Maxwell's equations are applied to thin rectangular loops in space. In particular you should note that the displacement current (introduced in this chapter) is necessary to obtain these waves.
   b. Learning the properties of plane electromagnetic waves. These are conveniently summarized at the end of Section 35-4.
   c. Learning the definition and the meaning of the Poynting vector $\underline{S}$. This is related to the energy transported by an electromagnetic wave.

35-1 Displacement Current

Previously we have applied Ampere's law, $\oint \underline{B}\cdot d\underline{\ell} = \mu_0 i$, to steady state situations. There is an inconsistency in this law for the non-steady state case. The cause of this can be seen in the following example.

Suppose a capacitor C is being charged by the current I as shown in Figure 35-1a.

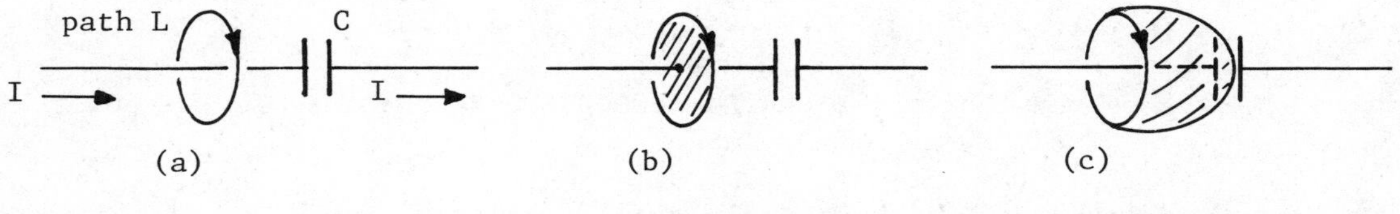

Figure 35-1. (a) Capacitor C is being charged by the current I. The left hand side of Ampere's law, $\oint \underline{B}\cdot d\underline{\ell}$, is to be applied to the closed path L. (b) The plane circular area is bounded by the path L. The current crossing this surface (to the right) is i = I. (c) The hemispherical surface is bounded by the path L. The current crossing this surface (to the right) is i = 0.

We may apply the left hand side of Ampere's law to the closed path "L" shown. On the right hand side of Ampere's law, "i" is the current crossing any open surface "S" whose boundary is the path L (referring to Chapter 29 for the sign convention.) If as in Figure 35-1b we choose a plane circular area for S, then i = I. On the other hand if we choose the hemispherical surface shown in Figure 35-1c for S, then i = 0 (for there is no current crossing this surface). This is a contradiction: the right hand side of Ampere's law should not depend upon which surface we choose for S (provided merely that S have the path L for its boundary).

Obviously, the trouble here is that the current is not continuous: it ends abruptly at the left plate of the capacitor and starts again at the right plate. To take care of the lack of current between the capacitor plates Ampere's law is modified by having an extra term added to its right hand side. This extra term is chosen to be $\mu_o i_d$ where

$$i_d = d(\Phi_E)/dt \quad . \tag{35-1}$$

Here i_d (whose unit is the ampere) is called the displacement current. Ampere's law now becomes $\int \underline{\underline{B}}\cdot d\underline{\underline{\ell}} = \mu_o(i + i_d)$ or

$$\oint \underline{\underline{B}}\cdot d\underline{\underline{\ell}} = \mu_o i + \mu_o\varepsilon_o\, d(\Phi_E)/dt \quad . \tag{35-2}$$

In the right hand side of (35-2), i is the current crossing the surface S and $\Phi_E = \int \underline{\underline{E}}\cdot d\underline{\underline{S}}$ is the flux of $\underline{\underline{E}}$ through this same surface S.

>>> Example 1. A parallel plate capacitor C is being charged by the current I as shown below. Show that the displacement current crossing the plane P_1 to the right is equal to the current crossing the plane P_2 to the right.

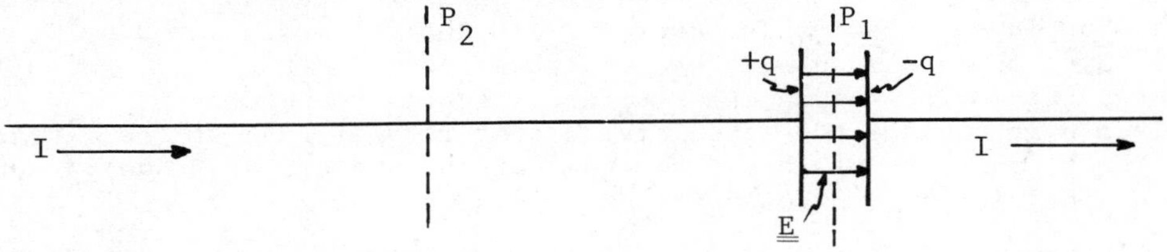

Using Gauss's law we find that the $\underline{\underline{E}}$ field between the plates is directed toward the right. Its magnitude is given by

$$|\underline{\underline{E}}| = \sigma/\varepsilon_o \quad , \qquad (\sigma = \text{charge per unit plate area})$$

$$= q/\varepsilon_o A \quad , \qquad (A = \text{area of each plate}) \quad .$$

The flux of $\underline{\underline{E}}$ through P_1 (to the right) is

$$\Phi_E = \int \underline{\underline{E}}\cdot d\underline{\underline{S}} = |\underline{\underline{E}}|A = q/\varepsilon_o \quad .$$

Here we have made use of the fact that

(a) $\underline{\underline{E}}$ is parallel to $d\underline{\underline{S}}$, and

(b) $\underline{\underline{E}}$ is uniform over the area A, zero outside this area.

The displacement current crossing P_1 (to the right) is

$$i_d = \varepsilon_o\, d(\Phi_E)/dt = \varepsilon_o\, d(q/\varepsilon_o)/dt = dq/dt \quad .$$

From the diagram we see that $dq/dt = + I$. Thus

$$i_d = I \quad .$$

Clearly I is also the current crossing P_2 (to the right).

Remark: The quantity $(i + i_d)$ is therefore continuous. Note that we obtain the correct result with the correct sign. This justifies the choice of sign in i_d = "+" $\varepsilon_o\, d(\Phi_E)/dt$.

<<<

35-2 Maxwell's Equations

We now have four integral laws for electricity and magnetism:

$$\oint \underline{\underline{E}}\cdot d\underline{\underline{S}} = q/\varepsilon_o \quad , \tag{35-3a}$$

$$\oint \underline{\underline{B}}\cdot d\underline{\underline{S}} = 0 \quad , \tag{35-3b}$$

$$\oint \underline{\underline{E}}\cdot d\underline{\underline{\ell}} = - d(\Phi_B)/dt \quad , \quad \text{where } \Phi_B = \int \underline{\underline{B}}\cdot d\underline{\underline{S}} \quad , \tag{35-3c}$$

$$\oint \underline{\underline{B}}\cdot d\underline{\underline{\ell}} = \mu_o i + \mu_o\varepsilon_o\, d(\Phi_E)/dt \quad , \quad \text{where } \Phi_E = \int \underline{\underline{E}}\cdot d\underline{\underline{S}} \quad . \tag{35-3d}$$

These laws, known as Maxwell's equations, are respectively: Gauss's law for electricity, Gauss's law for magnetism, Faraday's law, and Ampere's law (as modified above).

In the first pair of laws the left hand side deals with a closed surface S while the right hand side involves something contained in the volume enclosed by S. In the second pair of laws the left hand side deals with a closed path L while the right hand side involves something crossing an open surface which is bounded by L.

In the first pair of laws the sense of the vector $d\underline{\underline{S}}$ is that of the outward normal to S. In the second pair of laws the senses of the vectors $d\underline{\underline{\ell}}$ and $d\underline{\underline{S}}$ are related to each other by the sign convention as discussed in Chapter 29.

35-3 Review of Wave Motion

The remainder of this chapter is concerned with electromagnetic waves. The waves we will deal with are of the form

$$\psi(x,t) = A \sin (kx - \omega t) \quad . \tag{35-4}$$

This represents a traveling sinusoidal wave.* Note that ψ is a function of two independent variables, x and t. The important concepts are:

(a) A, the amplitude of the wave (maximum value of ψ) ;
(b) ω, the angular frequency (radians/time) ;
(c) ν (= $\omega/2\pi$) , the frequency (cycles/time);
(d) T (= $1/\nu$) , the period (time/cycle);
(e) k, the wave number (radians/length);
(f) λ (= $2\pi/k$) , the wavelength (length/cycle);
(g) v (= $\omega/k = \nu\lambda$) , the speed of the wave (length/time) .

The concepts (a) - (d) are features of SHM (they involve the dependence of ψ upon t, for fixed x).

*In general there could also be a phase constant term ϕ: $\psi = A \sin(kx - \omega t + \phi)$.

The wave (35-4) moves in the +x direction. If we had written $\psi = A \sin(kx + \omega t)$ instead, then the wave would move in the -x direction.

35-4 Traveling Waves and Maxwell's Equations

To investigate the possibility of electromagnetic waves the text assumes* that

$$\mathbf{E} = E(x,t)\,\mathbf{k} \qquad (\mathbf{k} = \text{unit vector in +z direction}) \; ; \qquad (35\text{-}5a)$$

$$\mathbf{B} = -\,B(x,t)\,\mathbf{j} \qquad (\mathbf{j} = \text{unit vector in +y direction}) \; . \qquad (35\text{-}5b)$$

This says that $\mathbf{E}$ is directed along the z axis and is a function only of x and t (not y or z). Similarly $\mathbf{B}$ is directed along the y axis and is also a function of only x and t. $\mathbf{E}$ and $\mathbf{B}$ will then describe plane waves.** We will consider these waves in free space only. Therefore in Maxwell's equations both "q" and "i" are zero. The Maxwell equations then reduce to

$$\oint \mathbf{E}\cdot d\mathbf{S} = 0 \quad , \quad \oint \mathbf{B}\cdot d\mathbf{S} = 0 \quad ,$$

$$\oint \mathbf{E}\cdot d\boldsymbol{\ell} = -\,d(\Phi_B)/dt \quad , \quad \oint \mathbf{B}\cdot d\boldsymbol{\ell} = \mu_o \varepsilon_o\, d(\Phi_E)/dt \quad .$$

The first two of these are automatically satisfied (since from 35-5) the lines of $\mathbf{E}$ and and $\mathbf{B}$ never terminate). The Maxwell equation $\oint \mathbf{E}\cdot d\boldsymbol{\ell} = -\,d(\Phi_B)/dt$ is then applied to the narrow rectangular path L_1 shown in Figure 35-2. This yields (see text)

$$\frac{\partial E}{\partial x} = -\,\frac{\partial B}{\partial t} \quad . \qquad (35\text{-}6)$$

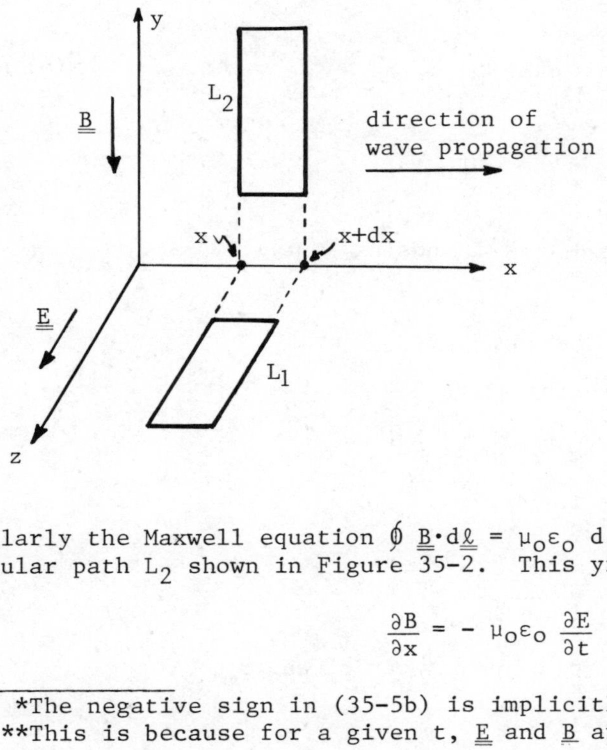

Figure 35-2. Paths L_1, L_2 for application of two of Maxwell's equations. Note that according to equations (35-5) there is a flux of $\mathbf{B}$ through the area bounded by L_1 and a flux of $\mathbf{E}$ through the area bounded by L_2.

Similarly the Maxwell equation $\oint \mathbf{B}\cdot d\boldsymbol{\ell} = \mu_o\varepsilon_o\, d(\Phi_E)/dt$ is applied to the narrow rectangular path L_2 shown in Figure 35-2. This yields (see text)

$$\frac{\partial B}{\partial x} = -\,\mu_o\varepsilon_o\,\frac{\partial E}{\partial t} \quad . \qquad (35\text{-}7)$$

---

*The negative sign in (35-5b) is implicitly assumed in the text figures.

**This is because for a given t, $\mathbf{E}$ and $\mathbf{B}$ are constant over any plane x = constant.

To solve equations (35-6) and (35-7) we assume traveling sinusoidal waves of the form

$$E(x,t) = E_m \sin(kx - \omega t) \tag{35-8a}$$

$$B(x,t) = B_m \sin(kx - \omega t) \ . \tag{35-8b}$$

Substituting these into (35-6) and (35-7) leads to an equality provided that $\omega/k = E_m/B_m = 1/\sqrt{\mu_o\varepsilon_o}$.

>>> Example 2. Show that $\omega/k = E_m/B_m = 1/\sqrt{\mu_o\varepsilon_o}$ by substituting (35-8) into (35-6) and (35-7).

From (35-8) we have

$$\frac{\partial E}{\partial x} = kE_m \cos(kx - \omega t) \ , \quad \frac{\partial E}{\partial t} = -\ \omega E_m \cos(kx - \omega t) \ ,$$

$$\frac{\partial B}{\partial x} = kB_m \cos(kx - \omega t) \ , \quad \frac{\partial B}{\partial t} = -\ \omega B_m \cos(kx - \omega t) \ .$$

Substituting into (35-6),

$$\frac{\partial E}{\partial x} = -\frac{\partial B}{\partial t}$$

$$kE_m \cos(kx - \omega t) = +\ \omega B_m \cos(kx - \omega t)$$

$$E_m = (\omega/k)\ B_m \ . \tag{a}$$

Similarly using (35-7),

$$\frac{\partial B}{\partial x} = -\ \mu_o\varepsilon_o \frac{\partial E}{\partial t}$$

$$kB_m \cos(kx - \omega t) = +\ \mu_o\varepsilon_o\omega E_m \cos(kx - \omega t)$$

$$kB_m = \mu_o\varepsilon_o\omega E_m \tag{b}$$

Substituting the expression (a) for E_m into (b) gives

$$kB_m = \mu_o\varepsilon_o\omega(\omega/k)\ B_m$$

$$(\omega/k)^2 = 1/(\mu_o\varepsilon_o) \ .$$

We take the positive square root:

$$\omega/k = 1/\sqrt{\mu_o\varepsilon_o} \ .$$

Substituting this back into (a) gives the relation between E_m and B_m,

$$E_m/B_m = 1/\sqrt{\mu_o\varepsilon_o} \ .$$

Remark: Note that the trigonometric term cos(kx - ωt) cancelled out. If E and B were <u>out</u> of phase (say E involving cos(kx - ωt) while B involved sin(kx - ωt)), then it would be impossible to satisfy equations (35-6) and (35-7). <<<

The speed of these waves, $v = \omega/k$, is then $1/\sqrt{\mu_0 \varepsilon_0} = 3 \times 10^8$ m/s. This value is identical with the speed of light (denoted by c). Thus the above results become

$$\omega = ck \tag{35-9a}$$

$$E_m = cB_m \tag{35-9b}$$

$$c = 1/\sqrt{\mu_0 \varepsilon_0} \quad . \tag{35-9c}$$

In summary, the important results regarding plane electromagnetic waves in free space are:

(a) Electromagnetic waves <u>travel with the speed of light</u> c.
(b) The speed of light c is related to the quantities μ_0, ε_0 by $c = 1/\sqrt{\mu_0 \varepsilon_0}$.
(c) The amplitudes of $\underline{E}$ and $\underline{B}$ are related by $E_m = cB_m$.
(d) The $\underline{E}$ and $\underline{B}$ fields are <u>in phase</u>: they are both proportional to the same function, sin(kx - ωt).
(e) The electromagnetic wave is <u>transverse</u>, i.e. $\underline{E}$ and $\underline{B}$ are both perpendicular to the direction of the wave propagation (in this case the x axis).
(f) $\underline{E}$ and $\underline{B}$ are perpendicular to each other.

\>>> Example 3. A plane electromagnetic wave in free space has a wavelength of 100 meters. The maximum electric field for this wave is 10^{-4} volt/meter. Write a possible equation for B as a function of x and t.

The desired equation is of the form $B = B_m \sin(kx - \omega t)$. The constants in this equation are given by

$$B_m = E_m/c = (10^{-4}\ \text{V/m})/(3 \times 10^8\ \text{m/s}) = 3.33 \times 10^{-13}\ \text{Wb/m}^2 \quad ,$$

$$k = 2\pi/\lambda = 2\pi/(10^2\ \text{m}) = 6.28 \times 10^{-2}\ \text{m}^{-1} \quad ,$$

$$\omega = ck = (3 \times 10^8\ \text{m/s})(6.28 \times 10^{-2}\ \text{m}^{-1}) = 1.88 \times 10^7\ \text{s}^{-1} \quad .$$

Therefore

$$B = (3.33 \times 10^{-13}\ \text{Wb/m}^2) \sin[(6.28 \times 10^{-2}\ \text{m}^{-1})x - (1.88 \times 10^7\ \text{s}^{-1})t].$$

<<<

35-5 Poynting Vector

Electromagnetic waves can transport energy. A useful quantity to describe this flow of energy is the <u>Poynting vector</u> $\underline{S}$. This is defined by

$$S = \frac{1}{\mu_0} \underline{E} \times \underline{B} \quad . \tag{35-10}$$

The physical significance of $\underline{S}$ is as follows. Imagine an area A perpendicular to $\underline{S}$ (i.e. the normal to A is parallel to $\underline{S}$). Then

(a) the magnitude of $\underline{S}$ is the power crossing A <u>per unit area</u>,
(b) the direction of $\underline{S}$ is the direction of this energy flow.

The units of $\underline{S}$ are those of power/area (W/m^2). In general both $\underline{E}$ and $\underline{B}$ are functions of location, $\underline{S}$ is therefore a <u>vector field</u>. The formula for the power crossing an arbitrary area is

$$P = \int \underline{\underline{S}} \cdot d\underline{\underline{A}} \qquad (35\text{-}11)$$

where, to avoid confusion, we use $d\underline{\underline{A}}$ rather than $d\underline{\underline{S}}$ to denote a vectorial surface area element.

For the special case of the plane wave (35-8), $\underline{\underline{S}}$ is in the +x direction. Its magnitude is

$$|\underline{\underline{S}}| = \frac{1}{\mu_0} E_m B_m \sin^2(kx - \omega t) \quad .$$

Since the average value of $\sin^2(kx - \omega t)$ is one half, the average value of the magnitude of this Poynting vector is

$$|\underline{\underline{S}}|_{av} = \tfrac{1}{2} |\underline{\underline{S}}|_{max} = \frac{1}{2\mu_0} E_m B_m \quad . \qquad (35\text{-}12)$$

>>> Example 4. A plane electromagnetic wave in free space has a maximum electric field of 10^{-4} V/m. A 1.0 cm^2 area is perpendicular to the direction of the wave propagation. Calculate the average power crossing this area.

$$|\underline{\underline{S}}|_{av} = \frac{1}{2\mu_0} E_m B_m = \frac{1}{2\mu_0 c} E_m^{\ 2} \qquad (\text{since } B_m = E_m/c)$$

$$|\underline{\underline{S}}|_{av} = \frac{(10^{-4}\ \text{V/m})^2}{(2)(4\pi \times 10^{-7}\ \text{Wb/A·m})(3 \times 10^8\ \text{m/s})} = 1.33 \times 10^{-11}\ \text{W/m}^2 \quad .$$

$$P_{av} = |\underline{\underline{S}}|_{av} A = (1.33 \times 10^{-11}\ \text{W/m}^2)(10^{-4}\ \text{m}^2)$$

$$P_{av} = 1.33 \times 10^{-15}\ \text{watt} \quad .$$

<<<

Chapter 36: NATURE AND PROPOGATION OF LIGHT

REVIEW AND PREVIEW

In the previous chapter you studied some basic concepts concerning electromagnetic waves. In this chapter you will study some further properties of electromagnetic waves.

GOALS AND GUIDELINES

In this brief chapter there are two goals:

1. Learning that an electromagnetic wave can transport momentum as well as energy. As a result such a wave can exert a pressure on an absorbing or reflecting surface.
2. Understanding the phenomenon called the Doppler effect. This is a shift in the frequency of the wave due to any relative motion between the source and the observer.

36-1 Momentum of an Electromagnetic Wave

An electromagnetic wave (e.g. light wave, radio wave, etc.) can transport momentum as well as energy. It is shown in the text that if an energy U is absorbed from the wave by an object, then an amount of momentum

$$p = U/c \qquad \text{(absorption)} \tag{36-1}$$

is transferred to the object. If the wave is perfectly reflected directly back instead of being abosrbed, then the momentum transferred is twice as large:

$$p = 2\,U/c \qquad \text{(reflection)} \quad . \tag{36-2}$$

36-2 Radiation Pressure

Suppose that an energy U is absorbed from an electromagnetic wave in a time t. From (36-1) the rate of momentum transfer to the absorber is

$$p/t = U/(ct) \quad .$$

Assuming that the wave is normally incident upon some surface of area A we have $U = |\underline{\underline{S}}|At$ where $\underline{\underline{S}}$ is the Poynting vector ($|\underline{\underline{S}}|A$ is the power, multiplication of this by the time gives the energy)*. Thus $p/t = (|\underline{\underline{S}}|At)/(ct)$ or

$$p/t = |\underline{\underline{S}}|A/c \quad .$$

The absorbing object will experience a force equal to this rate of momentum transfer, $F = p/t = |\underline{\underline{S}}|A/c$. Finally, the radiation pressure ($P_{rad} = F/A$) is the force per unit area exerted on the absorber:

$$P_{rad} = |\underline{\underline{S}}|/c \quad . \tag{36-3}$$

Of course if the wave were perfectly reflected, the radiation pressure would be twice as large.

---

* Here $|\underline{\underline{S}}|$ denotes the time average Poynting vector; see Section 35-5.

>>> Example 1. An intense beam of light has an average power per unit area of 100 watt/ cm^2 (= 10^6 W/m^2). Calculate the radiation pressure it would exert on an absorber. Express your answer in pounds per square inch.

$$P_{rad} = |\underline{S}|/c$$
$$= (10^6 \text{ W/m}^2)/(3 \times 10^8 \text{ m/s})$$
$$= (3.33 \times 10^{-3} \text{ N/m}^2)\left(\frac{1.45 \times 10^{-4} \text{ lb/in}^2}{\text{N/m}^2}\right)$$
$$= 4.8 \times 10^{-7} \text{ lb/in}^2 \quad .$$

Remark: As this example shows, the radiation pressure is very small in most cases. <<<

36-3 Speed of Light

Light waves (as well as any other electromagnetic wave) can travel in vacuum; there is no medium necessary for this wave propagation. The speed of light in vacuum, c, is an accurately measured quantity. Experimentally, this value is completely independent of any motion of either the source or the observer.

36-4 Doppler Effect

Suppose that a source of light "S" and an observer "O" are in relative motion along the straight line connecting them. Although both "S" and "O" will measure the same value c for the speed of light, they will measure different values for the frequency of the light. If u is the relative velocity of "S" and "O"* these frequencies are related by

$$\nu' = \nu \frac{1 - u/c}{\sqrt{1 - (u/c)^2}} \quad . \tag{36-4}$$

Here ν is the frequency as measured by the source "S", ν' the frequency as measured by the observer "O". Equation (36-4) is also valid if "S" and "O" are approaching each other; we need only use a negative value for "u".

If u is small compared with c, a useful approximation to (36-4) is

$$\nu' = \nu(1 - u/c) \quad ; \quad [|u| \ll c] \quad . \tag{36-5}$$

>>> Example 2. An observer "O" is approaching a stationary source "S". The relative velocity of approach is one third the speed of light. Calculate the ratio of the frequencies: ν'/ν.

The ratio of the frequencies is

$$\frac{\nu'}{\nu} = \frac{1 - u/c}{\sqrt{1 - (u/c)^2}} \quad .$$

Since "O" is approaching "S", u is negative: $u = - c/3$.

*This means the velocity of recession of "S" relative to "O", or equivalently the velocity of recession of "O" relative to "S".

$$\frac{\nu'}{\nu} = \frac{1 - (-\ 1/3)}{\sqrt{1 - (-\ 1/3)^2}} = \frac{(4/3)}{\sqrt{(8/9)}} = \sqrt{2} = 1.41 \quad .$$

Remark: The approximate formula (36-5) does not give too good an answer since u is not small compared with c. According to (36-5),

$$\nu'/\nu = 1 - (u/c) = 1 - (-1/3) = 4/3 = 1.33 \quad .$$

<<<

Chapter 37: REFLECTION AND REFRACTION -- PLANE SURFACES

REVIEW AND PREVIEW

The previous two chapters dealt with some of the wave properties of electromagnetic waves (e.g. light waves). There are many aspects of light, however, in which its wave properties are unimportant. One then is concerned with a ray of light and the path that this ray follows as various lenses, mirrors, prisms, etc. are encountered. This subject is called geometrical optics; it is the topic of this, and the next, chapter. This chapter in particular is concerned with the behavior of a ray of light as it encounters a plane surface.

GOALS AND GUIDELINES

In this chapter there are two major goals:

1. Learning the meaning of the index of refraction of a medium. Using this, you should be able to calculate what happens to the speed, frequency, and wavelength of light when it enters a medium.
2. Learning what happens to a light ray as it strikes a plane interface separating two media. This involves the laws of reflection and refraction.

You should study the definitions involved in the laws of reflection and refraction very carefully (Section 37-3). Figure 37-1, showing the incident, reflected, and refracted rays illustrates the meaning of many of these definitions.

37-1 Index of Refraction

The speed of light v in a transparent substance (hereafter called a medium) such as glass, water, etc., is generally less than the speed of light c in vacuum. The index of refraction n (a dimensionless property of the medium) is defined as the ratio of the two speeds:

$$n = c/v \quad . \tag{37-1}$$

Since v is less than c, n is greater than one. The speed of light in air is so close to c that we can take $n_{air} = 1$.

When light enters a medium, its frequency f does not change. Since the product of frequency times wavelength equals speed, the equations governing the light as it enters the medium are

$$f\lambda = c \quad ,$$

$$f\lambda_{med} = v \quad .$$

Here λ_{med} is the wavelength of the light in the medium; the symbol λ always means the wavelength in vacuum. Note that we have used the fact that the frequency does not change. Using equation (37-1), the above equations give λ_{med} in terms of λ.

$$\lambda_{med} = \lambda/n \quad . \tag{37-2}$$

In summary the frequency is unchanged and the wavelength is shorter (by a factor of n), compared with the corresponding values in vacuum.

>>> Example 1. Calculate the thickness of a layer of air ($n_{air} = 1.0003$) such that it will contain exactly one more wavelength of yellow (λ= 589 nm) light than the same thickness of vacuum.

Let the unknown thickness be t. The number of wavelengths in air, N_{air}, is

$$N_{air} = t/\lambda_{air} = t/(\lambda/n_{air}) = tn_{air}/\lambda \quad .$$

Similarly the number of wavelengths in the same thickness of vacuum, N_{vac}, is

$$N_{vac} = t/\lambda \quad .$$

We want $N_{air} - N_{vac} = 1$. Therefore

$$tn_{air}/\lambda - t/\lambda = 1$$

$$t = \lambda/(n_{air} - 1)$$

$$t = (589 \text{ nm})/(.0003) = 2 \times 10^6 \text{ nm} = 2 \text{ mm} \quad .$$

Remark: In this problem we cannot approximate the index of refraction of air, $n_{air} \simeq 1$. This is because we are dealing with the difference between air and vacuum: $n_{air} - 1 = 0.0003$.

<<<

37-2 Geometrical Optics

Of course light, an electromagnetic wave, is really a wave phenomenon. If the width of a beam of light is large compared with the wavelength of the light (the beam can still be quite narrow*) then the spreading of the beam due to its wave nature is negligible. The lack of spreading means that (for fixed n) light can be considered to travel in a straight line at constant speed. We can then speak of a ray of light, indicating this by a straight line on any diagram. This type of analysis, in which we deal with rays of light, is called geometrical optics.

37-3 Reflection and Refraction

Figure 37-1 shows a ray of light incident upon an interface separating two media. The media are characterized by their indices n_1 and n_2, the incident ray being in the first medium. The angle θ_1 between the incident ray and the normal to the interface is called the angle of incidence. Part of the incident light is reflected back at an angle of reflection θ_1', the remainder of the light is refracted into the second medium at an angle of refraction θ_2. Note that all three angles (θ_1, θ_1', θ_2) are measured with respect to the normal as shown.

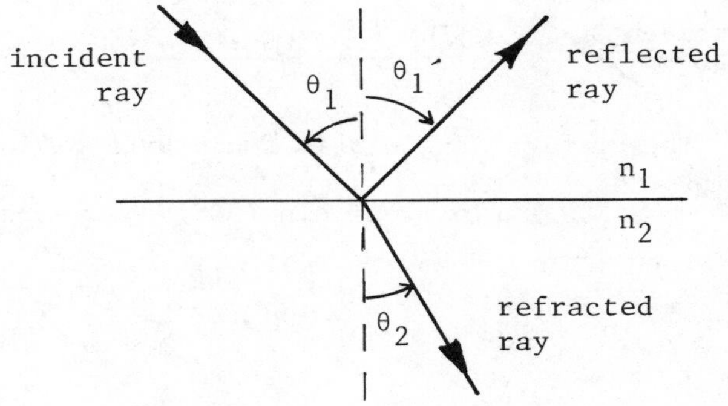

Figure 37-1. A ray of light incident upon an interface separating two media. The dashed line is normal to the interface.

*Recall that the wavelengths of visible light lie in the approximate range 400 - 800 nm = 4-8 × 10^{-4} mm.

The laws governing these rays (which can be derived from Maxwell's equations) are:

(a) The three rays (incident, reflected, refracted) as well as the normal to the interface are all coplanar. (37-3a)

(b) The angle of reflection equals the angle of incidence,

$$\theta_1' = \theta_1 \quad . \tag{37-3b}$$

(c) The angle of refraction is related to the angle of incidence by Snell's law,

$$n_1 \sin \theta_1 = n_2 \sin \theta_2 \quad . \tag{37-3c}$$

36-4 Total Internal Reflection

If n_1 is greater than n_2 there is some angle of incidence θ_c, called the critical angle, for which the angle of refraction is 90°. At larger angles of incidence ($\theta_1 > \theta_c$), refraction is not possible.* All the incident light will then be reflected; this is known as total internal reflection.

The critical angle may be found from Snell's law by setting $\theta_2 = 90°$. This yields

$$\theta_c = \sin^{-1} (n_2/n_1) \quad . \tag{37-4}$$

This expression clearly shows that in order for total internal reflection to occur, the incident light must be in the medium with the larger index of refraction.

>>> Example 2. Light is incident upon a 45-45-90° glass prism as shown in Figure 37-2. Calculate the angle of incidence at which total internal reflection begins to occur at the hypotenuse of the prism. Take the index of refraction of the glass to be 1.55.

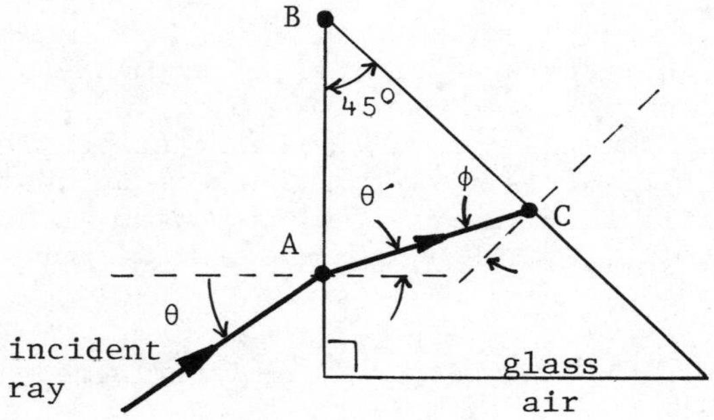

Figure 37-2. A ray of light incident upon one surface of a glass prism. The dashed lines are normal to the surfaces at points A and C.

The critical angle ϕ is found by applying Snell's law at point C in the figure.

$$n \sin \phi = 1 \sin (90°) \quad , \; n = \text{index of refraction of glass}$$

$$\sin \phi = 1/1.55 = 0.645$$

$$\phi = 40° \; 10' \quad .$$

*For $\theta_1 > \theta_c$, Snell's law would require that $\sin \theta_2$ be greater than one.

Now consider the sum of the three interior angles in triangle ABC.

$$(90^o - \theta') + (45^o) + (90^o - \phi) = 180^o$$

$$\theta' = 45^o - \phi = 45^o - 40^o\ 10' = 4^o\ 50' \quad .$$

Applying Snell's law at point A,

$$1 \sin\theta = n \sin\theta'$$

$$\sin\theta = (1.55)\sin(4^o\ 50') = (1.55)\cdot(0.0843) = 0.131$$

$$\theta = 7^o\ 30' \quad .$$

The required angle of incidence is 7° 30' below the normal.

Remark: The angle θ' (= 4° 50') happened to be positive. Had it turned out to be negative, the refracted ray in the prism (as it left point A) would be inclined downward from the normal. Note that the figure assumed this ray to be inclined upward from the normal; the positive calculated value of θ' means that this assumption is correct. <<<

37-5 Huygen's Principle

There is a geometrical construction for predicting the behavior of a ray of light. The method, called Huygen's principle, is as follows.

(a) Every point on the surface of a wavefront* acts as an emitter of light waves; these waves are called "secondary wavelets".
(b) The secondary wavelets travel with speed $v = c/n$; after a small time t they have traveled a distance $d = vt$.
(c) After this time t, a new wavefront exists. This new wavefront is that surface which is tangent to these secondary wavelets.

The laws of reflection and refraction can be derived using Huygen's principle.

Although Huygen's principle uses the wave nature of light somewhat (e.g. it deals with wavefronts rather than rays), it cannot replace Maxwell's equations. For example, Huygen's principle does not predict the proportion of the incident light which is reflected or refracted at an interface. This, as well as many other phenomenon (such as interference effects) are correctly predicted by Maxwell's equations. Nonetheless, Huygen's principle can be useful, especially when we do not seek too detailed information.

---

*Recall that a wavefront is a surface of constant phase. Wavefronts are perpendicular to the rays (provided that the speed of light is independent of its direction of propagation).

37-6 Programmed Problems

1. A light wave enters a glass (n = 1.60) block from air (n = 1.00). The quantity n is called the __________ of the medium.

   The speed of light v in a medium is related to the speed of light c in vacuum by the formula

   $$v = \_\_\_\_\_\_ .$$

   Calculate the speed of light in the glass.

index of refraction.

$v = c/n$ (this is the definition of n)

$$v = \frac{c}{n} = \frac{3.00 \times 10^8 \text{ m/s}}{1.60}$$

$$= 1.88 \times 10^8 \text{ m/s} .$$

---

2. Thus the light travels more slowly in glass than in air.

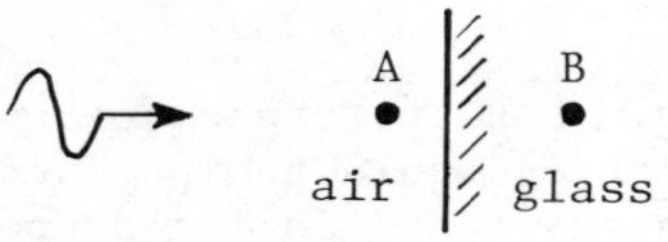

   Suppose that the frequency of the light at point A is 5.0×10^{14} Hz. That is, an observer at point A would measure 5.0×10^{14} oscillations in one second. What is the frequency of the light when it reaches point B?

It is the same as the frequency in the air, namely 5.0×10^{14} Hz. (If it were not the same as at point A then the number of cycles contained between points A and B would be changing with time. This is clearly absurd.)

---

3. For any wave there is a fundamental relation involving the frequency ν, the wavelength λ, and the speed of the wave v. What is this relation?

   Show that your answer is dimensionally correct. Remember that Hz = cycle/second.

$\nu\lambda = v$

$$\frac{\text{cycle}}{\text{second}} \cdot \frac{\text{meters}}{\text{cycle}} = \frac{\text{meters}}{\text{second}}$$

("cycle" is not really a unit but it is convenient to use in this type of problem).

---

4. Using the answer to the above frame calculate the wavelength of the light at points A and B.

   $$\lambda_A = \_\_\_\_\_\_ .$$

   $$\lambda_B = \_\_\_\_\_\_ .$$

$$\lambda_A = \frac{3.00 \times 10^8 \text{ m/s}}{5.0 \times 10^{14} \text{ s}^{-1}}$$

$$= 6.0 \times 10^{-7} \text{ m} .$$

$$\lambda_B = \frac{1.88 \times 10^8 \text{ m/s}}{5.0 \times 10^{14} \text{ s}^{-1}}$$

$$= 3.8 \times 10^{-7} \text{ m} .$$

Thus the wavelength in the glass is shorter than in the air.

5.

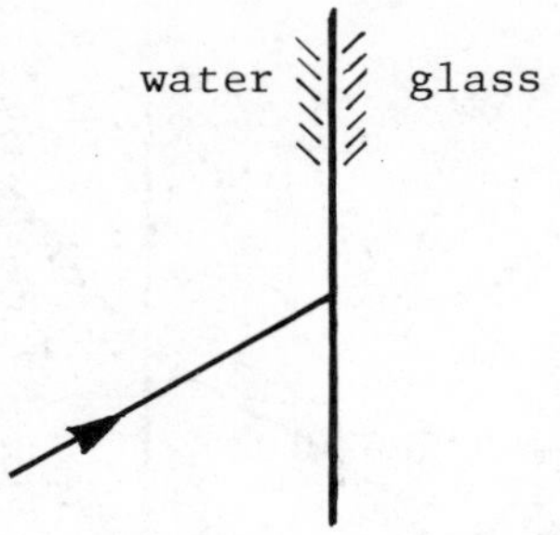

A ray of light in water ($n_1 = 1.33$) is incident upon a plane glass ($n_2 = 1.60$) surface as shown. Complete the diagram showing the reflected and refracted rays.

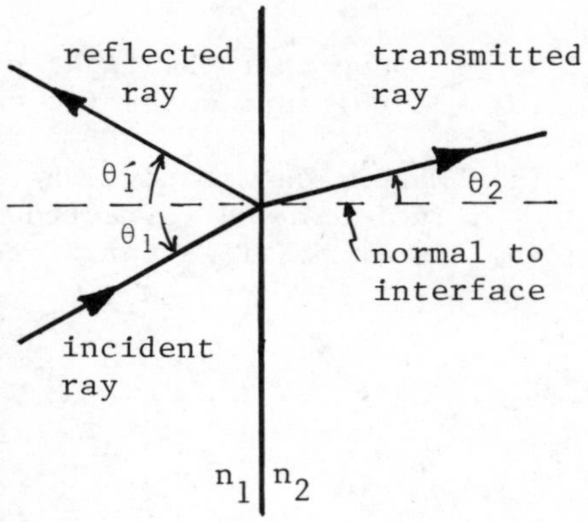

---

6. Refer to the answer to frame 5.
   (a) θ_1 is called the angle of __________ .
   (b) θ_1' is called the angle of __________ .
   (c) θ_2 is called the angle of __________ .
   (d) All three of these angles are measured from the __________ to the interface.
   (e) All three rays (incident, reflected, refracted) as well as the normal to the interface lie in the same __________ .

(a) incidence.
(b) reflection.
(c) refraction.
(d) normal.
(e) plane.

---

7. Again referring to the answer to frame 5,
   (a) How is the angle of reflection θ_1' related to the angle of incidence θ_1?
   (b) How is the angle of refraction θ_2 related to the angle of incidence θ_1?

(a) $\theta_1 = \theta_1'$ (law of reflection).
(b) $n_1 \sin\theta_1 = n_2 \sin\theta_2$

(law of refraction, known as Snell's law).

---

8. In the answer to frame 5, why is the angle of refraction θ_2 smaller than the angle of incidence θ_1?

From Snell's law

$$\sin\theta_2 = \frac{n_1}{n_2}\sin\theta_1 \quad .$$

In this case n_1 (= 1.33) is less than n_2 (= 1.60). Hence $\sin\theta_2$ is less than $\sin\theta_1$ and therefore θ_2 is smaller than θ_1.

---

9. In general then:
   (a) When light passes from a medium of smaller index of refraction n_1 into a medium of larger index n_2, the refracted ray is bent (toward, away from) the normal.
   (b) When light passes from a medium of larger index n_1 into a medium of smaller index n_2, the refracted ray is bent (toward, away from) the normal.

(a) toward

$n_1 < n_2$:

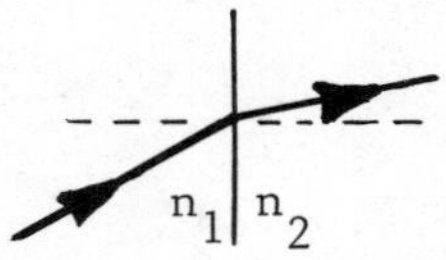

(b) away from

$n_1 < n_2$:

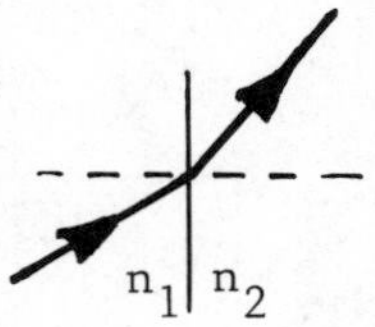

10. Now suppose that the light passes from glass (n_1 = 1.60) into water (n_2 = 1.33).

(a) Make a qualitative sketch showing the incident and refracted rays.
(b) Is it possible for θ_2 to be 90°?

(a)

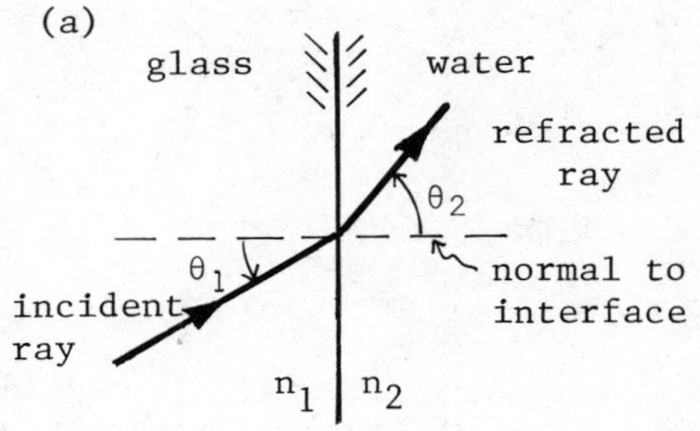

(b) Yes, because n_1 is greater than n_2.

11. (a) Calculate the value of θ_1 which makes θ_2 become 90°.
(b) This particular value of the angle of incidence is called ___________.

(a) 56°.

$$\sin \theta_1 = \frac{n_2}{n_1} \sin \theta_2$$

$$= \frac{1.33}{1.60} \sin (90^o)$$

$$= 0.83 \quad ,$$

$$\theta_1 = \sin^{-1}(0.83) = 56^o \quad .$$

(b) the critical angle (θ_c).

12. In frame 10, suppose that the angle of incidence θ_1 were made larger than the critical angle.
(a) Describe what would happen to the incident light.
(b) This phenomenon is called ___________.

(a) There would be no refracted ray, all the incident light would be reflected.

(b) total internal reflection.

Chapter 38: REFLECTION AND REFRACTION -- SPHERICAL SURFACES

REVIEW AND PREVIEW

In the previous chapter you studied the laws of reflection and refraction for a plane surface. In this chapter you will investigate the consequences of reflection and refraction at curved surfaces. You will see how to calculate the location of the image of a given object due to various optical devices (e.g. lenses, mirrors). As a result of studying this chapter you should have an understanding of how optical instruments (e.g. telescope, microscope) work.

GOALS AND GUIDELINES

In this chapter there are four major goals:

1. Understanding the behavior of a ray of light as it encounters a spherical reflecting surface, i.e. a spherical mirror (Section 38-3).
2. Understanding the behavior of a ray of light as it encounters a spherical refracting surface (Section 38-4). This serves primarily as background material for the next goal.
3. Understanding the behavior of a ray of light as it encounters a thin lens.

By far the most important thing you must learn from this chapter is how to calculate the location of the image of a given object due to a spherical mirror (goal number 1) or a thin lens (goal number 2). For each of these you must

- a. understand the pertinent definitions and their sign conventions;
- b. learn the formulas for the focal length and for the location of the image;
- c. be able to graphically find the location of the image.

You should study Examples 1 and 2 (spherical mirror) and Example 5 (thin lens) very carefully.

4. The final goal is to understand some terminology relating the image to the object. You should learn
   - a. whether an image is erect or inverted;
   - b. whether an image is real or virtual;
   - c. whether an object is real or virtual;
   - d. how to calculate the lateral magnification of the image.

The programmed problems illustrate some of the above goals. In addition, they show how to treat a "multi-element" (in this case a two lens) system.

38-1 Introduction

In this chapter the text applies geometrical optics to three types of optical systems:

(a) the spherical reflecting surface (mirror),
(b) the spherical refracting surface,
(c) the thin lens.

All the derivations are clearly presented in the text and will not be repeated here. What is important for problem applications is a thorough understanding of the results of these derivations. In order to correctly use these results the student must know the exact meaning of the various symbols in the equations as well as the sign conventions associated with their use.

38-2 Paraxial Rays

An example of an optical system is shown in Figure 38-1. Rays from an object O are bent by the two lenses so as to meet at the image I. All rays we will deal with are assumed to:

(a) make small angles with the axis of the optical system,
(b) make small angles with the normal to all optical surfaces (e.g. the four surfaces of the two lens system shown in the figure).

Such rays are called paraxial rays. Paraxial rays permit the use of small angle approximations ($\sin\theta \simeq \tan\theta \simeq \theta$).

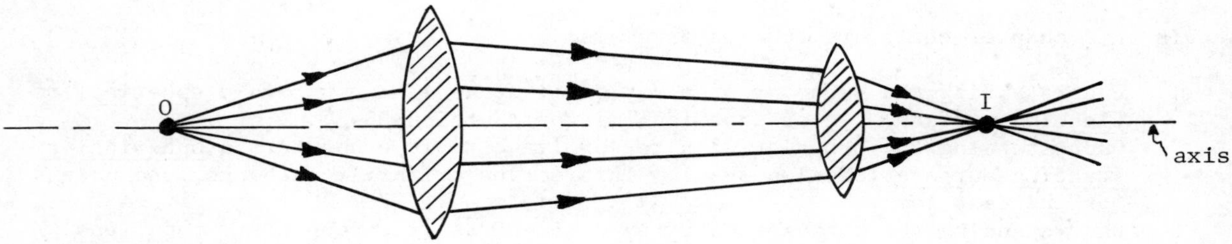

Figure 38-1. Rays from the object O are bent by the two lens optical system so as to pass through the image I.

38-3 Spherical Reflecting Surface

Figure 38-2 shows a concave spherical reflecting surface (mirror). Rays from an object O are reflected so as to pass through the image I. The symbols used to describe this situation are:

Points	Lengths
O = location of object	o = object distance
I = location of image	i = image distance
C = location of center of curvature of surface	r = radius of curvature of surface
F = location of focal point	f = focal length
V = location of vertex	

Note that all the distances o, i, r, f are measured from the vertex V. The equation for calculating the image location is

$$1/o + 1/i = 1/f \qquad \text{(spherical reflecting surface)} \quad . \qquad (38\text{-}1a)$$

Here the focal length f (a property of the mirror) is given by

$$f = r/2 \qquad \text{(spherical reflecting surface)} \quad . \qquad (38\text{-}1b)$$

In Figure 38-2 the incident light travels from left to right. In this case the distances o, i, r, f are considered positive if the corresponding points O, I, C, F are located with respect to the vertex as shown in the figure. Thus in order to remember the sign conventions, the student need merely remember the situation shown in Figure 38-2. If for example i were negative, it would mean that the image I is located on the other (in this case the right) side of the vertex. Similarly if the reflecting surface were convex, then r and f = r/2 would be negative; the points C and F being now located on the right side of the vertex.

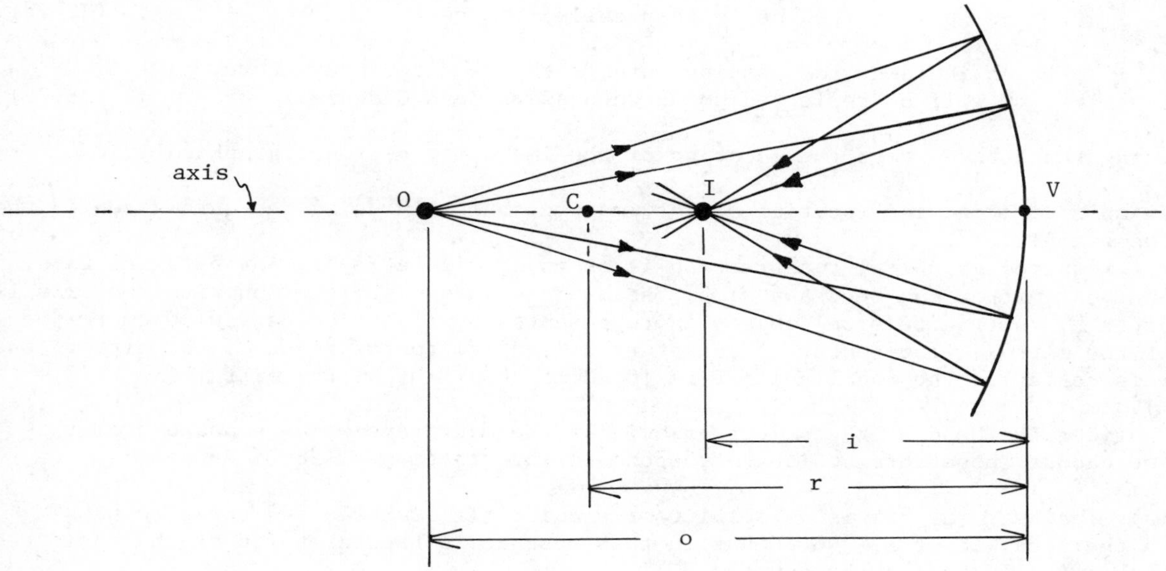

Figure 38-2. Rays from the object O are reflected by the spherical reflecting surface so as to pass through the image I.

>>> Example 1. An object is located 30 cm to the left of a convex spherical mirror whose radius of curvature is 20 cm. Calculate the location of the image.

The magnitude of r is given to be 20 cm. Since this mirror is convex, we see from Figure 38-2 that r is negative. Thus r = -20 cm, f = r/2 = -10 cm. Working in units of centimeters, the image location is found using (38-1a):

$$1/o + 1/i = 1/f$$

$$1/i = 1/f - 1/o = 1/(-10) - 1/(30) = -4/30$$

$$i = -30/4 = -7.5 \text{ cm} .$$

Referring to Figure 38-2 we see that the negative sign means that this image is located on the other side of the mirror from the object. Therefore the image is located 7.5 cm to the right of the mirror.

<<<

In Figure 38-2 the object is a point on the optical axis. Equations (38-1) are equally valid even if the object is not located on the axis (provided we consider only paraxial rays). Of the many rays which emanate from an object point located off the axis, there are three particularly simple ones to follow:

An incident ray parallel to the axis will be reflected so as to pass through the focal point F. (38-2a)

An incident ray passing through the focal point F will be reflected so as to be parallel to the axis. (38-2b)

An incident ray passing through the center of curvature C will be reflected so as to pass through C again. (38-2c)

By tracing these three rays the location of the image may be found graphically.

>>> Example 2. Find the location of the image graphically using the data given in the previous example.

The required graphical construction is shown in Figure 38-3. For paraxial rays, the spherical mirror can be drawn simply as a plane whose intersection with the axis is the vertex V. A (finite size) object is represented by a vertical arrow 30 cm to the left of the mirror. The center of curvature C is 20 cm to the right of the mirror (the mirror is convex). The focal point F is 10 cm to the right of the mirror ($f = r/2 = -10$ cm).

Consider the head of the object arrow. Of the many rays which emanate from this point we choose those three which correspond to the statements (38-2) above:

(a) Incident ray "a" is parallel to the axis. It is reflected so as to pass through the focal point F. Note that in this case it is the extension of the reflected ray which actually passes through F.

(b) Incident ray "b" passes through the focal point F. It is reflected so as to be parallel to the axis. Note that in this case it is the extension of the incident ray which actually passes through F.

(c) Incident ray "c" passes through the center of curvature C. It is reflected so as to pass through C again. Note that in this case it is the extension of the incident and reflected rays which actually pass through C.

These three reflected rays (when extended) have a common intersection point. This point is the image of the head of the object arrow. It is located 7.5 cm to the right of the mirror in agreement with the result of Example 1.

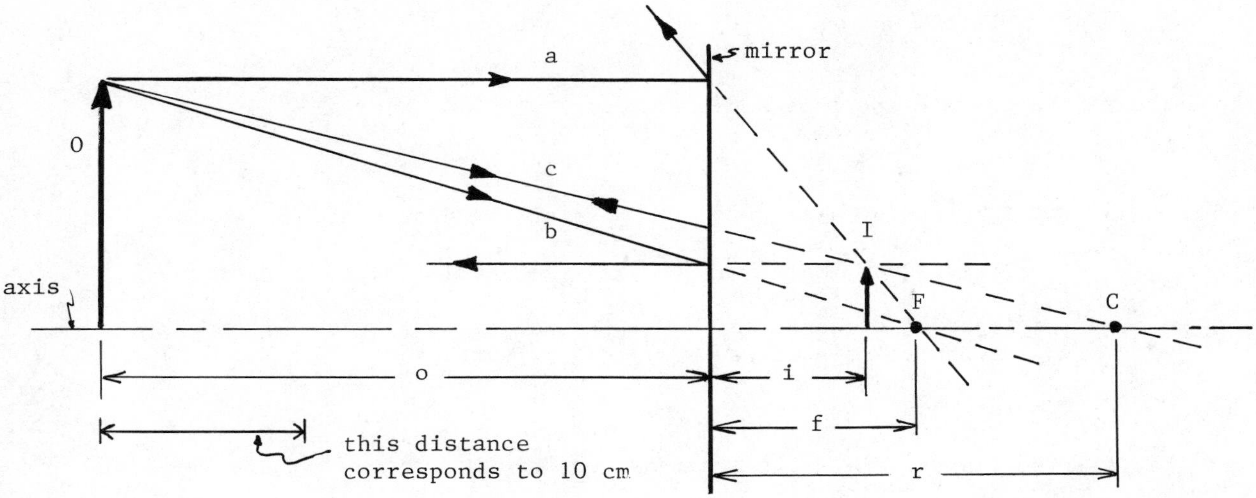

Figure 38-3. Graphical construction to find the image location for a convex spherical mirror.

Remarks:

1. Of course every point on the object arrow emits rays. Clearly the result of considering all the points on the object arrow would be the vertical image arrow shown in the figure.

2. As this example shows, the rules (38-2) must be interpreted so as to allow various rays to be <u>extended</u> if necessary for the graphical construction (even though no rays may actually exist in that region).

3. The vertical scale in the figure has been greatly exaggerated. In order to have paraxial rays, the object and image sizes must be small compared with the object and image distances respectively. By drawing the spherical surface as a plane, we may exaggerate the vertical scale for purposes of clarity; the graphical construction is then equivalent to equations (38-1). <<<

A <u>plane mirror</u> may be treated as a special case of a spherical reflecting surface. To do this we merely set $r = \infty$ in equations (38-1). This gives

$$i = -o \quad \text{(plane mirror)} \quad . \tag{38-3}$$

Image and object are thus equally distant from the mirror, the image being located on the opposite side of the mirror from the object. In the case of a plane mirror, the restriction to paraxial rays is not necessary: equation (38-3) is true for all rays.

38-4 Spherical Refracting Surface

Figure 38-4 shows a convex <u>spherical refracting surface</u>. Rays from an object O are refracted so as to pass through the image I.

The symbols used to describe this situation are:

Points	Lengths
O = location of object	o = object distance
I = location of image	i = image distance
C = location of center of curvature of surface	r = radius of curvature of surface
V = location of vertex	

Note that all the distances o, i, r are measured from the vertex V. The equation for calculating the image location is

$$n_1/o + n_2/i = (n_2 - n_1)/r \quad \text{(spherical refracting surface)} \quad . \qquad (38\text{-}4)$$

Here n_1, n_2 are the indices of refraction of the two media as shown. In Figure 38-4 the incident light travels from left to right. In this case the distances o, i, r are considered positive if the corresponding points O, I, C are located with respect to the vertex as shown in the figure.

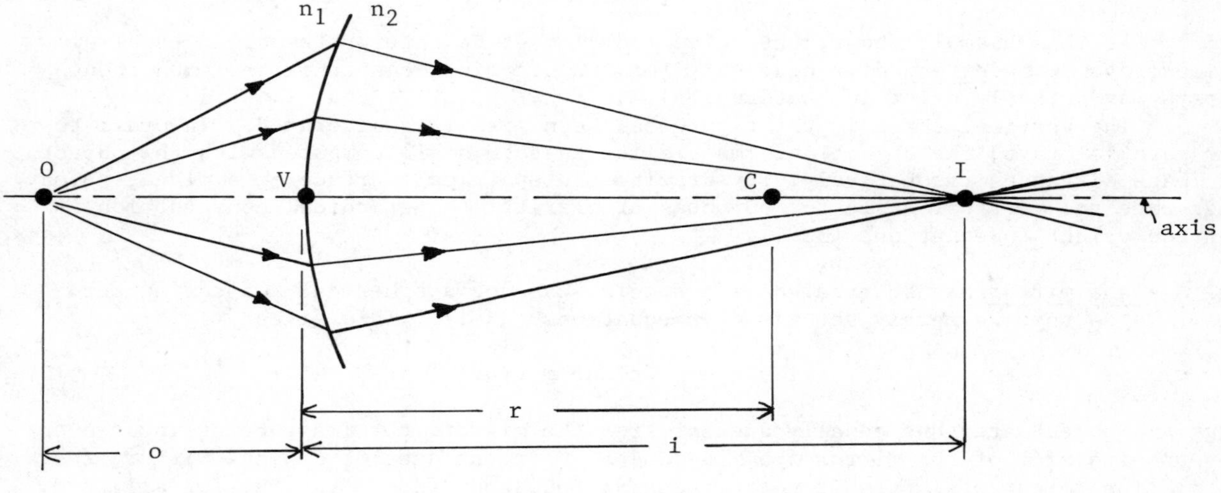

Figure 38-4. Rays from the object O are refracted by the spherical refracting surface so as to pass through the image I.

>>> Example 3. An object is at the bottom of a lake, 200 centimeters below the surface. An observer in a boat see an image of this object. Assuming paraxial rays, calculate the location of this image.

The plane surface of the lake may be considered to be a spherical refracting surface with an infinite radius of curvature. The given quantities are: $n_1 = 1.33$ (water), $n_2 = 1.00$ (air), $o = 200$ cm, $r = \infty$ (plane surface).

With $r = \infty$, equation (38-4) becomes

$$n_1/o + n_2/i = (n_2 - n_1)/r = 0$$

$$i = -(n_2/n_1)o = -(1.00/1.33)(200\text{ cm}) = -150\text{ cm} \quad .$$

Referring to Figure 38-4 we see that the negative value for i means that this image is located on the same side of the surface as the object. Therefore the image is located 150 cm below the surface of the lake.

Remark: In order to use only paraxial rays, the line from the eye of the observer to the object must be (very nearly) perpendicular to the surface of the lake. <<<

38-5 Thin Lens

Figure 38-5a shows a thin lens. This consists of a medium of index n (say glass) having two spherical surfaces, the distance between these surfaces being negligible. The medium surrounding the lens is assumed to have an index of unity (say air).

In Figure 38-5b rays from an object O are refracted by the thin lens so as to pass through the image I. The symbols used to describe this situation are:

Points	Lengths
O = location of object	o = object distance
I = location of image	i = image distance
C' = location of center of curvature of first surface	r' = radius of curvature of first surface
C'' = location of center of curvature of second surface	r'' = radius of curvature of second surface
F_1 = location of first focal point	f = focal length
F_2 = location of second focal point	f = focal length
C = location of center of lens	

Here the "first" surface means that surface which the incident light encounters first as it passes through the lens. Note that all the distances o, i, r', r'', f are measured from the center of the lens C. The equation for calculating the image location is

$$1/o + 1/i = 1/f \qquad \text{(thin lens)} \quad . \tag{38-5a}$$

Here the focal length f (a property of the lens) is given by

$$1/f = (n - 1)(1/r' - 1/r'') \quad \text{(thin lens)} \quad . \tag{38-5b}$$

In Figure 38-5 the incident light travels from left to right. In this case the distance o, i, r', r'', f are considered positive if the corresponding points O, I. C', C'', F_1, F_2 are located with respect to the center of the lens as shown in the figure.

Note that F_1, F_2 always lie on opposite sides of the lens at the same distance f from C; if f is negative then the locations of F_1, F_2 are interchanged.

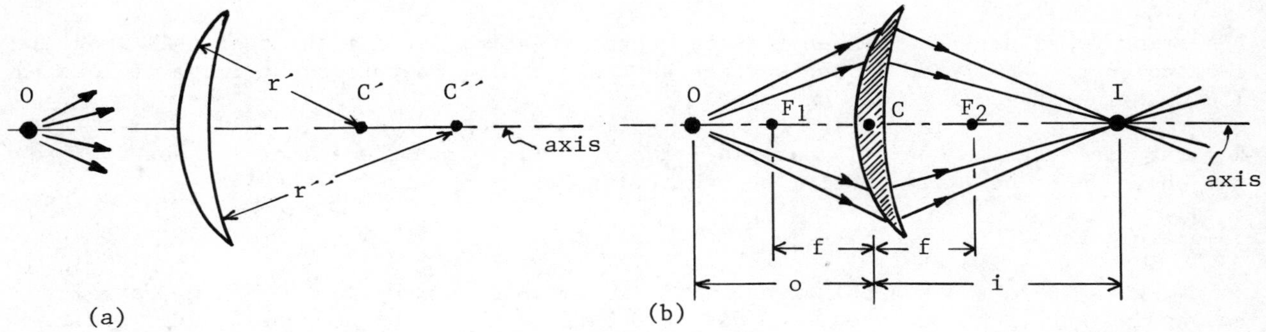

Figure 38-5. (a) A thin lens has two radii of curvature: r´, r´´. The incident light from the object O first strikes that surface whose radius of curvature is r´. (b) Rays from the object O are refracted by the thin lens so as to pass through the image I.

>>> Example 4. The diagram below gives the radius of curvature of the surfaces of four thin glass lenses. Calculate the focal length of each lens. Take the index of refraction of the glass to be 1.67.

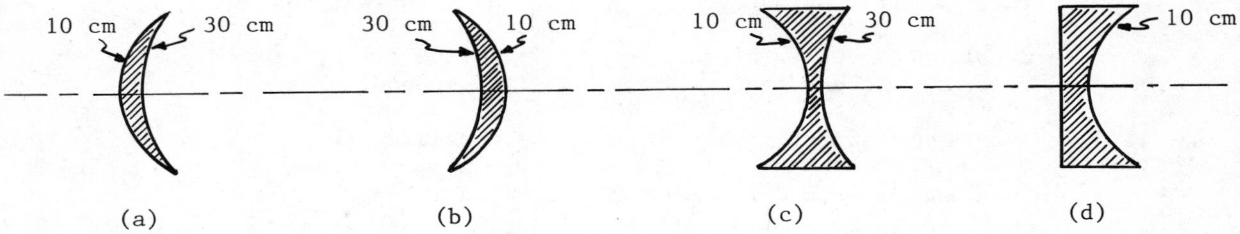

For convenience we imagine that the object is located to the left of each lens.

(a) Working in units of centimeters, we have from the diagram: $r' = +10$, $r'' = +30$. Equation (38-5b) gives

$$1/f = (n - 1)(1/r' - 1/r'') = (1.67 - 1)(1/10 - 1/30)$$

$$f = +22.4 \text{ cm} \quad .$$

Similarly for the other three lenses:

(b) $r' = -30$, $r'' = -10$, $f = +22.4$ cm;

(c) $r' = -10$, $r'' = +30$, $f = -11.2$ cm;

(d) $r' = \infty$, $r'' = +10$, $f = -14.9$ cm.

Remark: Lens (b) is the same as lens (a) except that it is "turned around". This does not affect its focal length.

In Figure 38-5 the object is a point on the optical axis. Equations (38-5) are equally valid even if the object is not located on the axis (provided we consider only paraxial rays). Of the many rays which emanate from an object point located off the axis, there are three particularly simple ones to follow:

An incident ray parallel to the axis will be refracted so as to pass through the second focal point F_2. (38-6a)

An incident ray passing through the first focal point F_1 will be refracted so as to be parallel to the axis. (38-6b)

An incident ray passing through the center of the lens C will be undeviated. (38-6c)

By tracing these three rays the location of the image may be found graphically.

>>> Example 5. An object is 10 centimeters in front of a thin lens whose focal length is 30 centimeters. Find the location of the image graphically and analytically.

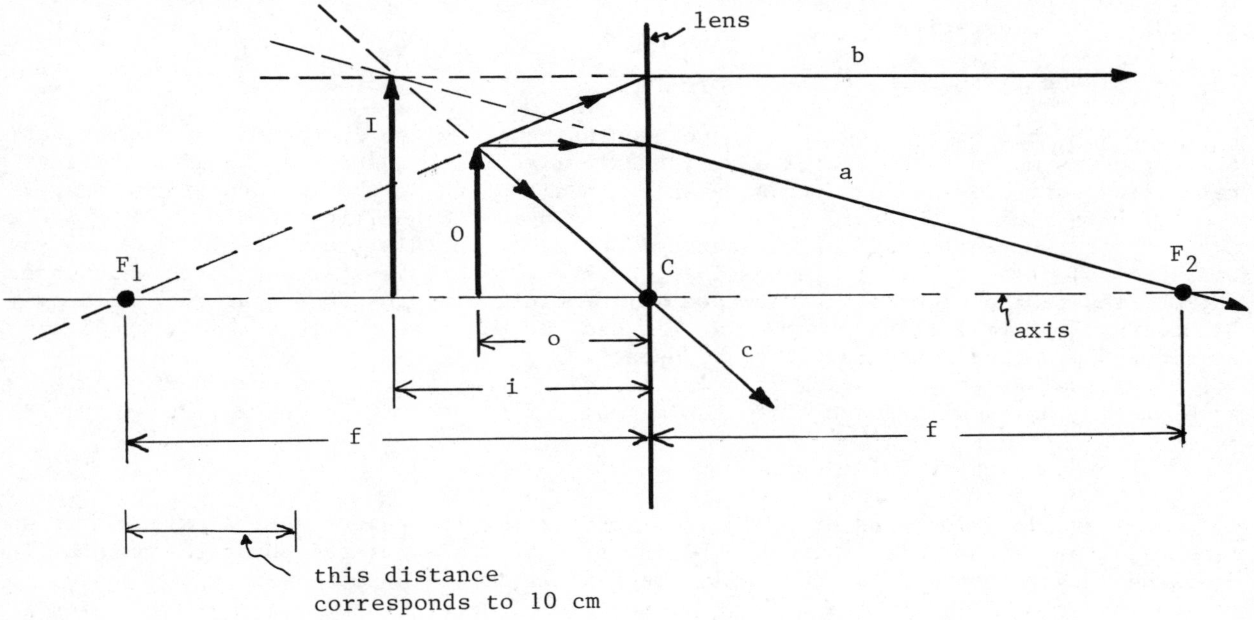

Figure 38-6. Graphical construction to find the image location for a thin lens

The required graphical construction is shown in Figure 38-6. For paraxial rays, the lens can be drawn simply as a <u>plane</u>. The object is represented by a vertical arrow 10 cm to the left of the lens. Since the focal length is positive (f = + 30 cm), F_1 is 30 cm to the <u>left</u> of the lens and F_2 is 30 cm to the <u>right</u> of the lens. In the figure, rays "a", "b", "c" correspond to the statements (38-6a), (38-6b), (38-6c) respectively.

The common intersection point of these three rays gives the location of the image: it is 15 cm to the left of the lens

The image location can be found analytically using (38-5a). Working in units of centimeters,

$$1/i = 1/f - 1/o = 1/30 - 1/10$$

$$i = -\ 15\ \text{cm} \quad .$$

In this case (refer to Figure 38-5b), the negative sign means that the image is located on the same side of the lens as the object. Therefore the image is located 15 cm to the left of the lens. This is in agreement with the above graphical construction.

Remark: As this example shows, the rules (38-6) must be interpreted so as to allow various rays to be extended if necessary for the graphical construction.

38-6 Real and Virtual Images

An image is said to be real if the rays of light actually intersect at the image location. A real image can be physically located on a screen placed at the image location. An image is said to be virtual if the rays of light only seem to be coming from the location of the image. In Figures 38-2,4,5 the image will be real if i is positive, virtual if i is negative.

38-7 Real and Virtual Objects

The sign conventions which we have been using are given in the figures which define the various symbols [see Figures 38-2 (spherical reflecting surface), 38-4 (spherical refracting surface) and 38-5 (thin lens)]. In all of these the incident light travels from left to right. Object distances, o, are considered positive if the object lies to the left of the (reflecting or refracting) surface. In this (the ususal) case, we say that the object is real. It is possible, however, to create a situation in which the object lies to the right of the surface while maintaining the fact that the incident light travels from left to right. In this case we say that the object is virtual and consider the object distance, o, to be negative.

A virtual object can arise in a multi-lens system. See the sequence of programmed problems beginning with frame 15 for an example of the occurrence of a virtual object.

38-8 Erect and Inverted Images

An image is said to be erect if it is oriented with respect to the axis in the same manner as the object. It is said to be inverted if it is rotated about the axis by 180° compared with the object, or generally speaking if it is "upside down" with respect to the object.

38-9 Lateral Magnification

The lateral magnification m is the ratio of the (transverse) image size to the (transverse) object size. For a spherical mirror or a thin lens, it is given by

$$m = -\ i/o \qquad \text{(spherical mirror or thin lens)} \quad . \tag{38-7}$$

A positive value for m means that the image is erect, a negative value means that the image is inverted.

>>> Example 6. Refer to the previous example. (a) Calculate the lateral magnification. (b) Is the image real or virtual? (c) Is the image erect or inverted?

(a) Using equation (38-7)

$$m = i/o = - (- 15 \text{ cm})/(10 \text{ cm}) = + 1.5 \quad .$$

(b) Since i is negative, the image is virtual. (Note that in Figure 38-6, the rays had to be extended back to the image. To an observer located to the right of the lens, the transmitted rays seem to be coming from the image.)

(c) Since m is positive, the image is erect. This agrees with the graphical construction in Figure 38-6.

<<<

38-10 Programmed Problems

1.

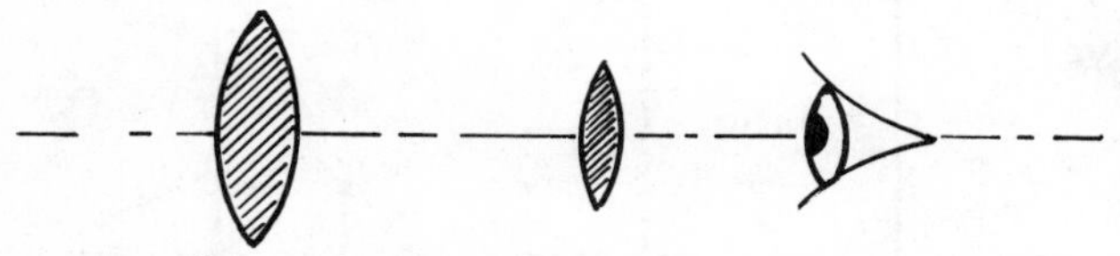

The diagram shows an "astronomical" telescope. It consists of a long focal length (f = 100 cm) objective lens and a short focal length (f = 2 cm) ocular lens. Suppose an object is located 100 meters to the left of the objective lens. Considering the effect of the objective lens only, what is the location of the image?

$$\frac{1}{o} + \frac{1}{i} = \frac{1}{f}$$

$$\frac{1}{i} = \frac{1}{f} - \frac{1}{o} = \frac{1}{100} - \frac{1}{10,000}$$

$$i = 101 \text{ cm} \quad .$$

The image is located 101 cm to the right of the objective lens.

---

2. In the above frame the object distance was much larger than the focal length. It turned out that the image distance was only slightly larger than the focal length. Verify this qualitatively by sketching three light rays in the following diagram.

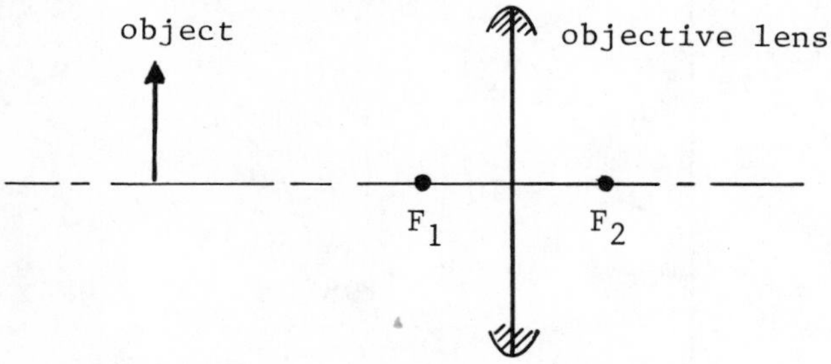

The three rays shown correspond to the three statements (38-6a, b,c).

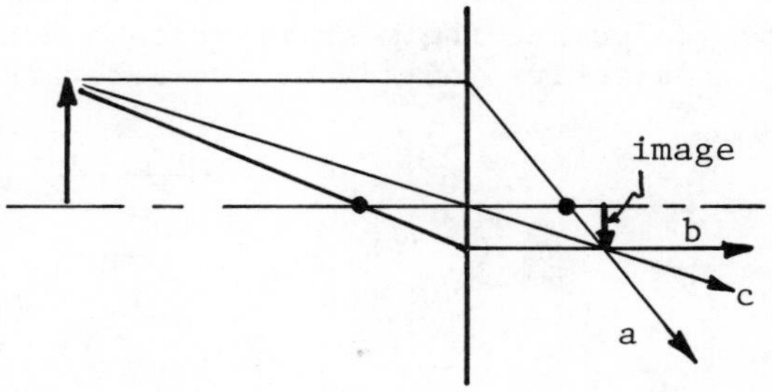

3. The above image will now act as an object for the second lens (the ocular). Suppose it is desired that the image of <u>this</u> object be located 200 cm to the left of the ocular. Where should the ocular be placed?

It is given that i = - 200 cm.

$$\frac{1}{o} + \frac{1}{i} = \frac{1}{f}$$

$$\frac{1}{o} = \frac{1}{f} - \frac{1}{i} = \frac{1}{2} - \frac{1}{-\ 200}$$

$$o = 1.98 \text{ cm} \quad .$$

The ocular must be located about 103 cm (101 cm + 1.98 cm) to the right of the objective.

4. In the above frame the object distance was only slightly smaller than the focal length. The image distance was negative, its magnitude being much larger than the focal length. Verify this qualitatively by sketching three light rays in the following diagram.

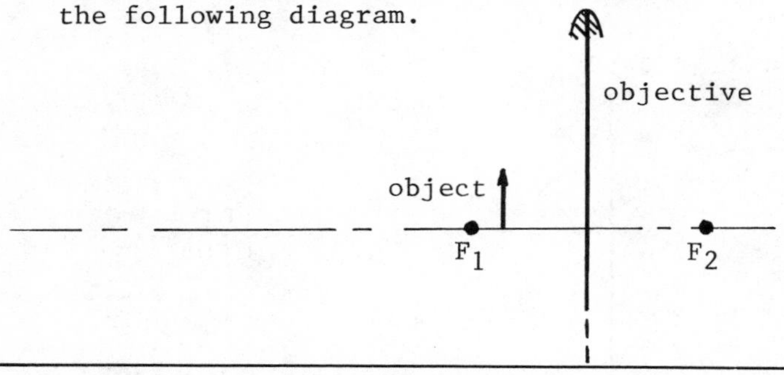

The three rays shown correspond to the three statements (38-6a, b,c).

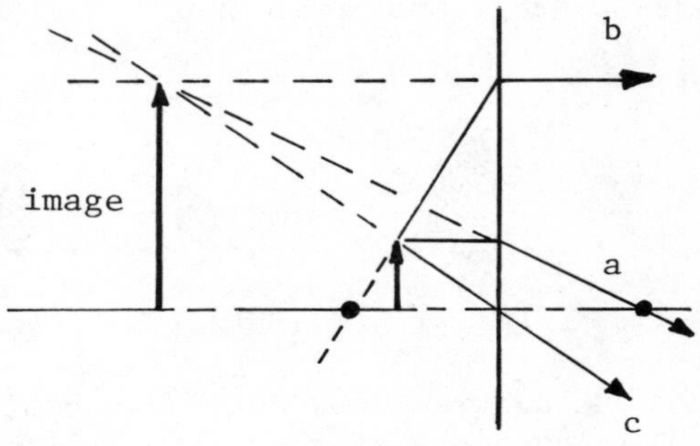

5. (a) What is the formula for the lateral magnification m due to a thin lens?

m = __________ .

(b) What is the physical meaning of m?

(c) What does the sign of m mean?

(a) $m = -\ i/o$

(b) The magnitude of m is the ratio of the image size to the object size ("size" means the lateral dimension, perpendicular to the axis).

(c) A positive value for m means that the image is erect, a negative value for m means that the image is inverted.

6. Calculate the magnification due to the objective and the ocular separately.

m_{obj} = __________ .

m_{oc} = __________ .

$$m_{obj} = -\ \frac{i}{o} = -\ \frac{101}{10{,}000}$$

$$\simeq -\ 10^{-2} \quad .$$

$$m_{oc} = -\ \frac{i}{o} = -\ \frac{-\ 200}{1.98}$$

$$\simeq 10^{2} \quad .$$

7. Let y = size of original object, y' = size of the image of this object due to the objective lens only, y'' = size of final image. The total lateral magnification m of the two lens system is defined by $m = \frac{y''}{y}$. Express this in terms of m_{obj} and m_{oc}.	$m_{obj} = \frac{y'}{y}$ $m_{oc} = \frac{y''}{y'}$ $(m_{obj})(m_{oc}) = (\frac{y'}{y})(\frac{y''}{y'})$ $= \frac{y''}{y} = m$. Therefore $m = m_{obj}m_{oc}$; m is simply the product of the individual lateral magnifications.
8. Using the answer to the above frame calculate the total lateral magnification m for the two lens system. m = ________ .	$m = (m_{obj})(m_{oc})$ $\simeq (-10^{-2})(10^{2}) = -1$. Therefore the final image is (very nearly) the same size as the object; the final image is also inverted.
9. The total lateral magnification is only about one. Also, its value depends strongly upon the 200 cm distance in frame 3. But this distance could really have been anything for which viewing was comfortable (say larger than 25 cm). In the case of a telescope the lateral magnification is not important. What do you think is an important measure of the "power" of a telescope?	Actually one should consider the angle subtended by the image as seen from the eye. When we say that a telescope "magnifies" we mean that the angle subtended by the image is larger than the angle subtended by the object.
10. Using the notation of frame 7 write an expression for (a) the angle θ_i subtended by the final image θ_i = ____________ ; (b) the angle θ_o subtended by the original object θ_o = ____________ . Note that the eye is essentially at the ocular (frame 1). Assume all angles are small.	(a) $\theta_i = \frac{y''}{200}$. (b) $\theta_o = \frac{y}{10,103} \simeq \frac{y}{10,000}$.

11. The angular magnification M is defined as the ratio of the above two angles:

$$M = \frac{\theta_i}{\theta_o} \; .$$

Calculate M for this telescope.

M = __________ .

$$M = \frac{\theta_i}{\theta_o} = \frac{(y''/200)}{(y/10{,}000)}$$

$$= \left(\frac{y''}{y}\right)(50) = (m)(50)$$

$$= (-1)(50) = -50 \; .$$

Since the image is inverted, the negative value makes sense. We say that the telescope is "50 power" (written 50 X).

12. In view of the above numerical answer ($M = -50$), can you guess at a general formula for M in terms of the focal lengths of the two lenses?

M = __________ .

$$M = -\frac{f_{obj}}{f_{oc}}$$

This formula is valid provided

(1) the original object is much further away from the objective lens than f_{obj} (10,000 >> 100), and

(2) the final image is much further away from the ocular than its focal length (200 >> 2).

13. The problem we have just completed shows us how to handle a "multi-element" (e.g. two lens) system. Given the location of the original object, we found the location of the final image. Without getting into the details of the formulas, describe in words how we attacked the two lens system.

First we treated only the objective lens (the first element). Given the location of the original object and the focal length of the lens we found the location of the image due to this lens only. This image then acted as the object for the ocular lens. Thus again we know an object distance and a focal length; we can solve for the image location.

14. The astronomical telescope shown in frame 1 produces an inverted image. This is alright if we wish to look at the moon. For more earthly situations we need a "terrestrial" telescope which produces an erect image. Look at the answer to frame 12 and see if you can figure out how to make a two lens terrestrial telescope.

Use a diverging ocular (say $f = -2$ cm). This makes M positive. It must be placed so that the image due to the objective is slightly more than 2 cm to the right of the ocular.

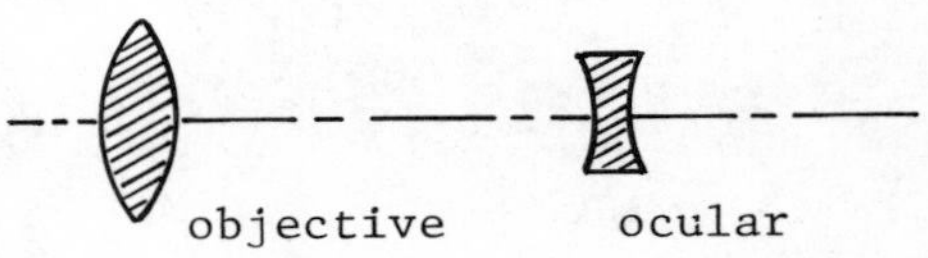

15. So far, all objects have been real. The following two-lens system will illustrate the concept of a virtual object.

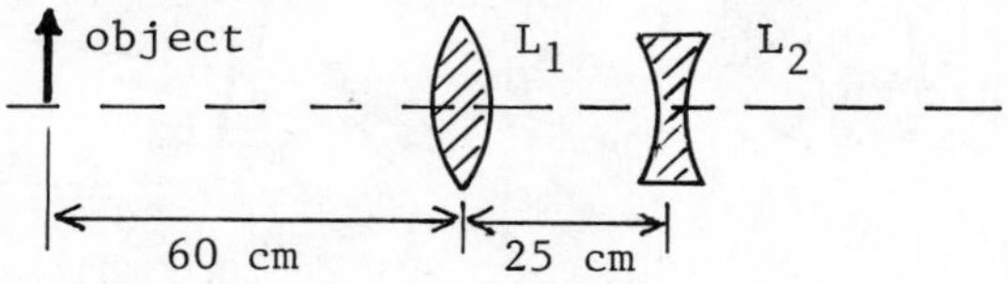

Lens L_1 has focal length $f_1 = + 20$ cm, lens L_2 has focal length $f_2 = - 10$ cm.

Calculate the location of the image treating only the first lens L_1.

$o_1 = 60$ cm, $f_1 = 20$ cm.

$$1/i_1 = 1/f_1 - 1/o_1$$

$$= 1/(20) - 1/(60) \ ,$$

$$i_1 = + 30 \text{ cm} \ .$$

The image is 30 cm to the right of lens L_1.

---

16. Now we shall treat the second lens L_2. Remembering that the incident light for L_2 travels from left to right, what is object distance o_2?

$-$ 5 cm.

The image due to L_1 serves as the object for L_2. It is 30 cm to the right of L_1, hence 5 cm to the right of L_2. From Figure 38-5 we see that this should be considered as a negative object distance. This is a virtual object!

---

17. Calculate the image distance i_2 due to lens L_2.

$o_2 = - 5$ cm, $f_2 = -10$ cm.

$$1/i_2 = 1/f_2 - 1/o_2$$

$$= 1/(-10) - 1/(-5) \ ,$$

$$i_2 = + 10 \text{ cm} \ .$$

---

18. (a) Describe the location of the final image.

(b) State whether the final image is real or virtual.

(c) Calculate the overall magnification.

(d) State whether the final image is erect or inverted.

(a) 10 cm to the right of L_2.

(b) Real (since the last image distance, i_2, is positive).

(c) $m = m_1 \cdot m_2$

$$= [-i_1/o_1] \cdot [-i_2/o_2]$$

$$= [-(30)/(60)] \cdot [-(10)/(-5)]$$

$$= - 1 \ .$$

(d) Inverted (since the overall magnification m is negative).

---

19. Calculate the distance between the original object and the final image.

95 cm. $[o_1 + 25 \text{ cm} + i_2 = 95 \text{ cm}]$
Don't forget the 25 cm between lenses.

Finally, here is an example illustrating virtual object, virtual image as well as a reversal of the sign conventions. Light from the object passes through the converging lens L_1, is reflected from the plane mirror M_2 and then passes through the diverging lens L_3.

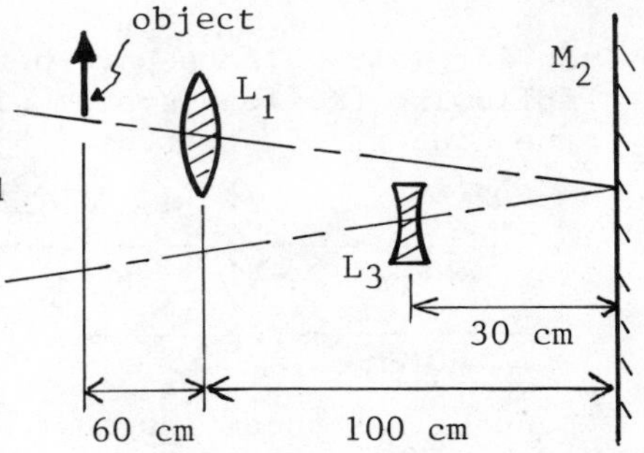

Lens L_1 has focal length f_1 = + 40 cm,
mirror M_2 has focal length f_2 = ∞,
lens L_3 has focal length f_3 = - 5 cm.

20.	First we treat lens L_1. Calculate the image distance i_1. [In this problem you can ignore the slight "tilt" in the axis of the optical system.]	o_1 = 60 cm, f_1 = 40 cm. $1/i_1 = 1/f_1 - 1/o_1$ $= 1/(40) - 1/(60)$ i_1 = + 120 cm . [<u>real image</u>]
21.	Now we treat the plane mirror M_2. (a) Where is the object for M_2? (b) What is the object distance o_2? (c) How are o and i related for a plane mirror? (d) Calculate the image distance i_2.	 (a) 120 cm to right of L_1. (b) o_2 = - 20 cm . [<u>virtual object</u>] (c) i = - o . [Eq. (38-5)] (d) $i_2 = - o_2 = - (- 20)$ = + 20 cm . [<u>real image</u>]
22.	Thus the plane mirror M_2 produced a real image of the virtual object. Now we treat lens L_3. Where is the object for this lens?	20 cm to the left of M_2. [Following the sign convention in Figure 38-2 we see that a positive i means that the image is to the left.]
23.	The incident light for lens L_3 is travelling (left to right/right to left).	Right to left. [It has been reflected from mirror M_2.]
24.	Therefore, all sign conventions are reversed (i.e. compared with those shown in Figure 38-5). Thus o_3 = + 10 cm (note that this object is to the right of L_3). Calculate the image distance i_3 of this <u>real object</u>.	o_3 = 10 cm, f_3 = - 5 cm. $1/i_3 = 1/f_3 - 1/o_3$ $= 1/(-5) - 1/(10)$ i_3 = - 3.33 cm .
25.	The final image is (real/virtual).	Virtual. [i_3 is negative]
26.	Where should one look to see this final image?	Place your eye to the left of lens L_3 and look to the right. It will appear to be located 3.33 cm behind this lens.

Chapter 39: INTERFERENCE

REVIEW AND PREVIEW

The previous two chapters concentrated on geometrical optics. In this and the next two chapters you will return to the wave aspect of light. The chief consequence of the wave nature of light is that interference can occur between two light waves. In this chapter you will study two different physical situations which show the phenomenon of interference:

a. Light passing through two slits is allowed to fall on a distant screen. Alternate bright and dark regions are observed.
b. Light passing through (or reflected from) a thin film may be either bright or dark according to the film thickness.

Both of these effects depend upon the wavelength of the light.

GOALS AND GUIDELINES

In this chapter there are three major goals:

1. Understanding why there is a phase difference between two beams of light which (coming from a common source) travel different path lengths and then recombine at some point (Section 39-2).
2. Understanding two slit interference (Section 39-3). You should be able to calculate the locations of the maxima and minima.
3. Understanding thin film interference (Section 39-4). Here, in addition to the phase difference due to the different path lengths, there may be a phase difference due to reflection.

Note that Equation 39-2 is the starting point for solving interference problems (Examples 1 and 2).

39-1 Wave Optics

The previous two chapters dealt with geometrical optics. Chapters 39-41 are concerned with the wave properties of light; this topic is called wave optics. Wave optics becomes important when certain physical dimensions of the apparatus are comparable with the wavelength of the light.

39-2 Interference of Light

Light is a wave phenomenon. If a beam of light is split into two (or more) beams and these beams are allowed to combine in some region of space, we expect interference to occur. Depending upon the phase difference between the two beams, the resultant amplitude may be more or less than the amplitude of each individual beam. If the two beams are in phase (or differ in phase by 2π, 4π, 6π, etc.) there will be constructive interference. On the other hand if they differ in phase by π (or 3π, 5π, 7π, etc.) there will be destructive interference.

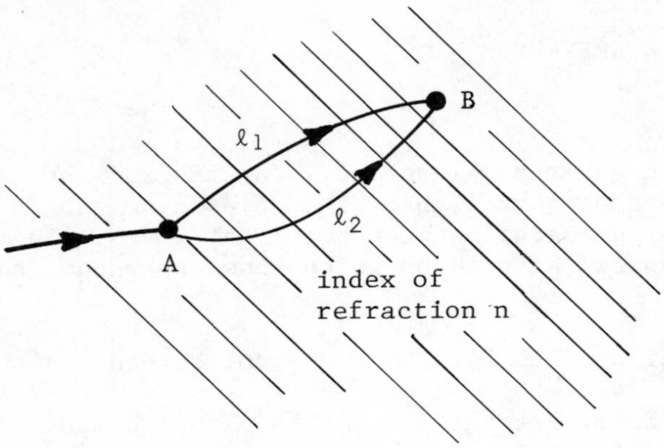

Figure 39-1. A beam of light is split into two beams at point A. The two beams combine at point B where interference occurs.

An example of the above situation is shown in Figure 39-1. A beam of light is split into two beams by some device at point A. These two beams are then allowed to combine at point B. Although the two beams were originally in phase at point A, there will be some phase difference between them at point B because they have traveled different distances. This phase difference, $\Delta\phi$, is proportional to the path length difference, $\Delta\ell = \ell_2 - \ell_1$:

$$\Delta\phi = (2\pi/\lambda_{med})\ \Delta\ell \quad . \tag{39-1}$$

This shows (for example) that for a path length difference of one wavelength, there is a phase difference of 2π . Note that we must use the wavelength, λ_{med}, in the medium in which the path length difference occurs. Recalling that $\lambda_{med} = \lambda/n$,

$$\Delta\phi = (2\pi n/\lambda)\ \Delta\ell \quad . \tag{39-2}$$

Here λ is the wavelength in vacuum and n is the index of refraction of the medium in which the path length difference occurs. Equation (39-2) is the starting point for solving many interference problems: it gives the phase difference $\Delta\phi$ associated with a path length difference $\Delta\ell$.

39-3 Two Slit Interference

Figure 39-2a shows the apparatus used to demonstrate two slit interference (known as Young's experiment). Light is incident normally upon the two slit system, the slits being separated by a distance d. Each slit is very narrow so that geometrical optics does not apply: light emerges outward from each slit in all directions. Point P lies on a screen a large distance D >> d from the slits; the location of P is specified by the angle θ shown.

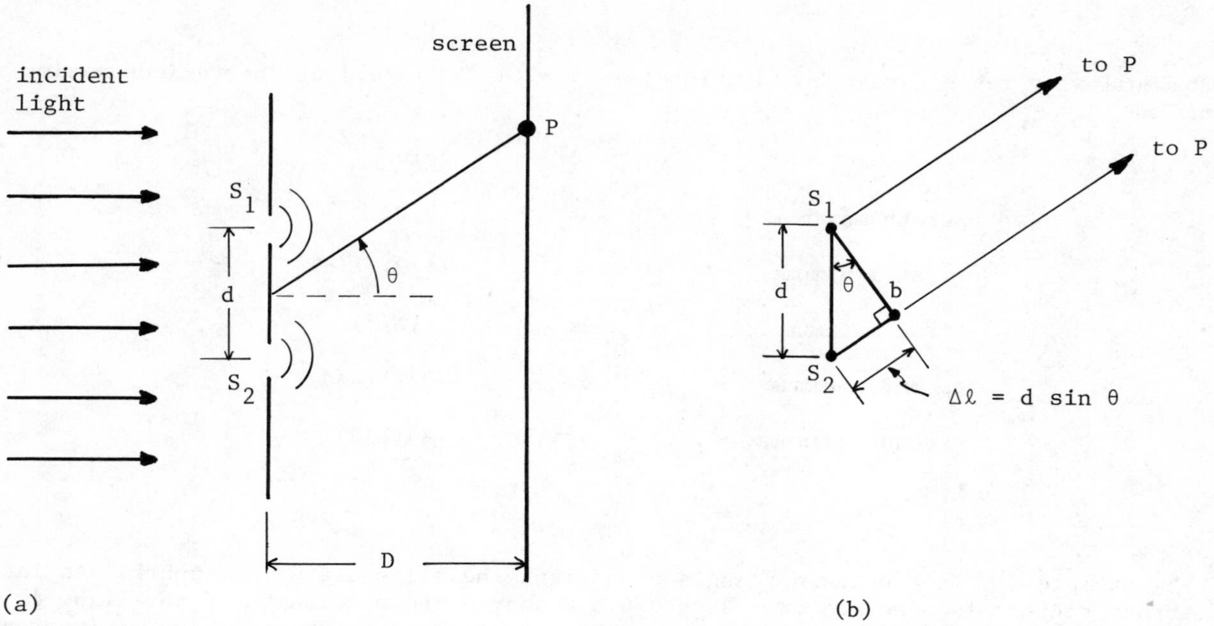

Figure 39-2. (a) Apparatus for Young's experiment. Light is incident normally upon the two slits (S_1, S_2). The interference pattern is observed on a screen a distance D away from the slits. (b) The path length difference $\Delta\ell = d \sin\theta$ is obtained from the right triangle S_1bS_2.

Young's experiment is an example of the situation shown in Figure 39-1 (with n = 1). Here the two slit system plays the role of point A, point P on the screen plays the role of point B. Light coming through the two slits will interfere at P. If there is constructive interference there will be a "maximum" (bright region) at P; similarly if there is destructive interference there will be a "minimum" (dark region) at P.

By construction the distances S_1P and bP in Figure 39-2b are equal. Since the screen is very far away from the slits we have approximately (see text):

(i) The angle "θ" in Figure 39-2b equals the angle "θ" in Figure 39-2a.

(ii) Triangle S_1bS_2 is a right triangle.

From the right triangle in Figure 39-2b, the path length difference ($\Delta\ell = S_2b$) is

$$\Delta\ell = d \sin\theta \qquad \text{(two slits)} \quad . \qquad (39\text{-}3)$$

Using equation (39-2) with n = 1, the phase difference associated with this path length difference is

$$\Delta\phi = (2\pi/\lambda)\, d \sin\theta \qquad \text{(two slits)} \quad . \qquad (39\text{-}4)$$

The student should be able to recall quickly this formula by simply sketching a diagram similar to Figure 39-2b.

On the axis ($\theta = 0$) of the system we have $\Delta\phi = 0$. The screen is bright at this point; this is called the "central (or zeroth) maximum". As we go up the screen from the central maximum, θ (and therefore $\Delta\phi = (2\pi/\lambda)(d \sin\theta)$ increases. Thus starting at the central maximum there are (in either direction) alternate bright and dark regions.

The following table gives the location [$\sin\theta = (\Delta\phi/2\pi)(\lambda/d)$] of the various maxima and minima.

Name	$\Delta\phi$	$\sin\theta$
zeroth maximum	0	0
first minimum	π	$(1/2)(\lambda/d)$
first maximum	2π	(λ/d)
second minimum	3π	$(3/2)(\lambda/d)$
second maximum	4π	$2\ (\lambda/d)$
etc.		

>>> Example 1. In a certain Young's experiment the slits are 0.2 mm apart. An interference pattern is observed on a screen 0.5 m away. The wavelength of the light is 500 nm. Calculate the distance between the central maximum and the third minimum on the screen.

The given quantities are

$$d = 2 \times 10^{-4}\ \text{m} \qquad \text{(slit separation)} \quad ,$$

$$\lambda = 5 \times 10^{-7}\ \text{m} \qquad \text{(wavelength)} \quad ,$$

$$D = 5 \times 10^{-1}\ \text{m} \qquad \text{(distance from slits to screen)} \quad .$$

The minima are given by $\Delta\phi = \pi, 3\pi, 5\pi, 7\pi$, etc. At the third minimum we then have $\Delta\phi = 5\pi$.

$$\Delta\phi = (2\pi/\lambda)\ \Delta\ell = (2\pi/\lambda)\ (d \sin\theta)$$

$$\sin\theta = \left(\frac{\Delta\phi}{2\pi}\right)\left(\frac{\lambda}{d}\right) = \left(\frac{5\pi}{2\pi}\right)\left(\frac{5 \times 10^{-7}\ \text{m}}{2 \times 10^{-4}\ \text{m}}\right)$$

$$\sin\theta = 6.25 \times 10^{-3} \quad .$$

The required distance on the screen is then

$$y = D \tan\theta$$

$$= D \sin\theta \qquad \text{(since } \theta \text{ is small)}$$

$$= (5 \times 10^{-1}\ \text{m})(6.25 \times 10^{-3}) = 3.1\ \text{mm} \quad .$$

<<<

39-4 Thin Film Interference

Figure 39-3 shows the various light paths involved in the phenomenon of thin film interference. Light is incident approximately normally upon a thin film of thickness t and index of refraction n. According to the location of the observer, there are two cases:

(i) If, as in Figure 39-3a, the observer is on the same side of the film as the incident light we say that it is being viewed "by reflection". Some of the incident light is reflected to the observer from the upper surface of the film. Light may also reach the observer by reflection from the lower surface of the film as shown. These two beams can then interfere with each other.*

(ii) If, as in Figure 39-3b, the observer is on the opposite side of the film from the incident light we say that it is being viewed "by transmission". Some of the incident light is transmitted to the observer directly through the film. Light may also reach the observer by being reflected first from the lower surface and then from the upper surface of the film as shown. These two beams can then interfere with each other.*

From Figure 39-3 we see that in either of these cases the path length difference is twice the thickness** of the film: $\Delta\ell = 2t$. Using equation (39-2), the phase difference due to this path length difference is

$$\Delta\phi = (2\pi n/\lambda)(2t) \quad \text{(due to path length difference)} \ . \qquad (39\text{-}5)$$

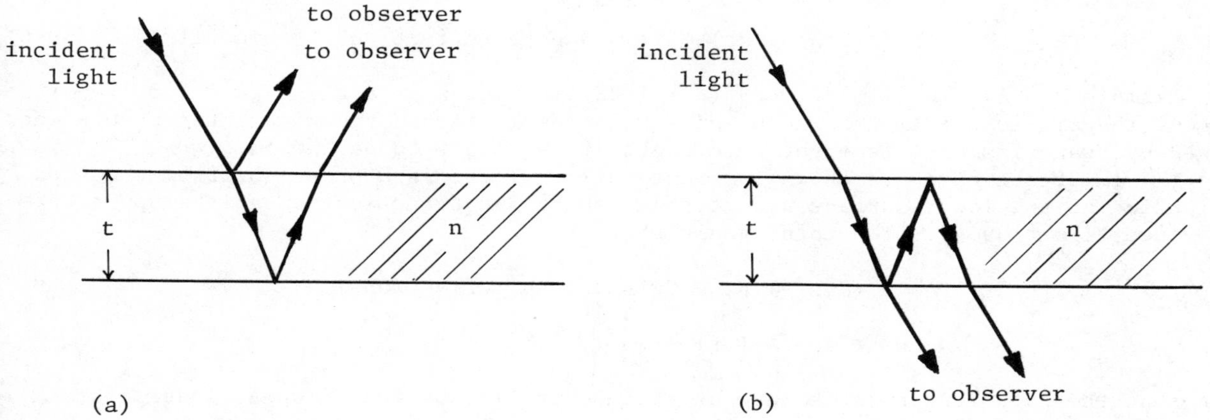

Figure 39-3. The important light paths for thin film interference. All rays are approximately normal to the film. (a) Thin film interference viewed by reflection. (b) Thin film interference viewed by transmission.

In addition to the phase difference (39-5) there is the phenomenon of a phase change due to reflection. The amount of this phase change is (for each reflection) either π or zero. In Figure 39-4a the light is being reflected from a medium whose index of refraction is greater than that of the incident medium; in this case the reflected light will undergo a phase change of π due to the reflection. In Figure 39-4b the light is being reflected from a medium whose index of refraction is less than that of the incident medium; in this case the reflected light will undergo no phase change due to the reflection. In neither case is there a phase change due to transmission.

---

*There are other possible beams due to multiple reflections within the film. The two beams shown in the figure are the most important ones.

**All rays are approximately normal to the film.

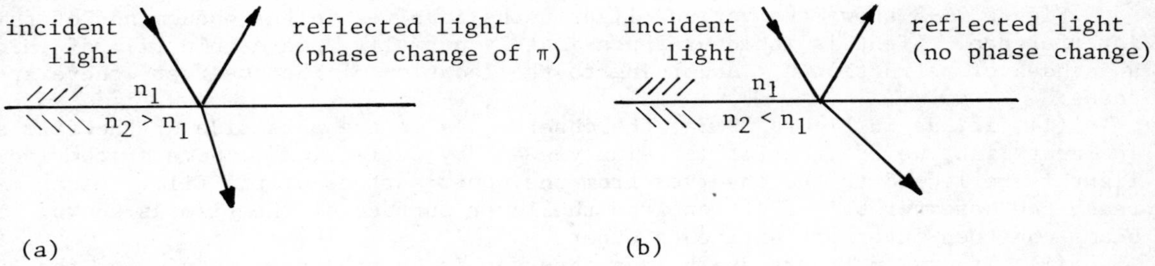

Figure 39-4. (a) With n_2 greater than n_1 the reflected light undergoes a phase change of π due to reflection. (b) With n_2 less than n_1 the reflected light undergoes no phase change due to reflection.

To solve thin film interference problems, one must first compute the phase change (39-5) due to the path length difference. Then every reflection involved must be compared with Figure 39-4 to determine if there are any phase changes due to reflection. The appropriate number of π's must be added* to the phase difference (39-5) to obtain the total phase difference,

$$\Delta\phi = (2\pi n/\lambda)(2t) + \text{"}\pi\text{'s due to reflection"} \quad \text{(thin film)} \qquad (39\text{-}6)$$

>>> Example 2. A thin film of MgF_2 ($n = 1.38$) is placed on a glass ($n = 1.50$) surface. What is the smallest film thickness which will result in constructive interference when viewed by transmission? Take the wavelength of the light to be 550 nm.

The important rays are shown in Figure 39-5. There is a phase shift of π due to reflection at the lower surface and no phase shift due to reflection at the upper surface (see Figure 39-4). The total phase shift is

$$\Delta\phi = (2\pi n/\lambda)\,\Delta\ell + \text{"}\pi\text{'s due to reflection"}$$

$$= (2\pi n_2/\lambda)(2t) + \pi \quad .$$

Note that the path length difference $\Delta\ell = 2t$ occurs in the medium whose index of refraction is n_2 (1.38). Solving for the thickness t,

$$t = (\lambda/4\pi n_2)(\Delta\phi - \pi) \quad .$$

For constructive interference we must have $\Delta\phi = 0, \pm 2\pi, \pm 4\pi$, etc. In this case the smallest (positive) thickness for constructive interference is given by $\Delta\phi = 2\pi$. Therefore

$$t = \frac{550\ \text{nm}}{4\pi(1.38)}\ (2\pi - \pi) = 100\ \text{nm} \quad .$$

---

*It does not matter whether one adds or subtracts these π's. Also, phase differences of 2π may be neglected.

Figure 39-5. A thin film of MgF_2 on a glass surface. The important rays are shown for the case where it is viewed by transmission.

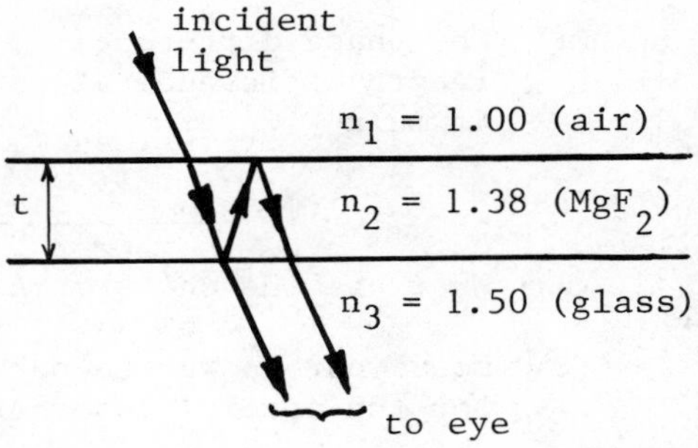

Remarks:

1. If we had subtracted the π due to reflection instead of adding it we would have had $t = (\lambda/4\pi n_2)(\Delta\phi + \pi)$. In this case $\Delta\phi = 0$ (instead of 2π) would give the smallest (positive) thickness for constructive interference. The answer ($t = 100$ nm) is the same.

2. The student should verify that constructive interference by transmission implies destructive interference by reflection. Thus the problem is really the same as that of a "non-reflecting" thin film. <<<

39-5 Programmed Problems

1. In the diagram below, light is incident normally upon the two slit system. At point P on the screen the two beams of light (one from each slit) will interfere. There is a certain phase difference $\Delta\phi$ between these two beams at point P. Why is there a phase difference?

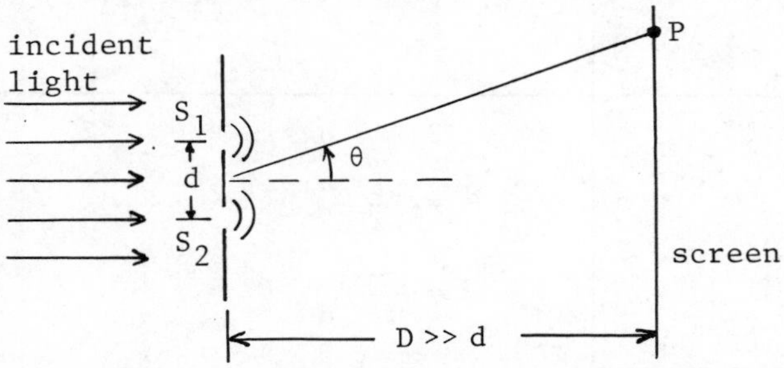

There is a phase difference because the two beams travel different distances to get to point P.

2. Let $\Delta\ell$ be the path length difference between the two beams.

(a) On the diagram below, show this path length difference $\Delta\ell$.

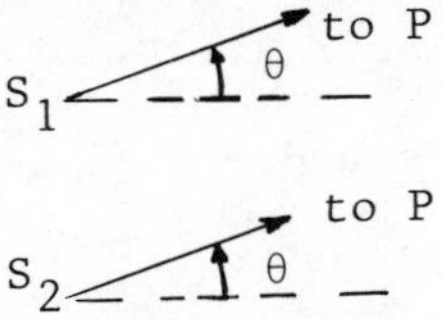

(b) Express $\Delta\ell$ in terms of the slit separation d and the angle θ.

$\Delta\ell$ = ______________ .

(a)

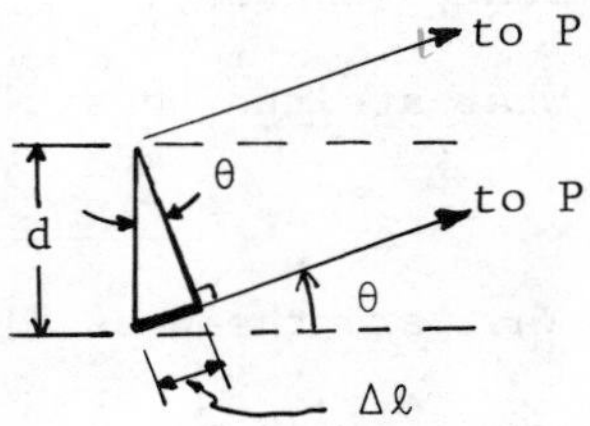

(b) $\Delta\ell = d \sin\theta$.

3. (a) The phase difference $\Delta\phi$ due to the path length difference $\Delta\ell$ is given by the formula

$$\Delta\phi = \_\_\_\_\_\_\_\_ .$$

(b) What are the units of $\Delta\phi$?

(c) Does your answer to part (a) make sense for the special cases $\Delta\ell = 0, \lambda/2, \lambda$?

(a) $\Delta\phi = (2\pi/\lambda)\ \Delta\ell$.

(b) radians

(c) $\Delta\ell = 0 \Rightarrow \Delta\phi = 0$

(constructive interference)

$\Delta\ell = \lambda/2 \Rightarrow \Delta\phi = \pi$

(destructive interference)

$\Delta\ell = \lambda \Rightarrow \Delta\phi = 2\pi$

(constructive interference)

---

4. Using the answers to frames 2 and 3, express $\Delta\phi$ in terms of d, λ, θ.

$$\Delta\phi = \_\_\_\_\_\_\_\_ .$$

$\Delta\phi = (2\pi/\lambda)\ d \sin\theta$

---

5. What are the possible values of $\Delta\phi$ for constructive interference (maxima)?

$$\Delta\phi = \_\_\_\_\_\_\_\_ .$$

$\Delta\phi = 0, 2\pi, 4\pi, 6\pi, \ldots$

$= 2\pi m \quad (m = 0, 1, 2, \ldots)$

---

6. Therefore what are the possible values of $\sin\theta$ for constructive interference?

$$\sin\theta = \_\_\_\_\_\_\_\_ .$$

$\Delta\phi = (2\pi/\lambda)\ d \sin\theta$

$$\sin\theta = \frac{\Delta\phi}{2\pi}\frac{\lambda}{d}$$

$$= \frac{2\pi m}{2\pi}\frac{\lambda}{d}$$

$$= m\frac{\lambda}{d} \quad (m = 0, 1, 2, \ldots)$$

---

7. Suppose we were interested in destructive interference (minima).

(a) What are the possible values of $\Delta\phi$?

$$\Delta\phi = \_\_\_\_\_\_\_\_ .$$

(b) What are the possible values of $\sin\theta$?

$$\sin\theta = \_\_\_\_\_\_\_\_ .$$

(a) $\Delta\phi = \pi, 3\pi, 5\pi, \ldots$

(b) $\sin\theta = \frac{1}{2}\frac{\lambda}{d}, \frac{3}{2}\frac{\lambda}{d}, \frac{5}{2}\frac{\lambda}{d}, \ldots$

8. Consider the case λ = 500 nm and d = 2000 nm. Make a sketch of the intensity on the screen as a function of sin θ.

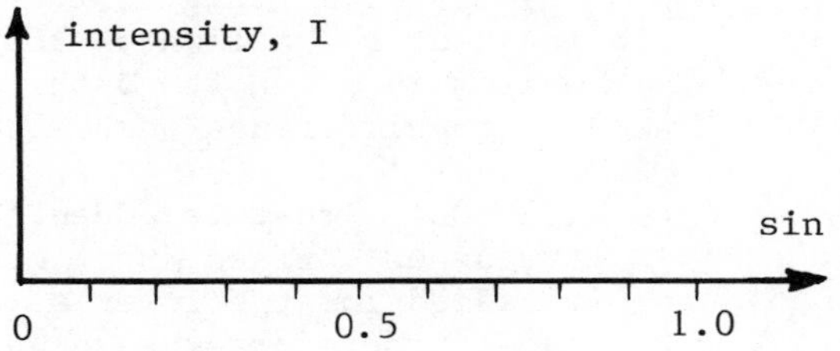

The maxima are located at

$\sin\theta = m\,\lambda/d = 0.25\,m$

$= 0,\ 0.25,\ 0.50,\ 0.75,\ 1.00.$

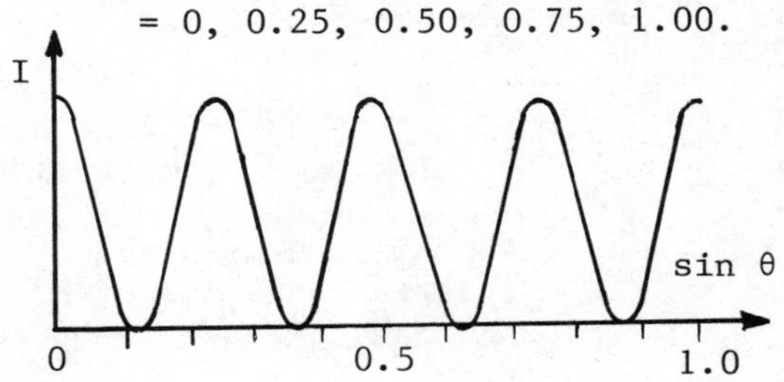

9. In the above sketch, where is the third order maximum?

The third order maximum is located at sin θ = 0.75. (Remember that the maximum at sin θ = 0 is called the zeroth order maximum.)

10. Another type of interference problem occurs when light is incident upon a thin film as shown below.

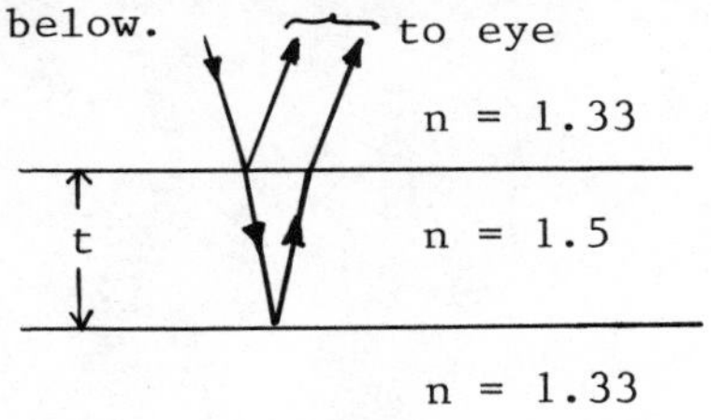

In this case, what is the path length difference $\Delta\ell$ (assume normal incidence)?

$\Delta\ell$ = __________ .

$\Delta\ell = 2t$.

Remember, the light goes back (distance t) and forth (distance t).

11. We would like to apply the formula used before:

" $\Delta\phi = (2\pi/\lambda)\ \Delta\ell$ ".

There are two modifications which must be made before this is correct. What are these modifications?

(i) We must use the wavelength λ_{med} in the medium in which the path length difference occurs. This is given by

$$\lambda_{med} = \lambda/n \ .$$

(ii) We must allow for possible phase changes of π due to reflection.

12. The correct formula is then

$$\Delta\phi = (2\pi/\lambda_{med})(\Delta\ell) + \text{"}\pi\text{'s due to reflection"}$$

$$= 4\pi nt/\lambda + \text{"}\pi\text{'s due to reflection} \quad .$$

(a) What is the meaning of λ?
(b) What is the value of n?
(c) How many π's are to be added?

(a) λ is the wavelength of the light in vacuum.

(b) $n = 1.5$ (this is the index of refraction of the medium in which the path length difference occurs).

(c) One π must be added for the upper surface, no π must be added for the lower surface: therefore the answer is one π.

Chapter 40: DIFFRACTION

REVIEW AND PREVIEW

In the previous chapter you studied two slit interference. Each slit was assumed to be so narrow that any phase difference due to light passing through different parts of the same slit could be ignored. In this chapter you will study single slit diffraction; that is, the interference pattern due to light passing through a single slit. Here the effect of the slit width cannot be ignored. The problem is attacked by dividing the single slit into many narrow strips. To systematically treat the addition of many waves (one from each strip), the concept of phasors is introduced.

GOALS AND GUIDELINES

In this chapter there are three major goals:

1. Understanding how to add waves (which differ in phase from one another) by the method of phasors.
2. Understanding the nature of single slit diffraction from the point of view of the phasor method. You should be able to calculate the locations of the maxima and minima as well as the relative intensities of the maxima.
3. Learning the definition of Rayleigh's criterion for resolving two sources by means of an optical instrument. This involves the diffraction due to a circular aperture.

The method of phasors has application not only to single slit problems but also to multiple slit problems (see next chapter). It is very important that you fully understand this geometrical technique for adding waves.

40-1 Phasors

The projection of a vector rotating with a constant angular velocity ω is a sinusoidal function of time. This fact permits us to add such functions using a simple geometrical construction. As an example, suppose we wish to find the sum $u = u_1 + u_2 + u_3$ where

$$u_1 = 4\sin(\omega t + 10^\circ) \quad ,$$

$$u_2 = 6\sin(\omega t + 45^\circ) \quad ,$$

$$u_3 = 3\sin(\omega t + 60^\circ) \quad .$$

The functions of u_i are of the form

$$u_i = A_i \sin(\omega t + a_i) \quad ,$$

i.e. they are all sinusoidal functions of time with the same (angular) frequency ω. As in this example, they may have different amplitudes A_i (4, 6, 3 respectively) and different phase constants a_i (10°, 45°, 60° respectively). For each of the functions u_i we draw an arrow in the x-y plane, called a phasor, such that:

(a) the length of the phasor represents (to some chosen scale) its amplitude A_i, (40-1a)

(b) the angle that the phasor makes with the x axis is its phase: $\omega t + a_i$. (40-1b)

Note that the y component of each phasor is the given function $u_i = A_i \sin(\omega t + a_i)$. We then add these phasors vectorially to obtain the "resultant phasor". The construction for this example is shown in Figure 40-1. In the figure the phasors are drawn "head to tail" to facilitate finding their sum.

It turns out that the desired sum u is also of the same form as the u_i: $u = R \sin(\omega t + b)$. Moreover, the resultant amplitude R and the phase constant b are related to the resultant phasor by the same rules (40-1) from which the individual phasors were drawn. Thus R and b can be easily determined from the figure.

Although the construction was made for some particular time t, as t increases all the phasors (including the resultant) merely rotate counterclockwise together at the common angular rate ω. Because of this, only the relative directions of the phasors are really important.

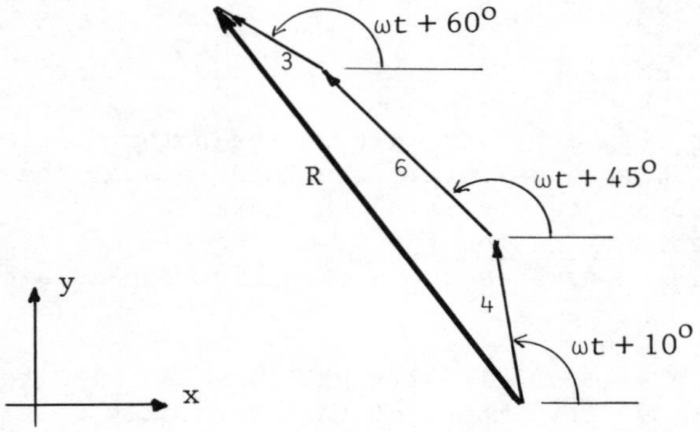

Figure 40-1. The phasor diagram for some time t. The lengths of the phasors are proportional to the amplitudes A_i. The angles $\omega t + a_i$ are measured up from the x axis. The resultant magnitude is approximately 12.3. As t increases the entire diagram rotates counterclockwise.

To simplify the construction we usually draw the first phasor along the x axis. The angle between the second and first phasors is then simply the phase difference ($45° - 10° = 35°$) between the corresponding functions u_2 and u_1. Similarly the angle between the third and second phasors is the phase difference ($60° - 45° = 15°$) between u_3 and u_2. This phasor diagram is shown in Figure 40-2. In general then the procedure for constructing a phasor diagram is to draw a phasor for each function u_i such that:

(a) its length corresponds to its amplitude A_i, (40-2a)

(b) the angle between any two phasors is the phase difference between them. (40-2b)

The length of the resultant phasor then corresponds to the resultant amplitude R.* Briefly, the point to remember about phasor diagrams is that lengths correspond to amplitudes and that angles mean phase differences.

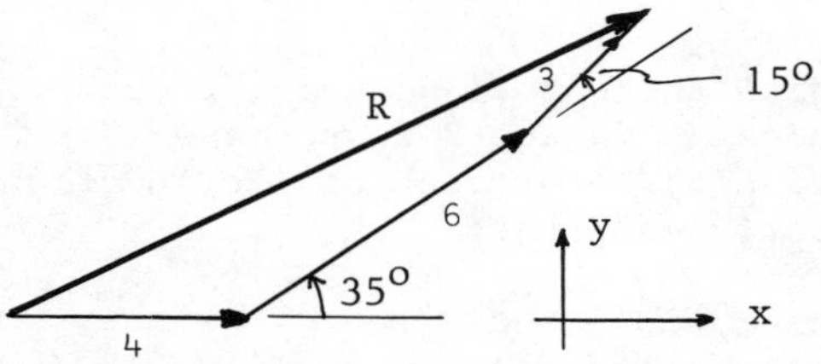

Figure 40-2. This is the same phasor diagram as in Figure 40-1 except that the first phasor is drawn along the x axis. The angles shown are the phase differences between successive phasors.

*Uusually we are interested only in the amplitude R. The phase constant b can be found by noting that the angle between the resultant phasor and the first phasor is $b - a_1$.

40-2 Single Slit Diffraction

When the width of a slit is not negligible we must take into account interference effects between light beams which come from different parts of the slit; this type of interference is known as diffraction.* In the single slit diffraction apparatus shown in Figure 40-3a, light is incident normally upon a slit of width a. Point P lies on a screen a distance D >> a away from the slit; the location of P is specified by the angle θ shown. The slit is imagined to be divided into a large number N of equally spaced strips of width $\Delta x = a/N$ as shown in Figure 40-3b. At point P the many beams of light (one from each strip) will interfere. In the limit $N \to \infty$ ($\Delta x \to 0$), we have infinitely many beams of light interfering at P.

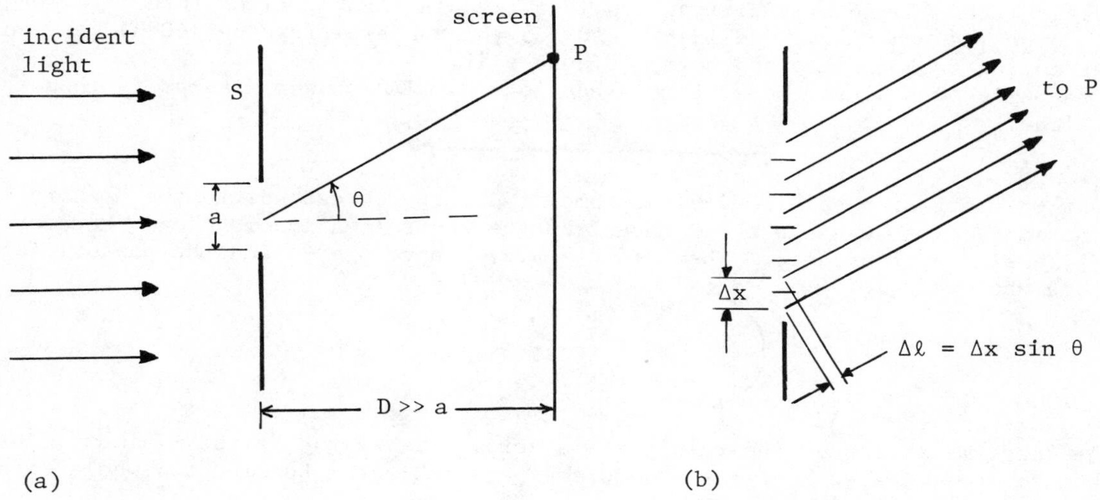

Figure 40-3. (a) Apparatus for single slit diffraction. Light is incident normally upon the slit S. The diffraction pattern is observed on a screen a distance D away from the slit. (b) The slit is imagined to be divided into strips of width Δx. The path length difference for adjacent strips is $\Delta\ell = \Delta x \sin\theta$.

The phasor diagram for single slit diffraction is shown in Figure 40-4. Since all the strips are of the same width (Δx), the individual phasors are all of equal length. Also, the angle between any two successive phasors, namely the phase difference $\Delta\phi = (2\pi/\lambda)(\Delta x \sin\theta)$, is the same for any pair of successive phasors. Because of these two facts the phasor diagram is an N sided (open) polygon; all sides having the same length and all adjacent angles being equal. In the limit $N \to \infty$ this becomes a circular arc. The angle φ between the first and last phasor in this phasor diagram is the phase difference between the light coming from the top and bottom of the slit, i.e.

$$\phi = (2\pi/\lambda)(a \sin\theta) \qquad \text{(angle between first and last phasors, single slit)} . \qquad (40\text{-}3)$$

*The interference of infinitely many beams is called diffraction.

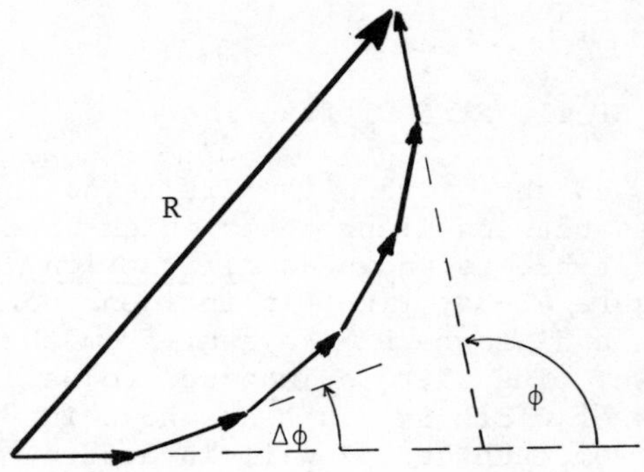

Figure 40-4. Phasor diagram for single slit diffraction. The diagram is an N sided (open) polygon. The angle between adjacent phasors is $\Delta\phi = (2\pi/\lambda)(\Delta x \sin\theta)$. The figure is drawn for N = 6. In the limit $N \to \infty$ the phasor diagram becomes a circular arc, the angle ϕ between the first and last phasor is $\phi = (2\pi/\lambda)(a \sin\theta)$.

The total arc length of this circular arc is the same for all values of θ (because this arc length is the sum of the amplitudes of the individual phasors). In summary the phasor diagram for single slit diffraction is a circular arc of fixed total arc length, the angle ϕ between the first and last phasor is the phase difference (40-3) between the light coming from the top and bottom of the slit.

On the axis ($\theta = 0$) of the system we have $\phi = 0$. The phasor diagram is then a straight line:

This gives the largest possible resultant amplitude; it corresponds to the "central (or zeroth) maximum". As we go up the screen from the central maximum, θ (and therefore $\phi = (2\pi/\lambda)(a \sin\theta)$) increases. When $\sin\theta = \lambda/a$ we have $\phi = 2\pi$ and the phasor diagram becomes one whole circle:

Since the diagram is closed, the resultant is zero; this is the "first minimum". Similarly when $\sin\theta = 2\lambda/a$ we have $\phi = 4\pi$ and the phasor diagram becomes two whole circles:

The resultant is again zero, this is the "second minimum". In general there will be a minimum every time the phasor diagram becomes a whole number of circles. This will occur whenever $\phi = 2\pi m$ ($m = 1, 2, 3, \ldots$). Since from (40-3) $\sin\theta = (\phi/2\pi)(\lambda/a)$, the minima are located at

$$\sin\theta = m\,\lambda/a \quad ; \quad m = 1, 2, 3, \ldots \quad \text{(minima, single slit).} \qquad (40\text{-}4)$$

Note that m = 0 is not included in (40-4). Between these minima there are weak "secondary maxima". These secondary maxima occur when the phasor diagram is (approximately, see text) 1½, 2½, 3½, ... circles. The first secondary maximum is only 0.045 as intense as the central maximum, the other secondary maxima are progressively less intense. Figure 40-5 shows the intensity as a function of $\sin\theta$ for single slit diffraction.

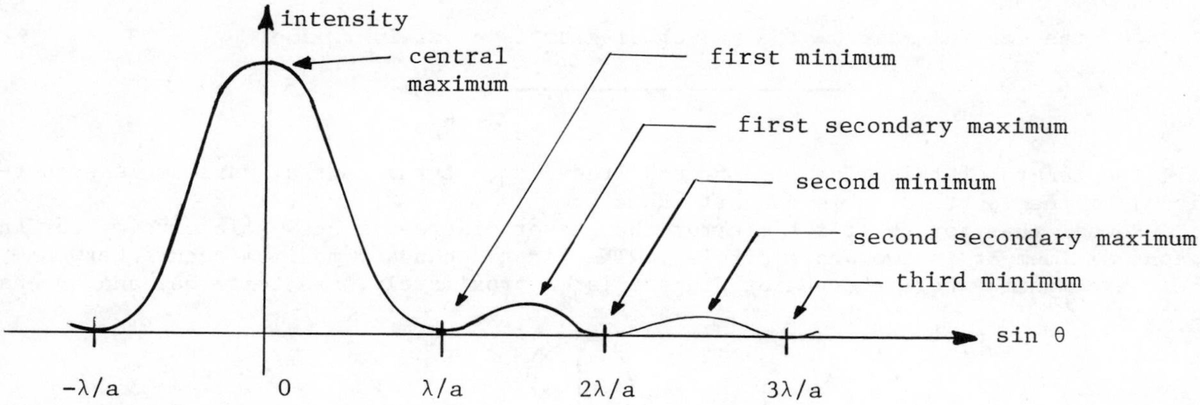

Figure 40-5. Single slit diffraction pattern.

>>> Example 1. Light of wavelength 550 nm is incident normally upon a single slit. A diffraction pattern is observed on a distant screen. The angle between the second minimum and the central maximum is 1°. Calculate the width of the slit.

The phasor diagram for the second minimum is two whole circles:

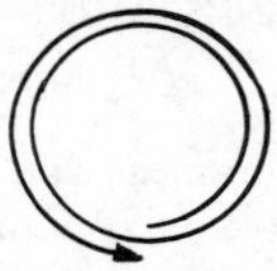

The angle ϕ between the first and last phasor is then 4π. Since this is also the phase difference between light coming from the top and bottom of the slit,

$$\phi = (2\pi/\lambda)(a \sin \theta) = 4\pi$$

where a is the width of the slit. Solving for a,

$$a = \frac{2\lambda}{\sin \theta} = \frac{2\ \lambda}{\sin\ (1^o)}$$

$$a = \frac{(2)\cdot(5.5 \times 10^{-7}\ m)}{1.75 \times 10^{-2}} = 6.3 \times 10^{-5}\ m = 0.063\ mm \quad .$$ <<<

40-3 Intensity and Amplitude

One property of a wave is its amplitude. For a wave on a string the amplitude is the maximum transverse displacement. For an electromagnetic wave (such as light) the amplitude could mean the maximum value of the electric field. Another property of a wave is its intensity. This is an energy related concept, for an electromagnetic wave it could be the average energy per unit area per unit time (i.e. the average Poynting vector). Regardless of exactly what is meant by amplitude or intensity, the intensity is proportional to the square of the amplitude. Thus for example if the ratio of two amplitudes is 10:1, the ratio of the corresponding intensities will be 100:1.

>>> Example 2. Calculate the ratio of the intensity of the first secondary maximum to that of the central maximum for single slit diffraction.

For the central maximum the phasor diagram is a straight line:

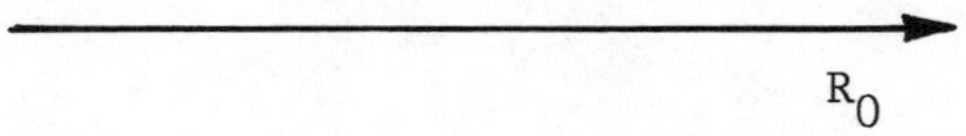

Call the length of this line L. The amplitude, R_0, of the central maximum is proportional to the length of this phasor: $R_0 \propto L$.

Recall that for the first minimum the phasor diagram is one whole circle, for the second minimum it is two whole circles. The first secondary maximum occurs between these two minima where the phasor diagram is (approximately, see text) one and one-half circles:

The total arc length of this diagram is of course L (total arc length is the same for all values of θ). The circumference (of one whole circle, i.e. 360°) is then 2/3 L. The resultant phasor is the diameter in the above diagram, the length of this diameter is (circumference)/π = $2L/3\pi$. The amplitude, R_1, of the first secondary maximum is proportional to the length of this resultant phasor: $R_1 \propto 2L/3\pi$. The ratio of the two amplitudes is

$$R_1/R_0 = (2L/3\pi)/L = 2/3\pi \quad .$$

Note that the proportionality constants have cancelled out. Since intensity is proportional to the <u>square</u> of the amplitude, the ratio of the two intensities is

$$I_1/I_0 = (R_1/R_0)^2 = (2/3\pi)^2 = 0.045 \quad .$$ <<<

40-4 Diffraction at a Circular Aperture

When light is incident upon a circular aperture, diffraction will occur. The diffraction pattern observed on a distant screen is qualitatively similar to that due to a single slit. Of course for the circular aperture this pattern will be circularly symmetric: it consists of a bright central disk surrounded by progressively weaker secondary rings.

For a single slit we see from (40-4) that the angle θ between the first minimum and the central maximum is given by $\sin\theta = \lambda/a$, where a is the slit width. A somewhat similar formula holds for the case of a circular aperture of <u>diameter</u> d:

$$\sin\theta = 1.22\,\lambda/d \qquad \text{(first minimum, circular aperture).} \tag{40-5}$$

40-5 Resolution of an Optical Instrument

Most optical instruments (e.g. a telescope) are circular in cross section. Incident light from a distant point source will produce a diffraction pattern characteristic of a circular aperture. Two such sources (e.g. two stars being viewed with a telescope) will produce two overlapping diffraction patterns. If these patterns overlap too much then we cannot tell them apart, i.e. we cannot <u>resolve</u> them. If, as in Figure 40-6, the two diffraction patterns are such that the central maximum of one falls on the first minimum of the other we say that they are just resolvable; this is known as <u>Rayleigh's criterion</u>.

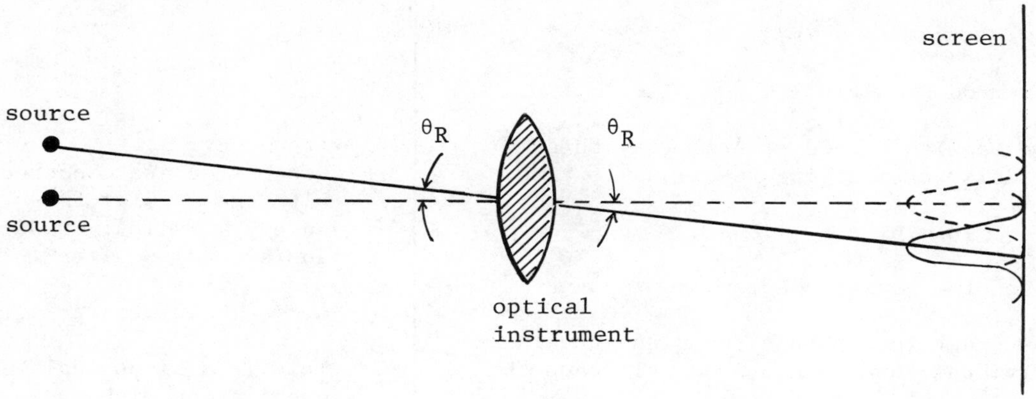

Figure 40-6. Two sources each produce a diffraction pattern shown graphed on the screen. The central maximum of one diffraction pattern falls on the first minimum of the other. Note that the angle θ_R between the first minimum and the central maximum of one diffraction pattern is the same as the angular separation of the two sources.

The angular separation, θ_R, of two sources which are just resolvable according to Rayleigh's criterion is then simply the angle (40-5) between the first minimum and the central maximum. In practice θ_R is quite small. Making the small angle approximation $\sin\theta_R \simeq \theta_R$, we have from (40-5)

$$\theta_R = 1.22\ \lambda/d \quad . \tag{40-6}$$

If the angular separation of two sources is larger than θ_R the sources can be resolved, if it is smaller than θ_R the sources cannot be resolved.

>>> Example 3. Two sources of light are one foot apart. They are viewed with a telescope at a distance of 10 miles. The objective lens of the telescope is 15 millimeters in diameter. Can the two sources be resolved? Assume a wavelength of 550 nm.

From (40-6) the angle θ_R between two sources which are just resolvable is

$$\theta_R \simeq 1.22\ \lambda/d$$

$$= (1.22)(5.5 \times 10^{-7}\ \text{m})/(1.5 \times 10^{-2}\ \text{m}) = 4.5 \times 10^{-5}\ \text{(radian)} \quad .$$

The actual angle between the sources is

$$\theta = \left(\frac{1\ \text{foot}}{10\ \text{mile}}\right)\left(\frac{1\ \text{mile}}{5.28 \times 10^{3}\ \text{foot}}\right) = 1.9 \times 10^{-5}\ \text{(radian)} \quad .$$

Since θ is less than θ_R the two sources cannot be resolved. <<<

40-6 Programmed Problems

1. Suppose we wish to add several (say three) sinusoidally oscillating functions of time,

$$u = u_1 + u_2 + u_3 \quad ,$$

by means of a phasor diagram.

(a) In order to use this technique the individual functions u_i must all have the same ________ .

(b) Each function u_i therefore has the mathematical form

$$u_i = \_\_\_\_\_ \; .$$

(a) frequency. Usually we are concerned with the angular frequency ω (units: radians/second).

(b) $u_i = A_i \sin(\omega t + a_i)$

This is a sinusoidal oscillation with angular frequency ω. Each function u_i must have the same ω.

---

2. We have:

$$u_1 = A_1 \sin(\omega t + a_1)$$

$$u_2 = A_2 \sin(\omega t + a_2)$$

$$u_3 = A_3 \sin(\omega t + a_3) \quad .$$

Suppose that the phasor diagram for adding these three functions is as shown below.

(a) What are the amplitudes A_2 and A_3 in comparison with A_1? (You will have to measure certain lengths in the diagram in order to answer this question.)

(b) What are the phase differences between the various functions u_i?

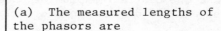

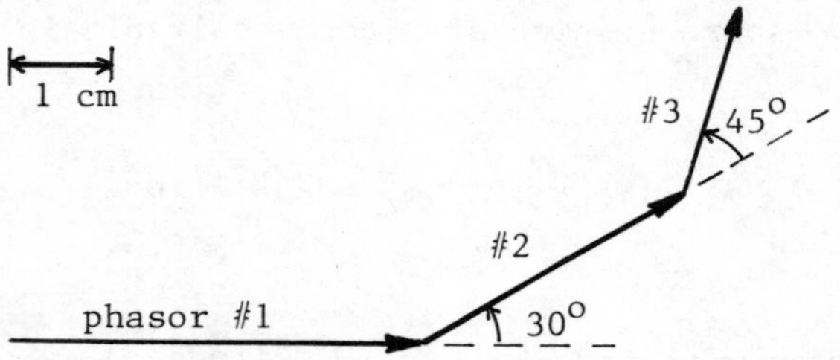

(a) The measured lengths of the phasors are

phasor #1: 4 cm
phasor #2: 3 cm
phasor #3: 2 cm .

Therefore

$A_2 = 0.75\ A_1$

$A_3 = 0.50\ A_1$.

(b) $a_2 - a_1 = 30^\circ$ (or $\pi/6$ radian)

$a_3 - a_2 = 45^\circ$ (or $\pi/4$ radian)

Also, $a_3 - a_1 = 75^\circ$.

---

3. The answers to the above frame were obtained by using the facts that

(a) The length of a phasor is proportional to the ________ of the corresponding function u_i.

(b) The angle between two phasors (say phasors #2 and #1) is equal to the ________ between the corresponding functions u_2 and u_1.

(a) amplitude A_i. (Note the word <u>proportional</u>; A_1 is not <u>equal</u> to 4 cm.)

(b) phase difference $a_2 - a_1$.

4. What is the mathematical form of the function $u = u_1 + u_2 + u_3$?

$u =$ ________ .

$u = R \sin(\omega t + b)$

It is of the same form as the individual functions u_i. The quantity R is the "resultant amplitude".

---

5. Using the phasor diagram shown in frame 2, estimate the resultant amplitude R in comparison with the amplitude A_1.

The "resultant phasor" is obtained by drawing an arrow from the tail of the first phasor (#1) to the head of the last phasor (#3). The measured values are

phasor #1: 4.0 cm
resultant phasor: 7.9 cm .

Therefore

$$R = (7.9 \text{ cm}/4.0 \text{ cm})\ A_1 = 1.98\ A_1 .$$

---

6. If all three functions u_i were in phase what would be the value of R in comparison with A_1?

In this case the phasor diagram would be a straight line:

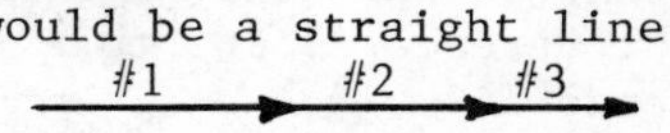

[half-scale]

The length of the resultant phasor would be 9.0 cm. Therefore

$$R = (9.0 \text{ cm}/4.0 \text{ cm})\ A_1 = 2.25\ A_1 .$$

---

7. We are now going to apply this technique of phasor addition to the problem of the diffraction of light by a single slit of width a. We first divide the slit into small equal strips of width Δx. Light coming from the various strips will interfere at point P on a distant screen. Why is there a phase difference between these various light beams?

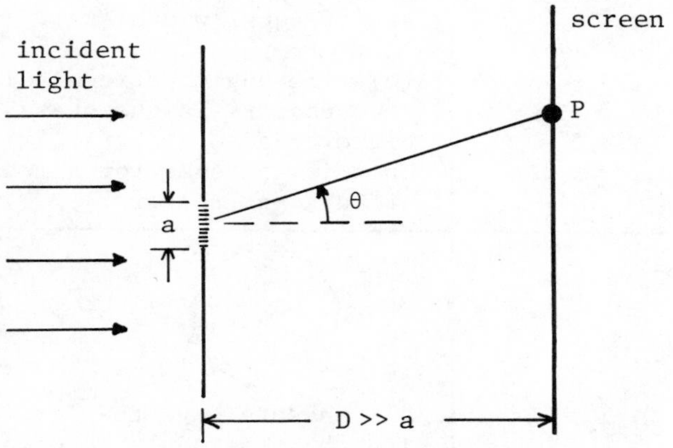

There is a phase difference because the light coming from different strips must travel different distances to get to point P.

8. Now consider two adjacent strips as shown below. Complete the diagram by showing the path length difference $\Delta\ell$ between the two light beams.

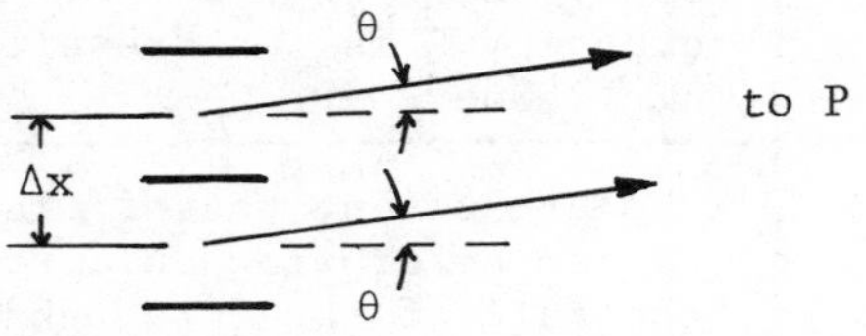

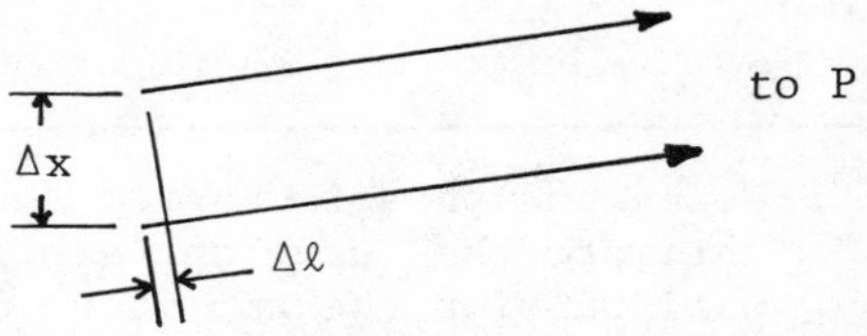

---

9. In frame 8 note that the distance between centers of the two adjacent strips is Δx. Express the path length difference $\Delta\ell$ in terms of Δx and the angle θ.

$\Delta\ell$ = ___________ .

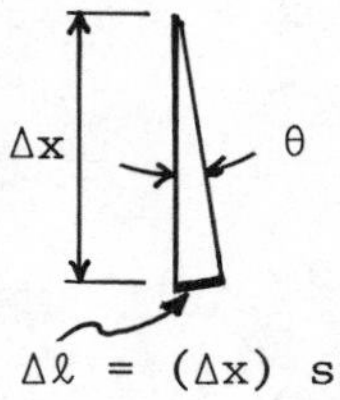

$\Delta\ell = (\Delta x) \sin\theta$

---

10. (a) What is the general formula for the phase difference $\Delta\phi$ due to a path length difference $\Delta\ell$?

$\Delta\phi$ = ___________ .

(b) Using the answer to frame 9 express $\Delta\ell$ in terms of Δx, λ, θ.

$\Delta\phi$ = ___________ .

(a) $\Delta\phi = (2\pi/\lambda)\ \Delta\ell$

(b) $\Delta\phi = (2\pi/\lambda)\ (\Delta x) \sin\theta$

---

11. Suppose that we have divided the original slit into 10 strips. We now want to use the phasor diagram technique to find the resultant amplitude.

(a) How many phasors will appear in the phasor diagram?
(b) What can you say about the relative lengths of these phasors?
(c) What can you say about the angle between adjacent phasors?

(a) 10. There will be one for each strip.

(b) They will all have the same length. The individual light beams (one from each strip) will all have the same amplitude because the strips are of equal width (Δx).

(c) The angle between adjacent phasors is the phase difference $\Delta\phi = (2\pi/\lambda)(\Delta x) \sin\theta$. This is the same for all pairs of adjacent phasors.

---

12. Sketch the phasor diagram assuming that $\Delta\phi = 30^{\circ}$.

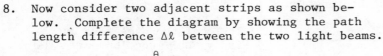

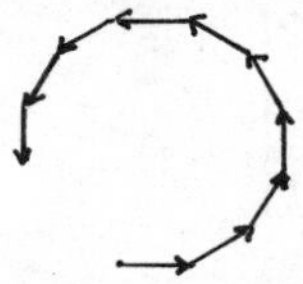

All phasors have the same length. The angle between adjacent phasors is 30°.

13. The above phasor diagram is not quite correct. The individual strips are of finite width; there will be interference between light coming from different parts of the same strip. To improve the situation we can take smaller strips. Suppose we take 20 strips instead of 10.
(a) The number of phasors in the phasor diagram will now be ________ .
(b) Each phasor will be ________ as long as the phasors in frame 12.
(c) The angle $\Delta\phi$ between adjacent phasors will be ________ as large as in frame 12.

(a) 20

(b) one half

(c) one half

---

14. As we continue this process of dividing the original slit into more and more strips the phasor diagram will approach a circular arc:

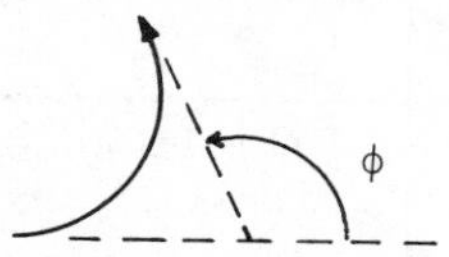

The angle ϕ between the first phasor and the last phasor is the phase difference corresponding to the path length difference

$\Delta\ell =$ __________ .

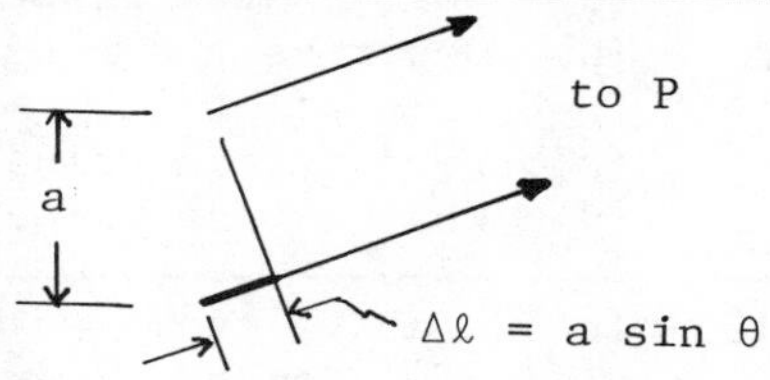

Note that ϕ is the angle between the first and last phasor while $\Delta\phi = (2\pi/\lambda)(\Delta x) \sin\theta$ is the angle between adjacent phasors.

---

15. Therefore the angle ϕ between the first phasor and the last phasor is

$\phi =$ __________ .

$\phi = (2\pi/\lambda)a \sin\theta$

---

16. The point P on the screen is specified by the angle θ. Suppose that P is directly on the axis of the system, i.e. $\theta = 0$.
(a) Sketch the corresponding phasor diagram.
(b) Describe the intensity at P.

(a)

The phasor diagram is a straight line.

(b) The intensity is very large. This is called the central maximum.

---

17. What happens to the phasor diagram as θ increases?

Since $\phi = (2\pi/\lambda)a \sin\theta$, the phasor diagram wraps around in a tighter circle. The total arc length of the circular arc remains constant.

18. The resultant phasor is drawn from the tail of the first phasor to the head of the last phasor. This is shown in the answer to frame 17. As θ increases $\phi = (2\pi/\lambda)a \sin \theta$ increases until eventually the phasor diagram becomes one whole circle:

For this condition,
(a) What is the resultant amplitude?
(b) Describe the intensity on the screen at this point P.
(c) What is the value of ϕ?
(d) What is the value of $\sin \theta$?

(a) zero

(b) dark. This is the first minimum.

(c) 2π (radians)

(d) $\sin \theta = (\phi/2\pi)(\lambda/a)$

$= (2\pi/2\pi)(\lambda/a)$

$= \lambda/a$

---

19. In the above frame we dealt with the first minimum. Now consider the second minimum.
(a) Describe the phasor diagram.
(b) What is the resultant amplitude?
(c) What is the value of ϕ?
(d) What is the value of $\sin \theta$?

(a) It consists of two whole circles.

(b) zero

(c) 4π

(d) $2\lambda/a$

---

20. Now generalize the above answers to the case of the mth minimum.
(a) Describe the phasor diagram.
(b) What is the resultant amplitude?
(c) What is the value of ϕ?
(d) What is the value of $\sin \theta$?
(e) What are the possible values of m?

(a) It consists of m whole circles.

(b) zero

(c) $2\pi m$

(d) $m\ \lambda/a$

(e) $m = 1, 2, 3, 4, \ldots$.

---

21. Why isn't $m = 0$ included in the above answer?

$m = 0$ corresponds to $\theta = 0$. This is not a minimum; it is the central maximum (see frame 16).

---

22. Make a sketch of the intensity as a function of $\sin \theta$. Indicate the values of $\sin \theta$ at which the minima occur.

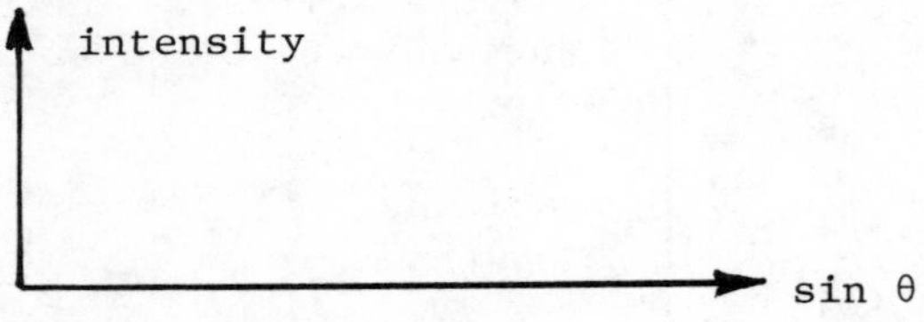

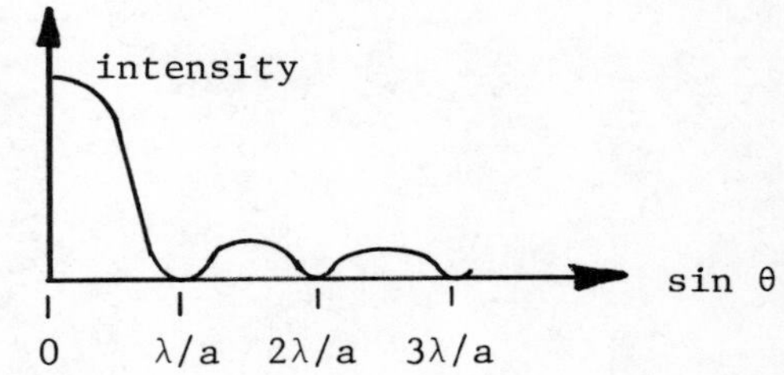

23. The first minimum occurs at

$$\sin\theta = \lambda/a$$

and the second minimum occurs at

$$\sin\theta = 2\lambda/a \quad .$$

Sketch the phasor diagram for

$$\sin\theta = (1.5)\lambda/a \quad .$$

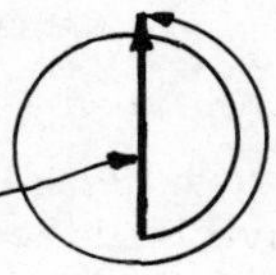

The phasor diagram is therefore 1½ circles.

24. Describe the intensity of the screen at point P for the above situation i.e. $\sin\theta = (1.5)\lambda/a$.

This is a secondary maximum. The resultant phasor is much smaller than that for the central maximum (see frame 16).

Chapter 41: GRATINGS

REVIEW AND PREVIEW

So far you have studied two slit interference and single slit diffraction. In this chapter you will study the interference pattern due to a diffraction grating (a diffraction grating consists of a large number of equally spaced parallel slits). The phasor technique will be used to explain many of the features of the interference pattern. Diffraction gratings permit us to measure the various wavelengths in a given source of light.

GOALS AND GUIDELINES

In this chapter there are four major goals:

1. Understanding the features of the interference pattern due to a diffraction grating (Section 41-1). You should be able to calculate the locations of the principle maxima as well as all the minima by means of the phasor diagram technique (Section 41-2).
2. Learning the definitions of
   a. dispersion: this is related to the angular width of the pattern (Section 41-3).
   b. resolving power: this concerns the ability to distinguish effects due to two adjacent wavelengths.
3. Understanding the effect of the individual slit width (Section 41-6). In particular you should be able to calculate the missing orders (Example 4).
4. Understanding the basic features of X-ray diffraction; in particular Bragg's law. You should be able to calculate the distance between lattice planes in a crystal from the X-ray diffraction data (Section 41-7).

41-1 Diffraction Grating

Figure 41-1 shows light incident normally upon a diffraction grating. This consists of a large number N of identical slits, the spacing between centers of adjacent slits is d. For now we will assume that the slits are so narrow that the effects due to individual slit width can be neglected. Point P lies on a screen a distance D >> Nd away from the grating; the location of P is specified by the angle θ shown.

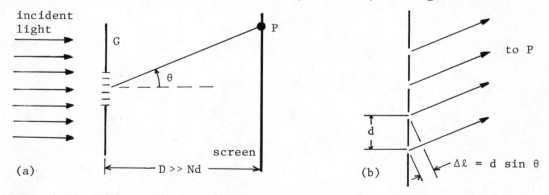

Figure 41-1. (a) Light is incident normally upon the diffraction grating G. The intensity pattern is observed on a screen a distance D away from the grating. (b) The grating consists of slits which are separated by a distance d. The path length difference for adjacent slits is $\Delta\ell = d \sin\theta$.

At point P the many beams of light (one from each slit) will interfere. An example of the intensity pattern obtained on the screen for the case N = 6 is shown in Figure 41-2. The chief features of this pattern are:

1. There are intense principle maxima located at $\sin\theta = m\lambda/d$ (m = 0, 1, 2, 3, ...). The principle maxima are narrow; for a given slit separation d, the width of these maxima decreases as the number of slits N is increased.
2. Between the principle maxima there are weak secondary maxima. The number of these increases as N is increased (there are N - 2 secondary maxima between adjacent principle maxima). Generally, the intensity of these secondary maxima (relative to that of the principle maxima) decreases as N is increased.

In practice N is usually so large (several thousand) that all one sees are the very narrow intense principle maxima; between these principle maxima the intensity is essentially zero.

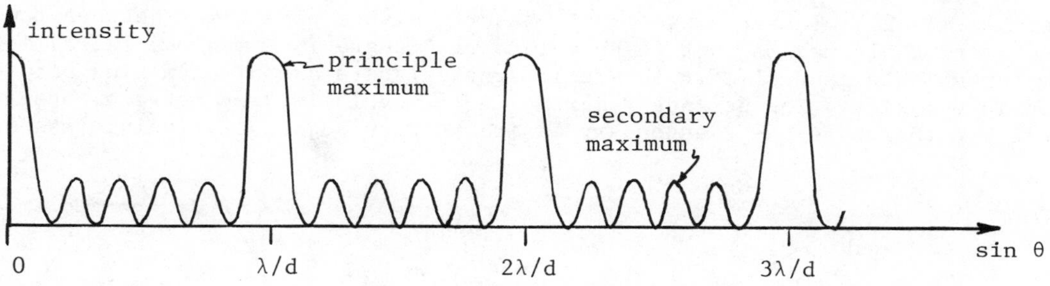

Figure 41-2. Intensity pattern for a six slit grating.

41-2 Phasor Analysis

The intensity patterns shown in Figure 41-2 can be explained by considering the associated phasor diagram. For a given point P on the screen (i.e. for a given value of θ) the path length difference $\Delta\ell$ for the light coming from adjacent slits is $\Delta\ell = d\sin\theta$. From equation (39-2) the phase difference $\Delta\phi$ due to this path length difference is

$$\Delta\phi = (2\pi/\lambda)\, d\sin\theta \quad \text{(phase difference for adjacent slits).} \qquad (41\text{-}1)$$

Note that this is the same for every pair of adjacent slits. The associated phasor diagram is therefore an N sided (open) polygon, the angle between successive phasors being the phase difference (41-1). Since all the slits are of the same width, the individual phasors are all of equal length. Figure 41-3a shows an example of such a phasor diagram for the case N = 6, $\Delta\phi = 75^o$. Note that in this example the magnitude of the resultant phasor is much smaller than the sum of the magnitudes of the individual phasors.

Generally, as in Figure 41-3a, the resultant phasor is small. An exception to this arises if the angle $\Delta\phi$ between successive phasors is a whole number of circles, i.e. if $\Delta\phi = 2\pi m$ (m = 0, 1, 2, 3, ...). In this case the phasor diagram becomes a straight line, the magnitude of the resultant phasor being the sum of the magnitudes of the individual phasors. Figure 41-3b shows an example of this situation for the case N = 6, $\Delta\phi = 4\pi$. The large resultant phaosr corresponds to a principle maximum.

The location of the principle maxima can therefore be obtained by substituting $\Delta\phi = 2\pi m$ into equation (41-1). The result is

$$\sin\theta = m\lambda/d \quad \text{(principle maxima).} \tag{41-2}$$

The integer m (m = 0, 1, 2, 3, ...) is called the <u>order</u> of the principle maximum.* The location of the mth principle maximum (as given by (41-2)) depends only upon the ratio of the wavelength λ to the slit separation d; it is independent of the number of slits N. Note that the light coming from <u>all</u> N slits interferes constructively at a principle maximum.

Now consider, for example, what happens to the phasor diagram between the first and second principle maxima for the case N = 6. At the first principle maximum (m = 1) we have $\Delta\phi = 360^\circ$. At some larger value of θ, $\Delta\phi = 360^\circ + 60^\circ = 420^\circ$. Here the phasor diagram becomes closed (it is a hexagon); the resultant phasor is then zero as shown in Figure 41-3c. Similarly the resultant phasor is zero for $\Delta\phi = 360^\circ + 120^\circ$, $360^\circ + 180^\circ$, $360^\circ + 240^\circ$, $360^\circ + 300^\circ$. For $\Delta\phi = 360^\circ + 360^\circ = 720^\circ$ we obtain the second principle maximum (m = 2). Thus the resultant phasor vanishes five times between the first and second principle maxima. At these values of θ there is a minimum (zero) intensity on the screen. In general, for an N slit grating there are N - 1 minima between adjacent principle maxima. The secondary maxima lie (roughly) midway between these minima. Therefore there are N - 2 secondary maxima between adjacent principle maxima.

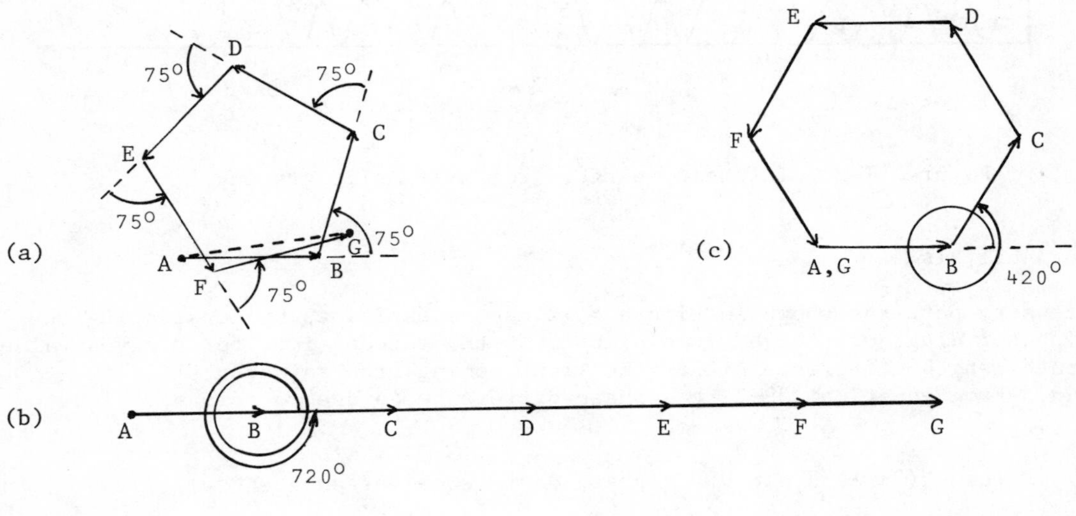

Figure 41-3. (a) Phasor diagram for N = 6, $\Delta\phi = 75^\circ$. The second phasor BC makes an angle of 75° with the first phasor AB. Similarly the angle between any two successive phasors is 75°. The resultant phasor AG is indicated by the dashed line. (b) Phasor diagram for N = 6, $\Delta\phi = 4\pi = 720^\circ$. The magnitude of the resultant phasor AG in this case is the sum of the magnitudes of the individual phasors. (c) Phasor diagram for N = 6, $\Delta\phi = 420^\circ$. The resultant phasor AG is zero in this case.

>>> Example 1. A diffraction grating has 2000 rulings (slits) per centimeter. It is used with light whose wavelength is 560 nm. (a) Calculate the angle θ at which the second order maximum occurs. (b) How many orders exist for this wavelength?

*The mth order principle maximum is usually simply called the mth order maximum.

The given quantities are

$$\lambda = 560 \text{ nm} = 5.6 \times 10^{-5} \text{ cm} \quad ,$$

$$d = (1/2000) \text{ cm} = 5 \times 10^{-4} \text{ cm} \quad .$$

(a) Using equation (41-2),

$$\sin\theta = m\lambda/d = (2)(5.6 \times 10^{-5} \text{ cm})/(5 \times 10^{-4} \text{ cm}) = 0.224$$

$$\theta = 12.9^{\circ} \quad .$$

(b) Solving equation (41-2) for m,

$$m = \frac{d}{\lambda}\sin\theta = \left(\frac{5 \times 10^{-4} \text{ cm}}{5.6 \times 10^{-5} \text{ cm}}\right)\sin\theta = 8.93 \sin\theta \quad .$$

Since the maximum value of sin θ is one, the largest possible <u>integer</u> value for m is m = 8. Thus the highest order for this wavelength is the <u>eighth</u>. Of course there are really seventeen orders on the screen: eight on either side of the zeroth order maximum plus the zeroth order maximum itself. <<<

41-3 Dispersion

Suppose that for some wavelength λ the mth order maximum is located at an angle θ (this angle could be computed using equation (41-2)). For a slightly longer wavelength, λ + dλ, the mth order maximum is located at a slightly larger angle θ + dθ. The ratio of the angular separation dθ of these maxima to the wavelength difference dλ is called the <u>dispersion</u> D.

$$D = d\theta/d\lambda \quad . \tag{41-3a}$$

By differentiating equation (41-2) we obtain an expression for D,

$$D = \frac{m}{(d)(\cos\theta)} \quad . \tag{41-3b}$$

In using (41-3b) the student must remember that the angle θ describes the location of the mth order maximum. The units of D are those of angle per length; for example if d is in nanometers then (41-3b) gives D in radian/nanometer.

>>> Example 2. A diffraction grating has 5000 rulings per centimeter. (a) Compute the dispersion for a wavelength of 500 nm in the second order. Express your answer in minutes/nm. (b) Estimate the angular separation between the second order maximum for the wavelengths 500 nm and 502 nm.

(a) The given quantities are

$$\lambda = 5 \times 10^{-5} \text{ cm} \quad ,$$

$$d = 2 \times 10^{-4} \text{ cm} \quad ,$$

$$m = 2 \quad .$$

The location of the desired principle maximum is

$$\sin\theta = m\left(\frac{\lambda}{d}\right) = (2)\left(\frac{5\times10^{-5}\text{ cm}}{2\times10^{-4}\text{ cm}}\right) = 0.500$$

Using equation (41-3b),

$$D = \frac{m}{d\cos\theta} = \frac{2}{(2\times10^{-4}\text{ cm})(\cos 30^\circ)} = 1.15\times10^4\text{ rad/cm}$$

$$D = \left(1.15\times10^4\,\frac{\text{rad}}{\text{cm}}\right)\left(\frac{1\text{ cm}}{10^7\text{ nm}}\right)\left(\frac{180^\circ}{\pi\text{ rad}}\right)\left(\frac{60'}{1^\circ}\right) \qquad [' \equiv \text{minute of arc}]$$

$$D = 4.0\ '/\text{nm} \quad .$$

This value for the dispersion applies to the second order maximum for wavelengths near 500 nm.

(b) Since the wavelength difference $\Delta\lambda = 2$ nm is small,

$$\Delta\theta \simeq \left(\frac{d\theta}{d\lambda}\right)(\Delta\lambda) = D(\Delta\lambda) = (4.0\ '/\text{nm})(2\text{ nm})$$

$$\Delta\theta = 8.0\text{ minutes} \quad .$$

<<<

41-4 Width of Principle Maxima

We can obtain an estimate of the (angular) width of a principle maximum by considering the associated phasor diagram. At the mth order principle maximum the phasor diagram is a straight line, the angle $\Delta\phi$ between successive phasors is $\Delta\phi = 2\pi m$. At the adjacent minimum the phasor diagram is an N sided closed polygon, the angle $\Delta\phi$ between successive phasors is $\Delta\phi = 2\pi m + 2\pi/N$ (e.g. for $N = 6$ the desired minimum occurs at $\Delta\phi = 2\pi m + 2\pi/6 = (360^\circ)\,m + 60^\circ$). The angle $\theta = \theta_m$ for the mth order maximum as given by (41-1) is

$$\sin\theta_m = \left(\frac{\Delta\phi}{2\pi}\right)\left(\frac{\lambda}{d}\right) = \left(\frac{2\pi m}{2\pi}\right)\left(\frac{\lambda}{d}\right) = m\left(\frac{\lambda}{d}\right) \quad .$$

At the adjacent minimum the angle θ has increased to $\theta = \theta_m + \Delta\theta_m$ where $\Delta\theta_m$ is the angular separation between the mth order maximum and the adjacent minimum. This angle θ as given by (41-1) is

$$\sin(\theta_m + \Delta\theta_m) = \left(\frac{\Delta\phi}{2\pi}\right)\left(\frac{\lambda}{d}\right) = \left[\frac{2\pi m + 2\pi/N}{2\pi}\right]\left(\frac{\lambda}{d}\right) = (m + 1/N)\left(\frac{\lambda}{d}\right) \quad .$$

Subtracting the previous equation from this one,

$$\sin(\theta_m + \Delta\theta_m) - \sin\theta_m = \frac{\lambda}{Nd} \quad .$$

The first term can be simplified by means of the trigonometric identity: $\sin(x + y) = (\sin x)(\cos y) + (\cos x)(\sin y)$.

$$\sin(\theta_m + \Delta\theta_m) = (\sin\theta_m)(\cos\Delta\theta_m) + (\cos\theta_m)(\sin\Delta\theta_m)$$
$$= \sin\theta_m + (\cos\theta_m)(\Delta\theta_m) \quad .$$

Here the small angle approximations, $\cos\Delta\theta_m \simeq 1$ and $\sin\Delta\theta_m \simeq \Delta\theta_m$, have been made.* Substituting into the above equation and solving for $\Delta\theta_m$ we obtain

$$\Delta\theta_m = \frac{\lambda}{Nd\cos\theta_m} \qquad \text{(width of mth order maximum).} \qquad (41\text{-}4)$$

As defined above, $\Delta\theta_m$ is the angular separation between the mth order principle maximum and the adjacent minimum; it therefore serves as an estimate of the angular width of this maximum. In using equation (41-4), if λ and d are expressed in the same units then $\Delta\theta_m$ will be in radians.

Note the presence of the product Nd in equation (41-4); this is the width of the entire diffraction grating. If we ignore the factor $\cos\theta_m$, equation (41-4) is similar to the equation (40-4) for the first minimum due to a single slit. The width $\Delta\theta_m$ of a principle maximum may therefore be interpreted as a single slit diffraction effect, the entire diffraction grating acting as a single slit of width Nd.

41-5 Resolving Power of a Grating

Suppose that for some wavelength λ the mth order maximum occurs at an angle $\theta = \theta_m$. For a slightly different wavelength $\lambda + \Delta\lambda$ the mth order maximum will occur at a slightly different value of θ. If the angular separation between these two principle maxima is too small, then they will not be resolvable due to their finite width $\Delta\theta_m$. The Rayleigh criterion (see Chapter 40) as applied to this situation says that these two maxima will be just resolvable if the maximum for the wavelength λ coincides with the adjacent minimum for the wavelength $\lambda + \Delta\lambda$. That is, the angular separation between the two mth order maxima should be $\Delta\theta_m$ (as given by 41-4) if they are to be just resolvable. The smallest wavelength difference $\Delta\lambda$ that is just resolvable is then

$$\Delta\lambda \simeq \left(\frac{d\lambda}{d\theta}\right)(\Delta\theta_m) = \left(\frac{1}{D}\right)(\Delta\theta_m) \quad .$$

Using equations (41-3b) and (41-4),

$$\Delta\lambda = \left(\frac{d\cos\theta}{m}\right)\left(\frac{\lambda}{Nd\cos\theta}\right) = \frac{\lambda}{Nm} \quad .$$

The ratio of the wavelength λ to the wavelength difference $\Delta\lambda$ is called the resolving power R,

$$R = \frac{\lambda}{\Delta\lambda} \quad . \qquad (41\text{-}5a)$$

Using the expression for $\Delta\lambda$ above, the resolving power becomes

$$R = Nm \quad . \qquad (41\text{-}5b)$$

The resolving power is a pure number. For example, a resolving power of 10,000 with m = 3 means that one could resolve two wavelengths which differed by one part in 10,000 by using the diffraction grating in the third order.

*In effect what we have done is to derive the relation $d(\sin\theta)/d\theta = \cos\theta$.

>>> Example 3. It is desired to design a diffraction grating for which the third order maximum for a wavelength 600 nm will occur at $\theta = 30^o$. The resolving power under these conditions must be such that a wavelength difference 0.05 nm can be resolved. Calculate: (a) the slit spacing, (b) the resolving power for the above conditions, (c) the minimum number of slits required, (d) the minimum width of the entire diffraction grating.

(a)
$$\sin\theta = m\lambda/d$$

$$d = m\lambda/\sin\theta = (3)(6 \times 10^{-5}\ \text{cm})/\sin 30^o = 3.6 \times 10^{-4}\ \text{cm} \quad .$$

(b)
$$R = \lambda/(\Delta\lambda) = 600\ \text{nm}/(0.05\ \text{nm}) = 12\ 000 \quad .$$

(c)
$$R = Nm$$

$$N = R/m = 12{,}000/3 = 4000 \quad .$$

(d)
$$\text{width} = Nd = (4000)(3.6 \times 10^{-4}\ \text{cm}) = 1.4\ \text{cm} \quad .$$
<<<

41-6 Effect of the Slit Width

So far we have ignored the effect of the individual slit width. If the width, a, of each individual slit is not negligible then the intensity pattern observed on the screen is the mathematical product of two terms:
(a) The intensity pattern obtained if we ignore the individual slit width (as in Figure 41-1).
(b) The intensity pattern associated with a single slit of width a (as in Figure 40-5).

The factor (a) accounts for the interference between light coming from different slits. The factor (b) accounts for the interference between light coming from different parts of any one slit, i.e. single slit diffraction.

It can turn out that the intensity of some principle maximum occurring in the term (a) gets multiplied by the zero intensity of a minimum occurring in the single slit diffraction term (b). In this case we say that we have a <u>missing order</u>.

>>> Example 4. A diffraction grating has 2000 rulings per centimeter. The individual slit width is 1000 nm. Sketch the intensity pattern obtained for a wavelength of 450 nm. What orders (if any) are missing?

The given quantities are

$$d = 5 \times 10^{-4}\ \text{cm} \qquad \text{(slit separation)} \quad ,$$

$$a = 10^{-4}\ \text{cm} \qquad \text{(slit width)} \quad ,$$

$$\lambda = 4.5 \times 10^{-5}\ \text{cm} \qquad \text{(wavelength)} \quad .$$

(a) We first neglect the effect of the individual slit width. The principle <u>maxima</u> are located at

$$\sin\theta = m\lambda/d = m\ (4.5 \times 10^{-5}\ \text{cm})/(5 \times 10^{-4}\ \text{cm}) = (0.09)m \quad , \quad m = 0, 1, 2, 3, \ \ldots$$

Assuming that N is large enough so that we need only consider the principle maxima, the resulting intensity pattern is sketched in Figure 41-4a. Note that the sketch must terminate at $\sin\theta = 1$.

(b) We now consider only the single slit diffraction pattern. From equation (40-4) the minima are located at

$$\sin\theta = m'\left(\frac{\lambda}{a}\right) = m'\left(\frac{4.5 \times 10^{-5}\ \text{cm}}{10^{-4}\ \text{cm}}\right) = (0.45)\,m' \quad , \quad m' = 1, 2, 3, \ldots \quad .$$

(Note that we use a different letter, m', to avoid confusion with the order m of the principle maxima.) This intensity pattern is sketched in Figure 41-4b.

(c) The desired intensity pattern is the product of the above two curves. The result is shown in Figure 41-4c.

(d) The missing orders occur when a principle maximum for the N slit grating coincides with a minimum for the single slit diffraction pattern. That is

$$\sin\theta = m\lambda/d = m'\lambda/a$$

$$m = m'd/a = m'(5 \times 10^{-4}\ \text{cm})/(10^{-4}\ \text{cm}) = 5\,m' \quad .$$

In the above equation, m and m' are restricted to

$$m = 0, 1, 2, 3, \ldots, 11 \quad \text{(since there are only 11 orders)}$$

$$m' = 1, 2, 3, \ldots \quad .$$

The only solutions are

$$m = 5 \quad \text{(corresponding to } m' = 1)$$

$$m = 10 \quad \text{(corresponding to } m' = 2) \quad .$$

Therefore the fifth and tenth orders are missing. This agrees with the sketch in Figure 41-4c.

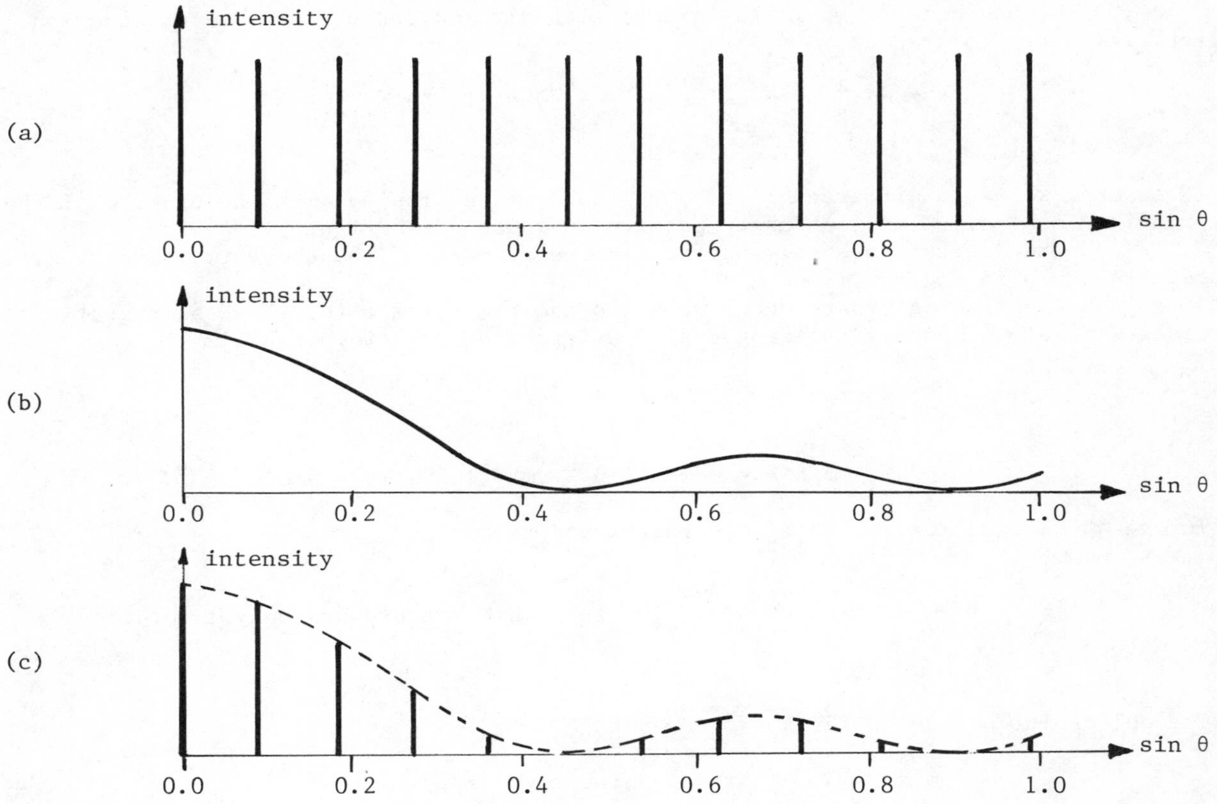

Figure 41-4. (a) Intensity pattern for diffraction grating neglecting slit width effects. The principle maxima occur at $\sin\theta = m\lambda/d = 0.09\ m$; $m = 0, 1, 2, \ldots$. For simplicity these narrow maxima are represented by a vertical line. (b) Intensity pattern for single slit. The minima occur at $\sin\theta = m'\lambda/a = 0.45\ m'$; $m' = 1, 2, \ldots$. (c) Intensity pattern for diffraction grating including slit width effects. The dashed curve is the envelope of the intensity pattern. Note that $m = 5$ and $m = 10$ orders are missing. <<<

41-7 X-ray Diffraction

The regularly spaced atoms in a crystalline solid can be used to produce interference effects in somewhat the same manner as the regularly spaced slits in a diffraction grating. The distance between adjacent atoms in a solid (a few Angstroms) is much smaller than the wavelength of light (4000-8000 Å). Therefore to obtain practical interference effects, X-rays (an electromagnetic radiation) whose wavelength is comparable with this interatomic spacing are used. Such interference phenomena associated with the use of X-rays are known as X-ray diffraction. [The unit Angstrom Å = 0.1 nm is used frequently in X-ray diffraction.]

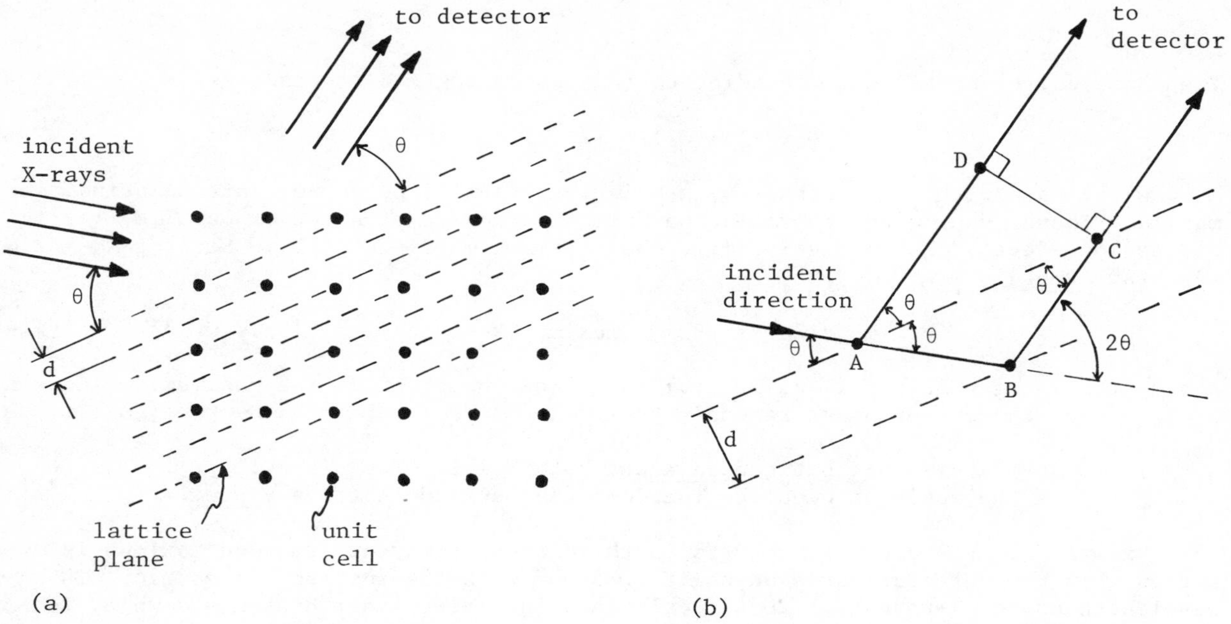

Figure 41-5. (a) Apparatus for X-ray diffraction. Each dot represents a unit cell. The law of reflection is obeyed by the lattice planes. (b) To calculate the path length difference $\Delta\ell$ = ABC − AD:

$$ABC = 2(AB)$$

$$AD = (AC)(\cos\theta) = (2\ AB\cos\theta)(\cos\theta) = 2\ AB\cos^2\theta$$

$$\Delta\ell = ABC - AD = 2\ AB(1 - \cos^2\theta) = 2\ AB\sin^2\theta$$

$$\Delta\ell = 2(d/\sin\theta)\sin^2\theta = 2d\sin\theta \quad .$$

Note that the angle between the detector and the incident direction is 2θ.

In Figure 41-5a, X-rays are incident upon a crystal. The dashed lines represent an edge view of a set of parallel lattice planes,* the separation between adjacent planes is d. It turns out that a lattice plane can be considered to act as a (partially reflecting) mirror in the sense that the law of reflection is obeyed: the angle of incidence equals the angle of reflection (even though the conditions for geometrical optics are certainly not satisfied). Figure 41-5b shows two of these adjacent parallel lattice planes. From this figure we see that the path length difference $\Delta\ell$ associated with these two planes is

$$\Delta\ell = 2d\sin\theta \quad .$$

Note that the angle θ (called the glancing angle) is measured from the plane itself rather than the normal to the plane. (The experimentally measurable angle is 2θ, the angle between the reflected and incident directions as shown in the figure.)

*A crystal is a periodic structure of unit cells, the unit cell in general consists of several atoms (see text for an example of a unit cell of NaCl). Imagine that each unit cell is drawn simply as a point. Then a lattice plane is a plane containing (at least) three non-colinear unit cells.

The phase difference $\Delta\phi$ corresponding to this path length difference is

$$\Delta\phi = (2\pi/\lambda)(\Delta\ell) = (2\pi/\lambda)(2d \sin \theta) \quad .$$

As was the case with the diffraction grating, there will be narrow intense principle maxima. These occur when the radiation from all members of a set of parallel lattice planes interferes constructively, i.e. when $\Delta\phi$ is a multiple of 2π. Setting $\Delta\phi = 2\pi m$, we obtain Bragg's law:

$$m\lambda = 2d \sin \theta \qquad \text{(maxima, X-ray diffraction).} \qquad (41\text{-}6)$$

Again, the integer m (m = 1, 2, 3, ...) is called the order of the maximum.* In using Bragg's law, the student must remember that:

1. d is the distance between adjacent parallel lattice planes.
2. θ is the angle between the incident beam and the lattice plane.

>>> Example 5. For a certain crystal a third order X-ray diffraction maximum is observed when the detector makes an angle of 50.0° with the incident direction. The wavelength of the X-rays is 1.20 Å. Calculate the interplanar distance d which is responsible for this maximum.

The given quantities are

$$2\theta = 50.0^\circ \quad \text{(therefore } \theta = 25.0^\circ \text{, see Figure 41-5b)}, \quad m = 3 \quad , \quad \lambda = 1.20\ \text{Å} \quad .$$

Using Bragg's law,

$$m\lambda = 2d \sin \theta$$

$$d = \frac{m\lambda}{2 \sin \theta} = \frac{(3)(1.20\ \text{Å})}{2 \sin (25.0^\circ)} = 4.26\ \text{Å}$$

<<<

>>> Example 6. Using a fixed X-ray wavelength, maxima are observed for a certain crystal when the angle between the detector and the incident X-ray beam is 16.2°, 31.0°, 32.8°, 50.0°, 64.6°. What reasonable interpretation can you attach to these values?

The observed angles represent values of 2θ. We first calculate θ and $\sin \theta$. These numbers are listed in the following table.

2θ	θ	$\sin \theta$
16.2°	8.1°	0.141
31.0°	15.5°	0.267
32.8°	16.4°	0.282
50.0°	25.0°	0.423
64.6°	32.3°	0.534

*The case m = 0 is usually omitted since if m = 0 then $\theta = 0$. The incident and reflected beams then coincide and cannot be distinguished experimentally.

Now we look for simple ratios in the last column. In this problem we have

$$(0.141):(0.282):(0.423) = 1:2:3$$

$$(0.267):(0.534) = 1:2 \quad .$$

From Bragg's law, $m\lambda = 2d \sin\theta$, we see that a reasonable interpretation of this data is:

1. The observed angles $2\theta = 16.2^o$, 32.8^o, 50.0^o are the first, second, and third order maxima for some interplanar distance d.
2. The observed angles $2\theta = 31.0^o$, 64.6^o are the first and second order maxima for some other interplanar distance d. <<<

41-8 Programmed Problems

1.

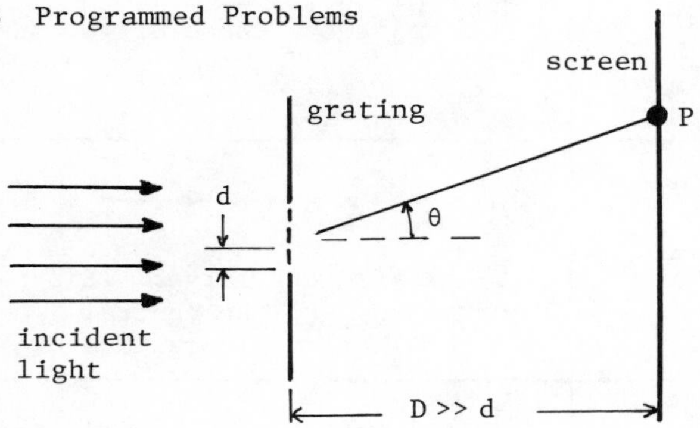

The diagram above shows light incident normally upon a "diffraction grating". Let d = separation between adjacent slits and N = number of slits. We would like to apply the technique of phasor addition to this problem. (For now, let's ignore the effects of the individual slit width.) How many phasors will appear in the phasor diagram?

N. One for each slit.

---

2. What is the path length difference $\Delta\ell$ for light coming from adjacent slits and arriving at point P?

$\Delta\ell$ = ____________ .

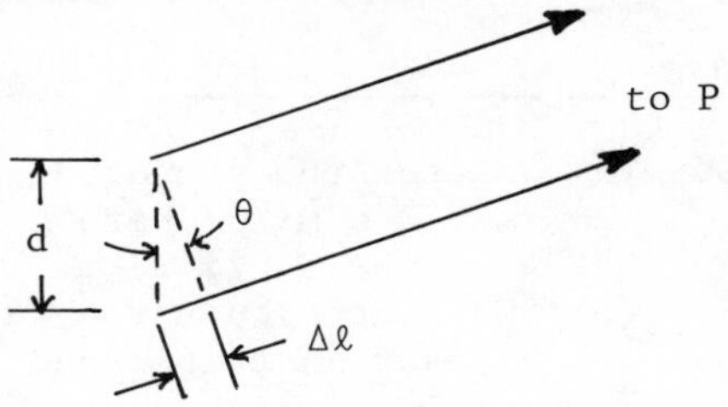

$\Delta\ell = d \sin\theta$

---

3. What is the general formula for the phase difference $\Delta\phi$ due to a path length difference $\Delta\ell$?

$\Delta\phi$ = ____________ .

$\Delta\phi = (\frac{2\pi}{\lambda})\ \Delta\ell$

4. Using the answer to frame 2, express the phase difference $\Delta\phi$ in terms of d, λ, θ.

$\Delta\phi = (\frac{2\pi}{\lambda})d \sin \theta$

---

5. $\Delta\phi$ is a phase difference. Which two light beams have this phase difference?

$\Delta\phi$ is the phase difference between light coming from any pair of adjacent slits and arriving at point P. (Note that d is the same for all pairs of adjacent slits.)

---

6. Describe the phasor diagram for this problem.

It is an N sided (open) polygon. All phasors have the same length. The angle between all pairs of adjacent phasors is the phase difference:

$$\Delta\phi = \frac{2\pi}{\lambda} d \sin \theta .$$

---

7. (a) Sketch the phasor diagram for the case in which P lies on the axis ($\theta = 0$). Take N = 4.

(b) What can you say about the resultant phasor in this case?

(a)

(b) It is very large. This corresponds to the central (zeroth) principle maximum.

---

8. (a) Sketch the phasor diagram for the case $\sin \theta = \lambda/d$. Take N = 4.
(b) What can you say about the resultant phasor in this case?

(a)

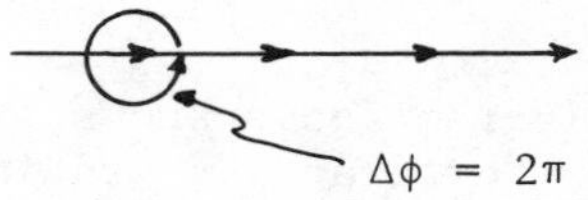

$$\Delta\phi = \frac{2\pi}{\lambda} d \sin \theta$$
$$= \frac{2\pi}{\lambda} d(\frac{\lambda}{d}) = 2\pi .$$

(b) It is very large again. This corresponds to the first principle maximum.

---

9. (a) Sketch the phasor diagram for the case $\sin \theta = m\lambda/d$ where m is an integer ($m = 0, 1, 2, 3, \ldots$). Take N = 4.
(b) What can you say about the resultant phasor in this case?

(a)

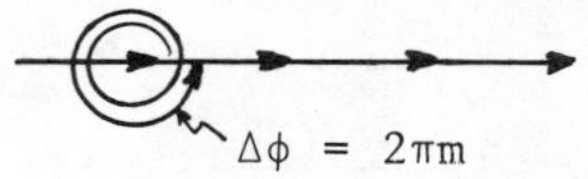

$$\Delta\phi = \frac{2\pi}{\lambda} d \sin \theta$$
$$= \frac{2\pi}{\lambda} d(m \frac{\lambda}{d}) = 2\pi m .$$

(b) It is very large. This corresponds to the mth principle maximum (or mth order maximum).

10. Are there any values of sin θ between the zeroth and first principle maxima for which the resultant phasor is zero? If so, what are these values of sin θ? (Again take N = 4.

Yes.

$\sin \theta = \frac{1}{4}\frac{\lambda}{d}, \frac{2}{4}\frac{\lambda}{d}, \frac{3}{4}\frac{\lambda}{d}$.

In these cases the phasor diagram becomes a closed polygon. These correspond to the minima.

---

11. In terms of N, how many such minima occur between adjacent maxima?

N - 1

For example, for N = 4 there were 3 such minima.

---

12. For N = 4, sketch the intensity on the screen as a function of sin θ. Show the location of the principle maxima.

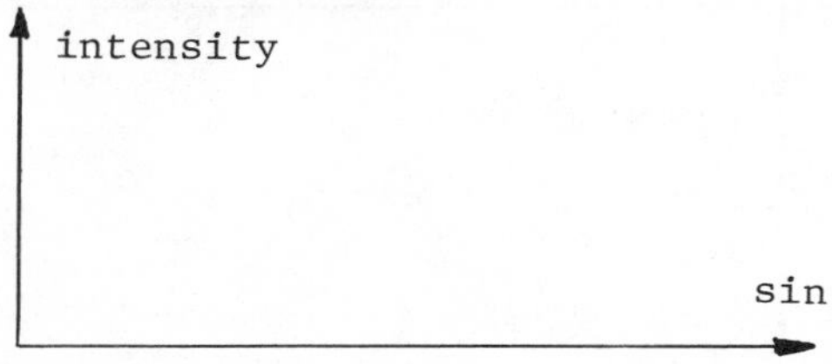

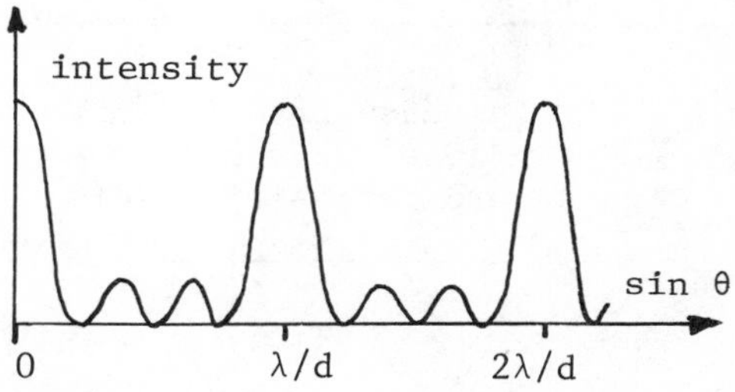

---

13. In the above frame the small maxima between the principle maxima are called __________ .

secondary maxima

---

14. Suppose the number of slits N is increased keeping the slit separation d fixed. Describe what happens to the answer to frame 12.

The location of the principle maxima remains the same ($\sin \theta = m\lambda/d$). The principle maxima become sharper (more narrow). Between the principle maxima the intensity is very small.

---

15. Now let's consider the effect of the individual slit width. Suppose each slit is of width a. Sketch the intensity due to one slit of width a as a function of sin θ. Indicate the location of the minima.

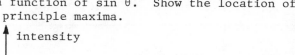

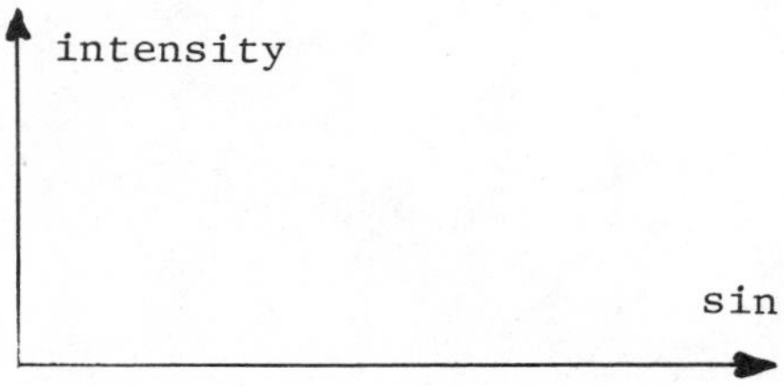

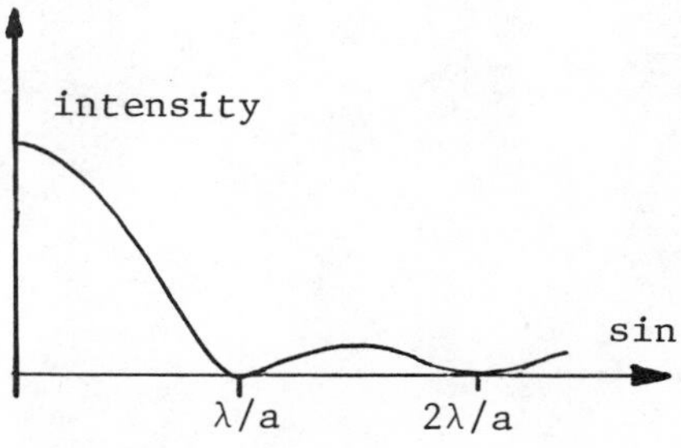

16. Describe how to use the above answer to obtain the intensity pattern due to the N slit grating.

The graph of frame 15 will be an envelope of the graph of frame 12. A typical result might be as shown below. Here the narrow principle maxima are shown simply as vertical lines.

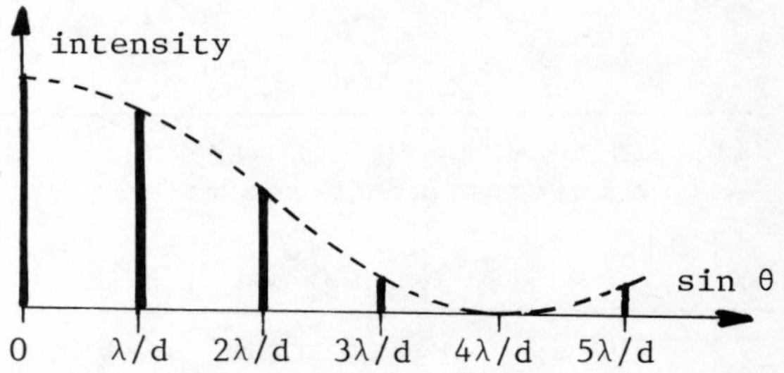

17. In the answer to frame 16, the fourth principle maximum has been reduced to zero intensity. This is called a ____________________ .

missing order

Chapter 42: POLARIZATION

REVIEW AND PREVIEW

In the previous three chapters you studied some of the consequences of the wave nature of light; these were all interference phenomena. Another aspect of the wave nature of light is that light is a transverse wave. A result of this is that light waves can be polarized. For example, light waves may be transmitted or absorbed by a Polaroid according to the state of polarization of the light.

GOALS AND GUIDELINES

The chief goal of this chapter is to gain an understanding of the polarization properties of light. You should learn about

a. plane polarized light (Section 42-2);
b. unpolarized light (Section 42-3);
c. Malus' law for the fraction of light transmitted by a Polaroid (Section 42-4);
d. double refraction (Section 42-6);
e. circular polarization (Section 42-7).

42-1 Polarization

The waves which we have studied so far in physics can be classified as being either longitudinal or transverse. For example:

(a) A sound wave in air is a longitudinal wave. Here the displacement of the air molecules is parallel to the direction of propagation of the wave.
(b) A wave on a string is a transverse wave. Here the displacement of each segment of string is perpendicular to the direction of propagation of the wave.

Consider waves which travel in the z direction. In the case of the longitudinal sound wave (a) there is only one possible direction for the particle displacement, namely parallel to the z axis. However for the case of the transverse wave on a string (b) there are two possible directions for the displacement of a particle, namely parallel to the x or y axes.* We then say that the transverse wave (b) can be polarized.

The electromagnetic waves which we have studied (Chapter 35) are transverse waves. In this case it is the $\underline{E}$ and $\underline{B}$ fields which are perpendicular to the direction of propagation of the wave. In particular, light waves are transverse waves; hence light waves can be polarized. This chapter deals with the polarization of light waves.

42-2 Plane Polarized Light

Suppose a light wave is traveling in the z direction. If the $\underline{E}$ field vectors are always in the x direction we say that the light is plane polarized, the plane of polarization being the x-z plane. (By convention, the plane of polarization is taken to be the plane containing the $\underline{E}$ field and the direction of propagation.)

---

*Any other direction of displacement in the x-y plane can be regarded as a superposition of a displacement in the x direction and a displacement in the y direction.

42-3 Unpolarized Light

Common sources of light (such as an incandescent solid) emit unpolarized light. This can be thought of as being similar to plane polarized light except that the plane of polarization is randomly changing with time. For example if an unpolarized light wave is traveling in the z direction then the $\underline{\underline{E}}$ field is in the x-y plane and is randomly changing its direction in this plane.

42-4 Polarizing Sheets

Certain sheets of material, called polarizing sheets (e.g. the familiar Polaroid sheets), have the property that they will transmit light which is plane polarized in a particular direction and absorb light which is plane polarized perpendicular to this direction. This particular direction in the sheet is called the polarizing direction; light which is plane polarized parallel to the polarizing direction is transmitted while light which is plane polarized perpendicular to the polarizing direction is absorbed. When unpolarized light is incident upon a polarizing sheet, one half of the incident intensity is transmitted.

Regardless of the nature of the incident light (plane polarized, unpolarized, etc.), the light which is transmitted through a polarizing sheet is always plane polarized parallel to the polarizing direction of the sheet. A polarizing sheet can therefore be used to produce plane polarized light. When used in this manner it is called a polarizer. By rotating a polarizing sheet it can also be used to determine the plane of polarization of a given beam of plane polarized light. When used in this manner it is called an analyzer.

Suppose that plane polarized light is incident upon a polarizing sheet; let the angle between the plane of polarization and the polarizing direction be θ. The $\underline{\underline{E}}$ field of the incident light can be resolved into two components: the component $|\underline{\underline{E}}| \cos \theta$ parallel to the polarizing direction is transmitted, the component $|\underline{\underline{E}}| \sin \theta$ perpendicular to the polarizing direction is absorbed. Since intensity is proportional to the square of the amplitude of the $\underline{\underline{E}}$ field, the ratio of the transmitted intensity I to the incident intensity I_o is $\cos^2 \theta$. This is known as Malus' law:

$$I = I_o \cos^2 \theta \quad . \tag{42-1}$$

>>> Example 1. Unpolarized light is incident upon a stack of three polarizing sheets. The polarizing direction of the second sheet makes an angle of 30° with that of the first sheet; the polarizing direction of the third sheet makes an angle of 45° with that of the second sheet. Calculate the fraction of the incident intensity which is transmitted.

One half the intensity of the incident unpolarized light is transmitted through the first sheet. This transmitted light is plane polarized parallel to the polarizing direction of the first sheet and therefore is plane polarized at an angle of 30° with respect to the polarizing direction of the second sheet. From Malus' law, the fraction $\cos^2 (30^o)$ of this intensity is transmitted through the second sheet. Similarly, the third sheet transmits the fraction $\cos^2 (45^o)$ of the intensity which is incident upon it. Therefore the fraction of the original intensity which is transmitted through the stack of three sheets is

$$I/I_o = (0.5)(\cos 30^o)^2(\cos 45^o)^2 = (0.5)(0.866)^2(0.707)^2 = 0.188 \quad .$$ <<<

41-5 Polarization by Reflection

One way to produce plane polarized light is to pass unpolarized light through a polarizing sheet. Another way is to use the phenomenon of polarization by reflection. When unpolarized light is incident upon a transparent medium the reflected light is partially polarized, the amount of this polarization depending upon the angle of incidence. For a certain angle of incidence (known as the polarizing angle θ_p) the reflected light is completely plane polarized, the plane of polarization being perpendicular to the plane of incidence (the plane of incidence is the plane containing the incident ray and the normal to the surface). The polarizing angle θ_p is given by Brewster's law:

$$\tan \theta_p = n_2/n_1 \quad . \tag{42-2}$$

Here n_1 is the index of refraction of the incident medium, n_2 is the index of refraction of the refracting medium, θ_p is the angle of incidence (i.e. the angle between the incident ray and the normal to the surface).

42-6 Double Refraction

Certain materials (such as calcite crystal) are called doubly refracting; they are characterized by two numbers, n_o and n_e, called principle indices of refraction. Doubly refracting materials possess a certain characteristic direction called the optic axis.*

First consider the case in which light is incident normally upon the surface of a doubly refracting material and let the optic axis be parallel to this surface. In this case the light traveling through the material is split into two parts:

(a) An "ordinary" ray (called the "o-wave") which travels through the material with a speed $v_o = c/n_o$. The ordinary ray is plane polarized perpendicular to the optic axis.

(b) An "extraordinary" ray (called the "e-wave") which travels through the material with a speed $v_e = c/n_e$. The extraordinary ray is plane polarized parallel to the optic axis.

In summary, the important property of a doubly refracting material in this case is that it splits the incident light into two parts: the part which is plane polarized perpendicular to the optic axis (o-wave) and the part which is plane polarized parallel to the optic axis (e-wave). These two parts travel through the material with different speeds (c/n_o and c/n_e respectively).

The more general case in which the direction of propagation is not perpendicular to the optic axis is much more complicated. The o-wave still travels through the material with speed c/n_o and obey's Snell's law at an interface. However the speed of the e-wave depends upon its direction of propagation; it can be any value between c/n_e and c/n_o. Also, the e-wave need not obey Snell's law at an interface. In general, for a given incident ray, there will be two different angles of refraction (one for the o-wave and one for the e-wave); this is known as double refraction.

42-7 Circular Polarization

Consider two plane polarized beams of light which are identical in all respects (intensity, frequency, direction of propagation) except that their planes of polarization are perpendicular to each other. For example, if the direction of propagation is along the +z axis, then for one beam of light the plane of polarization might be the x-z plane while for the other beam it would be the y-z plane.

---

*Of course this should not be confused with the optical axis of an optical system (e.g. the line through the centers of a system of lenses).

If the two beams of light are in phase (e.g.if they both involve a factor sin $(kz - \omega t)$) then they will combine to form plane polarized light, the plane of polarization making an angle of 45° with the x and y axes. However if the two beams of light are 90° out of phase (e.g. if one beam involves a factor sin $(kz - \omega t)$ while the other involves a factor cos $(kz - \omega t)$) then they will combine to form circularly polarized light. (See text figure.)

For plane polarized light the $\underline{E}$ field (at any fixed location) is always along the same direction and oscillates sinusoidally with time (e.g. according to sin ωt). On the other hand for circularly polarized light the $\underline{E}$ field is of constant magnitude; its direction rotates about the propagation direction of the wave, the tip of the $\underline{E}$ field vector executing uniform circular motion with angular frequency ω. Circularly polarized light may be classified as being either clockwise or counterclockwise according to the sense of rotation of the $\underline{E}$ field vector (the convention is made that the observer is facing the source of the light when classifying circularly polarized light in this manner.)

42-8 Quarter-wave Plate

Consider a slab of doubly refracting material whose optic axis is parallel to the surface of the slab. When plane polarized light is incident upon the slab it is split into two parts. One part (the o-wave, which is plane polarized perpendicular to the optic axis) travels with speed c/n_o. The other part (the e-wave, which is plane polarized parallel to the optic axis) travels with speed c/n_e. At the incident surface of the slab these two waves are in phase. Due to the fact that these waves travel with different speeds they will emerge from the slab with some phase difference between them; this phase difference will depend upon the thickness of the slab. If the slab thickness is such that this phase difference is 90°, the slab is called a quarter-wave plate.

>>> Example 2. The principle indices of refraction of calcite are $n_o = 1.658$ and $n_e = 1.486$. Calculate the thickness of a quarter-wave plate made of calcite. Assume that the wavelength of the incident light is 500 nm.

The number of o-wave wavelengths contained within a calcite slab of thickness t is

$$N_o = t/\lambda_o = n_o t/\lambda \quad .$$

Note that o-wave wavelength in the material is $\lambda_o = \lambda/n_o$. Here $\lambda = 5000$ A is the wavelength in vacuum. Similarly the number of e-wave wavelengths contained within the slab is

$$N_e = n_e t/\lambda \quad .$$

In order to have a quarter-wave plate we must have N_o to be greater than N_e by 1/4.

$$N_o - N_e = 1/4$$

$$n_o t/\lambda - n_e t/\lambda = 1/4$$

$$t = \frac{\lambda}{4(n_o - n_e)} = \frac{500 \text{ nm}}{4(1.658 - 1.486)} = 730 \text{ nm}$$

$$t = 7.3 \times 10^{-4} \text{ mm} \quad .$$

Remark: If n_o were less than n_e then N_o would be less than N_e. In this case we would set $N_e - N_o$ equal to 1/4. <<<

To study what happens when plane polarized light is incident upon a quarter-wave plate, three special cases will be considered.

(a) Plane of polarization of incident light perpendicular to optic axis: In this case the light traveling through the quarter-wave plate is entirely o-wave. The emerging light is plane polarized, the plane of polarization is the same as that of the incident light.

(b) Plane of polarization of incident light parallel to optic axis: In this case the light traveling through the quarter-wave plate is entirely e-wave. The emerging light is plane polarized, the plane of polarization is the same as that of the incident light.

(c) Plane of polarization of incident light at an angle of 45° with the optic axis: In this case the light traveling through the quarter-wave plate is partially o-wave and partially e-wave. As these two waves emerge they

    (i) have equal amplitudes (since $\sin 45^\circ = \cos 45^\circ$),
    (ii) are plane polarized perpendicularly to each other,
    (iii) differ in phase by 90° (due to the quarter-wave plate).

They therefore combine to produce circularly polarized light.

In summary, a quarter-wave plate can be used to change plane polarized light into circularly polarized light. In order to do this the quarter-wave plate must be oriented so that its optic axis makes an angle of 45° with the plane of polarization of the incident light. Similarly, circularly polarized light can be changed into plane polarized light by passing it through a quarter-wave plate.

Chapter 43: LIGHT AND QUANTUM PHYSICS

REVIEW AND PREVIEW

You have seen that light can exhibit wave-like properties. The purpose of this chapter is to introduce you to other, particle-like, properties of light. Evidence is presented to show that, in certain circumstances, light acts as if it were "quantized"; a quantum of light is called a photon.

GOALS AND GUIDELINES

The goal of this chapter is to become familiar with the following topics:

a. The distribution of wavelengths in the radiation emitted by a heated solid (Sections 43-1, 43-2 and 43-3).
b. The photoelectric effect: electrons can be emitted when light strikes a metal surface (43-4).
c. The Compton effect: a shift in the wavelength of a photon scattered by an electron (Section 43-5).
d. The Bohr model of the hydrogen atom: this accounts for the observed wavelengths of light emitted by hydrogen atoms (Sections 43-6 and 43-7).

You should try to understand why the latter three items are evidence that light is quantized.

43-1 Radiation from a Heated Solid

A heated solid emits electromagnetic radiation (heat, light, etc.) at all wavelengths. To describe this radiation two quantities are useful:

1. The spectral radiancy R_λ is the rate at which this energy is radiated per unit area per unit wavelength. That is, $R_\lambda d\lambda$ is the rate at which electromagnetic energy is radiated with wavelengths in the range λ to $\lambda + d\lambda$ (per unit area of the emitting surface).
2. The radiancy R is the total rate at which this energy is radiated per unit area (here "total" means without regard to wavelength). R is then the sum (integral) of all the $R_\lambda d\lambda$'s, i.e.

$$R = \int_0^\infty R_\lambda \, d\lambda \quad .$$

43-2 Cavity Radiation

In general the dependence of R_λ upon the wavelength and the temperature is very complicated and may vary from one material to another. There exists an "idealized heated solid" called a cavity radiator for which this dependence is relatively simple. A cavity radiator consists of a solid material at some temperature T, containing a cavity (i.e. a hollow region in the interior of the solid). A small hole permits the electromagnetic radiation, called cavity radiation, to escape. Cavity radiation is completely independent of the particular material which surrounds the cavity.

43-3 Planck's Radiation Formula

Planck derived an expression for the spectral radiancy $\mathcal{R}_\lambda$ in the case of cavity radiation:

$$\mathcal{R}_\lambda = \frac{c_1}{\lambda^5} \frac{1}{e^{c_2/\lambda T} - 1} \qquad (43\text{-}1a)$$

Here c_1, c_2 are constants. By integrating this expression the radiancy $\mathcal{R} = \int \mathcal{R}_\lambda \, d\lambda$ is found to be proportional to the fourth power of the temperature,

$$\mathcal{R} = \sigma T^4 \quad . \qquad (43\text{-}1b)$$

The proportionality constant $\sigma = 5.67 \times 10^{-8}$ W/m$^2 \cdot$K^4) is called the Stefan-Boltzmann constant. To derive equation (43-1a) Planck made two assumptions:

1. The atoms within a solid act as oscillators which can emit or absorb electromagnetic radiation. Planck assumed that such an oscillator (e.g. a mass and spring) cannot have any arbitrary energy, rather its energy is quantized according to the equation $E = nh\nu$. Here ν is the oscillator's frequency,* $n = 0, 1, 2, \ldots$ is an integer, and $h = 6.63 \times 10^{-34}$ J$\cdot$s is a fundamental constant (known as Planck's constant).
2. The above oscillator can radiate energy only in quanta; when n decreases by one an amount of energy $\Delta E = h\nu$ is radiated.

The important point here is that certain mechanical properties (in this case the energy of an oscillator) cannot have arbitrary values but rather are quantized to certain discrete values.

43-4 Photoelectric Effect

When light of frequency ν is incident upon a metal surface in a vacuum, electrons (called photoelectrons) may be emitted from the surface. The maximum kinetic energy K_{max} of these photoelectrons can be determined by allowing them to travel through a potential difference V. At a certain value $V = V_o$ (called the stopping potential) all the photoelectrons will be stopped. The maximum kinetic energy is then

$$K_{max} = eV_o \quad . \qquad (43\text{-}2)$$

Experimentally, the maximum kinetic energy K_{max} of the photoelectrons is related to the frequency of the incident light ν as follows:

$$h\nu = K_{max} + E_o \qquad (43\text{-}3)$$

where E_o (called the work function) is a constant characteristic of the particular material which is emitting the photoelectrons. Combining the above two equations we have

$$V_o = \frac{h}{e}\nu - \frac{E_o}{e} \quad . \qquad (43\text{-}4)$$

Using experimental data, the measured stopping potential V_o can be plotted as a function of the frequency of the incident light ν. According to (43-4) this results in a straight line whose slope is h/e and whose intercept is $- E_o/e$.

---

*Here ν is the frequency in cycles per time (not radians per time).

Einstein explained the photoelectric effect and equation (43-3) in the following manner.

1. The incident light exists in concentrated bundles called photons, the energy of a photon being given by $E = h\nu$.
2. When a photon strikes the metal surface it is absorbed as a unit. Part of its energy (E_o) is used to remove the photoelectron from the metal, the remaining energy ($h\nu - E_o$) is available as kinetic energy for the electron. Those electrons which suffer no collisions as they escape from the metal will have the largest kinetic energy, i.e. $K_{max} = h\nu - E_o$.

Since the smallest that K_{max} can be is zero, there is a smallest frequency ν_o (called the cutoff frequency) for which photoelectrons will just be emitted. Setting $K_{max} = 0$ in (43-3) gives an expression for the cutoff frequency ν_o:

$$h\nu_o = E_o \quad . \tag{43-5}$$

If the frequency of the incident light is greater than ν_o then photoelectrons will be emitted; an increase in the intensity of the light will cause an increase in the number of photoelectrons emitted per unit time. If the frequency of the incident light is less than ν_o then no photoelectrons will be emitted regardless of the intensity of the light.

In summary, the photoelectric effect provides evidence that light is quantized. The quantum of light, i.e. the photon, has energy $E = h\nu$.

>>> Example 1. The work function of a certain metal is 1.8 eV. Calculate the "threshold wavelength" λ_o. (The threshold wavelength is the longest possible wavelength of the incident light for which the photoelectric effect will occur for the given material.)

According to the definition above, the threshold wavelength λ_o is the wavelength corresponding to the cutoff frequency ν_o. Since a photon whose frequency is ν_o will have an energy equal to the work function of the material,

$$h\nu_o = E_o \quad , \quad \nu_o = E_o/h$$

$$\lambda_o = \frac{c}{\nu_o} = \frac{hc}{E_o}$$

$$\lambda_o = \frac{(6.63 \times 10^{-34}\ J\cdot s)(3.00 \times 10^8\ m/s)}{1.8\ eV} \times \frac{1\ eV}{1.6 \times 10^{-19}\ J}$$

$$\lambda_o = 6.9 \times 10^{-7}\ m = 690\ nm \quad .$$ <<<

43-5 Compton Effect

The Compton effect involves the "collision" of a photon (usually an X-ray) with a free electron. Figure 43-1a shows the situation before the collision: the incident photon with wavelength λ (= c/ν) travels in the +x direction, the electron is assumed to be initially at rest. After the collision the electron recoils at an angle θ with some velocity $\underline{v}$ and a photon with wavelength λ' (= c/ν') emerges at a "scattering angle" ϕ as shown in Figure 43-1b. In order that energy be conserved the emerging photon must have less energy than the incident photon (because the kinetic energy of the electron has been increased); therefore λ' must be larger than λ. The wavelength increase $\Delta\lambda = \lambda' - \lambda$ is called the Compton shift.

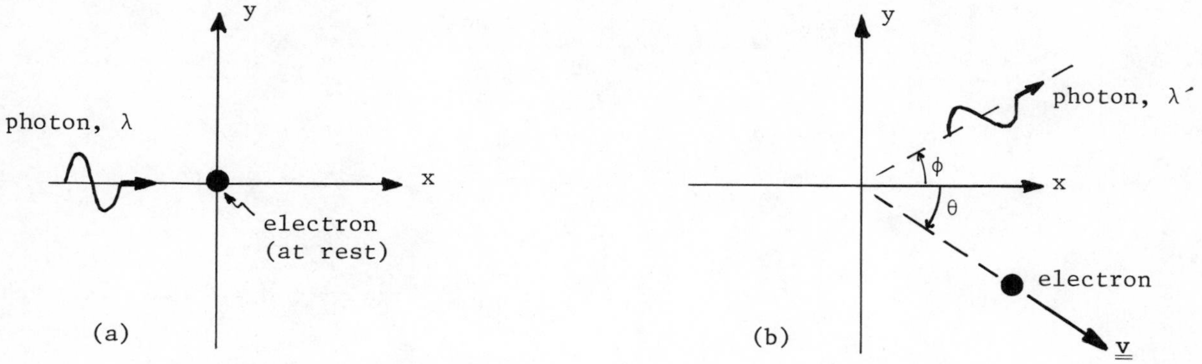

Figure 43-1. The Compton effect. (a) Before collision. (b) After collision.

Altogether there are five variables involved in the Compton effect: λ, λ', v, θ, ϕ. There are three conservation laws that can be applied: conservation of energy, conservation of momentum in each of the x and y directions. Usually the electron recoil speed v is so close to the speed of light that the correct relativistic expressions for the energy and momentum of the electron must be used. As for the photon, it is treated as a "particle" with energy hc/λ and momentum h/λ. See text for these details. Using these three conservation laws, the two variables v and θ can be eliminated leaving one equation relating λ, λ', ϕ. This relation is

$$\Delta\lambda = \lambda' - \lambda = \frac{h}{m_0 c}(1 - \cos\phi) \tag{43-6}$$

where m_0 is the (rest) mass of the electron. The maximum Compton shift occurs when the photon is "back scattered" (ϕ= 180°).

The Compton effect provides further evidence for the particle-like (quantum) nature of light.

>>> Example 2. In a certain Compton effect experiment the wavelength of the incident X-ray is 0.1 Å. If the scattering angle is 90°, what is the angle at which the recoil electron emerges?

Referring to Figure 43-1, the angle ϕ is 90° and the angle θ is the unknown angle at which the electron emerges. Let $\mathbf{p}_1$, $\mathbf{p}_2$ be the initial and final momentum of the photon and let $\mathbf{P}_1$, $\mathbf{P}_2$ be the initial and final momentum of the electron. The total momentum before the collision must equal the total momentum after the collision: $\mathbf{p}_1 + \mathbf{P}_1 = \mathbf{p}_2 + \mathbf{P}_2$. Since $\mathbf{P}_1 = 0$,

$$\mathbf{P}_2 = \mathbf{p}_1 - \mathbf{p}_2 \quad .$$

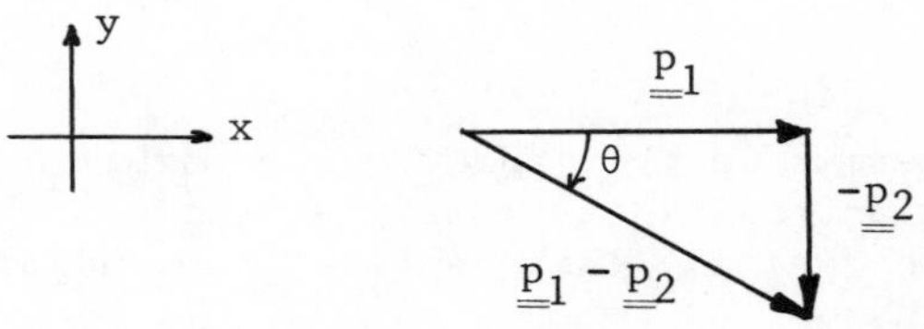

We also know that $\mathbf{p}_1$ is in the x direction and that $\mathbf{p}_2$ is in the y direction (since ϕ = 90°). The diagram shows the vector construction for determining $\mathbf{P}_2$. From it we see that the angle θ is given by

$$\tan\theta = \frac{|\underline{\underline{P}}_2|}{|\underline{\underline{P}}_1|} = \frac{hc/(\lambda + \Delta\lambda)}{hc/\lambda} = \frac{\lambda}{\lambda + \Delta\lambda} .$$

Now

$$\Delta\lambda = \frac{h}{m_o c}(1 - \cos\phi) = \frac{h}{m_o c} \qquad \text{since } \phi = 90^o$$

$$\Delta\lambda = \frac{6.63\times10^{-34}\ \text{J}\cdot\text{s}}{(9.11\times10^{-31}\ \text{kg})(3.00\times10^{8}\ \text{m/s})} = 2.4\times10^{-12}\ \text{m} = 0.024\ \text{Å} .$$

Thus

$$\tan\theta = \lambda/(\lambda+\Delta\lambda) = (0.1\ \text{Å})/(0.124\ \text{Å}) = 0.81$$

$$\theta = 39^o .$$

This is the direction of the final momentum $\underline{\underline{P}}_2$ of the electron. Since $\underline{\underline{P}} = m\underline{\underline{v}}$, the velocity has the same direction as the momentum. <<<

43-6 Line Spectra

Radiation from a heated solid (in particular a cavity radiator) exists at all wavelengths; this is known as a <u>continuous spectrum</u>. In certain other situations (e.g. a gas in an electric arc) the radiation is emitted at discrete wavelengths only; this is known as a <u>line spectrum</u>. The existence of line spectra is evidence that "something" in the atom is quantized.

43-7 Hydrogen Atom

The simplest line spectrum is that of hydrogen. As a model of the hydrogen atom, consider an electron in a circular orbit about a fixed proton. Using this model Bohr explained the observed line spectrum by making two chief assumptions:

1. The atom can exist only in certain "stationary states"; call the energies of these states E_k. Electromagnetic radiation (in the form of a single photon) is emitted only when the atom makes a transition from one of these states (E_k) to another (E_j); the emitted photon carries away the energy difference: $E_k - E_j$.
2. The angular momentum of the orbiting electron must be an integer multiple of $h/2\pi$. This is the "quantization" condition.

Since the electron is assumed to be in a circular orbit about the proton, application of F = ma gives

$$\frac{1}{4\pi\varepsilon_o}\frac{e^2}{r^2} = m\frac{v^2}{r} \qquad (43\text{-}7a)$$

Here r is the radius of the orbit, m is the mass of the electron, v is the speed of the electron. Note that the proton's charge is the same in magnitude as the electron's charge: e. The quantization condition for the angular momentum of the electron states that

$$mvr = nh/2\pi \qquad (43\text{-}7b)$$

where n = 1, 2, 3,... . Equations (43-7) may be solved simultaneously for r and v.

The results are

$$r = n^2 \frac{h^2 \varepsilon_0}{\pi m e^2} \qquad (43\text{-}8a)$$

$$v = \frac{e^2}{2nh\varepsilon_0} \quad . \qquad (43\text{-}8b)$$

The energy is the sum of the kinetic and potential energies,

$$E = \tfrac{1}{2} mv^2 - \frac{1}{4\pi\varepsilon_0} \frac{e^2}{r} \quad . \qquad (43\text{-}9)$$

Substituting equations (43-8) into (43-9) gives

$$E = - \frac{me^4}{8\varepsilon_0{}^2 h^2} \frac{1}{n^2} \quad , \quad n = 1, 2, 3, \ldots \quad . \qquad (43\text{-}10)$$

The value of the coefficient of $1/n^2$ is 13.6 eV. Thus

$$E = - (13.6 \text{ eV})(1/n^2) \quad . \qquad (43\text{-}11)$$

When the atom undergoes a transition, there is an initial value of n and a final value of n. If the final value of n is n = 1, the emitted photon is said to belong to the Lyman series. Similarly if the final value of n is n = 2 we obtain the Balmer series, and for n = 3 the Paschen series. As an example the lowest energy photon in the Balmer series is due to a transition from n = 3 to n = 2. The energy of this photon can be easily calculated using (43-11):

$$\Delta E = [-(13.6 \text{ eV})/3^2] - [-(13.6 \text{ eV})/2^2]$$

$$\Delta E = 1.9 \text{ eV} \; .$$

The allowed energy values (43-11) are frequently displayed by means of an energy level diagram as shown in Figure 2. Here the energy values are dispayed as horizontal lines with the amount of energy **increasing** (becoming less negative) as one moves vertically upward. Also shown are the values of the energy E as well as the corresponding quantum number n. A transition between energy levels is indicated by a vertical arrow.

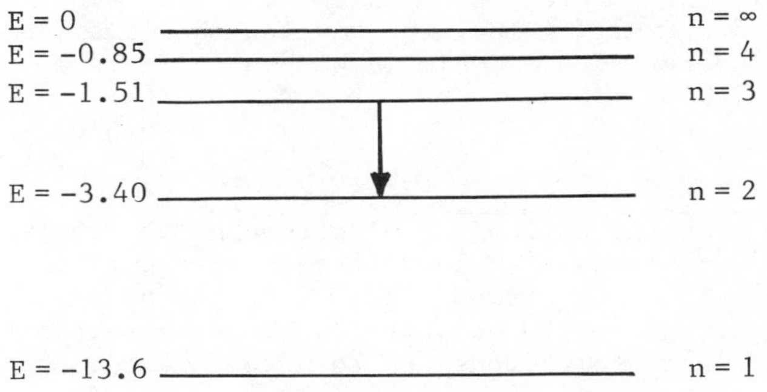

Figure 43-2. Energy level diagram for the hydrogen atom. The energy values increase in the vertical direction; numerical values for E are in electron volts (eV). The vertical arrow indicates a transition from the n = 3 state to the n = 2 state. This transition results in the emission of a photon whose energy is 1.9 eV and belongs to the Balmer series.

Chapter 44: WAVES AND PARTICLES

REVIEW AND PREVIEW

You have seen that light can have both wave (e.g. interference) and particle (e.g. photoelectric effect) properties. Although you probably regard matter as having only particle properties, in this chapter you will see that matter can also exhibit wave properties.

GOALS AND GUIDELINES

The chief goal of this chapter is to become familiar with some of the wave properties of matter. The starting point is the association of a wavelength with the momentum of the particle (Section 44-1). Using this concept you should be able to explain

a. electron diffraction (Section 44-2);
b. the Bohr model of the hydrogen atom (Section 44-3).

44-1 Matter Waves

We have seen that light can have both wave and particle properties. For example, the phenomena of interference and diffraction are explained using the wave properties of light; the photoelectric and Compton effects are best explained using a particle (photon) point of view. The energy E and momentum p of a photon are related to the frequency ν and wavelength λ of the corresponding electromagnetic wave by

$$E = h\nu \qquad (44\text{-}1a)$$

$$p = h/\lambda \ . \qquad (44\text{-}1b)$$

De Broglie hypothesized that matter can also exhibit both wave and particle properties; the frequency and wavelength of these <u>matter waves</u> being related to the particles's energy and momentum by the very same formulas (44-1) which apply to photons. However, matter waves are <u>not</u> electromagnetic waves and they do not travel with the velocity of light.*

Usually, the concept of the frequency of matter waves is relatively unimportant; however the concept of the wavelength of matter waves can be very important (recall that it is the wavelength which enters into the various formulas for interference and diffraction). From (44-1b), the wavelength (sometimes called the "de Broglie wavelength") of the matter wave associated with a particle of momentum p is

$$\lambda = h/p \quad \text{(de Broglie wavelength)} \ . \qquad (44\text{-}2)$$

---

*In the case of an electromagnetic wave we have $\nu\lambda = c$. Using this, equations (44-1) can then also be written as $E = hc/\lambda$ and $p = h\nu/c$. Although correct for photons, the latter two equations do <u>not</u> apply to matter waves since these waves do not travel with velocity c. To help recall the correct formulas (44-1) which apply <u>both</u> to photons and to matter waves, note that the constant "c" does not appear at all in these relations.

44-2 Electron Diffraction

The wave properties of matter can result in interference phenomena. A beam of electrons with kinetic energy K incident upon a crystal gives rise to an electron diffraction pattern as observed on a distant fluorescent screen. The analysis of this is quite similar to that of X-ray diffraction as discussed in Chapter 41. The lattice planes can be considered to act as partially reflecting mirrors. Electrons are found to be located at those points on the screen where constructive interference of the electron matter waves occurs. The location of these interference maxima are given by 41-6:

$$m\lambda = 2d \sin \theta \quad . \tag{44-3}$$

Here $\lambda = h/p = h/\sqrt{2m_eK}$ is the wavelength of the matter wave, d is the interplanar spacing in the crystal, θ is the angle between the incident beam and the lattice planes (recall that θ is also the angle between the diffracted beam and the lattice planes so that 2θ is the angle between the diffracted and incident beams), m = 1, 2, 3, ... is the order of the maximum. The kinetic energy K can be computed using K = eV where V is the potential difference through which the electrons were originally accelerated.

The diffraction of a beam of particles by a crystal is not limited to electrons. Any particle (charged or not) can be used in such an experiment. Of course the velocity of the particle must be such as to make its de Broglie wavelength λ comparable with the interplanar spacing d in order to yield an appreciable effect.

>>> Example 1. Electrons are accelerated from rest through a potential difference of 500 volts. The resulting beam of electrons is then used in an electron diffraction experiment. A second order maximum is observed where the angle between the diffracted and incident beams is 40°. (a) Calculate the wavelength of the electron matter waves. (b) Calculate the interplanar distance, d, of those lattice planes which are responsible for this maximum.

(a)

$$\lambda = \frac{h}{p} = \frac{h}{(2\ m_eK)^{1/2}} = \frac{h}{(2\ m_eeV)^{1/2}}$$

$$= \frac{6.63 \times 10^{-34}\ \text{J}\cdot\text{s}}{[(2)(9.11 \times 10^{-31}\ \text{kg})(1.60 \times 10^{-19}\ \text{C})(5.00 \times 10^{2}\ \text{V})]^{1/2}}$$

$$= 5.5 \times 10^{-11}\ \text{m} = 0.55\ \text{Å} \quad .$$

(b)

$$m\lambda = 2d \sin \theta$$

$$d = \frac{m\lambda}{2 \sin \theta} \quad .$$

In this problem we have m = 2 (second order) and $\theta = 20^\circ$ (since $2\theta = 40^\circ$). Substituting these values,

$$d = \frac{(2)(0.55\ \text{Å})}{(2) \sin (20^\circ)} = 1.6\ \text{Å} \quad .$$

<<<

44-3 Atomic Structure and Standing Waves

In the discussion of Bohr's explanation of the hydrogen atom (Chapter 43), one of the assumptions made was that the angular momentum must be a multiple of $h/2\pi$. That is,

$$mvr = nh/2\pi \quad ; \quad n = 1, 2, 3, \ldots \; .$$

Since the de Broglie wavelength of the electron is given by $\lambda = h/mv$, the above equation may be written as

$$n\lambda = 2\pi r \quad .$$

Thus an integral number of wavelengths must "fit" around the circumference of the orbit. This is the condition for a standing wave. Hence the Bohr quantization condition may be interpreted as demanding that the matter wave for the orbiting electron be a standing wave. The existence of discrete energies for an atom is thus analogous to the existence of discrete frequencies for a vibrating mechanical system (such as a string which is clamped at both ends).

44-4 Wave Mechanics

In classical mechanics we specify the location of a particle as a function of time. In wave mechanics (or quantum mechanics) we specify a wave function Ψ as a function of x, y, z (and time). For example in the case of a particle confined to be on the x-axis between $x = 0$ and $x = \ell$, the wave function turns out to be proportional to $\sin(n\pi x/\ell)$. This insures that Ψ vanish at the two extremities ($x = 0, \ell$).

>>> Example 2. A particle is confined to be on the x-axis between $x = 0$ and $x = \ell$. Obtain an expression for the possible energies.

In this case an integral number of half wavelengths must "fit" in the length ℓ. That is $n\lambda/2 = \ell$ or $\lambda = 2\ell/n$, $n = 1, 2, 3, \ldots$. The energy is then

$$E = \frac{p^2}{2m} = \frac{1}{2m}\left(\frac{h}{\lambda}\right)^2 = \frac{1}{2m}\left(\frac{h}{2\ell/n}\right)^2$$

$$E = n^2h^2/(8\,m\ell^2) \quad .$$ <<<

The interpretation of the wave function Ψ is the following. The square of Ψ, Ψ^2, is proportional to the probability that the particle will be found to be near the point at which Ψ is evaluated. For example, in a one dimensional problem, $\Psi^2\,dx$ is proportional to the probability that the particle will be found to be located between x and x + dx. Thus in wave mechanics we abandon the concept that a particle has a precise location; instead we speak of the probability that it will be found in various locations.

>>> Example 3. Consider the situation described in example 2. What is the wave function for the nth state?

The wave function will be of the form $\Psi = A_n \sin(2\pi x/\lambda)$ where A_n is a constant. In this problem $n(\lambda/2) = \ell$ so that $\lambda = 2\ell/n$. Thus $\Psi = A_n \sin(n\pi x/\ell)$. The constant A_n can be determined by demanding that the probability of finding the particle somewhere between $x = 0$ and $x = \ell$ be one. That is

$$\int_0^\ell \Psi^2\,dx = 1$$

$$A_n^2 \int_0^\ell \sin^2(n\pi x/\ell)\,dx = 1 \quad .$$

Using the trigonometric identity $\sin^2 \theta = \frac{1}{2}(1 - \cos(2\theta))$,

$$\frac{1}{2} A_n^2 \int_0^\ell (1 - \cos(2n\pi x/\ell))\, dx = 1$$

$$\frac{1}{2} A_n^2 \left[x - \frac{\ell}{2n\pi} \sin(2n\pi x/\ell)\right]_{x=0}^{x=\ell} = 1$$

$$\frac{1}{2} A_n^2 \ell = 1 \quad , \quad A_n = \sqrt{2/\ell} \quad .$$

The desired wave function is therefore

$$\Psi = \sqrt{2/\ell} \sin(n\pi x/\ell) \quad .$$

Remark: The fact that A_n turned out to be independent of n is an accident of this particular problem. <<<

43-5 The Uncertainty Principle

According to wave mechanics, it is impossible to simultaneously measure certain pairs of quantities to arbitrary accuracy. This is known as the uncertainty principle. In particular,

$$\Delta x \Delta p_x \geq h \quad . \tag{44-4a}$$

Here Δx is the uncertainty in the x coordinate of the particle and Δp_x is the uncertainty in the x component of momentum of the particle. Equation (44-4a) says that the product of these two uncertainties must be at least as large as Planck's constant h. Similarly,

$$\Delta y \Delta p_y \geq h \tag{44-4b}$$

$$\Delta z \Delta p_z \geq h \quad . \tag{44-4c}$$

The uncertainty principle does not prohibit very accurate simultaneous measurement of, say, x and p_y.

Another form of the uncertainty principle is

$$\Delta L_z \Delta \phi \geq h \quad . \tag{44-4d}$$

Here L_z is the z component of the (orbital) angular momentum $\underline{\underline{L}} = \underline{\underline{r}} \times \underline{\underline{p}}$ of a particle and ϕ is the angular coordinate of the particle in the x-y plane ($\phi = \tan^{-1}(y/x)$). Finally, a somewhat different form (in that it involves the duration of the measurement) of the uncertainty principle is

$$\Delta E \Delta t > h \quad . \tag{44-4e}$$

Here E is the energy of the particle and Δt is the duration of time used in measuring this energy E.

Chapter 45: THE ATOM

REVIEW AND PREVIEW

In the previous chapter the concept of a wave function $\Psi(x)$ for a particle was introduced. Here, we shall continue with a discussion of the description of a particle by its wave function (i.e. the use of quantum, rather than classical, mechanics). The system to which quantum mechanics will be applied is the atom: first to the one-electron atom (hydrogen, or hydrogen-like) and then, more qualitatively, to the many-electron atom. (Note: The text also discusses two other applications of quantum mechanics, namely (i) x rays and the historical numbering of the elements and (ii) the laser. These topics will not be treated here.)

GOALS AND GUIDELINES

Due to the complexity of this subject, the treatment omits most of the derivations. There are, however, a considerable number of new concepts and associated terminology to be learned. The major items are

1. Hydrogen atom:
   - (a) quantum numbers and the labelling of the states,
   - (b) total angular momentum,
   - (c) magnetic moments,
   - (d) application of external magnetic field (Zeeman effect).

2. Stern-Gerlach experiment.

3. Many-electron atoms:
   - (a) Pauli exclusion principle,
   - (b) periodic table.

45-1 Quantum Mechanics in More Than One Dimension

In Section 44-4 we saw that the problem of a particle confined to a "one-dimensional box" of length ℓ led to the wave function

$$\Psi(x) = A_n \sin(n\pi x/\ell), \quad n = 1,2,3,\ldots \quad . \tag{45-1a}$$

Since there is no potential energy in this problem, $E = p^2/2m$; the momentum p is related to the wavelength $\lambda = 2\ell/n$ by the de Boglie relation $p = h/\lambda$. Putting this all together gave

$$E = n^2h^2/(8\ m\ell^2) \tag{45-1b}$$

for the allowed energies in terms of the quantum number n.

The problem of an electron in an atom differs from the above problem in two significant ways (i) there is a potential energy (for the hydrogen atom this is the Coulomb potential energy between the electron and the proton) and (ii) this is a three-dimensional problem. To obtain a feeling for the effect (ii), consider a particle confined to a "two-dimensional box" of side length ℓ. Analogous to (45-1a) we now have

$$\Psi(x,y) = A_{n_x,n_y} \sin(n_x\pi x/\ell)\ \sin(n_y\pi y/\ell). \tag{45-2a}$$

Here the two quantum numbers, n_x and n_y, each take on the values 1,2,3,... separately. Eq. (45-2a) arises by simply fitting a sinusoidal wave into the box in each of the x and y directions. One may think of there being a wavelength $\lambda_x = 2\ell/n_x$ associated with the x coordinate and another wavelength $\lambda_y = 2\ell/n_y$ associated with the y coordinate. The energy E, which is all kinetic, is given by $E = p^2/2m = (p_x{}^2 + p_y{}^2)/2m$; each component of p being related to the corresponding wavelength by the de Broglie relation: $p_x = h/\lambda_x$ and $p_y = h/\lambda_y$. Putting this all together gives

$$E = (n_x{}^2 + n_y{}^2)h^2/(8\ m\ell^2) \quad . \tag{45-2b}$$

This expression illustrates two important features which were not present in the one dimensional problem. First, the energy depends upon two quantum numbers (n_x and n_y), rather than one. Second, a given energy may correspond to more than one set of quantum numbers. That is, several states may have the same energy; this is called degeneracy, such states themselves are called degenerate. For example, we may always interchange n_x and n_y; the state $(n_x,n_y) = (2,3)$ has the same energy as the state $(n_x,n_y) = (3,2)$ according to (45-2b). In additional to this obvious degeneracy associated with interchanging n_x and n_y, there is some "accidental" degeneracy; e.g. the states (1,7), (7,1) and (5,5) all have the same energy.

45-2 Hydrogen Atom

The quantum mechanical problem of an electron under the influence of the Coulomb potential due to the nucleus is a three dimensional one. Because of the spherical symmetry, it is treated in spherical coordinates (r,θ,ϕ) rather than in Cartesian coordinates (x,y,z). One would expect there to be three quantum numbers, one number associated with each of these three coordinates. In fact, however, there are four quantum numbers; the additional quantum number being associated with the spin of the electron. These quantum numbers are denoted by (n,ℓ,m_ℓ,m_s). The names and possible values of these four quantum numbers are listed below.

quantum number	name	possible values
n	principal quantum number	1,2,3,4,......
ℓ	orbital quantum number	0,1,2,..,(n-1)
m_ℓ	orbital magnetic quantum number	$-\ell,-\ell+1,\ldots,\ell-1,\ell$
m_s	spin magnetic quantum number	$\pm\frac{1}{2}$

In words: n can have any integral value starting with 1; for each n, ℓ can have any integral value from 0 to n-1; for each ℓ, m_ℓ can have any integral value from $-\ell$ to ℓ; m_s has only two possible values, $+\frac{1}{2}$ and $-\frac{1}{2}$. We shall now discuss the physical meaning of each of these four quantum numbers in detail.

Principal quantum number n: This is associated with the r coordinate of the spherical (r,θ,ϕ) coordinate system. For the hydrogen atom (ignoring a small effect due to the spin), it alone determines the energy

$$E = -(13.6\ \text{eV})/n^2 \quad . \tag{45-3}$$

All states of a given n are said to belong to the same shell. The shells are (for historical reasons) sometimes indicated by a letter code called a shell symbol as follows:

n	1	2	3	4	...
Shell symbol	K	L	M	N	...

<u>Orbital quantum number ℓ</u>: This is associated with the θ coordinate. This quantum number gives the quantized value of the magnitude of the orbital angular momentum

$$L = |\underline{\underline{L}}| = \sqrt{\ell(\ell+1)}\ \hbar \quad . \tag{45-4}$$

The quantity $\hbar$ (read "h-bar") is Planck's constant divided by 2π,

$$\hbar \equiv h/(2\pi) = 1.05 \times 10^{-34} \quad \text{J·s} \ ; \tag{45-5}$$

this serves as a natural unit of angular momentum in quantum mechanics. All states having the same values for the pair of quantum numbers (n,ℓ) are said to belong to the same <u>subshell</u>. The subshells are (for historical reasons) indicated by a letter code called a subshell symbol as follows:

ℓ	0	1	2	3	4	5	...
Subshell symbol	s	p	d	f	g	h	...

Note that the shell symbols are capital letters while the subshell symbols are lower case letters.

<u>Orbital magnetic quantum number m_ℓ</u>: This is associated with the ϕ coordinate. This quantum number gives the quantized value of the z-component of the orbital angular momentum

$$L_z = m_\ell \hbar \quad . \tag{45-6}$$

The word "magnetic" in the name of this quantum number has to do with the fact that certain effects of m_ℓ become apparant if the atom is placed in a magnetic field (see Section 45-5).

<u>Spin magnetic quantum number m_s</u>: The electron has a "spin" angular momentum $\underline{\underline{S}}$. This has a fixed magnitude

$$S \equiv |\underline{\underline{S}}| = \sqrt{\tfrac{1}{2}(\tfrac{1}{2}+1)}\ \hbar = (\sqrt{3}/2)\ \hbar \quad . \tag{45-7}$$

(Sometimes the symbol $s = \frac{1}{2}$ is used; s is called the spin quantum number. Then Eq. (45-7) with s is analogous to Eq. (45-4) with ℓ.) The quantum number m_s gives the quantized value of the z-component of the spin angular momentum

$$S_z = m_s \hbar \quad . \tag{45-8}$$

Note that this has only two possible values; $S_z = \pm\frac{1}{2}\ \hbar$. Ignoring a small effect due to the spin, the effect of this fourth quantum number is to double the number of states: for each set (n,ℓ,m_ℓ) there are two possible values of m_s.

>>> Example 1. An electron in a hydrogen atom is in the N shell. (a) How many states are there in each subshell? (b) How many states are there in this shell?

The N shell means $n = 4$. Since $\ell = 0,1,\ldots(n-1)$, the possible values of ℓ are $\ell = 0,1,2,3$. Let us tabulate the remaining quantum numbers, m_ℓ and m_s for each of these subshells.

subshell	m_ℓ	m_s	number of states
s ($\ell = 0$)	0	$-\frac{1}{2}, +\frac{1}{2}$	2
p ($\ell = 1$)	-1, 0, +1	$-\frac{1}{2}, +\frac{1}{2}$	6
d ($\ell = 2$)	-2, -1, 0, +1, +2	$-\frac{1}{2}, +\frac{1}{2}$	10
f ($\ell = 3$)	-3, -2, -1, 0, +1, +2, +3	$-\frac{1}{2}, +\frac{1}{2}$	14

The last column shows the number of states in each subshell; this was obtained by multiplying the number of different m_ℓ's by the number of different m_s's for each row in the chart. The total number of states in this N shell is

$$2 + 6 + 10 + 14 = 32 \quad .$$

<<<

>>> Example 2. An electron is in a d subshell. Calculate the possible values of L and L_z.

The letter d denotes $\ell = 2$. The magnitude of the orbital angular momentum is

$$L = \sqrt{\ell(\ell+1)}\ \hbar = \sqrt{2(2+1)}\ \hbar = \sqrt{6}\ \hbar = 2.45\ \hbar \quad .$$

The possible values of the z-component of the orbital angular momentum are

$$L_z = m_\ell \hbar = -2\ \hbar,\ -\hbar,\ 0,\ \hbar,\ 2\ \hbar$$

since $m_\ell = -2,-1,0,1,2$. Note that all the L_z's are smaller in magnitude than L, i.e. any component of $\underline{\underline{L}}$ must be smaller than L.

<<<

45-3 Total Angular Momentum

Instead of considering the orbital angular momentum $\underline{\underline{L}}$ and the spin angular momentum $\underline{\underline{S}}$ separately, it is sometimes more convenient to consider their (vector) sum

$$\underline{\underline{J}} = \underline{\underline{L}} + \underline{\underline{S}} \quad . \tag{45-9}$$

$\underline{\underline{J}}$ is called the total angular momentum. As an angular momentum, $\underline{\underline{J}}$ obeys rules similar to those of $\underline{\underline{L}}$ and $\underline{\underline{S}}$. The magnitude of $\underline{\underline{J}}$ is

$$J \equiv |\underline{\underline{J}}| = \sqrt{j(j+1)}\ \hbar \tag{45-10}$$

and the z-component of $\underline{\underline{J}}$ is

$$J_z = m_j \hbar \quad . \tag{45-11}$$

Here the quantum number j takes on the two possible values

$$j = \ell \pm \tfrac{1}{2} \quad , \tag{45-12a}$$

unless $\ell = 0$ in which case j has the single value $j = \frac{1}{2}$. The quantum number m_j takes on the possible values

$$m_j = -j, -j+1, \ldots, j-1, j \quad , \tag{45-12b}$$

that is, m_j can have any value between -j and j in integral steps.

So far, we have labelled the states with the quantum numbers (n, ℓ, m_ℓ, m_s). An alternate description is to label the states with the quantum numbers (n, ℓ, j, m_j).

To show that this relabelling scheme is consistent, consider the $n = 4$, $\ell = 3$ subshell. This subshell has 14 states (see Example 1). In the (n,ℓ,j,m_j) scheme, the possible values of j are $j = \ell + \frac{1}{2} = 7/2$ and $j = \ell - \frac{1}{2} = 5/2$. Associated with $j = 7/2$ are 8 states ($m_j = -7/2,-5/2,-3/2,-1/2,1/2,3/2,5/2,7/2$) and associated with $j = 5/2$ are 6 states ($m_j = -5/2,-3/2,-1/2,1/2,3/2,5/2$). Therefore the total number of states is 8 + 6 = 14. We thus obtain the same number of states (14 in this case) whether we use the four quantum numbers (n,ℓ,m_ℓ,m_s) or the four quantum numbers (n,ℓ,j,m_j).

So far, we have ignored any interaction between the orbital magnetic momentum $\mathbf{L}$ and the spin angular momentum $\mathbf{S}$. However, there is a magnetic moment $\boldsymbol{\mu}_\ell$ associated with $\mathbf{L}$ and a magnetic moment $\boldsymbol{\mu}_s$ associated with $\mathbf{S}$ (see Section 45-4). These two magnetic moments interact with each other via a term proportional to the dot product of the corresponding two angular momenta, $\mathbf{L}\cdot\mathbf{S}$; this is called the spin-orbit interaction. As a result of this interaction, m_ℓ and m_s are no longer "good" quantum numbers; however j and m_j remain "good" quantum numbers. For this reason, it is preferable to use the (n,ℓ,j,m_j) labelling scheme if spin-orbit interaction is to be taken into account. In the situation dealt with in the above paragraph ($n = 4$, $\ell = 3$), we would say that there are

eight $4f_{7/2}$

and

six $4f_{5/2}$

states. In this notation, the first number (4) is the value of n, the next letter (f) stand for the value of ℓ (i.e. $\ell = 3$) and the subscript (7/2 or 5/2) is the value of j. It turns out that the spin-orbit interaction among these fourteen degenerate states "split" these states into a set of eight degenerate states and another set of six degenerate states according to their value of j.

>>> Example 3. Show that the spin-orbit interaction (proportional to $\mathbf{L}\cdot\mathbf{S}$) depends only on j and ℓ. Hint: "Square" the equation $\mathbf{J} = \mathbf{L} + \mathbf{S}$.

We start with $\mathbf{J} = \mathbf{L} + \mathbf{S}$. To "square" this equation, we take the dot product of it with itself.

$$\mathbf{J}\cdot\mathbf{J} = (\mathbf{L} + \mathbf{S})\cdot(\mathbf{L} + \mathbf{S})$$

$$\mathbf{J}\cdot\mathbf{J} = \mathbf{L}\cdot\mathbf{L} + \mathbf{S}\cdot\mathbf{S} + 2\,\mathbf{L}\cdot\mathbf{S}$$

Solving for $\mathbf{L}\cdot\mathbf{S}$,

$$\mathbf{L}\cdot\mathbf{S} = (\mathbf{J}\cdot\mathbf{J} - \mathbf{L}\cdot\mathbf{L} - \mathbf{S}\cdot\mathbf{S})/2 \quad .$$

Now $\mathbf{J}\cdot\mathbf{J} = J^2 = j(j+1)\hbar^2$, $\mathbf{L}\cdot\mathbf{L} = L^2 = \ell(\ell+1)\hbar^2$ and $\mathbf{S}\cdot\mathbf{S} = S^2 = (3/4)\hbar^2$. Thus $\mathbf{L}\cdot\mathbf{S}$ depends only on j and ℓ. <<<

45-4 Magnetic Moments

In Section 32-5 we saw that an electron (in a circular orbit) constitutes a current loop whose orbital magnetic moment $\boldsymbol{\mu}_\ell$ is related to its orbital angular momentum $\mathbf{L}$ by

$$\boldsymbol{\mu}_\ell = -\left(\frac{e}{2m}\right)\mathbf{L} \qquad (45\text{-}13)$$

where e is the electron charge and m is the mass of the electron. The subscript ℓ reminds us that this magnetic moment is associated with the orbital angular momentum; the negative sign indicates that $\boldsymbol{\mu}_\ell$ and $\mathbf{L}$ are anti-parallel. Although (45-13) was derived using classical mechanics and circular orbits, it is completely valid in quantum mechanics. In addition to its orbital magnetic moment $\underline{\underline{\mu}}_\ell$ associated with its orbital angular momentum $\underline{\underline{L}}$, the electron also has a spin magnetic moment $\underline{\underline{\mu}}_s$ associated with its spin angular momentum $\underline{\underline{S}}$. This is given by

$$\underline{\underline{\mu}}_s = -\left(\frac{e}{m}\right)\underline{\underline{S}} \tag{45-14}$$

Again, this magnetic moment is anti-parallel to its associated angular momentum. Comparing Eqs. (45-13) and (45-14) we see that the spin angular momentum is twice as important as the orbital angular momentum with regard to producing a magnetic moment.

45-5 The Atom in a Magnetic Field

If a magnetic moment $\underline{\underline{\mu}}$ is in a region of magnetic field $\underline{\underline{B}}$, there is a potential energy given by $U = -\underline{\underline{\mu}}\cdot\underline{\underline{B}}$. For convenience, we take the $\underline{\underline{B}}$ field to be in the z direction. Then the potential energy associated with this magnetic moment in the magnetic field is

$$U = -\mu_z B \tag{45-15}$$

where μ_z is the z-component of $\underline{\underline{\mu}}$.

If there were only an orbital magnetic moment and no spin magnetic moment, then Eqs. (45-15) and (45-13) would give

$$U = -\mu_{\ell,z} B = -\left(-\frac{e}{2m}L_z\right) B = \frac{e}{2m}(m_\ell \hbar)\, B$$

$$U = m_\ell\, \mu_B\, B \tag{45-16}$$

where the quantity

$$\mu_B \equiv \frac{e\hbar}{2m} = 5.78\times 10^{-5}\ \text{eV/T} = 9.3\times 10^{-24}\ \text{J/T} \tag{45-17}$$

is called the Bohr magneton, a natural unit of magnetic moment for the electron. Similarly, if there were only a spin magnetic moment and no orbital magnetic moment, one would obtain $U = -\mu_z B$ or

$$U = 2\, m_s\, \mu_B\, B \tag{45-18}$$

for the potential energy. Note the additional factor of 2 in (45-18) when compared with (45-16); this arises because the spin angular momentum is twice as important as the orbital angular momentum with regard to producing a magnetic moment (see Section 45-4).

In reality, one has to deal with both types of magnetic moment. In addition, due to spin-orbit interaction, m_ℓ and m_s are no longer "good" quantum numbers. However, j and m_j remain as "good" quantum numbers. It turns out that the formula for the potential energy still resembles (45-16) and (45-18) except that it uses m_j,

$$U = g\, m_j\, \mu_B\, B \quad . \tag{45-19}$$

The quantity g appearing in this is called the (Landé) g-factor; it is given by

$$g = 1 + \frac{j(j+1) + s(s+1) - \ell(\ell+1)}{2\,j(j+1)} \quad . \tag{45-20}$$

Here j is related to the total angular momentum $\underline{\underline{J}}$, ℓ is related to the orbital angular momentum $\underline{\underline{L}}$ and $s = \frac{1}{2}$ is related to the spin angular momentum $\underline{\underline{S}}$.

>>> Example 4. Evaluate the g-factor for the following three states: $3p_{3/2}$, $3p_{1/2}$ and $3s_{1/2}$.

The fact that n = 3 is irrelevant. From Eq. (45-20),

For $3p_{3/2}$: $j = 3/2$ and $\ell = 1$. $g = 1 + \dfrac{(3/2)(3/2+1) + (1/2)(1/2+1) - (1)(1+1)}{(2)\ (3/2)(3/2+1)} = \dfrac{4}{3}$.

For $3p_{1/2}$: $j = 1/2$ and $\ell = 1$. $g = 1 + \dfrac{(1/2)(1/2+1) + (1/2)(1/2+1) - (1)(1+1)}{(2)\ (1/2)(1/2+1)} = \dfrac{2}{3}$.

For $3s_{1/2}$: $j = 1/2$ and $\ell = 0$. $g = 1 + \dfrac{(1/2)(1/2+1) + (1/2)(1/2+1) - (0)(0+1)}{(2)\ (1/2)(1/2+1)} = 2$.

Since $\ell = 0$ for the $3s_{1/2}$ state, Eq. (45-18) would apply from which it is obvious that g = 2 (note that m_s is the same as m_j if $\ell = 0$). <<<

The splitting of the energy levels due to an external magnetic field is called the Zeeman effect. According to Eqs. (45-19) and (45-20), states of a given ℓ and j will split into $2j+1$ equally spaced levels (the number $2j+1$ is the number of m_j's belonging to that j). The magnitude of the separation between adjacent levels is proportional to the magnetic field B, the proportionality constant being a complicated function of ℓ and j according to (45-20). We shall now consider the Zeeman effect for sodium in more detail. Sodium has 11 electrons; 10 of them fill the 1s, 2s and 2p levels. These 10 electrons have a total magnetic moment of zero. Thus we treat the 11th electron (the "valence" electron) as if it were in a hydrogen-like atom. The next shell is n = 3. For sodium (as opposed to hydrogen), the 3s level lies below the 3p level. When spin-orbit interaction is taken into account the 3p level splits into a $3p_{3/2}$ and a $3p_{1/2}$ level with the former being higher in energy. See Figure 45-1. As for the 3s level, there is no spin-orbit interaction ($\underline{L}\cdot\underline{S} = 0$ since $\ell = 0$); this state has $j = \frac{1}{2}$, thus it is now labelled $3s_{1/2}$. When the external magnetic field $\underline{B}$ is applied, the $3p_{3/2}$ state splits into four equally spaced levels corresponding to $m_j = -3/2, -1/2, 1/2, 3/2$. Similarly, the $3p_{1/2}$ and $3s_{1/2}$ states each split into two levels corresponding to $m_j = -1/2, 1/2$. The magnitude of the splitting of the $3p_{3/2}$ state is proportional to its g-factor (g = 4/3 from example 4). Similarly the slitting of the $3p_{1/2}$ state is proportional to 2/3 and that of the $3s_{1/2}$ state proportional to 2.

Figure 45-1. Some energy levels for sodium. The 3p state is split into a $3p_{3/2}$ and a $3p_{1/2}$ state by the spin-orbit interaction. Under an applied $\underline{B}$ field, these two and the $3s_{1/2}$ states split further according to their m_j values. Vertical lines indicate possible transitions which result in an emitted photon. (Since the photon also has an angular momentum of 1 unit of $\hbar$, only certain transitions are allowed: both $\Delta\ell = \pm 1$ and $\Delta m_j = 0, \pm 1$ must be obeyed; these are known as "selection rules".)

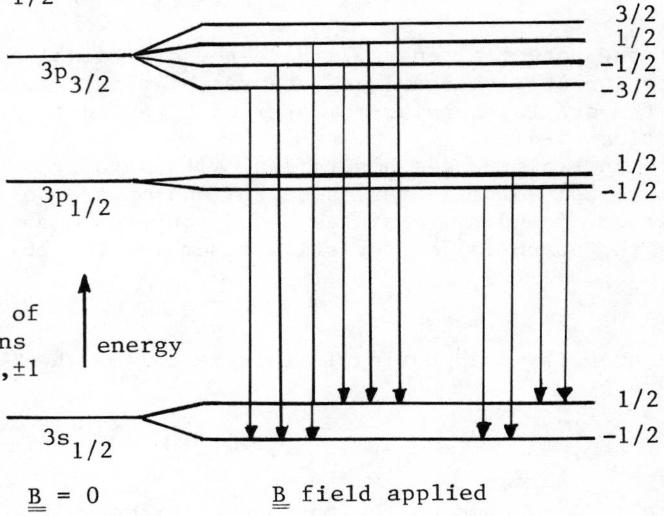

45-6 The Stern-Gerlach Experiment

In section 32-4 we saw that although there is no force exerted on a magnetic moment by a uniform magnetic field, there is a force exerted on a magnetic moment by a non-uniform magnetic field. To obtain the amount of this force we start with the potential energy of a magnetic moment $\underline{\underline{\mu}}$ in a magnetic field $\underline{\underline{B}}$, $U = -\underline{\underline{\mu}}\cdot\underline{\underline{B}} = -\mu_z B_z$ where, for convenience, we take $\underline{\underline{B}}$ to have only a z-component. The z-component of the force $\underline{\underline{F}}$ acting on this magnetic moment is, from Eq. (7-9),

$$F_z = -\,\partial U/\partial z = \mu_z\ \partial B_z/\partial z\quad . \tag{45-21}$$

As anticipated, the force depends on the non-uniformity of the magnetic field through the factor $\partial B_z/\partial z$. For the case of an atom, we can substitute the expression (45-19) for U into Eq. (45-21). This gives

$$F_z = g\ m_j\ \mu_B\ \partial B_z/\partial z \tag{45-22}$$

where $\mu_B = e\hbar/2m$ is the Bohr magneton. This relation takes into account the fact that there are both orbital and spin contributions to the effective magnetic moment.

In the actual experiment, a beam of atoms moving in the x-direction passes through a region in which the $\underline{\underline{B}}$ field is in the z-direction and is highly non-uniform. According to Eq. (45-22), the force F_z exerted on an atom by the magnetic field is quantized into $2j+1$ possible values according to the value of m_j. Thus the beam of atoms "splits" into $2j+1$ sub-beams as shown schematically in Figure 2. This provides direct evidence that the z-component of $\underline{\underline{J}}$, $J_z = m_j\hbar$, is quantized; this is known as "space quantization".

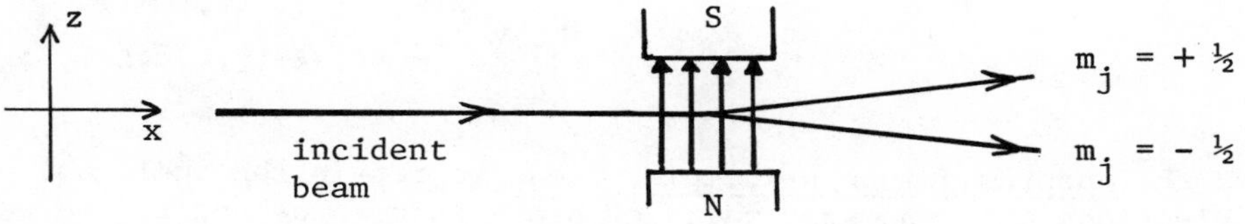

Figure 45-2. Stern-Gerlach experiment. The incident beam of atoms moves through a region of non-uniform $\underline{\underline{B}}$. The $\underline{\underline{B}}$ field is in the z-direction (from north pole N to south pole S). The pole faces of the magnet are shaped so as to make B_z strong at the S pole and weak at the N pole; thus $\partial B_z/\partial z$ is positive in this case. In this particular illustration, $j = \frac{1}{2}$ so that the beam splits into $2j+1 = 2$ sub-beams corresponding to $m_j = +\frac{1}{2}$ and $m_j = -\frac{1}{2}$. A beam of silver atoms, for example, would exhibit this behavior.

>>> Example 5. Estimate the angular separation which occurs in a typical Stern-Gerlach experiment. Take the mass of the atom to be $M = 1.8\times10^{-25}$ kg (silver), the velocity to be $v_x = 800$ m/s, $\partial B_z/\partial z = 1000$ T/m, the length of the magnetic field region (along the x-axis) to be X = 3 cm, g = 2 and $j = \frac{1}{2}$.

Each atom is in the magnetic field for a time $t = X/v_x = (3\times10^{-2}\text{ m})/(800\text{ m/s}) = 3.8\times10^{-5}$ s. During this time, there is an acceleration in the z-direction

$$\begin{aligned} a_z &= F_z/M = (g\ m_j\ \mu_B\ \partial B_z/\partial z)/M \\ &= (2)(\tfrac{1}{2})(9.3\times10^{-24}\text{ J/T})(10^3\text{ T/m})/(1.8\times10^{-25}\text{ kg}) = 5.2\times10^4\text{ m/s}^2\quad . \end{aligned}$$

The atom therefore acquires a velocity in the z-direction

$$v_z = a_z t = (5.2\times10^4\ m/s^2)(3.8\times10^{-5}\ s) = 2.0\ m/s\ .$$

Therefore the angle between this beam and the incident beam is (for small angles)

$$\theta = v_z/v_x = (2.0\ m/s)/(800\ m/s) = 0.0025\ radian$$

$$\theta = 0^o\ 9'\ .$$

Although small, this angular deviation of the beam can be seen in the experiment. <<<

45-7 The Pauli Exclusion Principle

The electron in a hydrogen atom is described by its four quantum numbers. These can be chosen as (n,ℓ,m_ℓ,m_s) if we ignore spin-orbit interaction or (n,ℓ,j,m_j) if spin-orbit interaction is taken into account. The important point here is that four quantum numbers are required to completely describe the state of the electron.

In a many-electron atom, the situation is more complicated. Not only does each electron interact with the positively charged nucleus but also each electron interacts with every other electron. Further, this electron-electron interaction is not only due to the Coulomb potential but is also due to the magnetic moments (both orbital and spin) of the electrons. However, it is still possible to assign each electron a set of four hydrogen-like quantum numbers. That is, we can associate with each electron the quantum numbers

$$(n,\ell,m_\ell,m_s) \quad \text{where:} \quad \begin{aligned} n &= 1,2,3,\ldots; \\ \ell &= 0,1,2,\ldots,(n-1); \\ m_\ell &= -\ell,-\ell+1,\ldots,\ell-1,\ell; \\ m_s &= \pm\tfrac{1}{2}\ . \end{aligned}$$

This is called the <u>quantum number principle</u>. We also retain the shell and subshell terminology: electrons with the same value of n are said to belong to the same <u>shell</u>; electrons with the same values of the pair (n,ℓ) are said to belong to the same <u>subshell</u>.

There is, however, one further restriction on the allowed quantum numbers which arises in the case of a many electron atom. In order to explain the observed filling of the shells and subshells (see Section 45-8), it is necessary that

"no two electrons in the same atom have the same set of four quantum numbers". (45-23)

This is the content of the <u>Pauli exclusion principle</u>.

>>> Example 6. What is the maximum number of electrons which may occupy the L shell?

The L shell means n = 2. For this value of n, the possible (ℓ,m_ℓ,m_s) values are: $(0,0,-\tfrac{1}{2})$, $(0,0,+\tfrac{1}{2})$, $(1,-1,-\tfrac{1}{2})$, $(1,-1,+\tfrac{1}{2})$, $(1,0,-\tfrac{1}{2})$, $(1,0,+\tfrac{1}{2})$, $(1,1,-\tfrac{1}{2})$ and $(1,1,+\tfrac{1}{2})$. According to the Pauli exclusion principle, there is a maximum of eight electrons in this shell. <<<

45-8 The Periodic Table

If we ignore the small effect of the spin-orbit interaction, the energy of an electron in a given state of the hydrogen atom depends only upon the principal quantum number n (see Eq. (45-3)). The energy increases (i.e. becomes less negative) with increasing n; the orbital quantum number ℓ (as well as the other two quantum numbers) does not enter into the formula for the energy.

For a many-electron atom, it turns out that the energy again depends upon n (energy increasing with increasing n) but also depends upon ℓ. In order to see how this ℓ dependence arises, consider the two orbits (belonging to the same value of n) shown in Figure 3. The electron in orbit #1 (a circular orbit) has a larger value of ℓ than the electron in orbit #2 (a highly elliptical orbit). The shaded region in the figure indicates the inner electrons in this many-electron atom. Since orbit #2 penetrates this shaded region more than orbit #1, we see that electron #1 is "shielded" from the full Coulomb force of the nucleus more than is electron #2. Consequently, electron #2 has a lower (more negative) energy than electron #1. We therefore have the following rule

$$\text{For a given principal quantum number n, the order of increasing energy is the order of increasing orbital quantum number } \ell. \qquad (45\text{-}24)$$

Figure 45-3. Two electron orbits belonging to the same value of n. The circular orbit (#1) has a larger value of ℓ than the highly elliptical orbit (#2). The electron in orbit #2 penetrates the inner electrons (shaded region) and is therefore less shielded from the nucleus than is the electron in orbit #1.

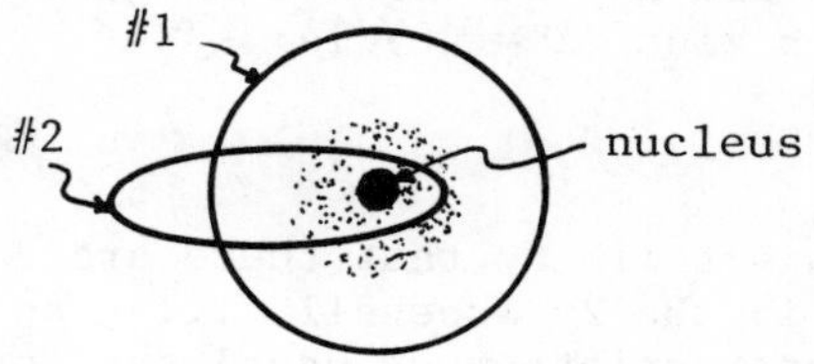

To form an atom of atomic number Z (i.e. whose nucleus contains Z protons), we must surround the nucleus with Z electrons. Because of the Pauli exclusion principal (45-23), no two electrons can have the same set of four quantum numbers. For any given subshell, i.e. a given n and ℓ, the maximum number of electrons is $2(2\ell+1)$. This is because there are $2\ell+1$ possible values of m_ℓ (from $-\ell$ to ℓ) and for each of these values of m_ℓ, there are two possible values of m_s ($-\frac{1}{2}$ and $+\frac{1}{2}$). Thus for a given subshell,

symbol	value of ℓ	maximum number of electrons
s	0	2
p	1	6
d	2	10
f	3	14

The subshells fill in order of increasing energy. Generally this means that those with lower values of n fill first and for a given n, from (45-24), those with lower values of ℓ. The actual order in which the subshells are filled is the following:

symbol	n	ℓ	maximum number of electrons
1s	1	0	2
2s	2	0	2
2p	2	1	6
3s	3	0	2
3p	3	1	6
4s	4	0	2
3d	3	2	10
4p	4	1	6
5s	5	0	2
4d	4	2	10
5p	5	1	6
6s	6	0	2
4f	4	3	14
5d	5	2	10
6p	6	1	6
7s	7	0	2
5f	5	3	14
6d	6	2	10

Note that the ℓ dependence (45-24) is so important that, e.g., the 4s subshell is filled before the 3d subshell. To describe the filling of the subshells in an atom we use an "electron configuration notation". For example, the electron configuration for zinc (Z = 30) is:

$$1s^2 2s^2 2p^6 3s^2 3p^6 4s^2 3d^{10} \quad .$$

This tells us that there are 2 electrons in the 1s subshell, 2 in the 2s subshell, 6 in the 2p subshell, ... , and finally 10 in the 3d subshell. The sum of the superscripts must equal the total number of electrons, i.e. the atomic number (Z = 30 in this case).

>>> Example 7. Write the electron configuration for carbon (Z = 6) and for potassium (Z = 19).

Following the above rules, we have

$$\text{for carbon: } 1s^2 2s^2 2p^2$$
$$\text{for potassium: } 1s^2 2s^2 2p^6 3s^2 3p^6 4s^1 \quad .$$ <<<

Although the above rules are remarkably successful, there are a few atoms for which the order of filling of the subshells is different (see text).

Chapter 46: ELECTRICAL CONDUCTIVITY IN SOLIDS

REVIEW AND PREVIEW

In the previous chapter we applied quantum mechanics to the electrons in an atom. This led to the concept of energy levels and the periodic table. In this chapter we apply quantum mechanics to the electrons in a solid as opposed to an isolated atom. We seek to explain why some solids are excellent insulators while others are excellent conductors. We shall also investigate some properties of semiconductors, the basis of today's solid state electronics.

GOALS AND GUIDELINES

Again, due to the complexity of the subject, the treatment omits most derivations. The major goal is to develop an understanding of why different solids can have such drastically different electrical conductivity properties. The main topics to be studied are

1. Free-electron model
   (a) state distribution function $n_s(E)$ and the Fermi energy E_F,
   (b) Fermi-Dirac probability function $p(E)$,
   (c) particle distribution function $n_p(E)$.

2. Band structure of solids
   (a) how band structure arises,
   (b) band structure for conductors,
   (c) band structure for insulators,
   (d) band structure for semiconductors.

3. Semiconductors
   (a) intrinsic semiconductors,
   (b) doping: n-type and p-type semiconductors.

4. Semiconducting devices
   (a) diode
   (b) MOSFET transistor

46-1 Free-electron Model

In the free-electron model of a metal, the conduction electrons are treated as if they were free (i.e. not under the influence of any potential energy). An obvious extension of Eq. (45-2b) to the case of a three dimensional box of side length ℓ gives

$$E = (n_x{}^2 + n_y{}^2 + n_z{}^2)\ h^2/(8m\ell^2) \qquad (46\text{-}1)$$

for the allowed energies of an electron in this "solid". Here the quantum numbers n_x, n_y and n_z each take on the values 1,2,3,4,... For a macroscopic sample ($\ell\to\infty$), the spacing between adjacent energy levels is extremely small. To describe this, we introduce the state distribution function $n_s(E)$:

$n_s(E)dE$ is the number of states (per unit volume of sample) whose energy lies in the range E to E + dE. (46-2)

For the free-electron model, it turns out that the state distribution function is given by

$$n_s(E) = (35.5\ m^{3/2}/h^3)\ E^{\frac{1}{2}} \quad . \qquad (46\text{-}3)$$

The state distribution function $n_s(E)$ increases with increasing energy according to $E^{\frac{1}{2}}$. The reason for this increase can be seen from Eq. (46-1): the number of states in a given energy range depends upon the number of combinations of the quantum numbers (n_x, n_y, n_z) which correspond to this energy range; as E increases there are many more such combinations. The following example illustrates this and qualitatively verifies the $E^{\frac{1}{2}}$ dependence.

>>> Example 1. Take $h^2/8m\ell^2 = 1$ in Eq. (46-1) or, equivalently, measure energy in units of $h^2/8m\ell^2$ for simplicity. (a) Calculate the number of states in the energy range $E = 25 \pm 5$. (b) Repeat for the energy range $E = 100 \pm 5$. (c) Compare the ratio of these two answers with the ratio of the square roots of the corresponding (nominal) energies.

(a) From Eq. (46-1) we see that the answer is the number of combinations of the three quantum numbers which satisfy $n_x^2 + n_y^2 + n_z^2 = 25 \pm 5$. We first list those solutions which obey $n_x \geq n_y \geq n_z$. These are

(3,3,2), (4,2,1), (4,3,1), (5,1,1),
(3,3,3), (4,2,2), (4,3,2), (5,2,1).

The total number of states is the number of distinct permutations of these. Noting that if the three quantum numbers are all different there are six such permutations, if two of the three quantum numbers are the same there are three such permutations and if all three quantum numbers are the same there is one permutation, we have

number of states = (6)(4 states) + (3)(3 states) + (1)(1 state) = 34 states.

(b) Similarly for $E = 100 \pm 5$, the solutions which obey $n_x \geq n_y \geq n_z$ are

(6,6,5), (7,7,1), (8,5,3), (8,6,2), (9,4,2),
(7,5,5), (7,7,2), (8,5,4), (9,3,3), (10,1,1),
(7,6,4), (8,4,4), (8,6,1), (9,4,1), (10,2,1).

Allowing for the distinct permutations, we have

number of states = (6)(8 states) + (3)(7 states) + (1)(0 states) = 69 states.

(c) The ratio of these two answers is 69/34 = 2.03. The ratio of the square root of the corresponding (nominal) energies is $(100/25)^{\frac{1}{2}} = 2.00$. We see that the proportionality between the number of states per unit energy range and $E^{\frac{1}{2}}$ is verified. <<<

Because of the Pauli exclusion principle, there cannot be more than one electron in each quantum state (provided m_s is included as one of the quantum numbers). To obtain the lowest energy state for the system, the individual electrons states are filled starting with the lowest energy state. Figure 1 shows a graph of the state distribution function $n_s(E)$ as a function of E (it is conventional to plot E on the vertical axis in these graphs). The total number of electrons per unit volume (without regard to energy) is the sum (integral) of the occupied individual electron states,

$$n_o = \int_0^{E_F} n_s(E)\, dE \,. \tag{46-4}$$

This can be interpreted as the area to the left of the curve in the figure, as indicated by the shaded region. The upper limit of integration, denoted by E_F, is called the Fermi energy. Note that the Fermi energy is the energy of the most energetic conduction electron when the system is in its lowest energy state.

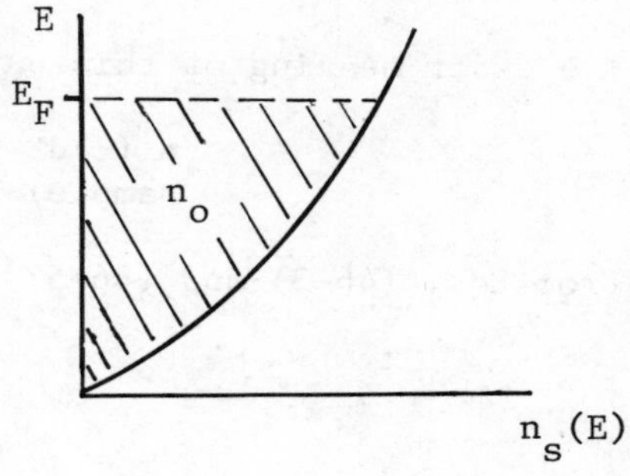

Figure 46-1. The state distribution function $n_s(E)$ as a function of the energy E. The total number of electrons per unit volume, n_o, is the shaded area. The assumption is that the states are filled from below consistent with the Pauli exclusion principle; this yields the lowest energy for the system. The particular energy E_F is called the Fermi energy.

The above analysis is appropriate for the lowest energy of the system, i.e. for a temperature T = 0 K. To treat the case of a finite temperature we need to know the probability that a given energy level is occupied. This probability is given by the Fermi-Dirac probability function p(E),

$$p(E) = \frac{1}{e^{(E - E_F)/kT} + 1} \quad . \tag{46-5}$$

Here T is the (absolute) temperature, k is Boltzmann's constant and E_F is the Fermi energy. Note that

$$\text{for } T \to 0,\ p(E) \to \begin{cases} 1, & \text{if } E < E_F \\ 0, & \text{if } E > E_F \end{cases} \quad . \tag{46-6}$$

Figure 2 shows the Fermi-Dirac probability function p(E) as a function of the energy E. In Fig. 2a we see p(E) for T = 0 while Fig. 2b shows p(E) for a finite temperature T. Note that the finite temperature case differs from the zero temperature case only in an energy region of about kT about the Fermi energy. Since a typical Fermi energy is several electron volts while kT is only a fraction of an electron volt (kT = 1/40 eV at room temperature), the two curves in the figure are almost identical.

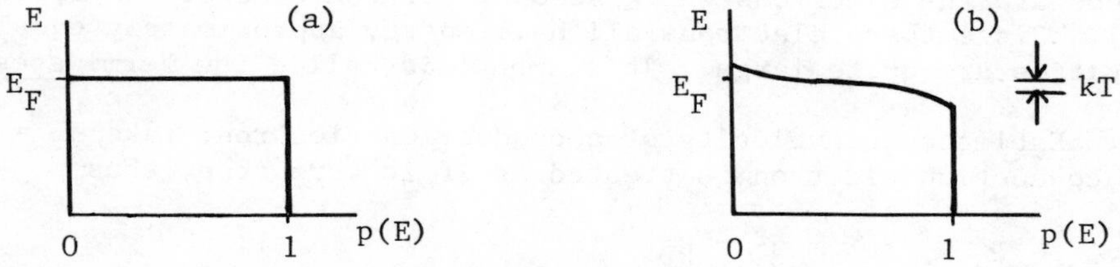

Figure 46-2. The Fermi-Dirac probability function p(E) as a function of the energy E. (a) The T = 0 case. (b) The finite temperature case.

The state distribution function, $n_s(E)$, gives the number of states (per unit volume of sample) per unit energy. The Fermi-Dirac probablilty function, p(E) gives the probability that a given energy state is occupied. If we multiply these two, we obtain the particle distribution function $n_p(E)$

$$n_p(E) = n_s(E)p(E) \quad . \tag{46-7}$$

The exact meaning of this particle distribution function is:

$n_p(E)dE$ is the number of occupied states (per unit volume of sample) whose energy lies in the range E to E + dE. (46-8)

From Eqs. (46-3) and (46-5) the particle distribution function is

$$n_p(E) = C \frac{E^{1/2}}{e^{(E - E_F)/kT} + 1} \tag{46-9}$$

where the constant $C = 6.78 \times 10^{27}\ m^{-3}\ eV^{-3/2}$. Figure 3 shows the particle distribution function $n_p(E)$ as a function of the energy E. Due to the nature of p(E), the particle distribution function essentially coincides with the state distribution function $n_s(E)$ for all energy up to within a small range (of the order kT) about the Fermi energy E_F where it rapidly falls to zero.

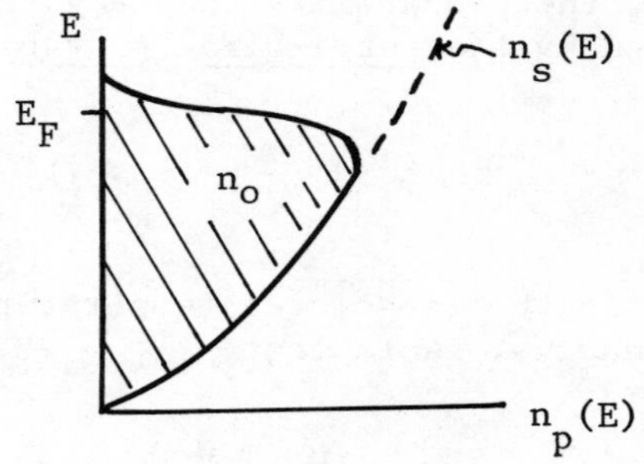

Figure 46-3. Particle distribution function $n_p(E)$ as a function of E. The dotted curve indicates the state distribution function $n_s(E)$. The shaded region indicates the total number of electrons per unit volume, n_o.

When an external electric field is applied to this sample, some electrons can acquire an additional energy which is typically of the order of a fraction of an eV (say 1 meV). This is extremely small on the scale of Figure 3 since a typical Fermi energy is several eV. Those electrons in states somewhat below the Fermi energy cannot absorb this additional dnergy since all the adjacent states are occupied. In effect, these electrons are "frozen" at their present energy levels. Only those electrons near the Fermi energy can find unoccupied states nearby in energy. It is these electrons which are affected by an external electric field; thus these are the electrons which account for the conduction properties of the material. Since these electrons all have energy approximately equal to E_F, their velocities are quite large. This amount is called the <u>Fermi speed</u>.

\>>> Example 2. Estimate the velocity of a conduction electron; take $E_F = 5$ eV.
Solution: The conduction electron is treated as if it were free, thus

$$\tfrac{1}{2} m v^2 = E_F$$

$$v = (2E_F/m)^{1/2}$$

$$= [(2)(5\ eV)(1.6 \times 10^{-19}\ J/eV)/(9.11 \times 10^{-31}\ kg)]^{1/2}$$

$$= 1.33 \times 10^6\ m/s \quad .$$

<<<

46-2 Band Structure of Solids

The free-electron model, discussed in the previous section, is a reasonable one for conductors. It does not, however, explain why certain solids are conductors while others are insulators. A better model, called the band theory model, can adequately predict the electrical conductivity properties of solids; this is because this band theory model takes into account the discrete energy levels of the constituent atoms.

As an example, consider an isolated atom of copper. The electron configuration for this atom (one exception to the rules of the previous chapter!) whose $Z = 29$ is $1s^2 2s^2 2p^6 3s^2 3p^6 3d^{10} 4s^1$. This is shown in Figure 46-4a where the electrons (solid circles) are shown occupying these energy states; similarly, unoccupied states are indicated with open circles. Note that the 4s state can accomodate two electrons but only one is actually occupied. When a large number, N, of these copper atoms are combined into a crystal there is an interaction between the atoms causing the individual energy levels to split into "bands". The N 4s levels, each of which can accomodate two electrons form a band which can accomodate 2N electrons; since there are only N 4s electrons (one from each atom) this band will be only half full as shown in Figure 46-4b. This furnishes the band model for a conductor: since there are many unoccupied energy levels adjacent to the occupied ones, electrons can easily gain energy due to an applied electric field.

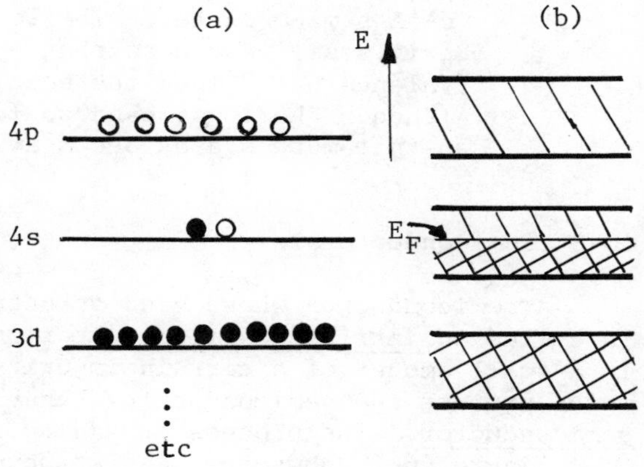

Figure 46-4. (a) Energy levels for an isolated copper atom. Solid circles indicate occupied states, open circles indicate unoccupied states. (b) The copper atoms combine to form a crystal. The energy levels combine to form "bands". The band corresponding to the 4s states is half full. Vacant (unoccupied) states and occupied states are indicated by shading as follows:

Other possible situations which can occur are shown in Figure 46-5. Fig. 46-5a shows a conductor (similar to that shown in the previous figure). Fig. 46-5b shows the band structure for an insulator. Here the highest occupied band is full; this is separated from the next band by an energy gap E_g. Since $E_g \gg kT$, electrons cannot be easily thermally excited into this higher band. In effect, the electrons are "frozen" into their present energy levels; hence the sample acts as an insulator. Fig. 46-5c shows the situation for a semiconductor. Here the energy gap E_g is smaller than that for an insulator. Some electrons can now be thermally excited into the upper band. These few electrons can participate in the electric conduction mechanism. Referring to the figure, we see that these electrons have been thermally excited from the "valence band" into the "conduction band". Finally, Fig. 46-5d shows another type of conductor. Here the bands "overlap" so that electrons can easily move from one band to the next.

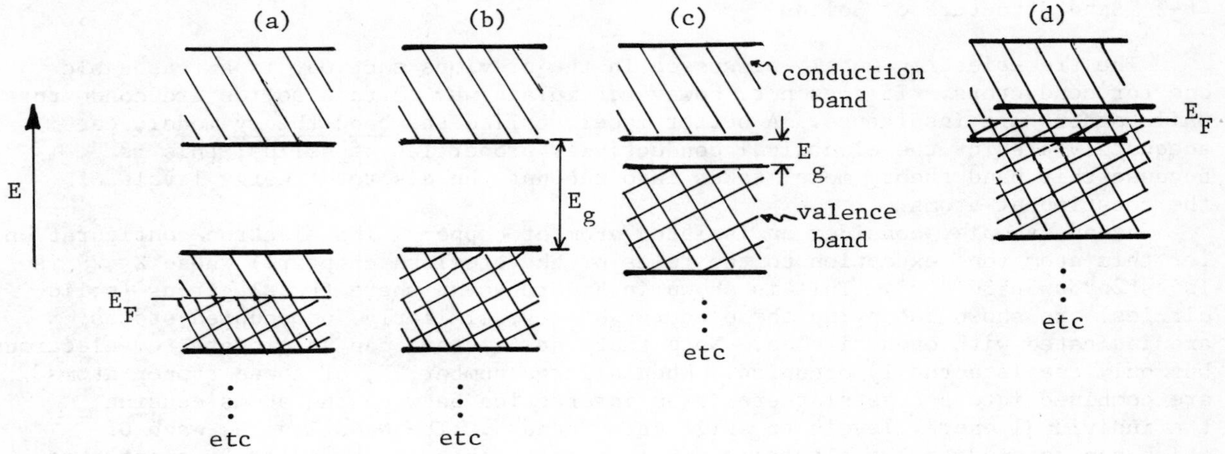

Figure 46-5. Some possible energy bands. Occupied states are indicated by double shading, unoccupied states are indicated by single shading. (a) A conductor. The lower band shown is half full. (b) An insulator. The lower band shown is completely full. The energy gap E_g obeys $E_g >> kT$. (c) A semiconductor. The lower band is completely full. The energy gap E_g is small enough so that some electrons are thermally excited from this "valence band" into the next "conduction band". (d) Another conductor. Although the lower band would normally be full, it overlaps the next band so that more states becom available.

46-3 Semiconductors

A semiconductor whose band structure is similar to that shown in Fig. 46-5 is called an intrinsic semiconductor; silicon is an example of such a material. If a small amount of a certain impurity is added, additional energy levels can appear which promote the semiconducting behavior; we call such a material an extrinsic semiconductor. The process of adding this impurity is called doping the semiconductor.

Figure 46-6 shows the band structure for two types of extrinsic semiconductors. In Fig. 46-6a a small amount of phosphorous has been added to silicon. The phosphorous contains one more valence electron than does the silicon. As a result there is an extra energy level immediately below the conduction band. This occupied energy level can readily "donate" its electron to the conduction band thus contributing to the semiconducting behavior of the material. We call this an n-type semiconductor since the doping material contributed an electron (a negative charge). The small energy difference between the "donor" level and the conduction band is denoted by E_d. In Fig. 46-6b a small amount of aluminum has been added to silicon. The aluminum contains one less valence electron than does the silicon. As a result there is an extra energy level immediately above the valence band. This unoccupied energy level can readily "accept" an electron from the valence band. This leaves behind an unoccupied state in the valence band; this unoccupied state, called a "hole", acts as a positive charge carrier in the valence band thus contributing to the semiconducting behavior of the material. We call this a p-type semiconductor since the doping material contributes a hole (a positive charge). The small energy difference between the "acceptor" level and the valence band is denoted by E_a.

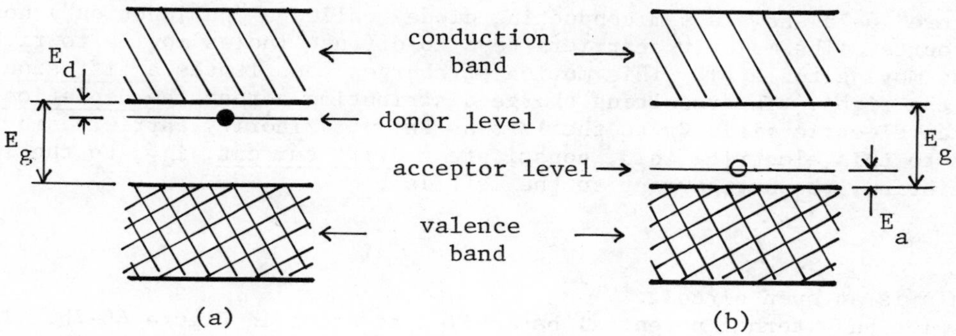

Figure 46-6. Two types of extrinsic semiconductors. (a) n-type semiconductor. Since E_d is small, the electron in the donor state can readily be donated to the conduction band. (b) p-type semiconductor. Since E_a is small, the acceptor state can readily accept an electron from the valence band; this leaves behind a positive hole in the valence band.

46-4 Semiconducting Diode

In this section we discuss the semiconducting diode. This consists of a junction of a p-type semiconductor with an n-type semiconductor. Each type has three different kinds of particles of interest as listed in the following table.

	Particle	Charge	Origin	Remarks	Excitation Energy
n-type	electron	-e	electron donated from donor level into conduction band	majority carrier	E_d
	hole	+e	electron excited from valence to conduction band, hole remains in valence band	minority carrier	E_g
	ion	+e	donor atom loses electron	not mobile	E_d
p-type	hole	+e	electron accepted from valence band into acceptor level, hole remains in valence band	majority carrier	E_a
	electron	-e	electron excited from valence to conduction band	minority carrier	E_g
	ion	-e	acceptor atom gains electron	not mobile	E_a

Figure 46-7a shows a semiconducting diode (called a "pn junction") not connected to any source. The majority carriers tend to diffuse (holes moving to right and electrons moving to left). This motion of charges constitutes a diffusion current, i_{df}, to the right. The resulting charge distribution across the depletion layer, d_o, creates an electric field E_o to the left as shown. Minority carriers, moving in response to this electric field, constitute a drift current, i_{dr}, to the left. At equilibrium, the net current to the left is

$$i_o = i_{dr} - i_{df} = 0$$

since this is an open circuit.

Now let an external potential be applied as shown in Figure 46-7b. The positive terminal of the external potential is connected to the "n" terminal of the junction; in this situation we say that the junction is "back biased". The effect of this is to increase the electric field across the junction to E_B. The diffusing charges (majority carriers) now have a higher potential barrier to cross. As a result the depletion layer enlarges to d_B and the diffusion current decreases compared to that in Fig. 46-7a. However, the drift current, i_{dr}, which depends upon the minority charges remains the same. The net current to the left is

$$i_B = i_{dr} - i_{df} < 0$$

and is rather small.

Now let the external potential be applied as shown in Figure 46-7c. The positive terminal of the external potential is connected to the "p" terminal of the junction; in this situation we say that the junction is "forward biased". The effect of this is to decrease the electric field across the junction to E_F. The diffusing charges (majority carriers) now have a lower potential barrier to cross. As a result the depletion later shrinks to d_F and the diffusion current increases compared to that in Fig. 46-7a. However, the drift current, i_{dr}, which depends upon the minority charges remains the same. The net current to the left is

$$i_F = i_{dr} - i_{df} > 0$$

and is rather large.

Note that, for a given amount of external potential, the current through the junction is large if the polarity corresponds to forward biasing and small if it corresponds to backward biasing; hence the junction acts as a diode rectifier.

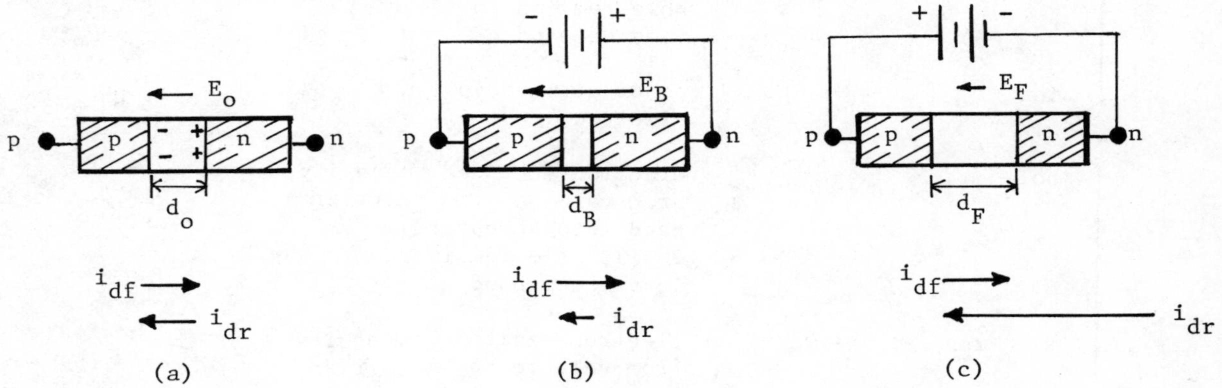

Figure 46-7. A pn junction. (a) no external source, (b) back biased, (c) forward biased.

46-5 The MOSFET Transistor

In this section we discuss a three terminal semiconducting device called the MOSFET (metal-oxide-semiconductor-field effect-transistor). This is shown in Figure 46-8. A p-type substrate contains a source S and a drain D, both of n-type material, connected by a thin channel of n-type material called the n-channel. An insulating layer of SiO_2 covers this. Metal contacts allow access to S and D; a thin metallic layer called the gate G is above the SiO_2 near the n-channel. A relatively small change in the gate potential can control the drain-to-source current i_{ds}. The mechanism for this is explained in the next paragraph.

Suppose that the potential of S as well as the substrate is 0 V and that of the drain is 10 V. The potential of the gate is taken to be 1 V. Then the gate is more negative than most of the n-channel and the electric field $\underline{\underline{E}}$ in the n-channel will be toward the gate G. Now if the gate potential is reduced to, say, 0 V this electric field will be larger in magnitude. The effect of this larger electric field is to repel conduction electrons out of the n-channel (toward S or D). This reduces the number of charge carriers in the n-channel and thus increases its resistance. Therefore the current i_{ds} decreases.

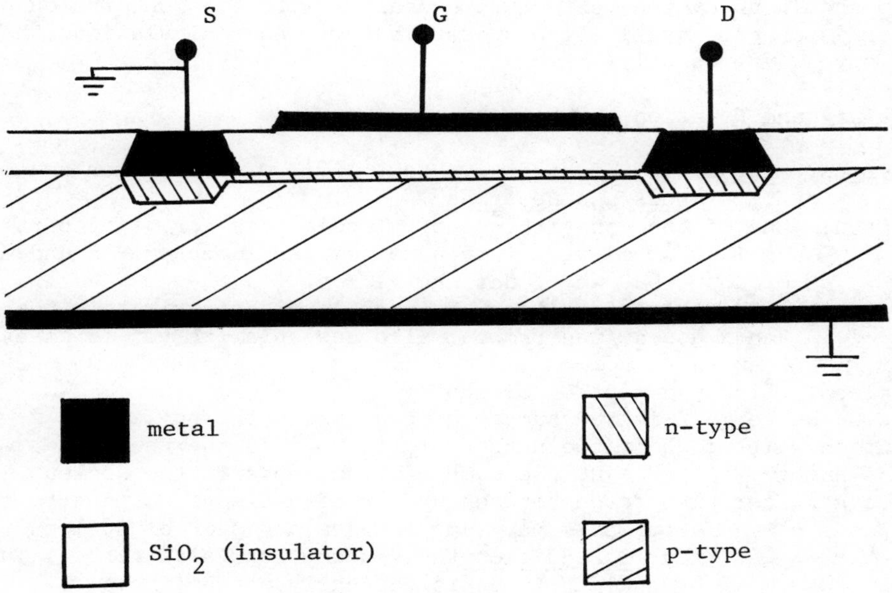

Figure 46-8. MOSFET

Chapter 47: NUCLEAR PHYSICS

REVIEW AND PREVIEW

In previous chapters you have studied a few properties of atoms. These properties are largely dictated by the electrons of the atom moving under the influence of the coulomb force. In this chapter you will study some properties of the nucleus of the atom. The coulomb force still plays a role, but a minor one in comparison with the much stronger nuclear force. You will study some basic properties of nuclei including reactions. This will finish with a look at nuclear energy.

GOALS AND GUIDELINES

In this chapter there are many things for you to learn but basically no new physical principles are involved.

1. Vocabulary: There will be many new names to learn. Section 47-1 starts you off with some basic ones.
2. You should study the binding energy section (47-2) with particular care since the ideas there are central to energy considerations in all that follows.
3. In Section 47-3 note the parallelism between decay and reaction energetics. You will need to learn the mechanics of Q-value calculations.
4. Section 47-4 deals with the use of nuclide charts.
5. Study the radioactive decay law in Section 47-5 with care.
6. You should study the last section dealing with fission, fusion and nuclear energy for good qualitative understanding. Any calculations there will largely involve Q-values only.

47-1 Vocabulary and Notation

Nucleus: The center of an atom, about 10^{-14} m in diameter and occupied by protons and neutrons.

Proton: One of the constituents of a nucleus; a very tiny particle ($r \sim 10^{-15}$ m) with intrinsic angular momentum $\frac{1}{2}\hbar$ and mass energy of about 940 MeV. Has charge + e.

Neutron: The other constituent of a nucleus; approximately the same size and mass as the proton; also has spin $\frac{1}{2}\hbar$ but is electrically neutral.

Nucleon: Either a neutron or proton.

Atomic Number: Z is the number of protons in the nucleus.

Neutron Number: N is the number of neutrons in the nucleus.

Mass Number: A = Z + N, the number of nucleons in the nucleus.

Strong Force: The force between a pair of nucleons not due to charge; apparently the same between any pair of nucleons. Protons have in addition the coulomb repulsive force. The strong or nuclear force is basically attractive.

Nuclides: The collection of all nuclei.

Isotope: Any of a set of nuclei having the same Z but different N and hence A.

Isotone: Any of a set of nuclei having the same N but different Z and hence A.

Isobar: Any of a set of nuclei having the same A but differing in N and Z.

A given nucleus is denoted by placing A as a left superscript to the chemical symbol used to denote the atomic number. (A neutral atom has Z electrons.) Example: C stands for carbon whose atomic number is 6. There are known isotopes of carbon from ^{10}C to ^{16}C.

In each case one finds N from N = A - 6.

A chart of all known nuclides may be prepared by graphing Z versus N as boxes.* A small section of such a chart is shown as Figure 47-1. This has been simplified by elimination of considerable information which is normally presented. Notice that <u>isotopes</u> occur on horizontal lines, that <u>isotones</u> occur as a vertical column. <u>Isobars</u> occur as diagonally adjacent boxes. The absence of an entry means that that nuclide neither occurs naturally nor has been produced in the laboratory.

A given nucleus can be thought of very roughly as a small sphere of radius of the order of 10^{-14} m. A more convenient unit of size here is the <u>fermi</u> (abbreviated fm or F) defined by

$$1 \text{ fermi} = 10^{-15} \text{ m} .$$

Z ↑ \ N →	3	4	5	6	7	8
8			^{13}O	^{14}O	^{15}O	^{16}O
7			^{12}N	^{13}N	^{14}N	^{15}N
6	^{9}C	^{10}C	^{11}C	^{12}C	^{13}C	^{14}C
5	^{8}B	^{9}B	^{10}B	^{11}B	^{12}B	^{13}B
4	^{7}Be	^{8}Be	^{9}Be	^{10}Be	^{11}Be	^{12}Be

Figure 47-1. Portion of Nuclide Chart

Because of quantum effects the nucleus does not have a sharp edge. The distribution of nucleons is roughly constant within some volume and then falls to zero in a rather fuzzy surface zone. One can, however, define a nuclear radius which is given by

$$R = R_o A^{1/3} \tag{47-1}$$

which says the nuclear volume $V \propto A$. Nucleons are tightly packed in the nucleus. This formula with $R_o \simeq 1.1$ fm applies for roughly $20 \lesssim A$.

47-2 Nuclear Masses and Binding Energies

Nuclei are either <u>stable</u> or <u>unstable</u>. Stable nuclei occur naturally. Some unstable nuclei also occur in nature, but all unstable nuclei decay ultimately into other nuclei, eventually into stable nuclei. Except for the lightest nuclides (small A) N > Z for stable nuclei because of coulomb repulsion.

For any nucleus the binding energy, E_B, is the energy necessary to break the nucleus into Z free protons and N free neutrons. Thus

$$E_B = Zm_p c^2 + Nm_n c^2 - m_A c^2 \tag{47-2a}$$

where

$$m_p = \text{proton mass}$$
$$m_n = \text{neutron mass}$$
$$m_A = \text{nuclear mass} .$$

Nuclear masses are difficult to measure but atomic masses are relatively easily measured to great precision. The basis of the atomic mass scale is the neutral atom ^{12}C whose mass is 12 <u>unified atomic mass units</u> (12 u).

*A complete chart has been prepared by D.T. Goldman and J.R. Roesser and is available from Educational Relations, General Electric Company, Schenectady, New York, 12305.

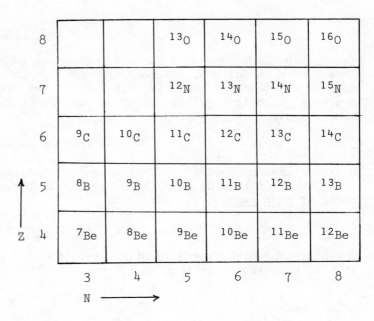

$$1\ u = 1.66 \times 10^{-27}\ kg \sim 931\ MeV\ .$$

The latter means $(1\ u)c^2 = 931$ MeV. The mass number A will be the nearest integer to the atomic mass in units u. One may add and subtract Zm_ec^2 (m_e = electron mass) to the right side of Eq. (47-2a). Now $Z(m_p + m_e)c^2$ differs from ZM_Hc^2 (m_H = atomic mass of the hydrogen atom) by only Z × 13.6 eV. This difference is ignored and one always works with atomic masses. Eq. (47-2a) becomes

$$E_B = ZM_Hc^2 + NM_nc^2 - M_Ac^2\ . \tag{47-2b}$$

$E_B > 0$ means the nucleus is stable against breakup into Z protons and N neutrons.

Tables of binding energies E_B are available. You may find these also tabulated in terms of <u>mass excess</u>, Δ, defined by

$$\Delta_A = (M_A - A)c^2\ . \tag{47-3}$$

Sometimes the mass excess is given in units of u as well. From Eq. (47-3) it is clear that E_B may also be written as

$$E_B = Z\Delta_H + N\Delta_n - \Delta_A \tag{47-2c}$$

since one may add and subtract Ac^2 and use A = N + Z in Eq. (47-2b). Table 47-1 gives the mass excess both in MeV and micro-u and binding energies for several nuclei.

Element	Z	A	mass excess (μ-u)	mass excess (MeV)	E_B (MeV)
n	0	1	8,665	8.071	-------
H	1	1	7,825	7.289	-------
D (or ^{2}H)	1	2	14,102	13.136	2.224
T (or ^{3}H)	1	3	16,050	14.950	8.482
He	2	4	2,603	2.425	28.296
Li	3	6	15,125	14.088	31.993
Li	3	7	16,004	14.907	39.245
C	6	12	0	0	92.163
C	6	14	3,242	3.020	105.286
N	7	14	3,074	2.864	104.659
N	7	15	108	.100	115.494
O	8	15	3,070	2.860	111.952
O	8	16	-5,085	-4.737	127.620
Ca	20	40	-37,411	-34.848	342.056
Th	90	234	43,583	40.596	1,777.700
U	92	238	50,770	47.291	1,801.726

Table 47-1. Mass excess and binding energies for some selected nuclei given in MeV (energy equivalent) or μ-u (10^{-6} u).

>>> Example 1. (a) Find the atomic masses of the neutron, H and ^{16}O. (b) Compute the binding energy of ^{4}He.

(a) From Table 47-1 and $M_A = A + \Delta_A$ in units u we have

$$M_n = (1 + 8{,}665 \times 10^{-6})\ u = 1.0087\ u$$

$$M_H = (1 + .0073)\ u = 1.0073\ u$$

$$M_{^{16}O} = (16 + (-\ .0051))\ u = 15.9949\ u\ .$$

(b) From Table 47-1 and Eq. (47-2c)

$$E_B = 2(7.289) + 2(8.071) - 2.425 = 28.295\ \text{MeV}\ .$$ <<<

47-3 Nuclear Decay, Nuclear Reactions

All nuclear decay processes may be written symbolically as

$$C \rightarrow Y + b \qquad \text{(Decay)} \qquad (47\text{-}4a)$$

where C is the original nucleus called the parent, Y is the new nucleus formed called the daughter and b stands for everything else. Nuclear reactions differ from this basic decay form in that C is an intermediate state of the protons and neutrons formed by the collision of nucleus X with projectile a. Then one writes

$$X + a \rightarrow Y + b \qquad \text{(Reaction)} \qquad (47\text{-}4b)$$

or sometimes

$$X(a,b)Y\ . \qquad \text{(Reaction)} \qquad (47\text{-}4c)$$

In the case of decay we define decay energy, Q, as

$$Q = (m_C - m_Y - m_b)c^2 \qquad \text{(Decay Q-value)} \qquad (47\text{-}5a)$$

whereas for reactions we define the reaction energy, Q, as

$$Q = (m_X + m_a - m_Y - m_b)c^2\ . \qquad \text{(Reaction Q-value)} \qquad (47\text{-}5b)$$

In both cases we speak of Q-value.

For decay to be energetically possible Q must be positive. Reactions with $Q > 0$ are exothermic and those with $Q < 0$ are endothermic. In the latter case, the reaction may still be energetically possible if X and a have some initial kinetic energy. Only the kinetic energy of relative motion is available to the reaction (see Section 9-3) so endothermic decay is not possible. A decay or reaction can proceed only if it is energetically possible and if

i) the number of nucleons is the same on both sides of the arrow

and

ii) the charge is the same on both sides of the arrow.

The masses in Eqs. (47-5) are nuclear masses so we always add and subtract enough electron masses to use atomic masses.

>>> Example 2. What is the Q value for $^{238}U \rightarrow {}^{234}Th + \alpha$? The symbol U means uranium and Z = 92, so N = 146. The symbol Th stands for Z = 90 or thorium, so N = 144. The alpha particle (α) is the nucleus of ^4He which has Z = 2.

We use Eq. (47-5a) with $92m_ec^2$ added and subtracted. Thus

$$Q = (m_{^{238}U} + 92m_e)c^2 - (m_{^{234}Th} + 90m_e)c^2 - (m_\alpha + 2m_e)c^2$$

or

$$Q = (M_{^{238}U} - M_{^{234}Th} - M_{^4He})c^2$$

which with Eq. (47-3) becomes

$$Q = (238 - 234 - 4)c^2 + \Delta_{^{238}U} - \Delta_{^{234}Th} - \Delta_{^4He} = 4.27\ \text{MeV}\ .$$

The decay is energetically possible and also satisfies i) and ii). The probability of this allowed decay is governed by other conditions. <<<

>>> Example 3. What is the Q value for ^{60}Ni (α,n) ^{63}Zn?

Note first that 60 + 4 = 63 + 1 and since Z of nickel is 28 while that of zinc is 30, 28 + 2 = 0 + 30 so ii) is also satisfied. Here we use Eq. (47-5b) and add and subtract $30m_ec^2$, so that

$$Q = \Delta_{^{60}Ni} + \Delta_{^4He} - \Delta_{^{63}Zn} - \Delta_n \; .$$

A table similar to 47-1 would show

$$\Delta_{^{60}Ni} = -\,64.471 \text{ MeV}$$

$$\Delta_{^{63}Zn} = -\,62.217 \text{ MeV} \; .$$

Thus

$$Q = -\,7.9 \text{ MeV} \; .$$ <<<

This endothermic reaction cannot proceed unless the incident projectile, the α, has a threshold kinetic energy. From Section 9-3 the kinetic energy of relative motion (assuming the ^{60}Ni is at rest) is

$$K_{rel} = K_\alpha \left(\frac{m_{^{60}Ni}}{m_{^{60}Ni} + m_\alpha} \right) \; .$$

To a close enough approximation $m_\alpha = 4$, $m_{^{60}Ni} = 60$ and so $K_{rel} = 0.94\ K_\alpha \gtrsim 7.9$ MeV. Thus $K_\alpha \gtrsim 8.4$ MeV; the threshold is 8.4 MeV.

Nuclear decays are classified according to the particle emitted. In addition to α-decay there is $\beta^\pm$ decay where the emitted particle is an electron (β^-, or e^-) or positron (β^+, or e^+). There is also γ-decay where electromagnetic radiation is emitted. All other types of decays would be a type of fission.

In $\beta^\pm$ decay, only a very few of the particles emitted from a decaying sample have $K_{\beta^\pm} = Q$; most have considerably less. The rest of the decay energy (that not carried off by motion of the daughter which involves very little energy) is carried off by a particle called the neutrino. The neutrino has zero mass and spin $\frac{1}{2}\hbar$. Neutrinos are very hard to detect since they almost don't interact with matter. For either $\beta^\pm$ decay

$$A_Z \rightarrow A_{(Z\pm1)} + e^\mp + \nu \; . \tag{47-6a}$$

To use Eq. (47-5a) we add and subtract Zm_ec^2 and have

$$Q = (M_{A_Z} - M_{A_{(Z+1)}})c^2 = \Delta_{A_Z} - \Delta_{A_{(Z+1)}} \qquad (\beta^- \text{ decay}) \tag{47-7a}$$

$$Q = (M_{A_Z} - M_{A_{(Z-1)}} - 2m_e)c^2 = \Delta_{A_Z} - \Delta_{A_{(Z-1)}} - 2m_ec^2 \; . \qquad (\beta^+ \text{ decay}) \tag{47-7b}$$

Eq. (47-6a) displays the basic equations $n \rightarrow p + e^- + \nu$ and $p \rightarrow n + e^+ + \nu$. One interesting process competes with β^+ decay for medium to heavy nuclei. The innermost electrons move very close to and sometimes inside the nucleus. There, they may be captured by a proton and the process $p + e^- \rightarrow n + \nu$ known as electron capture (EC) can occur, more generally we have

$$A_Z + e^- \rightarrow A_{(Z-1)} + \nu \qquad (\text{EC}) \tag{47-6b}$$

and

$$Q = \Delta_{A_Z} - \Delta_{A_{(Z-1)}} \quad . \quad \text{(EC)} \tag{47-7c}$$

>>> Example 4. Compute the Q values for (a) the β^- decay of ^{14}C and (b) the β^+ decay of ^{15}O.

(a) First of all, one must decide what ^{14}C decays to. It is probably easiest to lay out the result as below:

Parent		Daughter
$Z = 6$	$\xrightarrow{\beta^-}$	$Z = 7$
$N = 8$		$N = 7$

so the daughter is ^{14}N. Thus

$$Q = \Delta_{^{14}C} - \Delta_{^{14}N} = 3.020 - 2.864 = 0.156 \text{ MeV} .$$

(b) Since this is β^+ decay one has $p \rightarrow n$, so

Parent		Daughter
$Z = 8$	$\xrightarrow{\beta^+}$	$Z = 7$
$N = 7$		$N = 8$

so the daughter is ^{15}N. From Table 47-1 together with $m_ec^2 = 0.511$ MeV one has

$$Q = \Delta_{^{15}O} - \Delta_{^{15}N} - 2m_ec^2$$
$$= 2.860 - 0.100 - 2(0.511) = 1.738 \text{ MeV} .$$ <<<

The protons and neutrons of a nucleus are a quantum mechanical system just as are the electrons of an atom. As with atoms, nuclei have excited energy states which are at various allowed energies above the lowest energy state called the ground state. We denote that a nucleus is in some excited state by placing an asterisk as an upper right superscript to the symbol. For example, $^{238}U^*$ means some excited state of the nucleus uranium whose mass number is 238. Of course we'd have to give other information to specify just which excited state; this is just a general notation. If the energy of this state above the ground state is sufficiently high, the state may decay by emitting one or more particles from the nucleus. If it is not high enough in energy for the emission of any particles, it can only decay to lower energy states (and ultimately to the ground state) by the emission of γ-radiation. We would denote, for example,

$$^{238}U^* \rightarrow {}^{238}U + \gamma$$

which is decay to the ground state. Schematically the situation is like Figure 47-2. If the excited initial state is above the final state in energy by $E_i - E_f$ then the emitted γ-ray will have nearly this energy. Some very tiny fraction of the energy must go into the recoil of the nucleus with lower energy in

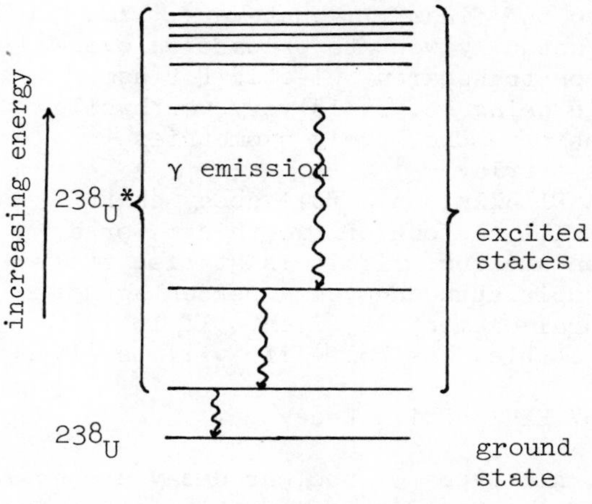

Figure 47-2. Excited States

order that momentum be conserved. Usually this is so small that to a good approximation

$$E_\gamma = \hbar\omega \simeq E_i - E_f \ .$$

47-4 Decay and Reactions - Use of the Nuclide Chart

The decay and reaction rules can be summarized by the typical reaction

$$^{A_1}_{Z_1}X + ^{A_2}_{Z_2}a \quad ^{A_3}_{Z_3}Y + ^{A_4}_{Z_4}b \ . \qquad (47\text{-}8)$$

This equation can also be used to represent a decay if we merely neglect a (take $A_2 = Z_2 = 0$). Furthermore, if b symbolizes the β and neutrino in $\beta^\pm$ decay, we would take $A_4 = 0$ (no nucleons) and $Z_4 = \pm$ for $\beta^\pm$ decay. If b symbolizes a γ-ray, then $A_4 = Z_4 = 0$. With this understanding the rules become

$$A_1 + A_2 = A_3 + A_4$$
$$Z_1 + Z_2 = Z_3 + Z_4 \ . \qquad (47\text{-}9)$$

In computing the Q value we need merely to remember to add $(Z_1 + Z_2)m_ec^2$ to $(m_{A_1} + m_{A_2})c^2$ and subtract it from $-(m_{A_3} + m_{A_4})c^2$. Each mass will then become an atomic mass, and there may be some m_ec^2 left over.

We also want to be able to follow the reaction or decay on the nuclide chart. This is easily done by using Eqs. (47-9). Since the chart is Z versus N we see, for example, that the reaction (n,γ) which means a = neutron, b = γ-ray will move us from the nucleus Z_1, $N_1 = A_1 - Z_1$ to the nucleus $Z_3 = Z_1$, $N_3 = A_3 - Z_3 = N_1 + 1$. That is, one box to the right along the row of constant Z. One can make a box chart to mentally overlay on the nuclide chart such as shown in Figure 47-3. The blank box is placed over the nuclide X with Z_1 and $N_1 = A_1 - Z_1$. Then any decay or reaction (represented on the overlay) can be easily followed. Only the decays $\beta^\pm$ and EC are indicated. All other decays lead to the same box where one finds (γ, whatever). You may actually want to prepare an overlay on transparent plastic for use. While doing so, it is very worthwhile to observe the lovely symmetries in this overlay.

			(α,n)	(α,γ)
	(p,n) β^-	(p,γ) (d,n)	(d,γ) (α,d)	(α,p)
	(γ,n) (p,d)		(n,γ) (d,p)	
(p,α)	(γ,d) (d,α)	(n,d) (γ,p)	(n,p) β^+,α	
(γ,α)	(n,α)			

Z ↑   N →

Figure 47-3. Chart Overlay

Finally, on a full decay chart you will see one of two things for a given nucleus. If it is stable, its isotopic abundance as a percentage of natural element is given. If it is not stable, its half-life will be given and its principal nodes of decay will be noted.

47-5 Radioactive Decay Law

All types of nuclear decay are collectively called <u>radioactive decay</u>. In a given sample of N radioactive nuclei, some fraction dN will decay in some time interval dt. The process for the sample is statistical. A given nucleus may, however, remain unchanged for billions of years and only when internal workings are just right does it decay. For a bulk sample, the decay rate $R = -\, dN/dt$ is proportional to N; the constant of proportionality, λ, is called the <u>decay constant</u> or <u>disintegration constant</u> and it depends upon the species of nuclei and the decay process.

$$R = -\frac{dN}{dt} = \lambda N \tag{47-10}$$

may be integrated to

$$N = N_o e^{-\lambda t} \tag{47-11}$$

where N_o is the number of nuclei in the sample at $t = 0$. One also has

$$R = R_o e^{-\lambda t} \tag{47-12a}$$

where the $t = 0$ decay rate is

$$R_o = \lambda N_o \ . \tag{47-12b}$$

The <u>half-life</u>, $t_{1/2}$, is that time at which $R = \frac{1}{2} R_o$, or $N = \frac{1}{2} N_o$. Thus $e^{-\lambda t_{1/2}} = 1/2$ or

$$t_{1/2} = \frac{\ln 2}{\lambda} = \frac{0.693}{\lambda} \ . \tag{47-13}$$

For $t_{1/2}$ of the order of seconds to a few days one measures R at various times and graphs log R vs time. From Eq. (47-12a) the slope determines λ. For long half-lives one must simultaneously measure R and N and use Eq. (47-10) to determine λ.

>>> Example 5. $^{32}P \xrightarrow[\beta-]{14.22\ d} {}^{32}S$ means ^{32}P decays to ^{32}S by β^- emission with a half life of 14.22 days. Suppose at the start of an experiment R = 500 disintegrations/hr. How long would one have to wait until R = 450 dis/hr?

We use Eq. (47-12a) and

$$\frac{450}{500} = \frac{R_2}{R_1} = \frac{R_o e^{-\lambda t_2}}{R_o e^{-\lambda t_1}} = e^{-\lambda(t_2 - t_1)} = e^{-\lambda \Delta t} \ .$$

Thus $\Delta t = 2.2$ days. <<<

A similar calculation for $^{14}C \xrightarrow[\beta-]{5730\ yr} {}^{14}N$ would have one waiting for 871 years! The use of ^{14}C in dating relies on the fact that living matter exchanges CO_2 with the atmosphere and some of this is ^{14}C. At death this exchange ceases, so if one knows R_o for a type of living matter, a measurement of the current R yields the time since death.

>>> Example 6. A type of living rosewood normally exhibits ^{14}C decay at the rate R_o = 10 dis/min/gram of carbon. Suppose an ancient cabinet of such rosewood has activity of R = 7.6 dis/min/gram of carbon. How old is the cabinet approximately?

We have

$$e^{-\lambda t} = \frac{7.6}{10}$$

so

$$t = \frac{1}{0.693} \left(\ln \frac{10}{7.6}\right) 5730 \text{ yr} = 2270 \text{ yr} \ .$$ <<<

Notice that this is equivalent to knowing R and N at t = 2270 yr.

47-6 Nuclear Energy; Fission, Fusion

There are two processes by which nuclear energy can be made available in reasonably large amounts. One is <u>fission</u> in which an <u>actinide</u> (thorium and upward in A) splits into two medium mass nuclei. The other is <u>fusion</u> in which two light nuclei fuse into one medium mass nucleus. Both are reactions because no spontaneous fission (decay) occurs with enough probability. The most important fission process involves <u>thermal neutrons</u> ($K \sim .04$ eV) and the nuclear material which will undergo fission by capture of thermal neutrons is called <u>fissile</u>. Among all the nuclides, ^{235}U is the only naturally occurring fissile nucleus. The basic process is

$$^{235}U + n \rightarrow {}^{236}U^* \rightarrow X + Y + bn \tag{47-14}$$

where X and Y are called <u>fission fragments</u> and there are b <u>emission</u> neutrons. They are any two medium mass nuclei such that

$$\begin{aligned} Z_X + Z_Y &= 92 \quad \text{(Z of uranium)} \\ A_X + A_Y + b &= 236 \,. \quad \text{(Number of nucleons)} \end{aligned} \tag{47-15}$$

The greatest probability of fission is for one of the masses near 90 and the other near 145. Some of the more common fissions are

$$\begin{aligned} {}^{235}_{92}U + {}^{1}_{0}n \rightarrow {}^{236}_{92}U^* &\rightarrow {}^{148}_{58}Ce + {}^{87}_{34}Se + {}^{1}_{0}n \\ \text{or } &\rightarrow {}^{144}_{56}Ba + {}^{89}_{36}Kr + 3\,{}^{1}_{0}n \\ \text{or } &\rightarrow {}^{143}_{55}Cs + {}^{90}_{37}Rb + 3\,{}^{1}_{0}n \\ \text{or } &\rightarrow {}^{140}_{54}Xe + {}^{94}_{38}Sr + 2\,{}^{1}_{0}n \,. \end{aligned} \tag{47-16}$$

There are many other modes of decay of $^{236}U^*$ (called the compound nucleus). For any process the Q-value is

$$Q = \Delta_{235U} - \Delta_X - \Delta_Y - (b - 1)\Delta_n \,.$$

In Eqs. (47-16) the left subscript is the Z number. In each of these fission <u>modes</u> the fission fragments are too neutron rich and decay by β^- emission. For example, in the last of these the subsequent decays are

$$^{140}Xe \xrightarrow{16\text{ s}} {}^{140}Cs \xrightarrow{66\text{ s}} {}^{140}Ba \xrightarrow{12.8\text{ d}} {}^{140}La \xrightarrow{40.2\text{ h}} {}^{140}Ce$$

$$^{94}Sr \xrightarrow{12\text{ m}} {}^{94}Y \xrightarrow{20\text{ m}} {}^{94}Zr \,.$$

One may as well say

$${}^{235}_{92}U + {}^{1}_{0}n \rightarrow {}^{140}_{58}Ce + {}^{94}_{40}Zr + 6\beta^- + 6\nu + 2\,{}^{1}_{0}n$$

and the <u>total</u> Q-value is

$$\begin{aligned} Q_{Total} &= \Delta_{235U} - \Delta_{140Ce} - \Delta_{94Zr} - \Delta_n \\ &= 208.23 \text{ MeV} \,. \end{aligned}$$

Of this 208 MeV, 177 MeV shows up promptly as kinetic energy of the fission fragments and emission neutrons. The balance shows up <u>later</u> as kinetic energy of the β^- electrons, daughter nuclei and neutrinos. In some modes the β^- decay is to an excited state of a daughter and one or more γ-rays are emitted. How much later this energy shows up depends upon the half-lives involved.

The fact that some modes of fission emit more than one neutron is what makes practical use of nuclear energy possible. A <u>chain</u> reaction is possible because the emitted neutrons can stimulate further fissions in the ^{235}U nuclear fuel.

In the first fission mode of Eq. (47-16), the ^{87}Se decays as follows

$$^{87}\text{Se} \xrightarrow{\beta^-} {}^{87}\text{Br} \begin{cases} \xrightarrow{\beta^-} {}^{87}\text{Kr}^* \longrightarrow {}^{86}\text{Kr} + n \\ \xrightarrow{\beta^-} {}^{87}\text{Kr} \xrightarrow{\beta^-} {}^{87}\text{Rb} \end{cases}$$

The β^- decay of ^{87}Bromine has high probability of decaying to an excited state of ^{87}Krypton which then decays to ^{86}Kr plus a neutron. Such neutrons are called <u>delayed neutrons</u>. The existence of these is vital to the control of the fission rate in a nuclear reactor.

Considering all possible modes of fission, the "average" reaction is

$$Q_{\text{Total}} \simeq 200 \text{ MeV}$$

$$K_{\text{fission fragments}} \simeq 170 \text{ MeV}$$

$$K_{\text{emitted neutrons}} \simeq 5 \text{ MeV}$$

$$K_{\beta^- \text{ and } \gamma\text{-rays}} \simeq 15 \text{ MeV}$$

$$\text{Energy of neutrinos} \simeq 10 \text{ MeV}$$

$$b \simeq 2.5 \quad \text{(neutrons released)} .$$

On the average then about 190 MeV is available since the neutrinos carry theirs away mostly without interaction. The fact that b is on the average > 1 makes practical use of this fuel source possible.

>>> Example 7. Suppose that an electrical power plant operates on the thermal energy derived from a nuclear reactor. Suppose that the power output (electrical) is 1200 MW, the effiency is 30%, and that the uranium fuel is enriched to 3% ^{235}U. How much fuel is used up in a month of operation?

First, we need the reactor thermal power, P_{reactor}, which we can find from

$$e = \frac{P}{P_{\text{reactor}}} = 0.3 = \frac{1200 \text{ MW}}{P_{\text{reactor}}}$$

So $P_{\text{reactor}} = 4000$ MW. (By the way, this means that 4000 - 1200 MW = 2800 MW of waste heat must be disposed of.) Next, we can calculate the fission rate R since Q ≃ 200 MeV/fission = 3.2×10^{-11} J/fission and

$$R = \frac{P_{\text{reactor}}}{Q} = 1.25 \times 10^{20} \text{ fissions/sec} .$$

Now we can compute the mass of ^{235}U used up in a month since

$$M = Rt \left(\frac{235 \times 10^{-3} \text{ kg/mol}}{6.0 \times 10^{23} \text{ atoms/mol}} \right) \leftarrow \text{mass of } ^{235}\text{U atom}$$

$$M = 1.25 \times 10^{20} \frac{\text{atoms}}{\text{sec}} \times 2.592 \times 10^{6} \text{ sec} \times 3.92 \times 10^{-25} \frac{\text{kg}}{\text{atom}}$$

$$M = 127 \text{ kg} .$$

Thus 127 kg of ^{235}U is used up in one month of operation. This 127 kg is 3% of the fuel used, so the amount of fuel expended is 127/.03 kg = 4230 kg.

One can go a bit further. Since $P_{reactor}$ = 4000 MW mass is converted to energy at the rate

$$\frac{dM}{dt} = \frac{dE}{dt}\frac{1}{c^2} = 4.4 \times 10^{-8} \text{ kg/sec}$$

so in one month (2.6×10^6 sec) the mass decrease is 0.1 kg. Therefore nearly all of the 4230 kg of fuel becomes waste. <<<

Generally, energy is released when any pair of light nuclei are crashed together with enough kinetic energy to overcome their coulomb repulsion. They fuse into a moderate mass nucleus and release energy. Four such reactions involve very light nuclei. These are

$$\begin{aligned}
d + d &\rightarrow {}^3He + n + 3.3 \text{ MeV} , \\
d + d &\rightarrow t + p + 4.0 \text{ MeV} , \\
d + t &\rightarrow \alpha + n + 17.6 \text{ MeV} , \\
d + {}^3He &\rightarrow \alpha + p + 18.0 \text{ MeV} .
\end{aligned} \tag{47-17}$$

Here d symbolizes the <u>deuteron</u>, the nucleus of ^2H whose name is <u>deuterium</u> and t symbolizes the <u>triton</u> which is the nucleus of tritium (^3H). The Q-values are as indicated. For the d - d process one finds that in order for the relative kinetic energy to equal the maximum coulomb repulsion energy, each deuteron must have about 200 MeV kinetic energy. To have sustained fusion in bulk matter therefore implies high temperatures; this is called <u>thermonuclear fusion</u>. Even in the sun's interior where temperatures are about 20 million Kelvin, the mean thermal kinetic energy of the deuterons is only about 2 MeV. Because of the Maxwellian distribution, a fair number have energies well in excess of this, and further, quantum mechanically there is a small (but non-zero) probability that the pair can come together with less energy. (This is known as tunneling through the coulomb energy barrier.) Stellar energy is therefore in some sense a giant confirmation of quantum mechanics.

The solar fusion cycle is called the <u>proton-proton cycle</u>; it proceeds in three steps:

p - p cycle, step 1, done twice	$p + p \rightarrow d + \beta^+ + \nu$	
p - p cycle, step 2, done twice	$d + p \rightarrow {}^3He$	(47-18)
p - p cycle, step 3	${}^3He + {}^3He \rightarrow \alpha + 2p$.	

Overall the process is equivalent to

$$4p \rightarrow \alpha + 2\beta^+ + 2\nu$$

and the Q-value is

$$Q_1 = (4m_p - m_\alpha - 2m_e)c^2 .$$

In addition, the $2\beta^+$ annihilate with two of the numerous free electrons (β^-) and release

$$Q_2 = 4m_ec^2 .$$

so

$$Q = Q_1 + Q_2 = (4m_p + 2m_e - m_\alpha)c^2$$

$$Q = 4\Delta_{^1H} - \Delta_{^4He}$$

$$Q = 26.7 \text{ MeV} .$$

About 0.5 MeV is carried off by the neutrinos so 26.2 MeV = 4.2×10^{-12} J is available from each proton cycle to contribute to the solar power output of about 4×10^{26} W. In addition, there is the carbon-cycle whose steps are

solar carbon cycle

$$p + {}^{12}C \rightarrow {}^{13}N \rightarrow {}^{13}C + \beta^+ + \nu$$
$$p + {}^{13}C \rightarrow {}^{14}N$$
$$p + {}^{14}N \rightarrow {}^{15}O \rightarrow {}^{15}N + \beta^+ + \nu$$
$$p + {}^{15}N \rightarrow {}^{12}C + \alpha .$$

The carbon is merely a catalyst and so the energy release is exactly the same as the p - p cycle. Some stars older than our sun and having much higher internal temperatures "burn" heavier fuel.

The p - p process almost doesn't happen. It is only practical in stars where there is an enormous supply of protons. For earth-bound fusion, one of (47-19) will have to do; the most promising is the d - t process. About .02% of the sea-water hydrogen is deuterium so its supply is almost unlimited. But tritium doesn't even occur in nature. It can easily be produced by

$${}^{6}Li + n \rightarrow t + \alpha + 4.8 \text{ MeV}$$

so, it is theoretically possible to design a fusion breeder reactor using

$$d + t \rightarrow n + \alpha + 17.6 \text{ MeV}$$
$$n + {}^{6}Li \rightarrow t + \alpha + 4.8 \text{ MeV}$$

which is equivalent to

$${}^{6}Li + d \rightarrow 2\alpha + 22.4 \text{ MeV} .$$

Since some of the neutrons would escape, this would merely use up the hard won initial supply of tritium. Fortunately natural lithium contains more ${}^{7}Li$ than ${}^{6}Li$ and

$${}^{7}Li + n \rightarrow {}^{6}Li + 2n$$

so one could recover the "lost" neutrons. As yet, however, successful sustained fusion reaction has eluded man's efforts because no design has been able to simultaneously satisfy the demands of high enough temperature and Lawson's criteria. For d - t one needs $T \sim 4 \times 10^{7}$ K. Lawson's number is the product of particle density, n, and time, τ, of confinement to this density at the required temperature. For the d - t process Lawson's number $n\tau$ must be $\gtrsim 10^{14}$ sec/cm^3.

Chapter 48: SPECIAL RELATIVITY

REVIEW AND PREVIEW

You probably began your study of physics with kinematics. There were some definitions to learn and some formal structure was built to organize your thinking. Nevertheless, things undoubtedly came out a great deal like you felt they ought to based upon your intuition. In this chapter you will touch on some basic kinematics and a little dynamics as described in the special theory of relativity. This is not a study of new physics, but rather a reformulation to make the laws valid in the case where the velocity is sufficiently high in comparison with the velocity of light. You have probably already used one of the dynamic equations, namely $E = mc^2$.

GOALS AND GUIDELINES

1. You should learn the two basic postulates of the theory although you will not see here any derivations from these first principles.
2. You should carefully study the Lorentz transformation equations and the time dilation and length contraction which follow directly from these.
3. You should learn how to use the relativistic velocity addition equation and the relativistic doppler shift formulas. Pay particular attention to the meaning of proper frame.
4. You should examine the structure of the relativistic definitions of momentum and kinetic energy and should note how these revert to the more familiar forms when v/c is small enough.
5. Throughout this chapter you should note the need to be very precise about what occurs where and how quantities are to be measured. Intuition is simply no longer a reliable guide. Most paradoxes that arise are the result of imprecise statements of the problem. Above all, enjoy your glimpse of this fascinating world which truly exists even though we do not often see it directly.

48-1 Basic Postulates

There are two basic postulates for the special theory of relativity. It is important to remember that both, and therefore all consequences of them, apply to <u>inertial frames</u>, that is, non-accelerated coordinate frames. Both postulates have repeatedly stood experimental test. They are:

i) The laws of physics are the same in all inertial coordinate frames of reference.

ii) The speed of light is the same in all inertial frames.

From these follow numerous modifications to the classical statements of physics. Some are very difficult to accept, but that is because our practical experience (the basis of our gut feelings) is limited to cases where speeds involved are much less than the velocity of light, $c = 3 \times 10^8$ m/s. In trying to master this material one must constantly ask the fundamental question: What is being measured and with respect to which coordinate frame?

This chapter is limited to a summary of some elements of kinematics and dynamics of mechanics as formulated in the special theory.

48-2 Lorentz Transformation Equations

In classical mechanics we write the equations relating coordinates in one frame to those in another (moving at constant velocity relative to the first) without much fuss. They seem "natural" because they jibe with our experience. In special relativity one must exercise more care.

There are to be two coordinate frames S and S′ whose axes are parallel. Frame S′ is moving along the x axis of S with a velocity v as seen by S. (We will use S or S′ both to label a frame and to label an observer in that frame.) If v is positive, S′ is moving to the right in S and if v is negative, to the left. Both S and S′ have clocks which were synchronized to read zero when the S and S′ origins coincide.

By an event one means a sharp occurrence of something which can be said to occur at a precise point in both space and time. Some famous physicists like to think of this as the bop on the head with a beer bottle! S and S′ observe the same event. S assigns space-time coordinates x, y, z, t and S′ assigns x′, y′, z′, t′.

In classical physics one relates these by the Galilean transformation equations

$$
\begin{aligned}
x' &= x - vt \\
y' &= y \\
z' &= z \\
t' &= t\ .
\end{aligned}
\qquad \text{Galilean transformation equations} \qquad (48\text{-}1)
$$

The correct equations according to the special theory are those of the Lorentz transformation

$$
\begin{aligned}
x' &= \frac{1}{\sqrt{1-(\frac{v}{c})^2}}(x - vt) \\
y' &= y \\
z' &= z \\
t' &= \frac{1}{\sqrt{1-(\frac{v}{c})^2}}\left(t - \frac{v}{c^2}x\right)\ .
\end{aligned}
\qquad \text{Lorentz transformation equations} \qquad (48\text{-}2a)
$$

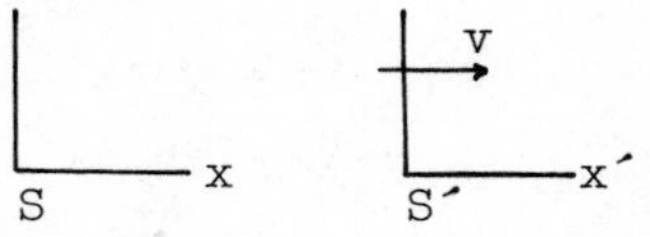

If $|v/c| << 1$ and is neglected, then Eqs. (48-2a) reduce to their classical counterparts. This is a good guiding principle you should use in any algebraic problem. Your answer may not be correct even if it does reduce to the correct classical result, but it is most assuredly wrong if it does not.

The ratio v/c occurs throughout the theory and so does the factor $1/\sqrt{1-(v/c)^2}$. It is convenient to label these as

$$\frac{v}{c} \equiv \beta \qquad \text{and} \qquad \frac{1}{\sqrt{1-(\frac{v}{c})^2}} \equiv \gamma\ . \qquad (48\text{-}3)$$

In the transformation equations we then use ct or ct′ so that these have the same dimensions as x (or x′). One then rewrites Eqs. (48-2a) as

$$
\begin{aligned}
x' &= \gamma(x - \beta ct) \\
y' &= y \\
z' &= z \\
ct' &= \gamma(ct - \beta x)\ .
\end{aligned}
\qquad (48\text{-}2b)
$$

S and S′ will record changes in coordinates and times. For example, S measures $\Delta x = x_2 - x_1$ and S′ measures correspondingly $\Delta x' = x_2' - x_1'$; if S′ measures $\Delta t' = t_2' - t_1'$, then S measures $\Delta t = t_2 - t_1$. What each will measure depends crucially on the circumstances which must be spelled out carefully. One sees, for example, that if $\Delta x = 0$ but $\Delta t \neq 0$, then not only is $\Delta t' \neq 0$, but neither is $\Delta x'$! Various circumstances are considered in Section 48-3. We conclude this section with the note that to obtain the inverse transformation (i.e., x, y, z, t in terms of x', y', z', t') one needs merely use Eqs. (48-2a or 2b) and interchange the primed and unprimed quantities and replace β by $-\beta$, i.e., replace v by $-v$. This is just the result of the fact that if S observes S´ moving with velocity v, then S´ observes S to move with velocity $-v$.

48-3 Time Dilation, Length Contraction, Simultaneous Events

Suppose S´ observes two events at different locations in the S´ system and notes that they occur simultaneously. S observes the same two events. They will record the same numbers for y and z so we ignore those. Thus S´ records $\Delta t' = 0$, $\Delta x' = x_2' - x_1'$ while S records $\Delta t = t_2 - t_1$ and $\Delta x = x_2 - x_1$. We use the inverse transformation equations to relate these two measurements

$$\begin{aligned} x &= \gamma(x' + \beta c t') \\ y &= y' \\ z &= z' \\ ct &= \gamma(ct' + \beta x') \ . \end{aligned} \qquad (48\text{-}2c)$$

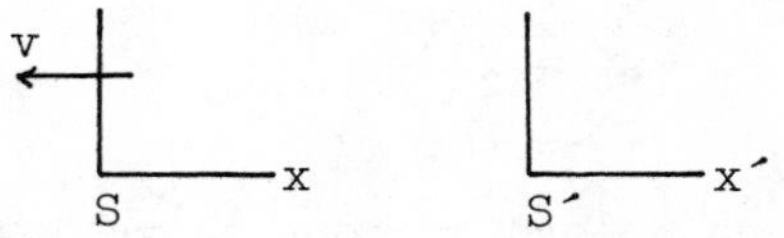

The meaning of $\Delta t' = 0$ is $t_2' = t_1'$. Thus

$$\Delta x = x_2 - x_1 = \gamma((x_2' - x_1') + \beta c(t_2' - t_1'))$$

$$c\,\Delta t = c(t_2 - t_1) = \gamma(c(t_2' - t_1') + \beta(x_2' - x_1'))$$

or

$$\begin{aligned} \Delta x &= \gamma\, \Delta x' \\ c\,\Delta t &= \gamma\beta\, \Delta x' \ . \end{aligned} \qquad \text{simultaneous events in S´}$$

So, not only does S observe the spacing between the events to be different from what S observes, but furthermore S does not see them as simultaneous.

Now, consider two events that occur at the same location in S´ but at different times. For example, the ticks of a clock at rest in S´. S´ records the time interval $\Delta t'$. This is called the proper time because it is recorded by a clock at rest. Often we denote $\Delta t'$ by T_0 to indicate the proper time. S will observe the ticks at different times and different places because the clock has velocity v relative to S; S records the interval $\Delta t = T$. From the last of Eqs. (48-2c) one has

$$\Delta t = \gamma\, \Delta t' \qquad \text{Time dilation}$$

or (48-4)

$$T = \gamma T_0 \ . \qquad \text{Clock at rest in S´}$$

Since $\gamma = 1/\sqrt{1 - \beta^2} \geq 1$ (equal only for $\beta = v/c = 0$) one says that the moving clock runs slow.

Next, consider a rod aligned along the $x - x'$ axes again at rest in S´. Since the rod is at rest, S´ can locate one end of it at some time (according to the S´ clock) and can locate the other end any other time he likes since the rod isn't going to move. From the point of view of S, however, the rod is moving. To sensibly measure its length, he must locate the position of each end simultaneously according to the S clock. Thus $\Delta t = 0$. We denote the proper length measured in the at rest frame by L_0; thus $\Delta x' = L_0$. The length according to S is $L = \Delta x$. From the first of Eqs. (48-2b)

$$L_0 = \gamma L$$

since $\Delta t = 0$, or

$$\Delta x = \frac{1}{\gamma}\, \Delta x'$$

or

$$L = \frac{1}{\gamma} L_0 . \qquad \text{Length contraction} \qquad (48\text{-}5)$$

Since $1/\gamma \leq 1$ (equal only for $v = 0$) the moving rod is measured to be shorter than its rest frame length.

The greatest difficulty students encounter with relativity problems is in deciding what goes in what frame of reference. The following example illustrates the ideas.

>>> Example 1. Sam Swift is an exceptionally fast jogger who can run at a speed $v = 0.866\ c$. Some referees have laid out a straight mile-long track to time Sam who is allowed a running start. Sam carries his jogging chronograph which is of the same make as those used by the referees. (a) What is the length of the track as measured by the referees? (b) What length does the track appear to have to Sam as he runs? (c) How long does it take for Sam to run the track according to the referees when they use their clocks? (d) How long a time interval for this run would Sam record on his chronograph? (e) If the referees observed Sam's clock, what interval would it record for them?

(a) The track is at rest with respect to the referees so its length is $L_0 = 1$ mile.

(b) From Sam's point of view, the track moves toward him at a speed 0.866 c. It is therefore a moving length and he would ascertain its length as

$$L = \frac{1}{\gamma} L_0 = \sqrt{1 - 3/4} \times 1 \text{ mile} = 0.5 \text{ mile} .$$

(c) If we call the referee's frame S and that of Sam S´, then the referees simply see Sam (S´) move one mile at a speed $0.866\ c = 0.866 \times 186{,}000$ miles/s. Thus they time him at

$$\frac{1 \text{ mile}}{161{,}000 \text{ miles/second}} = 6.2 \times 10^{-6} \text{ seconds} .$$

(d) This part is more tricky. From Sam's point of view the 0.5 mile long interval moves toward him at a speed $0.866\ c = 161{,}000$ miles/s. So he would record a time interval

$$\Delta t' = \frac{0.5 \text{ mile}}{161{,}000 \text{ mile/s}} = 3.1 \times 10^{-6} \text{ seconds} .$$

(e) Using Sam's clock which is moving relative to them, the referees record a time interval

$$\Delta t = \gamma\ \Delta t' = 2 \times 3.1 \times 10^{-6} \text{ s} = 6.2\ \mu\text{s} .$$

Notice that both Sam and the referees agree on his speed, but the subtle and important difference (at least to Sam) is that according to his proper time clock he has used up only 3.1 μs of his life while the referees have aged 6.2 μs! <<<

48-4 Velocity Addition, Doppler Effect

In the relativistic velocity addition there are three entities involved. There is an observer S, an observer S´ and an object which each observer watches. Observer S assigns a velocity u (along the x-axis) to the object while observer S´ assigns a velocity u´ (along the x´ axis). From the point of view of S, S´ moves along the x - x´ parallel axes with velocity v. We first note that

$$u = \frac{dx}{dt} \qquad \text{and} \qquad u' = \frac{dx'}{dt'} .$$

Then, from Eqs. (48-2c) one has

$$dx = \gamma(dx' + \beta c\ dt')$$

$$c\ dt = \gamma(c\ dt' + \beta\ dx')\ .$$

Divide the first by the second and one has the relationship sought

$$\frac{u}{c} = \frac{dx' + \beta c\ dt'}{c\ dt' + \beta\ dx'}$$

$$\frac{u}{c} = \frac{\frac{u'}{c} + \beta}{1 + \beta\,\frac{u'}{c}}$$

or finally

Velocity Addition

$$u = \frac{u' + v}{1 + \frac{u'v}{c^2}}\ . \qquad (48\text{-}6)$$

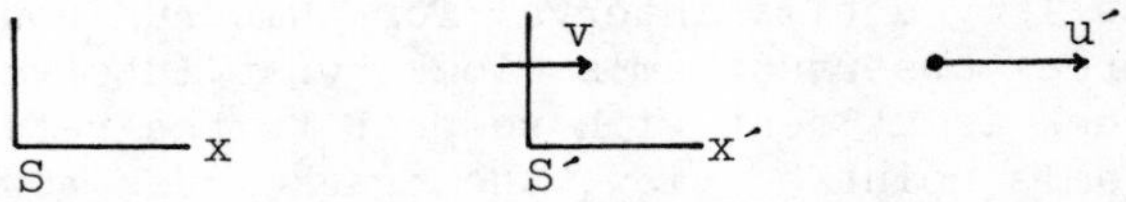

In using Eq. (48-6) one should clearly identify each entity and then use the given speed to solve for the unknown one. Which is assigned which role is immaterial as long as the velocity directions are kept straight as the next problem illustrates.

>>> Example 2. An observer O sees two space ships Alpha and Beta moving along the same line directly away from her. She notes the speeds as: Alpha 0.8c, Beta 0.9c. Beta is further away. With what speed would an observer on Alpha see Beta move?

First of all, it seems clear that an observer on Alpha will see Beta move away from Alpha. Consistent calculations will verify this in this case, but one should always check; intuition can often be wrong in relativity problems. We'll do this problem three ways:

i)

Entity	Frame or object	Velocity
O	S	$u = 0.9c$
Alpha	S'	$v = 0.8c$
Beta	object	$u' =\ ?$

From Eq. (48-6) we can solve algebraically for u' and find

$$u' = \frac{u - v}{1 - \frac{uv}{c^2}} = \frac{0.9c - 0.8c}{1 - \frac{(0.9c)(0.8c)}{c^2}} = 0.36\ c\ .$$

ii)

Entity	Frame or object	Velocity
O	S'	$v = -\ 0.8c$
Alpha	S	$u =\ ?$
Beta	object	$u' = 0.9c$

Notice that from the point of view of Alpha, O moves along the $-x$ direction, i.e., has velocity $-\ 0.8\,c$. Then

$$u = \frac{0.9c - 0.8c}{1 - \frac{(0.9c)(0.8c)}{c^2}} = 0.36\,c \;.$$

iii)	Entity	Frame or object	Velocity
	O	object	$u = -\,0.8c$
	Alpha	S	
	Beta	S´	$v = ?\;,\; u' = -\,0.9c$

Then

$$-\,0.8c = \frac{-\,0.9c + v}{1 - \frac{(0.9c)v}{c^2}}$$

or

$$-\,0.8c + \frac{(0.8c)(0.9c)v}{c^2} = -\,0.9c + v$$

or

$$v(1 - 0.72) = 0.1\,c$$

so

$$v = 0.36\,c \;.$$

<<<

This last method, although a little perverse, indicates the relative nature of all the equations of the special theory. Doing a problem more than one way serves as an excellent check.

It is sometimes loosely said that the relativistic doppler shift can be derived from the velocity addition result. While this is strictly speaking true one must involve the transverse velocity components which we have ignored here. Furthermore, one must keep carefully in mind that the electromagnetic waves travel with speed c relative to either S or S´. A simple derivation follows:

Let a source reside in S and imagine that it emits pulses at proper time intervals τ = one period. Each pulse travels with speed c. Now, suppose there is an observer in S´ moving along the x axis of S with velocity v (i.e., away from the source). Suppose that S´ is at x_0 when $t = 0$ and that the first pulse is sent out. It reaches S´ at time t_1 (as measured by S). The second pulse is sent out at time $t = \tau$ and reaches S´ at time t_2. Meanwhile S´ moves, being at positions x_1 at t_1 and x_2 at t_2. Since the pulses travel at speed c we have $x_1 = ct_1$. But also $x_1 = x_0 + vt_1$. Similarly $x_2 = c(t_2 - \tau)$ and $x_2 = x_0 + vt_2$. From these one has

$$t_2 - t_1 = \frac{c}{c - v}\,\tau \qquad , \qquad x_2 - x_1 = \frac{vc\tau}{c - v} \;.$$

Now we may use Eqs. (48-2b) to find $t_2' - t_1' = \tau'$, the period between successive pulses as observed in the moving frame S´. One has

$$c(t_2' - t_1') = c\tau' = \gamma(c(t_2 - t_1) - \beta(x_2 - x_1))$$

$$= \gamma\left(\frac{c^2}{c - v} - \frac{\beta vc}{c - v}\right)\tau$$

from which

$$\tau' = \sqrt{\frac{1 + \beta}{1 - \beta}}\,\tau$$

follows by the use of $\gamma = 1/\sqrt{1 - \beta^2}$, $\beta = v/c$. Since τ' is the period seen by a moving observer, $1/\tau'$ is the frequency seen by this observer; similarly $1/\tau$ is the frequency measured by an observer at rest with respect to the source. Let us use ν_0 = proper frequency (as measured by an observer at rest with respect to the source) and ν = frequency measured by a moving observer. Thus the doppler shift is

$$\nu = \sqrt{\frac{1 - \beta}{1 + \beta}}\,\nu_0 \, . \qquad \text{observer moves away from source}$$

Had we considered the source moving away from the observer we would encounter the same result. Furthermore, if the source and observer move toward one another with speed v then the sign of β is merely changed and we have in summary

$$\nu = \sqrt{\frac{1 \pm \beta}{1 \mp \beta}}\,\nu_o \, . \qquad (48\text{-}7a)$$

(ν = observer frequency; ν_o = proper frequency)

Relativistic Doppler Shift

Upper signs   O→   ←*S

Lower signs   ←O   *S→

Longitudinal

This is called the <u>longitudinal</u> doppler shift and occurs whenever source and observer move along the line joining each other. Qualitatively this agrees with the classical result (for mechanical waves) in that if source and observer come toward one another the frequency measured by the observer increases; if source and observer move away from one another, then the observed frequency is lower than the proper frequency. Qualitatively, the result is quite different. Relativistically, it makes no difference whether it is source or observer which moves; only the relative motion is important.

Since one often considers wavelength rather than frequency for electromagnetic waves, it is convenient to rewrite Eq. (48-7a) by using $\lambda\nu = \lambda_0\nu_0 = c$.

$$\lambda = \sqrt{\frac{1 \mp \beta}{1 \pm \beta}}\,\lambda_o \, . \qquad (48\text{-}7b)$$

(λ = observer wavelength; λ_o = proper wavelength)

Upper signs   O→   ←*S

Lower signs   ←O   *S→

For visible light, as source and observer <u>approach</u>, the light seen by the observer is <u>blue-shifted</u>; as source and observer <u>move apart</u> the light is <u>red-shifted</u> (i.e., toward longer wavelengths).

There is also a <u>transverse</u> doppler shift which is just a time dilation effect. This occurs when the source and observer move at right angles to the line joining them. It is given by

Transverse Doppler Shift

$$\nu = \sqrt{1 - \beta^2}\,\nu_o \, . \qquad (48\text{-}8)$$

O   ↑*S

>>> Example 3. The prominent red line of the hydrogen spectrum has proper wavelength 6563 A. If this line is observed at 7000 A from a stellar source, what is the velocity of the source? (Presume that both the earth and star are inertial frames and consider only a longitudinal shift.)

We have $\lambda_0 = 6563$ A and $\lambda = 7000$ A. This is a red shift, so in using Eq. (48-7b) we must use the lower signs. So

$$\sqrt{\frac{1+\beta}{1-\beta}} = \frac{7000}{6563}$$

or

$$\frac{1+\beta}{1-\beta} = 1.14 \ ,$$

so

$$\beta = 0.06 \ .$$

Thus $v = 0.06\,c = 1.9 \times 10^7$ m/s away from earth. <<<

>>> Example 4. The earth's orbit is roughly circular and its orbital velocity is nearly 3×10^4 m/s. By how much will the red line of hydrogen ($\lambda_0 = 6563$ A) from the solar spectra be shifted due to this motion?

From Eq. (48-8) and $\lambda\nu = \lambda_0\nu_0 = c$ one has

$$\lambda = \frac{1}{\sqrt{1-\beta^2}}\,\lambda_0$$

for the transverse shift. Thus since $\beta = 10^{-4}$

$$\lambda = 1.0001\,\lambda_0 = 6563.3 \text{ A} \ .$$

This is hardly noticeable! Indeed the spectral lines from the sun are doppler broadened much more than this. The broadening occurs because emission occurs from hydrogen atoms which are moving both toward and away from earth with a variety of velocities. Furthermore, the sun rotates. <<<

>>> Example 5. A radar speed detector depends upon the doppler shift. In principle, one could detect the difference between emitted and received frequencies at the assumed stationary patrol car. This difference will produce a beat frequency, ν_b. Use $\delta = \nu_b/\nu_0$ and determine an expression for β of the target car both for longitudinal and transverse shifts. (See Example 6 of Section 17-5 for comparison.)

First of all, the target car receives doppler-shifted waves and then becomes a moving source of these waves which are again doppler-shifted back at the patrol car. In contrast with the classical case however, the same formula applies to both.

Longitudinal case: The patrol car emits waves of frequency ν_0 and the car receives them at

$$\nu = \nu_0\sqrt{\frac{1 \mp \beta}{1 \pm \beta}} \ .$$

The moving car reemits waves of this frequency which are then received back at the patrol car at frequency

$$\nu' = \nu\sqrt{\frac{1 \mp \beta}{1 \pm \beta}} = \nu_0\left(\frac{1 \mp \beta}{1 \pm \beta}\right) \ .$$

Now we use $\beta \ll 1$ to rewrite this as

$$\begin{aligned}\nu' &= \nu_o(1 \mp \beta)(1 \mp \beta - \beta^2 + \ldots) \\ &= \nu_0(1 \mp 2\beta + \text{terms of order } \beta^3 \text{ and higher}) \ .\end{aligned}$$

Then with $\delta \equiv |\nu' - \nu_0|/\nu_0$ we have

$$\beta = \pm \frac{1}{2}\,\delta \ .$$

Since one is interested only in the speed, not direction,

$$v = \frac{1}{2}\,\delta\, c \ .$$

Transverse case: Now $\nu = \nu_0\sqrt{1 - \beta^2}$ and $\nu' = \nu\sqrt{1 - \beta^2}$, so $\nu' = \nu_0(1 - \beta^2)$, so with $\delta = |\nu' - \nu_0|/\nu_0$ we have

$$v = c\sqrt{\delta} \ .$$

Since $c \simeq 7 \times 10^8$ mph, at $v = 70$ mph one has

$$\delta_{\text{longitudinal}} = 2 \times 10^{-7}$$

$$\delta_{\text{transverse}} = 10^{-14} \ .$$

Since this is mostly longitudinal, the transverse effect is ignored in practical applications. <<<

48-5 Mass, Energy and Momentum

In order that linear momentum be preserved in collisions it is necessary to define momentum in the special theory as

$$\underline{\underline{p}} = \gamma m_0 \underline{\underline{v}} = \frac{1}{\sqrt{1 - \beta^2}}\, m_0 \underline{\underline{v}} \tag{48-9}$$

where m_0 is the mass of the particle measured in its proper frame. This is usually called the <u>rest mass</u>. Historically, Eq. (48-9) was interpreted as saying that the mass of a particle increased with speed according to

$$m = \gamma\, m_0 \ . \tag{48-10}$$

While it remains at times convenient to use this artifact, most modern physicists mean rest mass when they speak of the mass of a particle. This definition of relativistic mass as it was called arises out of the need on the part of some scientists to always write $\underline{\underline{p}} = m\underline{\underline{v}}$. Note that in the limit as $\beta \to 0$ (small velocities), $\underline{\underline{p}} \to m_0\underline{\underline{v}}$ as it must.

The correct form for the kinetic energy in the special theory is

$$K = m_0 c^2(\gamma - 1) = m_0 c^2 \left(\frac{1}{\sqrt{1 - \beta^2}} - 1 \right) \ . \tag{48-11}$$

Again $\lim_{\beta \to 0} K = \frac{1}{2} m_0 v^2$ as we expect. If one uses Eq. (48-10), then K has the form $K = mc^2 - m_0 c^2$. This suggested defining the total energy of a particle as

$$E_{\text{Total}} = m_0 \gamma\, c^2 \tag{48-12a}$$

so that

$$K = E_{\text{Total}} - E_0 \tag{48-12b}$$

and

$$E_0 = m_0\, c^2 \tag{48-12c}$$

is the <u>rest mass energy</u> or merely <u>rest energy</u> of the particle of mass m_0. E with no subscript means E_{Total}.

From Eqs. (48-12a) and (48-9) it follows that

$$E^2 = (pc)^2 + E_o^2 \ . \qquad (48\text{-}13)$$

This equation clearly shows that if $m_o = 0$ (as for the photon and probably the neutrino), then the total energy is

$$E = p\,c \qquad E_o = 0 \qquad (48\text{-}14)$$

and such particles must travel with speed $v = c$. Equation (48-14) is also useful for particles whose mass m_o is not zero. In that case, Eq. (48-14) is a good approximation when $\beta \simeq 1$. This is called the extreme relativistic approximation.

>>> Example 6. What speed must a particle of mass m_o have in order that $E = 10\ E_o$? In this case we have

$$K = \ E - E_o \ = 9\ E_o$$

from Eq. (48-12b). Then, from Eqs. (48-12c) and (48-11) we have

$$K = E_o(\gamma - 1) = 9\ E_o$$

so $\gamma - 1 = 9$ or $\gamma = 10$. Thus

$$\frac{1}{\sqrt{1 - \beta^2}} = 10$$

or

$$1 - \beta^2 = \frac{1}{100} \ ,$$

so

$$\beta = .995 \ .$$

<<<

APPENDIX

INTEGRATION

We want to examine the process of integration by making the ideas plausible but not by giving rigorous proofs. An application of integration occurs when we want the work done by a force, $F(x)$, whose value depends upon position, x. If the object upon which this work is done moves over some interval from say $x = a$ to $x = b$, the work done can be interpreted as the area under the curve $F(x)$ versus x between the limits a and b.

This then is the general problem: Given a function $f(x)$, what is the area bounded by: $f(x)$, the x-axis, and the lines $x = a$ and $x = b$? This area, A, is indicated in Fig. A-1a.

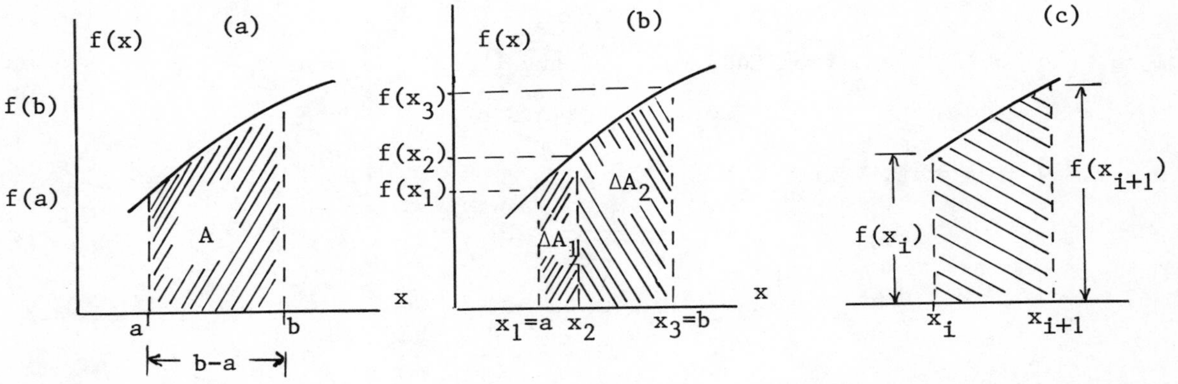

Figure A-1. (a) Area under a curve $f(x)$ between the limits a and b; (b) the area is divided into two segments ΔA_1 and ΔA_2 by the three points $x_1 = a$, x_2, $x_3 = b$; (c) a small segment of the curve between x_i and x_{i+1}.

A crude approximation to A would be either of the rectangular areas $f(a)(b - a)$ or $f(b)(b - a)$. Either would be a fair approximation if $f(x)$ were nearly constant between $x = a$ and $x = b$ but not if $f(x)$ varies considerably. We can improve the approximation by dividng the interval (a,b) into two pieces as indicated in Fig. A-1b and approximating $A = \Delta A_1 + \Delta A_2$ by $\Delta A_1 \simeq f(x_1)(x_2 - x_1)$, $\Delta A_2 \simeq f(x_2)(x_3 - x_2)$. In these smaller intervals $f(x)$ may vary less.

Clearly we can improve things further by dividing the interval (a,b) into many, say N, smaller segments by choosing points $x_1 = a, x_2, x_3 \ldots . x_{N+1} = b$ with $x_i < x_{i+1}$. In Fig. A-1c we show a typical segment x_i to x_{i+1}. Call $x_{i+1} - x_i = \Delta x_i$. Then ΔA_i lies in value near to $f(x_{i+1})\Delta x_i$ or $f(x_i)\Delta x_i$.

It seems reasonable (indeed it is true) that there is some point in the interval (a,b), say $x = \alpha$, such that

$$A = f(\alpha)(b - a) \quad . \tag{A-1}$$

Similarly in each interval there is some point α_i which depends upon the interval (x_i, x_{i+1}) such that ΔA_i is <u>exactly</u> given by

$$A_i = f(\alpha_i)\Delta x_i \quad . \tag{A-2}$$

The area A is then given exactly by

$$A = \sum_{i=1}^{N} \Delta A_i = \sum_{i=1}^{N} f(\alpha_i)\Delta x_i \quad . \qquad \text{(A-3)}$$

While Eq. (A-3) is exact, it is not very useful because we don't know the α_i. We do know, however, that if we shrink a Δx_i to zero, α_i approaches x_i. Also from Eq. (A-2) we see that in this limit $\Delta A_i/\Delta x_i$ goes to $f(x_i)$. If we make each segment Δx_i infinitesimally small however, the number of segments must become infinitely large because

$$\sum_{i=1}^{N} \Delta x_i = b - a \quad . \qquad \text{(A-4)}$$

Thus we have

$$\lim_{\substack{\Delta x_i \to 0 \\ N \to \infty}} \sum_{i=1}^{N} \Delta x_i = b - a \quad .$$

This limit of the sum is an example of what is known as a <u>definite integral</u>. We denote the integral by the sign $\int$ and write

$$\int_b^a dx \equiv \lim_{\substack{\Delta x_i \to 0 \\ N \to \infty}} \sum_{i=1}^{N} \Delta x_i = b - a \quad . \qquad \text{(A-5)}$$

Similarly we have the definite integral

$$\int_a^b f(x)\, dx \equiv \lim_{\substack{\Delta x_i \to 0 \\ N \to \infty}} \sum_{i=1}^{N} f(\alpha_i)\Delta x_i \quad . \qquad \text{(A-6)}$$

In Eq. (A-6), f(x) is called the <u>integrand</u>, a and b are called the <u>limits of integration</u> (lower and upper respectively).

Now, the right side of Eq. (A-6) is equal to the area A so Eq. (A-6) gives us a formal way to denote A but still doesn't say how to find it. Look at the left two members of Eq. (A-3) however. This becomes

$$A = \lim_{\substack{N \to \infty \\ \Delta A_i \to 0}} \sum_{i=1}^{N} \Delta A_i \equiv \int_?^? dA \quad . \qquad \text{(A-7)}$$

The question is, what do we put in for the limits of integration. From Eq. (A-5) we see that

$$\int_a^b dx = x\Big|_b - x\Big|_a$$

where $|_b$ means "evaluated at b" and similarly $|_a$ means "evaluated at a". In analogy then we could put $\int_0^A dA$ which is also formal. If now we look at Eq. (A-2) and remember that x_i is merely some point in the interval, we can conclude that there is some function of x, call it A(x), such that

$$dA(x)/dx = f(x) \quad . \qquad \text{(A-8)}$$

Then in analogy with Eq. (A-5) we would put A(a) for the lower limit and A(b) for the upper limit in Eq. (A-7).

This gives a prescription for evaluating $\int_a^b f(x)\,dx$.

1. Find a function, A(x), such that

$$dA(x)/dx = f(x) \quad ;$$

2. The integral is then A(b) - A(a) .

To see that this works out correctly in some simple cases consider first f(x) = 1 (as in Eq. (A-5)). Then A(x) (in this case) is x because dx/dx = 1. So

$$\int_a^b dx = x\Big|_a^b$$

where we use the notation $x\big|_a^b$ to mean $x\big|_b - x\big|_a$. As a second example, suppose f(x) = x. Since

$$d(\tfrac{1}{2}x^2)/dx = x$$

we have

$$\int_a^b x\,dx = \tfrac{1}{2}x^2\Big|_a^b = \tfrac{1}{2}b^2 - \tfrac{1}{2}a^2 \quad .$$

As a check, consider Fig. A-2.

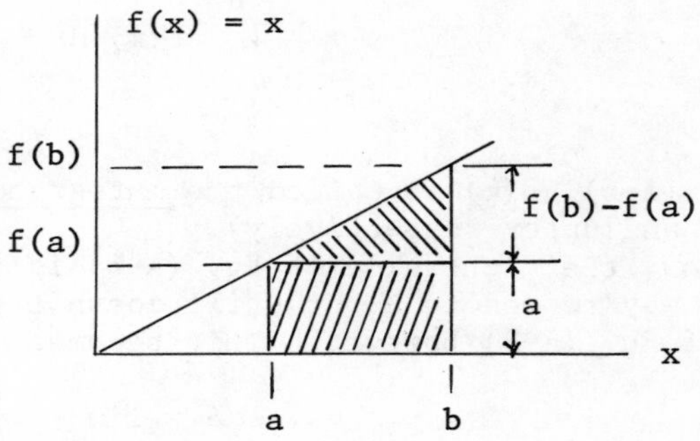

Figure A-2. Integral of f(x) = x from a to b shown as the area A under the curve.

The area under the curve f(x) = x is obviously

$$(b - a)f(a) + \tfrac{1}{2}[f(b) - f(a)](b - a) = (b - a)a + \tfrac{1}{2}(b - a)(b - a)$$

$$= \tfrac{1}{2}(b - a)(b + a) = \tfrac{1}{2}b^2 - \tfrac{1}{2}a^2$$

which agrees.

General properties of the definite integral follow from the definition. Some useful ones are:

$$\int_a^b cf(x)\,dx = c\int_a^b f(x)\,dx \quad ; \quad c = \text{constant}.$$

That is, a constant may be "factored out" of the integral. Also, the integral of the sum of two functions is the sum of their separate integrals:

$$\int_a^b [f_1(x) + f_2(x)]dx = \int_a^b f_1(x)\, dx + \int_a^b f_2(x)\, dx \quad .$$

Finally, the interval of integration may be broken up as for example,

$$\int_a^b f(x)\, dx = \int_a^{a'} f(x)\, dx + \int_{a'}^b f(x)\, dx \quad .$$

There also exists the concept of the <u>indefinite</u> integral which is what you will find in tables. This is just the $A(x)$ from before. That is, if

$$d(A(x))/dx = f(x)$$

then

$$\int f(x)\, dx = A(x) \quad .$$

Notice that no limits are given. Examples are given in an appendix to the main text. They are written as

$$d(x^m)/dx = mx^{m-1} \quad , \quad \int x^m\, dx = x^{m+1}/(m+1) \quad , \quad m \neq -1 \; .$$

One must note that in a sense this is incomplete because we could add a constant to $A(x)$. That is, suppose $B(x) = A(x) + \text{constant}$. Then

$$d(B(x))/dx = f(x)$$

so we could equally well write $\int f(x)\, dx = B(x)$. The point is, that to the particular integral [say the $x^m/(m+1)$ above] we can always add a constant. In a physical problem this constant is evaluated by the initial conditions.

>>> Example 1. Use the definite integral to derive the one dimensional, constant acceleration, kinematic equations.

Let x,v,a stand for the position, velocity, acceleration respectively. Then, since

$$dv/dt = a = \text{constant}$$

we have

$$dv = a\, dt \quad .$$

Now we integrate both sides and use the limits $v = v(0)$ when $t = 0$ and $v = v(t)$ when $t = t$. Thus

$$\int_{v(0)}^{v(t)} dv = \int_0^t a\, dt$$

or

$$v\Big|_{v(0)}^{v(t)} = at\Big|_0^t$$

or

$$v(t) - v(0) = at \quad .$$

Then from $dx/dt = v(t)$ we have

$$dx = v(t)\ dt = [v(0) + at]\ dt$$

so that

$$x(t) - x(0) = \int_{x(0)}^{x(t)} dx = \int_0^t [v(0) + at]\ dt$$

$$= v(0)t + \tfrac{1}{2} at^2 \quad .$$

We could also use the indefinite integral in this problem. Then we would have

$$dv/dt = a = \text{constant}$$

thus

$$v(t) = \int a\ dt = a \int dt = at + C$$

where C is a constant. But when $t = 0$, $v(t) = v(0)$, which is the initial condition, so $v(0) = a \cdot 0 + C$ and thus

$$v(t) = at + v(0) \quad .$$

You should be able to get $x(t)$ this way. Hint: $\int at\ dt = \tfrac{1}{2} at^2 + C$. <<<

INDEX

SUPPLEMENTARY INDEX*

*This Supplementary Index covers Study Guide chapters 45-48. These chapters pertain only to the "extended version" of "Fundamentals of Physics".

SUPPLEMENTARY INDEX (CONTINUED)*

*This Supplementary Index covers Study Guide chapters 45-48. These chapters pertain only to the "extended version" of "Fundamentals of Physics".

Notes

Notes

Notes

Notes

Notes

Notes

Notes

Notes

Notes